Immo Warntjes
The Munich Computus: Text and Translation

SUDHOFFS ARCHIV

Zeitschrift für Wissenschaftsgeschichte

Beihefte

Herausgegeben von
Klaus Bergdolt
Peter Dilg
Menso Folkerts
Gundolf Keil
Fritz Krafft

Heft 59

Immo Warntjes

The Munich Computus: Text and Translation

Irish computistics between Isidore of Seville and the Venerable Bede and its reception in Carolingian times

Franz Steiner Verlag Stuttgart 2010

Gedruckt mit freundlicher Unterstützung
der Arno-Borst-Stiftung

Bibliografische Information der Deutschen Nationalbibliothek
Die Deutsche Nationalbibliothek verzeichnet diese Publikation in der Deutschen Nationalbibliografie; detaillierte bibliografische Daten sind im Internet über <http://dnb.d-nb.de> abrufbar.

ISBN 978-3-515-09701-7

Gedruckt auf säurefreiem, alterungsbeständigem Papier.

Druck: AZ Druck und Datentechnik, Kempten
Printed in Germany

TO THE MEMORY OF

BRUNO KRUSCH

BARTHOLOMEW MAC CARTHY
EDUARD SCHWARTZ

&

ARNO BORST

CONTENTS

Abbreviations X

Foreword XIII

Introduction
The Munich Computus in modern times XV
The Munich Computus in the history of computistics XXX
The Munich Computus in the history of computistical textbooks LII

The history of the Munich Computus
The date LVII
The author LXII
The provenance LXXVII
The transmission XCVII

The context of the Munich Computus
Structure and sources CVII
Reception and influence CLIX

The edition
Editorial principles CCIII
The manuscript (Munich, Bayerische Staatsbibliothek, Clm 14456) CCXI

The Munich Computus: Text & Translation 1
I. De divisionibus temporum 2
II. De atomo 8
III. De momento 12
IIII. De minuto 14
V. De puncto 14
VI. De hora 16
VII. De quadrante 18
VIII. De diei nomine 20
VIIII. De nocte 28
X. De ebdomada 34
XI. De feriis 34
XII. De mensibus 44
XIII. De Ianuario 50
XIIII. De Februario 52
XV. De Martio 54

XVI.	De Aprelio mense	54
XVII.	De Maio	56
XVIII.	De mense Iunio	58
XVIIII.	De Iulio	60
XX.	De Augusto	60
XXI.	De Septimbrio	62
XXII.	De Octimbrio	62
XXIII.	De Nouimbrio et Decimbrio	62
XXIIII.	Continuatio: De mensibus	64
XXV.	De Kalendis	72
XXVI.	De nomine Nonarum	74
XXVII.	De nomine Iduarum	76
XXVIII.	De regulis mensium	78
XXVIIII.	De calculatione feriarum	82
XXX.	De tempore	86
XXXI.	De anno	88
XXXII.	De uerno	94
XXXIII.	De aestate	94
XXXIIII.	De autumno	94
XXXV.	De hieme	96
XXXVI.	Continuatio: De anno	96
XXXVII.	De sole	100
XXXVIII.	De divisione anni	100
XXXVIIII.	De augmento diei et noctis per annum	110
XL.	De V diebus superfluis	116
XLI.	De bissexto	120
XLII.	De aetate	140
XLIII.	De seculo	142
XLIIII.	De mundo	142
XLV.	De anno solis	154
XLVI.	De luna	154
XLVII.	De quattuor partibus mundi	160
XLVIII.	Continuatio: De luna	162
XLVIIII.	De aepactis	166
L.	De aetatibus lunaribus in Kalendis mensium	172
LI.	De loco et calculatione aepactarum	188
LII.	De anno resurrectionis in computatione Graecorum	194
LIII.	De anno lunae	202
LIIII.	De pascha	206
LV.	De regulis primi mensis	210
LVI.	De pascali subputatione	222
LVII.	De spatio annorum lunae	228
LVIII.	De initio quadragesimae	230
LVIIII.	De communibus et embolismis	240

LX. De inpletione inter solem et lunam per decennouennalem ciclum 256
LXI. De embolismis 266
LXII. De saltu 270
LXIII. De prohibitione celebrandi pascha in luna XIIII 290
LXIIII. De annis in temporibus 298
LXV. De comparatione variorum ciclorum cum ciclo Victorii 300
LXVI. De cursu temporum 304
LXVII. De ciclo Grecorum XCV annorum 310
LXVIII. De XII ciclis mundi 314

Appendices 319

Glossary 341

Bibliography 354

Indices 377

ABBREVIATIONS

ANGERS 477: Angers, Bibliothèque Municipale, 477 (461).
ARG. AQUENS.: 'Der Aachener Vorbehalt von 816', ed. Borst, *Schriften*, 1356–66.
ARN SERM.: 'Die Predigt Arns von Salzburg um 802', ed. Borst, *Schriften*, 833–84.
BC: Bobbio Computus, ed. in *PL* 129, 1275–372 (checked against the sole MS witness, Milan, Biblioteca Ambrosiana, H 150 inf).
CCCM: *Corpus Christianorum Continuatio Mediaevalis*.
CCSL: *Corpus Christianorum Series Latina*.
CE: *Computus Einsidlensis* (Einsiedeln, Stiftsbibliothek, 321 (647), p. 82–125).
CAP. COMP.: 'Das Aachener Verhör von 809', ed. Borst, *Schriften*, 1040–53.
COMP. COL.: 'Das Kölner Lehrbuch von 805', ed. Borst, *Schriften*, 891–950.
COMPUTUS COTTONIANUS: London, British Library, Cotton Caligula A XV, fol. 73r–80r.
CSEL: *Corpus Scriptorum Ecclesiasticorum Latinorum*
DCH: *Disputatio Chori et Praetextati* (cited from the Sirmond MS (Oxford, Bodleian Library, Bodley 309, fol. 101r–105v)).
DDT: *De divisionibus temporum*, ed. *PL* 90, 653–64.
DE ANNO: ed. Mountford, 'De mensium nominibus', 115–6.
DE BISSEXTO I–II: ps-Alcuin, *De bissexto I–II*, ed. *PL* 101, 993–9.
DE HEBDOMADIBUS: ed. Jones, *Bedae opera*, 394–5.
DE SALTU LUNAE I–VIII: ps-Alcuin, *De saltu lunae I–VIII*, ed. *PL* 101, 984–93.
DIAL. BURG.: 'Das burgundische Lehrgespräch von 727', ed. Borst, *Schriften*, 353–74.
DIAL. LANGOB.: 'Das langobardische Zwiegespräch um 750', ed. Borst, *Schriften*, 433–61.
DIAL. NEUSTR.: 'Das neustrische Streitgespräch von 737', ed. Borst, *Schriften*, 381–423.
DNR: Isidore, *De natura rerum*, ed. Fontaine, *Traité*, 163–327.
DRC: *De ratione conputandi*, ed. Ó Cróinín in Walsh & Ó Cróinín, *Cummian's letter*, 113–213.
DRP: Anatolius (?), *De ratione paschali*, ed. Mc Carthy & Breen, *De ratione paschali*, 44–53.
DT: Bede, *De temporibus*, ed. Jones, *Bedae opera*, 295–303 (including only the beginning of the chronicle).
DTR: Bede, *De temporum ratione*, ed. Jones, *Bedae opera*, 175–291 (not including the chronicle).
EPIST. RAT.: 'Der Regensburger Protestbrief von 809', ed. Borst, *Schriften*, 1027–53.
EPISTOLA CUMMIANI: ed. Walsh & Ó Cróinín, *Cummian's letter*, 55–97.

EPIT.: Virgilius Maro Grammaticus, *Epitomae*, ed. Löfstedt, *Virgilius Maro Grammaticus*, 103–245.

ETYM.: Isidore, *Etymologiae*, ed. Lindsay, *Isidori Hispalensis episcopi etymologiarum sive originum libri XX*.

EW: Bede, *Epistola ad Wicthedum*, ed. Jones, *Bedae opera*, 319–25.

HE: Bede, *Historia ecclesiastica gentis Anglorum*, ed. Plummer, *Baedae opera* 1, 1–360.

KAL.: The various recensions of the *Reichskalender* (Borst, *Reichskalender*, 399–1644).

LECT. COMP.: 'Die rheinische Anleitung von 760/792', ed. Borst, *Schriften*, 544–659.

LIB. ANN.: 'Das Veroneser Jahrbüchlein von 793', ed. Borst, *Schriften*, 676–722.

LIB. CALC.: 'Die Salzburger Enzyklopädie von 818', ed. Borst, *Schriften*, 1383–451.

LIB. COMP.: 'Die Aachener Enzyklopädie von 809', ed. Borst, *Schriften*, 1087–334.

LDA: Dicuil, *Liber de astronomia*, ed. Esposito, 'An unpublished astronomical treatise', 378–446.

M: Munich, Bayerische Staatsbibliothek, Clm 14456.

M*: M before correction.

MC: The original Munich Computus (cited from the edition by chapter and line), as well as its author.

MGH: *Monumenta Germaniae Historica*.
- *AA*: *Auctores antiquissimi*.
- *Epp.*: *Epistolae (in Quart)*.
- *Epp. sel.*: *Epistolae selectae*.
- *DD Mer.*: *Diplomata regum Francorum e stirpe Merovingica*.
- *LL*: *Leges (in Folio)*.
- *Poetae*: *Poetae Latini medii aevi*.
- *SS*: *Scriptores (in Folio)*.
- *SS rer. germ.*: *Scriptores rerum Germanicarum in usum scholarum separatim editi*.

PG: *Patrologiae cursus completus, series graeca*.

PL: *Patrologiae cursus completus, series latina*.

PP: *Pauca problesmata* ('The Irish reference Bible'), ed. Mac Ginty, *CCCM* 173.

PROL. AQUIT.: 'Das aquitanische Vorwort zur Ostertafel von 721', ed. Borst, *Schriften*, 337–47.

PV: Pacificus of Verona, *Computus*, ed. Meersseman & Adda, *Manuale di computo*, 53–166.

QUAEST. AUSTR.: 'Die austrasische Abhandlung von 764', ed. Borst, *Schriften*, 466–508.

QUAEST. LANGOB.: 'Die langobardische Abhandlung um 780', ed. Borst, *Schriften*, 514–26.

RM: Rabanus Maurus, *De computo*, ed. Stevens, *CCCM* 44, 199–321.

SAT.: Macrobius, *Saturnalia*, ed. Willis, *Saturnalia*.

SER. NOV.: 'Die ostfränkische Ahnentafel von 807', ed. Borst, *Schriften*, 971–1008.
SIRMOND MS: Oxford, Bodleian Library, Bodley 309.
VERS. TUR.: 'Das Tourer Lehrgedicht um 800', ed. Borst, *Schriften*, 804–19.

FOREWORD

The present study is a revised and extended version of my Ph.D. thesis, submitted and accepted at the Department of History of NUI, Galway, in September 2007. In the four years of working on this thesis, and the two years since, I was extremely fortunate in meeting many very supportive colleagues, who are not only fine scholars, but who also soon became close friends. This thesis profited greatly from the exchange of ideas with them – on a daily basis with the group of medievalists in Galway, and less frequently with scholars further afield. First and foremost I would like to thank my supervisor, Dáibhí Ó Cróinín, for inviting me to participate in the *Foundations of Irish culture project* that he set up in Galway in 2002 and which I entered a year later. The door of his office was always open for any discussion of computistica (as well as other fields of medieval Irish history), and he kindly allowed me to use his notes and unpublished material not only on the Munich Computus (which included a transcription), but also on other published and unpublished computistical texts; without his initiative, his belief in his student, and his tremendous support even before starting the Ph.D., this book would never have been written. Furthermore, I would like to particularly thank the two post-doctoral researchers of the project, Mark Stansbury and Rick Graff, for patiently discussing many of my questions, providing me with material whenever I was abroad, and assembling a fine microfilm collection that opened my eyes to the absolute necessity of studying the original texts, especially in a field like computistics where more texts remain unpublished than published. Anne Connon first introduced me to early medieval Irish history, and I am very grateful to her for setting me on that path. Great social support came, besides the above named, from the fellow Ph.D. students in the *Moore Institute* (formerly the *Centre for the Study of Human Settlement and Historical Change*), who created a very friendly, collegial atmosphere, from which I profited a great deal, both academically and personally. Among my fellow Ph.D. students I would like to thank Pádraic Moran and Jacopo Bisagni in particular for extremely interesting discussions of matters of shared interest, as well as a very enjoyable life outside of college. Outside of Galway, I especially benefited from many memorable afternoons and evenings discussing computistica (especially its technical dimensions) with Dan Mc Carthy in Dublin. All these people created a wonderfully kind, warm, and friendly atmosphere for a *peregrinus pro studio* in Ireland, and I very much miss them since being back in Germany. Additionally, ever since the memorable *First International Conference on the Science of Computus* in Galway in 2006, I have had the pleasure and privilege of many learned conversations, personal and via e-mail, with Leofranc Holford-Strevens. Closer to home, I would like to thank Kerstin Springsfeld for numerous stimulating discussions, as well as Karl-Heinz Spieß for giving me the chance to pursue an academic career. Menso Folkerts took a keen interest in my studies of early medieval science throughout and kindly accepted my thesis for the *Beihefte* to *Sudhoffs Archiv*, a series that has, at this stage, established a tradition of publishing studies of science in the so-called 'Dark Ages' (ca. AD 500–1100) and therefore appeared as the most obvious and best choice to print the present book. In the process of preparing my thesis for print, Dirk Schultze helped out in technical matters and Daniel Frisch did a great job when it came to indexing and layout.

It remains to thank certain institutions for their help and support. I gratefully acknowledge the financial support of the *Programme for Research in Third Level Institutions* (2003–6) and the *Arno-Borst-Foundation* (2007), the latter also for covering the publication costs of the present book. I am very grateful to various libraries for allowing me access to their manuscripts *in situ*, namely the Bayerische Staatsbibliothek in Munich, the Dombibliothek in Cologne, the Landesbibliothek in Kassel, the Herzog-August-Bibliothek in Wolfenbüttel, the Zentralbibliothek in Zürich, the Universitätsbibliothek in Basel, the Burgerbibliothek in Bern, and most of all the Stiftsbibliothek in Einsiedeln, especially its librarian, Odo Lang, OSB. Additionally, I want to thank the Staatsbibliothek in Berlin and especially the Staats- und Universitätsbibliothek of my *alma mater* in Göttingen for letting me use their outstanding research facilities. I am also grateful to the Bayerische Staatsbibliothek in Munich, the Staatsbibliothek in Berlin, the Stiftsbibliothek in Einsiedeln, the Dombibliothek in Cologne, and the Biblioteca Apostolica in the Vatican for granting me permission to reproduce in the Appendices photographic material under their copyright.

The thesis itself is dedicated to the four scholars who introduced the text here edited for the first time to modern scholarship and first established its context. First and foremost it is dedicated to Bruno Krusch, who discovered the Munich Computus when researching in Munich in 1878, being a Ph.D. student himself, though some five years younger than myself when I started the project. Bartholomew Mac Carthy and Eduard Schwartz (the latter studying this text when at my *alma mater*, the University of Göttingen, after a short spell in the late 1870s of studying at the university I am working in at present, the University of Greifswald) placed this text into its right geographical and chronological context. Of the scholars this book is dedicated to, its greatest debt is owed to Arno Borst, who first realized the Munich Computus' implications on Carolingian intellectual thought. I had the pleasure of meeting him twice in his home in Konstanz, where we discussed our work, as well as the more general developments in the field. He also very kindly asked me to proof-read his *Schriften zur Komputistik im Frankenreich*, and this work, as well as our correspondence about his texts, was extremely instructive and illuminating. His encouragement never ceased and his support continued after the publication of his monumental work with a kind offer of financial support for my project from his own *Arno-Borst-Foundation*. His death in April of 2007 marks the end of an outstanding career, and is a great loss not only to computistical studies, but to the study of the middle ages in general. It is a great pity that I was not able to discuss the final draft of my thesis with him, which would, no doubt, have very much profited from his comments. My thesis is dedicated to his memory and that of the other three fine scholars.

Greifswald, February 2010

INTRODUCTION

THE MUNICH COMPUTUS IN MODERN TIMES

Ever since Jean Mabillon, the founder of modern palaeography and diplomatics, studied the codex containing the Munich Computus in the monastery of St Emmeram in Regensburg as part of his travels through German and Swiss libraries in 1683,[1] it became well known for its unique transmission of the Regensburg annals (*Annales Ratisponensis*), which he subsequently edited in volume four of his *Veterum analectorum*.[2] Therefore, when this codex was transferred to the Königliche Hof- und Centralbibliothek (now Bayerische Staatsbibliothek) in Munich in 1812 as a result of the secularisation of Bavarian monasteries,[3] it received immediate attention because of these annals, particularly since these annals had been re-edited twice in the Benedictine Colomann Sanftl's handwritten catalogue of St Emmeram codices only three years earlier.[4] In 1819, the precursor to the *Monumenta Germaniae Historica* (*MGH*) was founded, with the primary object of editing all German sources of the medieval

1 For Mabillon's voyage to Switzerland and Germany see especially Mabillion's own account entitled *Iter Germanicum* in the fourth volume of his *Veterum analectorum*, where he gives a detailed description of his stay at St Emmeram in Regensburg from 20 to 25 August 1683 (Mabillon, *Veterum analectorum* 4, 3–92, the stay in Regensburg on p. 51–61; the *Iter Germanicum* was published separately in Germany in 1717, where Mabillon's stay in Regensburg can be found on p. 55–66). Cf. also Jadart, *Mabillon*, 31–3, 206–8 (the latter passage is a summarized itinerary of the voyage); Bergkamp, *Mabillon*, 55–7; Ruinart, *Mabillon*, 64–9; Leclercq, *Mabillon*, 200–30 (a very lively description of Mabillon's stay at St Emmeram on p. 220–1); Barret-Kriegel, *Mabillon*, 64–7 (Mabillon's stay in Regensburg just briefly noted on p. 66).

2 Mabillon, *Veterum analectorum* 4, 476–7 (without indication of the codex). For the codices used by Mabillion during his stay in St Emmeram cf. Bischoff, *Mittelalterliche Bibliothekskataloge* 4,1, 133–4.

3 For the transfer of manuscripts from the Bavarian monasteries to the Königliche Hof- und Centralbibliothek in Munich in the course of the secularisation in the early 19th century cf. especially Hauke, 'Bedeutung', 87–97 (the case of St Emmeram in Regensburg on p. 91). For the transfer of manuscripts from St Emmeram in Regensburg in particular cf. Docen, 'Anzeige', 425; Hemmerle, *Benediktinerklöster* I, 105; idem, *Benediktinerklöster* II, 242; Bezzel, *Bayerische Staatsbibliothek*, 12–3; Bischoff, *Mittelalterliche Bibliothekskataloge* 4,1, 138; Kellner & Spethmann, *Historische Kataloge*, 385.

4 Sanftl, *Catalogus* II, 934–6; IV, 443–4. In the section on mathematics (*mathesis*), Sanftl also mentions the Munich Computus and other computistica from this MS (Sanftl, *Catalogus* III, 1729). For Sanftl's catalogue cf. Kellner & Spethmann, *Historische Kataloge*, 388–9.

period, AD 500 to 1500.[5] On foot of this, Bernhard Joseph Docen, the Munich librarian, was contacted with a request for a list of all texts in the newly acquired Regensburg manuscripts that would be of special interest to the intended corpus of editions. In his list, published in 1820 in the first volume of the *MGH*'s (or rather its precursor's) just-founded journal, Docen mentioned the *Annales Ratisponenses*, but without any reference to the manuscript in which they are contained.[6] This led to a further inquiry from the *MGH* about the manuscript in question, which Docen answered by providing a catalogue description of this codex. In this description, the Munich librarian mentioned a *Computus S. Augustini –, S. Dionysii, S. Quirili Greciae et ceterorum* as the main text of this manuscript. Not finding the time to study and contextualize this computus further, Docen tentatively conjectured that this text might have been composed at the time of or even by Bede himself; but, as he explicitly stressed in the final sentence of his article, it was also possible, if not likely, that this text presents an otherwise unknown, unpublished, and important source for Christian time-reckoning.[7] Yet, whereas the *Annales Ratisponenses* were re-edited by the first president of the *MGH*, Georg Heinrich Pertz, in the *MGH*'s first volume of editions,[8] the computus did not receive any further attention for another fifty years. The reason for this neglect presumably was Docen's tentatively assumed connection between this text and the Anglo-Saxon scholar Bede, which placed it outside of the *MGH*'s interest.

It was only due to Bruno Krusch's non-national, chronological interest that the Munich Computus did not remain in obscurity any longer. As a 21-year old doctoral student he came across the Munich Computus in the Bayerische Staatsbibliothek in 1878, while working on Victorius' paschal cycle and its precursors, the 84-year Easter tables.[9] Krusch was exclusively interested in only

5 For the foundation of the *Gesellschaft für ältere deutsche Geschichtskunde* see Bresslau, 'Geschichte', 34–40 (p. 38: the principal object of editing all German medieval texts constitutes the first paragraph in the foundation statute); Fuhrmann, 'Goethe', 3; idem, *Gelehrtenleben*, 11–3; Schmitz, 'Entstehungsgeschichte', 503–7.

6 Docen, 'Anzeige', 425–9 (the *Annales Ratisponenses* are listed on p. 428). Docen worked directly from Colomann Sanftl's handwritten early 19th-century inventory of Regensburg manuscripts, in which Sanftl had re-edited the *Annales Ratisponenses* and had referred to the Munich Computus (cf. note 4). For Docen's career and his position and occupation in the Munich library at that time see especially Haller, *Bayerische Staatsbibliothek*, 121–2, 132.

7 Docen, 'Notizen', 515–9.

8 The edition in *MGH SS* 1, 91–3. For this first *MGH* volume of editions cf. Bresslau, 'Geschichte', 151–6; Fuhrmann, 'Goethe', 20–1; idem, *Gelehrtenleben*, 33; Wesche, 'Der erste Band', 17–21; Schmitz, 'Entstehungsgeschichte', 518–9.

9 For the date of Krusch's discovery see Krusch, *Studien* II, 58. For his scholarly occupation at that time see Krusch, *Studien* I, v; idem, *Studien* II, 5. For his early career and his chronological studies see Heymann, 'Bruno Krusch', 505–6. Cf. also Ó Cróinín, *Early Irish history*, 1. Krusch may have known the brief reference to the Munich Computus in Halm et al., *Catalogus*, 175, which was published only two years earlier, in AD 1876; it seems likely that this or Docen's earlier reference stimulated Krusch's initial interest in this text; in his first publication on this computus, Krusch (*Studien* I, 10) refers only to Docen.

one feature of this text, namely its frequent references to a *latercus*, which he correctly identified as an 84-year Easter table with 14-year *saltus*, having Easter lunar limits of 14 to 20. Most unfortunately for chronological studies to the present day, however, he connected the *latercus* information of the Munich Computus with the *laterculus* of Augustalis as transmitted in the *Computus Carthaginiensis*, because of the similarity in terminology. This resulted in his wrong reconstruction of the *laterculus* of Augustalis, which Krusch believed covered the years AD 213–312.[10] It was only some 20 to 25 years later that Bartholomew Mac Carthy as well as Eduard Schwartz proved that the Munich *latercus* did not in the least refer to the *laterculus* of Augustalis, and that Krusch's reconstruction was therefore obsolete.[11] Krusch himself accepted this view shortly before his death.[12] However, many historians of chronology to the present day refer to Krusch's theory of the *laterculus* of Augustalis as historically correct,[13] so that it cannot be overemphasized that the basis for Krusch's reconstruction, the Munich *latercus*, has nothing to do with Augustalis' table.[14]

Despite the faultiness of his theory, Krusch certainly deserves all due credit for rescuing the Munich text from obscurity and for highlighting its exceptional chronological value in its unique *latercus* references. Precisely this unique data attracted two of the leading chronologists of their time to the Munich Computus, the Reverend Bartholomew Mac Carthy in 1901 and the classicist Eduard Schwartz in 1905. Mac Carthy was the first to prove that the *latercus* mentioned in the Munich text refers, in fact, to the 84-year Easter cycle followed in some regions of Britain and Ireland until the eighth century. Since no Easter table of that reckoning was known to have survived, the Irish scholar realized the outstanding value of the Munich Computus' information about that reckoning; he

10 Krusch, *Studien* I, 5–19. The *Computus Carthaginiensis* is edited in Krusch, *Studien* I, 279–97.

11 Mac Carthy, *Annals of Ulster* 4, lxvi–vii, and especially Schwartz, 'Ostertafeln', 63–6. Cf. O'Connell, 'Easter cycles', 73–4; Wallis, *Bede*, xlv; Warntjes, '84 (14)-year Easter reckoning', 69–70.

12 Krusch, *Studien* II, 58.

13 Rühl, *Chronologie*, 122–4 (before the publication of Schwartz's correction); Schmid, *Osterfestberechnung in der abendländischen Kirche*, 19–20; Jones, *Bedae opera*, 15–6, 19; Cordoliani, 'Computistes insulaires', 6, 12; David, 'Saint Martin', 285; Strobel, *Ursprung*, 137, 161–2, 228, 273–4, 365, 384–6; Gougaud, *Christianity*, 186; Grumel, 'Problème', 167–8; Pedersen, 'Ecclesiastical calendar', 39–40; Stevens, 'Scientific instruction', 95; idem, 'Cycles of time', 37, 50; Blackburn & Holford-Strevens, *Companion to the year*, 806, 870, 872; Butzer & Butzer, 'Mathematics', 79; Lejbowicz, 'Computus', 160; idem, 'Tables paschales', 21, 44; Dekkers & Gaar, *Clavis patrum latinorum*, 725; Machielsen, *CCSL Clavis Patristica* 3A, 219; von den Brincken, *Chronologie*, 74; Holford-Strevens, *History of time*, 47.

14 The correct theory about this *laterculus* of Augustalis is Schwartz, 'Ostertafeln', 63–6. This is accepted by Gentz, 'Ostern', 1651; O'Connell, 'Easter cycles', 73–4; Wallis, *Bede*, xlv; Warntjes, '84 (14)-year Easter reckoning', 69–70. Ginzel, *Handbuch* 3, 243, only outlines Krusch's and Schwartz's theories, without stating any preference; Mc Carthy & Breen, *De ratione paschali*, 17 have their reservations about the Julian calendar and lunar limits attributed to this table by Krusch. See now also Mosshammer, *Easter Computus*, 224–8.

even attempted to reconstruct such a table from the data provided by the Munich text.[15] Subsequent to Mac Carthy's study, the Munich Computus was primarily analyzed for its references to this obscure and rather legendary Easter reckoning followed by the Irish and British in the early centuries of the middle ages. The most comprehensive analysis of the Munich *latercus* references was published by Schwartz, in his seminal study of the history of Easter tables, only four years after Mac Carthy's book had appeared in print,[16] which he may have known, even though he did not refer to it.[17] Being a very thorough and cautious scholar, Schwartz believed a reconstruction of the 84-year Easter table followed in Ireland and Britain based on the Munich data to be an impossible task.[18] Nevertheless, the Reverend D.J. O'Connell published another attempt at reconstructing such an Easter table on the basis of the Munich Computus in 1940,[19] an attempt that was refined by the Church historian Knut Schäferdiek in 1983.[20] However, only two years later an Easter table of that reckoning was discovered by Dáibhí Ó Cróinín in a Padua manuscript, which was subsequently reconstructed by Dan Mc Carthy.[21] In this reconstruction, the Munich Computus played a major part, since it transmits reliable and crucial information about the sequence of lunations underlying this table; the importance of this technical detail becomes immediately apparent from the fact that the reconstruction failed in the first place, precisely because this sequence of lunations was not consid-

15 Mac Carthy, *Annals of Ulster* 4, lxv–lxxxi.

16 Schwartz, 'Ostertafeln', 89–104.

17 Schwartz does not mention Mac Carthy's work anywhere in his study. Consequently, O'Connell, 'Easter cycles', 67 assumes that Schwartz was not familiar with Mac Carthy's book; likewise, Mc Carthy & Ó Cróinín, 'Easter table', 66. Mc Carthy, 'Easter principles', 223, however, convincingly argues that the parallels between Mac Carthy's and Schwartz's studies in the analysis of the *latercus* are so close that they must have been, in some way, dependent. In his opinion, the two scholars were likely to have collaborated, with Schwartz providing the source for Mac Carthy, since Schwartz's account of the *latercus* is the more detailed of the two, even though the publication dates speak against this hypothesis. I am inclined to think that Schwartz knew Mac Carthy's study and extended and corrected it. Note, however, that the Göttingen library (Schwartz wrote his study of Easter tables in his time at the University of Göttingen) did not acquire a copy of Mac Carthy's volume 4 of the *Annals of Ulster* before 1929 (I would like to thank the Göttingen librarian Helmut Rohlfing for providing me with this information). Therefore, if Schwartz did know Mac Carthy's work, he probably worked from his own copy. Dáibhí Ó Cróinín informed me that he could not find any reference to Mac Carthy in Schwartz's Nachlaß in Munich.

18 Schwartz, 'Ostertafeln', 102.

19 O'Connell, 'Easter cycles', 84–106.

20 Schäferdiek, 'Osterzyklus', 357–78.

21 For the date of the discovery of the Padua table see Ó Cróinín, *Early Irish history*, 4; Mc Carthy & Breen, *De ratione paschali*, 10. It was first analyzed and published in Mc Carthy & Ó Cróinín, 'Easter table', 58–75, but correctly reconstructed only in Mc Carthy, 'Easter principles', 204–24; cf. Mc Carthy & Breen, *De ratione paschali*, 10–1. A translation and concise summary of the technicalities underlying this table can be found in Blackburn & Holford-Strevens, *Companion to the year*, 870–5. A full facsimile is printed in Warntjes, '84 (14)-year Easter reckoning', 80–2.

ered.[22] After the discovery of the Padua table, the Munich Computus obviously lost its importance as the primary witness for this Easter reckoning. Only one further study of the Munich *latercus* followed, a detailed comparison between the Munich data and the Padua table, with the object of identifying the source underlying the information about the *latercus* in the Munich text, as well as analyzing its author's familiarity with this reckoning. Moreover, this study proved that a reform of this 84-year reckoning to prevent it from becoming increasingly inaccurate had never been executed, and neither did this reckoning include mechanisms that would have made it more accurate astronomically while abandoning its cyclic character at the same time.[23]

A different interest in the Munich Computus also existed, beyond the technical details of the 84-year Easter reckoning followed in Ireland and Britain, because of the few Old Irish words contained in this text. Generally, the incorporation of Old Irish words in the main body of a Latin text, as is the case in the Munich Computus, is a very rare phenomenon compared to the regular occurrence of Old Irish in interlinear or marginal glosses to other Latin texts. This phenomenon is yet to be fully explained, and any future study of it will need to rely on the evidence of the Munich text in particular, and of early Irish computistical material in general.[24] Moreover, any new discovery of Old Irish terms complements the comparatively small corpus of Old Irish vocabulary from this early period of the written Irish language. Mac Carthy drew attention to the occurrence in the Munich Computus of the bilingual term *dies cetene*,[25] Ó Cróinín to the Old Irish verb *tomel*.[26] This terminology, together with the additional occurrence of the curious term *noinaic*, and a few Old Irish numerals, have only recently been analyzed linguistically, and thoroughly discussed in the context of code-switching and code-mixing.[27]

22 For the problems occurring in the first attempt of reconstruction due to the application of the alternating sequence of lunations cf. Mc Carthy & Ó Cróinín, 'Easter table', 231–2. The non-alternating *latercus* sequence of lunations was then applied in Mc Carthy's definite reconstruction of the Padua table (Mc Carthy, 'Easter principles', 210–3); its importance for this reconstruction is subsequently stressed in Mc Carthy, 'The origin of the *latercus*', 25–6; Warntjes, '84 (14)-year Easter reckoning', 43.

23 Warntjes, '84 (14)-year Easter reckoning', 31–85.

24 Other computistical texts, in which Old Irish words occur in the main body of the Latin text, are the newly discovered *Computus Einsidlensis* and a lemmatized treatise on the Dionysiac and ps-Dionysiac *argumenta* in Padua, Biblioteca Antoniana, I 27, 77v–78r. For Old Irish in the *Computus Einsidlensis* cf. Warntjes, 'Computus Einsidlensis', 62 (note that one of the page references to the occurrence of Old Irish in this text has been cited wrongly due to a printing problem, which led to all '7' being substituted by '9' throughout the article; in note 7 – itself misprinted as 9 – it should read 97 instead of 99) and especially the full analysis of all Old Irish terms found in this text in Bisagni & Warntjes, 'Early Old Irish material', 77–105. For the occurrence of Old Irish in the Padua MS see Ó Cróinín, 'Dionysius Exiguus', 272.

25 Mac Carthy, *Annals of Ulster* 4, clxxx.

26 Ó Cróinín, 'Old Irish gloss', 131–2. Cf. also Ó Cróinín, 'Earliest Old Irish glosses', 16–7.

27 Bisagni & Warntjes, 'Latin and Old Irish', 1–33. For the Old Irish terms in the Munich Computus cf. also p. LXXV–LXXVI below.

Since the primary interest in the Munich Computus lies in the passages that deal with either these Old Irish words or the details of the *latercus*, the question remains whether these features present all of the text's originality and therefore constitute the only points of interest in this text. If so, would not an edition and detailed analysis of these passages, as now provided by the two most recent studies, suffice? In other words, is an edition of the entire text necessary and of any interest?

Docen's verdict would have been that if the Munich Computus proves to be an independent and unpublished text, as it surely does, then it certainly deserves to be edited in its entirety, as would any other text with these two characteristics.[28] However, the two German scholars who had worked most intensively on this text, Krusch and Schwartz, explicitly denied any value in editing the Munich Computus. In 1905, Schwartz wrote on this matter:[29]

> Sollte jemand auf den Einfall kommen den münchener Computus in ganzem Umfang abdrucken zu lassen, so würde der wesentliche Erfolg der sein, dass Bedas chronologisches Wissen und seine nüchterne, nie sich verwirren lassende Praecision sich von einem dunklen Beispiel occidentalischer Ignoranz mit wirklich Erfurcht gebietender Klarheit abheben.

Previously, in 1878, Krusch had only randomly studied the Munich Computus for his dissertation. After having read Schwartz's account of the Munich *latercus*, however, he returned to the text and transcribed it in full. His final verdict, formulated in 1937, was:[30]

> Ich habe den Computus zuerst entdeckt. Dann hat Schwartz die Hs. sich kommen lassen und ihn abgeschrieben. Um seine Ergebnisse nachzuprüfen, mußte ich sie mir wieder kommen lassen, und jetzt habe ich den Computus ganz abgeschrieben. Aber ich bin mit Schwartz der Ansicht, daß er den Druck nicht verdient. Die Hs. ist sehr fehlerhaft geschrieben. Schwartz hat vieles verbessert, aber noch mehr ist zu tun.

For these two scholars, then, three principal arguments spoke against an edition of this text. It was contemporaneous with Bede, so that in all probability most of the information given in the Munich Computus could be found in a clearer and more precise style in Bede's major computistical work, *De temporum ratione*. Therefore, the extremely time-consuming work of correcting this highly corrupted text would not prove worth the effort. Moreover, in some instances the Latin appears 'barbaric', to a degree that the sense of certain passages may never be fully understood.

Interestingly enough, in the 20th century it was particularly this last aspect, the 'barbaric' Latin, that attracted scholars to this text in its entirety. Schwartz, being one of the leading classicists of his time, showed little or no appreciation for non-classical Latin. Yet, in the last two decades of the 19th century, and especially with the appointment of Ludwig Traube to the newly created chair of

28 Cf. Docen, 'Notizen', 518–9.

29 Schwartz, 'Ostertafeln', 93.

30 Krusch, *Studien* II, 58. Cf. Ó Cróinín, 'A seventh-century Irish computus', 104.

medieval Latin in Munich in 1904, this general attitude changed.[31] Regional differences and characteristic features of medieval Latin became the focus of analysis, with Hiberno-Latin constituting one of these regional categories.[32] Mac Carthy, in a brief and rather uninspired analysis of the Munich Computist's orthography, was the first to hint at the potential of the Munich Computus for the study of Hiberno-Latin.[33] Traube's second successor, Bernhard Bischoff, arguably one of the most prolific scholars of Hiberno-Latin in the 20th century, referred to the Munich Computus only in passing.[34] A few more Hiberno-Latin aspects in this text have more recently been pointed out by Ó Cróinín,[35] but a comprehensive analysis of the Munich Computist's Latin as a whole has not been considered to the present day. It is hoped that the present edition provides the stimulus for such a study, especially since the study of the Latin of early medieval scientific texts (Hiberno-Latin or not) is a more general desideratum.

The true computistical value of the Munich Computus has only most recently been emphasized. Charles W. Jones, the author of the outstanding edition of Bede's computistical works, pointed to Bede's dependency on an Irish collection of computistical tracts, which he identified with a large section of the Sirmond manuscript (Oxford, Bodleian Library, Bodley 309).[36] With the exception of some minor pieces, the tracts themselves are, however, not of Irish ori-

31 Traube was one of three scholars who are regarded as the founding fathers of the study of medieval Latin; the other two are Traube's contemporaries Wilhelm Meyer in Göttingen and Paul von Winterfeld in Berlin. For the creation of the chair of medieval Latin in Munich, Traube's early career, his pioneering work and impact on the study of medieval Latin cf. Boll, 'Traube', XVIII–XXXI, XLI–VII; Silagi's notes to Traube, *Rückblick*, 3–9, 30–1; Lehner & Berschin, 'Nachwort', 243–4.

32 Seminal is Traube's 'Die lateinische Sprache des Mittelalters', which is published in his *Vorlesungen* 2 (the special place of Ireland and Britain in the development of medieval Latin on p. 39–41, 61–2, 91). For the subsequent development of the study of Hiberno-Latin in the 20th century and its results cf. Herren, 'Sprachliche Eigentümlichkeiten', 425–33; Stotz, *Handbuch zur lateinischen Sprache* 1, 85–6, 107–12. An overview of the literature on the subject prior to 1972 is provided by Bieler, 'Hiberno-Latin dictionary', 248–55, without reference to any scientific work. A detailed linguistic analysis of a Hiberno-Latin text, as well as a thorough (though sometimes outdated) discussion of Hiberno-Latin features, is provided in Bengt Löfstedt's dissertation on the Irish grammarian Malsachanus (Löfstedt, *Malsachanus*, 81–156), and also in his discussion of the language of the *Anonymus ad Cuimnanum* (Bischoff & Löfstedt, *Anonymus ad Cuimnanus*, xxiv–xxxviii). It is worth noting here that Traube also had an interest in computistical texts; his study of the *Computus* of Helperic of Auxerre (Traube, 'Computus Helperici') still is the best study of that text to date; cf. Borst, *Kalenderreform*, 140.

33 Mac Carthy, *Annals of Ulster* 4, clxxviii–ix; cf. also his discussion of the Hiberno-Latin term *singularis* in Mac Carthy, *Annals of Ulster* 4, lxix.

34 Bischoff, 'Das griechische Element', 250.

35 Ó Cróinín, 'Hiberno-Latin calcenterus', 56–7; idem, 'A seventh-century Irish computus', 104–7; Walsh & Ó Cróinín, *Cummian's letter*, 62–3, 182, 211.

36 Jones, 'Sirmond manuscript', 208–19; idem, *Bedae opera*, 105–13. For subsequent discussions of this manuscript and Bede's dependency on its material cf. Ó Cróinín, 'Irish provenance', 173–90; idem, 'Bede's Irish computus', 201–12; Wallis, *Bede*, lxxii–ix; Springsfeld, *Alkuins Einfluß*, 68–80; Graff, 'Recension of two Sirmond texts', 112–42.

gin.[37] Jones also mentioned the Munich Computus in this pre-Bedan Irish context, but he never discussed it and its relation to Bede's and other computistical texts in any detail.[38] It was not until the studies of Dáibhí Ó Cróinín in the 1980s that the genuinely Irish contribution in the field of computistics in the period between Isidore and Bede was placed on a solid footing. After having discovered a most original Irish computistical textbook of this period, *De ratione conputandi*, he compared selected passages of this new textbook with the Munich Computus in the article announcing the discovery, as well as in his subsequent edition of *De ratione conputandi*.[39] Unfortunately, in both studies the comparison was not systematic, so that many parallels between the two texts remained unnoticed. Nevertheless, in the company of *De ratione conputandi*, the Munich Computus was rightly considered as an extremely important witness to what may be termed as the Irish phase in the history of computistics, i.e. the period between the reception of Isidore and that of Bede. Yet, in Jones' and Ó Cróinín's studies, the Munich Computus was almost exclusively discussed in an Insular context. It is the merit of Arno Borst to have placed this computus in

37 Jones divided the allegedly pre-Bedan section of the Sirmond manuscript into two books; book one contains items 3 to 9 in his list, while book two consists of items 13–45 (cf. the references in the previous note). In this second book, evidently of Irish origin are sections of item 26 (published in Ó Cróinín, 'Bede's Irish computus', 209–10; the originally Irish bits are numbered VI and VIII–XII in Ó Cróinín's edition) and of items 35–36 (published in Ó Cróinín, 'Bede's Irish computus', 204–7; quite certainly of Irish origin are the unidentified pieces numbered IX and X by Ó Cróinín; note that the source references on p. 207 of Ó Cróinín's article are out of sequence: the Isidorian citation listed under V belongs to IV, and subsequently the source identifications VI to IX refer, in fact, to V to VIII; number IX is, therefore, unidentified); for parallels between these sections of the Sirmond manuscript and the Munich Computus cf. the following passages in the edition of the Munich Computus below (the edition is abbreviated as MC in the following, with references to chapter.lines): 41.107–110, 44.11–12, 50.22–28, 50.47–49, 59.38–69, 62.65–67, 64, and see also the discussion on p. LVII–LVIII and LXXX–LXXXII below; item 13 may also be of Irish origin, since it shows parallels to a heavily corrupted, apparently Irish tract in Padua, Biblioteca Antoniana, I 27, 77v–78r (cf. Ó Cróinín, 'Dionysius Exiguus', 272–4). In the first book, items 4–5, the tracts *De computo dialogus* and *De xiiii divisionibus temporum*, ultimately derive from Irish tracts, but here appear in a Carolingian recension (my reading of the evidence; for the controversy about place and time of these tracts cf. note 55); items 6 to 8, the ps-Alcuin tracts on the bissextile day and the *saltus lunae*, show many parallels to Irish texts (cf. MC 8.38–43, 24.12–14, 36.2–5, 41.7–8, 41.38–49, 41.80–88, 41.92–106, 46.16–20, 48.2–7, 55.6–12, 62.14–63, 62.68–72, 62.87–95, 62.111–117), but they may also have been (and in my opinion are) continental compositions drawing on Irish sources (again, their origin is highly disputed; cf. Jones, *Bedae opera*, 110; Cordoliani, 'Traités', 53; Dekkers & Gaar, *Clavis patrum latinorum*, 736; Borst, 'Alkuin', 59–61; Stevens, 'Rabani', 173–4; idem, 'Present sense', 18–20; Machielsen, *CCSL Clavis Patristica* 3A, 215; Springsfeld, *Alkuins Einfluß*, 77–9; Butzer & Butzer, 'Mathematics', 79).

38 Jones, 'Sirmond manuscript', 209–10, 213–4, and especially idem, *Bedae opera*, 110, where he simply states that Bede does not cite the Munich Computus.

39 Ó Cróinín, 'A seventh-century Irish computus', 104–7, 110–1, 118–9, 124–7; Walsh & Ó Cróinín, *Cummian's letter*, 23–4, 115–6, 134, 143–5, 156, 163, 165, 169, 172, 181, 204–5, 210. For the Munich Computus in the context of the reception of Virgilius Maro Grammaticus in computistical literature see Ó Cróinín, 'Virgilius Maro Grammaticus', 197–200.

a wider, Western European context. In his 2006 monumental corpus of editions of Frankish computistical texts, the Munich Computus plays an important part as a crucial source for Frankish computistica of the eighth century.[40] This demonstrates the influence of this Irish computistical textbook and highlights its role in the shaping of western medieval computistics.

Consequently, the principal value of the Munich Computus lies in the fact that it is a crucial text in the history of the most important science of the early middle ages, computistics, and as such essential to the understanding of the development of science in this period. Its scientific context, and especially its outstanding place in the formative period of medieval computistics, will be accentuated in detail in the following two chapters.

Terminology: Before proceeding to these, however, a note on the titles given to the Munich Computus in the studies outlined above is necessary. In general, the reference to Munich does not appear appropriate for a text that was evidently composed in Ireland and copied in Regensburg, where it was subsequently housed for almost a millenium before being transferred to Munich. Yet, ever since Mac Carthy's study of this text, which was the first such study published in English, this text is exclusively referred to as the 'Munich Computus' in English publications.[41] It appeared inappropriate, therefore, to change this terminology for the *editio princeps* of this computus, which would only lead to confusion about the text in question.

40 The fact that Borst begins the introduction to his corpus of editions with a quote from the Munich Computus illustrates the importance placed by him on this text. Borst, *Studien*, 1. On p. 134–7 he discusses this text in the context of pre-Bedan Irish computistics, there presented as one of the main foundations of Frankish computistics. The Munich Computus is referred to as *Comp. Hib.* throughout Borst's editions.

41 Mac Carthy, *Annals of Ulster* 4, lxvii–lxxv, ccxxviii–xxx; Kenney, *Sources*, 223; O'Connell, 'Easter cycles', 84–90; Lapidge & Sharpe, *Bibliography*, 95; Ó Cróinín, 'A seventh-century Irish computus', 102–27; idem, 'Old Irish gloss', 131–2; idem, 'Irish provenance', 183; idem, 'Virgilius Maro Grammaticus', 197; idem, 'Columbanus', 52; idem, 'Earliest Old Irish glosses', 16; idem, *Early medieval Ireland*, 188; idem, *Irish history*, 4–5; Walsh & Ó Cróinín, *Cummian's letter*, 258; Mc Carthy & Ó Cróinín, 'Easter table', 58–67; Mc Carthy, 'Easter principles', 205–24; idem, 'Origin', 49; Warntjes, '84 (14)-year Easter reckoning', 31–85; idem, 'Earliest occurrence', 96–105; Bisagni & Warntjes, 'Latin and Old Irish', 1–33; idem, 'Early Old Irish material', 77–91. The manuscript reference is preferred by some authors (Jones, 'Sirmond manuscript', 209; idem, *Bedae pseudepigrapha*, 48–9, 125; idem, *Bedae opera*, 110), and on very few occasions the lengthy heading of this text is referred to (Mac Ginty, 'Irish Augustine', 78). Note, however, that Jones, *Bedae pseudepigrapha*, 67, in his imprecise and vague treatment of the Munich MS, describes this text as 'the Irish Computus, composed AD 689'; he gives the heading of the Munich Computus as the incipit, but no explicit or folio number for the end; it appears from Jones, *CCSL* 123B, 351 that he regarded the entire MS from fol. 8r onwards as one recension of the now lost, hypothetical 'Irish computus'; hence, 'the Irish Computus, composed AD 689' was a description rather than a title, and referred to more than just the text from fol. 8r to 46r. For 'Munich Computus' denoting the MS as a whole, rather than the specific text of fol. 8r–46r, see note 54.

Unfortunately, anonymous computistical texts of the early middle ages are often referred to under various titles by modern commentators, depending on their personal preferences. This led to the bizarre situation that some of these texts are referred to by three or more different titles, with only their dates of composition providing definite clues about their identity. The wish to avoide such a scenario for the Munich Computus may serve here as a justification for retaining this rather inappropriate title. Some of the more prominent anonymous eighth-century computistical texts may illustrate the argument: The computistical anthology Milan, Biblioteca Ambrosiana, H 150 inf, is published under the title *Liber de computo* in volume 129 of the *Patrologia Latina*, and referred to by this name in some studies; in others, however, it appears as the Bobbio Computus (because of its provenance), the Milan Computus (because of its present location), or as one (if not the main or only) recension of a computistical compilation called *Computus Graecorum sive Latinorum*.[42] Similarly, evidently Frankish computistical texts in particular have received numerous different titles over the past century: The Frankish computus of AD 727 based on Victorian principles was first called according to the sole manuscript witness 'Berner Computus Nr. 611 von 727' by Krusch, and it was described as 'Komputus im Berner Codex n. 611 aus dem Jahre 727 n. Chr.' by Schmid; in Krusch's following *editio princeps*, however, he published it under the title 'Der merovingische Computus Paschalis vom Jahre 727 n. Chr.', so that it was subsequently referred to as 'Der merowingische Computus von 727', with the English equivalent 'Merovingian computus of 727', the French 'Comput Mérovingian de 727', the Latin *Computus paschalis merowingicus anni 727*; yet, Jones preferred to term it *Computus Victorianus*; in catalogues it is listed as *Computus paschalis a. 727*, or simply *Computus paschalis*, accompanied by additional reference to the manuscript, while it appeared as 'L'Anonyme de 727' in French literature; it has just recently been critically edited by Borst as 'Das burgundische Lehrgespräch von 727', with the Latin title *De ratione conpoti* and the abbreviation *Dial. Burg.*; an earlier publication by Borst makes it apparent

42 *Liber de computo*: *PL* 129, 1275–372; Cordoliani, 'Traités', 64; Jones, *Bedae pseudepigrapha*, 151; idem, 'Sirmond manuscript', 208; idem, *Bedae opera*, 111, 401; idem, *CCSL* 123A, XIII; idem, *CCSL* 123C, 777; Boschen, *Annales Prumiensis*, 246, 252; Rissel, *Rezeption*, 28–9; Walsh & Ó Cróinín, *Cummian's letter*, 115, 257; Dekkers & Gaar, *Clavis patrum latinorum*, 736; Machielsen, *CCSL Clavis Patristica* 3A, 198. Bobbio Computus (which is the title used in the present study, so that it is not confused with Rabanus Maurus' or Helperic's *Liber de computo* or other texts of the same title): Wallis, *Bede*, lxxii–iii, 451; Warntjes, '84 (14)-year Easter reckoning', 41–3. Milan Computus: Ó Cróinín, 'A seventh-century Irish computus', 105, and more often. *Computus Graecorum sive Latinorum*: Borst, 'Alkuin', 57; idem, *Plinius*, 119; idem, *Kalenderreform*, 181–2; idem, *Streit*, 143, 168; idem, *Studien*, XXVII (abbreviated as *Comp. Graec.* throughout Borst's corpus of editions); Kühnel, *End of time*, 102; Dekkers & Gaar, *Clavis patrum latinorum*, 736; Cordoliani, 'Traités', 59, 64; idem, 'Encyclopédie carolingienne', 237; idem, 'Contribution', 174; idem, 'Manuscrit de comput ecclesiastique', 20; Machielsen, *CCSL Clavis Patristica* 3A, 200–3; Lejbowicz, 'Tables paschales', 22; Germann, *De temporum ratione*, 44, and more often. Untitled: Wiesenbach, *Sigebert von Gembloux*, 59.

that the latter abbreviation stands for *Dialogus de computo Burgundiae*.[43] The Frankish computus of AD 737 based on Dionysiac principles was first referred to as *Tractatus de computo ecclesiastico* by Labbe, who was only concerned with the dating clause incorporated in the text; it was not given any title by Valentin Rose, who first described it in some detail in his catalogue of Berlin manuscripts, and there only tentatively characterized it as 'ein Schulbuch über den computus vom Jahre 737'; Krusch called this text more rigorously 'Das älteste fränkische Lehrbuch der dionysischen Zeitrechnung', and it was later referred to as simply 'Fränkisches Lehrbuch von 737', translated into English as 'Merovingian manual of 737'; Cordoliani refers to this text as 'Comput dionysien de 737', and, in accordance with Cordoliani's title, it appears as *Compotus Dionysii a. 737* or *Computus Dionysianus a. 737* in recent catalogues; Borst terms this text in the *editio princeps* 'Das neustrische Streitgespräch von 737', with the Latin title *De paschali racione aliique causis* and the abbreviation *Dial. Neustr.*; again, an earlier publication by Borst reveals that the latter abbreviation stands for *Dialogus de computo Neustriae*.[44] The Frankish computistical formulary of AD 793 has received less variation in its titles over the years; it has mostly been referred to as *Annalis libellus*, which was also the Latin title first preferred by Borst (with the abbreviation *Ann. lib.*), before he opted for changing the order of words to *Libellus annalis* with the corresponding abbreviation *Lib. ann.* in his recent edition; the German title employed there and earlier is 'Das Veroneser Jahrbüchlein von 793'; in a recent catalogue of computistical

43 For Krusch's titles see Krusch, 'Lehrbuch', 241; idem, *Studien* II, 53. 'Komputus im Berner Codex n. 611 aus dem Jahre 727 n. Chr.': Schmid, *Osterfestberechnung in der abendländischen Kirche*, 82. 'Der merowingische Computus von 727': Borst, 'Computus', 15; idem, *Computus*, 42, 152. 'Merovingian computus of 727': Thorndike & Kibre, *Catalogue of incipits*, 82; Walsh & Ó Cróinín, *Cummian's letter*, 118, and more often; Wallis, *Bede*, 13, and more often (it does not appear in the indices of either work). 'Comput mérovingien de 727': Cordoliani, 'Traités', 59 ; idem, 'Table pascal de Périgueux', 60 (with additional MS ascription). *Computus paschalis merowingicus anni 727*: Machielsen, *CCSL Clavis Patristica* 3A, 192. *Computus Victorianus*: Jones, *Bedae opera*, 400 ; idem, *CCSL* 123C, 735. *Computus paschalis a. 727*: Dekkers & Gaar, *Clavis patrum latinorum*, 732. '*Computus paschalis* in der Handschrift Bern 611': Frede, *Kirchenschriftsteller*, 78. 'L'Anonyme de 727': Lejbowicz, 'Computus', 159–61, 181. For Borst's titles see: Borst, *Schriften*, XXIX, 348, 353. *Dialogus de computo Burgundiae*: Borst, *Streit*, 84, 168. Pedersen, 'Ecclesiastical calendar', 55 describes it as 'a Merovingian *Computus Paschalis* from A.D. 727'.

44 *Tractatus de computo ecclesiastico*: Krusch, 'Einführung', 137; Schmid, *Osterfestberechnung in der abendländischen Kirche*, 83. For Rose's description see Rose, *Handschriften-Verzeichnisse*, 285–6; for Krusch's title Krusch, 'Lehrbuch', 232. 'Comput dionysien de 737': Cordoliani, 'Traités', 59; idem, 'Table pascale de Périgueux', 57. *Compotus Dionysii a. 737*: Thorndike & Kibre, *Catalogue of incipits*, 1249. *Computus Dionysianus a. 737*: Frede, *Kirchenschriftsteller*, 84 ; Dekkers & Gaar, *Clavis patrum latinorum*, 732; Machielsen, *CCSL Clavis Patristica* 3A, 282. 'Fränkisches Lehrbuch von 737': Borst, 'Alkuin', 55; idem, *Plinius*, 114–6. 'Merovingian manual of 737': Kühnel, *End of time*, 101. Borst's titles: Borst, *Schriften*, XXIX, 375, 381. *Dialogus de computo Neustriae*: Borst, *Streit*, 24, 169. Untitled: Jones, *Bedae opera*, 66; Rissel, *Rezeption*, 28; Stevens, 'Rabani', 170; Walsh & Ó Cróinín, *Cummian's letter*, 161; Declercq, *Anno Domini*, 162.

texts it has been simply termed *libellus computisticus*.[45] The two enormous Frankish computistical compendia of AD 809 and 818, however, have received a great variation of titles: The earlier one has been called 'astronomisch-komputistisches Lehrbuch', 'astronomisch-komputistisches Werk von 809', 'Seven-book computus' with the German equivalent '7-Bücher-Computus', 'une grande compilation d'astronomie et de comput de l'an 809', 'Aix-la-Chapelle encyclopedia', while it was termed by Borst in his recent edition and earlier as 'Die Aachener Enzyklopädie von 809' (which is translated into English as the 'Aachen encyclopaedia of 809'), with the Latin title *Libri computi* and the corresponding abbreviation *Lib. comp.*[46] The later one was named 'Three-book computus' with the German equivalent '3-Bücher-Computus', whereas Borst in his edition employed the German title 'Die Salzburger Enzyklopädie von 818' (while he earlier preferred 'Salzburger Kompilation') and the Latin *Liber calculationis* with the corresponding abbreviation *Lib. calc.*[47]

45 *Annalis libellus*: Borst, *Plinius*, 138–9, 375, 428; idem, *Kalenderreform*, 317 (but with the abbreviation *Lib. ann.*); Kühnel, *End of time*, 105–6; Springsfeld, *Alkuins Einfluß*, 12, 409. Borst's titles: Borst, *Computus*, 50, 154; idem, 'Alkuin', 61–2; idem, *Plinius*, 138, 144, 375, 428; idem, *Streit*, 41, 173; idem, *Schriften*, XLI, 660, 679. *Libellus computisticus*: Machielsen, *CCSL Clavis Patristica* 3A, 207. Untitled: Krusch, 'Lehrbuch', 233; Thorndike & Kibre, *Catalogue of incipits*, 1311. Rose, in his catalogue of the Phillipps MSS in the Berlin library (Rose, *Handschriften-Verzeichnisse*, 283), describes it as 'Werk kurzer Belehrungen über den Computus (Zeitrechnung), verfasst i. J. 793 (bzw. 776)'. Stevens, 'Present sense', 19 refers to it as *Liber annalis*.

46 'Astronomisch-komputistisches Lehrbuch': Köhler, *Karolingische Miniaturen* 3, 119–27; Mütherich, 'Buchmalerei', 50; idem, 'Erneuerung', 18; idem, 'Leidener Aratus', 150, and more often; Mütherich & Gaehde, *Buchmalerei*, 8, 12, 89. 'Astronomisch-komputistisches Werk von 809': Boschen, *Annales Prumiensis*, 13, 17–8, 24, 242–6. 'Seven-book computus': King, *Excerpts*, 3–27 (cited from Borst, *Plinius*, 171); Eastwood, 'Astronomy in Christian Latin Europe', 251; idem, 'Plinian astronomy', 201; idem, 'Plinian astronomical diagrams', 148, and more often; idem, 'Astronomies of Pliny', 164, and more often; Butzer, 'Scholars', 50 (confusing it with *Lib. calc.*); Butzer & Butzer, 'Mathematics', 80. '7-Bücher-Computus': Springsfeld, *Alkuins Einfluß*, 12, 409. 'Une grande compilation d'astronomie et de comput de l'an 809': Cordoliani, 'Encyclopédie carolingienne', 237; idem, 'Contribution', 174. 'Aix-la-Chapelle encyclopedia': Kühnel, *End of time*, 103, 107–10. 'Aachen encyclopaedia of 809': Wallis, *Bede*, xci–ii (with wrong manuscript ascriptions, since the Munich and Vienna MSS rather contain *Lib. calc.*); Butzer & Butzer, 'Mathematics', 80. Borst's titles: Borst, 'Alkuin', 71; idem, *Plinius*, 156, 382, 389; idem, *Kalenderreform*, 319; idem, *Streit*, 16, 173; idem, *Schriften*, XLI, 1054, 1056, 1087. Wilmart, *Codices*, 160, in his very detailed description of one of the MSS of this work, terms this text *De temporum ratione atque de rerum natura libri septem, seu chronologica et astronomica syllogia*. Untitled: Neuß, 'Kopie', 113–40. For the problem of terminology for this and the following work cf. also Germann, *De temporum ratione*, 30–1, 88–90.

47 'Three-book computus': Eastwood, 'Plinian astronomy', 201; idem, 'Plinian astronomical diagrams', 144, and more often; idem, 'Astronomies of Pliny', 163, and more often; Butzer, 'Scholars', 50 (confusing it with *Lib. comp.*); Butzer & Butzer, 'Mathematics', 80. '3-Bücher-Computus': Springsfeld, *Alkuins Einfluß*, 12, 409. Borst's titles: Borst, 'Alkuin', 73; idem, *Plinius*, 171, 381–2, 421; idem, *Kalenderreform*, 321; idem, *Streit*, 36, 173; idem, *Schriften*, XLI, 1367, 1369–70, 1383; Kühnel, *End of time*, 110. 'Salzburger Kompilation': Borst, *Plinius*, 171, 381–2, 421; Kühnel, *End of time*, 110. Cordoliani, 'Traités', 58; idem, 'Manuscrit de comput ecclesiastique', 26 merges the MSS of this Frankish encyclopaedia

These are only the most prominent examples of the widespread tendency of renaming anonymous computistical texts whenever previous titles appear inappropriate. Such a situation could be avoided for the Munich Computus by adopting the title unanimously given to this text in previous studies in English, though at the cost of a more suggestive and appropriate title. In German literature, however, the titles attributed to this text vary. Docen referred to it by its rather lengthy heading, *Compotus sancti Augustini, sancti Hieronimi, sancti Ysidori, sancti Dyonisii, sancti Quirilli Greciae, et ceterorum*, as do almost all catalogue entries.[48] Krusch did the same when introducing this text in his 1880 analysis, but in the following discussion he used the shorter 'Münchener Computus', which is the origin of the terminology applied in English studies of this text.[49] This terminology was adopted by Schwartz, and then in turn by Schäferdiek, who based his study on Schwartz's results.[50] Yet, in the summary of his article, Schäferdiek calls this text very precisely 'ein durch eine Münchener Handschrift überlieferter irischer Komputus aus dem Jahre 719'.[51] This phrase appears to rely on Krusch's one page note on this computus published shortly before his death, where, after Mac Carthy and Schwartz had demonstrated the Irish origin of this text, Krusch decided to coin it 'Der große irische Computus vom Jahre 719 n. Chr.'.[52] This then led Borst to the Latin title *Computus Hibernicus*, abbreviated as *Comp. Hib.*[53] Even though this terminology is more adequate, it is nevertheless more confusing than 'Munich Computus': In modern literature, only one text has been termed 'Munich Computus', namely the text edited here (Munich, Bayerische Staatsbibliothek, Clm. 14456, fol. 8r–46r), and

of AD 818 with those of the earlier encyclopaedia of AD 809, and terms the text 'Compilation d'astronomie et de comput (809)' (Cordoliani is more correct in other studies; cf. previous note); presumably based on Cordoliani's confused entry, Stevens, 'Present sense', 23–4, calls this text quite mistakenly *Compilatio computistica et astronomica AD DCCCVIIII* (he terms it simply *Compilatio DCCCVIIII* in idem, 'Karolingische Renovatio', 674), apparently inverting the dates and thus the chronological order of *Lib. comp.* and *Lib. calc.* (cf. Borst, *Schriften*, 1086); Stevens's terminology and mistakes were copied in Machielsen, *CCSL Clavis Patristica* 3A, 196–8, which is generally useless, since it appears not to be based on first-hand manuscript research; consequently, relying on secondary literature, it uncritically includes almost all of the numerous mistakes of previous scholars without qualifying them; McCluskey, 'Astronomies in the Latin West', 153 refers to this and the previous text as 'astronomical and computistical anthologies that emerged around the year 809', while in *Astronomies*, 135–9 he inverts the titles (and manuscript witnesses) of these two texts by referring to the earlier one as 'three-book computus', to the later one as 'seven-book computus'. Similar confusion in Butzer & Butzer, 'Mathematics', 50. Untitled: Neuß, 'Kopie', 118–40; Mütherich, 'Buchmalerei', 50; Stevens, 'Rabani', 170.

48 Docen, 'Notizen', 516. For the catalogue entries cf. Halm et al., *Catalogus*, 175; Cordoliani, 'Traités', 59; Thorndike & Kibre, *Catalogue of incipits*, 244; Machielsen, *CCSL Clavis Patristica* 3A, 188–9. Under this title also McGinty, 'Irish Augustine', 78; Stevens, 'Rabani', 170 (wrongly described as a collection of *argumenta*).

49 Krusch, *Studien* I, 10–6.

50 Schwartz, 'Ostertafeln', 89–102, especially 89; Schäferdiek, 'Osterzyklus', 360–77.

51 Schäferdiek, 'Osterzyklus', 378.

52 Krusch, *Studien* II, 58.

53 Cf. especially Borst, *Schriften*, XXVIII, 1.

therefore this title is unambiguous.[54] The same cannot be said about the title *Computus Hibernicus* (*Comp. Hib.*), since this term was already used by Jones in a different context, referring to a hypothetical, now lost 'Irish Computus'.[55]

54 Note, however, that Ó Cróinín, in some of his studies, applies the term 'Munich Computus' to the entire MS rather than specifically to the text on folios 8r–46r, although his argument is exclusively based on that text: Ó Cróinín, 'A seventh-century Irish computus', 102; idem, 'Virgilius Maro Grammaticus', 197; Walsh & Ó Cróinín, *Cummian's letter*, 258. Unfortunately, this transfer of title subsequently led, in some instances, to the transfer of the characteristics of the text on folios 8r to 46r to the entire MS, so that it is argued, quite mistakenly, that the whole MS is a copy of an Irish exemplar of 718: Ó Cróinín, 'Old Irish gloss', 131–2; idem, 'Earliest Old Irish glosses', 16; similarly Ohashi, 'Sexta aetas', 59. In only one of the studies dealing with material of this MS other than the text on folios 8r to 46r is the term 'Munich Computus' explicitly employed for the entire MS: Graff, 'Thirteenth figure', 321, 329. Concerning the extent of the text, only Thorndike & Kibre, *Catalogue of incipits*, 244 disagree with common opinion; they appear to regard the ps-Dionysiac *Argumentum XIV* which immediately follows the Munich Computus in the manuscript as part of that text, since they argue that this computus extends from fol. 8r to 47v rather than 46r; from the MS it is, however, perfectly clear that the ps-Dionysiac *argumentum* constitutes a separate treatise, because the last quarter of fol. 46r is left blank so that this independent text could start at the beginning of the following page.

55 Jones uses the term *Computus Hibernicus* (*Comp. Hib.*) only in his 1980 *CCSL* edition of Bede's computistical works, not in his earlier 1943 edition. In this 1980 *CCSL* edition he does not clearly define this term, neither in the *index auctorum* (Jones, *CCSL* 123C, 735), nor anywhere else in this edition. From the general introduction to this work it is, however, immediately apparent that the Munich Computus is certainly not meant by this term, since Jones mentions it without referring to it as *Computus Hibernicus* (Jones, *CCSL* 123A, XIII). An analysis of all source references to *Comp. Hib.* then shed light on what Jones associated with this title. Six of the eight references listed in the *index auctorum* of the 1980 *CCSL* edition correspond to cross-references to the appendix in the 1943 edition (Jones, *CCSL* 123B, 299–303; *CCSL* 123C, 587 versus Jones, *Bedae opera*, 195–7, 296). In this appendix, Jones published 'excerpts from the Irish computus', namely the preface and table of contents of a now lost computus from the Sirmond MS (Oxford, Bodleian Library, Bodley 309, fol. 62r–v) and a chapter of that computus headed *De Hebdomadibus* from Bern, Burgerbibliothek, 417, fol. 52v–53v. It is therefore clear that Jones denoted a hypothetical pre-Bedan Irish Computus (for which see also Jones, *Bedae opera*, 112) with the term *Computus Hibernicus*, parts of which survive in the Sirmond group of manuscripts. The exact contents of this lost computus obviously cannot be established, but it is apparent from Jones' two further references to *Comp. Hib.* (Jones, *CCSL* 123B, 310, 351) that he regarded an excerpt from the Sirmond MS (Oxford, Bodleian Library, Bodley 309, 73v) and the Anatolian (?) table for calculating the number of days from 1 January to any given Julian calendar date as part of it (his reference here is to Munich, Bayerische Staatsbibliothek, Clm 14456, 65v–67r, but it should only be to fol. 66v; cf. the table of contents of this MS p. CCXIII–CCXXI below; already in *Bedae opera*, 110, Jones argued that this MS contains parts of the Irish computus). For further clarification of Jones' *Comp. Hib.* references see Wallis, *Bede*, 32. Two problems with Jones' Irish computus need to be pointed out here: First, a distinction between the *Comp. Hib.* and *De divisionibus temporum* (which is repeated in Machielsen, *CCSL Clavis Patristica* 3A, 192–5, 236–8) appears not to be justified; a version of *DDT* apparently was, according to the table of contents published by Jones, part of the *Comp. Hib.*, and Jones should have included the *DDT* references among the *Comp. Hib.* ones (cf. Wallis, *Bede*, 34, where she identifies *DDT* as an 'Irish computus tract'). Second, the table of contents published by Jones from the Sirmond MS quite cer-

Accordingly, the term 'Munich Computus' is used in the present study for its unambiguity and for the sake of consistency with English literature on the subject, as well as the two authoritative German studies; the author of this text is consequently referred to as the 'Munich computist'.

tainly refers to a late eighth-century Frankish Computus based on Irish material rather than a pre-Bedan Irish text: If this table of contents is compared to the three pre-Bedan Irish computistical textbooks, it becomes immediately apparent that certain chapters listed in that table of contents were not part of Irish computistical teaching of ca. AD 700, namely the chapters dealing with the incarnation year, the indiction, the *cyclus lunaris*, the calculation of the lunar age and weekday of any given day of a year, the time of the day of the kindling of the moon, the length of moonlight per day, the rogation, as well as the astronomical chapters. In accordance with this, some scholars regard Jones' *Comp. Hib.* rather as a later Frankish compilation based on Irish material, called *Sententiae s. Augustini et Isidori in laude computi* (short *Sententiae*), which appears to have survived in numerous differing versions. Cf. especially Cordoliani, 'Encyclopédie carolingienne', 237–43; idem, 'Traités', 66; Borst, *Plinius*, 118–9; idem, *Kalenderreform*, 187–8; the references in the index of Borst, *Schriften*, 1487; and furthermore Frede, *Kirchenschriftsteller*, 91; Stevens, 'Rabani', 170–1, 179–80; Dekkers & Gaar, *Clavis patrum latinorum*, 735–6; Machielsen, *CCSL Clavis Patristica* 3A, 192–5; Springsfeld, *Alkuins Einfluß*, 77. It is, however, quite problematic that neither Jones' *Comp. Hib.*, nor the *Sententiae* are anywhere clearly defined, and in the end do not appear to be identical. The task of future studies will be to precisely define both texts and to identify the Irish kernel, as well as the Frankish additions. Cf. note 115.

THE MUNICH COMPUTUS IN THE HISTORY OF COMPUTISTICS

The period between the collapse of the Western Roman Empire in the fifth century and the arrival of Greek scientific texts through Arabic channels, as well as genuinely Arabic scientific tracts, from the eleventh century onwards is commonly regarded as a 'dark' period in the history of Western science.[56] The reason for this view is that the sciences taken into consideration are almost exclusively astronomy and fields of pure mathematics, especially geometry and number theory. In astronomy, the most important author, the second-century Alexandrian scholar Ptolemy (who had assembled the essentials of antique astronomical knowledge in one outstanding book, the Almagest) was not known in the West before Gerard of Cremona's (and a Sicilian anonymous' earlier) translation of this work in the twelfth century.[57] Western knowledge of astronomy in the early middle ages amounted only to random information from the works of Calcidius, Macrobius, Martianus Capella, Isidore, and Pliny.[58] In ge-

56 Historians of mathematics in particular show little appreciation for Western science in the early middle ages, at least in their overview accounts of the history of mathematics; note that computistics is not considered by these scholars, even though, in the early middle ages, it was primarily a mathematical science. Cantor, *Vorlesungen*, 821–47 is still the best and most balanced account for the early medieval period to date. Wußing et al., *Algebra*, 204 note that no medieval pre-twelfth-century text on algebra is known to have originated in the West; Linn, *Mathematics*, 13–24, heads the chapter on Western mathematics in the early middle ages with the title 'The West's asleep', stating that 'there can be little argument about the general lack of real intellectual and mathematical activity'; particularly harsh is Kline, *Mathematics*, 115: 'A people unacquainted with the rudiments of arithmetic could hardly be expected to advance mathematics. Actually history has no surprise for us in this instance. In no one of the civilisations that have contributed to our modern one did mathematical learning exist on as low a level as it did in medieval Europe. From the years 500 to 1400 there was no mathematician of note in the whole Christian world'. Singer begins his description of scientific development in the period AD 400 to 1000 in the following way (Singer, *History of scientific ideas*, 137): 'We now enter the last and longest phase of the Great Failure'. Cf. also Haskins, *Studies*, 3; Eves, *Mathematics*, 207–8; Burton, *History of mathematics*, 271–3.

57 For Ptolemy's Almagest, its transmission and importance, cf. especially Pedersen, *Survey* (p. 11–25 for the transmission and reception); Kunitzsch, *Almagest* (p. 6–112 for the transmission); and furthermore Stahl, 'Dominant traditions', 95–8, 123–4; idem, *Roman science*, 125–6; North, *History of astronomy*, 106–20; Pedersen, *Early physics*, 76–89; Dreyer, 'Medieval astronomy', 103, 108–20; Haskins, *Studies*, 14–5, 103–10, 157–64, 189–93; Pannekoek, *History of astronomy*, 146–62; McCluskey, *Astronomies*, 20–4, 188–90; Hoskins, *History of astronomy*, 16–24; Burton, *History of mathematics*, 190–2, 274–6.

58 For the history of astronomy in the West in the early middle ages cf. especially Eastwood, 'Astronomy in Christian Latin Europe', 235–53; McCluskey, 'Astronomies in the Latin West', 139–56; and furthermore Meier, 'Sieben freien Künste', 23–4; North, *History of astronomy*, 226–30; Pedersen, *Early physics*, 216–8; Dreyer, 'Medieval astronomy', 103–6; Stahl, 'Dominant traditions', 97–106, 111–9, 121–4; Bergmann, *Innovationen*, 18–9, 21–6; McCluskey, *Astronomies*, 114–49, 157–64; Hamel, *Geschichte der Astronomie*, 85–100; Hoskins & Gingerich, 'Medieval Latin astronomy', 68–72; Hoskins, *History of astronomy*, 29–32; Stevens, 'Karolingische Renovatio', 674–80; Eastwood, 'Plinian astronomy', 197–

ometry (as well as other mathematical disciplines) the West had to rely on a pseudo-Boethian text, a miscellany of excerpts from Boethius (and through Boethius of Euclid, but under omission of all proofs), Cassiodorus, Isidore, the Roman *agrimensores*, and others;[59] in number theory the most prominent text was Boethius' *De arithmetica*;[60] in arithmetic the calculation tables of Victorius of Aquitaine constituted the most widely used tool for mathematical operations.[61] The fundamental studies by Greek scientists, especially Euclid's Ele-

212; idem, 'Plinian astronomical diagrams', 141–3; idem, 'Dungal's letter', 119–21; idem, 'Astronomies of Pliny', 161–77; Eastwood & Graßhoff, *Planetary diagrams*, 1–10, 14–21. For a cultural perspective on astronomy in the early middle ages cf. Englisch, *Artes liberales*, 182–279; for astronomy in the early medieval classroom cf. Rissel, 'Hrabans Liber de computo', 149–51.

59 This text has never been published in full. For its content and transmission in the middle ages cf. especially the fundamental studies by Tannery, 'Notes', 39–50; Folkerts, 'Pseudo Boethian geometria', 189–90, 193–204; idem, 'Altercatio', 85–102; and furthermore Cantor, *Vorlesungen*, 580–9; Manitius, *Geschichte*, 28; Reindel, 'Beginn des Quadriviums', 518–9; Folkerts, *'Boethius' Geometrie II*, XI; idem, 'Geometry II', 1; idem, 'Euklidbearbeitungen', 5–6; Pingree, 'Boethius's geometry', 155–7; Zaitsev, 'Early medieval geometry', 523–35; Machielsen, *CCSL Clavis Patristica* 3A, 168–70; Butzer & Butzer, 'Mathematics', 81–2. For the history of early medieval geometry cf. epecially Folkerts, 'Development of mathematics', 1–6; for geometry in the Carolingian period cf. Stevens, 'Compotistica', 38–43; for a cultural perspective on early medieval geometry cf. Englisch, *Artes liberales*, 149–82.

60 Edited by Oosthout & Schilling, *CCSL* 94A. For this text and its reception in the early middle ages cf. especially Masi, *Boethian number theory*; White, 'Boethius', 163–5, 168–88 (for the reception of this work); and furthermore Cantor, *Vorlesungen*, 579–80; Manitius, *Geschichte*, 26; Stahl, *Roman science*, 198–9; Burton, *History of mathematics*, 238; Jones, 'Sirmond manuscript', 219; Cadwell, 'De institutione arithmetica', 137–9, 142; Stevens, 'Compotistica', 35–8; idem, 'Karolingische Renovatio', 670; Folkerts, 'Development of mathematics', 16–7; Machielsen, *CCSL Clavis Patristica* 3A, 182–3. For a cultural perspective on early medieval number theory cf. Englisch, *Artes liberales*, 91–149.

61 Arithmetic is here understood (according to modern mathematical understanding) as the study and application of elementary mathematical operations; it has to be clearly distinguished from number theory: arithmetic forms the basis of every calculation, so that basic knowledge of it was obviously as essential in everyday life as it was for any science; the theory of numbers, however, was a highly specialized theoretical field studied by only few intellectuals. Victorius' *Calculus* is edited by Peden, *Abbo of Fleury*, 1–62 as the basis of Abbo's commentary on this text; her edition supersedes the 19th-century editions by Christ, 'Argumentum calculandi', 132–6 (partial); Friedlein, 'Victorii calculus', 443–63; idem, 'Calculus des Victorius', 58–79. For this text cf. Christ, 'Argumentum calculandi', 100–32; Friedlein, 'Calculus des Victorius', 42–58; Jones, 'Sirmond manuscript', 218–9; idem, *Bedae pseudepigrapha*, 53; Rissel, 'Hrabans Liber de computo', 143–6 (including facsimiles of some multiplication tables); Bergmann, *Innovationen*, 34–6; Folkerts, 'Development of mathematics', 18; Stevens, 'Compotistica', 36, 38; idem, 'Cycles of time', 28; idem, 'Karolingische Renovatio', 667–70; Peden, *Abbo of Fleury*, xv–vii, xxxvi–xlix. For the execution of arithmetical operations in the early middle ages cf. Meier, 'Sieben freien Künste', 6–7; French Anderson, 'Arithmetical computations', 145–9; Rissel, 'Hrabans Liber de computo', 143–6 (underrating the calculation with fractions); Williams & Williams, 'Finger numbers', 590–3; Wallis, *Bede*, 254–63; Springsfeld, 'Rechnen', 224–31; Pillonel-Wyrsch, *Calcul*, 24–31.

ments, were not available before the late twelfth century,[62] and western scientists hardly improved on the few late antique tracts available to them. Therefore, the period roughly between 500 and 1000 is usually regarded as a time of scientific stagnation, ignorance, and disinterest. This point of view would quite certainly be valid if it was not for one science, computistics, which is too often neglected by scholars of the history of science, who traditionally focus on the quadrivium, and do not necessarily regard computistics as a medieval science in its own right.[63]

When the Christian church decided in the second century that Easter was neither to be celebrated on the same day as the Hebrew pasch, nor on a fixed Julian calendar date, but rather on the Sunday after the first full moon in spring, mathematical and astronomical methods needed to be applied for determining this date. Out of this necessity of calculating the most important feast in Christianity, a science developed with the primary object of calculating Easter, called computistics.[64] In the scientific centres of late antiquity, most notably Alexandria, computistics was, in its scientific relevance, little more than a subfield of applied astronomy and mathematics. However, the spread of Christianity to a less scientifically developed West led to an increase in the importance of computistics relative to the traditional sciences, especially since the bishops of Rome were eager to understand the calculation of Easter, so that they would not be dependent on Alexandria in that fundamental question. Two developments, then, accompaning the collapse of the Western Roman Empire had a profound impact on the study of science in the West in the early middle ages: Contact to the Greek speaking world was limited, so that availability of important scientific texts as well as personal expertise was severely restricted, to say the least. At the same time, Christian rather than secular institutions, most notably the

62 For Euclid's Elements, its transmission and importance, cf. Haskins, *Studies*, 24–5; Kline, *Mathematics*, 59–79; Eves, *Introduction*, 115–25; Burton, *History of mathematics*, 145–84, 273–8; Folkerts, 'Euclid', 1–49; idem, 'Development of mathematics', 6–11. For the parts of Euclid's work that were available in the West before the twelfth century see especially Folkerts, 'Euclid', 1–3, 19–25.

63 Pannekoek, *History of astronomy*, 173 calls computistics 'a thin rivulet of science'. Contrary to Wallis' opinion (*Bede*, xviii), computistics do 'seek to establish universal principles' (namely one universal system for the calculation of Easter) and do boast a scientific theory (a well defined luni-solar system based primarily on mathematical principles). Pedersen, *Early physics*, 216 rightly refers to computistics as 'a particular Medieval science', and similarly Cordoliani, 'Comput', 45 refers to computistics as 'cette science de comput ecclésiastique'.

64 For discussions and definitions of the term computistics and the various meanings of the Latin *computus* (as well as its equivalents in other languages) cf. especially the detailed studies by Borst, 'Computus', 10–51; idem, *Computus*, 9–56; Lejbowicz, 'Computus', 151–87; and furthermore Sickel, 'Lunarbuchstaben', 153; Bach, *Osterfestberechnung*, 3; Henel, *Studien*, 1–4; Jones, *Bedae pseudepigrapha*, 4; idem, *Bedae opera*, 75–6; idem, *CCSL* 123A, XII; Neugebauer, *Ethiopic astronomy*, 3, 68–9; Rissel, 'Hrabans Liber de computo', 138; Ó Cróinín, 'Irish provenance', 173; Stevens, 'Rabani', 167; idem, 'Cycles of time', 28–9; Wiesenbach, *Sigebert von Gembloux*, 31; Englisch, *Artes liberals*, 280–1, 283; Wallis, *Bede*, 425–6; Machielsen, *CCSL Clavis Patristica* 3A, 185; Blackburn & Holford-Strevens, *Companion to the year*, 801, 878.

monasteries, became the centres of science. Accordingly, the sciences were studied from a purely Christian perspective, in an environment suspicious of heathen authors, as well as of a study of science disconnected from religious objectives.[65] It does not surprise, therefore, that the traditional sciences were studied only rudimentarily, primarily as *Hilfswissenschaften* to the only 'truely' Christian science, computistics.[66]

Computistics, on the other hand, flourished to an exceptional degree, as the hundreds of manuscripts dealing with computistical matters surviving from the seventh to the tenth century impressively illustrate. It can hardly be doubted that almost every person educated in a monastic school was taught at least the basics of this science.[67] In this sense computistics was more than merely the applied science of calculating Easter. In the early sixth century, when Dionysius Exiguus made the Alexandrian Easter table available to the West, this science was rather narrowly defined as the means necessary for an understanding of this new table and the system underlying it. From the middle of the seventh century

65 For early medieval Christian attitudes towards science cf. especially Wallis, *Bede*, xxi–viii; McCluskey, *Astronomies*, 29–48; for Bede's case Stahl, *Roman science*, 229; for the case of geometry cf. Zaitsev, 'Early medieval geometry', 528–53. Cf. also note 119.

66 This is the role assigned to astronomy, at least in the pre-Carolingian period, if not thereafter, by Eastwood, 'Dungal's letter', 118; McCluskey, 'Astronomies in the Latin West', 146; to mathematics by Meier, 'Sieben freien Künste', 7; Struik, *History of mathematics*, 84; to arithmetic and astronomy by Rissel, 'Hrabans Liber de computo', 139–40; Bergmann, *Innovationen*, 20, 28, 36; Stevens, 'Cycles of time', 46; implicitly by Stahl, *Roman science*, 225.

67 The central document of the Carolingian educational reform, the *Admonitio generalis* of AD 789, lists c. 79 the *computus* among the subjects to be taught in monastic schools (*MGH LL* 1, 65): *Et ut scolae legentium puerorum fiant. Psalmos, notas, cantus, compotum, grammaticam per singula monasteria vel episcopia, et libros catholicos bene emendatos; quia saepe dum bene aliqui Deum rogare cupiunt, sed per inemendatos libros male rogant*. Likewise, a Frankish capitulary of AD 805 (*Capitulare duplex in Theodonis villa promulgatum*) states explicitly that all ecclesiastics should learn the *computus* (*MGH LL* 1, 132): *De computo, ut veraciter discant omnes*. This opinion is repeated as late as the 13th century by Guillaume Durand in his *Rationale divinorum officiorum* 8.1 (*CCSL* 140B, 131): *Quoniam, sicut ait beatus Augustinus, sacerdotes compotum scire tenentur, alioquin uix in eis sacerdotis nomen constabit, sub uerbo notitiam cursus temporis lunae ac kalendarii intelligimus quoniam compotus est scientia certificandi tempus secundum solis et lune progressum*. The most explicit statement comes from the Irish context, with the tenth-century *Saltair na Rann* ll. 261–268 arguing that every learned churchman should be able to correlate Julian calendar, lunar, and weekday data: (Stokes, *Saltair na Rann*, 4–5; the translation is cited from David Green's notes now in the Dublin Institute for Advanced Studies and available on the world wide web at:
http://www.celt.dias.ie/publications/online/saltair_na_rann/ (see Canto 001–010, p. 31)): *A coic cach*lae *d'fiss cenbrath / dlegair docachintliuchtach, / docachoen, cengláma gné, / bis fog*rada *ecailse. // Laa mis g*rene, *ésca aes, / rith mara cen immarbáes, / laa sechtmaine, feili noeb n-uag, / iarcertglaine con-imluad*. 'There are five things which a learned man should know about each day, everybody who is connected with the church, without appearance of censure. // The day of the solar month, the age of the moon, the running of the sea – without folly – the day of the week, of pure festivals, according to right clarity, with their variations.'

onwards, however, computistics developed into a synthesis of everything even remotely relevant to time reckoning in its most general form. Therefore, computistics covered every aspect of theoretical as well as applied science, including methods of finger-counting and other simple arithmetic, basic astronomical theory, tracts on the various units and divisions of time, descriptions of the history and workings of the lunar and solar calendar, algorithms for the calculation of chronological data (especially those connected to Easter), theories about and simple tools for the measuring of time, and the like. In short, science was almost synonymous with computistics in the Christian milieu of the early middle ages.[68]

Yet, the first and foremost aspect of computistics still was the calculation of Easter. Accordingly, a classification of the history of computistics into different periods quite naturally has to be based on the methods underlying this calculation. From the resurrection of Christ to the present day, four different phases can specified:[69]

1) The period from the resurrection of Christ to ca. the early second century, which is marked by the celebration of Easter on the fourteenth moon of the Hebrew first month, Nisan. The few sources available for this early period suggest that the earliest Christian communities in Asia Minor commemorated the passion of Christ at the same time as the Hebrew pasch, in accordance with John's Gospel.[70]

2) The period from ca. the early second century to the early ninth century, which is marked by the existence of competing systems, differing in their lunar calendars. At some point in the second century, most Christian communities agreed that Easter should commemorate the resurrection rather than the passion of Christ, and it should therefore be celebrated on a Sunday. The general rule was, as it still is to the present day, to celebrate Easter on the first Sunday after the first full moon (which is the fourteenth moon) in spring. Some communities in Asia Minor, however, rejected this reform, and stuck to the old practice of celebrating Easter on the fourteenth moon, irrespective of the weekday; they were soon condemned as heretics, called quartodecimans (from celebrating Easter on *luna quartodecima*).[71] More difficulties arose when the Hebrews

68 For similar conclusions cf. especially Pedersen, 'Ecclesiastical calendar', 60; Stevens, 'Scientific instruction', 83; and also idem, 'Compotistica', 49.

69 A similar classification of the history of computistics on the basis of competing systems can be found in Bach, *Osterfestberechnung*, 7–20.

70 For this early period cf. especially Hilgenfeld, *Paschastreit*, 160–214; and furthermore Ideler, *Handbuch*, 200–1; Bach, *Osterfestberechnung*, 7–8; Rühl, *Chronologie*, 110; Schwartz, 'Osterbetrachtungen', 2, 6; Ginzel, *Handbuch*, 210–1; Chaîne, *Chronologie*, 19–20; O'Connell, 'Easter cycles', 67–8; Gentz, 'Ostern', 1647–9; Jones, *Bedae opera*, 8–9; Declercq, *Anno Domini*, 50; Vogtherr, *Zeitrechnung*, 60–1; Holford-Strevens, *History of time*, 44–5.

71 For the developments in the second century and the quartodecimans cf. Ideler, *Handbuch*, 202–5; Hilgenfeld, *Paschastreit*, 216–320; Duchesne, 'Question', 6–16; Bach, *Osterfestberechnung*, 8–11; Rühl, *Chronologie*, 110–1; Schwartz, 'Osterbetrachtungen', 13–4; Ginzel, *Handbuch*, 212–6; Chaîne, *Chronologie*, 20–5, 42; O'Connell, 'Easter cycles', 68; Gentz, 'Ostern', 1649; Jones, *Bedae opera*, 9–10; Mohrmann, 'Conflit pascal', 154–71;

started to cyclically calculate their lunar calendar. In this calendar the first month, Nisan, and with it the Hebrew pasch on the fourteenth day of that month, could occur before the beginning of spring according to Christian reckoning;[72] accordingly, the Christian communities of Asia Minor who celebrated Easter on the Sunday after the Hebrew pasch commemorated Christ's resurrection earlier than the rest of Christendom; for that reason they were referred to as protopaschists.[73]

The main paschal conflict of late antiquity, however, developed from the late third century onwards between the church of Alexandria and that of Rome. Since Easter had to fall on the Sunday after the first full moon in spring, the obvious task was to create a working system by combining the solar with the lunar calendar. No difference existed between Alexandria and Rome in the choice of the solar calendar, since the Julian calendar was well-established throughout the Mediterranean world at that time. However, they differed in their lunar calendars: While Alexandrian computists applied a 19-year lunar cycle from the late third century onwards,[74] their Roman counterparts first tried an 8-year lunar cycle, which formed the basis for the Hippolytan 112-year Easter cycle;[75] soon thereafter they developed a more accurate 84-year lunar cycle with 12-year *saltus*. The period of 84-years was chosen because it constitutes a multiple of the 28-year solar cycle, so that, in fact, a luni-solar and there-

Pedersen, 'Ecclesiastical calendar', 24–30; Declercq, *Anno Domini*, 50–1; Blackburn & Holford-Strevens, *Companion to the year*, 791; Vogtherr, *Zeitrechnung*, 61–2. For quartodeciman practice cf. especially the detailed studies of Strobel, *Ursprung*, 17–69; and furthermore Dugmore, 'Quartodecimans', 411–21.

72 Cf. especially Grumel, 'Problème', 166–76.

73 For the protopaschists cf. Ideler, *Handbuch*, 206; Rühl, *Chronologie*, 110–1 (both with a wrong definition); Bach, *Osterfestberechnung*, 12–3; Ginzel, *Handbuch*, 212, 216–8; Declercq, *Anno Domini*, 51.

74 For the Alexandrian reckoning cf. Petavius, *De doctrina temporum* 1, 286–98; van der Hagen, *Dissertationes*, 208–47, 267–328; Rühl, *Chronologie*, 116–9; Mac Carthy, *Annals of Ulster* 4, l–lv; Schwartz, 'Ostertafeln', 8–29; Ginzel, *Handbuch*, 233–5; Chaîne, *Chronologie*, 40–2; Jones, *Bedae opera*, 29–33; Strobel, *Ursprung*, 133–7; Neugebauer, *Ethiopic astronomy*, 7–10, 56–63, 98–101 (through Ethiopic sources); Blackburn & Holford-Strevens, *Companion to the year*, 803–5; Holford-Strevens, *History of time*, 48–9, 132–4.

75 For the Hippolytan Easter table cf. especially Schwartz, 'Ostertafeln', 29–36; Neugebauer, *Ethiopic astronomy*, 85–7; Pedersen, 'Ecclesiastical calendar', 32–9 (with a facsimile of the original table on p. 33); Lejbowicz, 'Tables paschales', 13–7, 44–5, 48; most recently Mosshammer, *Easter Computus*, 116–27; and furthermore Petavius, *De doctrina temporum* 1, 110–1; Bucherius, *De doctrina temporum*, 133–5, 291–312; Ideler, *Handbuch*, 214–25; Hilgenfeld, *Paschastreit*, 331–40; Duchesne, 'Question', 17–9; Rühl, *Chronologie*, 119–22; Mac Carthy, *Annals of Ulster* 4, xxxii–xl, clxii–iii; Schwartz, 'Osterbetrachtungen', 17–8; Ginzel, *Handbuch*, 236–8; Chaîne, *Chronologie*, 43–4; O'Connell, 'Easter cycles', 689; Gentz, 'Ostern', 1650–1; Jones, *Bedae opera*, 11–2; van de Vyver, 'L'évolution', 7; Strobel, *Ursprung*, 122–4; Ferrari d'Occhieppo, 'Osterberechnung', 100–2; Stevens, 'Cycles of time', 35–6; Wallis, *Bede*, xxxvi–vii; Blackburn & Holford-Strevens, *Companion to the year*, 805; Mc Carthy & Breen, *De ratione paschali*, 16–7; Declercq, *Anno Domini*, 69–71. For its history, context, and subsequent development, cf. especially Richard, 'Comput de cent-douze ans', 257–77; idem, 'Comput pascal par octaétéris', 308–39 (arguing for the use of reformed *octaeterides* well into the fourth century).

fore Easter cycle of 84-years was created, called the *Supputatio Romana.*[76] But there were also other fundamental differences between the *Supputatio Romana* and the Alexandrian reckoning. The Romans were in favour of lunar limits 16 to 22 for Easter Sunday, since *luna* 16 occurred on the resurrection according to the Gospel of John. The Alexandrians, for their part, remained closer to Hebrew tradition by advocating lunar limits 15 to 21, which agreed with the Hebrew period of unleaven bread, as well as with the fact that according to the Synoptic Gospels Christ was cruxified on *luna* 15. Additionally, 21 March was strictly observed as the spring equinox (representing the beginning of spring) in the Alexandrian reckoning, so that the Easter full moon had to fall on or after this date. For the Romans, however, it was of paramount importance that the most important feast of Christianity did not coincide with the pagan celebrations commemorating the foundation of their city; therefore, Easter Sunday could not be celebrated later than 21 April, which meant that concessions had to be made to the lower Julian calendar limit in order to establish a working system, so that Easter could fall as early as 18 March in Rome. The conflict between Rome and Alexandria continued throughout the fourth and well into the fifth century.[77]

In the course of time it was realized in the West that the *Supputatio Romana* had become increasingly inaccurate. For this reason, various reforms of the 84-year cycle were implemented in the first half of the fifth century.[78] Only

76 For the *Supputatio Romana* cf. especially Schwartz, 'Ostertafeln', 40–50; most recently Mosshammer, *Easter Computus*, 204–13; and furthermore Bucherius, *De doctrina temporum*, 419–32; Noris, *Dissertationes*, 81–148; van der Hagen, *Observationes in anonymi cyclum LXXXIV annorum*, 247–89; Ideler, *Handbuch*, 239–53; Krusch, *Studien* I, 31–115 (which has to be read with caution, since only the discussion of what Krusch calls the 'younger' *Supputatio Romana* is correct; the same holds true for every subsequent study based on Krusch); Mac Carthy, *Annals of Ulster* 4, lxxxii–iii; Rühl, *Chronologie*, 124–5; Ginzel, *Handbuch*, 238–42; Jones, *Bedae opera*, 27–8; O'Connell, 'Easter cycles', 71–4; Cordoliani, 'Computistes insulaires', 6 (explicitly preferring Krusch's wrong theory to Schwartz's correction); Strobel, *Ursprung*, 225–33; Blackburn & Holford-Strevens, *Companion to the year*, 805–7.

77 For the history of the conflict between the Alexandrian and the Roman church concerning the date of Easter in the fourth and fifth centuries cf. especially Schwartz, 'Ostertafeln', 50–8; Schmid, *Osterfestberechnung in der abendländischen Kirche*, 1–29; and furthermore Ideler, *Handbuch*, 253–8; Hilgenfeld, *Paschastreit*, 369–72; Duchesne, 'Question', 22–42; Bach, *Osterfestberechnung*, 14–5; Chaîne, *Chronologie*, 48–60; van de Vyver, 'L'évolution', 16–8; Stevens, 'Cycles of time', 38–40; Declercq, *Anno Domini*, 72–9; Blackburn & Holford-Strevens, *Companion to the year*, 792–4, 807–8.

78 Besides the *latercus* or 84 (14), which will be discussed presently, four fifth-century modifications of the *Supputatio Romana* are known at present: 1) The *Laterculus* of Augustalis, which some scholars still believe to be an original third century table (cf. notes 13 and 14); only random information about this table is transmitted in the *Computus Carthaginiensis* (ed. Krusch, *Studien* I, 279–97); 2) two further tables described by the Carthaginian computist; cf. Krusch, *Studien* I, 164–88; Schwartz, 'Ostertafeln', 68–9; Schmid, *Osterfestberechnung in der abendländischen Kirche*, 15–7; Ginzel, *Handbuch*, 243–4; O'Connell, 'Easter cycles', 75; Strobel, *Ursprung*, 137, 271–3; Stevens, 'Cycles of time', 38; Declercq, *Anno Domini*, 79–80; Warntjes, '84 (14)-year Easter reckoning', 70–1; 3) the Zeitz table, first critically edited by Mommsen in a separate treatise ('Zeitzer Ostertafel', 541–8) and then in *MGH AA* 9, 507–10; Krusch, 'Bruchstücke', 996 managed to reconstruct an addi-

one of these reforms, however, had an impact on the following centuries, namely the *latercus* or 84 (14)-year Easter reckoning.[79] It appears that Sulpicius Severus created this 84-year Easter cycle with *saltus* in 14-year intervals in ca. AD 410. At the very least it can be regarded as fairly certain that the Easter cycle in question had its origin in the early fifth century.[80] Influenced by Eastern computistics, this cycle shows a number of considerable differences from the *Supputatio Romana*. Its Julian calendar and lunar limits for Easter Sunday were 26 March to 23 April and *luna* 14 to 20 respectively. On a more technical level, it applied a different sequence of lunations and the *saltus* were placed in every fourteenth year instead of every twelfth. If and for how long this cycle was used in Gaul and / or other parts of continental Europe must remain speculative. In the Insular remoteness of Ireland and Britain it became the predominant Easter reckoning at some stage of the fifth century and remained as such until it was gradually abandoned (depending on the area) between the early seventh and the late eighth or early ninth centuries.[81]

tional part at the beginning of the table; further fragments of the Zeitz table have just recently been rediscovered in the Stiftsbibliothek of Zeitz, and a full, though hardly readable facsimile of all fragments is published in the catalogue to the exhibition following the rediscovery (Overgaauw & Steving, *Zeitzer Ostertafel*, 13–24); in this catalogue, an upcoming new edition is announced (Overgaauw & Steving, *Zeitzer Ostertafel*, 30); for this table cf. also Mommsen, 'Zeitzer Ostertafel', 539–40, 549–66; idem in *MGH AA* 9, 503–6; Krusch, *Studien* I, 116–23; idem, 'Bruchstücke', 982–97; Schwartz, 'Ostertafeln', 71–2; Schmid, *Osterfestberechnung in der abendländischen Kirche*, 15–7; Rühl, *Chronologie*, 125; Ginzel, *Handbuch*, 244–5; O'Connell, 'Easter cycles', 74–5; Gentz, 'Ostern', 1651; Strobel, *Ursprung*, 270–1; Declercq, *Anno Domini*, 79, 84; Overgaauw & Steving, *Zeitzer Ostertafel*, 5–11; Warntjes, '84 (14)-year Easter reckoning', 71.

79 The construction of the 84 (14) has been the subject of debate for centuries. Only due to the discovery of an Easter table of this reckoning by Dáibhí Ó Cróinín in 1985 was it possible to convincingly reconstruct the 84 (14). For the discovery and the construction of the 84 (14) cf. the literature cited in note 21. All previous discussions are therefore outdated; for these cf. p. XVII–XIX above.

80 Aldhelm, *Epistula ad Geruntium* (*MGH AA* 15, 483) attributes this 84-year cycle to Sulpicius Severus. This identification has already been accepted by Ussher, *Antiquitates*, 173, 482, 514, but in the following centuries it was often rejected. Cf. van der Hagen, *Observationes in laterculum paschalem centum annorum*, 342–7 (based on outdated theories about the the 84 (14), but, as everything else from van der Hagen, certainly worth a read); Krusch, 'Einführung', 162; Schmid, *Osterfestberechnung auf den britischen Inseln*, 13–8 (a good summary of older opinions on this question, but useless in its technical discussion, as it is based on theories outdated since); Jones, *Bedae opera*, 101. Very recently Mc Carthy, 'Origin', 38–44, made a case for Sulpicius' authorship, which was judged unconvincing by Wallis, *Bede*, lvi, without outlining any argument to the contrary. Schwartz, 'Ostertafeln', 102; idem, 'Osterbetrachtung', 27 argues from the context of 84-year Easter tables that the 84 (14) was an invention of the fifth or sixth century (probably early fifth and pre-Victorius), a hypothesis that was confirmed by my own study (Warntjes, '84 (14)-year Easter reckoning', 34, 36–7), which has shown on technical grounds that an early fifth century origin, and thus Sulpicius' authorship, is most plausible.

81 For the abandonment of this reckoning in the various parts of Ireland and Britain cf. p. XXXIX–XL, LXXXIII–LXXXIX, XCIII–XCV, CLVI–CLVIII.

Interestingly enough, the creation of and adherence to this cycle appear to have been totally independent of Rome. The crisis between Alexandria and Rome concerning the correct date of Easter reached its height in the mid-fifth century. Pope Leo the Great, at that time, asked the reknown mathematician Victorius of Aquitaine to solve the differences between the two computistical centres. The Easter table created by Victorius, therefore, was designed to be a compromise between Roman and Alexandrian principles.[82] The basis of this table was a 19-year lunar cycle as used by the Alexandrians, though with slight technical differences, which had far-reaching consequences: The 19-year *saltus* was introduced 13 years before the Alexandrian one; the Julian calendar limits for Easter Sunday were 22 March to 24 April, so that the Alexandrian 25 April was not acceptable in the Victorian reckoning; moreover, the Roman lunar limits of *luna* 16 to 22 were kept, but alternatives were noted for cases in which Easter Sunday fell on the controversial lunar age 22. This Victorian system, problematic as it was in some of its details, nevertheless proved extremely successful, especially in Francia, where it was decreed as the definitive reckoning at the Council of Orleans in AD 541, and where it was only gradually abandoned in a slow process throughout the eighth century.[83] It also enjoyed a short spell of popularity in southern Ireland and Anglo-Saxon England.[84]

However, Victorius' ambiguity in listing double dates for certain years in particular, as well as the inclusion of the controversial lunar age 22, did not solve the disputes. For these reasons the papal curia decided in AD 525 to contact the monk Dionysius Exiguus, a famous canonist and translator from Greek to Latin. It appears that Alexandrian Easter tables and paschal tracts were available in Rome at that time, but only in Greek. Accordingly, Dionysius' principal task was to translate the available material, which he accomplished in style and

82 Victorius' computistical works are edited by Bucherius, *De doctrina temporum*, 2–10, 14–69; Mommsen, 'Victorii Aquitani cursus', 677–735; Krusch, *Studien* II, 17–52. For the Victorian reckoning cf. Bucherius, *De doctrina temporum*, 11–13, 145–81, 205–43; Noris, *Dissertationes*, 133–5; van der Hagen, *Observationes in veterum patrum prologos et epistolas*, 161–87; Ideler, *Handbuch* 2, 275–85; Rühl, *Chronologie*, 126–8; Mommsen, 'Victorii Aquitani cursus', 669–72; Bach, *Osterfestberechnung*, 17–8; Mac Carthy, *Annals of Ulster* 4, lxxxiii–ix; Schwartz, 'Ostertafeln', 72–80; Schmid, *Osterfestberechnung in der abendländischen Kirche*, 30–1; Ginzel, *Handbuch* 3, 245–7; Chaîne, *Chronologie*, 61–2; Krusch, *Studien* II, 4–5, 10–15; O'Connell, 'Easter cycles', 75–6; Jones, 'Victorian and Dionysiac Paschal tables', 409–13; Jones, *Bedae opera*, 61–8; Strobel, *Ursprung*, 138–9; Neugebauer, *Ethiopic astronomy*, 81–3; Pedersen, 'Ecclesiastical calendar', 47–9; Stevens, 'Scientific instruction', 92–3; idem, 'Cycles of time', 40–1; Wallis, *Bede*, l–lii; Declercq, *Anno Domini*, 82–95; idem, 'Dionysius Exiguus', 181–7; Blackburn & Holford-Strevens, *Companion to the year*, 793, 808–9; Holford-Strevens, *History of time*, 48–9; most recently Mosshammer, *Easter Computus*, 239–44.

83 For the decree of the Council of Orleans see *Concilium Aurelianense* 1 (*CCSL* 148A, 132), as well as the discussions in Bucherius, *De doctrina temporum*, 183–4; Noris, *Dissertationes*, 179; Krusch, 'Einführung', 125–6; Schmid, *Osterfestberechnung in der abendländischen Kirche*, 63–4. For the slow adoption of the Dionysiac reckoning cf. p. XXXIX–XLI.

84 Cf. p. XXXIX–XLI, LXXXIII–LXXXIX, CLVI–CLVIII.

therewith made the Alexandrian table, for the first time, readily accessible in the Latin West.[85] However, except for centres with Byzantine contact in the Mediterranean, it took some time for the adoption of this table throughout Western Europe. Rome itself appears to have finally decided to adhere to the Dionysiac reckoning in the 640s or slightly thereafter,[86] Anglo-Saxon England

85 Dionysius' computistical works are edited by Jan, *Historia cycli dionysiani*, 59–115 (repr. in *PL* 67, col. 483–520) and Krusch, *Studien* II, 59–86 (whose edition is in many aspects inferior to Jan's). Given the fact that Dionysius appears to have been a rather limited computist (cf. note 103), it seems that he reiterated or simply translated into Latin what his exemplars outlined in Greek, rather than that he would have executed many calculations himself. He may have tranferred Cyril's 95-year table to the subsequent 95-year period by using the manual he outlined in his prologue (cf. note 104) and his *argumenta*, but it seems more likely that this continuation of the Cyrillian table was already available in Greek. It is also likely that the conversion of the dates from the Alexandrian to the Julian calendar had already been accomplished in Dionysius' exemplar, or that Dionysius used a table synchronizing the dates of the Alexandrian calendar with the Julian one; Cyril's Easter table was the official one used in Alexandria, but it is possible, if not very likely, that Greek Easter tables of the Alexandrian reckoning adapted to the Julian calendar circulated in Dionysius' time. More importantly, Dionysius appears, in my opinion, not to have been capable of changing the era in his *argumenta* from the Diocletian to the Incarnation era (for the mathematical difficulties of that operation cf. Neugebauer, 'Computus paschalis', 293–301), so that the AD era must already have been the basis of the Greek *argumenta* that Dionysius then simply translated, while transferring the examples to his *annus praesens*. For different opinions on Dionysius' computistical and chronological skills and the task accomplished by him in AD 525 cf. especially Neugebauer, *Ethiopic astronomy*, 104–5; Ogg, 'Hippolytus', 2–3, 17; Declercq, *Anno Domini*, 99–147; idem, 'Dionysius Exiguus', 187–246; Holford-Strevens, *History of time*, 49–50; Mc Carthy, 'Emergence of anno domini', 32–8, 51–3; most recently Mosshammer, *Easter Computus*, 59–106; and furthermore Piper, *Kalendarium*, 87–9; Ginzel, *Handbuch* 3, 247–8; Jülicher, 'Dionysius Exiguus', 998–9; Krusch, 'Einführung', 106–8; idem, *Studien* II, 59; Jones, *Bedae opera*, 68–73; Mordek, 'Dionysius Exiguus', 1090–1; Teres, 'Time computations', 177–87; Pedersen, 'Ecclesiastical calendar', 49–54; Stevens, 'Scientific instruction', 90; idem, 'Cycles of time', 41; Wiesenbach, *Sigebert von Gembloux*, 47–8; Borst, *Kalenderreform*, 177; McCluskey, *Astronomies*, 87.

86 In general, a detailed history of the Easter controversy in the early middle ages, substituting the outdated accounts by Krusch, Schmid, Jones, and others, is one of the main desiderata in the field. Some scholars believe that Rome had adopted the Dionysiac reckoning already in the sixth century; cf. Krusch, 'Einführung', 110–4; Schmid, *Osterfestberechnung auf den britischen Inseln*, 50–60; Betten, 'Adoption', 487; O'Connell, 'Easter cycles', 76. The crucial evidence for the time of uncompromising adoption of this reckoning in Rome comes from the British Isles. Cummian, in his letter to the Iona abbot Ségéne and a certain Béccán, relates that the southern Irish clergy had sent an embassy to Rome, which apparently advocated the Victorian reckoning on its return (cf. note 238). Therefore, the Victorian system appears to have still been followed in Rome at this time (this is denied by Krusch, 'Einführung', 152; Schmid, *Osterfestberechnung auf den britischen Inseln*, 27; O'Connell, 'Easter cycles', 80; Jones, 'Paschal tables', 417–8; idem, *Bedae opera*, 90–1, who argue, quite unconvincingly, that the southern Irish converted to the Victorian principles not because of papal authority, but because the embassy had obtained Victorian tables in Gaul on their return to Ireland, or because they made no clear distinction between the Victorian and the Dionysiac systems). The next piece of evidence for Rome's conversion to the Dionysiac system is Bede's description of the papal letter to the northern Irish

followed around 664,[87] while it was unanimously accepted in Pictland in AD 710,[88] throughout the *regiones Scottorum* by the 720s,[89] in at least the greater part of Wales by 768.[90] In Francia, the process of adopting the Dionysiac reck-

clergy of AD 640. In this account, Bede (*Historia ecclesiastica gentis Anglorum* 2.19, ed. by Plummer, *Bedae opera* 1, 122–4; hereafter *HE*) argues that the pope-elect instructed the Irish to celebrate Easter Sunday from *luna* 15 to *luna* 21 (which are the Dionysiac lunar limits), but in the subsequent quote from this letter omits any reference to the lunar limits for Easter Sunday. The question now is whether Bede omitted this reference (if it was, in fact, part of the original letter) because he had already mentioned it, or because the lunar limits advocated by the papal curia were, in fact, still the Victorian. The former interpretation is obviously preferred by scholars arguing for a sixth-century change in Rome; cf. especially Jones, 'Paschal tables', 417–9; and furthermore Krusch, 'Einführung', 150; Schmid, *Osterfestberechnung auf den britischen Inseln*, 44; contradictory Gougaud, *Christianity*, 191, 194. I find the latter more likely, as does Poole, 'Earliest use', 59–60. It appears, then, very probable that Rome officially switched from the Victorian to the Dionysiac reckoning in the 640s or 650s, certainly before the Synod of Whitby of AD 664. Cf. Kenney, *Sources*, 215; Gougaud, *Christianity*, 191; Blackburn & Holford-Strevens, *Companion to the year*, 794; and the following note. Ó Cróinín, 'New heresy', 96 argues for this switch having taken place in the 630s. Jones, 'Legend of St. Pachomius', 205 quite unconvincingly argues that Rome remained uncommitted until the ninth century; similarly Stevens, 'Scientific instruction', 90; idem, 'Cycles of time', 41; idem, 'Karolingische Renovatio', 671; idem, 'Present sense', 16.

87 The first disputes in the Anglo-Saxon church concerning the date of Easter originated in the differences between the Victorian reckoning and the 84 (14). The Dionysiac system, for its part, appears to have been first accepted by Benedict Bishop and / or Wilfrid, who had learned the details of this reckoning in Rome in the 650s (where the papal curia must then have just recently adopted it; cf. previous note) and then fervently advocated it on their return to England. The crucial decision in favour of the Dionysiac reckoning was then taken at the Synod of Whitby in AD 664. Cf. *HE* 3.25, 5.19; Stephen of Ripon, *Vita sancti Wilfrithi* 5, 7, 10, 47 (Colgrave, *Life of bishop Wilfrid*, 12, 14–6, 20–2, 98); and the various differing opinions of Krusch, 'Einführung', 153–8; Schmid, *Osterfestberechnung auf den britischen Inseln*, 49–61; Poole, 'Earliest use', 60–1; Gougaud, *Christianity*, 194–7; Betten, 'Adoption', 489–90; O'Connell, 'Easter cycles', 84; Jones, 'Paschal tables', 413, 419; idem, *Bedae opera*, 103–4 (Jones's opinion is that the Dionysiac system was known and taught since the beginning of the Canterbury mission, and that no real distinction was made between the Victorian and the Dionysiac reckoning, both tables being used side-by-side, until the synod of Whitby); Grosjean, 'Recherches', 231–2 (following Jones' point of view with a stronger emphasis on the Dionysiac reckoning); Stevens, 'Scientific instruction', 97–8; Harrison, 'Easter cycles', 1–2; Evans, 'Celtic church', 222–3; Mc Carthy, 'Lunar and Paschal tables', 175–6; Smyth, *Understanding the universe*, 147; Wallis, *Bede*, lxi–iii; Declercq, *Anno Domini*, 155; Ohashi, 'Sexta aetas', 60; Blackburn & Holford-Strevens, *Companion to the year*, 795; Holford-Strevens, *History of time*, 52–4; idem, 'Marital discord'.

88 The source for the conversion of the Picts is Ceolfrith's letter to the Pictish king Nechtan, and Bede's additional information, in *HE* 5.21. For the conversion of the Picts cf. Ussher, *Antiquitates*, 366–7, 541; Krusch, 'Einführung', 163–5; Schmid, *Osterfestberechnung auf den britischen Inseln*, 65–8; Gougaud, *Christianity*, 199–200; Betten, 'Adoption', 493–4; Jones, *Bedae opera*, 104; Declercq, *Anno Domini*, 156.

89 For the developments in Ireland cf. p. LXXXIII–LXXXIX, XCIII–XCV, CLVI–CLVIII.

90 *Annales Cambriae s.a.* 768 (Morris, *Nennius*, 88): *Pasca commutatur apud Brittones †super dominicam diem† emendante Elbodugo homine Dei*. Cf. Krusch, 'Einführung', 166;

oning (as the Latin translation of the Alexandrian system is forthwith called) stretched over decades, if not centuries. That region must have made contact with this reckoning through Italian and Spanish channels in the seventh century, and the Franks were then more immediately confronted with it by Irish and Anglo-Saxon missionaries in the late seventh and throughout the eighth century; the numerous Irishmen remain anonymous, while the most prominent Anglo-Saxons are represented by Willibrord, Boniface, and Alcuin. By the early ninth century, every region in Francia seems to have converted to the Dionysiac reckoning.[91]

3) The period from the early ninth century to 1582, which is marked by the unanimous acceptance of the Alexandrian / Dionysiac reckoning throughout Christendom. Whereas the Easter controversy had been an integral part of society in the early middles ages with three fundamentally different systems (the 84 (14), the Victorian, and the Dionysiac reckoning) competing with each other, no such conflict existed from the early ninth century to the end of the middle ages, from the adoption of the Dionysiac reckoning in the last region of Western Europe to the Gregorian calendar reform.[92]

4) The period from 1582 to the present day, which is again marked by the existence of competing systems, differing in their solar and / or lunar calendars. Differences between competing systems and controversies about the correct calculation of Easter resumed with the introduction of the Gregorian calendar among Catholic churches in 1582. The inaccuracy of the Julian calendar had been discovered by the fact that the astronomical vernal equinox, the day of equal length of day and night, did not happened to fall on 21 March, which is regarded as the vernal equinox in the Dionysiac reckoning, but ten days earlier by the time of pope Gregory XIII.[93] This inaccuracy of ten days had accumu-

Schmid, *Osterfestberechnung auf den britischen Inseln*, 87–8; Gougaud, *Christianity*, 200–1; Betten, 'Adoption', 497–8; Hughes, 'Annales Cambriae', 235; Evans, 'Celtic church', 223–4; Declercq, *Anno Domini*, 156; Blackburn & Holford-Strevens, *Companion to the year*, 796.

91 For the conversion from the Victorian to the Dionysiac reckoning in Francia see the unsatisfactory accounts by Krusch, 'Einführung', 129–31, 136–41; Schmid, *Osterfestberechnung in der abendländischen Kirche*, 83–4; O'Connell, 'Easter cycles', 77; Declercq, *Anno Domini*, 160–4; and furthermore Ohashi, 'Sexta aetas', 61.

92 To be sure, certain chronological elements of the Victorian reckoning, like the *annus passionis*, can still be found in later chronicles; cf. Krusch, 'Einführung', 166. Note, however, that a fundamental difference exists between the observance of an Easter table and the use of its chronological elements: If these elements proved useful and suggestive for the compilation of chronicles, for the composition of an unambiguous linear timeline, then they could outlast by decades (if not centuries) the Easter tables in which they occurred. Accordingly, the fact that such chronological tools are found in chronicles does not allow for conclusions about the Easter reckoning followed at that time.

93 Critique of the Julian calendar and the need for reform was repeatedly articulated by various scholars since the twelfth century. Cf. especially the seminal study of Kaltenbrunner, 'Vorgeschichte', 293–411; and furthermore Wislicenus, *Kalender*, 18–9; Ginzel, *Handbuch*, 252–7; Ferrari d'Occhieppo, 'Osterberechnung', 105–6; McCluskey, *Astronomies*, 198–202; von den Brincken, *Chronologie*, 30–1; Blackburn & Holford-Strevens, *Companion to the year*, 682–3; Holford-Strevens, *History of time*, 33–5. For the discussions and

lated over centuries because a tropical solar year was shorter (the mean tropical year consists of ca. 365.24219, the vernal equinox year of ca. 365.2424 days) than the 365.25 days fixed by Julius Caesar; the Gregorian reformers, more accurately than Caesar, took the tropical solar year to consist of 365.2425 days, which amounted to a difference of 0.0075 days per year, or one day in 133 ⅓ years, if compared to the length of the Julian calendar year. For this reason, two reforms were implemented which turned the Julian into the Gregorian calendar:[94] Ten days were eliminated between 4 October and 15 October 1582, so that 4 October was followed by 15 October in that year. Moreover, the calendar was rendered more accurate by subduing three bissextile days in 400 years, namely the ones in centennial years not being divisible by four (i.e. no bissextile day was to be implemented in 1700, 1800, and 1900, but it was in 1600 and 2000). For the correct calculation of Easter, the same mechanism had to be applied to the lunar calendar, since every solar bissextile day corresponded to a lunar bissextile day; accordingly, whenever a bissextile day was subdued in the solar calendar, it had also to be subdued in the lunar calendar. Finally, the 19-year lunar cycle developed by the Alexandrians from Greek or Babylonian precursors was not absolutely accurate, either. After 310 years, the calculated moon lagged one lunar day behind the astronomical moon. For this reason, the calendar reformers decided to add three lunar days at once, and that in the future eight lunar days should be added in the course of 2500 year in 300-year intervals except for the last interval, which was extended to 400 years; the first such addition was to be implemented in 1800.

At the present day, the difference between the calculation of Easter in the Orthodox Church on the one hand, and Catholic and Protestant churches on the other, is that Orthodox Christians still apply the medieval practice, i.e. the Dionysiac reckoning with its Julian calendar and Alexandrian 19-year lunar cycle, while Catholics and Protestants follow the rules outlined in the Gregorian calendar reform. Yet, this modern unity between Catholics and Protestants in this question is a fairly recent consensus, achieved in 1867, following centuries of disputes about the correct method. Protestant regions were slow in adopting (and at the beginning vigorously opposed to) the Gregorian calendar introduced by the pope in 1582. In 1700, however, most Protestant countries accepted the Gregorian calendar, though not the Gregorian method of reckoning Easter.[95]

disputes immediately preceding the reform cf. Schmid, 'Zur Gregorianischen Kalenderreform', 389–407.

94 For the technicalities of the Gregorian calendar reform cf. especially Wislicenus, *Kalender*, 19–20, 46–51; Bach, *Osterfestberechnung*, 29–32; Ginzel, *Handbuch*, 257–64; Blackburn & Holford-Strevens, *Companion to the year*, 683, 817–9; Holford-Strevens, *History of time*, 35–6, 57–8; and furthermore von den Brincken, *Chronologie*, 31–2.

95 For the introduction of the Gregorian calendar in certain regions between 1582 and 1700 cf. Wislicenus, *Kalender*, 20–1; Ginzel, *Handbuch*, 266–71; Bach, *Osterfestberechnung*, 19–20 (a concise list of all places and dates of change); von den Brincken, *Chronologie*, 32–4; Blackburn & Holford-Strevens, *Companion to the year*, 683–5. For the adoption of the Gregorian calendar in most Protestant countries in AD 1700 and the difference in the calculation of Easter cf. Wislicenus, *Kalender*, 21–2; Goldscheider, 'Einführung', 4–7; Lind-

Incited by the recent astronomical successes of Tycho Brahe and Johannes Kepler, and eager to make a clear statement of independence from papal authority, they preferred to determine the Easter full moon, i.e. the first full moon after the spring equinox, on the basis of astronomical observations rather than cyclic calculations. It was only due to the intervention of Frederick the Great in 1775 that most Protestant countries accepted the Gregorian mathematical method.[96] Fourteen years earlier, England had given way to the Gregorian calendar (though in a different format) as well as the Gregorian mathematical calculation of Easter.[97]

The most turbulent history of adopting a suitable method for the calculation of Easter is that of Sweden.[98] There, the attempt was made to introduce the Gregorian reform gradually over a period of forty years from 1700 to 1740 by subduing the eleven bissextile days of this period. Accordingly, the Julian bissextile day of 1700 was subdued, the two subsequent ones, however, in 1704 and 1708, were not. This led to the bizarre situation that, since 1700, Easter Sunday was recorded a calendar and lunar day later compared to the Julian calendar date and lunar age calculated on Dionysiac principles. In years in which the increase of the lunar age by one led to a transgression of the lunar limit of *luna* 21 (i.e. in Sweden this became *luna* 22), Easter Sunday fell a week (or, more precisely, six days) earlier than in the Dionysiac reckoning. But this situation did not continue for long, since in 1712 the Julian calendar was restored by intercalating two days in February, namely the bissextile day of that year and the bissextile day that was subdued in 1700. Twenty-eight years later, in 1740, the next and equally unique reform was implemented in Sweden, when it was decided that the astronomically calculated Easter full moons should be observed in the Julian calendar, which resulted in Easter Sunday occurring as early as 14 March (25 March in the Gregorian calendar) in 1742. The Gregorian calendar was not introduced until 1753. From that year to 1823 the Swedes reckoned Easter by means of the Gregorian calendar and astronomically calculated full moons. In 1823, then, the Swedish king decreed that the Gregorian Easter was to be followed in the subsequent two years (1824 and 1828) in which the astronomical Easter would differ from the Gregorian one; the official change to the Gregorian mathematically calculated full moons was finally introduced in 1844, and with this act conformity with the Catholics and almost all Protestants was

hagen, 'Der schwedische Kalender', 1–2; Ginzel, *Handbuch*, 272–2; Bach, *Osterfestberechnung*, 20; von den Brincken, *Chronologie*, 34; Blackburn & Holford-Strevens, *Companion to the year*, 685, 798; Holford-Strevens, *History of time*, 60–1.

96 Cf. Wislicenus, *Kalender*, 22; Goldscheider, 'Einführung', 7; Ginzel, *Handbuch*, 273–4; Bach, *Osterfestberechnung*, 20; Blackburn & Holford-Strevens, *Companion to the year*, 798–9; Holford-Strevens, *History of time*, 61.

97 For the case of England and its colonies cf. Ginzel, *Handbuch*, 275; Blackburn & Holford-Strevens, *Companion to the year*, 685–7; Holford-Strevens, *History of time*, 62.

98 For the case of Sweden cf. especially Goldscheider, 'Einführung', 10–37 (34–7 for the post-1700 period); Lindhagen, 'Der schwedische Kalender', 2–4; and furthermore Ginzel, *Handbuch*, 275–6; Blackburn & Holford-Strevens, *Companion to the year*, 687, 798–9; Holford-Strevens, *History of time*, 61–2.

ensured. Only Finland, conquered from Sweden by Russia in 1809, observed the astronomically calculated Easter full moons until as late as 1867.[99]

Yet, the existence of rivalling systems can only be a rough guide for the classification of different periods in the history of computistics. On a more specific level, scientific knowledge and competence has quite naturally also to be taken into consideration, as these constitute the fundamental criteria in the assessment of the history of any science. In terms of computistics, the mere use of an Easter table has a profoundly different quality to the understanding of the technical system underlying such a table;[100] furthermore, the ability to construct a new system obviously lies far beyond the understanding of an existing system. This is especially important for the second phase in the history of computistics outlined above, from the early second to the early ninth century. By the mid-fifth century, the composition of Easter tables had stopped.[101] It may not be a coincidence that this end to the creation of Easter tables coincided with the general decline of scientific knowledge in the West due to the political upheavals of that time. In the decades, and even centuries, after the collapse of the Western Roman Empire, the scientific expertise necessary for the creation of a working luni-solar system appears to have been lacking. In fact, even the understanding of the existing systems seems to have been regressive. Two of the three reckonings followed in the West from the sixth to the early ninth century, the Victorian one and the 84 (14), were transmitted only in tables: On the one hand, no additional technical information that would explain the system underlying his table is given by Victorius in the prologue to his Easter table; on the other hand, the 84 (14) was not, to our present knowledge, accompanied by any explanatory text. Without such guidelines at hand, a full understanding of these reckonings must have been difficult.[102] The situation was slightly better concerning the

99 For the Finnish case cf. Blackburn & Holford-Strevens, *Companion to the year*, 799; Holford-Strevens, *History of time*, 61.

100 Cf. Wallis, *Bede*, lxii.

101 Note that Dionysius, in the early sixth century, did not construct an Easter reckoning, but merely translated an existing one, the Alexandrian reckoning, for western usage. For his computistical skills cf. notes 85 and 103.

102 Concerning the 84 (14), the only known Easter table of that reckoning is not accompanied by any explanatory text in the manuscript (Padua, Biblioteca Antoniana, I 27, fol. 76r–77v; this table is preceded by Anatolius (?)'s *De ratione paschali*, and followed by a lemmatized explanation of certain features of the Dionysiac and ps-Dionysiac *argumenta*). The only text that supplies a reasonable amount of technical details about the 84 (14) is the Munich Computus itself (for the passages cf. note 253). Victorius' computistical work consists of the prologue to his Easter table and the Easter table proper (Krusch, *Studien* II, 17–52); the prologue itself does not explain any constructional detail of this reckoning. Later textbooks describing the Victorian reckoning are extremely rare. Only three texts from the seventh and eighth centuries deal with this reckoning exclusively, namely *Dial. Burg.*, *Quaest. Austr.*, while the earliest and most interesting still remains unpublished in Bern, Burgerbibliothek, 645, fol. 41r–71v. But technical information about this reckoning can also be found in the few texts that compare the Victorian with the Dionysiac reckoning (cf. notes 184–5, 239, 252).

third reckoning, the Alexandrian, termed the Dionysiac in its Latin adaptation. Despite being a rather untalented computist himself,[103] Dionysius provided numerous interesting technical details about this reckoning when translating and applying the Greek material available to him. In his prologue, he gives detailed instructions for converting an expiring 95-year Easter table into one that covers the subsequent 95-year period.[104] Additionally, he attached a number of computistical algorithms to his Easter table, designed to calculate and check the data listed in that table.[105] Finally, the only object of Dionysius' letter to the papal

103 It appears that Dionysius' Greek sources rather than his own computistical ability must have been outstanding; cf. note 85 above. His computistical incompetence is revealed by his treatment of the cyclic character of the 19-year cycle: Being unable to prove this cyclic character himself, he looked for sources that would provide him with the essential proof. Yet, he could only find a proof of the cyclic character of the eight-year lunar cycle underlying the Hippolytan table in Quintus Julius Hilarianus' writings. Since the 19-year cycle is divided into *ogdoas* and *hendecas*, i.e. into periods of eight and eleven years, he used Hilarianus' proof to show that the *ogdoas* is truly cyclic. This left him with the impossible task of proving that the *hendecas* was truly cyclic as well. It appears that he never realized the simple mathematical fact that eight and eleven years each could only be truly cyclic if one year was truly cyclic, since the only common divisor of eight and eleven is one; however, a lunar year, as Dionysius himself very well knew, had eleven days less than a Julian calendar year, so that one year could not constitute a lunar cycle. Since no attention has been drawn to Dionysius' problematic discussion of the 19-year cycle since the 18th century (cf. Noris, *Dissertationes*, 225; van der Hagen, *Observationes in veterum patrum prologos et epistolas*, 216), and since these early modern scholars were bewildered by Dionysius' misconception rather than able to provide an explanation (being unaware of Dionysius' source), it may be appropriate to cite here at length the relevant passage of Dionysius' letter to Boniface and Bonus, as well as his source. *Epistola Dionysii* (Krusch, *Studien* II, 83): *In ogduade diximus V annos esse communes, tres embolismos. Quinquies ergo trecenteni quinquageni quaterni fiunt I̅DCCLXX et ter trecenteni octuageni quaterni, I̅CLII, ac per hoc simul fiunt I̅I̅DCCCCXXII. Similiter octo anni solares, si in summam redigantur, id est, octies trecenteni sexageni quini et quadrantes, faciunt simul I̅I̅DCCCCXXII. Simili modo et endicadis annos, qui sunt communes VII et quattuor embolismi, si in summam ea, qua diximus, supputatione congesseris, tantundem paene repperies, quantum XI solares anni conficiunt hoc est I̅I̅I̅I̅XIIII. Haec est igitur embolismorum, sicut praediximus, ratio, ut incrementis suis communium annorum detrimenta compensent*. Quintus Julius Hilarianus, *Expositum de die paschae et mensis* 13 (*PL* 13, 1113B; the corrections are mine): *Inde per annos singulos hac diversitate dies annorum lunae cum diebus solis, qui in trecentis sexaginta quinque et quadran<te> conficit annum, non sibi concordare videntur. Proinde etiam hoc ostendimus, ut appareat omnibus aequales eos invicem dies habere, et ab initio cursus eorum isto ordine cucurrisse, et damna communium annorum eadem luna in annis embolysmis compensasse. Octo annorum rationem, quae perfecta fore videtur, in medium proferamus. In octo scilicet annos luna quinque annos habet communes et tres embolysmos. Ergo quinquies CCCLIV faciunt <M>DCCLXX, et ter CCCLXXXIV faciunt MCLII; fiunt simul dies duo millia DCCCCXXII. Sic deinde et solis octo annorum summam in unum colligamus octies CCCLXV quadran. duo millia DCCCCXXII. Ita igitur congregavit solis ac lunae cursus cum diebus suprascriptis*.

104 Krusch, *Studien* II, 64. For this manual and its application cf. especially van der Hagen, *Observationes in veterum patrum prologos et epistolas*, 68–71, 194–6; Declercq, *Anno Domini*, 101, 105–6; idem, 'Dionysius Exiguus', 192, 198–9.

105 For the Dionysiac *argumenta* see especially Warntjes, 'Argumenta' and now also Mosshammer, *Easter Computus*, 97–106.

magistrates Boniface and Bonus was to explain the 19-year cycle. Yet, his technical explanations are not always correct, and numerous crucial technicalities remained unconsidered in his works.[106] It cannot be doubted that better computists than Dionysius existed in the western Mediterranean in the sixth century, especially in centres with Byzantine connections,[107] but the further these tables and texts travelled in time and space, the less likely it was that computists knew more about these reckonings than what they found in these tables and tracts.

In the seventh century, Insular computists, like almost all computists of Western Europe, were therefore faced with the situation that the material available to them did not provide for a full understanding of the computistical systems. When confronted with the Dionysiac reckoning, they were basically left with the challenging task of reconstructing the system underlying it. This task was successfully accomplished by Insular computists by the early eighth century, so that computists of regions subsequently converting to the Dionysiac reckoning, most notably Francia, could study all important details of this reckoning in Insular texts. Then, after the Dionysiac system had become the undisputed reckoning throughout almost all of Western Europe in the late eighth and early ninth centuries, the basic preoccupation of computists was the invention of new or better methods of calculating calendrical data within this system. The system itself was not questioned for another two hundred years, for various reasons: Its century-long tradition, going back to the legendary Alexandrian computists, as well as being connected to the decrees of the council of Nicaea of 325, gave it the utmost authority. The workings of this system were now completely understood and appreciated, with its errors being only marginal. And, most importantly, for the first time in the history of Christianity, unity was achieved in the observance of Easter, the most important Christian feast, due to the unanimous acceptance of this reckoning. This situation only changed with the improvement of scientific knowledge due to contacts with the Arabic world from the very late tenth century onwards. The introduction of fundamental

106 For the most obvious of Dionysius' mistakes cf. note 103 above. Crucial information about the construction of the 19-year cycle are not mentioned by Dionysius, most importantly the exact placements of the *saltus* and of the embolisms within the Julian calendar; moreover, he apparently was not aware of the existence of lunar bissextile days. This led to confusion and differing customs among seventh- and eighth-century computists. Cf. note 111.

107 The anonymous authors of the additions to Dionysius' *argumenta* were able computists (cf. Warntjes, 'Argumenta'). Slightly later, in AD 640/1 (for the date cf. van der Hagen, *Observationes in Maximi monachi computum paschalem*, 2–4), Maximus Confessor's *Computus ecclesiasticus* demonstrates that there must have been a tradition of good computists among the Greek-speaking population in the Western Mediterranean in the sixth century; it must be presumed that Dionysius himself relied on their studies. Unfortunately, Maximus Confessor's computus has not received much attention among modern scholars (as so often, the most detailed discussion of this work still is van der Hagen's 18th-century *Observationes in Maximi monachi computum paschalem*; it is also discussed in Schwartz's excellent study, 'Ostertafeln', 81–8) and this most interesting work still awaits a critical modern edition (the latest edition appeared in *PG* 19, 1217–80, a reprint of Dionysius Petavius's early 17th-century edition and Latin translation). For other Greek paschal texts from the fourth and early seventh century cf. Schissel & Ellend, 'Berechnung', 150–1.

Greek and Arabic texts, as well as astronomical instruments, led to a rapid improvement of scientific expertise, especially in astronomical observations. Even though the Dionysiac reckoning was not abandoned anywhere for another five hundred years, its faults became increasingly apparent, the critique of this system increasingly louder, until it finally culminated in the Gregorian calendar reform.[108]

This tour de force through the history of computistics provides the context for an understanding of the particular importance of the Munich Computus. The seventh and early eighth century transpired as an exceptionally significant period in the history of computistics, since it saw the final and possibly most vigorous controversy about the correct method of calculating Easter before the Gregorian reform, a controversy that continued in Wales and Francia right into the early ninth century. Moreover, the reconstruction of the Alexandrian / Dionysiac system in this formative period set this influential and long-lasting (in the Orthodox Church to the present day) reckoning for the first time on a solid footing in Western Europe. In modern literature on computistics, one man alone is usually credited with this exceptional achievement, the Anglo-Saxon scholar Bede, so that the formation of medieval computistics became synonymous with Bedan computistics.[109] Hardly ever is his work discussed in context, and it has

108 Cf. the references in note 93.

109 The only serious attempt to place Bedan computistics into context was undertaken by Jones in his first edition of *De temporum ratione* (*DTR*) (Jones, *Bedae opera*; this is the edition used in the present study), in which he frequently referred to texts like the Bobbio Computus, the pseudo-Alcuin tracts on the *bissextus* and *saltus*, *De divisionibus temporum*, *Dial. Burg*., and some minor treatises. However, unpublished comprehensive textbooks that had already been identified some decades before Jones's publication, like the Munich Computus and *Dial. Neustr*., were not consulted. Since Jones' day, studies of Bede's computistics were not based on more than Jones' evidence, even though new texts had come to light; in fact, most scholars of Bedan computistics did not even consider the same material as Jones did. The fact that most of these new texts remained unpublished appears to have been the reason for their neglect, which led to a totally unbalanced view of Bede's computistical achievement. But even those texts which received a critical modern edition, like *De ratione conputandi* (*DRC*), were never compared in detail to Bedan thought by Bedan scholars. The tendency of present scholarship is to analyze and explain Bedan computistics in isolation rather than against the background of the compustical literature immediately preceding and following Bede; this tendency is impressively illustrated by the latest commentary on *DTR* (Pillonel-Wyrsch, *Calcul*), in which no computistical text between Isidore and Rabanus Maurus is seriously considered. If scholars want to do Bede and his contemporaries justice, they will need to place Bedan computistics in the context of immediately preceding Irish and immediately following Frankish computistics; this means that, additional to the texts referred to by Jones, at least the three known Irish computistical textbooks (cf. p. LIV–LV; for the other two Irish computistical textbooks besides the Munich Computus, see especially p. CXXXIII–CLII for the *Computus Einsidlensis*, p. CXCI–CCI for *De ratione conputandi*) and the 20 Frankish texts just published by Borst, *Studien*, *always* need to be included in the analysis. For the common overrating of Bede's computistical achievement cf. Schwartz, 'Ostertafeln', 93 (comparing Bede's with the Munich computist's achievement); Krusch, *Studien* II, 58 (a statement approved of by Jones, *Bedae opera*, 138); Cordoliani, 'Comput', 46; Rissel, *Rezeption*, 26–7; idem, 'Hrabans Liber de

almost always been overlooked that Bede stands at the end of a tradition of intense scholarly research in computistical questions undertaken in Britain and Ireland, most prominently in the *regiones Scottorum*, in the seventh century. More importantly, Bede's works certainly did not solve every computistical question of his day, or of the centuries to come.[110] Certain fundamental issues, like the beginning of the Lenten fast, were not even touched upon in his texts, and for others he only recorded his own, regional customs, sometimes not even very clearly.[111] Accordingly, computistics of the centuries after Bede cannot

computo', 138–9; Pedersen, 'Ecclesiastical calendar', 56–9; Stevens, 'Bede's scientific achievements', 18–9; Bergmann, *Innovationen*, 16–20; Borst, 'Computus', 16; Englisch, *Artes liberales*, 281, 475; Wiesenbach, *Sigebert von Gembloux*, 52–3; von den Brincken, *Chronologie*, 10, 49, 74; Germann, *De temporum ratione*, 32; and also the following note.

110 The belief that Bede's work solved every computistical problem of its time is still common. Cf. Manitius, *Geschichte*, 78–9; Englisch, *Artes liberales*, 281; Germann, *De temporum ratione*, 34, 78; see also the previous note.

111 Walsh & Ó Cróinín, *Cummian's letter*, 102–3, have quite rightly observed that Carolingian computists writing after the reception of Bede's works clearly felt that the Northumbrian scholar had not solved every computistical problem of their day, and modern scholars should take this feeling seriously. E.g., Bede does not discuss the *initium quadragesimae* anywhere in his computistical works, even though it needs the same technical explanations as Easter Sunday, on which it totally depends. (Note that a thorough discussion of the *initium* given by Rabanus Maurus (*De computo* 83), who is often accused of not providing any information beyond Bede, constitutes one of the main differences between his and Bede's work; yet, Rabanus Maurus is only the redactor, not the original author of this chapter, which is largely based on chapter 28 of a still unpublished and originally unfinished Fulda computus of AD 789 (Basel, Universitätsbibliothek, F III 15k, fol. 36r–49r: fol. 43v–44r); this unfinished, but still extremely interesting text presumably has to be connected to the *Admonitio generalis* of the same year, and still awaits a thorough sudy and especially a critical edition). The placements of the embolisms listed by Bede (*DTR* 45.35–38) reflect only his own regional custom; since Dionysius had never defined these placements (cf. note 106), every computistical centre had its own theory about them; from the ninth century onwards Bede's placements became standard in most regions, but before AD 800 countless different opinions on them exist; these have never been studied in detail (only Sickel, 'Lunarbuchstaben', 174–80 notes that different customs existed, but ascribes the main variations to the late rather than the early middle ages; cf. now Holford-Strevens, 'Lunar calendars', 201), but it may suffice here to say that hardly ever do two texts written before the reception of Bede agree on the placements of the embolism (especially the tables listing the lunar age of the calends of every month for all 19 years vary considerably, reflecting different placements of the embolisms and the *saltus*); cf. p. CLXV. Bede (*DTR* 41) places special emphasis on the lunar bissextile day, and rightly so, since the application of lunar bissextile days was crucial to an understanding of the 19-year cycle; in this chapter, Bede argues that the lunar bissextile day should be placed at the end of the February lunation; he does not mention that a different (presumably Irish) custom exists, according to which the lunar bissextile day accompanies the extra Julian calendar day (this custom is deducible from the *Computus Einsidlensis* (*CE*) (Einsiedeln, Stiftsbibliothek, 321 (647), p. 82–125) p. 119–122; the Munich Computus (MC) c. 58; *De ratione conputandi* (*DRC*) c. 100–102). Concerning another feature not specified by Dionysius, the exact placement of the *saltus* in the Julian calendar, Bede mentions three options, namely the end of the July, November, or March lunation (*DTR* 42.55–63); as it is obvious from his calculations, he preferred and applied the November placement (cf. especially *DTR* 20.41–45); yet, he never explicitly states this preference, which confused modern commentators more than

entirely be explained or understood by reference to his works alone, since many aspects simply lie outside of Bedan tradition. This is even more so the case for the decades and even centuries preceding Bede. Even though the Easter controversy plays an immensely important part in his *Historia ecclesiastica gentis Anglorum*, he is remarkably silent in his computistical works about the technical details of the reckonings in question other than the Dionysiac one. Being an ardent supporter of the Dionysiac system, he wanted to draw his readers away from the older methods that he despised. Consequently, only very few details about the 84 (14) and the Victorian reckoning can be found in Bede's texts,[112] and he is equally silent about the technical arguments discussed in the Easter controversy (the theological ones are outlined in his *Historia ecclesiastica gentis Anglorum*). A study of the technical details of the 84 (14) and the Victorian reckoning in general, and of the technical issues of the Easter controversy in particular, in order to fully grasp und understand pre-Bedan computistics and especially the intellectual foundation of the Easter controversy, cannot be based on Bede, who deliberately obscures rather than illuminates the reckonings he condemned.[113]

The most important key to an understanding of computistical issues neglected or differently interpreted by Bede but prominent in seventh- and eighth-century and later discourse, as well as of the context of Bede's own work, is supplied by Irish computistics of the period between the reception of the writings of Isidore of Seville and those of Bede.[114] Only three Irish texts of this pe-

medieval computists; it is a widely accepted (and absolutely wrong) modern assumption that Bede favoured the March placement, an assumption based on Bede's phrasing in *DTR* 42 rather than his more significant calculations (cf. Noris, *Dissertationes*, 237; Wallis, *Bede*, 327–8; Borst, *Streit*, 48; idem, *Schriften*, 408; already van der Hagen, *Observationes in veterum patrum prologos et epistolas*, 287–8 criticizes this opinion, and then (p. 352–3) gives the proof that Bede followed the November placement; also correct is Springsfeld, *Alkuins Einfluß*, 138–42). For the *saltus* see p. CLXVI–CLXVII; for Irish computistical features of the late seventh, early eighth centuries see p. LXIX–LXXIII, CIII, CXXXIII–CLII, CLIX–CLXVIII, CXCV–CC below.

112 The only details about the Victorian reckoning mentioned by Bede are the placement of the *saltus* in the sixth year of the *ogdoas*, leading to an epactal change from 4 to 16 on 1 January (*DTR* 43.32–34), as well as the lunar limits for Easter Sunday of *luna* 16 to 22, which he does not explicitly attribute to Victorius (*DTR* 59.26–28); likewise, in *DTR* 62.12–38 Bede discusses the one year in the Victorian 19-year cycle in which the Easter full moon occurred before the equinox of 21 March, without any reference to Victorius; on the other hand, in *DTR* 51 he condemns the Julian calendar limits for the Easter new and full moon ascribed to the Latins in Victorius' prologue without specifying that Victorius, in fact, did not follow these limits in his Easter table; thereby, he gives medieval computists, as well as modern commentators, the totally misleading idea that these limits were Victorius' own, whereas Victorius only referred to them as the limits of the *Supputatio Romana*. Concerning the 84 (14), Bede only mentions the lunar limits for Easter Sunday, *luna* 14 to 20 (*DTR* 51.36–44).

113 Besides the three Irish computistical textbooks discussed below, the fundamental texts dealing with these two reckonings are listed in note 102.

114 For previous assessments of Irish computistics of this period cf. Ó Cróinín, 'A seventh-century Irish computus', 99–127; Borst, *Schriften*, 73, 134–8.

riod are known,[115] of which only one, *De ratione conputandi* (*DRC*), has been edited before the publication of the present book.[116] With the edition of the other two, the Munich Computus and the newly discovered *Computus Einsidlensis* (*CE*), this formative period in the history of computistics and its implications will be more readily understood.[117] Among these three Irish texts, the Munich Computus holds a special position, since it is the only one that is securely datable. It therefore must constitute the basis for the contextualisation of the other two texts.[118] Standing at the end of the Easter controversy in the *re-*

115 A note is necessary here on two texts that are often included in the corpus of Irish texts from this period, namely the Bobbio Computus (for its other names cf. p. XXIV above) and the work called the 'Irish computus', *De ratione temporum uel de compoto annali*, or *Sententiae in laude compoti*. The first of these is a computistical anthology compiled in Bobbio in the first quarter of the ninth century; it certainly contains numerous original Irish tracts, but likewise Anglo-Saxon and especially Frankish material, some of which were composed as late as AD 827; therefore, the Bobbio Computus in its entirety cannot be regarded as representative of Irish computistics, nor of Irish computistical knowledge in the seventh and early eighth centuries; it has to be carefully analyzed with the object of disentangling the various strata; for this task, the three computistical Irish textbooks will certainly be of special importance. The second text presents a very similar case; the extent of this text has never been clearly defined, but it is quite obvious that it is a Frankish text based, but not exclusively, on Irish material (cf. note 55); again, first the text, and then the various strata need to be carefully specified. More fruitful would be the reconstruction of the archetype of the text *De divisionibus temporum*, of which a later recension is published in *PL* 90, 653–64; it appears that the archetype of this text dealt exclusively with the fourteen divisions of time; since the first half of the Munich Computus is based on such a text, it is quite obvious that the archetype originated in seventh-century Ireland (for the numerous unedited versions of this text cf. especially Jones, *Bedae pseudepigrapha*, 48–51 and see also note 332). Besides these rather problematic texts, note should be taken of the 'Langobardische Zwiegespräch' (*Dial. Langob.*) just recently published by Borst, *Schriften*, 433–61; Borst, *Schriften*, 424–4 argues that this text was written by an Irishman in Bobbio at around AD 750; the detailed discussion of the indiction in the final chapter of this work (*Dial. Langob.* 26) certainly points to a continental origin of this text; on the other hand, there are numerous parallels between this text and the Munich Computus in otherwise rarely-attested computistical features; moreover, except for the chapter on the indiction, this text appears to reflect nothing but Irish computistical thought of the early eighth century; for these reasons I find it as likely that *Dial. Langob.* was composed in the *regiones Scottorum* early in the eighth century (cf. the more detailed discussion of this text on p. CLXXIV–CLXXIX below). Finally, the present study has unearthed another Irish treatise, to be termed *De comparatione epactarum Dionysii et Victorii* and to be dated to AD 689; it is discussed and edited below p. CLII–CLVIII and Appendix 2 respectively. An Irish computus based on Victorian principles, now lost, was also compiled in this year; cf. p. CXXIV–CXXVI below.

116 Walsh & Ó Cróinín, *Cummian's letter*, 113–213.

117 For the *Computus Einsidlensis* (Einsiedeln, Stiftsbibliothek, 321 (647), p. 82–125) cf. Warntjes, 'Computus Einsidlensis', 61–4; Bisagni & Warntjes, 'Early Old Irish material', 77–105; and p. CXXXIII–CLII below.

118 For an analysis of the chronological order of these three texts, with the securely datable Munich Computus at its heart, see the discussions of the *Computus Einsidlensis* (p. CXXXIII–CLII) and *De ratione conputandi* (p. CXCI–CCI) below; for the date of the Munich Computus itself see p. LVII–LXI below.

giones Scottorum, and having a scholarly rational and rather unbiased focus,[119] these texts, and especially the Munich Computus, provide a unique insight into the older reckonings[120] and the technical questions posed and arguments discussed in the debates of the seventh century.[121] On the other hand, documenting an earlier phase in the reception of the Dionysiac reckoning than Bede, they show a clearer picture of the difficulties faced at the earliest attempts of first analyzing and then reconstructing this new system. In this process, specifically regional customs, to be vaguely connoted as Irish, developed prior to and contemporaneous with Bede's writings. The computistics particularly of the eighth to the tenth centuries depended as much on these Irish customs as on Bedan theories.

119 Scholars of medieval mentalité have quite recently started to analyze scientific attitudes and rationality in computistics as the foremost science of this period. In these studies, the focus in the early middle ages lies exclusively on Bede. This narrow focus leads to the conclusion that early medieval science is dominated by Christian beliefs; a strictly scientific attitude did not exist, but science was rather steeped in theological explanations and justifications; computistics was a means of understanding God's creation, a *doctrina christiana*; cf. Englisch, *Artes liberales*, 301–6, 393, 475–6; Wallis, *Bede*, xxi–viii (conceding p. lxxxviii–xcvi that this attitude changed in Carolingian times); Pillonel-Wyrsch, *Calcul*, 4; Germann, *De temporum ratione*, 41–4. It should be noted, however, that Bede stands at the beginning of (or isolated in?) this attitude towards computistics. Computistics turned into a *doctrina christiana* only at a time when one system for the calculation of Easter was unanimously accepted. Admittedly, the arguments prevailing in the Easter controversy of the seventh century certainly were primarily of a theological and dogmatic nature. On the other hand, this controversy also shaped a scientific mentality that would even meet modern standards, in which differing systems and customs were compared in detail and explained on a purely technical, non-theological level. An outstanding example of this early medieval scientific attitude is the *Computus Einsidlensis*, which certainly deserves a detailed study in this respect, comparing its strictly scientific methodology with Bede's more theological and dogmatic approach.

120 Cf. the passages listed in notes 251–3.

121 In this respect, chapter 52 of the Munich Computus is particularly noteworthy. It shows that followers of the Victorian reckoning must have vehemently criticized the Dionysiac reckoning for not recording accurate data for the *annus passionis* (the year of Christ's passion); in the year regarded as the *annus passionis* by Victorius and his followers, AD 28, the lunar age of Easter Sunday was 15 in the Dionysiac reckoning rather than *luna* 17, which would have been the lunar age suggested by certain authoritative texts and the Synoptic Gospels. Adherents of the Dionysiac reckoning tried to counter this accusation by a technical trick, namely by correlating the epacts of the Dionysiac *ogdoas* with those of the Victorian *hendecas*; this led to the desired data occurring in the second year of this comparison, which is the equivalent to the Victorian *annus passionis*. Only two other texts deal with the same question, namely the rather obscure *Comp. Col.* 5.4–6, and a more detailed and comprehensible, but to the present day unedited and even unnoticed tract in Cologne, Dombibliothek, 83², fol. 176v–178r, here termed *De comparatione epactarum Dionysii et Victorii*, discussed and edited for the first time below on p. CLII–CLVIII and Appendix 2 respectively.

THE MUNICH COMPUTUS IN THE HISTORY OF COMPUTISTICAL TEXTBOOKS

This formative period in the history of medieval computistics, roughly AD 650 to 750, also witnessed the origin of a new literary genre, the computistical textbook. In late antiquity, western Latin computistical writings, as far as can be judged from the surviving texts, consisted almost exclusively of only one type of text (neglecting Easter tables and their prologues here), namely computistical letters discussing the Easter dates of certain years, or more generally the theological arguments for and against certain solar and lunar dates connected to this feast.[122] This situation changed with the computistical works of Dionysius Exiguus, who appended a body of *argumenta*, of mathematical algorithms for the calculation of calendrical data, to his Easter table. Such a strictly defined body of algorithms, often incorporating mathematical tools (like multiplication tables) and additional texts designed for an understanding of these algorithms, can be characterized as a computistical formulary. The origin of the computistical formulary written in Latin, therefore, lies in Dionysius' writings.[123] In the ca. 150 years after its initial composition, this body was only marginally extended. Yet, from the last quarter of the seventh century, this genre started to flourish, primarily due to the acceptance of the Dionysiac reckoning in certain regions. By the eighth century, computistical formularies were well-established, constantly growing in size with markedly differing structures.[124] Parallel to this development was the creation of computistical anthologies, large collections of computistical material of all sorts. It appears that the first such collections were established in seventh-century Spain (though no Spanish anthology of the seventh century has survived), but soon they became the primary means of transmission of computistical texts, formularies included, throughout the Latin West.[125] The computistical textbook, for its part, had a similar, but more specific and structured purpose. It was designed to supply its reader with a comprehensive and well-structured introduction and guide to all aspects deemed necessary for a thorough understanding of this subject.

The first author to assemble in a structured way a great variety of basic but essential information on the reckoning of time in general, and the calculation of

122 Cf. Wallis, *Bede*, xvii. There are two noteworthy exceptions, namely the paschal tract of Anatolius (?) (ed. Mc Carthy & Breen, *De ratione paschali*, 44–53; *DRP* hereafter) and the Carthaginian Computist (ed. Krusch, *Studien* I, 279–97).

123 Dionysius explicitly states that he worked from 'Egyptian' sources when compiling the *argumenta* (Krusch, *Studien* II, 67), so that it can be deduced that computistical formularies written in Greek had a longer tradition. Cf. especially Piper, *Kalendarium*, 144–5; Declercq, 'Dionysius Exiguus', 200–2; and also notes 85, 107. See also Neugebauer's (*Ethiopic astronomy*, 70–6) discussion of a Coptic formulary.

124 For this development cf. especially Warntjes, 'Argumenta'; and furthermore Piper, *Kalendarium*, 145–8.

125 For computistical anthologies cf. Jones, *Bedae opera*, 75–7; Rissel, *Rezeption*, 22–4; Borst, *Schriften*, 146–51; Germann, *De temporum ratione*, 14–5.

Easter in particular, was the Spanish bishop Isidore of Seville († AD 636). Yet, this information was scattered in three sections throughout his great encyclopaedia, the *Etymologiae*. In book three, chapters 24 to 71, he discusses astronomical theory; in book five, chapters 28 to 39, the various divisions of time are outlined; book 6 chapter 17, then, specifically deals with the calculation of Easter. Seventh-century computists were generally not too concerned with astronomical theory. To them, computistics was a mathematical rather than an astronomical science. Basic lunar and solar theory for explaining the structure of the solar and lunar calendar was all they were concerned with in astronomical terms. Astronomy was granted a more important place in computistics only from Bedan times.[126] Accordingly, when computists in the seventh century thought about the creation of a concise computistical textbook, detailed astronomical theory was not part of their consideration. In my opinion, the first computistical textbook consisted merely of a combination of book 5, chapter 28 to 39, and book 6, chapter 17 of Isidore's *Etymologiae*.[127] This combination must, however, soon have been regarded as too basic and not comprehensive enough, so that it was reshaped with additional information from Macrobius, Dionysius Exiguus, Victorius of Aquitaine, Isidore's *De natura rerum*, and other texts.

The exact date of the first of these comprehensive computistical textbooks cannot be determined. The earliest securely datable computistical textbook is Bede's *De temporibus* of AD 703 (*DT*), which was, however, far from comprehensive.[128] Precisely because the scope of this work was felt to be too narrow

126 Astronomical theory like the course of the sun and the moon through the zodiac is not discussed in any of the Irish computistical textbooks of the late seventh and early eighth centuries; Bede introduced this theory into his *De temporum ratione* (c. 16–19), and ever since it held a prominent place in the study of computistical theory (the Bedan chapters, e.g., are cited in *Lib. ann.* 69; *Lib. calc.* 36–38; RM 39–42; PV §§320–336; note that there is no sign of such discussion in Frankish computistics before the second half of the eighth century). Cf. also Eastwood, 'Dungal's letter', 118; McCluskey, 'Astronomies in the Latin West', 144 (arguing that 'geometrical astronomy' (dealing with the course of the sun and the moon through the zodiac and other theories) became an integral part of computistics only from Bedan times onwards).

127 Similarly, Jones, *Bedae opera*, 130–1 argues that Northumbrian schools before the compostion of Bede's *DT* in AD 703 relied predominantly on books 5 and 6 of Isidore's *Etym.* (as well as the texts transmitted in the Sirmond MS) for computistical theory. Problematic for the interpretation outlined above, however, is the fact that *Etym.* 5.28–39 and 6.17 are not transmitted as a separate tract in any MS to my knowledge. Note, however, that Isidore's chapters on the various divisions of time, his chronicle, his chapters on astronomy and on cosmology constitute an independent text in Paris, Bibliothèque Nationale, Lat. 5239, fol. 145r–162r; Paris, Bibliothèque Nationale, Lat. 5543, fol. 148–157v. Moreover, in the Bobbio Computus (*BC*) c. 140, Isidore's discussion of the calculation of Easter (*Etym.* 6.17.10–32, i.e. *Etym.* 6.17 without the history of paschal cycles and the Easter table) is transmitted as a separate text, denoted as a letter of Cyril; strangely enough, neither Jones nor Borst realized its true content: Jones, *Bedae opera*, 97 regards this text as 'a late corruption, probably Irish'; Borst, *Schriften*, 939 as a genuine Cyrillian letter.

128 In the present study, Jones' first edition of Bede's *De temporibusm* is used (Jones, *Bedae opera*, 293–303); this edition is preferable to Jones's second edition in *CCSL* 123C, since

by the brethren of Bede's monastery, he compiled a more comprehensive text, *De temporum ratione* (*DTR*),[129] which was to become the standard work in computistics for centuries. Bede composed this work in AD 725,[130] but six years earlier, in AD 719, an equally comprehensive textbook originated, the Munich Computus. Consequently, the Munich Computus is, contrary to modern belief, the *earliest securely datable comprehensive* computistical textbook known at present, and as such deserves all due attention.[131] The only comprehensive computistical textbooks known at present that may be earlier, but which do not incorporate dating clauses, are the other two Irish texts of that genre, *De ratione conputandi* and the *Computus Einsidlensis*. Both texts can be safely placed in the period between the reception in Ireland of Isidore on the one hand, and that of Bede on the other, that is to say roughly between AD 650 and 750.[132] Moreover, both texts advocate the Dionysiac reckoning, so that it must be presumed that they were not compiled before the unanimous acceptance of that reckoning at leastin certain parts of the *regiones Scottorum*. Consequently, it seems unlikely to me that they were compiled pre-ca. AD 670.[133] Whether or not they were composed at any stage prior to the Munich Computus can only be determined by a detailed comparison of these texts, which is provided further below, the results of which may be briefly summarized here: When the textual details of the *Computus Einsidlensis* and *De ratione conputandi* are compared

it has better readings and provides valuable commentary; but note that only the *CCSL* edition includes the chronicle and glosses. In this work, Bede explicitly refers to AD 703 as his *annus praesens* in the computistical algorithms of chapter 14.

129 In the present study, Jones' first edition of Bede's *De temporum ratione* is used (Jones, *Bedae opera*, 173–291; this edition is preferable to Jones' second edition in *CCSL* 123B, since it has better readings and provides valuable commentary; but note that only the *CCSL* edition includes the chronicle and glosses). Bede outlines the reasons for compiling *De temporum ratione* in the preface to this work. For commentaries on this passage and Bede's reasons for writing this textbook cf. Wallis, *Bede*, 253–4; Pillonel-Wyrsch, *Calcul*, 6–13.

130 Bede mentions AD 725 twice as the *annus praesens* in his work: *DTR* 49.4, 52.3; the date is given once without reference to the *annus praesens*: *DTR* 54.

131 For the false belief that Bede's *De temporum ratione* is the first comprehensive treatment of computistics cf. Cantor, *Vorlesungen*, 828; Rissel, *Rezeption*, 26–7; Englisch, *Artes liberales*, 475; Wallis, *Bede*, xvi–vii; Germann, *De temporum ratione*, 45, 78.

132 The reception of Isidore in Ireland is heavily disputed: AD 650 constitutes the earliest possible date, and therefore this is chosen here; cf. Anspach, 'Fortleben', 327–9, 337–8 (dating the first citation of Isidore's work in Irish texts to the first half of the seventh century); Bischoff, 'Europäische Verbreitung', 180–3 (certainly late seventh century); Löfstedt, *Malsachanus*, 50–1 (mid-seventh century); Smyth, 'Isidore of Seville', 69–102 (late seventh century); Herren, 'Earliest Irish acquaintance', 243–50 (mid-seventh century); Ó Cróinín, *Early medieval Ireland*, 213–4; idem, 'Hiberno-Latin literature', 390 (mid-seventh century). In the end, the Isidorian reception in Ireland depends very much on the dating of the Irish computistical texts here discussed. The reception of Bede's computistical works is equally difficult to establish; AD 750, 25 years after his composition of *DTR*, only serves as a rough guide here.

133 For the acceptance of the Dionysiac reckoning in Ireland cf. p. XXXIX–XL, LXXXIII–LXXXIX, XCIII–XCV, CLVI–CLVIII.

to those of the Munich Computus, it transpires that the *Computus Einsidlensis* must have been written before the Munich text, i.e. pre-AD 719, while *De ratione conputandi* shows dependencies on both of these two texts and therefore appears to have been compiled post-AD 719. Further analysis of these two texts then suggests that the *Computus Einsidlensis* was composed ca. AD 689–719, *De ratione conputandi* ca. AD 719–727.[134] Accordingly, my reading of the evidence is that these three Irish computistical textbooks originated roughly in the period AD 689–727, the chronological order being: *Computus Einsidlensis*, Munich Computus, *De ratione conputandi*.

The concept of a computistical textbook, which originated in this Insular context of the late seventh century, certainly shaped the study of that science in the early middle ages, and even thereafter. The earliest Frankish computistical texts, written before 750, appeared in this or related formats, though the structure may have been different and at times it is debatable whether these texts should be considered as comprehensive texbooks or rather treatises dealing with specific chronological issues.[135] In the ninth century, Pacificus of Verona, Rabanus Maurus, and, most influentially, Helperic of Auxerre at the end of that century, decided to arrange computistical knowledge in textbooks.[136] The tenth

134 See the detailed analyses p. CXXXIII–CLII and CXCI–CCI below. For the dating of the *Computus Einsidlensis* see now also Bisagni & Warntjes, 'Early Old Irish material', 81–91 (an earlier view, superseded by the just mentioned article, is presented in Warntjes, 'Computus Einsidlensis', 62–3). Prior to the present study, Ó Cróinín dated *De ratione conputandi* to ca. AD 650 (Ó Cróinín, 'A seventh century Irish computus', 121), and this dating is generally accepted (not, however, by Smyth, 'Isidore of Seville', 94, 100; eadem, *Understanding the universe*, 156). The main argument for this dating is that three sources used by this computist, namely Jerome's *Commentarium in Aggaeum*, Ambrosiaster's *Liber Quaestionum*, and an obscure tract attributed to Origen, are only cited in one other text, Cummian's letter (Ó Cróinín, 'A seventh century Irish computus', 111–3); the inference is that place and time of *De ratione conputandi* and Cummian's letter must have been reasonably close; this, however, does not exclude the possibility that *De ratione conputandi* was indeed written in the same monastic school as Cummian's letter by a person taught in the same tradition as Cummian and using the same library, but some decades after Cummian's death. It should be noted here that the citation from ps-Origen also occurs in Bede, *In Marci evangelium* 4.14 (*CCSL* 120, 604–5; cited by Sedulius Scottus, *In Mat-thaeum* (Löfstedt, *Sedulius Scottus*, 560)); *In Lucae evangelium* 6.22 (*CCSL* 120, 373); *Homiliae subdititiae, Homilia 53* (*PL* 94, 389C); ps-Bede, *De officiis* (*PL* 94, 536D); Smaragdus, *Collectiones* (*PL* 102, 174D–175A); Haymo of Halberstadt, *Homiliae de tempore, Homilia 66* (*PL* 118, 393A).

135 'Das burgundische Lehrgespräch von 727' (*Dial. Burg.*) (Borst, *Schriften*, 353–74); 'Das neustrische Streitgespräch von 737' (*Dial. Neustr.*) (Borst, *Schriften*, 381–423); 'Das langobardische Zwiegespräch um 750' (*Dial. Langob.*) (Borst, *Schriften*, 433–61); for the possibility that the last-mentioned text was composed in Ireland before AD 750 cf. note 115 and especially p. CLXXIV–CLXXIX below; for its structure and character see p. CXI below.

136 Pacificus' *Computus* is edited in Meersseman & Adda, *Manuale di computo*, 82–137; Rabanus' *De computo* by Stevens in *CCCM* 44, 199–321; Helperic's *De computo* in *PL* 137, 17–48, which simply is a reprint of Pez's early 18th-century edition from a single, rather late MS witness (of more than 80 known today); a critical modern edition of this most influential computistical text is desperately needed. For the role of these three text-

century was dominated by Helperic's work,[137] while new computistical studies of that century, by Notker Labeo and others, were of a more specific, treatise-like character dealing with some specific computistical aspects rather than constituting comprehensive textbooks. When the genre of the textbook became more popular again in the eleventh century, the character of these was fundamentally different compared to early medieval textbooks, because of the more general change in science due to contacts with the Arabic world. Any study of comprehensive computistical textbooks in the early medieval period, i.e. from the origin of this type of text in the late seventh century to the reception of Helperic's work in the tenth, must, quite naturally, start with its earliest dateable witness, the Munich Computus.

books in the computistical discourse of the ninth century cf. also Wiesenbach, *Sigebert von Gembloux*, 55–7, as well as p. CX–CXII below.

137 For this text and its various later recensions cf. especially Traube, 'Computus Helperici', 128–52; for a classification of the English manuscripts McGurk, 'Computus Helperici', 1–5.

THE HISTORY OF THE MUNICH COMPUTUS

THE DATE

The Munich Computus was first dated by Bruno Krusch to AD 689 on the basis of two passages mentioning the year of the consuls *Bero et Barbua*; in one of these instances this year is explicitly referred to as *annus praesens*.[138] Krusch correctly identified this year as the 130th year of the Victorian Easter table, under which the consuls Verus II and Bradua are listed.[139] Victorius designed his

138 MC 41.107–110 (*annus praesens*), 62.65–67. Krusch, *Studien* I, 10; Krusch, 'Einführung', 162–3 repeats AD 689 as the date of composition, as does Mommsen in *MGH AA* 9, 34, 696–7. For discussions of this dating clause see also Mac Carthy, *Annals of Ulster* 4, lxx–i; Schwartz, 'Ostertafeln', 89–91, who then correctly argue that AD 689 should not be regarded as the date of composition of the Munich Comutus, but of an older layer in this text. Note that Krusch later, in 1938 (*Studien* II, 58), accepted Schwartz's dating of this text to AD 719. Nevertheless, Jones (*Bedae opera*, 110) repeats Krusch's original dating of AD 689, even though he had earlier ('Sirmond manuscript', 209) accepted Mac Carthy's date of AD 718, and was even more accurate in *Bedae pseudepigrapha*, 125, giving AD 718 as the date of composition of this computus, AD 689 as the date of one of its exemplars (but in *Bedae pseudepigrapha*, 48 he argues that the *manuscript* derives from an exemplar of AD 718, while some part of it was written in AD 689; on p. 67 he also only mentions the AD 689 date; note that Jones indicated that he had not consulted the MS himself); even less understandable is the reference to AD 689 as the date of composition of the Munich Computus in Thorndike & Kibre, *Catalogue of incipits*, 244 (here, that date is followed by the date of the MS, AD 817–24) as well as in the 2003 publication Machielsen, *CCSL Clavis Patristica* 3A, 188, 287. In all other studies subsequent to Mac Carthy and Schwartz, the AD 689 date is only mentioned (and rightly so) as the date of an older layer within the Munich Computus; cf. Kenney, *Sources*, 223; O'Connell, 'Easter cycles', 84; Gentz, 'Ostern', 1652 (wrongly associating this layer with the *latercus*); Schäferdiek, 'Osterzyklus', 361, 377; Lapidge & Sharpe, *Bibliography*, 95; Ó Cróinín, 'A seventh-century Irish computus', 102; idem, 'Old Irish gloss', 131; idem, 'Virgilius Maro Grammaticus', 197; idem, 'Columbanus', 52; idem, 'Earliest Old Irish glosses', 16; Mc Carthy & Ó Cróinín, 'Irish 84-year Easter table', 66; Ohashi, 'Sexta aetas', 59; Warntjes, '84 (14)-year Easter reckoning', 61, 74, 77; Borst, *Schriften*, 1.

139 Among the Victorian Easter tables known at present, only three contain the full consul list (cf. Mommsen in *MGH AA* 9, 672–4; Krusch, *Studien* II, 6–8): Gotha, Forschungsbibliothek, 75, fol. 77v–106r; Leiden, Universiteitsbibliotheek, Scaliger 28, fol. 3r–21r; and the Sirmond manuscript, which was identified with Oxford, Bodleian Library, Bodley 309 (the Victorian Easter table on fol. 113r–120r) by Jones ('Sirmond manuscript') at the same time as Krusch published his edition of Victorius' computistica, and was therefore unknown to the German scholar. The Victorian Easter table in the Bobbio Computus (Milan, Biblioteca Ambrosiana, H 150 inf, fol. 130r–132r) also includes the consul list, but only lists the years AP 1–120, 144–154 (these final eleven years being denoted as AP 121–131) and therefore does not transmit the year in question here. The consuls of the year in question are spelled *Vero II et Bradua* in both the Gotha MS, fol. 84v, and the Leiden MS, fol. 6r (cf. *MGH AA* 9, 696; Krusch, *Studien* II, 33), but *Ivro et Bardua* in the Sirmond MS, fol.

532-year Easter table in such a way that it began with the year of the passion of Christ (*annus passionis*=AP) rather than with the year of the incarnation (AD), which was introduced into western Easter tables by Dionysius Exiguus only some seventy years after Victorius' composition of AD 457. The construction is such that AP 1=AD 28.[140] Therefore, the year in question is the 532-year cyclic equivalent to AD 157, i.e. AD 157, 689, 1221, and so on. Since the Victorian Easter table did not exist in AD 157 and the manuscript incorporating the Munich Computus was written in the early ninth century, there cannot be any doubt that AD 689 was meant by the reference to *Bero et Barbua*. Both passages give more chronological details concerning this year, but these details are to be reconstructed from the established fact that AD 689 is the year in question, rather than that they would provide independent evidence.

It was, however, pointed out by Mac Carthy and four years later by Schwartz that the Munich Computus is a compilation with two datable layers. The first of these is the one established by Krusch, to be dated, as we have seen, to AD 689. It is worth noting here that the dating clause itself reveals a Victorian bias of this layer, so that the later author of the Munich Computus obviously incorporated excerpts from a Victorian computus of AD 689 into his textbook.[141] The second dating clause, then, refers to a later date, which undoubtedly reflects the date of composition of this text. Since Mac Carthy and Schwartz slightly differ in their interpretations of the passage in question (the Irish scholar identifying AD 718 as the year of composition, the German AD 719),[142] it is worth discussing it in detail here.

115r. Because of the correspondence between MC and the Sirmond MS in the spelling of the second name, it seems that the author of the Munich computist's exemplar of AD 689 worked from a Victorian Easter table similar to the one that survives in the Sirmond MS, possibly its exemplar (in which the corruption of the first name to *Ivro* had not yet occurred). Note that Bucherius, in his edition of the Victorian Easter table (*De doctrina temporum*, 27), which was based on the Sirmond MS (cf. Mommsen in *MGH AA* 9, 673, 676; Krusch, *Studien* I, 210; idem, *Studien* II, 8), gives the same reading as the Gotha and Leiden MSS; therefore, Mommsen and Krusch, taking Bucherius' edition as witness for the Sirmond MS, did not note any variant for the Sirmond MS in their editions; Mommsen (*MGH AA* 9, 696–7), however, with his characteristic thoroughness, listed the variants from MC.

140 For the construction of the Victorian Easter table cf. the literature cited in note 82.

141 For the Victorian layer of AD 689 underlying parts of the Munich Computus cf. Schwartz, 'Ostertafeln', 102; Kenney, *Sources*, 223; Warntjes, '84 (14)-year Easter reckoning', 61, 74. See also p. LXXX–LXXXII and especially CXXIV–CXXVI.

142 Mac Carthy, *Annals of Ulster* 4, lxx; Schwartz, 'Ostertafeln', 91. Mac Carthy's dating is accepted by Kenney, *Sources*, 223; O'Connell, 'Easter cycles', 84 (arguing that Schwartz advocates the same date); Jones, 'Sirmond manuscript', 209; idem, *Bedae pseudepigrapha*, 125; Bischoff, 'Das griechische Element', 250 (through Kenney); Harrison, 'Luni-solar cycles', 73; Lapidge & Sharpe, *Bibliography*, 95; Ó Cróinín, 'A seventh-century Irish computus', 102–3, 126 (attributing this date to Schwartz); idem, 'Old Irish gloss', 131; idem, 'Virgilius Maro Grammaticus', 197; idem, *Early medieval Ireland*, 188; idem, 'Earliest Old Irish glosses', 16; idem, 'Columbanus', 52; Mc Carthy & Ó Cróinín, 'Easter table', 60, 66 (arguing that Schwartz advocates the same date); Mc Carthy, 'Easter principles', 221; Borst, *Kalenderreform*, 187; Ohashi, 'Sexta aetas', 59; Lejbowicz, 'Tables pas-

Chapter 56 outlines a method for determining the Julian calendar date and the lunar age of Easter Sunday:[143] The preconditions are the Julian calendar date of the Easter full moon (which is listed for every year of the 19-year cycle in the passage immediately preceding this calculation and then calculated from the lunar age of 1 January in the immediately following passage[144]) and the weekday of 1 January; with this information the weekday of the Easter full moon is calculated; the Julian calendar date and lunar age of Easter Sunday is, then, determined by simply counting forward from the weekday of the Easter full moon to the following Sunday. This calculation is illustrated by three examples, which contain the following chronological data:

example	bissextile or not	weekday of 1 January	Julian calendar date of Easter full moon	Julian calendar date of Easter Sunday	lunar age of Easter Sunday
1st year	non-bissextile	*dominicum* (Sunday)	*v Idus Aprilis* (9 April)	*xvi Kalendas Maii* (16 April)	21
2nd year	bissextile	*ii feria* (Monday)	*iiii Kalendas* Aprilis (29 March)	*ii Kalendas Aprilis* (31 March)	16
3rd year	non-bissextile	*iiii feria* (Wednesday)	*xv Kalendas Maii* (17 April)	*xii Kalendas Maii* (20 April)	17

This information is, of course, more than enough to date this sequence of years: The second and third columns alone identify these years as the 28th, 1st and 2nd year of the solar cycle, the fourth column as the 17th, 18th, and 19th year of the Dionysiac 19-year cycle respectively.[145] Such a combination is obviously

chales', 21. Schwartz is followed by Gentz, 'Ostern', 1652; Cordoliani, 'Traités', 59; Warntjes, '84 (14)-year Easter reckoning', 33; idem, 'Computus Einsidlensis', 63 (the date is incorrectly given as AD 919 instead of 719 due to printing problems); Borst, *Schriften*, 1, 261, 1022. Note especially that Krusch himself (*Studien* II, 58) later accepted Schwartz's dating. Characteristically careful and accurate is Schäferdiek, 'Osterzyklus', 360–1, stating that the text was arguably written 718 / early 719, though on p. 378 he only refers to AD 719.

143 MC 56.15–43.

144 MC 55.61–68, 56.44–57.

145 The solar cycle is designed in a way that its first year, corresponding to 9 BC and recurring in 28-year intervals, is bissextile with 1 January falling on a Monday. Since the second example has these criteria, it agrees with the first year of the solar cycle. For the solar cycle cf. Ideler, *Handbuch* 2, 185–9; Rühl, *Chronologie*, 63–72; Wislicenus, *Kalender*, 34–6; Ginzel, *Chronologie*, 124–34; von den Brincken, *Chronologie*, 57–8. In all these accounts of medieval chronology, the years of the solar cycle are characterized by dominical letters. Yet, these letters do not become standard before the early eleventh century, so that the calculation with them is quite anachronistic for the period in question here. In the early middle ages, the solar characteristic for a year was the weekday of 24 March (*concurrentes*) or

unique in a 532 (28×19)-year luni-solar cycle (since 28 and 19 do not have a common divisor other than one), so that these three examples undoubtedly refer to AD 719, 720 and 721.[146] The first of these examples, i.e. AD 719, is denoted in this passage as *annus inminens*, literally the imminent year.[147] It seems, then, that Mac Carthy thought in terms of calendar years beginning with 1 January, so that in his opinion the *annus inminens* started on 1 January 719, the author consequently writing in AD 718; Schwartz, however, when arguing that the Munich Computus was composed immediately before Easter of AD 719, appears to have rather thought in terms of Dionysiac lunar years extending from *luna* 15 of the paschal month to the Easter full moon of the following calendar year.[148] The fact that the computist's reference to the imminent year follows immediately after the description of the method for calculating the data of Easter Sunday certainly suggests that it was the Easter Sunday in question (i.e. of AD 719) that was imminent rather than the year itself. Therefore, I am inclined to follow Schwartz's argument; with certainty, however, it can only be argued that the Munich Computus was composed after Easter Sunday of AD 718 and before that of AD 719.

The exact same dating clause recurs slightly later in the text, in a passage that outlines a method for the calculation of the Julian calendar date and the

1 January, and it is therefore helpful that not only the dominical letter, but also the *concurrentes* is given for each year of the solar cycle in Rühl, *Chronologie*, 142, 144–5; Ginzel, *Chronologie*, 144; Blackburn & Holford-Strevens, *Companion to the year*, 821. The translation of the weekday of 1 January to the *concurrentes* is achieved by subtracting two in common, one in bissextile years; if the result is 0 or negative, seven has to be added. Note that in pre-Bedan Irish computistics the solar characteristic of a year always was the weekday of 1 January, and all calculations were executed from that datum. In fact, in the three Irish computistical textbooks, the *concurrentes* are discussed only once (MC 67), but without using the technical term (cf. note 188). As for the Dionysiac 19-year cycle (*cyclus decemnovennalis*), a list of the 19 full moons of that cycle is given in the Easter table composed by Dionysius himself (cf. Krusch, *Studien* II, 69–74), and they are also separately listed in MC 55.61–68, as in many other computistical texts (for such a list in modern literature on chronology cf. Ideler, *Handbuch* 2, 199; Rühl, *Chronologie*, 153; Blackburn & Holford-Strevens, *Companion to the year*, 821). The Easter full moons mentioned in the three examples occur in 17th to 19th place there. Generally, the first year of the Dionysiac 19-year cycle corresponds to 1 BC and recurs in 19-year intervals. For this cycle cf. Ideler, *Handbuch* 2, 197–9; Rühl, *Chronologie*, 133–42; Wislicenus, *Kalender*, 45–6; Ginzel, *Chronologie*, 134–43 (note again that these accounts are based on later medieval practice).

146 In the incarnation era, the first correspondence of the 17th year of the Dionysiac 19-year cycle with the 28th year of the solar cycle is AD 187, and it then recurs every 532 years. Since AD 187 is obviously too early for the composition of this text, and since the MS was written in the early ninth century, the only possible date for the *annus inminens* is AD 719.

147 MC 56.19.

148 Note that both interpretations can be supported by the evidence of MC: On the one hand, a year is generally characterized by the weekday and lunar age of 1 January throughout this text, and all calculations are based on that data (cf. MC 29.2–12, 36.2 (and cross-references there); 49.14–32, 50, 52, 56.15–55, 58.33–74, 61.11–27); on the other hand a lunar year is defined as extending from the Easter month of one Julian calendar year to the Easter month of the following one (cf. MC 36.3–5, 53.5–11; and the cross-references in these two passages).

lunar age of the *initium quadragesimae*, i.e. the beginning of the Lenten period.[149] Because of the proximity within the text of this passage to the just discussed calculation of Easter Sunday, as well as the similarity in the methods applied, the computist used the same three years as examples here, though this time without any reference to an *annus praesens* or *annus inminens*. In consequence, the chronological data supplied in this second passage does not differ from the first passage.

All other technical examples outlined in this computus are evidently chosen for reasons other than being the *annus praesens* of the Munich computist or any of his sources, and therefore they do not provide any additional evidence for possible dating-clauses.[150]

149 MC 58.33–80. Cf. Schwartz, 'Ostertafeln', 91.

150 In chapter 50, the years chosen are the first two years of the Dionysiac 19-year cycle, the Victorian and alleged *latercus* equivalents to that first year, as well as exceptions to the rule outlined in that chapter (cf. Warntjes, '84 (14)-year Easter reckoning', 46–51); in chapters 51 and 59, the three groups of three examples each refer to the first three years of the Dionysiac 19-year cycle; the first years of a certain cycle were generally chosen as fitting theoretical examples rather than references to the time of composition.

THE AUTHOR

As is the case with most early medieval computistical texts, the author of the Munich Computus remains anonymous. Nevertheless, the text itself provides enough indications for identifying at least the nationality of its author. In fact, the evidence for an Irish authorship of this text is so overwhelming that it has never been questioned since Bruno Krusch introduced this computus into modern scholarship. Krusch himself had discovered an explicit reference to the *Scotti* in the text, which led him to the assumption that it was composed in Britain;[151] in the imperial usage of his time, the term Britain quite naturally referred to the British Isles. This identification was further specified by Mac Carthy, who left no room for doubt that the author himself was a *Scottus*, an Irishman. His analysis, the most thorough investigation of the authorship of this text to date, focused on two criteria, language and sources. Concerning the language of the author, Mac Carthy pointed to the use of two extraordinary terms: 1) *singularis*, which he regarded as the Hiberno-Latin equivalent to Old Irish *uathad* (oneness or singularity), in context to be translated as 'single digit' (i.e. denoting numbers from one to nine);[152] 2) *dies cetene*, which he identified as a bilingual term consisting of Latin *dies* and the Old Irish genitive singular *cétaíne*; it appears in the text with the meaning of Wednesday, being the equivalent to Latin *dies Mercurii* (literally 'the day of Mercury') and to Old Irish *día cétaíne* (literally 'the day of the first fast').[153] Additionally, Mac Carthy listed orthographic features that he believed to have originated from Irish phonetics.[154] Concerning sources, Mac Carthy drew attention to the fact that three texts which had been identified as Irish forgeries were cited by the Munich computist, namely the so-called Acts of the Council of Ceasarea, the paschal tract of Anatolius (?) (*De ratione paschali*), and the letter of ps-Cyril (*Epistola Cyrilli*), as well as one other evidently Irish text, *De mirabilibus sacrae scripturae*.[155] All subsequent studies of the Munich Computus, from Schwartz to O'Connell to Schäferdiek, provided no additional evidence on this matter.[156] In fact, since Mac Carthy's analysis no need for further investigation of this question was felt. Nevertheless, in the 1980s, Dáibhí Ó Cróinín discovered two more elements that highlight the Irish authorship of this text, one in each of the two categories outlined by Mac Carthy. Concerning the language of the author, Ó Cróinín pointed out that the Munich computist used the Old Irish verb *to·mel* (Classical

151 Krusch, *Studien* I, 13; idem, 'Einführung', 162–3. This was accepted by Mommsen in *MGH AA* 9, 34.

152 Mac Carthy, *Annals of Ulster* 4, lxix.

153 Mac Carthy, *Annals of Ulster* 4, clxxx. Cf. p. XIX and especially the discussion on p. LXXV–LXXVI.

154 Mac Carthy, *Annals of Ulster* 4, clxxviii–ix.

155 Mac Carthy, *Annals of Ulster* 4, lxxix–lxxx. For the 'Irish forgeries' cf. note 162. For *De mirabilibus* cf. p. LXXVIII–LXXX.

156 Cf. Schwartz, 'Ostertafeln', 92, 100–1; Krusch, *Studien* II, 58; O'Connell, 'Easter cycles', 84; Schäferdiek, 'Osterzyklus', 361.

Old Irish *do·meil*, literally 'consumes', 'uses up') instead of a Latin equivalent (like *consumare* or *uti*) in the main body of the text, which clearly indicates that the author must have been an Irishman writing for an Irish audience.[157] Concerning sources, Ó Cróinín identified a phrase attributed to *Iohannes consilarius* with a fragment of the papal letter to the Northern Irish clergy of AD 640, which would not have had a wide circulation outside of the papal curia and the territory of the addressees.[158]

In general, therefore, there are various criteria to narrow down the identity of an author, some more convincing than others. In the following analysis, any discussion of the author's Latin is omitted. The study of Hiberno-Latin still stands very much at its beginning,[159] and the value of the Munich Computus rather lies in providing another source for the investigation of the Latin of an early eighth-century Irish author, after this authorship has been identified beyond any reasonable doubt. The Irish authorship of the Munich Computus will be discussed in different categories in the following, which are listed in descending order according to their significance for this question (from the least to the most significant).

Script and abbreviations: Almost every early medieval text is only transmitted through later copies, not through the original hand of its author. Therefore, the script of these copies hardly ever provides any indication about the author. The Munich Computus is a good example of this, since it can be dated to AD 719, and was evidently written by an Irishman (as will be seen presently), but it is transmitted in an early ninth-century manuscript from Regensburg, written in Carolingian script. However, despite a change in script according to regional customs, later copyists sometimes preserved the original and distinctive abbreviations of the original text.[160] Unfortunately, this is not the case with the Munich Computus, in which all abbreviations meet Carolingian standards. Yet, some readings in the Munich text can only be explained as confusions of certain textual elements with Insular abbreviations, or as misinterpretations of such abbreviations by continental copyists.[161] Therefore, it is obvious that the Munich Computus contained Insular abbreviations at some point in time. Whether

157 The reference in question was first published in Ó Cróinín, 'Old Irish gloss', 131. Cf. also Ó Cróinín, 'Earliest Old Irish glosses', 17; and p. XIX, LXXV.

158 This source was first identified in Ó Cróinín, 'A seventh-century Irish computus', 103–4, 126. Cf. also Ó Cróinín, 'New heresy', 89, 93–4; and p. LXXXII–LXXXVII.

159 Cf. note 32.

160 The other two textbooks in the corpus of Irish pre-Bedan computistica, *De ratione conputandi* and the *Computus Einsidlensis*, still show specifically Irish abbreviations in their ninth-century continental copies; cf. Ó Cróinín, 'A seventh-century Irish computus', 101; Warntjes, 'Computus Einsidlensis', 62.

161 Cf. the *apparatus criticus* in the edition (especially noteworthy is the confusion of the numerals *l* and *i* with the Insular abbreviations *ł* and *.i.* for *vel* and *id est* in MC 12.36, 51.32, 58.59, as well as the misinterpretation of the Insular abbreviation for *autem* as *hoc* in MC 38.37).

this Insular phase stands at the beginning or in an intermediary position in the transmission of the text can obviously not be determined by these features.

Sources: Among the internal evidences, the sources used by the author are the most controversial aspect in determining his identity. There are two main approaches, namely the identification of a) a group of texts predominantly read and cited in a certain region, or b) individual characteristic sources that would place the text into a more local context. Both approaches pose difficulties, since the identification of a body of sources that was primarily read in a certain region, as well as of a regional origin of individual sources, are quite often problematic, the arguments equally often circular. In the case of the Munich Computus, a body of sources that has been referred to, and that illustrates the problems of both approaches, are the so-called 'Irish forgeries',[162] which consist of Anatolius (?)' *De ratione paschali*,[163] ps-Cyril's letter (*Epistola Cyrilli*),[164] Mori-

162 The grouping of these texts under the generic term 'Irish forgery' developed in the following stages: Mac Carthy, *Annals of Ulster* 4, cxv–xxvii discussed the so-called Acts of the Council of Caesarea, ps-Athanasius' and ps-Anatolius' (as he thought) tracts as native (i.e. Irish) fabrications, and then referred to ps-Cyril's letter as another Irish fabrication later in his work (Mac Carthy, *Annals of Ulster* 4, cxxiv–v). Kenney, *Sources*, 217 then listed the same four works in the same order under the heading 'Paschal fabrications'; apparently, he overlooked Mac Carthy's reference to 'the Irish fictitious Morianus' (Mac Carthy, *Annals of Ulster* 4, cxl). O'Connell, 'Easter cycles', 77–9 then added this text attributed to a certain Morinus to the list, accepting the forgery theory for all of them. Jones, 'Sirmond manuscript', 216; idem, *Bedae opera*, 34 consequently referred to this group of texts, but made immediately clear that he only considered the texts attributed to Morinus and Anatholius as Irish forgeries. Almost at the same time and independent from Jones, Cordoliani discussed all five texts in what must be regarded as the seminal article on 'Irish forgeries' (Cordoliani, 'Computistes insulaires', 5–34). Ó Cróinín, 'Hiberno-Latin literature', 375 also refers to this group of texts, but stresses that the classification as 'Irish forgeries' is wrong and misleading. Wallis, *Bede*, lxxv–vi, in her description of the Sirmond MS, lists only *De ratione paschali*, the *Disputatio Morini*, and the Acts of the Council of Ceasarea under the heading 'Irish forgeries' for the simple reason that she was only interested in dividing the contents of this manuscript into suggestive groups of texts; yet, the reason why she excluded the *Epistola Cyrilli*, which immediately precedes this group in the MS, is not clear; the *Tractatus Adthanasi* is consequently not included, as it does not appear in this MS.

163 *De ratione paschali* is edited by Bucherius, *De doctrina temporum*, 439–49 (paralleling the passage found in Eusebius Greek *Historia Ecclesiastica*; repr. in *PG* 10, 209–22); Krusch, *Studien* I, 316–27; and most recently in the critical edition by Mc Carthy & Breen, *De ratione paschali*, 44–53.

164 The *Epistola Cyrilli* is edited by Petavius, *De doctrina temporum* 2, 503–4 (with notes p. 508); Bucherius, *De doctrina temporum*, 72–4; Krusch, *Studien* I, 344–9. This letter consists of a genuinely Cyrillian but slightly altered first part, followed by the main part, added in AD 607. It probably is the least clear, and therefore most controversial, of the 'Irish forgeries'. Its manuscript tradition shows strong Insular connections. For this text cf. Bucherius, *De doctrina temporum*, 71–2; van der Hagen, *Observationes in veterum patrum prologos et epistolas*, 92–110; Krusch, *Studien* I, 101–9; Mac Carthy, *Annals of Ulster* 4, cxxxiv–v; Kenney, *Sources*, 217; O'Connell, 'Easter cycles', 79; Jones, 'Sirmond manuscript', 215–6; idem, *Bedae opera*, 93–7; Cordoliani, 'Computistes insulaires', 25–8; idem,

nus' *Disputatio* (*Disputatio Morini*),[165] ps-Athanasius' Easter treatise (*Tractatus Adthanasi*),[166] and the so-called Acts of the Council of Caesarea.[167] Taking the first-mentioned tract as an example, it was for a long time assumed that this text was not an original work of Anatolius of Laodicaea, but rather an Irish

'Traités', 60; Grosjean, 'Recherches', 226–8, 238–9; Stevens, 'Scientific instruction', 91, 94; Harrison, 'Luni-solar cycle', 74; Wallis, *Bede*, lix–lx.

165 The *Disputatio Morini* is edited by Du Cange, *Chronicon paschale*, 480–1 (under the title *De paschate Judæorum*); Pitra, *Spicilegium Solesmense* 1, 14–5; as part of the Bobbio Computus in *PL* 129, 1357–8 (a reprint of Muratori's 18th-century edition); by Cordoliani, 'Computistes insulaires', 30–4; and most recently by Graff, 'Recension of two Sirmond texts', 140–2. This work is referred to by name in Cummian's letter (*Epistola Cummiani* l. 213; Walsh & Ó Cróinín, *Cummian's letter*, 86). Its manuscript tradition shows strong Insular connections. For this text cf. especially Strobel, *Texte*, 116, 118–20; Graff, 'Recension of two Sirmond texts', 112–3, 125–36; and furthermore van der Hagen, *Dissertationes*, 165–6; Pitra, *Spicilegium Solesmense* 1, XII–IV; Mac Carthy, *Annals of Ulster* 4, cxl; O'Connell, 'Easter cycles', 79; Jones, 'Sirmond manuscript', 216; idem, *Bedae opera*, 34, 97–8; Cordoliani, 'Computistes insulaires', 28–30; idem, 'Traités', 64; Grosjean, 'Recherches', 225–6, 238–9; Wallis, *Bede*, lxxv–vi.

166 The *Tractatus Adthanasi* is edited among the works of Athanasius in *PG* 28, 1605–10 (a reprint from Montfaucon's 18th-century edition; Migne draws attention to the fact that this work also appears in the Bobbio Computus and regarded it as spurious); and as ps-Athanasius in Krusch, *Studien* I, 329–36. Barlow, *Opera omnia*, 265–7 has shown that this text is a recension of Martin of Braga's *De pascha*. The manuscript tradition of this recension shows strong Insular connections. For both texts cf. van der Hagen, *Observationes in laterculum paschalem centum annorum*, 328–36; Krusch, *Studien* I, 328–9; Mac Carthy, *Annals of Ulster* 4, cvii–viii; Kenney, *Sources*, 217; O'Connell, 'Easter cycles', 78; Jones, 'Sirmond manuscript', 212–3; idem, *Bedae opera*, 51–2; Cordoliani, 'Computistes insulaires', 21–4; idem, 'Traités', 54; Barlow, *Opera omnia*, 259–75; David, 'Saint Martin', 283–4, 289–99; Strobel, *Texte*, 110–15; Ó Cróinín, 'Echternach manuscript fragments', 135, 141 (drawing attention to the fact that this text is attributed to Sulpicius Severus in an early Echternach fragment); Stevens, 'Scientific instruction', 91–2.

167 The so-called Acts of the Council of Caesarea exist in four recensions. Recension A is edited by Baluze, *Nova collectio conciliorum*, 13–5 (which I have not seen; cited from Cordoliani, 'Traités', 57); recension B is edited in *PL* 90, 607–10 (a reprint of Noviamagnus 16th-century edition) under the title *De ordinatione feriarum paschalium* and as *Epistola Philippi de pascha* by Bucherius, *De doctrina temporum*, 469–71; recension C, which is a combination of the first two, is edited as part of the Bobbio Computus (*PL* 129, 1350–3, a reprint of Muratori's 18th-century edition); and separately by Krusch, *Studien* I, 306–10, reprinted in Wilmart, 'Nouveau texte', 20–7; recension D, which is close to A, is edited by Wilmart, 'Nouveau texte', 20–7. The recension of interest for the present study is B (*De ordinatione feriarum paschalium*), since it is this version that is cited in MC 11.37–40; 55.13–27; 63.3–23; as well as *DTR* 47.87–94; note that this recension consequently was certainly composed before AD 719. For this text and its recensions cf. van der Hagen, *Observationes in laterculum paschalem centum annorum*, 328–36; Krusch, *Studien* I, 303–6; Mac Carthy, *Annals of Ulster* 4, cxv–vii; Wilmart, 'Nouveau texte', 19–20; Kenney, *Sources*, 217; O'Connell, 'Easter cycles', 77–8; Cordoliani, 'Computistes insulaires', 24; idem, 'Traités', 57–8; idem, 'Manuscrit de comput ecclesiastique', 31–2; Jones, 'Sirmond manuscript', 216; idem, *Bedae pseudepigrapha*, 44–5; idem, *Bedae opera*, 52, 87–9; Strobel, *Texte*, 84–95; Lapidge & Sharpe, *Bibliography*, 90; Ó Cróinín, 'Echternach manuscript fragments', 138; Stevens, 'Scientific instruction', 86–90; idem, 'Cycles of time', 38; Wallis, *Bede*, lxxvi.

(or British) composition of the sixth or seventh century; this hypothesis was based on various arguments, one of the main being that this tract was almost exclusively cited in Insular texts, so that it appeared that it was known only in Britain and Ireland.[168] The alleged Irish authorship of this text has, however, recently been seriously questioned,[169] while the assumption of an almost exclusive circulation in Ireland and Britain is quite simply false (an *early* circulation would be more to the point). Anatolius (?)' theory of the increase and decrease of daylight, outlined in chapter 13 of his work, was extremely popular throughout the early middle ages, especially in continental sources, a fact overlooked because most of these sources remained unpublished.[170] As long as the recep-

168 For this argumentation, with a list and discussion of sources, cf. van der Hagen, *Dissertationes*, 137–40 (explicitly stating that *hujus libelli nulla (quantum mihi constat) mentio occurrit, nisi apud Britones, Scotos, & Hibernos*; van der Hagen's full argument extends from p. 115–41); Krusch, *Studien* I, 312–6; Jones, *Bedae opera*, 34, 82–4, and following him Strobel, *Texte*, 1 provide a better argument by pointing out that the manuscript tradition can be traced back to an Irish source (for this cf. also Ó Cróinín, 'Rath Melsigi', 154). It is accepted as an Insular forgery by Anscombe, 'Paschal canon', 515–8, 526–34 (dating it to 457); Mac Carthy, *Annals of Ulster* 4, cxviii–xxvii (dating it to AD 556); Kenney, *Sources*, 217 (referring to Mac Carthy's date); O'Connell, 'Easter cycles', 78–9 (rejecting Mac Carthy's date); Cordoliani, 'Computistes insulaires', 7–21 (assigning the date 457 to the text); idem, 'Traités', 54; idem, 'Comput', 46; van de Vyver, 'L'évolution', 8; Stevens, 'Scientific instruction', 87–8; Harrison, 'Epacts', 22–3; Pedersen, 'Ecclesiastical calendar', 55, 73; Walsh & Ó Cróinín, *Cummian's letter*, 32–5; Ó Cróinín, 'Echternach manuscript fragments', 134; McCluskey, 'Astronomies in the Latin West', 140; Wallis, *Bede*, lvii–ix; German, *De ratione temporum*, 36 (the latter after the publication of Mc Carthy & Breen's recent opposite opinion; cf. the following note). Nicklin, 'Date and origin', 137–50 proposes that Rufinus composed this work ca. AD 410 by translating and seriously emending and revising an Anatolian original.

169 Mc Carthy & Breen, *De ratione paschali*, 19–24, 115–43. In fact, various scholars over the past four centuries have argued for *De ratione paschali* being an original composition by Anatolius. Cf. Bucherius, *De doctrina temporum*, 434, 450–2, 465–6; the detailed review of previous opinions in Mc Carthy & Breen, *De ratione paschali*, 19–23; and Lejbowicz, 'Tables paschales', 24 (accepting Mc Carthy & Breen's theory). Strobel, *Texte*, 13–42 appears to have followed the most balanced and fruitful approach by attempting to disentangle the later emendations and additions to the original tract (a thorough review of the literature on p. 13–20, the textual analysis primarily based on technical considerations on p. 20–42); a new theory is most recently advanced by Mosshammer, *Easter Computus*, 135–61. I have not studied the text well enough to arrive at a personal conclusion; therefore, I will refer to this text as Anatolius (?) throughout this study.

170 In *DRP* 13, Anatolius (?) outlines the theory that daylight increases between the winter and the summer solstice by the ratio of 2⅔ *momenta* per day, and equally decreases between the summer and the winter solstice by the same ratio; moreover, it was deduced from this chapter that the bissextile increment per year, i.e. a quarter-day, arose from a daily increase of 1⅓ *momenta* per day between the winter solstice and the spring equinox. This theory was extremely popular, not only in Irish texts (cf. MC 39; *CE* p. 99–103; *DRC* 50), but also in just recently published continental sources (cf. *Dial. Neustr.* 28, 28aA; *Dial. Langob.* 19C, 28; *Quaest. Austr.* 2.9B–C (the theory in question is only hinted at, but with explicit reference to Anatolius); *Lect. comp.* 7.1, 7.1a; *Lib. ann.* 54; *Comp. Col.* 3.2C (explicitly referring to Anatolius); *Lib. comp.* 3.14; for the intermediary sources cf. Borst's edition of these texts) and it occurs in many unpublished tracts still hidden in manuscripts. Cf.

tion of this work in the middle ages has not been seriously studied, arguments about its reception are quite premature. The same holds true for the other texts mentioned. Therefore, none of these can individually be taken as proof that a certain text, in which any of them is cited, belongs to an Insular background, as long as their own provenance and circulation remain doubtful.

On the other hand, in terms of a group of texts predominantly read and cited in a certain region, the argument can only be a statistical one: the more sources are cited which had a wide (not exclusive, since this argument cannot be sustained as long as so many computistical tracts remain unpublished) reception and / or origin and / or history of transmission in a certain region, the likelier it is that the text citing these works was composed in that very same region. In the case of the Munich Computus the statistical argument can certainly be applied, since there are numerous sources cited by the Munich computist which match at least one of these criteria for the British Isles. Besides the so-called 'Irish forgeries' mentioned above and the two texts discussed below, these are the Macrobian excerpt known as *Disputatio Chori et Praetextati*,[171] Theophilus'[172] and Cyril'[173] prologues, the letter of Proterius (*Epistola Proterii*),[174] ps-Jerome's or

note 477. Likewise, Anatolius (?)'s table for calculating the number of days from 1 January to any given Julian calendar date outlined in *DRP* 10 can be found in numerous MSS (for some of these cf. Borst's edition of *Lect. comp.* 1.11). It is quite obvious that these theories and tables would more often have derived from intermediary sources rather than directly from Anatolius (?)'s work (the mentioned table, e.g., through Bede's *DTR* 22.18–29). Yet, the fact that the Anatolian (?) tract was well known among eighth-century continental computists is well demonstrated by two texts in which *De ratione paschali* is cited directly, namely *Prol. Aquit.* 4A and *Comp. Col.* 4.6. Only a thorough analysis of all (including unpublished) texts dealing with theories that can be traced back to *De ratione paschali* and / or including direct citations from this work will allow conclusions to be drawn about the reception of Anatolius (?) in the early middle ages.

171 These Macrobian excerpts (*Saturnalia* 1.12.2–1.15.20) still remain unedited as a separate text (the *editio princeps* of the *Disputatio Chori et Praetextati* is in preparation by Leofranc Holford-Strevens), so that the edition of Macrobius' *Saturnalia* (Willis, *Saturnalia*, 54–73) has to be consulted. For the *Disputatio* cf. especially Arweiler, 'Text und Überlieferung', 45–57 (who is not aware of all manuscript witnesses; cf. the following reference); Ó Cróinín, 'Dionysius Exiguus', 263–4; and furthermore Jones, 'Sirmond manuscript', 213, 217–8; idem, *Bedae opera*, 108, 348; Stahl, *Roman science*, 231; Ó Cróinín, 'A seventh-century Irish computus', 115; idem, 'Irish as mediators', 44–5; Mc Carthy & Ó Cróinín, 'Easter table', 61; Wallis, *Bede*, lxxvii, 46. Its manuscript tradition shows strong Irish connections. Note that various unpublished recensions of this text still remain hidden in the manuscripts; a thorough study of the transmission and reception of this text is one of the great desiderata of the field.

172 The *Prologus Theophili* is edited by Petavius, *De doctrina temporum* 2, 501–2 (with notes p. 508); Bucherius, *De doctrina temporum*, 471–3; Krusch, *Studien* I, 220–6. Its manuscript transmission shows strong Insular connections. For this text cf. van der Hagen, *Observationes in veterum patrum prologos et epistolas*, 1–16; Krusch, *Studien* I, 84–7; Ginzel, *Handbuch*, 233; Jones, 'Sirmond manuscript', 217; idem, *Bedae opera*, 29–31; Cordoliani, 'Traités', 66; Wallis, *Bede*, xl.

173 The *Prologus Cyrilli* is edited by Petavius, *De doctrina temporum* 2, 502–3; Bucherius, *De doctrina temporum*, 481–4; Krusch, *Studien* I, 337–43. Its manuscript tradition shows strong Insular connections. For this text cf. van der Hagen, *Observationes in veterum*

ps-Columbanus' *De sollemnitatibus*,[175] as well as Virgilius Maro Grammaticus'

patrum prologos et epistolas, 41–91; Krusch, *Studien* I, 88–98; Jones, 'Sirmond manuscript', 217; idem, *Bedae opera*, 38–53; Cordoliani, 'Traités', 60; David, 'Saint Martin', 293; Strobel, *Ursprung*, 253–69. Note that it is the *addendum* to this text in particular that is cited by the Munich computist.

174 Proterius' original Greek letter (composed in AD 455, now lost) was translated into Latin by Dionysius Exiguus in AD 525, who used it as his main source for his description of the 19-year cycle. It is is edited by Petavius, *De doctrina temporum* 2, 497–8 (with notes p. 506); Bucherius, *De doctrina temporum*, 82–8; Jan, *Historia cycli dionysiani*, 95–104; Krusch, *Studien* I, 269–78. The manuscript tradition of this text shows strong Insular connections. For this text cf. Bucherius, *De doctrina temporum*, 88; van der Hagen, *Observationes in veterum patrum prologos et epistolas*, 131–9; Piper, *Kalendarium*, 113; Krusch, *Studien* I, 266–9; Jones, 'Sirmond manuscript', 215; idem, *Bedae opera*, 57–9; Cordoliani, 'Traités', 65; Declercq, *Anno Domini*, 78.

175 This tract was first edited as letter 149 among the corpus of Jerome's works, to whom this text is attributed in three MSS, in *PL* 22, 1220 (a reprint of Vallarsi's 18th-century edition), and by Hilberg, *CSEL* 56, 357–63 (noting that this text appears not have been originally composed by Jerome). It is edited as anonymous by Pitra, *Specilegium Sollesmense* 1, 9–13. When studying in the Bibliothèque Nationale in Paris in 1884, Krusch came across this text in a MS from that library and thought he had discovered an unknown text. He edited it in the following year (Krusch, 'Chronologisches', 84–8), tentatively identifying it as a composition of Columbanus. This identification was accepted by Gundlach (who subsequently published it among the letters of Columbanus (*MGH Epp.* 3, 177–80)), but vehemently rejected by Seebass, 'Handschriften', 257–8, who also drew attention to the previous edition of this tract by Vallarsi (Seebass, 'Columba der Jüngere', 246). Krusch (*MGH SS rer. germ.* 37, 31–2) then accepted Seebass' argument. Walker, being aware of the pre-Krusch editions and the controversy subsequent to Krusch's tentative ascription to Columbanus, published this text among Columbanus' spuria (Walker, *Sancti Columbani opera*, 198–206). Note that chapter 2 of this text appears as a separate tract as c. 150 of the Bobbio Computus (*PL* 129, 1361–3). The text itself is written in favour of the Dionysiac lunar limits for Easter Sunday of *luna* 15 to 21 and against celebrating that feast on *luna* 14 (*De sollemnitatibus* c. 2 (Walker, *Sancti Columbani opera*, 200–2): *Et hoc etiam intueri debemus, quod non in decima quarta die ad vesperum, ut lex praecipit, ille agnus Dei, qui tollit peccatum mundi, et Pascha nostrum immolatus est Christus, sed decima quinta die, in quo manifestum est diem festum Iudeorum cum suo sacrificio a Domino esse solutum. ... Et hoc tantum observare dignatus est Dominus, ut in primo mense post decimam quartam diem Paschalem festivitatem, praecedente una sabbatorum, celebrari sine ulla ambiguitate censuerit*); it is argued that the practice favoured here is generally accepted in almost all parts of Christendom, and particularly by the apostolic see (*De sollemnitatibus* c. 2 (Walker, *Sancti Columbani opera*, 202): *quod nunc maxime ecclesia, auctoritatem sedis apostolicae sequens, observat*); consequently, the situation described in this text agrees only with the Insular context of the late seventh century: the papal curia accepted the Dionysiac reckoning only in the 640s (cf. note 86), while the only regions in the Latin West in which the Dionysiac reckoning competed with a system that allowed the celebration of Easter on *luna* 14 were parts of Ireland and Britain; therefore, I am inclined to think that this text was composed in some part of Ireland or Britain ca. AD 650–720 (by an Irish monastic teacher known by the name of Jerome; cf. p. CXXXI above). Scholarly opinions, however, range between the second and the seventh century as the date of composition of this text with equally diverse ascriptions of provenance; cf. Pitra, *Specilegium Sollesmense* 1, XI–II; Krusch, 'Chronologisches', 88; Gundelach, 'Columban-Briefe', 499–500, 503; idem, 'Zu den Columban-Briefen', 426–9; idem, 'Columbani epistolae', 154, 177; Schmid, *Osterfestberechnung in der abendländischen Kirche*, 78–80; Kenney, *Sources*, 193; Jones, 'Sir-

Epitomae.[176] But even more convincing than the analysis of a group of texts predominantly read and cited in a certain region is, in the case of the Munich Computus, the individual evidence of two rather rare sources, namely the Munich computist' quotes from *De mirabilibus sacrae scripturae* and from the papal letter of AD 640.[177] *De mirabilibus* was evidently composed in the *regiones Scottorum*, presumably in southern Ireland, in AD 654, and it generally appears not to have enjoyed a wide reception before the twelfth century, particularly not outside of Ireland and Britain. The same holds true for the papal letter of AD 640, addressed to the northern Irish clergy, of which fragments survived only in Bede's *Historia ecclesiastica* and the Munich Computus.

Technicalities and technical terminology: An analysis of the technical terminology and content of a computistical text appears more promising than the source analysis for the determination of the background of its author, but it also poses similar difficulties. The main obstacle lies in the fact that not enough computistical material has been published and evaluated to the present day. Concerning specifically Irish features in computistical literature, the task to be accomplished would obviously be to identify and edit evidently Irish texts and compare their information with the computistica of other regions. Since the Munich Computus is an excellent starting point for this task (being evidently Irish (as we will see) *and* securely datable, namely to AD 719), and since a thorough comparison of this text with contemporary

mond manuscript', 215; idem, *Bedae opera*, 108–10; Walker, *Sancti Columbani opera*, lx–i; Strobel, *Texte*, 68, 73–9; Ó Cróinín, 'Computistical works', 49–53; Borst, *Schriften*, 945.

176 The *Epitomae* are edited by Löfstedt, *Virgilus Maro Grammaticus*, 103–245. The fact that this work received an early reception in Irish computistical works is demonstrated by Ó Cróinín, 'A seventh-century Irish computus', 109–10, 123; idem, 'Virgilius Maro Grammaticus', 192–200. Note that the list of *Epitomae* citations in computistical literature given in Ó Cróinín, 'Virgilius Maro Grammaticus', 198–200 relates only to Irish texts or texts with Irish connections; chapter 11 of the *Epitomae* is also frequently cited in Bede's *DTR* (but not noted in Jones's edition or Wallis' translation; *DTR* 2.2, 5.2–3, 7.2–3, 11.1, 35.63–7, 36.1; and presumably more often), the Frankish *Dial. Burg.* 6A, 8A–B, 9A, 10A–B, 11A–B; *Comp. Col.* 2.1A; the Irish *Dial. Langob.* 2B, 6, 11A (*Comp. Col.* and *Dial. Langob.* less direct), and, without doubt, in more unpublished computistical tracts. The identity of Virgilius and consequently the provenance of the *Epitomae* is well disputed (the three main theories are that he was either Gaulish, or Spanish, or Irish). Cf. Manitius, *Geschichte*, 119–27; Anspach, 'Fortleben', 335; Herren, *Hisperica famina*, 27–32; Roger, *L'enseignement*, 110–26 (especially 110–1); Löfstedt, 'Spät- und Vulgärlateinisches', 121–6; idem, 'Wortschatz', 99, 110; Herren, 'New light', 28–9, 35–42 (a neat overview of previous opinions), 47–69; idem, 'Earliest Irish acquaintance', 246–7; idem, 'Pseudonymous tradition', 125–6; Law, *Wisdom*, 2–3; eadem, 'Grammar', 224; Ó Cróinín, *Early medieval Ireland*, 212; additionally, Traube, 'Virgilius Maro grammaticus', 158 stresses the dependence of Irish grammarians on Virgilius Maro; Löfstedt, *Malsachanus*, 20 adds that the manuscript tradition is exclusively Irish; Holtz, *Donat*, 315 places him in an Irish context, while Smyth, 'Isidore of Seville' 91–4 and Ó Cróinín, 'Hiberno-Latin literature', 388–9 treat this text as a seventh-century Hiberno-Latin composition.

177 For these two texts as well as parallels between the Munich Computus and certain parts of the Sirmond MS, cf. p. LXXVIII–LXXXVII below.

computistical material is provided in the present edition, some of the results are briefly outlined in the following. In addition to shedding light on the technical training of the Munich computist and its Irish context, it is hoped that this discussion may prove helpful for future investigation of and search for Irish computistica.

At first, computistical technicalities found in Irish texts of the seventh and early eighth centuries need to be classified in two principal categories: 1) Computistical techniques and features that were soon outdated due to the acceptance of different reckonings or the invention of different methods of calculations; 2) computistical techniques and features that remained unaffected by the developments in this medieval science. It is obvious that technical concepts and regional customs belonging to the first category are more significant than the ones of the second category, since they were less likely to have been adopted in other regions.

A good example of the second category, and therefore less indicative, is the concept of not attributing the name of a Julian calendar month to lunations that fall between two calends. The Munich computist only describes these lunations, while they are referred to by the technical term *luna abortiva* in *De ratione conputandi* and in the Sirmond MS.[178] This phenomenon of lunations falling between two calends occurred in the lunar calendar of every Easter reckoning. Moreover, the concept of leaving such lunations unnamed (in order to be faithful to the rule of naming a lunation according to the Julian calendar month in which it ends) constituted an attractive alternative to the rival method, in which only the embolisms remained unnamed.[179] Accordingly, it is hardly surprising

178 MC 59.38–69; *DRC* 70, 73–4; the relevant passage in the Sirmond MS is transcribed in Walsh & Ó Cróinín, *Cummian's letter*, 178–9; Ó Cróinín, 'Columbanus', 55; idem, 'Bede's Irish Computus', 206–7. For the *luna abortiva* cf. Walsh & Ó Cróinín, *Cummian's letter*, 177–8; Ó Cróinín, 'A seventh-century Irish computus', 125–6; idem, 'Columbanus', 54–5.

179 In the concept of *lunae abortivae*, no exceptions to the rule of naming a lunation according to the Julian calendar month in which it ends were necessary. In the rival method, however, in which only the embolisms remain unnamed, quite a few exceptions to that rule had to be conceded (so that, e.g., an April lunation would end in May), depending on the placements of the embolisms. For this rival method, which was by far the more popular one, cf. Bede, who appears to have directed his own statement quite specifically against the concept of the *lunae abortivae*: *DTR* 45.38–44 (cited in *Lib. calc.* 85; PV §§270–1): *solertissime quantum potuere curantes ut cuiuscumque aetatis luna in kalendis occurrisset, ipsa eiusdem mensis luna diceretur esse putanda. Quod tamen non usquequaque valuerunt obtinere propter praefixum paschalis lunae cursum, cuius cum extrema nonnumquam in kal. mai. incurrant et secundum quoque vel etiam tertium post kalendas die teneant, non tantum maii mensis sed potius aprilis lunae, sicut semper esse dicenda est.* It is also discussed in the context of Bede's placements of the embolisms by van der Hagen, *Observationes in veterum patrum prologos et epistolas*, 359–95. The popularity of the method of leaving the embolisms unnamed derived from the fact that in this concept the Easter lunation was without exception referred to as the April lunation. In the concept of the *lunae abortivae*, however, the Easter lunation was in some instances referred to as the May lunation, leaving the impression that Easter was celebrated in the second rather than the first month, which was regarded as uncanonical.

to find this concept, which presumably originated in seventh- or early eighth-century Ireland, also in later, non-Irish, but certainly Irish-influenced texts.[180] There are various other examples of the second category,[181] but since the implications of these are limited, we should rather turn to the more interesting first category.

In all three Irish computistical textbooks of the seventh and early eighth century the interference of the *saltus* and the lunar bissextile day with the relation between the lunar age of Easter Sunday and that of the *initium quadragesimae* (the beginning of the Lenten period) is discussed in great detail.[182] Discussions of this problem soon became obsolete, however, since it was later preferred to implement these technical devices (the *saltus* and the lunar bissextile day) outside of the quadragesimal period, to avoid any such interference: The *saltus* was placed in November or July rather than in March, the lunar bissextile day at the end of the February lunation rather than on 24 February. Therefore, lengthy discussions of this matter can only be found in Irish and Irish-influenced texts before Bede's works received a wider distribution and reception.[183] The same holds true for the extensive comparisons between the Victorian and the Dionysiac 19-year cycles found in the *Computus Einsidlensis* and the Munich Computus.[184] Both texts favoured the Dionysiac system, but since they marked the end of a long period of comparing these two rivalling concepts, it was only natural that they would incorporate the general results of this comparison. A few decades

180 *DRC* 70, which defines the concept in question, was copied in ninth-century Carolingian computistics as a separate text entitled *De duabus lunationibus in uno mense et de luna abortiva* edited in Appendix 6; additionally, this chapter of *DRC* is partially copied in St Dunstan's classbook (Oxford, Bodleian Library, Auct. F 4 32, fol. 22v). An independent definition of the term *luna abortiva* is given in Oxford, Bodleian library, Digby 63, fol. 33v (transcribed in Walsh & Ó Cróinín, *Cummian's letter*, 178): *De abortive luna: Abortiua luna est que non incipt in Kalendis in caput mensis, sed incipit imedio mense (recte in medio mensis), et finitur in medio. Hoc est abortiva luna.* In this passage, *medium mensis* denotes the entire period of days between the calends of two months, excluding these. *DRC* 73, which provides a full discussion of all *lunae abortivae* in the Dionysiac 19-year cycle, is copied with minor ellipses in St Dunstan's classbook (Oxford, Bodleian Library, Auct. F 4 32, fol. 22r–v); it then appears in an altered, ninth-century continental version under the title *De contrari aetatibus VII cycli lunaris*, also edited in Appendix 6. Moreover, the term *luna abortiva* is also applied to a lunation falling between two calends in Zürich, Zentralbibliothek, Car C 180, fol. 104v–106v. Cf. also p. CLXV–CLXVI below.

181 Prominent examples are the placement of the Dionysiac epacts on 1 January (accompanied by a detailed outline of the underlying transfer of Dionysius' original placement on 22 March to 1 January; cf. MC 49.11–15 and the references there); the placement of the Dionysiac *saltus* on 22 March (MC 62.68–72); the division of the 24 hours of a *saltus* by 235 lunations (MC 62.14–61); and many more minor features. Cf. also note 111.

182 *CE* p. 119–122; MC 58.8–32; *DRC* 100–102; cf. p. CLXVII–CLXVIII.

183 Discussions of this kind can be found in *Dial. Langob.* 24B and c. 31 of the unfinished Fulda computus of AD 789 in Basel, Universitätsbibliothek, F III 15k, fol. 44v–45r (cited in Paris, Bibliothèque Nationale, Nouvelle acquisition latine 1615, fol. 150r–v and as part of a computus of 825/6 in Bern, Burgerbibliothek, 417, fol. 14v–15r).

184 *CE* p. 108–11, 119; MC 49.16–32, 50, 52, 55.46–52, 58.2–7, 62.68–110.

later, however, when the Dionysiac reckoning was well established and unrivalled, such comparisons lost their currency, at least in the *regiones Scottorum*. On the Continent, however, where the transition from the Victorian to the Dionysiac was not fully finalized before the late eighth century, interest in Irish opinion on this matter still existed.[185] From the early ninth century onwards, then, those comparisons had lost their actuality and factual value almost everywhere.

The most significant technical indication of an Irish authorship of ca. AD 700 is, however, provided by two other features, one expected, the other rather unexpected. The expected feature are references to the *latercus*, i.e. the 84 (14)-year Easter reckoning followed in the *regiones Scottorum*. This reckoning was abandoned by the Southern Irish clergy in AD 632, by the Anglo-Saxons of Northumbria in AD 664, by the community of Iona as late as AD 716, and even later in Wales. On the Continent, it was only followed in Columban houses for a short period after their foundation.[186] The technical term for this reckoning, *latercus*, is only attested in the three Irish computistical textbooks of the seventh and early eighth centuries.[187] The rather unexpected feature is the characteristic lack of mathematical algorithms in the calendrical calculations of these textbooks. Usually, chronological calculations were executed by means of brief formulae ever since Dionysius had appended nine computistical *argumenta* to his Easter table in AD 525. From the late-seventh century onwards, the corpus of computistical *argumenta* was significantly extended, so that secific algorithms existed for the calculation of almost every feature of the lunar and solar calendars. The Irish computistical texts, including *Dial. Langob.*, stand, however, outside of this Dionysiac tradition, even though they all promote the Dionysiac reckoning. In fact, not a single Dionysiac *argumentum* is mentioned in these texts. On the contrary, it seems that Irish computists were suspicious of all those chronological innovations that the Dionysiac reckoning imparted, especially the epacts of 22 March, the *concurrentes* (weekday) of 24 March, and the

185 Most of the eighth- and early ninth-century continental texts that incorporate comparisons of the Victorian with the Dionysiac reckoning show dependency on an Irish school of thought. Cf. the comparison of the Easter full moons in *Lib. comp.* 2.22A; the table Berlin, Staatsbibliothek, Phillipps 1831, 128r; Paris, Bibliothèque Nationale, Lat. 7296, 111v (I have not seen the Paris MS; cf. Jones, *Bedae opera*, 24); and especially the long discussions in *Comp. Col.*, books 4 and 5, which are clearly dependent on MC.

186 For these facts cf. p. XXXVII–XXXIX, LXXXIII–LXXXIX, XCIII–XCV, CLVI–CLVIII.

187 *CE* p. 119; MC 50.67–76, 52.3–26, 58.6–7, 62.118–22; *DRC* 99. This technical term is not spelled out in the rubric of the only surviving Easter table of this reckoning, the Padua *latercus*, where it appears as *L T R C S* (or *L T RC S* ?). For a discussion of the *latercus*-references see especially Warntjes, '84 (14)-year Easter reckoning', 31–85. A clear distinction must be made between this term denoting the 84 (14)-year Easter reckoning and its appearance in different contexts, independent of this reckoning. Such a different context is the reference to *latercus* as the alleged Egyptian word for 'number' or 'conpotus' in *CE* p. 83; *DRC* 3; BC 66; *Dial. Burg.* 1B; *Comp. Col.* 6.6; gloss to *DTR* (*CCSL* 123B, 268). Moreover, an etymology of *latercus* is given in *CE* p. 84; *DRC* 4; and in Braulio's letter of AD 640 it is used with the more general meaning of 'Easter table' (cf. Ó Cróinín, 'New heresy', 95).

AD date as the main chronological features of a year; in fact, the AD reckoning, which was the precondition for every Dionysiac *argumentum*, is only mentioned in one of these four texts (*DRC*), and there only once, rather in passing, as one of the columns of the Dionysiac Easter table.[188] Instead, Irish computists of that period preferred to stick to methods of calendrical calculation they were presumably familiar with for centuries, and which they had probably already applied to the 84 (14) and the Victorian reckoning. These methods consisted of nothing but simple finger-counting from the weekday and lunar data of 1 January.[189] Since only these Irish texts composed before the middle of the eighth century apply these rather lengthy, non-algorithmic calculations based on finger-counting and few very basic calendrical rules, such methods of calculation are clear indicators of pre- AD 750 Irish provenance.

References to nationality: In computistical literature, references to nationalities other than 'the Greeks' and 'the Romans' or 'the Latins', which usually are synonyms for the Dionysiac and the Victorian reckoning respectively, are extremely rare. Whenever they occur, directly or indirectly, they naturally provide clues about the background of the author. There are two principal contexts in which they appear, either in chronological references to a regnal year or the obit of a king, or in the description of regional computistical customs. The former is the case, e.g., in a passage of the Sirmond MS, which notes the number of years (631=AD 658) from the passion of Christ to 'the Easter of Suibine mac Commáin', a member of the Déisi from east Munster; therefore, this passage was in all likelihood composed in Suibine's sphere of influence.[190] Very similarly, the

188 In MC it is explicitly mentioned that 1 January is preferred to 22 March as the *sedes epactarum*, and the Dionysiac epacts of 22 March are transferred to 1 January; cf. MC 49.11–18, 51 and note 181; likewise *CE* p. 109 ; *Dial. Langob.* 23 uses the epacts of 1 January, while the ones of 22 March are not even mentioned anywhere in this text; in *DRC*, the *sedes epactarum* of 22 March is only twice briefly referred to, namely as a column of the Dionysiac Easter table (*DRC* 103) and as a characteristic of the Dionysiac 19-year cycle (*DRC* 104). The *concurrentes* of 24 March are discussed only once in MC (c. 67), without reference to the technical term itself; they are nowhere alluded to in *CE*, *DRC*, and *Dial. Langob.*; in *DRC* 103 they even appear to have been deliberately omitted in the discussion of the columns of the Dionysiac Easter table (Walsh & Ó Cróinín, *Cummian's letter*, 207 think that this omission 'is possibly the result of a lacuna in the exemplar'; in that case, then, the omission would have been executed by an earlier computist rather than the author of *DRC*, but it would probably still have been deliberate). This discussion in *DRC* is the only time that the incarnation era (AD) is mentioned in any of the three Irish computistical textbooks; in *Dial. Langob.* a reference to AD only occurs in an addition to that text of AD 818 (c. 27a).

189 For these calculations see *CE* p. 97–8, 106, 112, 120–1; MC 29, 56.15–57, 58.33–80; *DRC* 33–35, 73. Only the general principles applied in such simple weekday-calculations can be found in (the Irish) *Dial. Langob.* 16 and *Quaest. Austr.* 2.17 (which is, it will be noted, a Victorian computus of Frankish origin, and the only non-Irish text I am aware of that shows traces of the mentioned calculations). Cf. also p. CLX–CLXI below.

190 Sirmond MS (Oxford, Bodleian Library, Bodley 309, fol. 95v–95*bis*r). For this passage see Ó Cróinín, 'Early Irish annals', 83–4; idem, 'Irish provenance', 177–81; idem, 'The works

Bobbio Computus records, in an often overlooked passage immediately preceding the Victorian Easter table, the number of years from the creation of the world to the sixteenth regnal year of the Merovingian king Chlothar III of Neustria (5874=AD 673).[191] Whenever these dating clauses with local references appear in anthologies incorporating material of various provenances, as in the case of the two examples just given, the difficulty obviously lies in determining the textual limits of the passage for which this clause is indicative. This problem does not exist for well-defined textbooks like the 'neustrische Streitgespräch von 737' (*Dial. Neustr.*), which was connected to Neustria by its modern editor because of a reference to the number of years from the incarnation to the death of *Teudericus rex Francorum*, the Merovingian king Theuderic IV.[192]

The other context in which national ascriptions occur, namely the description of regional computistical customs, is well illustrated by Dicuil's discussion of the different customs concerning the placement of the Dionysiac *saltus*. In this discussion, Dicuil does not leave any room for doubt about his own national identity:[193]

> Etsi lunarem saltum in vigesimo quarto die mensis Novembris, secundum Anglos, complere volueris. […] Sed si secundum Grecorum ac Latinorum, quam mea gens in Hibernia in hac ratione semper custodit, praedictum saltum in vigesimo quarto die mensis Martii sequentis, iuxta primum tempus creationis lunae rationabiliter observabis.

A comparable reference is included in the Munich Computus, in which the author mentions the 'Roman' and Irish habit of recording the epact on 1 January as against the 22 March of the 'Greeks':[194]

> Communes sunt totius anni dies aepactis, sed tamen Greci in XI Kalendas Aprilis legitime, Romani uero in Kalendis Ianuarii epactas enumerant. Unde Romani nec minus Scotti in Kalendis Ianuarii Grecorum obseruationes ępactas rimantur. Hinc ab VIII singulari in Kalendis Ianuarii ępactas incipimus.

The fact that the author identifies himself and his audience with the custom attributed to the Irish by using a verb in the first person plural when turning to the

of Virgilius Maro Grammaticus'; 193–6; idem, 'Bede's Irish computus', 209–10. Cf. also p. LXXX below.

191 Milan, Biblioteca Ambrosiana, H 150 inf, 127v. It is transcribed by Krusch, 'Einführung', 132 and Mommsen in *MGH AA* 9, 674. First attention to this dating clause was drawn by Reifferscheid, *Bibliotheca*, 93 (in a discussion of Victorius' works in the Bobbio Computus), who, however, omitted the crucial part of this passage. A comparison between this and the dating-clause in the Sirmond MS would be a fruitful study and would provide further evidence for scholarly contact between the Frankish west and the Irish south in the second half of the seventh century.

192 *Dial. Neustr.* 30C. Cf. Borst, *Schriften*, 375, 378; Krusch, 'Lehrbuch', 232, 242; for arguments for a different, Austrasian provenance of this text cf. p. CII–CIII, CLXXI–CLXXIII.

193 Dicuil, *Liber de astronomia* 1.5.1–2 (Esposito, 'Astronomical treatise', 388).

194 MC 49.11–17. Cf. Mac Carthy, *Annals of Ulster* 4, lxviii–ix; Schwartz, 'Ostertafeln', 92; Krusch, *Studien* I, 13; idem, 'Einführung', 162–3; Harrison, 'Luni-solar cycle', 73; Walsh & Ó Cróinín, *Cummian's letter*, 84.

application of this practice in the following sentence is a clear statement about his own national affiliation, as well as that of his audience.

Bilingualism / Trilingualism: The most convincing proof for the Irish authorship of the Munich Computus is the use of the vernacular. There are three passages in which the author uses his mother-tongue, Old Irish.[195] In the first passage, which was discovered by Mac Carthy, the Munich computist compares the week of creation with the week of the Exodus from Egypt as well as the week of the passion of Christ, using the bilingual term *dies cetene* (consisting of Latin *dies* and Old Irish *cétaíne*) whenever he refers to Wednesday.[196] It has been argued that this isolated application of Old Irish weekday terminology was motivated by a statement of Augustine in his *Enarrationes in psalmos*.[197] Concerning the title of psalm 93, *In quarta sabbati*, Augustine argues that the heathen term *dies Mercurii* for Wednesday has to be avoided, and should be substituted by a Christian equivalent, if such was available in the vernacular of any commentator.[198] This plea was then met by the Munich computist three centuries later, as it was another three centuries later by the St Gall monk Notker Labeo when translating psalm 93 into Old High German, creating the word *mittauuechun* (literally 'mid-week', still current in modern German as *Mittwoch*) to avoid any reference to a pagan god.[199] The second passage, discovered by Ó Cróinín, contains the Old Irish verb *tomel* (Classical Old Irish *do·meil*, 'to consume').[200] This verb appears in a discussion of lunar theory and it was apparently applied for reasons of precision, for an exact description of a complicated concept.[201] The third and last passage has only recently been noted.[202] When classifying the Julian calendar months, the Munich computist uses the curious term *noinaic* (to be understood as a nominative plural adjective *nonaic[h]* with the meaning 'having nones'), the Old Irish numeral *trī* for three, and the mixed numeral *noi decem* for nineteen.

195 A comprehensive analysis of Old Irish in the Munich Computus is now available in Bisagni & Warntjes, 'Latin and Old Irish', 1–33. In the following edition, all words in the vernacular (Old Irish and Old English) are underlined.

196 MC 44.13–81. Cf. p. XIX and Mac Carthy, *Annals of Ulster* 4, clxxx; Ó Cróinín, 'Old Irish gloss', 131–2; idem, 'Earliest Old Irish glosses', 16–7; Bisagni & Warntjes, 'Latin and Old Irish', 8–12.

197 Bisagni & Warntjes, 'Latin and Old Irish', 10–2.

198 Augustine, *Enarrationes in psalmos, Enarratio in psalmum 93, sermon 3* (*CCSL* 39, 1302–3).

199 Notker Labeo, *Translatio barbarica psalterii, psalmus 93* (Tax, *Notker*, 346). Cf. Borst, *Computus*, 68. Kalesch, *Tag und Jahr*, 41 denies that any Church influence was responsible for the change in the denomination of Wednesday from attribution to the Germanic god Wodan to 'mid-week' in Old High German; but then he seems not to be aware of the Augustinian passage and its impact.

200 MC 48.13–18. Cf. p. XIX above; Ó Cróinín, 'Old Irish gloss', 132; idem, 'Earliest Old Irish glosses', 17; and the following note.

201 Cf. Bisagni & Warntjes, 'Latin and Old Irish', 12–7.

202 MC 28.15–23. Cf. p. XIX above; Bisagni & Warntjes, 'Latin and Old Irish', 17–28.

Additional to these Old Irish words, the Munich computist also applied an Old English term, namely *gerīm* in the sense of 'numbers of days'.[203] It appears, therefore, that technical vocabulary of the undoubtedly Irish author (and, it must be presumed, of his audience, i.e. of the other scholars and pupils of the monastic school he was working in) also included Old English words, an impressive witness to intercultural contact among seventh- and early eighth-century scholars.

The fact that all these instances of the vernacular appear to have been deliberately applied and incorporated in an otherwise Latin text indicates that the author switched to his own idiom whenever he felt it was appropriate or didactically necessary, clearly expecting this switch in language to clarify rather than to obscure matters. Consequently, it can hardly be doubted that the Munich Computus was written by an Irishman for an Irish-speaking audience, and was therefore composed in the *regiones Scottorum*, rather than in a culturally mixed community on the Continent.

203 MC 51.6–8. Cf. now Warntjes, 'Earliest occurrence', 91–105.

THE PROVENANCE

But is it possible to be more specific than *regiones Scottorum* concerning the provenance of the Munich Computus? Most scholars who worked on this text were not concerned with this question. Either the identification of the author as Irish was sufficient to them, or they did not believe that further specification was possible. Mac Carthy, however, may have implied a connection between this text and Iona in the following statement:[204]

> The main object of this work was to recommend and explain the Alexandrine 19-year cycle, accepted finally in Iona and Ireland two years previously.

The Canadian scholar James F. Kenney, working from Mac Carthy's account, is slightly more direct in his admirable *Sources for the History of Ireland*:[205]

> The date of composition was 718: some Victorian paschal computations of 689 are incorporated. The date is not without suggestiveness. It was in 716, according to the Annals of Ulster, that the paschal system was changed in Iona.

This connection between the Munich Computus and Iona was then strongly advocated and often repeated by Dáibhí Ó Cróinín. His argument, formulated in his 1982 article announcing the discovery of *De ratione conputandi*, can be divided into two categories, the sources cited and the historical context.[206] Among the sources, the papal letter to the northern Irish clergy of AD 640 is regarded as most indicative. Since the citation from this letter given in the Munich Computus is unique to this text, and since all addressees of this letter belonged to the northern Irish clergy, Ó Cróinín assumes that the author of the Munich Computus must have written in the northern parts of the *regiones Scottorum*. Following Kenney's argument, Ó Cróinín argues that the relative temporal proximity between the date of compilation (AD 718/9) and the date of conversion of Iona from the *latercus* – mentioned in the Munich Computus – to the Dionysiac reckoning (AD 716) – clearly the reckoning favoured by the Munich computist – in particular strongly suggests Iona as the place of composition of this text. Knowledge about the Dionysiac and the Victorian reckoning came through southern Irish channels to the more northerly regions, and this 'northward flow of texts' also explains why the evidently southern Irish text *De mirabilibus sacrae scripturae* was cited in a northern Irish compilation; it also is the reason for the textual parallels between the supposedly older, southern-Irish *De ratione conputandi*, and the later, northern Irish Munich Computus. Thus far goes Ó Cróinín's argument, which is, it will be noted, very cautious, especially concerning the identification of Iona as the place of composition of the Munich

204 Mac Carthy, *Annals of Ulster* 4, lxxi.
205 Kenney, *Sources*, 223.
206 Ó Cróinín, 'A seventh-century Irish computus', 126–7.

Computus. In subsequent years, Ó Cróinín abandoned this caution and simply stated, without supplying further arguments, that the Munich Computus was composed in Iona.[207] In the following, this question will be reassessed in the two categories of argumentation considered by previous scholars, the sources used by the Munich computist and the historical background of this text.

Sources: As already outlined in the discussion of the Irish authorship of the Munich Computus, the evidence of the sources, as long as it does not unanimously point in one direction, will always remain controversial. Any source analysis serving the purpose of identifying a specific region, or even one single monastic centre, as the place of composition of a text can obviously only be based on very few sources of a highly indicative quality. These sources, by their very nature, must appear unlikely to have spread to other regions, and therefore must have had only a very limited regional reception. In the case of the Munich Computus, the use of one southern Irish source, and parallels to another tract from that region, point in a southern Irish direction, while only one citation appears to indicate a northern Irish background. The only evidently southern Irish text cited in the Munich Computus is *De mirabilibus sacrae scripturae*.[208] From internal information *De mirabilibus* can be securely dated to AD 654,[209] which

207 Ó Cróinín, 'Virgilius Maro Grammaticus', 197; idem, *Irish history*, 4, 103. Ó Cróinín's view is repeated by Lejbowicz, 'Tables paschales', 21.

208 See MC 66; 68; *De mirabilibus sacrae scripturae* is edited in *PL* 35, 2149–200.

209 Scholars vary in their opinion on the exact date of composition of this text. The dating clause reads (*De mirabilibus sacrae scripturae* 2.4 (*PL* 35, 2175–6)): *donec decimus inde oriens nonagesimo secundo anno post passionem Salvatoris, Alia et Sparsa consulibus, peractis cursibus consummatur. Post quem undecimus a consulatu Paterni et Torquati ad nostra usque tempora decurrens, extremo anno Hiberniensium moriente Manichaeo inter caeteros sapientes, peragitur. Et duodecimus nunc tertium annum agens ad futurorum scientiam se praestans, a nobis qualem finem sit habiturus ignoratur.* Two methods of calculating the date of composition from this passage have been advanced, one by AM (*anni mundi*) reckoning, the other by reference to the Victorian Easter table. First, the number of years from the beginning of the world to the *annus praesens* of the author is explicitly given as 11 full cycles of 532 years plus 3 more years, i.e. AM 11×532+3=5855. But how does this year relate to AD reckoning? Since the whole text, and especially this passage, is obviously based on the Victorian reckoning, Victorius' works have to be consulted. The Aquitanian, however, is ambiguous in this question: He refers to the *annus passionis*, the first year of his Easter table, as AM 5228 (Krusch, *Studien* II, 24), and his Easter table makes it perfectly clear that this corresponds to AD 28, which results in AM−5200=AD; on the other hand, he gives his *annus praesens* as AM 5658, and the chronological data in his Easter table reveals that this year corresponds to AD 457, which leads to AM−5201=AD. Accordingly, only the context can reveal which of these two relations was preferred. The daing clause mentions the death of a certain Manichaeus in the final year of the eleventh 532-year cycle, i.e. AM 5852, three years before the composition of the text in question. This person has been identified as Manchianus of Min Droichit, whose death is recorded for AD 652 in the Irish annals (cf. following note). Consequently, the 5200 year difference between AM and AD is regarded as correct here, which leads to AD 655 as the date of composition. This date is today almost unanimously accepted; cf. especially Ussher, *Antiquitates*, 502–3 (hence the dating is as old as the 17th century); Reeves, 'On Augustin', 516–7; Mac Carthy, *Codex*, 365–8; Esposito, 'De mirabilibus', 196–8; Stevens, 'Bede's

is referred to as three years after the death of a certain Manichaeus. This Manichaeus has been identified with Manchéne (Manchianus), abbot of Min Droichit in modern county Laois, whose death is recorded for ca. AD 652 in the Annals of Ulster and the Annals of Tigernach.[210] The only other person mentioned by name, a certain Bathanus, has more tentatively been identified with Báetán moccu Cormaic, abbot of Clonmacnois from ca. AD 652 to 664.[211] From the evidence of these references, this text appears to have been written in the southern Midlands, just within the area of influence defined as southern Irish; the use of the synonym Carthage for the author's monastery may point,

scientific achievements', 12; Ó Cróinín, 'Early Irish annals', 82–3; and furthermore Mac Carthy, *Annals of Ulster* 4, lxx; Kenney, *Sources*, 276–7; Grosjean, 'Quelques exègetes irlandais', 74; Bischoff, 'Wendepunkte', 222, 268; Herren, 'Pseudonymous tradition', 127; Simonetti, 'De mirabilibus', 247; Mac Ginty, 'Irish Augustine', 71–2; idem, 'Irish provenance', 183; Harrison, 'Epacts', 27; Ó Cróinín, *Early medieval Ireland*, 188; idem, 'Virgilius Maro Grammaticus', 195; idem, 'Hiberno-Latin literature', 380. Yet, it should be noted that the chronology of the Irish annals, on which this argument is primarily based, is uncertain for the period in question (cf. Mc Carthy, 'Chronology', 228–41 and now idem, *Irish Annals*, 139–51). Moreover, in the seventh and early eighth centuries, the 5201-year difference is usually the preferred one (cf. the dating clauses published in Krusch, 'Einführung', 132–3; Ó Cróinín, 'Irish provenance', 177–81; *Dial. Burg.* 17), and only from the mid-eighth century does the 5200-year difference become fashionable. Most conclusive, however, are the consul references. They make it absolutely clear that the author worked from a Victorian Easter table in which the death of Manichaeus was noted in the year of the consuls Alia and Sparsa, alias Aviola and Pansa (cf. *MGH AA* 9, 694–5; Ó Cróinín, 'Early Irish annals', 81–3; idem, 'Irish provenance', 183); this year corresponds without doubt to AD 119 and its 532-year equivalent, AD 651. The above outlined argument that the obit of Manichaeus corresponds to AD 652 presupposes a defective Victorian Easter table here, which is extremely unlikely. Hence, the author rather used the 5201-year difference between AM and AD in line with his contemporaries, and the year of composition is AD 654 (the Irish annals should be dated accordingly from this independent source). On the evidence of the Victorian Easter table this date was already established by Noris, *Dissertationes*, 180–2; Krusch, 'Einführung', 158–9; Mommsen in *MGH AA* 9, 694–5. Undecided between the two dates is Smyth, 'Isidore of Seville', 72 (but she opts for common opinion, AD 655, on p. 77); idem, *Understanding the universe*, 12 (but she follows the 655 dating when discussing the dating clause on p. 168).

210 AU 651.1 (Mac Airt & Mac Niocaill, *Annals of Ulster*, 128); ATig [AU 651] (Stokes, *Annals of Tigernach*, 152). For the identification see Ussher, *Antiquitates*, 503; Reeves, 'On Augustin', 516; Mac Carthy, *Annals of Ulster* 4, lxx; Esposito, 'De mirabilibus', 196–8, 202; Kenney, *Sources*, 276; Mac Ginty, 'Irish Augustine', 74–6 (with reservations); Ó Cróinín, 'Early Irish annals', 82–3; idem, 'Irish provenance', 180–3; idem, 'Virgilius Maro Grammaticus', 195. A different identification (St Manchán of Líath Mancháin) is proposed by Grosjean, 'Quelques exègetes irlandais', 74, 84–92, 96. Krusch, 'Einführung', 159 wrongly identifies this Manichaeus with Áedán, abbot of Lindisfarne.

211 Báetán's obit is recorded in AU 663.5 (Mac Airt & Mac Niocaill, *Annals of Ulster*, 134), ATig [AU 663] (Stokes, *Annals of Tigernach*, 158). For the identification see Reeves, 'On Augustin', 515; Esposito, 'De mirabilibus', 196–8, 202; Kenney, *Sources*, 276; Mac Ginty, 'Irish Augustine', 74–6 (reviewing the readings, but not making a statement concerning identity). For Báetán moccu Cormaic see Ryan, 'Abbatial succession', 496; Byrne, *History of Ireland* 9, 246; Kehnel, *Clonmacnois*, 248. A different identification (a certain Banban of the Déisi) is proposed by Grosjean, 'Quelques exègetes irlandais', 78–80, 85, 96.

however, further south to a foundation by St Carthach, most notably Lismore or Rathan in south-east Munster, where a circle of exegetical scholars with a considerable interest in computistics appears to have been active in the seventh century.[212] Yet, the reception of this text was not restricted to southern Ireland. Especially the fact that it was cited by Alcuin in the second half of the eighth century[213] suggests that it could practically have been available anywhere in the British Isles by the early eighth century, the time when the Munich Computus was compiled.

Still, the Munich Computus can more conclusively be connected to a southern Irish region due to some interesting and unique parallels with the Sirmond MS. This MS incorporates a dating clause for AD 658[214] with reference to a certain Suibinus filius Commannus, whose obit the southern Irish Annals of Inisfallen independently record for precisely the year in question, AD 658.[215] Ó Cróinín has convincingly identified this Suibne mac Commáin as a member of a kindred from the Waterford area in east Munster,[216] which is in striking proximity to the provenance proposed for *De mirabilibus sacrae scripturae*. Whether this date and provenance is representative for a large section of this MS or only for the tract in which the dating clause occurs (item 26 in Jones' inventory) is a matter of debate,[217] and need not immediately concern us here,

212 This is Grosjean's argument ('Quelques exègetes irlandais', 70–2, 96); he is followed in this view by Herren, 'Pseudonymous tradition', 127–8; Simonetti, 'De mirabilibus', 247–51 (tentatively, being more concerned with the cultural background of the author); Smyth, *Understanding the universe*, 11–2; Ó Cróinín, 'Hiberno-Latin literature', 380. For the circle of exegetical scholars (with Cummian and the author of *De mirabilius sacrae scripturae* being representative for considerable computistical interest) in south-eastern Munster cf. especially Grosjean, 'Quelques exègetes irlandais', 67–97; Ó Cróinín, 'Irish provenance', 178–85; idem, 'Hiberno-Latin literature', 377–81; and furthermore Ó Cróinín, *Early medieval Ireland*, 187–8; Herren, 'Pseudonymous tradition', 130.

213 Alcuin, *Interrogationes et responsiones in Genesin* (*PL* 100, 530C–D). The relevant passages are paralleled in Esposito, 'De mirabilibus', 200–1; cf. also Mac Ginty, 'Irish Augustine', 79.

214 Sirmond MS (Oxford, Bodleian Library, Bodley 309, fol. 95v–95*bis*r). The passage in question is printed in Ó Cróinín, 'Early Irish annals', 83; idem, 'Irish provenance', 177; idem, 'Virgilius Maro Grammaticus', 194; idem, 'Bede's Irish computus', 209. In this dating clause it is argued that 631 years have elapsed from the passion of Christ to the year of the Easter of Suibne mac Commáin. The era applied evidently is the Victorian AP, in which AP 1=AD 28. Hence, the year in question is AD 631+27=658. Cf. also p. LXXIII–LXXIV above.

215 Annals of Inisfallen [658] (Mac Airt, *Annals of Inisfallen*, 94).

216 Ó Cróinín, 'Irish provenance', 179–82. Cf. Smyth, 'Isidore of Seville', 97–8.

217 The discoverer of the dating clause, Dáibhí Ó Cróinín, argues ('Irish provenance', 177; 'Virgilius Maro Grammaticus', 193–7 (in answer to Smyth's critique; cf. further below in this note); 'Bede's Irish computus', 202–3) that the date applies to all those parts of the manuscript that Jones regarded as the Irish computus used by Bede, i.e. items 3–9 and 13–45 of Jones' inventory of the Sirmond MS (note that Jones, 'Sirmond manuscript', 209 believes that the first book, i.e. items 3–9, were not part of Bede's exemplar; he changed his opinion later in Jones, *Bedae opera*, 106, 110). Ó Cróinín's view is accepted by Walsh, 'Some remarks', 225; Wallis, *Bede*, lxxvii; but challenged by Smyth, 'Isidore of Seville', 96–99; eadem, *Understanding the universe*, 156–7, who suggests that the dating clause

since two of the parallels between the Munich text and this MS appear in this very tract. Here, the number of years from the Exodus from Egypt to the resurrection of Christ are counted, a calculation that is, to my knowledge, unique to the Munich Computus and this item of the Sirmond MS.[218] Less indicative, but still extremely interesting, is the count of 133 bissextile days in the 532-year Victorian Easter cycle; it occurs not only in the Munich text and this item 26, but also in Cummian's famous letter of AD 632, which, it will be noted, also originated in the southern *regiones Scottorum*; however, Bede later gives a similar calculation.[219] Additionally, a textual parallel exists between the Victo-

may not refer to the *annus praesens* of the compiler, and that the clause itself may only be representative for a limited part of the MS. Springsfeld (*Alkuins Einfluß*, 77), discussing Jones' arguments rather than Ó Cróinín's, makes a case for the tracts of the first book (items 3 to 9), at least in their present form, being composed on the Continent in the late eighth century. Cf. also notes 37, 55.

218 MC 64; Sirmond MS (Oxford, Bodleian Library, Bodley 309, fol. 95v; printed in Ó Cróinín, 'Early Irish annals', 83; idem, 'Irish provenance', 177; idem, 'Virgilius Maro Grammaticus', 194; idem, 'Bede's Irish computus', 209).

219 MC 41.108; Sirmond MS (Oxford, Bodleian Library, Bodley 309, fol. 95v; edited in Ó Cróinín, 'Bede's Irish computus', 209); *Epistola Cummiani* 225–6 (Walsh & Ó Cróinín, *Cummian's letter*, 88; note, however, that this passage is reconstructed; for this passage cf. also Krusch, 'Einführung', 151–2; Schmid, *Osterfestberechnung auf den britischen Inseln*, 26–7); *DTR* 65.2–7, which is cited in *Lib. calc.* 98 and Rabanus Maurus' *De computo* 94.2–9. The date of Cummian's letter is inferred from internal evidence as well as Bede's *HE* (cf. Ussher, *Antiquitates*, 537–8 (dating the embassy correctly, but then wrongly assuming that 634 saw the beginning of a new 84-year cycle, which he regarded as the stimulus for Cummian's letter); Krusch, 'Einführung', 149–50 (wrongly connecting the Irish embassy and Cummian's letter with the papal letter of AD 640); Plummer, *Bedae opera* 2, 125, 352; Poole, 'Earliest use', 59–60 (adopting Krusch's wrong chronology); Mac Carthy, *Annals of Ulster* 4, cxl–i; Schmid, *Osterfestberechnung auf den britischen Inseln*, 33–42; Kenney, *Sources*, 220–1; Gougaud, *Christianity*, 192–4; Oulton, 'Epistle of Cummian', 128, 133; Betten, 'Adoption', 488–9; O'Connell, 'Easter cycles', 80–3; Jones, 'Paschal tables', 417; idem, *Bedae opera*, 98 (wrongly giving AD 631 to 633 as the three-year journey of the embassy); Stevens, 'Scientific instruction', 83–5; Harrison, 'Epacts', 22; Walsh & Ó Cróinín, *Cummian's letter*, 3–7; Wallis, *Bede*, lix): The only clearly identifiable addressee of the letter, the Iona abbot Ségéne, held that position from ca. AD 624 to 652; Bede (*HE* 3.3) implies that the southern Irish had converted before Áedán's Northumbrian mission of AD 634; moreover, it can be inferred that the initial impetus for conversion came from a letter sent by pope Honorius, the bishop of Rome from AD 625 to 638 (*HE* 2.19). Being provided with these chronological ranges, a more precise dating becomes possible by Cummian's statement that the southern Irish delegation sent to the pope for advice spent Easter in Rome in a year, in which the date of this feast followed by the papal curia differed from their own reckoning by a whole month (*Epistola Cummiani* 281–283; Walsh & Ó Cróinín, *Cummian's letter*, 94). This year has been identified as AD 631 on the basis of reconstructions of the Easter reckoning followed by the Irish which have now been proven wrong by the discovery of the Padua table (for the Padua table cf. p. XVII–XIX). After that discovery only Wallis, *Bede*, lix has briefly commented on its implications on the date of Cummian's letter, so it might be appropriate to confirm the AD 631 date on that basis here: The Victorian reckoning followed in Rome and the 84 (14) observed by the Irish differed by a whole month (four weeks) in the date of Easter Sunday only in AD 631 and 642 (24 March and 21 April respectively in both cases) in the period

rian Easter table found in a later part of the Sirmond MS (item 45) and two references to a certain year of that table in the Munich textbook.[220] It is striking that all parallels between the Munich Computus and the Sirmond MS deal, in one way or another, with Victorius' computistical writings. Since the chronicle copied and extended from *De mirabilibus sacrae scripturae* in the final chapter of the Munich Computus is also based on Victorian calculations, it must be presumed that all this material was already part of the Munich computist's Victorian *Vorlage* of AD 689. Accordingly, it is this Victorian computus of AD 689 whose provenance can be identified with southern Ireland, presumably eastern Munster.[221]

In Ó Cróinín's theory, this Victorian computus would then have travelled north to Iona, where selected passages from it were incorporated into the final redaction of the Munich Computus. The northern provenance is suggested by only one source, the papal letter to the northern Irish clergy of AD 640. Its citation can be regarded as highly indicative for a northern Irish provenance of the Munich Computus only if it can be convincingly demonstrated that this letter is unlikely to have received a southern Irish reception at any stage before the compilation of the Munich Computus. Yet, the reception as well as the context of this letter rather implies that the southern Irish had copies of it, presumably even as early as AD 640. That the reception of this letter went beyond, at least, the charmed circles of the addressees is impressively illustrated by the fact that the largest part of this letter (and at the same time the only citation of this letter other than the one found in the Munich Computus) survives in the *Historia ecclesiatsica gentis Anglorum* of the Northumbrian monk Bede.[222] At the time of composition, AD 640, the Northumbrian church was, however, within the northern Irish ecclesiastic sphere of influence, being dependent on the monastery of Iona.[223] Since the abbot of Iona, Ségéne, was one of the addresses of the

AD 620 to 650; since from Bede's evidence the 642 date appears too late, AD 631 must be the date in question. The letter itself was then composed after the return of this embassy, which spent three years on their journey (*Epistola Cummiani* 277–280; Walsh & Ó Cróinín, *Cummian's letter*, 92), with the stay in Rome obviously happening right in the middle of it; hence the AD 632 date for the letter. The identification of Cummian is heavily disputed; cf. Mac Carthy, *Annals of Ulster* 4, cxli–ii; Schmid, *Osterfestberechnung auf den britischen Inseln*, 69; Kenney, *Sources*, 220; O'Connell, 'Easter cycles', 81; Jones, 'Paschal tables', 417; idem, *Bedae opera*, 90; Stevens, 'Scientific instruction', 85; Walsh & Ó Cróinín, *Cummian's letter*, 7–15; Ó Cróinín, *Early medieval Ireland*, 187; idem, 'Hiberno-Latin literature', 377–8; Blackburn & Holford-Strevens, *Companion to the year*, 795.

220 The textual parallel lies in the misspelled consul name *Bradua* instead of *Bardua*. MC 41.108–10, 62.66–8; Sirmond MS (Oxford, Bodleian Library, Bodley 309, fol. 115r). Cf. note 139.

221 For this Victorian computus of AD 689 used by the Munich computist see especially p. CXXIV–CXXVI below.

222 *HE* 2.19.

223 This letter can be securely dated by the fact that its author John refers to himself as pope-elect. Therefore, it can only have been written between the death of his predecessor Severinus and his own succession, which narrows the time-span to between August and December AD 640. Cf. Krusch, 'Einführung', 149–50; Plummer, *Bedae opera* 2, 112; Schmid,

letter, it would only have been natural that copies of this letter were sent from the mother-house to its Northumbrian daughter-foundations. Especially Lindisfarne, the first and principal daughter-house of Iona on Northumbrian soil, would have constantly been supplied with important church literature, including this papal paschal letter. Therefore, Bede's citation can well be explained by the circulation of texts among monasteries dependent on Iona.[224]

On the other hand, any papal letter received in Irelad or Britain, which was an extremely rare phenomenon, would have aroused the interest of every church within this region. This is especially true for southern Irish churches, which had looked to Rome for guidance ever since the paschal controversy of the late 620s.[225] Therefore, a papal letter exclusively concerned with the correct celebration of Easter would have been of utmost interest to southern Irish ecclesiasts, and for this reason alone it seems likely that copies of this letter were available in southern Ireland soon after AD 640. Additionally, the context of the letter may suggest that copies of it were sent by the papal curia not only to the northern Irish addressees, but also to their southern neighbours.

It has been quite recently suggested that the papal letter of AD 640 was a reply to an earlier letter written by northern Irish churchmen to inform the papal curia that they would celebrate Easter in AD 641 on the date denoted as Greek in the Victorian Easter table (which was the same date as the one listed in the 84 (14)).[226] Two main arguments speak, in my opinion, strongly against this hypothesis. As mentioned above, Ségéne, the abbot of Iona, appears among the addressees of the papal letter. It is well known that Iona followed the 84 (14)-year Easter reckoning until adopting the Dionysiac method in AD 716.[227] The

Osterfestberechnung auf den britischen Inseln, 44; Kenney, *Sources*, 221–2; Jones, *Bedae opera*, 98; Harrison, 'Letter', 227. For the history of the Northumbrian church at this time, which is dominated by the Iona monk Áedán as the informal head of this church in his position as abbot of Lindisfarne, cf. *HE* 3.3, 3.5, 3.14–17.

224 Note, however, that Bede states in his prologue to *HE* (Plummer, *Bedae opera* 1, 6) that he received copies of papal letters directly from the papal curia through Nothelm. For this fact cf. especially Mommsen, 'Papstbriefe', 387–92; and furthermore Manitius, *Geschichte*, 72. Whether or not the letter in question here was among Nothelm's corpus is impossible to establish. Plummer, *Bedae opera* 2, 2; Kenney, *Sources*, 221 believe it was. Note that the evidence concerning Nothelm suggests that southern Irish ecclesiasts could also have copied letters in the archive of the papal curia around AD 700 or earlier.

225 See especially the southern Irish embassy of AD 630–632 sent to the papal curia to ask for guidance in the paschal question (*Epistola Cummiani* 273–285; Walsh & Ó Cróinín, *Cummian's letter*, 92, 94). Cf. Gougaud, *Christianity*, 213.

226 Ó Cróinín, 'New heresy', 90–6. Discussions of this letter (without addressing the crucial question concerning the date of composition of the initial letter by the northern Irish clergy) can be found in Mac Carthy, *Annals of Ulster* 4, cxlviii–ix; Schmid, *Osterfestberechnung auf den britischen Inseln*, 44; Poole, 'Earliest use', 59–60; Kenney, *Sources*, 221–3; O'Connell, 'Easter cycles', 83; Jones, *Bedae opera*, 91–2, 98–9; Harrison, 'Letter', 227–9; Wallis, *Bede*, lxi (arguing that the initial letter was composed by the southern Irish clergy to alert the papal curia that the northern Irish had still not adopted the 'canonical' reckoning).

227 AU 715.4 (Mac Airt & Mac Niocaill, *Annals of Ulster*, 172). *HE* 3.4, 5.22, 5.24. For Bede's chronology of this question see Plummer, *Bedae opera* 2, 335–6. For Iona's con-

Victorian reckoning, therefore, was never followed in that monastery, especially not in AD 640. Even presuming that the 84 (14) was compared with the Victorian Easter table in Iona and other computistical centres in the northern *regiones Scottorum* around AD 640, the only Victorian Easter table known to have circulated in the British Isles, namely the one found in the Sirmond MS, does not incorporate double dates. In the years for which continental Victorian Easter tables list two dates, the Sirmond table only records the dates denoted (explicitly or implicitly) as Latin in those continental tables.[228] More specifi-

version cf. Ussher, *Antiquitates*, 366–7, 541; Krusch, 'Einführung', 165; Schmid, *Osterfestberechnung auf den britischen Inseln*, 71–6; Gougaud, *Christianity*, 198–9; Betten, 'Adoption', 494–7; Jones, *Bedae opera*, 104; Declercq, *Anno Domini*, 155.

228 Not all of the known Victorian Easter tables contain the mentioned double dates (and a detailed study of the Victorian doubles dates and especially of the implications of their abandonment is a pressing desideratum). In fact, double dates are recorded in only three of the surviving Victorian Easter tables, namely Gotha, Forschungsbibliothek, 75, fol. 77v–106r, the Bobbio Computus (Milan, Biblioteca Ambrosiana, H 150 inf., fol. 130r–132r), and the *Paschale Campanum* (ed. by Mommsen in *MGH AA* 9, 744–50). All of these Victorian Easter tables have a continental background: The provenance of the Gotha MS is Murbach (Krusch, *Studien* II, 6) and internal evidence in the Victorian Easter table renders it likely that it was originally composed in this Burgundian monastery (Krusch, 'Handschrift', 271, 276–9; Mommsen in *MGH AA* 9, 672–3); in the Bobbio Computus the Victorian Easter table is preceded by a calculation of the number of years from the creation of the world to the sixteenth regnal year of the Merovingian king Chlothar III of Neustria (Milan, Biblioteca Ambrosiana, H 150 inf., fol. 129v; cf. p. LXXIV), so a Neustrian provenance of this table is apparent; the *Paschale Campanum*, judging from the annal entries in the Easter table, is based on an exemplar from Illyria and originated in Campania (Mommsen in *MGH AA* 9, 744). Four Victorian Easter tables do not record a single double date: Sirmond MS (Oxford, Bodleian library, Bodley 309, fol. 113r–120v); Leiden, Universiteitsbibliotheek, Scaliger 28, fol. 3r–21r (Victorian table only on the recto pages; the Dionysiac table appears on the verso pages); Bern, Burgerbibliothek, 645, fol. 41r–48v, 59r–67v; and the Victorian Easter table in *Dial. Burg.* 16 (Borst, *Schriften*, 368–72). The Bern and Leiden tables vary between recording either the Greek or the Latin dates. The Sirmond MS and *Dial. Burg.* invariably record the dates explicitly or implicitly denoted as 'Latin' (note that for certain years the date with *luna* 22 (usually the 'Latin' lunar age) is marked as 'Greek', and therefore the 'non-Greek' date with *luna* 15 is recorded in these tables). The table transmitted in Paris, Bibliothèque Nationale, Lat. 4860, fol. 147v–148r and Vatican, Biblioteca Apostolica, Reg. Lat. 586, fol. 9r–10v constitutes a special case: It covers AP 141–212 (=AD 700–771) and is divided into *ogdoades* and *hendecades*, modelled on the Dionysiac table; it does not record any double dates at the beginning, invariably listing the Latin date, but from AP 181 (=AD 740) it frequently also mentions the Greek dates as alternatives. Except for two additional columns for denoting the years as common or embolismic and for recording the AP, this table is arranged in exactly the same way as the Padua *latercus*, the Easter table of the 84 (14) (cf. the facsimile in Warntjes, '84 (14)-year Easter reckoning', 80–2), listing the weekday and moon of 1 January, followed by the Julian calendar date and lunar age of the *initium quadragesimae* and then of Easter Sunday. Hence, it appears that this table was modelled on the *latercus*, and that the archetype of it was composed in Ireland; the division of the Victorian table into *ogoades* and *hendecades* was seemingly also employed in Ireland, since the Munich Computus and *CE* bear witness that such divisions were first undertaken by Irish computists for reasons of comparison with the Dionysiac reckoning (cf. MC 49.25–29, 52, 55.46–52, 62.87–95; *CE* p. 109). In this respect, it is interesting to note that the first years of this table only record

cally, in the year in question, AD 641, the Sirmond table only lists the Latin date, 8 April, *luna* 22.[229] Accordingly, it appears more likely that Irish churches in AD 641 would have followed the Latin rather than the Greek date of the Victorian Easter table, and that northern Irish churches would only have noted the (at this point very frequent) divergence of one week between the 84 (14) and the Victorian date.

From this evidence, the following interpretation of these events seems more suggestive: Between the acceptance of the Victorian reckoning by the southern Irish church in AD 632 and AD 640, only one year, AD 634, contained data with which the northern Irish clergy, following the 84 (14) with Easter lunar limits 14 to 20, would formally disagree, namely the celebration of Easter Sunday on *luna* 22.[230] In AD 641 and 642, however, two features occurred in suc-

the Latin date. I suspect, therefore, that the Irish archetype only or mostly recorded the Latin dates similar to the Sirmond MS, and that the Greek dates were then added on the Continent. Unfortunately, since this table begins with AP 141 (=AD 700), it has no immediate relevance for the investigation of the year in question, AD 641. The same apparently holds true for the Victorian Easter table in Turin, Biblioteca Nazionale, F VI 2 (cf. Cordoliani, 'Contribution', 169), which I have not examined.
It appears that the southern Irish abandoned the double dates to counter one of the major critiques brought forward against the Victorian reckoning by northern Irish computists, namely the inconclusiveness of certain years; for this critique see especially Columbanus, *Epistola* 2.7 (Walker, *Sancti Columbani opera*, 18); as papal consent was obviously supposed for the 'Latin' rather than the 'Greek' dates, abandoning the 'Greek' dates was only logical.

229 Sirmond MS (Oxford, Bodleian library, Bodley 309, fol. 114r). Cf. Krusch, *Studien* II, 31. Note that the other two Victorian Easter tables that only list a single date for this year (Leiden, Universiteitsbibliotheek, Scaliger 28, fol. 6r; Bern, Burgerbibliothek, 645, fol. 66r; cf. previous note) both also record the Latin date.

230 For the data of these years in the Victorian Easter table cf. generally Krusch, *Studien* II, 30–1, but more specifically the Victorian Easter table in the Sirmond MS (Oxford, Bodleian library, Bodley 309, fol. 114r) as the one (or at least similar to the one) presumably being followed by the southern Irish clergy. Note that the Victorian Easter table in the Sirmond MS consistently differs from Krusch's edition in the lunar age of Easter Sunday in years with epact 27; Krusch's data implies that an embolism ends in April of such a year; this is, however, not the case in the Sirmond table, where the lunation ending in April is hollow, a fact which consequently leads to a lunar age of Easter Sunday increased by one in the Sirmond table; accordingly, Krusch gives *luna* 21 for Easter Sunday in the year in question (AD 634; epact 26 is a misprint of 27 in Krusch's edition), the Sirmond table *luna* 22. In general, the Sirmond table contains some very curious data, not the least being an unparalleled placement of the *saltus*, applying it in the tenth rather than the sixth year of the Victorian 19-year cycle (the tenth year, interestingly enough, coinciding with the year of Christ's passion), i.e. four years later than Victorius' original placement; only a detailed study of this table while reveal whether or not this table was ever actually used or rather constituted a (unfinished) theoretical construct (in that case presumably designed to solve theological differences or questions, like the chronological data of Christ's resurrection, as a placement of the *saltus* in the tenth rather than the sixth year of the Victorian 19-year cycle meant that the lunar age on Christ's resurrection increased from the Victorian 16 to the canonical 17 – though this change is not noted in the table!). In the end, a detailed study not only of this but of all Victorian Easter tables is desperately needed (cf. also the previous two notes as well as note 139). For the Easter principles followed by the northern Irish

cessive years of the Victorian Easter table that were totally unacceptable for followers of the 84 (14): In AD 641, the Latin date, which appears to have been the one followed by the southern Irish, fell as late as *luna* 22; AD 642 was the 16th year of the Victorian 19-year lunar cycle, in which the Easter full moon always fell on 20 March, and therefore before the equinox of 21 March.[231] Celebrating Easter on the 22nd moon, and observing the Easter full moon before the equinox, constituted breaches of more generally recognized rules for the observance of Easter, since they were unacceptable not only to adherents of the 84 (14), but also to followers of the Dionysiac reckoning. Consequently, the northern Irish clergy took this opportunity of two successive obvious breaches of commonly acknowledged Easter rules to vigorously and officially complain about the Victorian reckoning followed by their southern Irish counterparts.[232]

The argument and tone of this letter to the pope, on whose authority the southern Irish had converted in the first place, must have been very similar to, and as fierce as, in Columbanus' letter to pope Gregory the Great.[233] The papal

and the construction of their table, the 84 (14), cf. especially Mc Carthy, 'Lunar cycle', 204–24; idem, 'Origin', 25–49; Blackburn & Holford-Strevens, *Companion to the year*, 870–5; Warntjes, '84 (14)-year Easter reckoning', 31–85.

231 The 16th year of the Victorian 19-year cycle has epact 25 on 1 January; accordingly, *luna* 25 also falls on 1 March, and therefore the next full moon, *luna* 14, occurs 19 days later, on 20 March (for the relation between epact and Easter full moon cf. also Krusch, *Studien* II, 15); note that the earliest Easter Sunday possible in such a year of the Victorian reckoning consequently was 22 March, *luna* 16, and thus fell after the equinox of 21 March; the following full moon, which occurred 29 days later, on 18 April, was totally unacceptable as Easter full moon in the Victorian reckoning, as it could lead to Easter Sunday as late as 26 April. In the year in question, AD 641, Victorius lists 24 March, *luna* 18 for Easter Sunday; note that the Victorian table in the Sirmond MS (Oxford, Bodleian library, Bodley 309, fol. 114r) records a wrong date for Easter Sunday in this year: *VIIII Kl* has been corrupted to *Non-* (a misinterpretation of the numeral).

232 Mac Carthy, *Annals of Ulster* 4, cxlviii; Kenney, *Sources*, 222; Gougaud, *Christianity*, 194; Jones, *Bedae opera*, 98–9, also argue that the northern Irish clergy wrote the initial letter in justification of their own practise, i.e. of the 84 (14); Wallis, *Bede*, lxi, on the other hand, believes that the initial letter was composed by the southern Irish clergy to alert the papal curia that the northern Irish still clung to the 84 (14). Cf. also Blackburn & Holford-Strevens, *Companion to the year*, 795.

233 Columbanus complained about Victorius' practice of allowing *luna* 21 and 22 for Easter Sunday and not observing the equinox in *Epistolae* 1.3–4, 2.5 (Walker, *Sancti Columbani opera*, 2–6, 14–16). Concerning *luna* 21 and *luna* 22, Columbanus there argues that these are *extra ius lucis*; it appears that the northern Irish clergy wrote something very similar to the papal curia, if they did not indeed quote Columbanus, and that the passage of the papal reply preserved in the Munich Computus (cf. MC 63.26–28) constitutes a direct response to this accusation, arguing that it is rather *luna* 14 that belongs to the shadows. For Columbanus' criticism of Victorius cf. also Krusch, *Studien* I, 313–4; idem, 'Einführung', 144–8; Mac Carthy, *Annals of Ulster* 4, cxxvii–xxxi; Schmid, *Osterfestberechnung auf den britischen Inseln*, 8–9, 25; idem, *Osterfestberechnung in der abendländischen Kirche*, 68–76; Kenney, *Sources*, 191–2; Gougaud, *Christianity*, 189–91; O'Connell, 'Easter cycles', 79–80; Jones, 'Paschal tables', 419–20; idem, *Bedae opera*, 78–80; Cordoliani, 'Computistes insulaires', 8–9; Harrison, 'Epacts', 22; Walsh & Ó Cróinín, *Cummian's letter*, 20–1; Ó Cróinín, 'Mo Sinu maccu Min', 35–6; idem, 'Computistical works', 48–9; Mc Carthy,

response to this attack from the northern Irish clergy, therefore, concerned the Irish church as a whole. Southern Irish delegates had travelled to Rome only nine years earlier to obtain papal authority for their Easter customs.[234] Some of the officials in the papal curia who had dealt with this delegation would still have been alive and in office nine years later, and very well informed about the conflicts within the Irish church. It must have been as much their concern to reassure the southern Irish clergy of the correctness of their methods (or to inform them about a possible change in reckonings that had taken place in Rome in the meantime) as it was to rebuke the northern Irish for following an outdated Easter table based on uncanonical principles. Hence, it seems reasonable that copies of the papal response were not only sent to the northern Irish addressees, but also to the southern Irish clergy; and even if the southern Irish clergy did not receive copies of this letter directly from the papal curia, interest in this document dealing with the celebration of the most important Christian feast, as well as with the fundamental issue of controversy between the two Irish churches, must have been enormous, and presumably led to further debates between the two churches and to the southern Irish acquiring copies from their northern neighbours.

In summary, the question of a specific regional provenance of the Munich Computus cannot be conclusively established by an analysis of the sources cited by its author. Only the Victorian computus of AD 689 used by the Munich computist can, with some degree of certainty, be connected to southern Ireland, presumably eastern Munster. It may well be that this southern Irish computistical text of AD 689, based on the Victorian reckoning, travelled north and was there incorporated into a new compilation thirty years later. It appears, however, just as likely that the Victorian material of AD 689 was re-used for the composition of a new computistical textbook in the same, southern Irish monastery, which was also in possession of a copy of the papal letter of AD 640 addressed to the northern Irish clergy.

Historical context: The historical context may prove more revealing in this question. In previous studies, the only aspects of the Munich Computus that has been placed into historical context to shed more light on the provenance of this text are its date of composition, AD 719, combined with its Dionysiac bias and mentioning of the *latercus*. Since the *latercus*, i.e. the 84 (14)-year Easter reck-

'Lunar and Paschal tables', 149–50; McCluskey, *Astronomies*, 88–9; Holford-Strevens, *History of time*, 51–2. Eighty years later, Bede vehemently argues against the Victorian usage of observing the Easter full moon before the equinox (*DTR* 51, cited in *Lib. comp.* 4.21), and, interestingly enough, discusses the 16th year of Victorian 19-year cycle in detail in this context (*DTR* 62.24–38); the technical argument of the northern Irish clergy in AD 640, complaining about the Victorian Easter full moon falling before the equinox in AD 642, may have been very similar to this Bedan passage. For Bede's criticism on Victorius cf. Bucherius, *De doctrina temporum*, 194–201; van der Hagen, *Observationes in veterum patrum prologos et epistolas*, 308–12, 323–5; Jones, *Bedae opera*, 385–6, 389–90; Ohashi, 'Sexta aetas', 59–60; Wallis, *Bede*, 341, 348–9; Pillonel-Wyrsch, *Calcul*, 325–8, 357–9.

234 Cf. note 225.

oning, was replaced by the Dionysiac reckoning in Iona only three years previously, in AD 716, this centre was quite naturally regarded as the possible place of composition of the Munich Computus. However, the fact that the Victorian reckoning has a far more prominent place than the *latercus* in the Munich Computus has been passed over in silence. If a recent conversion to the Dionysiac reckoning motivated the compilation of the Munich Computus, then a conversion from the Victorian reckoning rather than the *latercus* cannot readily be ruled out, but appears, in fact, more likely.

Generally, the conversion to the Dionysiac reckoning is explicitly recorded for only one monastery in the *regiones Scottorum*, Iona.[235] Additionally, Bede suggests that Adomnán, the abbot of Iona from AD 679 to 704, converted some monasteries in the northern regions of the *regiones Scottorum* (though not his own) to the Dionysiac reckoning in the last twenty years of his abbacy.[236] But this is all the information that has survived.[237] It is quite likely that some northern monasteries converted from the 84 (14) to the Dionysiac reckoning in the three decades or so before Iona; others possibly only slowly followed Iona's example.

But what about the southern Irish clergy? The only trustworthy information is their conversion from the 84 (14) to the Victorian reckoning in AD 632, recorded in Cummian's famous letter.[238] After having taken this decisive step,

235 Cf. note 227.

236 *HE* 5.15, 5.21. Bede's statement that Adomnán converted many Irish churches to the correct (undoubtedly Dionysiac) Easter reckoning, but did not succede in the monastery over which he presided, Iona, is well discussed; cf. Ussher, *Antiquitates*, 366, 381, 541; Krusch, 'Einführung', 161; Schmid, *Osterfestberechnung auf den britischen Inseln*, 45–7, 70–1, 78–9; Gougaud, *Christianity*, 198–9; Betten, 'Adoption', 490–2; O'Connell, 'Easter cycles', 84; Jones, *Bedae opera*, 104; Wallis, *Bede*, lxiii; Blackburn & Holford-Strevens, *Companion to the year*, 795.

237 It has been argued that Bangor had adopted the Dionysiac reckoning in around AD 610, i.e. 75 or so years before Adomnán's conversion. This theory is based on a Würzburg Gospel fragment (Würzburg, Universitätsbibliothek, M.p.th.f. 61, inserted fol. 29r; transcription and translation in Ó Cróinín, 'Mo Sinu maccu Min', 37; facsimile Ó Cróinín, *Early Irish history*, 214) which states that Mo Sinnu maccu Min, the abbot of Bangor, was the first to learn *compotem* from a Greek. Mac Carthy, *Annals of Ulster* 4, cxxxiii–iv (accepted by Kenney, *Sources*, 218) and Jones, 'Legend of St. Pachomius', 203–5 thought that the Alexandrian / Dionysiac 19-year cycle was meant by *compotem* (in the form of the poem *Nonae Aprilis* or otherwise; this is accepted as a possibility by Stevens, 'Scientific instruction', 92, 94–5, who connects this person with the author of the *Disputatio Morini*). On this basis, Grosjean, 'Recherches', 215–43 vehemently advocated the hypothesis that Bangor had accepted the Dionysiac reckoning in around AD 610. Cf. also Harrison, 'Epacts', 22. However, Ó Cróinín, 'Mo Sinu maccu Min', 36–47 has demonstrated that *compotem* refers to a list of the numerical values of the letters of the Greek alphabet rather than to the Dionysiac reckoning in any form. Moreover, a representative of Bangor appears to have been among the addresses of the papal letter to the northern Irish clergy of AD 640, a fact clearly indicating that Bangor still adhered to the 84 (14) at that time.

238 Cummian's letter gives unambigious evidence that it was the Victorian reckoning which was adopted by the southern Irish clergy in AD 632, since Cummian speaks of a cycle of 532 years and lunar limits for Easter Sunday of 16 to 22, which are Victorian characteristics: *Epistola Cummiani* 10–11, 75–85 (Walsh & Ó Cróinín, *Cummian's letter*, 56, 66, 68).

they would quite certainly not have abandoned this reckoning immediately or too soon thereafter. Moreover, it seems that the southern Irish compared the Victorian with the Dionysiac method for quite some time before decisively and unreservedly adopting the Dionysiac system.[239] It is at least possible that the outright acceptance of the Dionysiac reckoning in the south was triggered by developments among the northern clergy. The eventual conversion of Iona may have led the southern Irish clergy to reconsider its position and to adopt the Dionysiac system soon after Iona. In that case, the Munich Computus could as well have been written in a southern monastery that had converted just after AD 716. However, the developments in Rome and in Anglo-Saxon England, of which the southern Irish must have been informed, render it more likely that the southern Irish monasteries had converted to the Dionysiac reckoning by the end of the seventh century.[240]

Cf. Krusch, 'Einführung', 150–1; Schmid, *Osterfestberechnung auf den britischen Inseln*, 25–7; Poole, 'Earliest use', 59–60; Gougaud, *Christianity*, 190; Harrison, 'Epacts', 22; Walsh & Ó Cróinín, *Cummian's letter*, 18–9, 22; Blackburn & Holford-Strevens, *Companion to the year*, 795. Betten, 'Adoption', 488–9 implies, quite mistakenly, that the southern Irish adopted the Dionysiac reckoning in the early AD 630s. Jones, *Bedae opera*, 90–1 argues quite unconvincingly that Cummian confused Victorian and Dionysiac principles. In the same vein, Stevens, 'Scientific instruction', 83–4 argues that it is not known which system was adopted by the southern Irish clergy at that time. Walsh & Ó Cróinín, *Cummian's letter*, 46, followed by Wallis, *Bede*, lx, and hesitantly by Declercq, *Anno Domini*, 153–5 outline the hypothesis that the southern Irish from Cummian's time followed a mixed Victorian-Dionysiac system (this theory appears to be based on Jones' similar statements concerning the Easter reckonings followed in Anglo-Saxon England; cf. note 87); I find this quite unlikely, since no such mixed Easter table has survived; more likely, in my opinion, is the adoption of a Victorian Easter table without double dates (cf. notes 228–230) and comparison of this table with a Dionysiac one in the subsequent decades (cf. the following note).

239 Both the Munich Computus and the *Computus Einsidlensis* impressively demonstrate that the Victorian and the Dionysiac reckonings were compared in detail in seventh-century Ireland. For the passages in the Munich Computus cf. note 252; the comparative material of the *Computus Einsidlensis* can be found in Einsiedeln, Stiftsbibliothek, 321 (647), p. 109–11, 119, 123–4. Less comparative material can be found in *DRC* 72, 86, 90, 98–9, an indication of its later date of composition (for which cf. especially the detailed discussion p. CXCI–CCI below). The fact that Victorian and Dionysiac Easter tables were used in parallel for a considerable period of time in seventh-century Britain and Ireland is stressed (but exaggerated) by Jones, *Bedae opera*, 103; Declercq, *Anno Domini*, 154.

240 The papal curia changed to the Dionysiac reckoning in the 640s or 650s, and in Northumbria the decisive step to the adoption of the Dionysiac reckoning was taken in AD 664 at the Synod of Whitby (cf. notes 86–87). Twenty years later, Adomnán adopted the Dionysiac reckoning and spread word about it in Ireland (cf. note 236). Influenced by these developments, I presume that the southern Irish would have converted from the Victorian to the Dionysiac reckoning between the 650s and the 690s, probably closer to the end of this period, as the Victorian computus of AD 689 cited by the Munich computist illustrates that the Victorian reckoning still enjoyed considerable popularity at that time (cf. now also Bisagni & Warntjes, 'Early Old Irish material', 90–1). Kenney, *Sources*, 215 argues that the southern Irish converted from the Dionysiac to the Victorian reckoning in the aftermath of the Synod of Whitby; Harrison, 'Luni-solar cycles', 72–3 argues that 'in the seventh century the Easter tables of Dionysius Exiguus gradually gained favour, more particularly first

But this hypothesis that the southern Irish clergy had adopted the Dionysiac reckoning some two or more decades before the composition of the Munich Computus does not rule the southern *regiones Scottorum* out as the provenance of this text. After all, there is no obvious reason why the date of composition, AD 719, should necessarily be connected to the date of conversion to the Dionysiac reckoning of the monastery in which this text originated. In fact, there is a far more likely explanation for this specific date. The Dionysiac Easter table known to the Munich computist was one of 95-years. Dionysius himself had composed a 95-year table extending from AD 532 to 626.[241] A certain Felix Gillitanus, presumably among other computists, recalculated the necessary data for the following 95-year period, from AD 627 to 721.[242] In AD 719, then, two

in the south'; Mc Carthy, 'Lunar and Paschal tables', 176 implies that the change had occurred in southern Ireland before Whitby; Wallis, *Bede*, lx–iii rather confusingly argues that the southern Irish had adopted the Victorian reckoning or a 'hybrid Victorian-Dionysian stystem' at the time of Cummian, that they 'converted to the Alexandrian reckoning' (which appears to refer to the mentioned hybrid system) before 640, but then that the southern Irish still followed the Victorian reckoning at the time of Bede. For the unconvincing theory of a hybrid system cf. note 238. Quite balanced, Blackburn & Holford-Strevens, *Companion to the year*, 795 state that 'Victorius was generally preferred to Dionysius in Ireland for many generations'.

241 Ed. by Krusch, *Studien* II, 64, 70–6.

242 Felix Gillitanus' prologue to his Easter table is published in Krusch, *Studien* II, 87. Cf. van der Hagen, *Dissertationes*, 38–41; Piper, *Kalendarium*, 90; Krusch, 'Einführung', 114–5; idem, *Studien* II, 60; Poole, 'Earliest use', 61; Jones, *Bedae opera*, 73–4; Cordoliani, 'Traités', 61–2; Wiesenbach, *Sigebert von Gembloux*, 50–1; Declercq, *Anno Domini*, 130–1, 152–3; idem, 'Dionysius Exiguus', 230. The table itself has not survived, and neither has any other Dionysiac Easter table covering the entire period AD 626–721. Isidore's 95-year Easter table (*Etym.* 6.17) as printed by Lindsay is a curious construction; it is based on Dionysiac principles, and, in fact, the data given matches the data of Dionysius' table from AD 532–626 exactly, except for (only!) the Julian calendar dates of Easter Sunday (but not the lunar data, nor the list of bissextile years) of the first 19-years, which are recalculated to match the first 19 years of the following 95-year period (in the 17th year it should simply read *Idibus Aprilis* instead of *XI Idus Aprilis*, a date that does not exist in the Julian calendar). For this Isidorian table cf. especially Krusch, 'Einführung', 117–9; and furthermore Bucherius, *De doctrina temporum*, 192; Noris, *Dissertationes*, 182; van der Hagen, *Dissertationes*, 35–7; idem, *Observationes in veterum patrum prologos et epistolas*, 241; Piper, *Kalendarium*, 85, 90–1; Anspach, 'Fortleben', 340; Jones, 'Paschal tables', 415; idem, *Bedae opera*, 74; Cordoliani, 'Contribution', 171; Declercq, *Anno Domini*, 151–2; Borst, *Schriften*, 330–1. According to Krusch, the MSS of *Etym.* vary considerably in the data given in the Easter table, and only a critical edition of this table based on all MSS can shed more light on Isidore's original construction and, more importantly, on the seventh- and eighth-century corrections or recalculations of it; concerning the original Isidorian table, Krusch's hypothesis that it covered the same period as the one compiled by Dionysius, AD 532–626, is very convincing. Any future discussion of the Isidorian table must also take note of its later reception independent of *Etym.*; two strands of this reception are most notable in the present context: first, the Isidorian table covering the years AD 532–626 apparently experienced a revival in the eleventh century, when it regained its currency at the end of the underlying Easter cycle of 532 years (e.g., Zürich, Zentralbibliothek, Rh. 83, fol. 14r–17r); second, the Easter table in the ninth-century MS London, British Library, Harley 3017, fol. 50r–51r is Isidorian in its structure; as Jones, 'Two Easter tables', 204 found out,

years before the expiration of this table, the need was felt to recalculate the data to match the following 95-year period, covering AD 722 to 816. Even though Bede's abbot Ceolfrith wrote in AD 710 that many computists understood how to create a cyclic 532-year table on Dionysiac principles, and Bede himself composed such a table,[243] this knowledge certainly was not universal. The Munich computist, for his part, without doubt regarded the Dionysiac table as consisting of 95 years. He discussed the non-cyclic character of the 95-year Diony-

it covers the period AD 627–721, but Jones did not realize the Isidorian connection; he probably would have been aware of this context, had he known that the same table can be found in the related (and earlier) MS Paris, Bibliothèque Nationale, Lat. 5543, fol. 145r–v, where it is embedded in an abridged version of Isidore's chapter on Easter caculation (*Etym.* 6.17; this Paris version does not contain all 32 paragraphs, but ends with §20). Besides this recension of the Isidorian Easter table transmitted through ninth-century Fleury or Fleury connected MSS, the only surviving Dionysiac Easter table known at present that covers almost the entire 95-year period of AD 627–721, i.e. the direct continuation of Dionysius' original table, is the Périgueux table. This table consists of only the calendar dates for Easter Sunday from AD 631 to 721 noted by consecutive numbering of the days of the months. It was engraved on a marble plaque which used to be incorporated in the right side of the main altar of the Church of Saint-Étienne in Périgueux. Scholars thought for a long time that this table had disappeared when the church was destroyed in 1575, but it was, in fact, placed in the southern wall in the subsequent rebuilding of the church. The Périgueux table is edited (with mistakes) in Mai, *Scriptorum veterum nova collectio* 5, 69, and a photograph is published in Cordoliani, 'Table pascal de Périgueux', facing p. 57. For this table cf. especially Cordoliani, 'Table pascal de Périgueux', 57–60; and furthermore Krusch, 'Einführung', 129–31; Schmid, *Osterfestberechnung in der abendländischen Kirche*, 83. Stevens, 'Rabani', 170 mistakenly argues that this table shows Victorian data. The only other two Dionysiac tables covering at least parts of the 95-year period between AD 627 and 721 are connected to Echternach, and through Echternach ultimately to Rath Melsigi in Ireland: The first is Willibrord's (Paris, Bibliothèque Nationale, Lat. 10837, fol. 40v–41v, 43r–44r), covering AD 684 to 797, while it appears to have been composed in four separate parts and stages: 1) AD 684–702 (fol. 44r); 2) AD 703–721 (fol. 40v); 3) AD 722–759 (fol. 41r–v); 4) AD 760–797 (fol. 43r–v); cf. especially Lowe, *Codices Latini Antiquiores* 5, 26; Ó Cróinín, 'Rath Melsigi', 155–6; idem, 'Echternach manuscript fragments', 137; Obrist, 'Astronomical sundial', 74; Declercq, *Anno Domini*, 160–1; Freise, 'Grundformen', 517; and furthermore Cordoliani, 'Contribution', 172; Ferrari, *Sancti Willibrordi*, 10. The second can be found in Paris, Bibliothèque Nationale, Lat. 9528, fol. 201 (not in Lat. 10399, as Ó Cróinín, 'Rath Melsigi', 155 mistakenly states; for the MS cf. Lowe, *Codices Latini Antiquiores* 5, 20 (No. 585)), which I have not seen; according to Ó Cróinín, 'Rath Melsigi', 153; Ferrari, *Sancti Willibrordi*, 10 it covers the period AD 706–778; but since the fol. in question is a fragment, the original table quite certainly covered a longer period; if that period was AD 703–816, then this table would exactly represent a continuation of the older 95-year table as could have been expected to be composed in AD 719 or 720 by adding the newly calculated 95-year table from AD 722 to 816 to the last 19-year cycle of the expiring 95-year table (AD 703–721) in the same way as Dionysius had included the final 19-year cycle of the Cyrillian table.

243 *HE* 5.21; *DTR* 65. For Ceolfrith's statement and Bede's construction of a 532-year Easter table based on Dionysiac principles cf. Noris, *Dissertationes*, 157–61; van der Hagen, *Dissertationes*, 19–21; idem, *Observationes in veterum patrum prologos et epistolas*, 251–2, 325–6; Piper, *Kalendarium*, 91–2; Krusch, 'Einführung', 166–7; Jones, *Bedae opera*, 391; Englisch, *Artes liberales*, 294–5; Wallis, *Bede*, 352–3; Declercq, *Anno Domini*, 157–9; Ohashi, 'Sexta aetas', 60; Pillonel-Wyrsch, *Calcul*, 368–9.

siac table compared to the cyclic 532-year table of Victorius without a word about the fact that the cyclic 532-year period could also easily be applied to the Dionysiac system.[244] More revealingly, the Munich computist compares the *concurrentes*, i.e. the weekdays of 24 March, of successive 95-year periods, apparently to prove that a 95-year period is not cyclic concerning its solar data.[245] Moreover, whenever the Munich computist used his present year, AD 719, as an example for a calculation, he always followed this up by the examples of the subsequent two years, AD 720 and 721, so that his series of examples went right to the end of the current 95-year Easter table.[246] Whether the Munich computist had the ability to translate the expiring 95-year Easter table into a subsequent one must remain speculative. Still, it is noteworthy in this respect that he does not refer to Dionysius' manual for this task.[247] It appears, therefore, that the Munich Computus was compiled in AD 719 to collect all the material available for a thorough understanding of the Dionysiac reckoning, so that this knowledge would enable the translation of the 95-year table expiring in AD 721 to the subsequent 95-year period, covering AD 722 to 816.

The underlying hypothesis that Dionysiac Easter tables composed at this time would not always cover the 532-year period from AD 532 to 1063, as Bede's statement may imply, but would still be calculated for 95-year periods in some regions where the 532-year cyclic character was still not adopted for the Dionysiac reckoning, is confirmed by some of the surviving Easter tables. Even though these tables do not start in AD 722 (for the simple reason that the years or 19-year periods preceding the time of copying in the second half of the eighth century were not deemed relevant), they end in AD 816.[248] Consequently, they represent only the second part of 95-year

244 MC 65.20–26.

245 MC 67.

246 MC 56.15–55, 58.33–80.

247 Dionysius gives quite a detailed manual for the translation of an expiring 95-year Easter table to a subsequent one in his prologue; cf. note 104. In the end, only the solar data of every bissextile year of the previous 95-year table had to be recalculated.

248 Piper, *Kalendarium*, 32–5 (commentary on p. 83–7, 93) published a Dionysiac Easter table extending from AD 779 to 816 from the so-called Godeshalc gospel (Paris, Bibliothèque Nationale, Nouvelle acquisition latine 1203, fol. 124v–126r). It is a very extensive table, including columns for the weekday and lunar age of 1 January, as well as the Julian calendar date of the *initium quadragesimae*. Since the Munich computist explicitly stresses that the Irish prefer to note the epact on 1 January instead of 22 March, and since in all of his calculations the lunar and solar data on 1 January are his point of orientation, it can be argued that this Easter table derives from an Irish exemplar. Moreover, the column of the Julian calendar date of the *initium quadragesimae* was a characteristic feature of the 84-year Easter table (cf. Mc Carthy & Ó Cróinín, 'Irish 84-year Easter table', 61–2; Mc Carthy, 'Easter principles', 218–9; Blackburn, Holford-Strevens, *Companion to the year*, 873–5; facsimile in Warntjes, '84 (14)-year Easter reckoning', 80–2), and for this reason Irish computists may have introduced this column also into Dionysiac Easter tables (as they may have done in Victorian tables; cf. note 228). A Dionysiac Easter table similar in extent, and possibly a continuation of Godeshalc's Easter table, can be found for the period AD 798 to 854 in Vatican, Biblioteca Apostolica, Reg. Lat. 1260, fol. 112r–114r; Vatican, Biblioteca Apostolica, Pal. Lat. 1447, fol. 19v–22r; Vatican, Biblioteca Apostolica, Pal.

Easter tables, originally covering the period AD 722 to 816 and being presumably composed in Ireland.

Hence, the date of composition does not necessarily imply that the Munich Computus was written in a centre just recently converted to the Dionysiac reckoning. The context of the reckonings referred to in this text may nevertheless provide the key to a more specific location of its provenance. An analysis of Easter reckonings generally allows for only a very broad division between the northern and the southern Irish churches, as these two areas had distinctively different histories. As outlined above,[249] the southern Irish church converted to the Victorian reckoning in AD 632. For some time in the second half of the seventh century, Victorian and Dionysiac Easter tables appear to have been compared and used in parallel, before the Dionysiac reckoning was finally accepted towards the end of that century. In the northern regions of the *regiones Scottorum*, the Victorian reckoning was never used, as far as can be established, for the determination of Easter Sunday. Some monasteries seem to have converted from the 84 (14) to the Dionysiac by the end of the seventh century, while Iona, and others following Iona's example, held out until AD 716, or slightly thereafter. Now, the general belief is that a text incorporating information about the *latercus* as late as AD 719 can only have been composed in the northern *regiones Scottorum*. But would not the very mentioning of that reckoning have been regarded as heretic so shortly after its abandonment? Moreover, why would the Victorian reckoning have such a prominent place in a text written in a monastery that never followed that reckoning? The argument here might be that this reckoning would have provided some valuable information about the workings of the 19-year lunar cycle, which underlay not only the Dionysiac, but also the Victorian reckoning. Yet, the Victorian reckoning appears to have been despised by the northern Irish clergy throughout the seventh century.[250] It is therefore hardly conceivable that this reckoning would appear so prominently in a northern text, especially since there was no immediate necessity to do so. The Dionysiac reckoning and its 19-year cycle could be understood without reference to the Victorian version. In fact, it was rather the Victorian reckoning and its 19-year cycle that was only fully understood after all the details of the Dionysiac system had been systematically analyzed and then applied to the Victorian one.

If, now, the Munich computist's statements concerning these three reckonings are investigated in detail, the following picture presents itself. This text deals almost exclusively with the Victorian and Dionysiac reckonings, while the latter is more prominent and clearly the one favoured by the author.[251] Quite

Lat. 1448 13v–16r, which, it will be noted, additionally includes the lunar data for the *initium quadragesimae* (see the facsimile in Appendix 8); only one 19-year period of this table (AD 817–832) is transmitted in St Dunstan's classbook (Oxford, Bodleian Library, Auct. F 4 32, fol. 21r).

249 For this and the following cf. p. LXXXVIII–LXXXIX.

250 Cf. notes 232–3.

251 Whenever certain methods of calculation are illustrated by examples, these examples invariably come from the Dionysiac reckoning: MC 51.22–34, 56.15–57, 58.33–80, 59.38–

a few passages outline detailed comparisons of these two reckonings, some of which being based on the earlier *Computus Einsidlensis*.[252] The 84 (14), by its technical term *latercus*, is only randomly introduced by the Munich computist into these earlier discussions, with referrences to it appearing invariably as additions to the comparisons of the Dionysiac with the Victorian reckoning.[253] The apparent reason for the inclusion of the *latercus* information is to make this textbook and its comparative data as comprehensive as possible by adding the few features known about the 84 (14) ad to gain a third perspective on certain problems and questions. In this respect it is of primary importance to be aware of the Munich computist's familiarity with these reckonings: He is perfectly acquainted with the Victorian reckoning, he records very peculiar, otherwise (that is later) rejected features of the Dionysiac one (even though this is the reckoning he clearly advocates),[254] while he is quite unfamiliar with the *later-*

105. Moreover, the Easter new and full moons, as well as the years of the 19-year cycle, the detailed relation between the lunar age of the *initium quadragesimae* and Easter Sunday, and the embolisms are only listed for the Dionysiac reckoning: MC 55.53–69, 57, 58.81–94, 61.11–27. The general Easter rules outlined are the Dionysiac ones: MC 56.2–14. On the other hand, the passages that only show a Victorian bias are almost exclusively of a chronological rather than of a strictly speaking computistical nature, since they deal with cyclic and linear concepts and recordings of time rather than with the calculation of Easter: MC 41.107–110, 62.65–67, 64, 65, 68.

252 Cf. MC 49.16–32, 50.2–67, 52, 55.46–52, 58.2–6, 62.68–110; for the *Computus Einsidlensis* being the source for some of these passages compare the *apparatus fontium* to these passages in the edition; cf. also note 239.

253 Cf. MC 50.67–76, 52.3–26, 58.2–7, 62.118–122. This reckoning is also referred to, without mentioning of the technical term *latercus*, in MC 65.27–33. All these passages dealing with the 84 (14) are discussed in Warntjes, '84 (14)-year Easter reckoning', 31–85, with facsimiles of them on p. 83–5.

254 The sequence of lunations attributed to the Dionysiac reckoning by the Munich computist and consistently applied throughout this textbook is very idiosyncratic, since it records full lunations for March and April, hollow lunations for May and June, instead of a sequence alternating between full and hollow lunar months; cf. MC 50.7–21, 50.29–39, 61.17–23, 62.97–110 (but note that this sequence of lunations is not applied in 50.22–28, and that the Munich computist in one instance (52.36–40) even argues against it). The only other text (besides the fragmentary passage of the Harvard fragment, which is directly related to the Munich Computus; cf. p. CXC–CXCI), which also applies this peculiar sequence of lunations is *CE* p. 109–10 (but note that the Dionysiac sequence of lunations subsequently outlined on p. 111 is the alternating one; I suspect that the original was corrected to the alternating one by a later copyist), on which the Munich computist' account is clearly based. This theory presumably originated in the fact that the lunations ending in April in the 8th, 11th, and 19th year of the Dionysiac 19-year cycle consist of 30 days, a fact that immediately became apparent to seventh century Irish computists when comparing the lunar data of Easter Sunday with the Dionysiac epacts of 1 January or 22 March; usually, these lunations were regarded as embolisms or deferred March lunations; seventh-century Irish computist, however, did not consider these options, but rather presumed that all April lunations were full, and consequently constructed this peculiar sequence of lunations, which is also mentioned in *DRC* 72; note, however, that the author of that Irish textbook immediately rejects this theory in favour of the common alternating sequence of lunations, explaining in detail the three exceptions of the 8th, 11th, and 19th year (which is one of the technical proofs that *DRC* must have been composed later than MC); cf. p. CXCIX–CC below.

cus, misapplying the *latercus* data available to him and mistakenly projecting features of the Dionysiac and Victorian reckonings onto this older 84-year cycle.[255] Particularly the misconceptions introduced by the Munich computist concerning the *latercus* (apparently not having an Easter table of that reckoning at hand) renders it unlikely that this text was written in a centre that had only recently abandoned this reckoning and therefore would still have had all the expertise in it.

From this evidence it is obvious that the Munich Computus was presumably written in a monastery with a tradition in following the Victorian reckoning, as well as in comparing the Victorian with the Dionysiac system; moreover, very basic information about the *latercus* was available in that computistical centre in AD 719, but not the expertise to apply this data properly. These facts point, in my opinion, to a southern Irish rather than a northern Irish monastery, as only in southern Irish monasteries was the Victorian reckoning in use for some time in the seventh century, while the *latercus* was abandoned there in the early 630s.[256] Judging, very cautiously, from the evidence of the source material, i.e. from *De mirabilibus sacrae scripturae* and the Sirmond MS, an east Munster monastery may be a candidate (not only for the composition of the underlying Victoran computus of AD 689, but als for the Munich Computus proper). The author himself appears to have been a person who was taught in the Victorian reckoning, who followed the slow conversion over decades to the Dionysiac reckoning by comparison of the two systems (with not all details of the Dionysiac reckoning being fully understood in the course of this process), and who also observed the controversy between the two Irish churches, especially its technical arguments, from which he knew the basic facts about the 84 (14). As an aged man, he then appears to have been commissioned by the *seniores* of his monastery to use his knowledge for the composition of a thorough and com-

255 The Munich computist mistakenly compares the first year of the Victorian table with the first year of the *latercus* (MC 52.3–26), he wrongly states that the *saltus* of the *latercus* is placed in November (MC 62.119–120), and he gives the wrong lunar limits for the *initium quadragesimae* (MC 58.6–7). A full discussion of these misconceptions can be found in Warntjes, '84 (14)-year Easter cycle', 50–4, 62–9, 74–8.

256 Yet, the possibility that the Munich Computus was composed in a northern Irish monastery can, of course, not totally be ruled out. The passage of this text that appears to point most prominently in a northern direction for the origin of this textbook is the chapter that includes the citation from the papal letter of AD 640 (MC 63). This chapter is primarily concerned with the condemnation of celebrating Easter on *luna* 14. It could be argued that such a chapter would have had a special relevance in an area that just abandoned the 84 (14), according to which the celebration of Easter Sunday on *luna* 14 was legitimate. Yet, a collection of authoritative citations against the celebration of Easter Sunday on *luna* 14 may as well have been assembled in a southern Irish monastery in the course of the Easter controversy, to counter arguments of the northern clergy. It would therefore have been quite natural to include such a list in a comprehensive textbook that reviewed the computistical developments of the past decades. In fact, it appears very likely that this chapter already constituted part of the now lost Victorian computus of AD 689 (cf. p. CXX–CXXI, CXXIV–CXXVI).

prehensive textbook of computistics, at a time when the need for a detailed understanding of the Dionysiac reckoning became pressing, due to the expiration of the current 95-year Dionysiac Easter table.

THE TRANSMISSION

In the previous chapters, the following facts have been established concerning the transmission of the Munich Computus: 1) It was originally composed in the *regiones Scottorum*, presumably in southern Ireland, possibly in east Munster, in the early months of AD 719. 2) It has been copied in St Emmeram in Regensburg in the first two decades of the ninth century, and this is the only copy of this text known to have survived to the present day. 3) In 1812, the Regensburg manuscript that contains the Munich Computus was transferred to the Bayerische Staatsbibliothek (formerly the Königliche Hof- und Centralbibliothek) in Munich due to the secularisation of Bavarian monasteries; there, the shelf-mark from the St Emmeram library was adopted (Em. E 79), which was later changed to Clm 14456.

The presence of the codex containing the Munich Computus in the library of St Emmeram between its compilation in the early ninth century and its transfer to Munich in 1812 can only be traced from 1500 onwards. It may have been part of the impressive section on computistics documented in the earliest surviving catalogue of this Benedictine monastery, dating to the second half of the tenth century, which lists 17 books on computus.[257] The 14th- and 15th-century catalogues do not record a computistical section, nor do they mention the Munich Computus or its codex explicitly.[258] One of the reasons for this may have been that computistical codices were kept separately, either privately, or on the shelves of the monastic school. In this respect the first explicit reference to the Munich Computus in any of the St Emmeram catalogues is extremely noteworthy. The catalogue in question was compiled by Dionysius Menger in 1500 and 1501. Yet, since Menger had only access to the codices that were on the library shelves at that time, and not to those that were in private possession, he had to add various manuscripts to his catalogue in the course the following 20 years.[259] The largest section of additions was appended to the original catalogue in 1504, when the private collection of the subprior Erasmus Dawn was transferred to the library after the latter's death.[260] In this section occurs the entry of the Munich Computus: *Item compotus sancti Augustini, Ieronimi, Ysidori, Dionisii etc.*,

257 *Adbreviatio librorum sancti Emmerammi* (Bischoff, *Mittelalterliche Bibliothekskataloge* 4,1, 144–6: 145): *De compoto XVII libri.*

258 *Libri ecclesie sancti Emmerammi Ratisbonensis* (Bischoff, *Mittelalterliche Bibliothekskataloge* 4,1, 154–61); *Repertorium patris Conradi Pleystayner prioris super Bibliothecam* (Bischoff, *Mittelalterliche Bibliothekskataloge* 4,1, 164–76); for the latter catalogue cf. Kellner & Spethmann, *Historische Kataloge*, 385.

259 For Menger's catalogue see Bischoff, *Mittelalterliche Bibliothekskataloge* 4,1, 185–8; idem, 'St. Emmeram im Spätmittelalter', 143–6; Ziegler, *St. Emmeram*, 190–2. It is edited under the title *Registrum sive inventarium librorum bibliothece monasterii Sancti Emmerammi episcopi et martyris* in Bischoff, *Mittelalterliche Bibliothekskataloge* 4,1, 188–385.

260 Cf. Bischoff, *Mittelalterliche Bibliothekskataloge* 4,1, 185.

et incipit: 'Tempus quid est' etc. In fine varias figuras habens etc.[261] Then, in the 17th century, Jean Mabillon consulted this codex during his stay at St Emmeram between 20 and 25 August 1683 and transcribed the annals found therein. Between Mabillon's days and Docen's detailed description of this codex in 1825, the principal focus of interest lay in the St Emmeram annals transmitted at the end of this manuscript, as Colomann Sanftl's enormous handwritten catalogue of St Emmeram codices of 1809 impressively illustrates, in which the annals are transcribed twice; it will be noted, however, that Sanftl also mentions the Munich Computus and the following computistica in the mathematical (*mathesis*) section of his inventory.[262]

With the history of the codex containing the Munich Computus being relatively clear, the principal task concerning the transmission of the Munich Computus lies in the reconstruction of the route it took and the copying stages it underwent from its compilation in Ireland in AD 719 to its early ninth-century Regensburg copy. This problem can best be approached backwards, from the Regensburg copy to the original compilation of the Munich Computus. Accordingly, the first step obviously is to trace the ecclesiastical centre from which this text reached Regensburg. From the outset, the most promising approach to this question appears to be an analysis of the background of the other items incorporated in the manuscript containing the Munich Computus, since they may already have accompanied that text at an earlier stage of transmission. Among those items, the easiest to compare with other manuscripts are the images of folios 70v–73r. These have been studied most recently by Barbara Obrist, with the result that the closest correspondence to the Regensburg images can be found in a Fulda manuscript (Basel, Universitätsbibliothek, F III 15a).[263] Drawing attention to the fact that Baturich, bishop and abbot of St Emmeram from AD 817 to 847, had been educated in Fulda, she implies that the images may have been brought to Regensburg by Baturich himself.[264] In that case Baturich

261 *Registrum sive inventarium librorum bibliothece monasterii Sancti Emmerammi episcopi et martyris* (Bischoff, *Mittelalterliche Bibliothekskataloge* 4,1, 268).

262 For Mabillon's, Sanftl's, and Docen's treatment of this codex cf. p. XV–XVI.

263 Obrist, 'La représentation', 9–11, 21; eadem, 'Astronomical Sundial', 80, 97.The similarity between the images in the Regensburg and the Fulda MS had already been pointed out by Bischoff, *Schreibschulen*, 196.

264 Obrist, 'La représentation', 9. The hypothesis that Baturich had studied and worked as a teacher in Fulda is based on Rabanus Maurus' poem *Versus ad Baturicum episcopum ex persona Isanberti* (*MGH Poetae* 2, 173–4), in which he calls himself pupil (*alumnus*) of Baturich's (p. 173, l. 17), as well as on the explicit reference to a Baturich being a Fulda monk in a letter fragment from Fulda (*MGH Epp.* 5, 517): *Baturicus, monachus Fuldensis, a Ludovico surrogatus est Adelvino*. Cf. especially Janner, *Geschichte*, 162–5; Fleckenstein, *Hofkapelle*, 169; Störmer, 'Adelsgruppe', 4–5, 25, 30–1; idem, *Früher Adel*, 335; and furthermore Lehmann, 'Alte Klosterbibliothek', 224; Hauck, *Kirchengeschichte* 2, 629; Hausberger, *Geschichte*, 36; Fried, *Weg in die Geschichte*, 278; Bischoff, 'Literarisches und künstlerisches Leben', 77. For the connection between Baturich and Fulda cf. also Bischoff, *Mittelalterliche Bibliothekskataloge* 4,1, 102; Spilling, 'Frühe Phase', 280–1. Freund, *Von den Agilolfingern*, 258–99 rejects the opinion that Baturich

must have left Fulda for Regensburg some years before his abbacy, since the Regensburg images were drawn in the years preceding Baturich's abbacy (which started in 817) according to Bischoff.[265] Other items physically closer to the Munich Computus in the manuscript, however, do not show the same Fulda background: The calendar on folios 48r–53v was, according to Borst, a copy of the Lorsch prototype of the Carolingian *Reichskalender*, which also shows parallels to a Mainz calendar.[266] The recension of the ps-Dionysiac *Argumentum XIV* immediately following the Munich Computus on folios 46v–47v gives a distinctive sequence of *concurrentes* (5, 6, 7) in its three examples, matching the *concurrentes* (i.e. the weekday of 24 March) of AD 684 to 686; the only two other manuscript, to my present knowledge, that contain the same recension of this *argumentum* originated in Reichenau and Fleury, while an abbreviated version occurs in a Cologne manuscript.[267] In summary, then, the items accompanying the Munich Computus in the Regensburg manuscript do not point unanimously to one particular ecclesiastical centre whence they may have travelled to Regensburg as a pre-existing body of texts, tables, and images.

Consequently, the approach for determining the route of transmission and the copying stages of the Munich Computus can only be textual, analyzing the traces that this textbook left in other eighth- and early ninth-century computistical texts. The first witness to be considered is the Harvard fragment (Harvard, Houghton Library, MS Type 613, fol. 7r–v), just recently (re-)discovered by Dáibhí Ó Cróinín.[268] This fragment shows otherwise unattested parallels to the Munich text which suggest that the larger part of this fragment is based on a

had been part of the Fulda community at any stage of his life; he rather believes that Baturich was, for some time, a student of Alcuin's in Tours.

265 Bischoff, *Schreibschulen*, 176–7, 195–6. In fact, Baturich, as archpresbyter, signs a legal contract on behalf of St Emmeram as early as AD 814; this contract is published in Widemann, *Traditionen*, 11–2; for the context cf. Janner, *Geschichte*, 157; Freund, *Von den Agilolfingern*, 246–7.

266 Borst, *Reichskalender*, 58, 75. In his *Schriften*, 1021–2, Borst, in fact, suggests that the Munich Computus may have come from Peronne to Regensburg together with *Epist. Rat.*; the present analysis, however, suggests a different route of transmission.

267 The chronological data given in this version agrees with AD 589–591 and AD 684–686; the high productivity in computistical literature in the AD 680s (cf. p. CLVII–CLVIII below) suggests that the latter rather than the former was intended here; note that these three years coincide with the first three years of a 19-year cycle, so that they do not necessarily need to have been composed in those years, but as likely a few years later. The Reichenau MS in question is Paris, Bibliothèque Nationale, Lat. 4860, fol. 151r–v, the Fleury MS Vatican, Biblioteca Apostolica, Reg. Lat. 586, fol. 117r–v; the first example only of this recension can be found in Cologne, Dombibliothek, 103, 187v–188r. Cf. Springsfeld, *Alkuins Einfluß*, 181–2, who is not aware that a copy of this recension of *Argumentum XIV* can also be found in the Munich and the Vatican MSS; generally, her list of MSS witnesses of *Argumentum XIV* is far from complete, and a thorough study of the history, transmission, and development of *Argumentum XIV* still is a desideratum.

268 Dáibhí Ó Cróinín came across this fragment on a research trip to the United States at the end of 2006, shortly before the completion of my thesis; cf. p. CLXXXVIII below. An article on this fragment is in preparation.

copy of the Munich Computus.[269] According to Bischoff, the fragment (being part of a collection of fragments now dispersed in three different libraries, namely the Houghton Library in Harvard (formerly in the Phillipps Collection of Cheltenham), the Bayerische Staatsbibliothek in Munich, and the G.A. Plimpton collection housed at Columbia University, New York), may have been composed as early as AD 800, but in all likelihood not before that date.[270] The provenance, however, cannot be established with certainty; Lowe and Bischoff would not be more precise than arguing for a scriptorium on the Continent with Irish connections.[271] Even though the script of this fragment does not show parallels to any other St Emmeram manuscript, it cannot be ruled out that it was written there; the collection of fragments, including the Harvard fragment in question here, had evidently been used for bindings of St Emmeram codices in the 15th century,[272] and the unique parallels to the Munich Computus itself render it likely that St Emmeram was indeed the place of composition of the Harvard text.[273] Now, the fact that some of the readings of the Harvard fragment are better than the ones in the Regensburg copy of the Munich Computus implies that the author of the fragment did not use this, but rather an earlier, less corrupted copy of that text. This evidence, then, suggests that the Harvard fragment was written in the first two decades of the ninth century, drawing on an earlier copy of the Munich Computus, presumably the one that had reached St Emmeram in the first place.

If this hypothesis is accepted, then some statements can be made about the state of the Munich Computus when arriving in Regensburg, at least about the few passages of this text that formed the basis for the Harvard fragment, primarily taken from chapters 50, 53, and 55. Accordingly, the following two mistakes appear to have already been incorporated in the copy of the Munich Computus that came to Regensburg, since they are shared by the Harvard fragment: 1) In c. 50, l. 67 it already read *k- ian-* instead of the correct *k- men-* (i.e. *Kalendis mensium*). 2) In the list of the Julian calendar dates of the 19 Easter full moons (c. 55, ll. 62–69), only one misreading occurs in the Regensburg copy of the Munich Computus, namely *V Nonas Aprilis* instead of the correct *II Nonas Aprilis*; the same list appears in the Harvard fragment, and when the scribe faced the date in question, he wrote *V*, then erased that number, and continued to note the correct *II Nonas Aprilis*; it seems, therefore, that the author of the Harvard fragment worked from a copy of the Munich Computus in which the numeral *II* had already been mistaken for *V*.

On the other hand, the comparison between the Harvard fragment and the Regensburg copy of the Munich Computus reveals numerous mistakes that crept into the text only at the final copying stage: When discussing a year of the

269 The parallels are listed on p. CLXXXVIII–CXC below.

270 Bischoff, *Schreibschulen*, 257.

271 Lowe, *Codices Latini Antiquiores* 2, 8 (No. 144); Bischoff, *Schreibschulen*, 257.

272 Cf. Lowe, *Codices Latini Antiquiores* 2, 8 (No. 144); Bischoff, *Schreibschulen*, 257.

273 Sotheby & Co, *Bibliotheca Phillippica* 6, 14 are more inclined than Bischoff and Lowe to ascribe this fragment directly to the St Emmeram scriptorium; cf. also p. CXCI below.

latercus with epact 8 on 1 January (c. 50, ll. 67–71), the Regensburg copy lists lunar ages 18, 19, and 20 for 1 November, 1 December, and the following 1 January respectively; the Harvard fragment, however, preserves the correct lunar ages 17, 18, and 19.[274] The corruption *Salinorum* for *Sabinorum* in c. 53, l. 14 only occurs in the Regensburg copy, while the Harvard fragment has the original reading; likewise in the following passage (c. 53, l. 19), the Regensburg copy gives the corrupt *anni quoddum* instead of the correct *annis usque* as transmitted in the Harvard fragment.

But whence, then, did the Munich Computus reach Regensburg? The suggested possibility (based on the evidence of the images in the St Emmeram codex) that it may have come from Fulda is undermined by the fact that the Munich Computus leaves no traces in any computistical text composed in that monastery (particularly not in the unfinished Fulda Computus of AD 789 and Rabanus Maurus' *De computo*[275]). Contrary to the evidence of Fulda manuscripts, the Munich Computus left many traces in or shows distinct parallels to certain texts assembled in the early ninth-century codex Cologne, Dombibliothek, 83^2, which strongly suggests that it came to Regensburg via Cologne. One concept in particular can only be found in the Munich Computus and two texts preserved uniquely in the Cologne codex. In the disputes of the seventh century concerning the correct calculation of Easter, the Dionysiac reckoning was discredited by the fact that it did not transmit chronological data for the year of Christ's passion in accordance with the evidence of the Gospels;[276] Victorius' data, on the other hand, was in agreement with John's Gospel. Facing this accusation from followers of Victorius, adherents of the Dionysiac system invented a pseudo-proof that would solve their dilemma:[277] They argued that the Dionysiac and the Victorian lunar data should be compared in a way that the Dionysiac *ogdoas* was to be correlated with the Victorian *hendecas*, not the Victorian *ogdoas* as would have been correct. In the Victorian reckoning the year of Christ's passion occurred in the second year of the Victorian *hendecas*, having epact 19 on 1 January; the corresponding Dionysiac year in this correlation of the Dionysiac *ogdoas* with the Victorian *hendecas* was the second year of the Dionysiac *ogdoas*, having epact 20 on 1 January. Accordingly, the Dionysiac lunar age was incremented by one compared to the Victorian lunar age on any day of this year, particularly on the days of Christ's passion and his resurrection. Since the Victorian tables recorded *luna* 14 for Christ's passion and 16 for his resurrection in accordance with John's Gospel, the Dionysiac data according to this comparison was *luna* 15 on the passion, *luna* 17 on the resurrection in accordance with the chronology derived from the Synoptic Gospels.

274 The Harvard fragment, therefore, confirms my discussion in Warntjes, '84 (14)-year Easter reckoning', 48–50.

275 For the unfinished Fulda Computus of AD 789 cf. n. 111 above; Rabanus Maurus' *De computo* is edited by Wesley M. Stevens in *CCCM* 44, 198–321.

276 This problem also troubled early eighth-century Anglo-Saxon computists, as Bede's discussion of this matter in *DTR* 47, 61 impressively illustrates.

277 For the details of this method cf. Schwartz, 'Ostertafeln', 94–6; Warntjes, '84 (14)-year Easter reckoning', 62–9.

Now, an entire chapter is devoted to this computistical trick in the Munich Computus,[278] but it is also discussed in two further texts (and in *no other text* to my knowledge), both only to be found in the Cologne codex: The first (to be termed *De comparatione epactarum Dionysii et Victorii*, a hitherto totally neglected text) provides the most detailed discussion of the comparison in question and its implications on the chronology of the *annus resurrectionis*; it is argued below that this tract served the Munich computist as his principal source when compiling the relevant chapter, and that it therefore appears to have originated in a seventh-century Irish context; moreover, it seems likely that this text accompanied the Munich Computus as one of its essential sources, and consequently reached Cologne in the same codex as the Munich textbook.[279] The second, *Comp. Col.*, was presumably composed in the monastic school of Cologne in AD 805 according to its editor, Arno Borst.[280] *Comp. Col.* shows parallels in content to the Munich Computus not only in this peculiar comparison of the Victorian with the Dionysiac epacts, but also in various other passages; e.g., the disruptions of the course of time mentioned in the Bible are discussed in only three known early medieval texts, namely *De mirabilibus sacrae scripturae*, the Munich Computus, and *Comp. Col.*[281] In this, as in few other passages, the phrasing of *Comp. Col.* even shows direct dependency on the Munich text, so that it can hardly be doubted that the Munich Computus was one of the major sources used by the author of *Comp. Col.*[282]

It is apparent from this evidence that the Munich Computus was studied in detail in the monastic school of Cologne in the late eighth century at the latest. Concerning the state of the text at this Cologne stage of transmission, *Comp. Col.*'s discussion of the increment of the *saltus lunae* per lunation (which is primarily based on the Munich text) suggests that the Munich Computus still had the correct division of 1/12 of a *momentum* into 47 parts at the end of the calculation instead of 48 parts as corruptly transmitted in the Regensburg copy;[283] consequently, this mistake appears to have crept into the text in the final (Regensburg) copying stage. On the other hand, it is quite striking that *Comp. Col.* does not continue to distribute in detail those $(5\times47)/(12\times47)=235/(12\times47)$ parts of a *momentum* among the 235 lunations of the 19-year cycle; the reason for this may be that *Comp. Col.*'s exemplar, the Munich Computus, was already corrupt in this final part of the calculation and hence could not be relied on.

278 MC 52; the method of correlation is also outlined in MC 49.16–32.

279 Cf. the discussion of *De comparatione epactarum Dionysii et Victorii* p. CLII–CLVIII below; this tract is edited (for the first time) in Appendix 2.

280 For the provenance and date of this text cf. Borst, *Schriften*, 885. The chronology of the *annus resurrectionis* and its reflection in the two reckonings on the basis of the described correlation of epacts is discussed in *Comp. Col.* 5.4–5.

281 *De mirabilibus sacrae scripturae* 1.8, 2.4, 2.28, 3.13 (*PL* 35, col. 2160, 2175–6, 2187–8, 2199); MC 66; *Comp. Col.* 5.7, 6.1B.

282 Cf. the discussion of *Comp. Col.*'s dependency on MC on p. CLXXIX–CLXXXIII below.

283 MC 62.14–61; *Comp. Col.* 4.10B; see also p. CLXVI below. Borst, *Schriften*, 922–3, does not record MC as a source for this calculation; cf. note 577 below.

From the evidence of an earlier Frankish text, *Dial. Neustr.*, it appears that the Munich Computus had arrived in the Cologne region as early as the AD 730s. *Dial. Neustr.* was clearly written in a region loyal to the Merovingian puppet-king Teuderic IV (or rather his *major domus* and de facto regent, Charles Martell) in AD 737, since the dating clause at the end of the text refers to the year of composition as AD 737, the year in which Teuderic, king of the Franks, died.[284] Arno Borst, the editor of this text, suggests that *Dial. Neustr.* may have originated in Boniface's circle.[285] In Boniface's circle, however, one would expect extensive use of at least Bede's *De temporibus*, which is not cited at all.[286] On the contrary, the text is heavily influenced by Irish computistical thought: *Dial. Neustr.* starts with a discussion of the concept of numbers (as do *De ratione conputandi* and especially the *Computus Einsidlensis*),[287] while Irish influence is particularly obvious in the concepts of the linear increase and decrease of daylight,[288] the linear bissextile increment in the first quarter of a year,[289] the division by 90 of the days exceeding 90 in the four quarters of a year and their equal distribution within the quarter in which they occur,[290] the various options for the placement of the *saltus*[291] and for the solar bissextile day,[292] the calculation of the lunar increment of the *saltus* per lunation,[293] as well as the terminology for the *saltus*[294]. On a more detailed level, the direct dependency of *Dial. Neustr.* on the Munich Computus is readily apparent, since some theories and explanations occur in no other than these two texts (except for some being repeated in *Comp. Col.*, which evidently made use both texts):[295] Only MC and *Dial. Neustr.* divide the years into four quarters by using the 'Greek' equinoxes and solstices as marker days, while explicitly noting the number of days in each of these quarters (90, 91, 92, 92);[296] the comparison with a pregnant woman of the bissextile increment completing a full day in the course of four years can only be found in MC, *Dial. Neustr.*, the verbatim copy of MC's passage among the Angers glosses, and the later *Quaest. Austr.*, so that *Dial. Neustr.*'s account is clearly dependent on MC;[297] the comparison outlined

284 *Dial Neustr.* 30C. For the dating clause cf. p. LXXIV above; for its implications on the place of composition note 539 below.

285 Borst, *Schriften*, 375. Krusch, in his article about *Dial. Neustr.* ('Lehrbuch'), does not discuss the provenance of this text.

286 Cf. p. CLXXII–CLXXIII below.

287 *Dial. Neustr.* 1A–C; cf. *CE* p. 82–85; *DRC* 2–12.

288 *Dial. Neustr.* 28; cf. *CE* p. 99–103; MC 39.

289 *Dial. Neustr.* 28aA; cf. *CE* p. 99; MC 39.11–13; *DRC* 50.

290 *Dial. Neustr.* 29aA (cited in *Comp. Col.* 2.3B); cf. *CE* p. 103–104; MC 40; *DRC* 50.

291 *Dial. Neustr.* 25; cf. *CE* p. 123–4; MC 62.68–110.

292 *Dial. Neustr.* 27aA; cf. *CE* p. 105–6; MC 41.43–49; *DRC* 51.

293 *Dial. Neustr.* 26B; cf. *CE* p. 122–3; MC 62.14–61; *DRC* 108.

294 *Dial. Neustr.* 26A (cited in *Comp. Col.* 4.9A); cf. MC 62.111–117; *DRC* 110.

295 Besides *Dial. Neustr.*'s dependency on MC in the concepts discussed here, cf. also the direct textual dependency outlined p. CLXXIII–CLXXIV below.

296 MC 38.17–22; *Dial. Neustr.* 29B.

297 MC 41.35–37; *Dial. Neustr.* 28aC; *Quaest. Austr.* 2.9C; for the Angers citation cf. p. CLXXXVI below.

in *Dial. Neustr.* of the full moon at the time of creation with those occurring at the Exodus from Egypt and at the Last Supper clearly derives from MC's detailed comparison of the chronological data of these years, and later also found its way into *Comp. Col.*[298] Now, since *Dial. Neustr.* is most prominently cited in a Cologne text (*Comp. Col.*[299]), it may be assumed that *Dial. Neustr.* itself originated at the Middle Rhine, if not in Cologne itself, rather than in Bonifatius' direct sphere of influence (like Fulda or Mainz). This suggests, then, that the Munich Computus, and Irish computistical thought beyond that text, was known in Cologne or its vicinities by AD 737.

In the period between the composition of the Munich Computus and *Dial. Neustr.* (AD 719 to 737), only one text is known to have been influenced by the Munich Computus, namely *Dial. Burg.*; in fact, quite a few passages of *Dial. Burg.* show direct textual dependency on the Munich text.[300] The provenance of *Dial. Burg.*, however, remains uncertain. Based on the sources used by its author, which reveal strong Irish connections, Borst convincingly suggests the Columban foundation of Luxeuil or one of its daughter-houses, like Corbie, as the most likely place of composition of *Dial. Burg.*[301]

The fact that the Munich Computus appears to have been studied and cited in a Columban house less than ten years after its composition may then provide the context for the original route of this text from southern Ireland to the Continent. Jonas, the biographer of Columbanus, provides valuable details about the travel-routes taken by the Irish saint. The route of primary importance for the present study is the one chosen by the saint, his followers, and guards when he was commanded by the Frankish king Theuderic II not only to abandon his foundation in Luxeuil, but to leave the Frankish kingdoms altogether and to return to his native Ireland.[302] Hence, the Frankish king would have wanted to make sure that Columbanus travelled on the fasted possible way, so that it must be presumed that the route described was one of the most convenient, and in subsequent decades and centuries one of the principal routes for travelling from Ireland into the Frankish heartland in general, to Luxeuil in particular. Since the person carrying the Munich Computus to Francia travelled from Ireland to the Continent, while Columbanus travelled in the opposite direction, the route described by Jonas is here reversed. As it was common in the early middle ages, the main part of this trip was undertaken by boat, which was the most comfort-

298 MC 44.13–80; *Dial. Neustr.* 15A–B; *Comp. Col.* 5.1.

299 Cf. especially *Comp. Col.* 1.1, 1.2, 2.1A, 2.3B, 2.4, 2.5A, 3.1A–B, 3.5B, 4.9A, 4.10B, 5.12, as well as Borst, *Schriften*, 885, 891.

300 Cf. p. CLXXI–CLXXII below.

301 Borst, *Schriften*, 218, 348–9. Krusch, in his *editio princeps* of *Dial. Burg.* (*Studien* II, 53), does not discuss the provenance of this text.

302 Jonas, *Vitae Columbani* 1.20–2 (*MGH SS rer. germ.* 37, 197–203). The route described by Jonas (and under discussion here) includes a detour from Autun via Avallon and Auxerre to Nevers; note that this detour corresponds to the *Itinerarium Antonini* (cf. Krusch, *Ionae vitae sanctorum*, 197), and literary dependency rather than historical fact may therefore have been the reason for its inclusion in Jonas' work (cf. Schäferdiek, 'Columbans Wirken', 368).

able and fastest means of travelling. Setting out from Ireland, the boat trip would go around the coast of Brittany as far as the mouth of the river Loire near Nantes (or on a land route through Brittany from its north-coast to the Loire[303]), and would then continue on the Loire, passing the cities of Tours and Orleans, as far as Nevers. From Nevers, a direct land route past Autun and Besançon would lead the traveller straight to Luxeuil.

Whether or not this was the route taken by the person who first brought the Munich Computus to the Continent must obviously remain speculative. Yet, it is extremely noteworthy in this context that the Munich Computus leaves most and sometimes unique traces in a manuscript originating on this route, namely in glosses to Bede's scientific works in Angers, Bibliothèque Municipale, 477 (461).[304] The codex itself appears to have been compiled in the Breton monastery of Landévennec in the late ninth century.[305] Accordingly, it could be argued that a) the glosses, if not b) the Munich Computus itself, came from any of the above mentioned centres (Luxeuil, Cologne, or the like) to Brittany in the course of the ninth century. However, it has to be noted concerning a) that no other MS containing these glosses is known at present, which makes it more likely that they were originally composed in Brittany;[306] concerning b), it appears unlikely that a text like the Munich Computus, outdated and replaced by more appropriate texts by the ninth century, would have been studied and memorized in such a detail (as the glossing suggests) if introduced that late. It seems more plausible that the Munich Computus was one of the first computistical textbooks known in this Breton monastery, and for that reason alone one of the standard texts on this medieval science for a certain period of time, probably the third, fourth, and fifth decades of the eighth century; after the introduction of Bedan computistica, then, the Bedan textbooks were glossed with the computistical knowledge previously acquired and memorized from the Munich and other, predominantly Irish, texts, and later copied into the codex that has survived to the present day.

Further confirmation for this theory that the Munich Computus travelled through Brittany and the Loire valley to the Continent (or that a copy of this text reached Brittany at a very early stage from a centre at the Loire) comes

303 In his *Vitae Columbani* (1.4–5; *MGH SS rer. germ.* 37, 160–1), Jonas relates that Columbanus, when first travelling to Francia, landed at the coast of Brittany (*sinus Brittanici*). Accordingly, an Irishman following Columbanus in subsequent decades may as well have travelled to Brittany, whence taking the land-route to a city at the Loire, and continuing the trip by boat to Nevers.

304 See p. CLXXXIV–CLXXXVII below.

305 For provenance and date of this MS see especially Fleuriot, *Dictionnaire*, 8–11. A very brief catalgue description of this MS is given by Molinier, 'Manuscrits', 349.

306 This hypothesis is confirmed by the fact that some glosses of this MS were written in Breton, which clearly speaks for a Breton provenance of them originally; cf. Fleuriot, *Dictionnaire*, 49–330; Lambert, 'Commentaires Celtique I', 119–39; idem, 'Commentaires Celtique II', 185–201; idem, 'Thirty', 29–43. Note, however, Ó Cróinín's suggestion (in Lambert, 'Commentaires Celtique II', 202) that the Angers MS may have been copied from an Irish exemplar, which would then, undoubtedly, already have had included the glosses copied or derived from the Munich Computus.

from one of the other Irish computistical textbooks, *De ratione conputandi*. Like the Munich Computus, it leaves distinctive traces among the Angers glosses just discussed.[307] Moreover, one of the two surviving copies of this text originated in the Loire valley,[308] while it was used as a source for the composition of *Dial. Burg.*, which may have been written in Luxeuil or one of its daughter-houses.[309] Consequently, the most prominent traces left by this Irish computistical textbook are on exactly the same route as the one outlined for the Munich Computus above.

In summary, this evidence suggests that the Munich Computus came to the Continent on the same or a very similar route as the one described by Jonas in his *Vitae Columbani*, having left early traces in Brittany and having reached Luxeuil or one of its daughter-houses by AD 727. By AD 737, it appears to have made its way to the Cologne region, either directly from the Columban monastery in question, or from any computistical centre in central France (especially the Loire valley) which the text may have passed on its way to Luxeuil or its vicinities. Cologne, then, was the place where this text enjoyed its greatest popularity and received its most extensive study. From Cologne the Munich Computus came to St Emmeram in Regensburg, presumably in the late eighth century. In this Bavarian monastery the only surving copy of this text was compiled in the second decade of the ninth century, and was finally transferred to Munich a millenium later.

307 The direct quotations from *De ratione conputandi* in the Angers glosses are numerous. Cf., e.g., *DRC* 107.7–8: *Item Augustinus dicit: Saltus est euentus famosus in luna, ut bissextus in sole habetur.* with Angers, Bibliothèque Municipale, 477 (461), fol. 20v: *Saltus euentus famosus in luna, ut bissextus in sole habetur.* and *DRC* 111.1–3: *Sciendum nobis cur saltus dicitur. Augustinus ostendit dicens: Saltus, quasi quoddam saliendi signum, quia conpotus transilit unum diem. Quamuis non hic saltus sed geminatio uideatur.* with Angers, Bibliothèque Municipale, 477 (461), fol. 20v: *Saltus saliendo dicitur. Augustus: Saltus quasi saliendi quoddam signum, quando compotus transilit diem. Quamuis non hic saltus sed gemminatio uideat numeri.*

308 Only two copies of *De ratione conputandi* have survived: A long version in Brussels, Bibliothèque Royale, 5413–22, fol. 77v–107v; the origin of this part of the MS is believed to be Northern France (Ó Cróinín, 'A seventh-century Irish computus', 129; Walsh & Ó Cróinín, *Cummian's letter*, 106; Bischoff, *Katalog* 1, 154–5; Borst, *Schriften*, 220–1), and consequently even this copy may have been written in close geographical proximity to the Loire valley. The shorter version is transmitted in Vatican, Biblioteca Apostolica, Reg. Lat. 1260, fol. 87r–99v. According to Bischoff (cf. Walsh & Ó Cróinín, *Cummian's letter*, 107) this MS was composed in the Loire valley, while Borst (*Reichskalender*, 147) argues more specifically for Auxerre or Fleury, Jones (*Pseudepigrapha*, 137) for Fleury. In a different publication, Borst ('Alkuin', 57) points out that the first 13 chapters of *DRC* are also to be found in Paris, Bibliothèque Nationale, Nouvelle acquisition latine 456, fol. 189v–190v, a tenth-century MS from Auch (which also includes Abbo of Fleury's recension of Helperic's *De computo*), and therefore the beginning of *DRC* may also have ultimately reached Auch from the Loire valley; for this MS see Springsfeld, *Alkuins Einfluß*, 117–8; Borst, *Reichskalender*, 187–8; idem, *Schriften*, 280–1.

309 For *Dial. Burg.* originating in Luxeuil or one of its daughter-houses cf. n. 301 above; for *De ratione conputandi* being cited by the author of *Dial. Burg.* cf. p. CC below.

THE CONTEXT OF THE MUNICH COMPUTUS

STRUCTURE AND SOURCES

All three known Irish computistical textbooks from the late seventh and early eighth century (the *Computus Einsidlensis*, the Munich Computus, and *De ratione conputandi*) show, save for only minor variations, a very similar general structure, which is a clear indication of their interdependency.[310] The basic framework of these textbooks consists of four sections, organized in the following order: 1) A discussion of the various divisions of time from the smallest units to the all-embracing, rather vaguely defined concept of *mundus*; within this general structure, the Julian calendar (as part of the analysis of the year) and issues related to the Julian calendar are discussed like the bissextile day, weekday calculations, the divisions of a year between astronomical marker-days (the equinoxes and solstices), as well as the theory of increase and decrease of daylight between these marker-days, etc.[311] 2) Solar theory is then followed by lunar theory, discussing astronomical characteristics of the moon,

310 For the interdependency of these texts cf. especially p. CXLVIII–CL. The similarity of their structure is already pointed out by Warntjes, 'Computus Einsidlensis', 63. Of these three Irish computistical textbooks, only the structure of *De ratione conputandi* has been discussed in some detail. Ó Cróinín, when announcing the discovery of this text ('A seventh-century Irish computus', 102) argued that its structure is based on *De computo dialogus* and *De divisionibus temporum* taken together. In his subsequent edition (Walsh & Ó Cróinín, *Cummian's letter*, 103) he analyzed *De ratione conputandi*'s structure in more detail, stating that the basic framework of at least parts of this text was set by the 14 divisions of time; he divided the text into two main sections: the first dealing with the concept of numbers (c. 1–12) and the divisions of time (c. 13–61), including various lengthy accounts of calendrical matters; the second main section (c. 63–115), then, comprising lunar theory, the mechanisms of the 19-year cycle and its technical features, theological and mathematical theories concerning Easter and the *initium quadragesimae* (ending in a discussion of the columns of the Dionysiac Easter table), a section on the *saltus* (but not on the 19-year cycle as Ó Cróinín states), and a history of the ecclesiastical calendar. Most recently, in a talk entitled 'Insular Paschal controversy and the Irish study of the computus' given in Clonmacnoise on 13 October 2007, Dan Mc Carthy presented the following outline of *De ratione conputandi* (I would like to thank Dan Mc Carthy for providing me with a print-out of his talk): 'On the names of the sun in Hebrew, Greek, Latin, Syriac, and with philosophers (1 section); On numbers (12 sections); On the fourteen divisions of time (14 sections); On the Julian calendar (8 sections); On the Roman, Macedonian and Hebrew months (6 sections); On the year, equinoctes, solstices, bissextile, zodiac (20 sections); On Victorius' date of the Passion (1 section); On the moon (5 sections); On the lunar cycles and their computation (16 sections); On the Pasch and its history (4 sections); On different Paschal computations – the *latercus*, Victorius, Dionysius (9 sections); On the Dionysiac Paschal table (15 sections); On the ecclesiastical calendar (2 sections).'

311 *CE* p. 85–108; MC 1–45; *DRC* 13–62.

especially its relation to the sun, but also, and more importantly, the mathematical construction of a lunar year, epactal theory, as well as the formal differences between the Dionysiac and the Victorian lunar data.[312] 3) These two major sections set the ground for the combination of the two, for the central theme of computistical textbooks, namely the reckoning of Easter proper with its numerous aspects and nuances, particularly the basic elements necessary for the various modes of calculation.[313] 4) The fourth and final section deals with the construction of the 19-year cycle, ending in a detailed analysis of the *saltus lunae*;[314] since the discussion of the 19-year cycle was not immediately relevant for a thorough understanding of the calculation of Easter, it seems that the basic theory was covered by the first three sections, with section four being designed for the more advanced student eager to learn more about the details of the system underlying the outlined methods.

Additional to this common framework, each Irish computistical textbook has its own preferences, which are either discussed as front matters, or appended as addenda of special interest. On the one hand, the authors of both the *Computus Einsidlensis* and *De ratione conputandi* thought it essential and necessary to first outline the concept and history of numbers, as well as the etymologies of the most important terms, as a basis for the following, more detailed and complicated analyses.[315] On the other hand, the discussion of the structure of the 19-year cycle at the end of his textbook lead the author of *De ratione conputandi* to a brief outline of the history of Easter cycles,[316] while the Munich computist added to the main part of his text few chapters of various contents, apparently all taken from the now lost Victorian computus of AD 689, including a short chronicle to mark the end of the work.[317]

The structure of these Irish computistical textbooks, particularly of its main part, certainly influenced subsequent computistical texts of the same genre. Yet, this structure was altered in a way that the entire textbook, not only its first main part, was pressed into the scheme of the divisions of time (which, in this instance, included cycles). The first to have followed this model was the Northumbrian scholar Bede.[318] In his early work, *De temporibus* of AD 703,

312 *CE* p. 108–11; MC 46–53; *DRC* 63–83.

313 *CE* p. 111–2; MC 54–58; *DRC* 84–103.

314 *CE* p. 113–25; MC 59–62; *DRC* 104–13. Note that many of the topics discussed by *CE* and MC in this section on the construction of the 19-year cycle are part of the section on lunar theory (section 2) in *DRC*; moreover, the discussion of the *initium quadragesimae*, which occurs in this section in *CE*, has more appropiately been placed in the section on the reckoning of Easter (section 3) by MC and *DRC*.

315 *CE* p. 82–5; *DRC* 1–12.

316 *DRC* 114–115.

317 MC 63–68. For the argument that these chapters stem from the Victorian computus of AD 689 cf. p. CXXIV–CXXVI below.

318 Jones, *Bedae opera*, 137–8 has already observed that Bede used the concept of the divisions of time as the basic framework for his computistical textbooks. Jones claims that Bede adopted this structure from the 'Irish computus', a term he never defined properly (cf. note 55); it is apparent from his discussion that what he had in mind in this instance was a combination of *De computo dialogus* (*PL* 90, 647–652) and *De divisionibus temporum* (*PL*

Bede adopted a structure that appears to have been influenced by, or rather modelled on a passage outlined at the beginning of the *Computus Einsidlensis* and later *De ratione conputandi* (as well as in other minor, still unedited tracts[319]): First, he discussed the units of time from *momentum* to the year (c. 1–

90, 653–664) in its original, now lost form. Note that Jones was unacquainted with the three Irish computistical textbooks; only the Munich Computus was known to him, but he had never studied that text; the other two textbooks were discovered after his death. A different approach of defining the structure of Bede's computistical works is taken by Wallis. Concerning *DT*, she (Wallis, *Bede*, lxv–vi) follows Jones in arguing that Bede was somewhat influenced by Isidore's *Etym.* and Irish texts (*De computo dialogus* and *De divisionibus temporum*), but that he introduced analyses of the calendar and the Dionysiac Easter table to their discussions; she is not aware of the fact that in some Irish texts (cf. the following note) one interpretation of the divisions of time includes cycles, so that the calendar is discussed as a concept related to the year, the Easter table in the context of the Easter cycle. Concerning *DTR*, she (Wallis, *Bede*, vii–ix, 5–7) divides her translation into six major sections without any further explanation: 1. Technical preparation; 2. The Julian calendar; 3. Anomalies of lunar reckoning; 4. The Paschal table; 5. The world chronicle; 6. Future time and the end of time; this structure, however, does not reflect Bede's approach. Yet another (and even more unconvincing) view of the structure of *DTR* is expressed by Pillonel-Wyrsch, *Calcul*, 13–17: c. 1–4 are considered as an introduction leading into c. 5–40, which constitute a discussion of the divisions of time and related material; c. 41–43 deal with lunar phenomena, c. 44–47 with the 19-year cycle, c. 48–49 with the indiction, c. 50–56 with the epacts, c. 56–60 with the lunar cycle, c. 61–65 with lunar calculations; c. 66, then, contains the chronicle, c. 67–71 deals with the ages to come and the last judgement. Cf. also Germann, *De temporum ratione*, 38.

319 *CE* p. 85: *Quomodo crescunt numeri in nominationibus specialibus? A momento in minutum, a minuto in punctum, a puncto in hora, ab hora in quadrante, a quadrante in diem, a die in septimanam, a septimana in mensem, a mense in tempus, a tempore in annum. Ab anno vero in circulum XIX annorum secundum lunę, cursum ratione saltus; si secundum solis cursum in XXVIII annos ratione bissexti. Communi itaque ordine solis scilicet et lunę, hoc est in die mensis et in die septimanę et in etate lunę, in circulum annorum DXXXII numerorum series crescit. Et hoc modo in eundem circulum peracto cursu, quo graditur circulus solis et lunę reuoluitur. DRC* 14: *Sciendum nobis quomodo diuisiones temporis crescunt. Ita: a momento in minutum, a minuto in punctum, a puncto in horam, ab hora in quadrantem, a quadrante in diem, a die in ebdomadam, ab ebdomada in mensem, a mense in uicissitudinem triformis, a uicissitudine triformis in annum, ab anno in ciclum XXVIII annorum secundum solem; secundum lunam XVIIII annorum; DXXXII secundum solem et lunam, quando fit integra reuersio*. Similar passages clearly related to or even copied from these two Irish accounts (or the now lost *De divisionibus temporum*) can also be found in numerous still unedited texts. Of particular interest is the fact that it appears in a prologue to Bede's *De temporibus* in dialogue form, clearly indicating that a later copyist and commentator of this text understood the importance of this passage for Bede's work and especially its structure; this prologue is preserved in Vatican, Biblioteca Apostolica, Pal. Lat. 1448, fol. 1v–2r; Vatican, Biblioteca Apostolica, Pal. Lat. 1447, fol. 3r; Schaffhausen, Stadtbibliothek, 61, fol. 22r–v, while the passage in question is almost verbatim to *DRC* (here cited from my unpublished edition): *D.: Quomodo hae diuisiones temporis crescunt? M.: Ita etiam ab athomo in momentum, a momento in minutum, a minuto in punctum, a puncto in horam, ab hora in quadrantem, a quadrante in diem, a die in ebdomadam, ab ebdomada in mensem, a mense in uicissitudinem triformen, a uicissitudine triformi in annum, ab anno in ciclum: XXVIII annorum secundum solem,*

6, 8–9), before turning to the theory of cycles, particularly the 19-year lunar cycle and the 532-year Easter cycle (c. 11, 13–15); the latter discussion is principally based on the various columns of the Dionysiac Easter table and the algorithms for calculating that data, ending in a reflection on the symbolism of Easter. Within this framework, he cleverly inserted several chapters at appropriate places: First, he placed a chapter on the solstices and the equinoxes (c. 7) between those concerned with the Julian calendar months and the seasons of a year, while devoting a chapter each to the two technical devices that deserved special attention, to the bissextile day (c. 10) in the context of the Julian calendar year on the one hand, to the *saltus lunae* (c. 12) in the context of the 19-year lunar cycle on the other; finally, he added a brief chronicle in place of a discussion of the largest unit of time, *aetas*, which is not referred to in the mentioned passage of the *Computus Einsidlensis* and *De ratione conputandi*, but which features prominently in the detailed discussions of the divisons of time in the Munich Computus and *De ratione conputandi*.[320]

In his later and more detailed work, *De temporum ratione* of AD 725, Bede kept this principal structure, though it is less easy to detect, since Bede, compared to his earlier work, inserted into this general framework many more chapters on topics not directly concerned with the divisions of time, but nevertheless relevant for time-reckoning in general: Like the *Computus Einsidlensis* and *De ratione conputandi*,[321] Bede prefixed a chapter on number theory, in his case on finger-counting (c. 1). Afterwards, he introduced the concept of time in general terms (c. 2), like the Munich Computus and *Dial. Langob.*,[322] before discussing the divisions of time from the smallest fractions to the unit of the month (c. 3–15). Between the discussion of the months and that of the next unit in line, the seasons, Bede, quite competently and consistently, introduced several chapters of various themes: the discussion of the twelve months led to an analysis of the twelve zodiac signs, which brought him to an outline of the course of the moon through the zodiac and the calculation thereof (c. 16–19); this outline of mathematical methods is then followed by a set of algorithms for weekday and especially lunar calculations (c. 20–24), which leads into more general lunar theory, including a discussion of the tides (c. 25–29); subsequently, he turned to the equinoxes and solstices (as an introduction to the following treatment of the seasons) as well as those parts of solar theory that are dependent on the intervals set by these marker-days (c. 30–34). After this long digression, Bede returned to the units of time, continuing with the seasons and the year (c. 35–37). The Julian calendar year, then, compelled the next excursus, dealing first with the bissextile day as a solar, but also as a lunar phenomenon, which quite naturally led to a discussion of the only other such technical device, the *saltus lunae*, and the consequences and problems that these, for mathematical reasons artifi-

XVIIII annorum secundum lunam, DXXXII annorum secundum solem et lunam, quando sit integra reuersio.

320 MC 1, 42; *DRC* 13, 59.

321 Cf. note 315 above.

322 Cf. following page.

cially implemented features may cause (c. 38–43). Having thus dealt with the divisions of time and related excursi up to the unit of the year, Bede turned to the theory of cycles, first discussing the 19-year lunar cycle (c. 44–46), then, as the culmination of his book, the Easter cycle, structured on the basis of the columns of the Dionysiac Easter table (c. 47–65). Finally, as in the case of his earlier work, Bede ended *De temporum ratione* with a chronicle, i.e. with an analysis of the largest unit of time, *aetas*, followed by theological theory about the end of time (c. 66–71).[323]

Interestingly enough, the earliest Frankish computistical textbooks, *Dial. Burg.* and *Dial. Neustr.*, do not apply the same or a related structure. Moreover, the predominantly Irish *Dial. Langob.* shows considerable structural differences to its Irish precursors. It begins, like the Munich Computus and, in fact, partially based on that text, with a general discussion of the term and concept of time (c. 1–2), followed by an outline of the various divisions of time (c. 3–4). This outline, however, does not lead into a detailed analysis of each individual unit from the smallest to the greatest as in the other Irish textbooks; on the contrary, the year is taken as the starting point from which the various elements of a year are discussed in descending order, namely the seasons, the months, the days of a month, followed by the principles of weekday calculations (c. 5–16). Subsequently, the four marker-days of a year, the equinoxes and solstices, are analyzed, as well as the theories directly dependent on these days (c. 17–19). Finally, the author of this text deals with the technical devices artificially implemented, for mathematical reasons, in lunar and solar cycles, namely the embolisms, the bissextile day, and the *saltus lunae* (c. 20–25).[324] Quite remarkably, the text avoids any reference to the calculation of Easter, and in this sense should not be regarded as a comprehensive computistical textbook, but rather as a text concerned with specific issues of time-reckoning, especially the artifical devices implemented in mathematically conctructed systems.

Generally, by the AD 740s or so, the production of comprehensive computistical textbooks ceased, probably due to the reception of Bede's *De temporum ratione*.[325] Computists rather concentrated on the collection of computistical formulae[326] and on scholarly debates of central chronological themes,[327] while

323 These final chapters of Bede's *DTR* do not feature in Jones' 1943 edition (a fact he very much regretted himself; cf. Jones, *Bedae opera*, 138), but they were edited separately by Mommsen in *MGH AA* 13, 233–327, whence they were republished in Jones' 1980 edition of *DTR* (*CCSL* 123B, 461–544).

324 It is argued p. CLXXVI–CLXXVII below that the final chapter in the edition of *Dial. Langob.*, chapter 26 dealing with the indiction, is a later, continental addition (which survives as a separate treatise in other MSS); it is obvious from the discussion above that an analysis of the indiction does not fit into the structure of the work, since it does not constitute a technical device, but rather a cycle (of 15 years) in itself and there is no logic to this cycle (rather than the 19-year lunar cycle or the 532-year Easter cycle) forming the end of this text; cf. also n. 115.

325 Cf. the discussion of the history of the computistical textbook p. LV–LVI above.

326 Good examples from the second half of the eighth century are *Lect. comp.* of AD 760, the *Computus Rhenanus* of AD 775 (cf. n. 574), and *Lib. ann.* of AD 793.

the early ninth century, then, saw the establishment of a new literary genre, the computistical encyclopediae.[328] It was not until AD 820 that a new computistical textbook saw the light of day, composed by the Fulda abbot Rabanus Maurus. In principal, this textbook was primarily based on Bede's *De temporum ratione*, and consequently adopted the main structure of Bede's work outlined above. Yet, Rabanus appears to have been also aware of Irish computistical concepts, which additionally influenced the structure of his work: Like the *Computus Einsidlensis* and *De ratione conputandi*, Rabanus begins his text with a detailed description of the concepts of numbers, considering aspects far beyond Bede's treatment of this topic (c. 1–8), before addressing the general question of the nature of time based on *DTR* (c. 9–10). Most intriguingly, Rabanus outlines 14 divisions of time (c. 10); the number is exactly the same as given in both the Munich Computus and *De ratione conputandi*; the details, however, are slightly different, since Rabanus omits the units 'week' and the rather vaguely defined *mundus*, but introduces instead *ostentum* as the second smallest unit and *partes* as the fourth smallest, placed between *momentum* and *minutum*. Nevertheless, Rabanus, more emphatically than Bede, leaves no room for doubt that these divisions of time serve as a framework for his work as a whole, impressively illustrated by the fact that the following chapter (c. 11) deals with the smallest of these units, *atomos*, while the final two chapters (c. 95–96) are concerned with the largest, *seculum* and *aetas*.[329]

From the composition of Rabanus Maurus' *De computo* until the end of the ninth century, no computistical textbook is known at present, a fact that may only be due to a lack of interest in ninth-century computistics by modern scholarship. In the year AD 903, then, a new textbook under the title *Liber de computo*, commonly ascribed to Helperic of Auxerre, was composed which proved to be extremely influential right into the 13th century.[330] Even though this work

327 Cf. especially *Quaest. Austr.*, *Quest. Langob.*, *Comp. Col.*, and texts like ps-Alcuin's *De saltu* and *De bissexto* (*PL* 101, 984–99).

328 *Lib. comp.* and *Lib. calc.*

329 Previous commentators on the structure of Rabanus' *De computo* do not consider the divisions of time as the framework of this text (with sections on the concept of numers and the concept of time prefixed). Rissel, in her discussion of Rabanus' sources (*Rezeption*, 30–40), divides this text into five sections, dealing with arithmetic (c. 1–10), divisions of time (c. 11–36), astronomy (c. 37–53), Easter calculation (c. 54–94), and a world-cronicle (c. 95–96) respectively; cf. also Rissel, 'Hrabans Liber de computo', 140. Stevens, when editing this text ('Rabani', 178–88), adopts Rissel's division in slight variation: 1) c. 1–8, dealing with the concept of numbers, 2) c. 9–36, being concerned with 'the discussion of time itself and its terminology', 3) c. 37–53, discussing celestial bodies and their relation, 4) c. 54–92, introducing the calculation of Easter and all its related features, especially those outlined in Easter tables, 5) c. 93–96, stressing the significance and redemptive character of Easter. For Rabanus' divisions of time cf. also p. CLIX–CLX below.

330 Helperic's *Liber de computo* exists only in an early 18th-century edition by Bernard Pez (based on only one – and a very late one – of the more than 80 manuscript witnesses, where it appears in a recension of AD 1090), reprinted in *PL* 137, 15–48. A critical edition (not only of the earliest witness of AD 903, but also of its main recensions) is desperately

abandoned the concept of the 14 divisons of time altogether, it nevertheless imitated, to a certain degree, the general structure employed in Irish textbooks of the late seventh, early eighth century by being divided into four major sections, dealing with: 1) lunar theory (c. 1–9), 2) solar theory (c. 10–22), 3) the combination of the two, i.e. the reckoning of Easter, including the Dionysiac algorithms for calculating the data found in the columns of Easter tables (c. 23–35), 4) noteworthy exceptions and problems in lunar reckoning, followed by an epilog (c. 36–38).

In summary, then, the concept of the divisons of time, which structured the first part of Irish computistical textbooks, was subsequently, by Bede and Rabanus Maurus, employed as the overriding theme for a computistical textbook as a whole, while the general Irish structure of 1) solar theory, 2) lunar theory, 3) the combination of the two as Easter calculations, and 4) more complicated chronological features for advanced students, was prominently adopted by Helperic of Auxerre, though he abandoned the concept of the divisions of time.

Having thus outlined the context of the structure followed by the Munich computist, we may now turn to the details and sources of each part of the Munich Computus. This text is divided into six major sections:

1) After very briefly introducing the concept of time in general (c. 1, ll. 2–3), the basic framework of chapters 1 to 45 is set by the 14 divisions of time outlined in the first chapter as *atomos*, *momentum*, *minutum*, *punctum*, the hour, the quarter-day, the day, the week, the month, the season, the year, *saeculum*, *aetas*, and *mundus*.[331] At appropriate places, several excursi are inserted into this framework, e.g. a discussion of the structure of the Julian calendar and of basic algorithms for weekday calculations. The source for this classification of the units of time into 14 categories probably was a seventh-century Irish tract called *De divisionibus temporum*. Unfortunately, this text is only transmitted in a great variety of versions in later Carolingian computi, in which it regularly breaks off at the discussion of the year.[332] In fact, the Munich Computus consti-

needed. For the date of composition, the author, as well as the reception of this work cf. the references in notes 136–137 above.

331 Note that the following mentioning of the indiction appears, for several reasons, to have been a gloss that crept into the text at a later copying stage: a) it would have constituted a fifteenth division, though the Munich computist explicitly announced that only 14 were to be discussed; b) it does not fit into the context in which it appears; and, most conclusively, c) the indiction does not feature in the following detailed discussions of each individual division.

332 As it is the case in the Munich Computus itself, the discussion of the unit 'year' quite naturally leads to a very detailed and lengthy discussion of the Julian calendar; at this point, then, most later Carolingian versions of *De divisionibus temporum* abandon the overriding theme of the 14 divisions of time, since they usually do not return to the greater units *saeculum*, *aetas*, and *mundus*; this fact is impressively illustrated by the only published version of this text from a now lost MS in *PL* 90, 653–64. To my present knowledge, only the Sirmond MS (Oxford, Bodleian Library, Bodley 309, fol. 64v–73v) and the related versions of *DDT* found in Basel, Universitätsbibliothek, F III 15k, fol. 24v–35v and Bern, Burgerbibliothek, 417, fol. 47r–61v, 33r–v (the discussion of the greater units of time is

tutes the earliest datable text containing a full discussion of these 14 divisions. Accordingly, as long as the original Irish kernel of the Carolingian versions is not established, the dependency of the Munich Computus on this undoubtedly seventh-century Irish tract cannot be analyzed in detail. The primary sources for *De divisionibus temporum* and this part of the Munich Computus are Isidore's *Etymologiae* and *De natura rerum*, but many other authors and texts are considered for more specific questions.

Following the outline given in chapter 1, the Munich computist remains faithful to his list for the first nine divisions of time, discussing each of these in turn in subsequent chapters: His discussion of *atomos* (c. 2) constitutes the chapter of the entire text that draws most on grammarians, which are quoted in order to complement Isidore's account of this matter; it may be worth noting here that the study of grammar and computistics was well connected in early medieval monastic schools, so that computists would quite naturally have been very well aquainted with grammatical texts.[333] In the descriptions of the following eight units of time, from *momentum* to the week (c. 3–10), the sources used are, as long as they can be traced, almost exclusively the two Isidorian works, the *Etymologiae* and *De natura rerum*.

At this point the Munich computist diverts from the list of his 14 divisions of time by turning from the week to a subsection of that unit of time, namely the weekdays (c. 11), with the principal aim of introducing the three main weekday-terminologies found in early medieval literature, i.e. *feriae*, *sabbati*, and the Roman planetary weekdays. Again, Isidore provides almost all of the source material, though the citation of the B-version of the so-called Acts of the Council of Cesarea, which should more appropriately be referred to by its title *De ordinatione feriarum paschalium*, is already noteworthy at this early stage of the text; it is cited at more length in chapters 55 and 63.[334]

With chapter 12 the Munich computist then returns to his list of the 14 divisions of time by defining the term *mensis* (month) and explaining the historical development of the Julian calendar months. While the etymologies of *mensis* derive from Isidore as well as the enigmatic Virgilius Maro,[335] the history of the Julian calendar is based on excerpts from Macrobius' *Saturnalia*, which circu-

clearly misplaced here and originally rather belongs to the end of the discussion of the divisions of time, as in the Basel MS) do return to the greater units, and therefore these versions appear to be the closest to the original, now lost seventh-century Irish text (but note that none of these versions represents the original, since they certainly include Carolingian additions). For *De divisionibus temporum* cf. also notes 55, 115 above and Graff, 'Recension of two Sirmond texts', 112–25.

333 The early medieval connection between grammatical and computistical studies still is a desideratum in modern scholarship. It may suffice here to draw attention to the fact that, e.g., Walahfrid Strabo's famous *vademecum* (St. Gallen, Stiftsbibliothek, 878) consists exclusively of grammatical and computistical texts; cf. Scherrer, *Verzeichnis*, 307–9; facsimiles and descriptions of this codex can be found online at: http://www.e-codices.unifr.ch/de/list/one/csg/0878.

334 For *De ordinatione feriarum paschalium* cf. especially note 167 above; for it being quoted in these later chapters of the Munich Computus see p. CXIX, CXXI below.

335 For Virgilius Maro's *Epitomae* cf. especially note 176 above.

lated under the title *Disputatio Chori et Praetextati* in early medieval manuscripts;[336] these excerpts constitute the ultimate source for this topic, for medieval computists as well as modern historians. In the following 11 chapters (c. 13–23) the etymologies of every single Julian calendar month are discussed in detail, primarily based on Macrobius and Isidore, but also on some interesting chronicle material in the discussion of January. The author then returns to the general discussion of the unit 'month' (c. 24), asking very original questions about the order of months and, more specifically, the various traditions concerning what is to be regarded as the first month, as well as the development of the lengths of the months and their origin. In this chapter, the most noteworthy of the very few sources are Jerome's commentary on Ezechiel and the *addendum* to the Cyrillian Prologue. As in his discussion of the week, the Munich computist continues with three chapters on the days of the Julian calendar months, or, to be more precise, on the three marker-days, the Calends, Nones, and Ides (c. 25–27), again primarily based on the authority of calendar theory, Macrobius, as well as on Isidore. Having introduced these marker-days, the author continues to outline the arrangement of the calends, ides, and nones in each of the Julian calendar months, and with this effectively explains the structure of the Julian calendar (c. 28). At this point, the computist deemed it important to introduce basic rules of weekday calculations, first for specific dates in the first third of a year which directly relate to the calculation of Easter, then more generally for any given Julian calendar date (c. 29). Both of these chapters (c. 28–29) are principally based on the *Computus Einsidlensis*.[337]

Following these excursi, the Munich computist again returns to his basic framework for this section, the 14 divisions of time. After listing the etymologies for 'season' (*tempus*) and 'year' (*annus*), as well as classifying the latter unit (c. 30–31), he proceeds to discuss the individual seasons in turn (c. 32–35), apparently as sub-divisions of a year, before returning to more general questions about the year (c. 36). This approach is in line with the one followed in the discussion of the week and the month, where the days of the week and the Julian calendar dates respectively are treated as sub-sections. In chapters 30 to 35, the principal source is again Isidore, with very few additions from Macrobius and Virgilius Maro; chapter 36, then, contains very original ideas on questions that appear to reflect the Munich computist's own interest, discussing the various traditions concerning the beginning of the year, the exact number of weeks and *momenta* per year, and the place of the year of creation in the 19-year cycle.

This is then followed by a block of five chapters (c. 37–41) which have no immediate relation to the 14 divisions of time, but all of which are, in some way, related to the discussion of the year. Chapter 37 deals with the names for and etymologies of 'sun' (*sol*), with most of the information being taken from Cassidorus' *Exposito in psalmos* and Virgilius Maro's *Epitomae*; it appears that

336 For the *Disputatio Chori et Praetextati* cf. note 171.

337 For the *Computus Einsidlensis* and the Munich computist's dependency on that text cf. especially p. CXXIII–CLII below.

this excursus on the sun is introduced here, immediately following the discussion of the year, since a year is defined as the period of time that the sun takes to travel through the zodiac. The following three chapters (c. 38–40), which primarily derive from the *Computus Einsidlensis*, relate, in one way or another, to the division of the year into four quarters between the solstices and equinoxes; this division of the year differs from the division into seasons, as the Munich computist argues himself,[338] and therefore had to be treated separately. Chapter 38 introduces this division, which then provides the basis for the theory of the increase and decrease of daylight outlined in chapter 39, whose principles are ultimately based on Anatolius (?)'s *De ratione paschali*.[339] One of the problems of the division of a year between the equinoxes and solstices was, however, that it was not equal, but rather resulted in quarters of differing lengths, namely 90, 91, 92, and 92 days respectively, which contradicted the theory of a linear increase and decrease of daylight by a value common to every day of the Julian calendar year. In chapter 40, the Munich computist addresses this problem by first analyzing the effect that the neglect of the days exceeding 90 in three of the four quarters (5 days in total) would have on this theory, while then dividing these days equally among the 90 days of the quarter in which they occur (with this implying that these five days are not to be regarded as full days and could therefore be left unconsidered for the general theory). Before returning to the list of the 14 divisons of time, the Munich computist felt the need of explaining one further feature related to the year, or rather to the construction of the Julian calendar year, namely the bissextile day (c. 41). The length of this chapter and the originality of large parts of it impressively illustrate the interest that seventh- and early eighth-century Irish computist had in this technical device, but also the difficulties they faced in understanding and applying it.

Chapters 42 to 44, then, deal with the final three divisions of time in the Munich computist's list, *aetas*, *saeculum*, and *mundus* respectively. While the chapter on *aetas* (c. 42) is exclusively based on Isidore's *Etymologiae*, no source is apparent for the etymology and definition of *saeculum* (c. 43). The subsequent discussion of *mundus* (c. 44) consists of three parts, the definition and etymology of the term (again exclusively taken from Isidore's *Etymologiae*), a reference to the number of years from the beginning of the world to the passion of Christ based on the prologue of Victorius of Aquitaine, and a lengthy comparison of the chronology of the first week of the three most important years in Christian history, namely the year of creation, the year of the Exodus from Egypt, and the year of the passion of Christ; the stimulus for the composition of this third part comes from Victorius' prologue, the content, however, is a very original study of the scripture and other authoritative texts, leading to unparalleled results.

Finally, chapter 45 supplies a proof for the number of days of the Julian calendar year, which may have been designed as a round-off of the preceding discussion of the various units of time smaller than the solar year as well as of

338 MC 38.39–43.

339 For Anatolius (?)'s *De ratione paschali* cf. p. LXV–LXVII above.

the elements of the Julian calendar. In that case, however, one would rather have expected it to occur in the context of the discussion of a year. Other possibilities are that it should mark the end of solar theory, or that it was, originally, a marginal commentary that crept into the text at a later copying stage.

As a cursive overview, the relation between the Munich computist's discussion of the 14 divisions of time on the one hand, and the various subsections and excursi on the other, is illustrated in the following table:

Chapter	Discussions of the 14 division of time	Subsections and excursi
1	*De divisionibus temporum*	
2	*De atomo*	
3	*De momento*	
4	*De minuto*	
5	*De puncto*	
6	*De hora*	
7	*De quadrante*	
8	*De diei nomine*	
9	*De nocte*	
10	*De ebdomada*	
11		*De feriis*
12	*De mensibus*	
13		*De Ianuario*
14		*De Februario*
15		*De Martio*
16		*De Aprelio mense*
17		*De Maio*
18		*De mense Iunio*
19		*De Iulio*
20		*De Augusto*
21		*De Septimbrio*
22		*De Octimbrio*
23		*De Nouimbrio et Decimbrio*
24	*Continuatio: De mensibus*	
25		*De Kalendis*
26		*De nomine Nonarum*
27		*De nomine Iduarum*
28		*De ordine et regularibus dierum mensium*
29		*De calculatione feriarum*
30	*De tempore*	
31	*De anno*	
32		*De uerno*
33		*De aestate*
34		*De autumno*
35		*De hieme*

36	*Continuatio: De anno*	
37		*De sole*
38		*De divisione anni*
39		*De augmento diei et noctis per annum*
40		*De V diebus superfluis*
41		*De bissexto*
42	*De aetate*	
43	*De seculo*	
44	*De mundo*	
45		*De anno solis*

2) The second major section of the Munich Computus consists of chapters 46 to 53, introducing the lunar year as well as lunar reckoning. Chapter 46 lists the etymologies of 'moon' (*luna*) and discusses its relation to the sun, being almost exclusively based on Isidore's *Etymologiae*. The specific question in which directions of the sky the moon may rise then led the Munich computist to a general excursus about the four directions of the world (c. 47), with the etymologies being again copied from Isidore's *Etymologiae*, while the definition of the directions by means of the positions of the sun in the sky appears to have been the Munich computist's own contribution. Additionally, this excursus serves as the basis for the following, very original continuation of the discussion of the moon (c. 48), dealing with the place and time of the kindling of the moon according to conflicting traditions, the course of the moon in the sky, as well as the question about the divisions of time in lunar reckoning; the originality of this chapter is further accentuated by the use of Old Irish.[340]

The general question about the units of time in lunar reckoning leads to a discussion of the most important feature in that reckoning, namely the epact, which also constitutes the principal characteristic of every lunar year (c. 49); this chapter, being primarily based on the *Computus Einsidlensis*, discusses the definitions of the term, but more prominently the various placements of the epacts in the Julian calendar year, as well as differing methods of comparing the 19 Dionysiac epacts with the 19 Victorian ones. The comparative approach is then followed up in the next chapter, in which the lunar age of the Calends of each month as well as the sequence of lunations are analyzed, comparing Dionysiac with Victorian and *latercus* data; such a comparison can already be found for the Dionysiac and the Victorian reckoning in the *Computus Einsidlensis*, with the introduction of the *latercus* information being the Munich computist's original contribution. In chapter 51, the Munich computist returns to the discussion of the epacts by defining that the eleven-day difference between the lunar and the solar year corresponds to the eleven Julian calendar days from 11 to 21 March; moreover, he analyzes the relationship between the lunar ages of 22 March and 1 January and 11 March of the following Julian calendar year; the basic concept and the main part of this chapter derives, again, from the *Computus Einsidlensis*, but the Munich computist also introduces con-

340 For the use of the Old Irish verb *do·meil* in this discussion cf. p. LXXV above.

siderable variations and new ideas. Chapter 52, then, addresses the question of the epact of 1 January of one specific year, namely the year of the passion of Christ, the most important year in Christian chronology; the problem faced by seventh-century computists was that a comparison of the Dionysiac with the Victorian reckoning resulted in the fact that the Dionysiac reckoning had epact 18 (and consequently *luna* 15 on Easter Sunday) in what Victorius had designated as the *annus passionis*; this result was contrary to the information supplied by the Gospels, as well as later authoritative literature; having established this fact, it appears that adherents of the Victorian reckoning accused followers of the Dionysiac one of heresy; in order to counter this accusation, Dionysiac advocates then constructed a (unhistorical) method of comparing these two lists of 19 epacts that would provide for the necessary results; it appears that the method in question was first formulated in a tract transmitted in the famous Cologne codex 83², to be termed *De comparatione epactarum Dionysii et Victorii*, which probably was the Munich computist's source for this chapter.[341] Finally, the last chapter of this section (c. 53) deals more generally with the lunar year, its definition, origin, and beginning; it appears to be designed to conclude the discussion of lunar theory, and, being without apparent source, reveals the Munich computist's own personal interests.

3) Having introduced the basics of solar and lunar theory, the Munich computist turns to the combination of the two, the principal object of a computistical textbook, the calculation of Easter (c. 54–58). First of all, chapter 54 introduces the term Easter (*pascha*) itself, primarily based on Isidore's *Etymologiae*, but some information about the pronounciation and the inflection of the word appear to have been derived from Jerome and the grammarian Pompeius (Maurus). Chapter 55, then, outlines the essential rules for the calculation of Easter; citing Anatolius (?), the Munich computist first discusses the beginning of the first month, i.e. the Easter month, according to different traditions, before introducing the Julian calendar and lunar limits for Easter Sunday, a discussion primarily based on the B-version of the so-called Acts of the Council of Caesarea (*De ordinatione feriarum paschalium*) and the ps-Jerome or ps-Columbanus text *De sollemnitatibus*, with both texts having an Insular background;[342] he then continues to discuss the Julian calendar limits for the Easter new and full moon, quoting the prologue of Victorius of Aquitaine, before comparing the Easter full moons of the Victorian and the Dionysiac reckoning; finally, he lists the Dionysiac Easter new and full moons, leaving no doubt about the reckoning favoured by himself. A detailed description of the calculation of Easter Sunday proper is then given in chapter 56, following a summary of the rules that deter-

341 For a discussion of this chapter and the method underlying it cf. Schwartz, 'Ostertafeln', 94–6; Warntjes, '84-year Easter reckoning', 62–9; for *De comparatione epactarum Dionysii et Victorii* being one of MC's sources cf. p. CLII–CLVIII below; it is edited (for the first time) as Appendix 2.

342 *De ordinatione feriarum* is also quoted in c. 11 and at length in c. 63 (cf. p. CXIV, CXXI); for this text see especially n. 167. For *De sollemnitatibus* cf. especially n. 175.

mine this calculation, loosely based on Martin of Braga's *De pascha*,[343] *De ordinatione feriarum paschalium*, and Anatolius (?)'s *De ratione paschali*; the calculation itself is illustrated by the examples of the three successive years AD 719–721; it uses the Julian calendar date of the Easter full moon and the weekday of 1 January as preconditions, a method taken from the *Computus Einsidlensis*; following this calculation, the Munich computist outlines an original method for calculating the Julian calendar date of the Easter full moon from the epact of 1 January, which allows for the calculation of Easter Sunday from the weekday and lunar data of 1 January when combined with the previous rule. Since, in his opinion, a lunar year extends from the Easter new moon to *luna* 30 preceding the Easter new moon of the following year, and therefore is directly dependent on Easter calculations, the computist includes a chapter on the Julian calendar intervals of the so reckoned 19 lunar years here (c. 57). Finally, the relation between Easter and the beginning of Lent is analyzed in great detail in the final chapter of this section on Easter (c. 58), relying once more on the even more detailed discussion of this question in the *Computus Einsidlensis*.

4) Chapters 59–62 are concerned with the theory of the 19-year cycle (*cyclus decemnovenalis*). It should be noted that a thorough understanding of the construction of the 19-year cycle is not immediately essential for the calculation of Easter and therefore there was no apparent need for this section to precede the discussion of Easter; it rather follows the discussion of Easter as an *addendum* for advanced students who have already successfully studied the basics outlined in the previous three sections.[344] The first chapter of this section (c. 59) deals with the concept of common and embolismic years, the general introduction and definitions being based on Dionysius' prologue and Isidore's *Etymologiae*, while the following detailed discussion of the sequence of common and embolismic years is partly taken from the *Computus Einsidlensis*. This is followed by one of the most curious chapters of the entire text (c. 60), which essentially summarizes the attempts made by seventh-century Irish computists to prove the cyclic nature of the 19-year cycle; the chapter as a whole derives from the *Computus Einsidlensis*. In chapter 61 the Munich computist returns to the theory of embolismic years, discussing their characteristics before outlining their relation to the epacts of 1 January and their placements within the 19-year cycle, an outline that is again based on the *Computus Einsidlensis*. The final chapter of this section on the 19-year cycle (c. 62), then, deals with the most difficult and controversial feature of this construction, the *saltus lunae*, i.e. the omission of one lunar day every 19 years in order to bring the total number of lunar days in line with the number of solar days of this period. Some parts of this long chapter are based on the *Computus Einsidlensis*, but most of it is the Munich computist's own composition.

5) The first four sections have supplied a good overview of early medieval time-reckoning, including detailed discussion of the various units of time, comprehensive descriptions of all aspects of the Julian calendar, a thorough outline

343 For Martin of Braga's *De pascha* cf. especially n. 166.

344 Cf. p. CVIII above.

of lunar reckoning (including comparisons between the three principal systems) and weekday calculations, a substantial analysis of the 19-year cycle, and, most of all, a full discussion of the technical details of the calculation of Easter and the beginning of Lent. Accordingly, the chapters of section five (c. 63–67) have to be regarded as appendices of special interest to the Munich computist, which is confirmed by the facts that a) the *Computus Einsidlensis*, the Munich Computus' main source, breaks off after the discussion of the 19-year cycle, and b) almost all of these chapters have no parallels, to my knowledge, in any other surviving computistical text or manuscript. Further below, the theory will be advanced that these chapters were copied from the now lost Victorian computus of AD 689.[345] Chapter 63 appears to be a remnant from the height of the Easter controversy, listing authoritative statements against the celebration of Easter Sunday on *luna* 14 from texts that were prominent in the Insular Easter debates of the seventh century, such as *De ordinatione feriarum paschalium*,[346] the letter of ps-Cyril (*Epistola Cyrilli*),[347] the letter of Proterius,[348] and Dionysius' prologue; note that this chapter also contains some chronicle material, presumably derived from Jerome. This is followed by a brief calculation of the number of years from the creation of the world to the Exodus from Egypt, as well as from the latter event to the *annus passionis*, based on Victorius' prologue (c. 64). In chapter 65 the Munich computist (or his unkown source, probably the lost Victorian computus of AD 689) tests all Easter tables known to him according to their congruence with the 19-year lunar cycle, the 28-year solar cycle, and the 532-year Victorian Easter cycle. Chapter 66 assembles all Biblical passages that may hint at a disruption of the course of time and discusses their impact on the reckoning of time; besides the Vulgate, the principal source for this chapter is the Irish mid-seventh-century text *De mirabilibus sacrae scripturae*.[349] The most original of the appendices is the final one, which provides an unparalleled proof for a 95-year Easter table not being truly cyclic in respect of its weekday data (since, it will be noted, 95 does not constistute a multiple of the 28-year solar cycle).

6) As is the case with other eighth- and ninth-century computistical textbooks[350] (but, it will be noted, no other Irish text of that genre), the Munich Computus ends with a chronicle (c. 68); this chapter should be regarded as a separate section, even though it probably was one of the chapters copied from the Victorian computus of AD 689 as addenda to the textbook proper. The Munich computist's chronicle differs considerably from all other chronicles attached to computistical textbooks in that its purpose is not to simply provide for a linear structure based either on Septuagint or Vulgate chronology, but rather

345 Cf. p. CXXIV–CXXVI below.

346 *De ordinatione feriarum paschalium* is also quoted in c. 11 and 55 (cf. p. CXIV, CXIX above).

347 For the *Epistola Cyrilli* cf. n. 164 above.

348 For the *Epistola Proterii* cf. n. 174 above.

349 For *De mirabilibus sacrae scripturae* cf. especially p. LXXVIII–LXXX above; it also served as the basis for the chronicle at the end of MC (c. 68).

350 *DT* 16; *DTR* 66–71; RM 96. Cf. p. CVIII–CXII above.

attempts to correlate the linear chronological data from the creation of the world to the *annus praesens* of the author with the concept of 532-year Easter cycles, in this following *De mirabilibus sacrae scripturae*.

Having analyzed the structure applied and the principal sources used by the Munich Computus, special attention needs to be drawn to seven of the Munich computist's sources that have never been discussed in detail (with the exception of the *latercus*) or even been identified before, and which for this reason alone deserve a separate treatment here.

Latercus: In a recent article, I have discussed all of the Munich computist's passages dealing with the *latercus* (i.e. the 84 (14)-year Easter reckoning followed in various regions of Ireland and Britain right into the eighth century) in great detail,[351] so that I restrict myself here to simply outlining the results of that analysis. It appears that the Munich computist himself was rather unfamiliar with this Easter reckoning (leading to the conclusion that he was a member of the southern Irish clergy that had abandoned the 84 (14) almost 90 years before the compilation of the Munich Computus[352]) and that he had relatively little information about it at hand. After having had the chance of examining the *Computus Einsidlensis* (of whose existence I was not aware when writing the mentioned article), I now believe that it was this text that triggered the Munich computist's interest in the 84 (14).[353] On the basis of this presumption, the following appears to be the most likely scenario: The Munich computist copied the *latercus* lunar limits for Easter Sunday as well as for the *initium quadragesimae* from the Einsiedeln text (the *initium* data having been wrongly reconstructed by the Einsiedeln computist) and on foot of this reference he seems to have made his own inquiries about the 84 (14). These inquiries (or his own memory) led him to the facts that the initial epact of the *latercus* was 19, that the *saltus* was applied in 14-year intervals, and that the lunar calendar of this reckoning was based on a peculiar sequence of lunations, which may still have been memorized in his own monastery.[354] Every other information referring to the *latercus* in the Munich text was then, in most cases wrongly, constructed by the Munich computist himself.[355]

351 Warntjes, '84 (14)-year Easter reckoning', which includes facsimiles of all relevant passages on p. 83–5. In the following edition, the passages referring to the *latercus* are MC 50.67–71, 50.72–76, 52.3–26, 58.2–7, 62.118–122. One further passage in MC (65.27–31) relates to the 84 (14), though not under its terminus technicus *latercus*; for this passage cf. p. CXXIII below.

352 Cf. p. LXXXVIII–XCVI above.

353 For the crucial passage cf. p. CXLVII–CL below.

354 In this context it should be remembered that an Irish mnemonic dealing with this peculiar sequence of lunations was well known as late as the 19th century! The relevant poem was discovered by Dáibhí Ó Cróinín; cf. Mc Carthy, 'Easter principles', 212–3; Warntjes, '84 (14)-year Easter reckoning', 43, 77.

355 The passages dealing with the *latercus* and their constructions are the following: As mentioned above, MC 58.2–7 dealing with the lunar limits of Easter Sunday and the *initium* is

Note, however, that yet another passage dealing with the 84 (14) appears in the Munich Computus, albeit not under its terminus technicus *latercus*.[356] In this passage, the computist (or rather his source) compares all Easter tables known to him according to their congruence or incongruence with the 19-year lunar cycle, the 28-year solar cycle, and the favoured 532-year Victorian Easter cycle. One of the tables examined consists of 100 years with the *saltus* occurring in 14-year intervals. Since no Easter table with 14-year *saltus* other than the 84 (14) is known at present (and if this reference was not to the 84 (14), then this reckoning that was so prominent in the seventh-century Irish context would not have been mentioned at all in this comprehensive list of Easter tables!), it can hardly be doubted that this Easter reckoning is referred to here. The context, namely the comparison with the 532-year table of Victorius of Aquitaine, shows a clear bias for the Victorian reckoning. It must therefore be concluded that the Munich computist copied this passage about the 84 (14) from an exemplar primarily concerned with the Victorian reckoning, most probably from the Victorian computus of AD 689 to be discussed below.

In summary, it now appears that the Munich computist took and copied basic information about the 84 (14) from two independent sources, a passage each from the *Computus Einsidlensis* on the one hand, the Victorian computus of AD 689 on the other. Additionally, he remembered or acquired basic information about this reckoning, namely the facts that the initial epact was 19, that the *saltus* occurred in 14-year intervals, that it constituted an 84-year Easter cycle, and that the basis of its lunar calendar was a peculiar (but very clever and suggestive) sequence of lunations; if this information was assembled rather than remembered by the Munich computist, the sources are most likely to have been oral, but the basic structure of this reckoning (like its 84-year cyclic character)

directly based on the *Computus Einsidlensis*. MC 50.72–76 correctly describes the *latercus*' peculiar sequence of lunations, which the Munich computist may have acquired from an oral source; with this sequence at hand, he managed to reconstruct the lunar data of the Calends of each month of a year with epact 8 on 1 January, which was considered to be the equivalent to the first year of the Dionysiac 19-year cycle (MC 50.67–71). The fact that the *latercus* started with epact 19 on 1 January and that the *saltus* were applied in 14-year intervals made it possible to reconstruct the entire list of all 84 *latercus* epacts, from which, then, the epactal changes of the *saltus* years could easily be noted (MC 62.118–122); moreover, the first 19 epacts of this list were also compared to the epacts of the Dionysiac and the Victorian reckoning (MC 52.3–26), though in a wrong (i.e. unhistorical) way which lead the Munich computist to numerous false conclusions: a) that the *latercus* data corresponded to the Victorian data in what Victorius had denoted as the *annus passionis* only because their epacts allegedly agreed in that year (MC 52.3–26); b) that the *saltus* of the *latercus* occurred (regularly) in the fourth year of the (Victorian) *ogdoas*, a statement that, besides the wrong alignment of epacts in the first place, disrespects the structural difference of a 19-year lunar cycle as against an 84-year one (MC 62.118–122); c) that the *saltus* of the *latercus* was implemented in November, which simply is a translation of Victorian usage to the *latercus*, as the *latercus*' hollow November lunation could hardly have been reduced to 28 days (MC 62.118–122).

356 MC 65.27–33. For this passage and its context cf. also Schwartz, 'Ostertafeln', 101–2; Warntjes, '84 (14)-year Easter reckoning', 55–6, 61; wrong conclusions are drawn by Mosshammer, *Easter Computus*, 219–22.

may also have been known from technical information randomly mentioned in the literature originating in the paschal controversy of the seventh century. All further information about the *latercus* given in the Munich Computus was constructed (often wrongly) from this data by the computist himself. Therefore, the Munich computist did not work from a single cohesive source about the *latercus*, neither an Easter table of that reckoning, nor a computus or a treatise on the 84 (14).[357]

Victorian computus of AD 689: The two references to the year of the consuls *Vero* and *Bradua* clearly reveal that the author of the Munich Computus used a text composed in AD 689 that was concerned with and in favour of the Victorian reckoning.[358] Since this source has not survived, it cannot be established whether this text dealt with selected aspects of this reckoning only, or whether it constituted a full scale textbook; for this reason it will simply be termed the Victorian computus of AD 689 here.[359] Methodologically, it seems appropriate to attribute to this now lost text all passages of the Munich Computus based on and favourable to the Victorian reckoning (the Munich computist himself was a follower of the Dionysiac reckoning) for which no other source can be traced. On this basis, it certainly included a discussion of the bissextile day, consisting at least of three calculations: a) of the number of years that contained 365 bissextile days, i.e. 'a full year of bissextile days' (or, in the phrasing of the author of this passage, 'a year full of bissextile days') consisting of 4×365=1460 years; b) of the number of bissextile days in the Victorian Easter cycle, i.e. 532/4=133 days; and c) of the number of 'years full of bissextile days' from the creation to the *annus praesens* of the author, based on the Eusebian-Victorian *annus mundi* era.[360] Similar calculations were executed in this lost computus for the 19-year *saltus lunae* of the Victorian reckoning, counting the number of *saltus* in an interval of 5100 years, as well as the number of *saltus* from the beginning of the world to the year of the consuls *Vero* and *Bradua*.[361] Since these are the two passages containing the dating clause for AD 689, they were evidently copied from the Victorian computus of AD 689.

Additionally, the entire fifth section of the Munich Computus, comprising chapters 63 to 67, as well as the following brief chronicle (c. 68), appear to have been taken from this Victorian exemplar. As has been established above,

357 For this argument cf. especially Warntjes, '84 (14)-year Easter reckoning', 75–7.

358 MC 41.107–110, 62.65–67. For these passages and their dating clause cf. p. LVII–LVIII above. For the context of the composition of this text in AD 689 cf. p. CLIV–CLVI above.

359 Even though modern scholars of computistics have widely acknowledge the fact that the Munich computist worked from a Victorian exemplar of AD 689 (cf. note 141 above), no attempt has been made to define the layer of the Munich Computus that ultimately derives from this Victorian exemplar. Besides the passages containing the Victorian dating clause (MC 41.107–110, 62.65–68), only c. 65 has been previously associated with the Victorian computus of AD 689 (cf. Schwartz, 'Ostertafeln', 101–2; Warntjes, '84 (14)-year Easter reckoning', 61, 74).

360 MC 41.107–110.

361 MC 62.65–68.

the Munich Computus presents a coherent and well structured textbook on the reckoning of time up to chapter 62, ending with a treatment of the *saltus*, like the *Computus Einsidlensis* (which seems to have been the immediate precursor to the Munich Computus); the following six chapters, then, can be regarded as addenda of special interest to the Munich computist.[362] These addenda may have been copied from one and the same source, namely the Victorian computus in question here. Only the first of these chapters (c. 63) does not show a direct Victorian bias; however, its principal purpose, listing arguments against celebrating Easter on *luna* 14, is evidently directed against the 84 (14) and may therefore have been written by any adversary of this reckoning, Victorian or Dionysiac alike. It would therefore be absolutely plausible to find this discussion in a Victorian text composed in AD 689, a time when southern Irish adherents of the Victorian reckoning had struggled for more than 50 years against the practice still followed by northern Irish churches. All remaining chapters of section 5 (c. 64–67) are clearly based on and / or in favour of the Victorian reckoning: chapter 64 calculates the number of years from the creation of the world to Moses' celebration of the first pasch, and then from this event to the resurrection of Christ, on the basis of the Eusebian-Victorian *annus mundi* chronology; chapter 65 compares all known Easter tables with the Victorian 532-year cycle; chapter 66 discusses the Biblical disruptions of time, a discussion stimulated by the southern Irish tract *De mirabilibus sacrae scripturae*, which in turn is based on Victorius' writings; chapter 67 demonstrates that the Dionysiac 95-year Easter table is not cyclic concerning its solar data, and the initial statement that this table revolves five times in the cycle of Victorius leaves no room for doubt that this chapter was written by an adherent of the Victorian reckoning. Moreover, the brief chronicle of chapter 68, then, is again based on *De mirabilibus sacrae scripturae*.

Having thus established the passages of the Munich Computus taken from the lost Victorian computus of AD 689, it is now also possible to determine the origin of this text. Since it appears to have been only cited in the Munich Computus, this suggests that the Victorian computus itself was an Irish production. The very fact that the Easter reckoning underlying this text is the Victorian one, then, suggests southern Ireland as its place of composition, since this reckoning appears not to have been used in any other region of the *regiones Scottorum*.[363] More conclusively, the passages from chapters 41 and 62, as well as chapter 64 as a whole, show parallels to items 26 and 45 of the Sirmond MS,[364] while chapters 66 and 68 are based on *De mirabilibus*, as has already been pointed out above; both *De mirabilibus* and item 26 of the Sirmond MS have convincingly been located to the east Munster region of southern Ireland, while both have been written at roughly the same time, the former being datable to AD 654, the latter to AD 658.[365] On the basis of this textual dependency on these two texts,

362 Cf. p. CXXC–XXI above.

363 Cf. especially p. LXXXVIII–LXXXIX and XCIII–XCV above.

364 Cf. p. LXXX–LXXXII above and the relevant passages in the edition.

365 Cf. p. LXXVIII–LXXXII.

it is highly probable that the lost Victorian computus of AD 689 was composed in east Munster.

Irish Isidore / Augustine / Jerome: In the Munich Computus, there are frequent ascriptions to Isidore, Augustine, and Jerome, even though the passages, sentences, or phrases in question cannot be found in any of the three saints' works. Generally, these ascriptions can be classified in two main categories: a) confusion by the Munich Computist when citing a different, but known author; and b) passages without any apparent source (at least not part of the surving body of texts of these Church fathers), but often shared by other computistical (primarily Irish) texts.

Category a) may stem from two distinctively different confusions, namely the Munich computist's own, or that of his source. E.g. passages quoted from Virgilius Maro are frequently attributed to Augustine (and to no other author, particularly not to Virgilius Maro himself),[366] and accordingly it has to be assumed that Virgilius Maro's *Epitomae* circulated under the name Augustine; at least the copy used by the Munich computist appears to have had this attribution. On the other hand, the Munich computist sometimes ascribes Isidorian passages to Augustine,[367] etc., which presumably is due to the fact that this computist recited many of these passages from memory, but confused the author in this process.

As for the more important category b), the obvious question is whether these late seventh- and early eighth-century Irish computists worked from now lost texts attributed to Isidore, Augustine, or Jerome. In most instances, I do not believe that this was the case. It appears more likely to me that early medieval computists used lists of excerpts from these authors classified according to content, i.e. assembled under suggestive headings like *De atomo*, *De embolismis*, etc. Such lists must have been frequently added to, revised, and updated, so that certain new definitions and concepts, originally glosses or commentaries to well-known citations, crept into these list under the name of any of the church-fathers. These lists then circulated among Irish computists, and this would explain the common ascription to certain authors in various texts. Most of the category b) ascriptions can, in my opinion, be explained by this pattern, but particularly the ascriptions to Augustine may rather point to an Irish teacher of that name, as the following discussion will show.

Irish Isidore: Unsurprisingly, if otherwise unknown passages are attributed to an author, the attribution most frequently is to Isidore. In the following, a list is provided of all passages attributed to Isidore in Irish computistical textbooks, but not found in any of the bishop of Seville's works. If these passages occur in more than one Irish (or later Frankish) computistical textbook, they are tabulated in parallel, with the differences in the citation proper being highlighted in bold font.

366 MC 12.3–7, 37.6–7.
367 MC 8.16–17, 31.5–7, 41.13–16.

Passage 1:

MC 4.2–4:	***DDT*** **4A16–B1:**
	Isidorus dicit:
Minutum a minuendo dictum, ***eo quod IIII*** *momenta* ***in lucem crescentem*** *de tenebris minu**at** ab XI Kalendas* ***Ianuarii*** *in XII Kalendas* ***Aprilis****.*	*Minutum dictum a minuendo,* ***quia duo*** *momenta* ***et duae partes tertii momenti*** *minu**untur** de tenebris* ***in unoquoque die****, ab undecimo Kalendarum* ***Aprilis****, usque in duodecimum Kalendarum* ***Julii****.*

Passage 2:

MC 4.6–8:	***DDT*** **4B3–5:**	***DRC*** **18.6–7:**	**BC 96A4–6:**	**RM 15.4–6:**
Ita autem definitur secundum Ysidorum dicentem:	*Isidorus diffinivit dicens:*	*Isidorus dicit:*	*Iterum*	*D. Vnde dicitur minutum?* *M. A minore interuallo quasi*
Minutum ***est*** *uelut minus momentum, quia minus numerat* ***quod*** *maius implet.*	*Minutum* ***dicitur*** *velut minus momentum, quia minus numerat,* ***sed*** *majus implet.*	*Minutum* ***dicitur*** *uelut minus momentum, quia minus numerat* ***quod*** *maius implet.*	*minutum uelut minus momentum, quia minus numerat* ***id,*** ***quod*** *maius implet.*	*minus momentum, quia minus numerat* ***quod*** *maius implet.*

Passage 3:

MC 5.3–5:	***CE*** **p. 88:**	***DDT*** **5B9–11:**	***DRC*** **18.6–8:**	**RM 16.5–7:**
Quod ita diuisit Ysidorus:	*Isidorus ita ut alibi definiuit:*	*Isidorus ostendit, cum dixit:*	*Isidorus dicit:*	
*Punctus a pungendo, eo quod quibusdam punctionibus certe designatio**nis** in horologiis* ***sit designatus****.*	*Punctus a pungendo* ***dictus est****, eo quod quibusdam punctionibus certe designatio**nis** in horologi<i>s* ***sit designatus****.*	*Punctus a pungendo* ***dictus est****, eo quod quibusdam punctionibus certae designatio**nis** in horologiis* ***designantur****.*	*Punctus a pungendo* ***dictus****, eo quod quibusdam conpunctionibus certe designatio**nis** in horologiis* ***sit designatus****.*	*Punctus* ***quippe*** *a pungendo* ***dictus est****, eo quod quibusdam punction**ibus** certe designation**ibus** in horologiis* ***designatur****.*

Passage 4:

MC 8.2–3	**Angers 477, fol. 19r:**	***CE* p. 89:**	***DRC* 22.7:**	***PP* I §94:**
		Aliter		*inde* ***dicitur***
Dies ***a*** *diuidendo lucem a tenebris* ***dictus est****, dicente Ysidoro.*	*<Dies>* ***a*** *diuidendo lucem a tenebris.*	*dies diuidendo lucem a tenebris.*	*dies diuidendo lucem a tenebris* ***dicta****.*	*dies diuidendo lucem a tenebris.*

This passage has received some variation in *DDT*, from where it was copied into *Dial. Burg.*

***DDT* 8A2–3:**	***Dial. Burg.* 11A:**
D. Dies unde nomen accepit? M. A divisione, eo quod dividat lucem a tenebris.	*Dies dicitur ab eo, quod dividat inter lucem et tenebras.*

Passage 5:

MC 63.28–29:	***DRC* 98.21–2:**
Ysidorus ait:	*Et Isidorus dicit:*
Quicumque ante XV luna pascha celebrari iubet, transgreditur mandatum.	*Quicumque ante XV lunam pascha celebrari iubet, transgreditur mandatum* ***domini****.*

Note that this last passage has a profoundly different quality compared to the previous ones, since the first four passages deal with etymologies of individual divisions of time, while this fifth passage is concerned with Easter rules. Moreover, it is noteworthy that the passages in question almost exclusively appear in Irish texts, which supports the hypothesis that they circulated in Irish lists of excerpts. Among later texts, only the Frankish *Dial. Burg.* of AD 727 and Rabanus Maurus' *De computo* of AD 820 cite any of these passages, both apparently working from some version of the Irish *De divisionibus temporum*. In particular, this highlights the influence of Irish computistical thought on Rabanus Maurus' work.

Additional to these passages it should be noted that the authentic Isidorian citations in Irish computistical texts often share readings that are not to be found in the Isidorian original. This again is proof that these Irish computist worked from very similar exemplars.

Irish Augustine: Turning to citations attributed to Augustine, but not to be found in any of Augustine of Hippo's works, it needs to be pointed out at first that the author of *De mirabilibus sacrae scripturae* refers to himself as Augustine (and is quoted under that name by the Munich computist[368]), and also that citations from Virgilius Maro's *Epitomae* are frequently attributed to Augustine in the Munich text.[369] Accordingly, there are two independent texts quoted by the Munich computist under Augustine's name that were certainly not composed by Augustine of Hippo, but rather show an Irish background.[370] Besides citations from these two texts, there are five further passages ascribed to Augustine in the Munich Computus and other Irish computistical texts.

Passage 1:

MC 41.6–7:	***DRC* 57.27–28:**
Hinc Augustinus ait: *Bissextus* ***famosus*** *solaris* ***euentus****, cui nulla uis est* ***secundum lunam****.*	*ut Isidorus dicit:* *Bissextus* ***est effectus quidam*** *solaris, cui nulla uis est* ***cum tenebris****.*

Passage 2:

MC 59.22–23:
ut Augustinus ait:
Seriunctus annus, qui communium annorum <damna> supplet.

Passage 3:

MC 59.65–66:
Augustinus ait: Omnis ratio embolismorum: In fine numeramus.

Passage 4:

MC 62.3–5:	***DRC* 111.1–2:**	**Angers 477, fol. 20v:**
Tamen Augustinus ait: *Saltus quasi saliend**o** quoddam signum,* ***quia*** *co**m**potus transilit* ***unum*** *diem.*	*Augustinus ostendit dicens:* *Saltus, quasi quoddam saliend**i** signum,* ***quia*** *co**n**potus transilit* ***unum*** *diem.*	*Augustus:* *Saltus quasi saliend**i** quoddam signum,* ***quando*** *co**m**potus transilit diem.*

368 MC 66.15.

369 Cf. the references in note 366 above.

370 For *De mirabilibus sacrae scripturae* being composed in southern Ireland cf. p. LXXVIII–LXXX above; for the controversy about Virgilius Maro's origin cf. note 176 above.

Passage 5:

MC 62.9–10:	***DRC* 107.7–8:**	**Angers 477, fol. 20v:**
	Item Augustinus dicit:	
Saltus euentus famosus in luna, ***cui nulla uis est secundum solem****.*	*Saltus* ***est*** *euentus famosus in luna,* ***ut bissextus in sole habetur****.*	*Saltus euentus famosus in luna,* ***ut bissextus in sole habetur****.*

These passages attributed to Augustine have a strikingly different quality compared to the Isidorian ones. First, they appear only in the Munich Computus, *De ratione conputandi*, and the Angers glosses; the Angers glosses were clearly copied from both the Munich Computus and *De ratione conputandi*, while the latter two are evidently dependent.[371] Accordingly, passages 1, 4, and 5 may either have been later additions to the lists of excerpts circulating among Irish computists (and therefore not known to the earlier authors of the now lost *De divisionibus temporum* and the *Computus Ensidlensis*), or they were introduced by the Munich computist and then directly copied and revised by the author of *De ratione conputandi*. Second, while almost all passages attributed to Isidore mentioned above (except for passage 5) deal with etymologies of certain divisions of time, the Augustian passages here are essentially definitions of technical terms, with one noteworthy exception: passage 3 refers to a method for determining whether a lunar year is common or embolismic. Since Augustine of Hippo is not known to have composed a technical treatise on the 19-year cycle, and since the described method, namely the count of the number of lunations of a lunar year, is unique to the Munich Computus and follows the patterns of seventh-century Irish calendrical calculations, the Augustine referred to here, as well as in all other passages listed above, should be identified as a teacher of the Munich computist rather than Augustine of Hippo. This hypothesis is further supported by the fact that the author of *De mirabilibus sacrae scripturae* called himself Augustine, so that this name seems not to have been uncommon in the scientific intellectual milieu of seventh-century Irish monasticism. Consequently, all five passages attributed to Augustine but stemming neither from Augustine of Hippo, nor from *De mirablibus sacrae scripturae*, nor from Virgilius Maro, may effectively go back to the Munich computist's teaching under a certain Irish monastic teacher known by the name of Augustine. Note that it is hardly likely that this teacher of the Munich computist is the same person as the author of *De mirabilis sacrae scripturae*, since the former favoured the Dionysiac, the latter the Victorian reckoning.

Concerning ascriptions to Augustine, another passage found in the Munich Computus and the Irish *Dial. Langob.*, but in no other computistical text (at least not in the same wording), is of particular interest.

371 For the Angers glosses cf. p. CV, CLXXXIV–CLXXXVII; for the relation between the Munich Computus and *De ratione conputandi* cf. p. CXCI–CCI below.

MC 1.2–3:	***Dial. Langob.* 1B:**
Interrogatio: Tempus, quid est? Responsio: Tempus est spatium tendens de principio usque in finem.	*Interrogatio: Quid est tempus? Responsio: Augustinus dicit: Tempus proprie dicitur spatium extendens ab initio usque ad finem.*

As can be seen in the quote, the attribution to Augustine occurs only in *Dial. Langob.*, not in the Munich Computus. It is possible that both texts cited the same source, in which this definition of time was ascribed to Augustine, with only the author of *Dial. Langob.* copying this ascription; or, and more likely, *Dial. Langob.* worked directly from the Munich Computus, in which case the attribution to Augustine appears to stem from a connection drawn between the general title of the Munich Computus (*Computus sancti Augustini* ...) and the definition of time immediately following this title.[372]

Irish Jerome: The passages attributed to Jerome in the Munich Computus fall into three categories: a) citations from the original works of Jerome, b) a single quotation from the tract *De sollemnitatibus*;[373] and c) two passages of unknown origin to be discussed presently.

De sollemnitatibus is also cited under Jerome's name in *De ratione conputandi* (which appears to be a direct quotation from the Munich Computus),[374] and at least three of the seven extant manuscripts containing this text refer to Jerome as the author in the incipit, all of which showing strong Insular connections.[375] Consequently, it can hardly be doubted that seventh- and eighth-century Insular computists took a certain Jerome for the author of this tract. In this respect it is noteworthy that only two, and the latest two, of the five witnesses (the Munich Computus, *De ratione conputandi*, and the three manuscripts), namely the ninth-century Geneva and the twelfth-century Vatican manuscript (and not, it will be noted, the oldest of the three manuscripts), refer to this Jerome as *Sanctus Hieronymus*; therefore, we may assume that *De sollemnitatibus* was composed by an Insular monastic teacher calling himself after the fifth-century saint, and that the authorship was attributed to Saint Jerome only by later continental copyists. This assumption is further strengthened by internal evidence (particularly chapter 2 of this tract), which suggests that *De sollemnitatibus* was composed ca. AD 650–720 in an Insular context.[376]

The two passages that cannot be traced to any extant text are the following:

372 Cf. the discussion of the relation between the Munich Computus and *Dial. Langob.* p. CLXXIV–CLXXIX below.

373 MC 55.28–33.

374 *DRC* 98; cf. p. CXCIII below.

375 For the MS transmission of *De sollemnitatibus* cf. especially Ó Cróinín, 'Computistical works', 51, and furthermore Walker, *Sancti Colmbani opera*, lx. The three MSS attributing this work to Jerome are Cologne, Dombibliothek, 83², fol. 201r; Geneva, Bibliothèque Publique et Universitaire, 50, fol. 121r; Vatican, Biblioteca Apostolica, Vat. Lat. 642, fol. 89r; I have not seen the Tours MS.

376 Cf. note 175.

Passage 1:

MC 8.41–43:	***DRC* 26.5–7:**	***DDT* 8(657)B9–11:**	**BC 70C1–4:**	**Reg. Lat. 49[377]:**	***De bissexto* I 995C12–14:**
Inde sanctus Hieronimus ait:	*sequentes Hieronimum dicentem:*	*propter illam auctoritatem Hieronymi, qua dixit:*	*sequentes illud quod scriptum est:*		
		Quia in		***quia in***	***quia in***
*Media nocte factus est mundus et **in** media nocte **iterum** destruitur.*	*Media nocte factus est mundus, et **in** media nocte **iterum** distruetur.*	*media nocte factus est mundus, et **in** media nocte **iterum** destruetur.*	*media nocte factus est mundus et media nocte **iterum** distruetur.*	*media nocte factus est mundus et **in** media nocte distruetur.*	*media nocte factu<s> est mundus, et **in** media nocte **iterum** destruetur.*

Passage 2:

MC 14.3–5:	**Angers 477, fol. 19v:**
Ut ait sanctus Hieronimus:	*ut*
Septem luperculis pauit inopem, quia omnia febricitant in eo.	*VII lupicinis satiauit (?) inopem.*

Concerning passage 1, its citation in the *Catéchèses celtiques* (forming part of MS Vatican, Biblioteca Apostolica, Reg. Lat. 49) suggests that it may have had its origin in an Insular Biblical commentary attributed to Jerome, and was for that reason well-known in the Irish monastic intellectual milieu; at least Irish computists of the seventh and early eighth centuries appear to have been extremely fond of this passage as an explanation for the reckoning of a day beginning of at midnight, since it is cited not only in the Munich Computus, but also in *De ratione conputandi*, *De divisionibus temporum*, and the Bobbio Computus; its quotation in the presumably late eighth-century Frankish *De bissexto* is one of the many indications that this text was composed by a person trained in Irish computistical thought.[378]

The second passage is more problematic, primarily because its content remains obscure. Even though the parallel passage in the Angers glosses shows great variation, there cannot be any doubt that the Munich Computus was its direct source, since the Munich Computus is cited elsewhere in the Angers glosses and since this passage, or anything similar, cannot be found in any other extant text. Still, the meaning of the passage, and the reason for its attribution to Jerome, cannot be determined.

377 Wilmart, 'Catéchèses celtiques', 105 (edited from Vatican, Biblioteca Apostolica, Reg. Lat. 49).

378 For this text, particularly the differing opinions on the time and place of its composition, cf. note 37 above.

Computus Einsidlensis (Einsiedeln, Stiftsbibliothek, 321 (647), p. 82–125): Since the *Computus Einsidlensis* was only discovered in early 2006, no detailed study of this text (nor an edition) exists to the present day.[379] Yet, even a random analysis of this text immediately allows for some general statements about its provenance and date of composition. The use of Old Irish terminology and phrases is clear proof that it was written by an Irishman for an Irish-speaking audience.[380] Concerning the date of composition, the latest datable author cited is Isidore of Seville (†636), while no trace of Bede's computistical works can be found.[381] These facts securely place the *Computus Einsidlensis* in the *regiones Scottorum* between the reception of Isidore and that of Bede. As in the case of the Munich Computus, the fact that in a number of passages the Victorian reckoning is compared with the Dionysiac one suggest that the *Computus Einsidlensis* was written in an area where the Victorian reckoning was followed at some point in time, and was then gradually replaced by the Dionysiac reckoning in a process of comparing the theories behind these two reckonings and their application; since the Victorian reckoning appears to have never been followed among the northern Irish clergy, the implication is that this text was composed in the southern *regiones Scottorum*.[382]

In fact, the detailed comparisons between the two mentioned reckonings provide some of the most conclusive evidence that the Munich Computus and the *Computus Einsidlensis* are dependent: These two texts (and, to my present knowledge, *only* these two texts) compare the Victorian and Dionysiac epacts of 1 January according to two different methods[383] as well as the lunar age of the Calends of each month for certain years of the respective 19-year cycles (in which an otherwise unattested Dionysiac sequence of lunations is applied).[384] Yet, the similarities between the two texts are far more substantial than this. As a matter of fact, the parallels cover the greater part of both textbooks, so that a full list of parallel passages appears inappropriate here (especially since they can easily be checked in the edition below). Nevertheless, certain passages, mainly in the more technical second half of both texts, are particulary noteworthy since they transmit seventh-century Irish computistical concepts not to be found in *any other* computistical text known at present; moreover, they impressively highlight the technical problems faced by seventh-century computists due to the darth of written source-material available to them.

379 For the discovery of and a brief introduction to this text see Warntjes, 'Computus Einsidlensis', 61–4.

380 *CE* p. 90, 93, 97, 123; these passage are now analyzed in detail in Bisagni & Warntjes, 'Early Old Irish material', 91–105; cf. also Warntjes, 'Computus Einsidlensis', 62 (where the reference to p. 97 has been misprinted as 99).

381 Cf. Warntjes, 'Computus Einsidlensis', 62.

382 For this argument in respect of the Munich Computus cf. p. LXXXVIII–XCVI above; for the passages with comparative character in the *Computus Einsidlensis* cf. the following discussion.

383 MC 49.16–32; *CE* p. 109.

384 MC 50 (which, quite artificially, also introduces the *latercus* into this comparison); *CE* p. 109–110.

Most of these difficulties arose from the need to understand the construction of the 19-year cycle.[385] The Dionysiac 19-year cycle employed a specific sequence of common and embolismic years, first outlined in Latin (from Greek sources) in Dionysius' letter to Boniface and Bonus.[386] Unfortunately, no proof of this sequence can be found in any pre-mid-seventh-century computistical text. Consequently, seventh- and early eighth-century Irish computists felt the need to verify this sequence. Based on Dionysius' statement that a lunar year extended from *luna* 15 of the Easter lunation to the Easter full moon (*luna* 14) of the following year, and that a common year has 354, an embolismic year 384 days, both the Munich and the Einsiedeln computist explicitly count the number of days in the mentioned interval for the first three years of the Dionysiac 19-year cycle;[387] the result is that the first two years are common years of 354 days, the third an embolismic one of 384 days; it is then argued that the whole sequence can be verfied by this method. Most strikingly, both computists use the exact same (and otherwise unattested) method of counting the number of days between *luna* 15 of the Easter lunation to the Easter full moon of the following year: in order to simplify the calculation, they fixed an interval of 325 days that would, according to definition, lie within the range of every lunar year, namely between 2 May and 22 March of the following year (the latter date being the earliest possible date of the Easter full moon); having fixed this interval, they only had to add the number of days between *luna* 15 of the Easter lunation and 2 May, as well as the number of days between 22 March and the Easter full moon of the following year.

More problematic was the proof of the cyclic character of the 19-year cycle. As mentioned earlier,[388] the authority on the construction of the 19-year cycle, Dionysius, had given a false proof of this, which led to confusion rather than clarity: From Quintus Julius Hilarianus' *Expositum de die paschae et mensis* (who discussed the Hippolytan 112-year Easter cycle based on an 8-year lunar cycle) Dionysius copied the proof that eight lunar years are truly cyclic, and then, quite expectedly, struggled to prove the same for the remaining eleven years; obviously, if an eight-year period is cyclic, an eleven-year period can also be cyclic only if one year is cyclic, since 8 and 11 do not share a common

385 There is the very unfortunate tendency among some modern historians of computistics (especially those, it will be noted, not being trained in science) to argue that as long as a scientific system provides reliable data, there was no need to understand the system itself (cf. most recently Englisch, 'Karolingische Reformkalender', 245–6). This opinion appears to be nothing but a projection of a very regrettable but common mentality of modern non-scientists to the mentality of medieval scientists; at no point in time (not even in the so-called 'dark ages') did scientists feel comfortable using the data of a system that they did not understand. Moreover, in an age in which scribal errors were common and could lead to utter confusion, it was of paramount importance to be able to check the data in question.

386 *Epistola Dionysii* (Krusch, *Studien* II, 82–3); this sequence also constitutes a column in the Isidorian Easter table (*Etym.* 6.17), and is outlined in *CE* p. 113 and MC 59.28–35 (where references to further texts can be found).

387 *Epistola Dionysii* (Krusch, *Studien* II, 84–5); MC 59.70–105; *CE* p. 113–4.

388 Cf. note 103.

divisor other than 1; yet, one lunar year does not have the same number of days as one Julian calendar year. The main difference in construction between an 8-year lunar cycle and a 19-year one is that the former does not allow for lunar bissextile days, while the latter does.[389] It was a long and painstaking process for seventh-century computists to detect this difference and the reason causing it. The actuality and importance of this question still in the first half of the eighth century is impressively illustrated by the fact that Bede devotes two chapters of his *De temporum ratione* to make a case for the application of lunar bissextile days.[390] And Bede, for his part, stands only at the end of a decade-long analysis of this question by seventh-century computists.[391]

In fact, earlier (no doubt seventh-century) attempts of solving this problem, particularly those that ultimately failed, are only transmitted in the Munich Computus and the *Computus Einsidlensis*.[392] Moreover, the methods underlying these attempts (again, particularly the ones that ultimately failed) are the same in both of these textbooks: At first, a proof is given for the cyclic character of the first eight years of the 19-year cycle (the *ogdoas*) by compensation of the remaining two lunar days (8×354+3×30=2922 lunar days vs 8×365=2920 solar days) by eight bissextile solar quarter-days (i.e. the two solar bissextile days of this eight-year period), which is identical with Dionysius' account in the result,

389 8 solar years consist of 8×365¼=2922 days, which equals the number of days of 8 common lunar years of 354 days plus three embolisms of 30 days: 8×354+3×30=2832+90=2922 days; consequently, the construction of the 8-year lunar cycle does not allow for lunar bissextile days. On the other hand, lunar bissextile days are absolutely vital for the construction of the 19-year lunar cycle: 19 solar years consist of 19×365¼=6939¾ days; the same number is gained by 19 common lunar years of 354¼ days plus 7 embolisms of 30 days minus the *saltus lunae*: 19×354¼+7×30−1=6730¾+210−1=6939¾ days. The difficulty of explaining the lunar bissextile day astronomically (as against the comparatively simple case of the solar bissextile day) is neatly expressed in a gloss to Bede' *DTR* (*CCSL* 123B, 405–6): *Simul enim praeparantur per IIIIor annos, id est quadrans solaris et quadrans lunaris. Sicut enim quadrans solaris ex tarditate solaris cursus per quattuor annos praeparatur, sic et quadrans lunaris ex tarditate lunaris cursus simul per eosdem quattuor annos praeparaur, tamen quia luna multo uelocius sole suum complet cursus, non facile patet tarditas sui cursus, sicuti solis cursus facit. Sol enim in anno semel firmamentum circuit, tardius (Gl.: Id est unoquoque anno.) tamen semper in VI horis. Luna autem in anno tredecies firmamentum circuit, et in ultimo circuitu tardius in VI horis peruenit ad illum punctum caeli ubi in antecedenti anno fuit, et ideo quia tredecies in anno firmamentum circuit, non facile patet sua tarditas*.

390 In *DTR* 41, Bede makes an explicit case for the application of lunar bissextile days, while he discusses their implications on the theory of the 19-year lunar cycle in *DTR* 46. Cf. the commentaries to these chapters of *DTR* in Jones, *Bedae opera*, 374–5, 380–1; Wallis, *Bede*, 325–6, 335–6; Pillonel-Wyrsch, *Calcul*, 287–9, 308–11; none of these scholars is aware that Dionysius, the main authority on the structure of the 19-year cycle, was the source of confusion, and likewise they totally neglect Irish pre-Bedan evidence.

391 This is explicitly denied by Wallis, *Bede*, 325–6, who is not aware of the Irish evidence discussed in the following.

392 MC 60; *CE* p. 114–7. A different, very comprehensive, and far more sophisticated proof of the cyclic character of the 19-year cycle is outlined in *DRC* 79, which illustrates that this text was composed at a later, more advanced stage (cf. p. CXCVII below).

but strikingly different in the modes of calculation. This is followed by a most curious, but quite intelligent (though false) pseudo-proof of the cyclic character of the remaining eleven years (the *hendecas*): It is argued that the one remaining solar day (11×354+4×30=4014 lunar days vs 11×365=4015 solar day) is compensated for by the *saltus lunae*, and the additional eleven bissextile solar quarter-days (which amount to almost three full days) by the final day (*luna* 30) of three embolisms.[393] Yet, seventh-century Irish computists appear to have soon realized that those constructions were either false or connected to a different cycle, and that bissextile days are as much a lunar phenomenon as they are a solar one, so that the bissextile increment was therefore irrelevant for the calculation (as it adds the exact same amount to the lunar as to the solar count). Again, this criticism, like the foregoing theories it refers to, cannot be found in any other text, and it is expressed in the same way in both the Munich and the Einsiedeln textbooks. Finally, both texts provide a correct proof of the cyclic character of the 19-year cycle; the method applied by both computists alike, however, is again rather curious and otherwise unattested.[394]

It should generally be kept in mind that in an age like the seventh and early eighth centuries, computists copied ideas and technical concepts, but tended to reproduce them in their own words, often applying their own (and different) modes of calculation. Accordingly, many parallels between two texts can be established even though there may not be any parallels in wording or even in the modes of calculation.[395] In the case of the two Irish computistical textbooks in question here, however, the dependency is absolutely evident not only in the methods applied (as has just been illustrated), but also in terminology and phrasing. There are, again, too many parallels than can be listed here (and again all of them can be checked in the edition below), so that few striking examples may suffice.

First, the most detailed and longest calculation found in both the Munich Computus and the *Computus Einsidlensis* is the division of the 24 hours of a *saltus* by the 235 lunations of a 19-year lunar cycle. This division was extremely challenging for early medieval computists since it involved calculations with great fractions (up to 1/564). Obviously, the result could be established by various modes of calculation; likewise, the terminology of these rather uncom-

393 The pseudo-logic of this proof lies in the following two observations: 1) The *saltus lunae*, being by definition the subtraction rather than addition of a lunar day, could obviously not compensate for an extra solar day; however, this subtraction of one lunar day due to the *saltus* led to an increase of the epact by one, which would appear like an extra lunar day to the untrained computist; this fact would have justified the connection between the *saltus* and the extra solar day. 2) Seventh-century computists regarded a lunar month as consisting of 29½ days; in order to avoid calculations with fractions, lunar months of alternately 30 and 29 days were applied; this balance, however, was disturbed by the introduction of seven embolismic months of 30 days; accordingly, it appeared as if the seven embolismic months introduced 3½ lunar days too many; in this respect, the eleven solar bissextile quarter-days were considered as compensating for at least part of this difference.

394 For this see further below (p. CXLIV–CXLV).

395 Note, however, that the opposite is true for computistical *formulae*.

mon fractions varied considerably.[396] Accordingly, the use of the same modes of calculation and terminology is a clear indication of the dependency of two texts. Both features are shared by the Munich Computus and the *Computus Einsidlensis*, as can be seen from the following correlation of the relevant passages; note, moreover, that only one further text exists using the same modes of calculation, namely *Comp. Col.*, which is evidently dependent on the Munich Computus.[397]

MC 62.14–61:

Sciendum, quot lunis preparatus. Id est CCXXXV. Quomodo? Id est IIII momenta uniuscuisque lunę usque ad caput anni XVIIII, et XII pars momenti <et VII pars XLma XIImae partis momenti>. Quomodo fit? Quando dicis accensionem lunae, id est precederunt IIII momenta pro uelocitate accensionis lunę. In capite XVIIII annorum haec IIII momenta I diem efficiunt horarum XXIIII. Sciendum quomodo crescit. Id est per octa <momenta inter> II lunas, XVI momenta inter IIII lunas, XX momenta inter V lunas, XL momenta inter X lunas, LXXX momenta inter XX lunas. Id est II horas coniunge, et fiunt IIII horae in fine lunarum XL. Adde X lunas et fiunt CC momenta in luna L et V horae. CCCC momenta in luna <C> et fiunt XX horae in fine lunarum CC. Adde XXX lunas, per <X> quorum XL momenta crescunt, unam horam coniunge, horae et fiunt XXIII. Adde V lunas, per quas XX momenta fiunt, id est dimidia hora.

Deest alia dimidia, id est XX momenta. Tolle I momentum de his XX, diuide illud in XII partes. Sic diuide aliud momentum in XII partes. Coniunge bis XII et efficiunt XXIIII partes. Adde alia duo momenta et efficiunt XXIIII. Coniunge hoc et fiunt XLVIII inter IIII momenta. Remanent tamen XVI momenta. Adde I momentum de XVI, ac diuide illud in XII partes, ad XLVIII partes; <id est adde II> et fiunt L, adde X et fiunt LX inter V momenta. Supersunt XV momenta. Sic inter V alia momenta LX fiunt partes. Sic in V aliis LX

***CE* p. 122–3:**

Incrementum utique lunare per decennouenalem ciclum crescit, hoc est IIII momentis et XII parte momenti et XLma et VIIma parte XIImę partis momenti per singulas lunas, quę sunt CCXXXV decennouennali ciclo. Itaque probemus: IIII momenta per singulas lunas. Veluti est IIII momenta ab una luna, et alia IIII momenta ab altera luna, quę simul VIII sunt. Adde item VIII momenta a duabus lunis, quę XVI momenta faciunt. Item adde ab una luna IIII momenta, quę omnia simul iuncta XX momenta fiunt a lunis V. Item adde ab aliis V lunis XX momenta, quę coniuncta reliquis XL momenta faciunt. Hinc est hora a X lunis. Cum autem perduxeris CC lunas, XX horas fecisti. Adde XXX et V lunas, quę simul faciunt horas XXIII et dimidia horam.

Hinc dimidia hora deest, hoc <est> XX momenta. Quibus conpudandis XII lunas iunge, a quibus duces unum momentum, hoc est XIImam partem momenti a singulis lunis. Item ab aliis XII lunis momentum accipe. Hinc II momenta a lunis XXIIII. A lunis vero XLVIII IIII momenta sunt. A lunis autem LX V momenta adduc. Hinc a lunis LX quater ductis XX momenta duces. Si vero sunt lunę LX quater, totam inpleuisti diem. Igitur conputa CCCXXXV (recte CCXXXV) lunas, ut scias, utrum sexagesimum numerum enumerare

396 Cf. the references to MC 62.14–61 in the edition.

397 For *Comp. Col.*'s dependency on the Munich Computus cf. p. CII, CLXXIX–CLXXXIII.

partes. Id est LX quater in XX momentis. Tolle X quater de IIII LXtis, et remanent IIII L L L L, id est CC. Cum his CC partibus CCas lunas impleui. Coniunge X quater, quas de <IIII> LXtis tollisti, et efficiunt XL partes. Diuide hanc XL et tolle XXXV. Nunc impleui CCas lunas et XXXV.

Sunt autem partes V, quas de <IIII> LXtis tulisti. ⌐ Diuide <CC in> IIII L L L L. Trahe X de unoquoque L et coniunge <IIII> X et efficiunt <XL> partes. Inde fiunt XL XL XL XL. ¬ Diuide V in hunc numerum, id <est> I I I I I. Tolle I de his et diuide in partes XLVII. Sic diuide unum aliud in XLVII; sic diuide I aliud in XLVII; sic diuide I aliud in XLVII; sic diuide I aliud in XLVII. <Id est XLVII quinquies. Tolle VII quinquies de his XLVII quinquies, et remanent XL quinquies.> Sic diuide I de his XLtis et da X quater super XL quater et fiunt IIII L L L L, id est CC. <Supersunt tamen VII quinquies partes. Tolle II quinquies de his, et remanent V quinquies.> Coniunge V quinquies et fiunt XXV. <Supersunt tamen II quinquies partes.> Coniunge II quinquies et fiunt X. Diuide hoc X in bis quinque. Aliud V super XXV et fit XXX. <Aliud V super XXX et fit XXXV.> Sic diuiditur VII pars XLma XIImae partis momenti uniuscuiusque lunae de lunis CCtis XXXV.

poteris quater, hoc modo: CC diuide in quinquagenarium numerum, fiuntque L quaterna diuisione. His adde XXX, fiuntque LX ter. Adiece V lunas super L quaternum.

V autem lunę desunt. Et V partes, quas diximus esse XII particulas momenti, desunt, id est maá triun laigu leut*. Quere igitur diuisionem a lunis CCXXXV. Hoc est XLmam VIImam particulam XIImę partis momenti. Hoc modo diuide lunas: CC, quę faciunt XL quinquies. Adde XXXV lunas, quę diuise fiunt VII quinquies. Hinc singulę partes XLVII sunt. Deinde IIII momenta et XIIma pars momenti et XLma VIIma pars partis XIImę momenti. Ab hanc diem perficiendam a singulis lunis accipiuntur, et hoc modo lunam uelocius accensam dicimus, quam putamus.*

The dependency between the two texts is even more and particularly obvious in a discussion of weekday calculations; some of the methods described here were later copied into other texts, but either not with the same amount of detail,[398] or in a different order.[399]

398 Cf. *Dial. Langob.* 12B; *Quaest. Austr.* 2.7A–B.

399 *DRC* discusses the days of a months with identical weekday data per class of months, i.e. the final section in *CE*'s and MC's discussions, first (as c. 33), followed by the general rules for the calculation of the weekdays on the Calends (c. 34), which also constitutes the middle part in *CE*'s and MC's discussions, and finally it outlines the weekday calculation for Easter (c. 35), which is the first section in *CE* and MC.

MC 29:	***CE* p. 97–8:**
Quomodo mensium initia ebdomatis diebus intelleguntur? Dies in quo Kalendae Ianuarii fit, quarto sequenti lumine Kalendae Februarii et Martii et VIII et Idus et XI et IIII Kalendas Aprilis, nisi fuerit bissextus. *Si autem contigerit, quinto die post Kalendas <Ianuarii> VIII Idus Martii et relicum erit; ita et Kalendae Martii, si in VI Kalendas eius bissextus esse fulserit.*	*Quocumque die Kalendae Ianuarii fuerint, quarto die postea Kalendae Februarii erint et Kalendae Martii et VIII Idus Martii et Idus Martii et XI Kalendas Aprilis et IIII Kalendas Aprilis, si non fuerit bissextus.* *Si autem fuerit bissextus et si in Februario: Quocumque die Kalendae Ianuarii fuerint, IIII die Kalendae Februarii post Kalendas Ianuarii erint, et quinto die Kalendae Martii et VIII Idus Matii et Idus Martii et XI Kalendas Aprilis et IIII Kalendas Aprilis fiet. Si uero in Martio bissextus: Quocumque die Kalendae Ianuarii fuerint, quarto die Kalendae Februarii et Kalendae Martii post Kalendas Ianuarii erint, VIII Idus autem Martii et Idus Martii et XI Kalendas Aprilis et IIII Kalendas Aprilis quinto die erit post Kalendas Ianuarii.*
Dies in quo Kalendae Ianuarii innotescit, priori die Kalendae Aprilis esse fulgebit, nisi bissextus sit. Nam si ipse euenerit, uno eodemque die Kalendae Ianuarii et Aprilis copulari apparebit.	*Item Kalendae Ianuarii quocumque die fuerint, pridie Kalendae Aprilis erit, si non fuerit bissextus. Si autem fuerit bissextus, eodem die erit Kalendae Ianuarii et Kalendae Aprilis***.**
	Item Kalendae Ianuarii quocumque die fuerint, IIII subsequentibus diebus principia IIII temporum erunt, si non fuerit bissextus. Si autem fuerit bissextus, IIII temporum initia uno interciden(d)te die IIII diebus subsequentibus post Kalendas Ianuarii erunt.
Omnis mensis XXXI: dies in quo <Kalendae eius> esse contigerit, quarto die Kalendae sequentis mensis erit. In triginta uero mense, IIIo die <Kalendae sequentis mensis> inueniri ardebit.	*Item mensis XXXI dies habens IIIIto die post Kalendas eius sequentis mensis Kalendae erit. Mensis autem XXX dies habens III die post Kalendas huius alterius mensis Kalendae erit. Mensis XXVIII dies habens eodem die Kalendae eius et sequentis mensis erit.*
Dies in quo fit Kalendae Ianuarii, sequenti die Kalendae Maii, nisi fuerit bissextus. Si enim bissextus fit, tertio die erit. Sequenti uero lumine in quo Kalendae Maii uiget, Augustus initium habet; in postero die Nouimbrio initium fulget; cui secundo die Kalendae Februarii conuenit. Dies in quo Kalendae Maii fieri innotescit, in eo VIII Kalendas Ianuarii sequentis et Kalendae eius indubitanter euenire florebit.	*Item Kalendae Maii quolibet die fuerit eodem die VIII Kalendas Ianuarii et Kalendae Ianuarii erit, si bissextus fuerit an non.*

Omnes menses trigesimi singuli, IIII Nonę: uno die inueniuntur Kalendae eorum, et VI Idus, et XVIII Kalendas, et XI Kalendas, et IIII Kalendas. Cuncti uero menses XXXmi, quattuor Nonae: in eodem die ebdomatis fiunt Kalendę eorum, et VI Idus, et XVII Kalendas, et X Kalendas, et III Kalendas. Omnes menses XXXI, sex Nonae: eodem die ebdomatis Kalendae eorum, et VIII Idus, et Idus, et XI Kalendas, et IIII Kalendas repperiuntur. Februarius XXVIII, IIII Nonae: in eodem die Kalendae eius, et VI Idus, et XV Kalendas, et VIII Kalendas, et Kalendae mensis sequentis, id est mensis Martii.

De epdomadibus regulandis in diebus mensis. Quicumque mensis trigenos singulos dies habuerit, et IIIItas Nonas, uno eodemque die Kalendarum ipsius, et VI Idus, et XVIII Kalendas, et XI Kalendas, et IIII Kalendas sequentis mensis erit, ueluti Ianuarius, Augustus, et Decimber. Itemque quicumque mensis trigenos possiderit dies, et IIItas, uno eodemque die Kalendarum ipsius et VI Idus, et XVII Kalendas, et X Kalendas, et III Kalendas mensis sequentis erit, ueluti Aprilis, Iunius, Septimber, et Nouimber. Itemque quicumque mensis trigenos singulos dies habuerit, et VItas Nonas, eodemque die ipsius Kalendarum, et VIII Idus, et Idus, et XI Kalendas, et IIII Kalendas sequentis mensis erit, ueluti Martius, Maius, Iulius, et Octimber. Item mensis dies XXVIII habens, eodem die Kalendarum eius, et VI Idus, et XV Kalendas, et VIII Kalendas, et Kalendarum sequentis mensis erit, ueluti Februarius.

Concerning direct textual dependency between these two Irish textbooks (which, it will be noted, is rather less common), the following passages may suffice:

Munich Computus	***Computus Einsidlensis***
9.16–20: *Vesperum* [...] . *Unde haec stella a maioribus Romam appetentibus uesperum uocata est, quae noctis initium trans Hesperiam oriebatur. Itaque Isperia ab Espero rege dicta, quę nunc Italia ab Italo duce dicitur.*	p. 91: *Vesperum quidam cum adspiratione hesperum dicunt a prouintia Hesperea, que ab Hespero Romanorum duce nomen sortita est. Hesperus ab Hesperea ciuitate, quam construxit Hespera filia Romuli. Tale uero nomen quod est Hesperus Macedonii appellauerunt, eo quod eis a Macedonia ad Ęgyptum nauigantibus haec stella oriebatur trans Hesperiam, quae nunc Italia ab Italo siculorum rege nominata est.*
11.2–3: *Feria a fando dicta est, uel a fiendo, dicente Domino: Fiat lux et facta est lux.*	p. 92: *Feria a fando dicta est, quasi faria, hoc est opera VII dierum fando. Aliter feria a fiendo dicta est, eo quod dixit Deus: Fiat lux et reliqua.*

12.15–19: *Hii sunt menses Latinorum: Ianuarius, Februarius, Martius, Aprilis, Maius, Iunius, Iulius, Augustus, Septimber, Octimber, Nouimber, Decimber. Nomina mensuum IIII rebus assumpta sunt: A diis, a regibus, a numeris, a rebus.*	p. 93: *Quibus modis menses uocabula sumpserunt? Hoc est a quattuor: a diis et regibus, a rebus et numeris. Quot menses sunt anni? Hoc est XII: Ianuarius, Februarius, Martius, Aprilis, Maius, Iunius, Iulius, Augustus, Septimber, Octimber, Nouimber, Decimber.*
24.40–41: *Quot sunt principales dies de nominibus dierum mensium? Id est III: Kalende, Nonae et Idus.*	p. 95: *Quod sunt principales dies, quibus menses conputantur? II uel III, id est Kalendę, Nonę, Idus.*

On the basis of this evidence (and much more is to be found in the edition below), it cannot be doubted that these two texts are dependent. The question that remains to be solved is which of these texts is to be regarded as the source for the other, i.e. which is the older text? Unfortunately, this question cannot be solved on purely chronological grounds, since contrary to the Munich Computus, the *Computus Einsidlensis* does not contain a dating clause of the year of composition. Only two examples of some calendrical calculation are given in the text; the first, however, is explicitly referred to as a theoretical example and as not representing the *annus praesens* of the author,[400] while the chronological data of the second agrees with the years AD 568 and 570, too early to coincide with the composition of this textbook.[401]

There are, however, numerous indications that the *Computus Einsidlensis* is the older of the two texts, and that it was, in fact, the Munich computist principal source. One of the main features of Irish computistical textbooks of the seventh and early eighth centuries is the discussion of the various units of time as an introduction to terminology as well as a basis for the subsequent analyses.

400 Cf. further below (p. CXLVI–CXLVII).

401 In *CE* p. 120–1, the Einsiedeln computist outlines an example each for the interference with the quadragesimal period of the bissextile day on the one hand, of the *saltus lunae* on the other. The essential chronological data given in the first example are: *feria* 1, *luna* 16 for 1 January of a bissextile year. The essential data of the second example are: *feria* 4, *luna* 8 for 1 January of a non-bissextile year. These chronological details reveal that the reckoning underlying the calculations is the Dionysiac one; between the year of Dionysius' writing, i.e. AD 525, and the second half of the ninth century, i.e. the time of composition of the MS incorporating the *Computus Einsidlensis*, only the year AD 568 agrees with the data given in the first example; the data of the second example, then, matches the next occurrence of a *saltus*, namely AD 570. In the second half of the sixth century, the Dionysiac reckoning was certainly not accepted anywhere in the *regiones Scottorum*, presumably not even known; more conclusively, as the Einsiedeln computist quotes Isidore, he cannot have written in the sixth century. In a previous publication (Warntjes, 'Computus Einsidlensis', 62), I have suggested that AD 663 may have been the year in question, for the following reasons: In the period in which the *Computus Einsidlensis* was certainly composed, i.e. ca. AD 650 to 750, only one year shows the chronological data of *feria* 1 and *luna* 16 on 1 January, namely AD 663; yet, this year is not bissextile. I assumed, therefore, that the author of the *Computus Einsidlensis* may have taken his own *annus praesens* as an example and constructed the hypothetical impact of a bissextile day on the respective calendrical calculations in this year, had a bissextile day occurred. It now appears more likely to me that the Einsiedeln computist copied this example from an earlier (non-Irish) exemplar.

Typically, in these Irish texts the discussion centred around 14 such divisions.[402] Just how this list of divisions developed must remain speculative. Since most of the definitions and etymologies of these units are based on Isidore's works, it appears that Isidore's list of eight divisions of time provided the basis and the impetus for subsequent lists and discussions composed in Ireland.[403] Isidore's original list consisted of *momenta*, hours, days, months, years, *lustra*, *saeculi*, and *aetates*.[404] The *lustra*, defined as five-year periods,[405] were soon regarded as an inappropriate unit of time, so that it fell out of the list at a very early stage. On the other hand, Isidore himself had discussed more divisons of time than the ones outlined in his list; some of these additional units were soon added to Isidore's original list, namely *atomos*, the quarter-day (*quadrans*), the week, the season, and *mundus*.[406] Other units, like *minutum* and *punctum*, were invented for the simplification of calculations, i.e. when it was deemed appropriate to invent a specific term for a frequently recurring multiple or fraction of a known unit.[407]

As for the development of the Irish list of 14 divisions of time, it appears, in fact, that the *Computus Einsidlensis* stands right at the beginning of this process. Clearly, the author of this text was not aware of any division of time into 14 units. His own core of divisions comprised *momenta*, *minuta*, *puncta*, hours, quarter-days, days, weeks, months, seasons, and years; these are the units that

402 Cf. MC 1.4–8; *DRC* 13; *DDT* 1 (*PL* 90, 653; note that *DDT* explicitly states that there are 14 divisions of time, but continues to list 15, adding the cycle; this, among other evidence, reveals that this text as printed in *PL* 90 is a later Carolingian reworking, not the original seventh-century Irish text); the order given in *DRC* and *DDT* differs slightly from MC's account, since *saeculum* and *aetas* are switched. Note that there are numerous short tracts on the divisions of time which are still unedited, most of which referring to 14 divisions, some, however, expanded this or an altered list to 15 or 16. Cf. the discussions p. CVII–CXVIII, CLIX–CLX.

403 This is also Ó Cróinín's (*Cummian's letter*, 103) verdict: 'The classification by fourteen divisions of time may have been an Irish development (of Isidore's seven).' Cf. also Borst, *Kalenderreform*, 566–8, and now Graff, 'Recension of two Sirmond texts', 120–1.

404 *Etym.* 5.29.1. The list is followed by detailed discussions of these units in *Etym.* 5.29–32, 5.36–37.

405 *Etym.* 5.37.2.

406 For Isidore's discussion of *atomos* see *Etym.* 13.2; *quadrans* (in the context of weights) *Etym.* 16.25.17; the week *Etym.* 5.32, *DNR* 3; the season *Etym.* 5.35, *DNR* 7; *mundus Etym.* 3.29, 13.1, *DNR* 9.

407 Interestingly and revealingly enough, the etymologies composed for these invented units were attributed to Isidore (cf. p. CXXVII above). *Minutum* was invented as a term for the increase and decrease of daylight per day, which was calculated as 2⅔ or 4 *momenta*, depending whether the bissextile increase of 1⅓ per day was included or not (which in itself was sometimes, though rarely, also defined as *minutum*; e.g., all three definitions can be found in *CE* p. 86; 4 *momenta* in MC 1.9, 4.2–4 and BC 33; 2⅔ *momenta* in *DDT* 4; explicitly pro 2⅔ and contra 4 *momenta* in *DRC* 18). *Punctum* was invented as a subdivision of an hour, marked by physical points on horologues; differences existed between 4 or 5 *puncta* completing an hour (4 *puncta* would have been the most common, as in MC 1.10, 6.8; for 5 *puncta*, ascribed exclusively to lunar theory, cf., e.g., *DTR* 3.10–12; *DDT* 1B19; *RM* 17.20–22).

he not only lists, but also discusses in detail in the following chapters.[408] Additionally, he mentions the three cycles that form the basis for Easter calculations according to the Dionysiac (and the Victorian) method, namely the 19-year lunar cycle, the 28-year solar cycle, and the combination of these two, i.e. the 532-year Easter cycle. More importantly, he precedes his list by a statement that some people regard, and quite rightly so, *atomos* as the smallest unit of time; furthermore, some of the greater units (*saeculum* and *aetas*) appear in the *Computus Einsidlensis* when its author cites Isidore's list at the end of this chapter on the divisions of time.[409] Taken together, then, the Einsiedeln computist mentions 13 of the 14 units of time that are listed and discussed in the other Irish texts; on the other hand, he did not combine these into one list, but rather discusses *atomos*, the list from *momentum* to the three cycles (possibly his own invention), and the Isidorian list separately. It appears, therefore, that this chapter of the *Computus Einsidlensis* was composed before the concept of the 14 divisons of time became a well-defined standard, and that it was, in fact, this chapter that provided the original stimulus for the compilation of the typically Irish list of 14 divisions of time, which included all 13 divisions mentioned by the Einsiedeln computist (the cycles not included) plus the addition of *mundus* as the all-embracing, greatest unit.[410]

Similarly, the contextualisation of the Munich and the Einsiedeln computists' accounts of the most complicated technical device in the Julian calendar, the bissextile day, may provide further evidence concerning the chronological order of these two texts. Generally, the authority on the Julian calendar throughout the middle ages was a rather lenthy excerpt from Macrobius' *Saturnalia*, which circulated under the title *Disputatio Chori and Praetextati*.[411] According to this text, the bissextile day is to be placed *ante quinque ultimos Februarii mensis dies*, 'before the five last days of February', i.e. before *VI Kalendas Martii*, 24 February in modern notation.[412] In a rather obscure passage, Isidore, however, connects the bissextile day with *VI Nonas Martii*, i.e. 2 March.[413] A specifically Irish tradition, on the other hand, connects the bissextile day with the beginning of the world, which, according to this tradition, was believed to have occurred on *XII Kalendas Aprilis*, i.e. 21 March. But which of these dates was the one to be followed as the placement of the bissextile day? Discussions about this question had stopped by the time of Bede in favour of

408 The list in *CE* p. 85 (and transcribed in note 319 above), with each of these units being subsequently discussed in detail in *CE* p. 85–95.

409 *CE* p. 85.

410 It is noteworthy, in this respect, that the unpublished prologue to Bede's *DT* transmitted in Vatican, Biblioteca Apostolica, Pal. Lat. 1448, fol. 1v–2r; Vatican, Biblioteca Apostolica, Pal. Lat. 1447, fol. 3r; Schaffhausen, Stadtbibliothek, 61, fol. 22r–v (cf. note 319) lists 13 rather than the typical 14 divisions of time, namely the 13 mentioned in *CE*, not considering the rather vague *mundus*.

411 For the *Disputatio Chori et Praetextati* cf. note 171.

412 *Sat*. 1.14.6.

413 *Etym*. 6.17.27.

Macrobius' authoritative statement.[414] In the seventh century, however, debates were still lively, with all three dates being treated as equal options and being tested for their applicability within the construction of the Julian calendar, as well as within the liturgical year. In the *Computus Einsidlensis*, all three placements of the bissextile day are not only discussed at some length without any obvious preference, the impact of the bissextile day on weekday calculation within a Julian calendar year is also outlined in equal detail for all three dates.[415] The Munich Computus, however, appears to belong to a later stage in this discussion. Even though all three placements are still mentioned, 21 March is immediately ruled out to be an appropriate placement of the bissextile day, as it would interfere with Easter calculations.[416] The remaining two dates, however, 24 February and 2 March, are still treated as equal options.[417]

Yet, the course of discussion of these technical questions in the seventh and eighth centuries must obviously remain speculative in character, so that it does not allow for definite statements about the chronology of the *Computus Einsidlensis* and the Munich Computus. Fortunately, additional evidence is provided by even more technical chapters. As mentioned above, both the Munich Computus and the *Computus Einsidlensis* provide a proof for the cyclic character of the 19-year cycle.[418] The method applied is as curious as it is unique to these two texts: The first lunar year of this cycle starts with *luna* 15 of the Easter lunation of that year, which happens to fall on 18 April; this date is taken as the point of orientation; then, the lunar day difference between the end of a lunar year (occurring on the Easter full moon) and 18 April is calculated for every single year of the 19-year cycle, with the result being that the eighth lunar year ends a day later than eight solar years (proving that the *ogdoas* is not truly cyclic); the end of the 19th lunar year, however, coincided with the end of 19 solar years. In the *Computus Einsidlensis*, this detailed discussion extends over 2½ manuscript pages; in the Munich Computus, however, it covers only 1½ manuscript pages. The reason for this discrepancy is that the Munich computist gives

414 In *DT* 10 and *DTR* 38, Bede mentions besides 24 February an Egyptian placement of the bissextile day on 29 August, which he, however, does not consider as an alternative; *DTR* 40, then, is fully devoted to 24 February as the right placement of this technical device in the Julian calendar. Bede's influence in this question is illustrated by the fact that the passage in question from *DT* 10 is cited in RM 55, the passage from *DTR* 38 in RM 54; PV §301; *Lib. comp.* 3.6; *Lib. calc.* 41, while parts of *DTR* 40 are cited in *Lib. calc.* 42B. The Frankish *Dial. Neustr.* of AD 737, working from Irish sources, still lists all three placements of the solar bissextile day in c. 27a. Various placements can also be found in *De bissexto* I (*PL* 101, 997) and *Comp. Col.* 3.3B, 3.5A, 3.8A, but for antiquarian reasons rather than scholarly interest. For *DRC* cf. p. CXCVIII below.

415 *CE* p. 104–6.

416 MC 41.43–49.

417 This is reveal by the fact that in MC 41.2–12 the Munich computist, when introducing the etymology for *bissextus*, argues that this term can stand for both *bis sex Kalendas*, if the bissextile day falls in February, or *bis sex Nonas*, if it is alternatively placed in March. Cf. also the passage on weekday calculations (MC 29.2–12), which is set in parallel to its source, a passage from the *Computus Einsidlensis*, on p. CXXXIX above.

418 MC 60.52–107; *CE* 116–8. Cf. p. CXXXIV–CXXXVI above.

a rather abbreviated version of this proof, not outlining the details for every single year. Accordingly, the assumption must be that the Einsiedeln computist, introducing this method for the first time, felt the need to outline it in as much detail as possible to provide for a full understanding among his readers; the Munich computist, then, worked from the full version, but at a time when this method was already well established and understood, so that he felt that an abridged version would suffice.

Another technical passage provides an even clearer indication that the Munich computist worked from the Einsiedeln text. The passage in question (chapter 40 of the Munich Computus) stands in direct connection to and immediately follows the discussion of the assumed linear increase and decrease of daylight throughout a calendar year (chapter 39). In this theory (generally based on chapter 13 of Anatolius (?)'s *De ratione paschali*), daylight increases or decreases by 2⅔ *momenta* per day; the year is divided into four quarters, so that daylight increases by this ratio in the first two quarters between the winter and the summer solstices, and decreases by the same ratio in the final two quarters. Every quarter is reckoned as consisting of 90 days to provide for this linear distribution. This, however, leads to a total of 360 days of a year, with five days remaining unconsidered. How are those five days, then, supposed to be dealt with? Even though the theory of the increase and decrease of daylight was quite popular in the early middle ages,[419] the Munich Computus and the *Computus Einsidlensis* are the only texts discussing the question of the five remaining days in detail.[420] They argue that those five days should be distributed equally among the 90 days of the quarters in which they occur; apparently, the logic appears to be that once those days are split up, they are not reckoned as full days anymore and can accordingly be neglected for the theory of the *daily* increase and decrease of daylight.

Both texts then continue to explain the impact that those five days would have had, if they were taken into account, namely that they would provide for an additional increase of 16 *momenta* per year, resulting in 24 hours in the course of 60 years (16×60=960 *momenta*=24 hours). Now, the composition of this additional increase is extremely revealing. All information about its composition comes from the *Computus Einsidlensis*, whereas the Munich computist did not feel the need to explain it any further (apparently, as will become obvious in the following, because he did not understand the theory himself). The first of these five days occurs in the second quarter (22 March to 20 June), another two in the third quarter (21 June to 20 September), and the remaining two in the fourth quarter (21 September to 20 December). According to the Einsiedeln computist, the regular daily increment of 2⅔ *momenta* had to be attributed to three of these five days only; the two days occurring in one of the final

419 Cf. the references to MC 39, and note that many tracts dealing with this question still remain unedited in the MSS.

420 MC 40; *CE* p. 103.

two quarters,[421] however, were supposed to have an increment of 4 *momenta* each had they not been distributed equally among the 90 days of their quarter (3×2⅔+2×4=16 *momenta*[422]). This difference in the increment was due to the rather obscure theory of a *nubs pallidae* ('a pale cloud', apparently part of the bissextile increment) outlined by the Einsiedeln computist, which adds an extra 1⅓ *momenta* per day to the regular increment (resulting in a total of 4 *momenta*, i.e. 2⅔+1⅓) to account for the periods of dawn and twilight that could neither be attributed to day nor night. It was this theory, then, that lead to a decrease of daylight by 4 rather than 2 ⅔ *momenta* in one of the final two quarters.[423] In the Munich computist's account of the increase and decrease of daylight, on the other hand, an increment of 4 *momenta* occurred *only* in the first quarter (21 December to 21 March; this period, it will be noted, does not contain more than 90 days), but in none of the following three; in fact, the idea that an increment of 1⅓ *momenta* of a nocturnal *nubs* is to be added to the regular increment of 2⅔ *momenta* in either of the final two quarters is explicitly rejected.[424] Consequently, the Munich computist's discussion of the five days exceeding 4×90(=360) in the course of a Julian calendar year is incongruent with his analysis of the increase and decrease of daylight. The only possible explanation for this incongruence is that the Munich computist worked from the *Computus Einsidlensis*, altered the latter's theory about the increase and decrease of daylight, but subsequently did not realize the implications of this alteration on the discussion of the five day exceeding the 360 days of the four 90-day quarters of a year.

In three further passages, then, the dependency of the Munich Computus on the *Computus Einsidlensis* becomes apparent beyond doubt, since the Munich computist evidently takes up ideas from his Einsiedeln precursor. First, following his general introduction to the calculcation of Easter, the Einsiedeln computist illustrates the theory by various examples.[425] Taking the Julian calendar dates of the Easter full moon and the weekday of 1 January as preconditions, he calculates the weekday of the Easter full moon, and from this the Julian calendar date of Easter Sunday. It appears that the examples are hypothetical rather than referring to the *annus praesens* of the author, since the chosen data are the

421 It is not quite clear which of the two quarters having 2 days in excess of a total of 90 days (i.e. the final two quarters of a Julian calendar year, from 21 June to 20 September and from 21 September to 20 December respectively) was supposed to have an increment of 4 rather than 2⅔ *momenta* per day; the discussion of the increase and decrease of daylight (*CE* p. 101–2) would suggest the first of these two (21 June to 20 September), the discussion of the five extra days (*CE* p. 103) rather the second (21 September to 20 December); cf. note 423.

422 Note that in this curious calculation, no distinction is made between increments referring to the increase of daylight and those referring to its decrease.

423 The *nubs pallidae* is discussed in the context of the first of these final two quarters (*CE* p. 101–2), but not in the context of the second, so I am inclined to think that the increment of 4 *momenta* relates to the first of these final two quarters; the discussion of the five extra days, however, rather suggests the second; cf. note 421.

424 Cf. MC 39.

425 *CE* p. 112.

most apparent ones if those were to be constructed, namely the Easter full moons of the first five years of the Dionysiac 19-year cycle and the weekdays of 1 January from *feria* 1 to 4, followed by *feria* 6 in the fifth year due to a bissextile day occurring after four years. This assumption is confirmed by the computist's own words: *Hęc autem exempli causa ostendimus*. Important for the present purpose, however, is the immediately following sentence: *Sed si quis in decennouenali ciclo huius numeri ueritatem cognoscere uoluerit, presenti sibi anno indesinenter inuestiget*. ('But if somebody wants to know the truth (concerning Easter Sunday) of his year in the 19-year cycle, he should himself investigate the present year without delay.') This call for investigating his own *annus praesens* was then directly taken up the Munich computist:[426] The method of calculating the Julian calendar date of Easter Sunday applied by him is exactly the same as the one outlined by the Einsiedeln computist; the three successive examples, however, refer to historical rather than hypothetical years, starting with the *annus inminens*, AD 719.[427] Consequently, there can hardly be any doubt that the Munich computist worked directly from the Einsiedeln text in this instances.

Concerning the second passage, one of the very curious features of early eighth-century Irish computistics was the equation of the two terms *quadrans* and *dodrans*. Classically, these two terms are complementary rather than identical, with *quadrans* defining a quarter, *dodrans* three quarters, adding up to one when combined. The contemporary of both the Einsiedeln and the Munich computist, the Anglo-Saxon scholar Bede, was totally aware of these classical definitions, as were computists and mathematicians before and after him.[428] *Only* Irish or Irish influenced texts of the early eighth century, the Munich Computus being among them, misinterpret *dodrans* and equate it with *quadrans*.[429] But where did this mistake originate? It is noteworthy that the only Irish text that does not explicitly mention this mistaken equation is the *Computus Einsidlensis*. In fact, it is quite apparent that the mistake arose from a misreading of the relevant passage in the Einsiedeln text. When discussing the characteristics of a quarter-day, the Einsiedeln computist makes the following statement:[430]

> Quadrans, quo sensu hoc nomen intellegitur? Aut dubium quod III significationes dicitur, hoc est: Quadras, ut dicitur gigas, gigantis; et quadrans participium a uerbo quadro, et nomen est quadrans; Grece autem integritas doras uel dodras, dodrantis.

Hence, one of three possible grammatical explanations of the word *quadrans* envisaged by the Einsiedeln computist is the theory that it ultimately is a Greek term, showing the same grammatical structure as *doras* or *dodras*, gen. *dodrantis*, the Greek term for Latin *integritas*. This statement, then, was mis-

426 MC 56.15–43.

427 For the chronological details of these years cf. p. LVIII–LXI above.

428 Cf., e.g., *DTR* 4.36–37 (cited in RM 8.35–36; PV §41); RM 18.6–7; Helperic, *De computo* 1 (*PL* 137, 21).

429 MC 7.2–3 (which is directly copied in the Angers glosses to *DTR*: Angers, Bibliothèque Municipale, 477 (461), fol. 49v); *DRC* 21.2–4; *DDT* 7D11–12; BC 97C10–13.

430 *CE* p. 89.

read to the effect that *dodras* was, in fact, considered as the Greek term for Latin *quadrans*. Since no other text is as close to the *Computus Einsidlensis* as the Munich Computus, it appears reasonable to assume that the mistaken equation of *quadrans* with *dodrans* originated in the Munich computist's misinterpretation / -reading of this Einsiedeln passage.

Probably the most revealing passage in this context, however, is concerned with a characteristically Irish concept (or rather a reckoning predominantly used by the Irish), the *latercus*; in fact, the chapter of the *Computus Einsidlensis* in which this passage occurs is the *only* part of this text referring to the *latercus*.[431] Interestingly enough, the parallel passage in *De ratione conputandi* also is the *only* part of that text dealing with the *latercus*;[432] only the Munich Computus contains additional references to this Easter reckoning in different contexts.[433] When discussing the *initium quadragesimae*, the beginning of the Lenten fast, the Einsiedeln computist compares the evidence of all three reckonings followed in the *regiones Scottorum* at different periods of time, namely the Dionysiac (clearly the one favoured), the Victorian, and the *latercus*, i.e. the 84 (14)-year Easter reckoning. Since the *initium quadragesimae* is totally dependent on Easter Sunday, the discussion begins with a comparison of the lunar limits for Easter Sunday of these three reckonings, namely *luna* 15 to 21, *luna* 16 to 22, and *luna* 14 to 20 respectively. From these, the lunar limits for the *initium quadragesimae* are then calculated as *luna* 3 to 9, *luna* 4 to 10, and *luna* 2 to 8 respectively.

Now, the interesting aspect of this passage is the fact that it presents an obvious error. Whoever composed this passage was not aware of the fact that the *latercus* applied a sequence of lunations different to the alternating one used in the Dionysiac and Victorian reckonings.[434] This difference had serious implications on the lunar relation between Easter Sunday and the *initium quadragesimae*, as the lunation that ends within this interval, namely the March lunation, is hollow (i.e. it consists of only 29 days) in the *latercus* sequence, but full (i.e. of 30 days) in the alternating sequences applied in the Dionysiac and Victorian reckoning. Since the *initium quadragesimae* was generally accepted to be placed exactly six weeks, i.e. 42 days, before Easter Sunday, this leads to a lunar day difference of 12 days between the two dates in the Victorian and the Dionysiac reckoning (42−30=12); according to the *latercus*, however, it would lead to a difference of 13 days (42−29=13). Consequently, the mistake made in the passage in question is that 12 was subtracted from the lunar age of Easter Sunday in order to calculate the lunar age of the *initium quadragesimae* not only for the Dionysiac and the Victorian

431 *CE* p. 119; right at the beginning of his textbook (*CE* p. 83–4), the Einsiedeln computist also mentions the term *latercus*, though not as the terminus technicus for a specific Easter reckoning, but rather as the alleged Egyptian term for number, or as a Latin term according to Isidore, followed by the latter's etymology. Cf. note 187.

432 *DRC* 99; also like *CE* (and in exactly the same context as *CE*; cf. previous note), *DRC* mentions the term *latercus* at the beginning of this texbook (c. 3–4), but not as the terminus technicus for a specific Easter reckoning.

433 Cf. p. CXXII–CXXIV above.

434 For a more detailed discussion of the implications of this difference in the sequence of lunations on the *initium* lunar data of the *latercus* see Warntjes, '84 (14)-year Easter reckoning', 51–4.

reckoning, but also for the *latercus*, though in that case 13 should have been subtracted, resulting in *luna* 1 to 7 as the lunar limits for the *initium quadragesimae* of the *latercus*.

The fact that this mistake occurs in all three Irish computistical textbooks (and in no other text) impressively highlights the interdependency of these three texts. In this respect, the most decisive question is which of these three texts introduced the error, and consequently appears to be the oldest of the three. Note that the Munich Computus appears as the least likely candidate from the outset, as its author is the only one among the three computists that could and should have known better: Only the Munich Computus transmits the *latercus* sequence of lunations, so if its author had calculated the three different lunar limits for the *initium quadragesimae*, he would, at least could have been aware of the difference in calculation.[435] In the end, both the Munich Computus and *De ratione conputandi* treat this error as fact, as if they had copied it from an authoritative text and therefore not feeling the need for any further comment. The Einsiedeln computist, on the other hand, expresses some doubts about the correctness of his statements; indeed, his phrasing strongly suggests that he was the author of the construction and therefore the origin of the mistake. I give the relevant passage of the three texts in parallel, with the crucial phrases of the Einsiedeln computist highlighted in bold font:

***CE* p. 119:**	**MC 58.2–7:**	***DRC* 99:**
De paschae figuris. Quot sunt figurę paschę? III sunt: A XIIII luna in XXmam secundum latercum, a XVmam (!) luna in XX\I/ secundum Grecos, a XVI luna in XXIIdam secundum Romanos. ***Dum autem pro certo nouimus has III formas in pascha esse, aeque III formas in initio fieri oppinamur. Et hoc uerum esse fatemur,*** *hoc est secundum latercum initium fit a IIda luna VIIIuam lunam, secundum Grecum (!) a III luna in IXnam lunam, secundum Romanos a IIII luna in X lunam.*	*Septem aetates paschae, quas prediximus, hae sunt: Secundum Grecos a XV luna usque in XXI, initii uero a III singulari in VIIII extenduntur. Secundum autem Victorium aetates pasche a XVI luna in XXII, initii a IIII singulari in Xam. Iuxta uero latercum a XIIII luna in XX, et initii a II luna in VIII singularem.*	*Sciendum nobis quod, quemadmodum hi uiri in septem paschae aetatibus discrepant, <in septem aetatibus initii idcirco discrepant>. Laterci enim sectatores, qui <a> XIIII luna usque ad XX septem aetates paschae numerant, a II luna usque ad VIII lunam septem aetates inítii conputant. Latini uero, septem aetates paschae a XVI luna usque ad XXII conputantes, a IIII luna usque in X lunam septem initii aetates numerant. Greci uero rationalibus, quos nos sequimur, a XV luna usque in XXI lunam septem aetates paschae numerantes, a III luna usque in VIIII lunam septem aetates initii conputant.*

435 The *latercus* sequence of lunations can be found in MC 50.73–77, preceded by a correct application of it when outlining a year of the *latercus* with epact 8 on 1 January (MC 50.68–72); yet, note also that the Munich computist failed to realize the implications of this sequence in chapter 52.

After introducing the lunar limits for Easter Sunday according to the three reckonings, the Einsiedeln computist states that he knows for certain (*pro certo*) that these are the lunar limits for Easter Sunday of the three reckonings in question; concerning the three lunar limits of the *initium quadragesimae*, however, he argues that he believes (*oppinamur*) that they happen to be in like manner (*aeque*), and that he therefore grants the following to be true, listing the three lunar limits in question. This clearly indicates that he worked from a source that provided him with the details of the lunar limits for Easter Sunday, from which he had to construct the lunar limits for the *initium quadragesimae*, not being entirely sure whether his method was correct for each of the three reckonings in question; in fact, his doubts were quite justified, but not shared by subsequent texts, i.e. the Munich Computus and *De ratione conputandi*, since these texts relied on the authority of their ultimate model for this passage, the *Computus Einsidlensis*.

Accordingly, this passage in particular reveals that the *Computus Einsidlensis* is older than the other two Irish computistical textbooks, and combined with all the other evidence discussed above it can hardly be doubted that this text was the Munich computist's principal source. Since the Munich Computus is datable, this result leads to AD 719 as a *terminus ante quem* for the composition of the Einsiedeln text. But is it possible to be more precise than that?

As mentioned above, the only technical dating clause found in the *Computus Einsidlensis* (AD 568 and 570) can hardly refer to the year of composition of this text.[436] A different passage of this textbook, however, may shed more light on this question. At the end of his discussion of the bissextile day, the Einsiedeln computist gives a curious mathematical explanation for this technical device:[437] If the 8760 hours of the 365 days of a year (the bissextile increment itself not being included) are divided by the seven days of a week, this division results in a remainder of three hours (8760mod7=3), which constitutes the bissextile increment per year (the bissextile day being reckoned as consisting of the 12 hours of daytime). For the present purpose, the interesting aspect of this explanation is not its strange argumentation, but the ascription of this theory to a certain Theodore. This name is as rare in computistical literature as in the more general early medieval, particularly the seventh- and early eighth-century, Insular context. If it belongs to this context rather than being a remnant of older computistical literature, the Theodore that immediately springs to mind is Theodore of Tharsus, the archbishop of Canterbury from AD 669 to 690. Aldhelm relates that many Irishmen studied in Theodore's cathedral school in Canterbury,[438] and computistics was, no doubt, one of the subjects taught there.[439]

436 Cf. note 401 above.

437 *CE* p. 108. For this calculation cf. the apparatus to MC 41.92–106 in the edition below and see Springsfeld, *Alkuins Einfluß*, 204–5, and now eadem, 'Beschreibung', 220–1.

438 Aldhelm, *Epistola ad Eahfridum* (*MGH AA* 15, 493): *etiamsi [beatae memoriae] Theodorus summi sacerdotii gubernacula regens Hiberniensium globo discipulorum, ceu aper truculentus molosorum catasta rigente vallatus, stipetur, limato perniciter grammatico dente iactura dispendii carens rebelles falanges discutit*. Cf. also Stubbs, 'Theodorus', 931; Stenton, *Anglo-Saxon England*, 180–1.

Consequently, the historical context certainly justifies the identification of a Theodore mentioned in an Irish computistical text belonging to the pre-AD 719 period with the archbishop of Canterbury. This identification is further strengthened by the parallel passages found in other contemporary computistical texts, most prominently the Munich Computus itself, where the same technical explanation of the annual bissextile increment can be found in exactly the same place, at the end of the chapter on the bissextile day.[440] As has been established above, the Munich computist worked from the Einsiedeln text, and the dependency of the two texts is strikingly apparent also in this passage, since both computists used the exact same modes of calculation:

***CE* p. 108:**	**MC 41.92–106:**
De bissexto, quem secundum Theodorum formamus hoc modo: ut in numero horarum diei per totum annum septenum ponamus numerum et quodcumque superfuerit septenum numerum, hoc fiat materia bissexti. Cuius probatione hęc est: Annus diebus CCCLXV conputatur. C uero dies horas habent $\overline{I}$ CC; ita et CCC dies habent horas $\overline{III}$ DC; LXta uero dies habent horas DCCXX; V autem dies horas LXta habent; que simul faciunt horas $\overline{IIII}$ CCCLXXX. Ita et nox eundem numerum horarum per totum annum habet. Hinc conpotus est horarum totius anni $\overline{VIII}$ DCCLX. Hoc septeno numero diuide VII milia et DCC, remanent $\overline{I}$ et LXta. Remitte itidem DCC, remanent CCLX. Partire iterum hunc numerum in quinquaginta, hoc est septies quinquaginta, remanent X. Deduc VII ex hoc numerum, remanent III horę, quę septenum numerum supersunt, et quę per quadriennium diem bissexti licet abusiuum faciunt et non diem proprium.	*Aliter Grecorum bissextus preparari artificiose intellegitur. Greci autem anni horas rimari sollicitant. Quas per VII diuidunt in figuram VII dierum per quos mundi circulus inuoluitur. Quod itaque extra VII sentiunt superesse, ex eo diem bissexti faciunt. Hoc modo horas anni numerabis: Nam X horę per X dies duabus horis relictis sunt uniuscuiusque diei, CXX efficiunt. Ac deinde C C C diebus $\overline{III}$ numerantur et duabus horis relictis DC. LX diebus DCCXX horae rimantur. Et de V diebus LX horę sunt. Hic est totus numerus, quem diximus, $\overline{IIII}$CCCLXXX. Quibus parem horarum noctium numerum coniungas, $\overline{VIII}$DCCLX utrumque inuenire scias. Inde per VII diuisas III horas superesse noscas. Quibus per quadriennium XII horas effici intellegas. Quas peracto IIII annorum curso in diem bissexti unari putant.*

As can be seen from this collation, the main difference between the two texts lies in the ascription of the concept: While the Einsiedeln computist attributes it to Theodore, his Munich successor ascribes it to the 'Greeks'. The conclusion, therefore, must be that the Munich computist, working from the Einsiedeln passage, regarded the Theodore referred to in the Einsiedeln text as a

439 For the subjects taught at the school of Canterbury see especially Bede's account in his *HE* 4.2: *Et quia litteris sacris simul et saecularibus, ut diximus, abundanter ambo errant instructi, congregate discipulorum caterua, scientiae salutaris cotidie flumina inrigandis eorum cordibus emanabant; ita ut etiam metricae artis, astronomiae, et arithmeticae ecclesiasticae disciplinam inter sacrorum apicum uolumina suis auditoribus contraderent.* For Theodore's teaching cf. Lapidge, 'Theodore', 444–6; idem, 'Theodoros', 636.

440 MC 41.92–106.

representative of the 'Greeks',[441] a fact certainly supporting the identification of this person with the archbishop of Canterbury, who was a Byzantine by birth as well as in his scholarly training.

This identification, then, suggests AD 669, the beginning of Theodore's episcopy, as the *terminus post quem* for the composition of the *Computus Einsidlensis*. In fact, the earliest datable occurrence of this curious explanation of the annual bissextile increment (attributed to Theodore in the Einsiedeln text) can be found as part of the ps-Dionysiac *Argumentum XVI* in the *Computus Digbaeanus* of AD 675.[442] Yet, the *Computus Einsidlensis* does not necessarily have to have been composed in Theodore's lifetime. Another important clue may be provided by the now lost Victorian computus of AD 689 used by the Munich computist.[443] Generally it appears that this Victorian text, the *Computus Einsidlensis*, and the Munich Computus were composed in relative close geographical proximity. At least it can be stated with some degree of certainty that all three of them were written in an area of the *regiones Scottorum* that had followed the Victorian reckoning at some point in time, i.e. in the southern part of Ireland.[444] The Victorian computus of AD 689 may, in this respect, indicate that the conversion to the Dionysiac reckoning (i.e. the one favoured by both the Einsiedeln and the Munich computists) had not yet taken place at that time. On this basis, I regard the period between AD 689 and 719 as the most likely date of composition of the *Computus Einsidlensis*.[445]

De comparatione epactarum Dionysii et Victorii (Cologne, Dombibliothek, 83², fol. 176v–178r):[446] This text, which is edited for the first time in Appendix 2 below, has, to the present day, not been discussed in print.[447] In fact, not even its general content has been correctly identified. The text itself consists of two parts, the second part being introduced by the heading *In nomine Dei summi incipit* written in red ink (fol. 178r) and ending a few lines before the explicit on

441 The vague ascription to the 'Greeks' of this technical explanation of the annual bissextile increment then found its way into the Irish influenced *De bissexto* I (*PL* 101, 994), and from there into *Lib. comp.* 1.3f (?). The same explanation is found, though without any attribution, in BC 39, 40; *Dial. Burg.* 14A; cf. also the following note.

442 Krusch, *Studien* II, 80–1. The *Computus Digbaeanus* of AD 675 is published under Dionysius Exiguus' name in both Jan, *Historia cycli dionysiani*, 79–94 (repr. in *PL* 67, 497–508) and Krusch, *Studien* II, 75–81. For this text cf. especially Warntjes, 'Argumenta'; note that in this article (p. 45) attention is drawn to the fact that *Argumentum XVI* may not necessarily have formed part of the *Computus Digbaeanus* of AD 675.

443 For this text cf. p. CXXIV–CXXVI above.

444 Cf. the argument p. LXXVIII–XCVI, CXXV–CXXVI above.

445 A more concise and therefore less detailed version of this argument can be found in Bisagni & Warntjes, 'Early Old Irish material', 81–91.

446 The title is my own invention based on the content of the tract; the tract itself has no heading (cf. Cologne, Dombibliothek, 83², fol. 176v; a facsimile of the entire MS is readily available on the world wide web at: http://www.ceec.uni-koeln.de).

447 It is only briefly referred to in Warntjes, '84 (14)-year Easter reckoning', 66, where the folio reference is wrong; it should read fol. 176v–178r instead of fol. 211r–213r.

the same folio.[448] In modern catalogues, this second part (right up to the explicit) is treated as a separate text, but content and phrasing reveal that it is closely connected to the first part.[449] Both parts primarily deal with the question of which year in the Dionysiac 19-year cycle coincides with the year of Christ's passion. The era applied in the Dionysiac Easter table leaves no room for doubt about the year that was regarded as the *annus incarnationis*; the *annus passionis*, however, was not referred to in any way. On the other hand, Victorius, in his Easter table, reckoned the years from the *annus passionis*. Accordingly, seventh-century computists thought that a comparison of the Victorian with the Dionysiac reckoning, or rather of their epacts, would at least reveal the Dionysiac equivalent to what Victorius had designated as the *annus passionis*.

The historically correct comparison, however, led to chronological data for the Dionysiac equivalent of Victorius' *annus passionis* that disagreed with the evidence of the Synoptic Gospels and other authoritative texts: The resurrection of Christ took place on *luna* 15, and consequently the passion on *luna* 13, whereas the Synoptic Gospels give *luna* 17 for the resurrection, *luna* 15 for the passion. For this reason, a different method of comparison was invented for the two sets of epacts, namely by starting the collation with the first year of the

448 The content of the final two passages before the explicit reveals that these passages originally did not belong to the main text: The first passage deals with the equinox (after mentioning the number of years of an Easter cycle (532) without any obvious reason), the second is the beginning of a calculation of the number of lunar days of the *hendecas*:
DXXXII. Aequinoctium est XIIII Kalendas Aprilis, et huius transitus est XII Kalendas Aprilis. Item aequinoctium est XII Kalendas Aprilis, et huius transitus XI Kalendas Aprilis. Rursumque aequinoctium est VIII Kalendas Aprilis, et huius transitus est VII Kalendas Aprilis.
Septies autem tricenteni quinquageni quaterni fiunt simul $\overline{II}$*CCCCLXXXVIII (recte* $\overline{II}$*CCCCLXXVIII), et quarter tricenteni octogeni quaterni, ut est CCC CCC CCC reliqua. Explicit.*

449 The entire text under discussion here is silently passed over in the 1874 catalogue of Cologne manuscripts (Jaffé & Wattenbach, *Coloniensis Codices*, 30). The first scholar who explicitly referred to it was Krusch, *Studien* I, 203; he divides the text into three parts, the table of fol. 176v, the following text up to the incipit *In nomine dei summi incipt* in red ink (fol. 177r–178r), and the text confined by this incipit and the explicit on the same fol. 178r; concerning the content of these sections, Krusch rightly argues that that the table shows a comparison of the epacts and the Easter full moons of the Victorian with those of the Dionysiac reckoning; however, the following statement by Krusch that this comparison is 'extremely faulty' reveals that he did not understand the logic behind it, namely that the Dionysiac *ogdoas* was correlated with Victorian *hendecas*; moreover, his argument that the following texts was written for the purpose of rejecting the Victorian reckoning is equally vague and misleading; it was rather written by adherents of the Dionysiac reckoning as a reaction to harsh accusations made by their Victorian counterparts, who claimed that the Dionysiac reckoning gives data for the *annus resurrectionis* contrary to the evidence of the Gospels. In the most recent description of this MS, van Euw, 'Kompendium der Zeitrechung', 154 divides this section into two separate texts (fol. 176v–178r up to the incipit; fol. 178r from the incipit to the explicit), but only repeats Krusch's wrong comments on the content. The fact alone that the phrase *qui Grecam magis regulam sequitur* occurs before and after the incipit is a clear indication that these two parts are evidently related, and therefore they are considered as one text here.

Dionysiac *ogdoas* on the one hand as against the first year of the Victorian *hendecas* on the other. In this purely constructed comparison, the Dionysiac equivalent to the second year of the Victorian *hendecas* (i.e. the lunar year of the Victorian *annus passionis*) is the second year of the Dionysiac *ogdoas*. The epact (of 1 January) of this second year of the Dionysiac *ogdoas* is 20, that of the second year of the Victorian *hendecas* 19. Consequently, the Dionysiac lunar age of every Julian calendar date in this year is incremented by one compared to the Victorian lunar age. In particular, the lunar age of the resurrection of Christ in the Victorian *annus passionis* is 16 in the Victorian, but 17 in the Dionysiac reckoning (and the passion consequently falls on *luna* 15) according to this unhistorical comparison, therewith obtaining data for the Dionysiac reckoning in total agreement with the evidence of the Synoptic Gospels and other authoritative texts.

Now, the principal purpose of the Cologne text is to explain this unhistorical comparison in all its details, and to promote it against the historically correct one. The reason for this preference becomes perfectly apparent when analyzing the attribution given to each of these methods: Whereas the historically correct comparison is ascribed to followers of the Victorian reckoning, the unhistorical one (aligning the Dionysiac *ogdoas* with the Victorian *hendecas*), which is the one favoured by the author himself, is attributed to advocates of the Dionysiac reckoning. These ascriptions quite clearly reveal the background to the composition of this text: It appears that followers of Victorius compared the Victorian reckoning with the Dionysiac one. This comparison led to the conclusion that the Dionysiac lunar data for what has been denoted as the *annus passionis* in the Victorian Easter table disagreed with the evidence of the Synoptic Gospels, as well as John's Gospel, the latter providing the basis for the Victorian data. Consequently, in the disputes about the correct system for the calculation of Easter, it seems that ardent supporters of the Victorian reckoning argued against their Dionysiac counterparts that Dionysius' system could not be followed, as it contained non-Biblical (and therefore heretical) lunar data for the passion and resurrection of Christ. Faced by this accusation, Dionysiac supporters then invented the comparison outlined in the Cologne text, which was supposed to serve as a technical justification of the Dionysiac system, constructing a comparison of the two reckonings that would provide the data transmitted in the Synoptic Gospels for the Dionysiac equivalent to the Victorian *annus passionis*.

Except for a table in the *Computus Einsidlensis* correlating the Dionysiac and Victorian epacts according to both methods, and a rather obscure account of this problem in *Comp. Col.*, the only text dealing with this question is the Munich Computus.[450] Accordingly, the Munich Computus and the Cologne tract are the only extant texts that discuss, in a structured and comprehensive way, the equivalent of the Victorian *annus passionis* in the Dionysiac 19-year cycle on the basis of the two correlations of epacts. This fact alone strongly suggests that the two texts are dependent. As in the case of the *Computus Einsidlensis*,

450 *CE* p. 109; *Comp. Col.* 5.4–5; MC 52. Cf. also p. CI–CII above.

the question that remains to be solved is whether the Munich computist worked from the Cologne text or vice versa.

Two aspects in particular reveal, in my opinion, that the Munich computist worked from the Cologne text. First, the bias and the fervour of the argument as expressed in the Cologne tract are not shared by the Munich computist: The latter simply outlines the two comparisons, pointing out that both appear problematic for different reasons; moreover, in the Munich Computus the methods of correlation are not attributed to any group of people, particularly not to opposing parties. It seems, therefore, that the Cologne treatise was written at the height of the controversy between adherents of the Victorian reckoning and those of the Dionysiac one, while the Munich textbook was composed a few decades later, when the disputes had been settled and the Victorian reckoning had lost its currency. Second, only the Munich computist's account of this problem includes references to the *latercus*. Considering the fact that the Munich computist rather artificially introduced *latercus* data into analyses that derive from the *Computus Einsidlensis*,[451] the same appears to be the case here. Generally, the impression is that the Munich computist could have easily written his discussion on the basis of the Cologne tract, while it is difficult to imagine that the Cologne computist would have been able to composed his thorough analysis from the Munich computist's less detailed chapter, while at the same time omitting all references to the *latercus* that feature so prominently in the Munich text. Consequently, the Munich computist seems to have worked from the Cologne text or a related treatise, and, having been pointed to the *latercus* in a different context by the Einsiedeln computist,[452] he tried to introduce the *latercus* information available to him into this discussion of the lunar data of the *annus resurrectionis* to gain a different perspective on this problem, an attempt that ultimately failed.[453]

Again, the question must be addressed whether a date more precise than pre-AD 719 (i.e. prior to the composition of the Munich Computus) can be established for the Cologne treatise. Only one passage illustrates the theory by use of examples, and the data given in these examples may, in fact, have been taken from years close to the date of composition. In the unhistorical comparison of the epacts of the two reckonings, the Dionysiac lunar age is incremented by one compared to the Victorian lunar age in 17 out of 19 years (or a few months less than that). The fundamental difference between the two reckonings in the lunar limits of Easter Sunday (*luna* 16 to 22 in the Victorian, *luna* 15 to 21 in the Dionysiac reckoning) leads to the facts that, according to this unhistorical comparison, the Dionysiac Easter Sunday would fall a week earlier than the Victorian one a) in each of the 19 years (with certain exceptions) whenever the Victorian Easter Sunday falls on *luna* 22, and b) additionally in each of the 17 years mentioned above whenever the Victorian Easter Sunday falls on *luna* 21,

451 Cf. especially MC 50.

452 Cf. p. CXXII above.

453 For an analysis of the *latercus* information in this chapter of the Munich Computus see Warntjes, '84 (14)-year Easter reckoning', 62–9, 74–8.

since this then corresponds to the unacceptable *luna* 22 in the Dionysiac reckoning. This phenomenon is discussed in the Cologne treatise and illustrated by some examples. It should be noted that an example could have been easily constructed for any year of the Victorian 19-year cycle; accordingly, the first year of the Victorian *ogdoas* or *hendecas*, or of the Victorian Easter table proper, would have been the most obvious choices for a hypothetical example. Yet, it seems that the author of the Cologne text browsed through the Victorian Easter table for a year or years that would be suitable for illustrating his argument. In this respect, it would only be natural for the author to look for those examples close to his *annus praesens*, especially when considering the fact that early medieval computists often only had a section of, but not the full Victorian Easter table avaible to them.[454] Now, the first year chosen by the Cologne computist has *feria* 6 and *luna* 4 on 1 January, resulting in Easter Sunday on 18 April, *luna* 22; the second year has *feria* 6 and *luna* 3 on 1 January, resulting in Easter Sunday on 18 April, *luna* 21.[455] In the almost three hundred years between Dionysius' writings (AD 525) and the composition of the Cologne MS (AD 805), two years with the mentioned data occurred only twice in the Victorian Easter table, namely in AD 594 and 583, and in AD 689 and 678.[456] The Cologne computist's account itself suggests that, if any of the two years discussed by him is chronologically close to his year of composition (and the solar data in particular do not render it likely that these examples were purely constructed), it would be the first one; the second year was then chosen for mathematical reasons, since the facts that it had identical solar information as and an epact of exactly one less than the first example made it possible to use most of the data of the first example. Therefore, either AD 594 or 689 may have been close to the *annus praesens* of the author of the Cologne text. Taking the fact into consideration that a very similar discussion can *only* be found in the Munich Computus, it appears reasonable to argue that these two texts originated in close geographical proximity, pointing to the *regiones Scottorum*, and more specifically to southern Ireland, as the place of composition of the Cologne treatise. From this perspective, AD 689 or thereabouts actually seems extremely likely as the date of composition of the Cologne text: The southern Irish clergy had converted to the Victorian reckoning in AD 632, while the following conversion from the Victorian to the Dionysiac reckoning took place in a rather slow process over decades, presumably in the second half of the seventh century.[457] Within this process, some churches probably converted to the Dionysiac method sooner than others, leading to controversies between computists following different systems. The Cologne treatise certainly would fit perfectly in the context of such a controversy, representing a direct reaction from adherents of the Dionysiac system against accusations made by their Victorian rivals.

454 Cf. the discussion of the Victorian Easter tables in notes 139, 228 above.

455 *De comparatione epactarum Dionysii et Victorii* 1.107–113 (see Appendix 2).

456 Cf. Krusch, *Studien* II, 27–38, 51–2.

457 Cf. especially p. LXXVIII–XCVI above.

In fact, the date AD 689 itself is very suggestive. It will be remembered that the now lost Victorian computus, from which the Munich computist copied considerable parts, was also composed in that year.[458] This fact does not, however, allow for the conclusion that the Cologne treatise originally constituted a part of that Victorian computus, since the latter appears to have been written from a purely Victorian bias, while the former vigorously took side for the Dionysiac reckoning against Victorian computists. Anyway, AD 689 certainly was a year of increased computistical interest, since it also saw the compilation of the *Computus Cottonianus*, which in all likelihood was composed in an Anglo-Saxon centre in the *regiones Scottorum*, namely in Rath Melsigi or an affiliated monastery for the Willibrordian mission.[459] There may have been various reasons for this increase in computistical activity, among which the need for providing missionaries with the basic methods of Easter calculation as well as working Easter tables appears to have been only of rather minor importance. A more prominent reason was the conversion of the Iona abbot Adomnán, the most influential churchman in the *regiones Scottorum*, to the Dionysiac reckoning. Bede relates that Adomnán, on one of his visits to Northumbria, discussed computistics with Ceolfrith, abbot of Wearmouth and Jarrow, and doubtlessly other *seniores* of that monastery; this scholarly conversation ultimately led to Adomnán's conversion from the 84 (14)-year Easter reckoning to the Dionysiac one.[460] Since Bede, in both accounts, argues that Adomnán's conversion took place when the Irishman was on a mission from his own people, this event should be connected to an entry in the *Annals of Ulster* for AD 687, which states that Adomnán brought 60 former captives back to Ireland.[461] If this was the case, then Adomnán travelled to Ireland as a convert to the Dionysiac reckoning in AD 687, and his presence and conversion will certainly have incited debates, if not controversies, among Irish computists. According to Bede's narrative, Adomnán did not manage to convince the brethren of his own monastery to follow his lead, but he was more successful in Ireland. In fact, Bede explicitly argues that the greater part of the Irish in Ireland itself (*plurima pars Scottorum in Hibernia*) converted to the correct Easter observation (*rationale paschalis observantiae tempus*). With this statement Bede certainly wants to imply that most of the Irish converted to the Dionysiac reckoning soon after Adomnán and on the Iona abbot's initiative. The *Annals of Ulster* also relate that Adomnán travelled (from Iona) to Ireland in AD 692,[462] and, combining Bede's statement with this evidence, it seems that Adomnán's main success in converting a considerable part of the Irish clergy to the Dionysiac reckoning happened in AD 692 and the subsequent years. Hence, it appears that Adomnán's visit to Ireland in AD 687 shortly after his conversion has stimulated discussions, if not

458 For this text cf. p. CXXII–CXXIV, for the dating clause p. LVII–LVIII.

459 This argument will be outlined in detail in an article forthcoming in the proceedings to the second conference on the science of computus.

460 *HE* 5.15, 5.21.

461 AU 686.5 (Mac Airt & Mac Niocaill, *Annals of Ulster*, 150–1).

462 AU 691.1 (Mac Airt & Mac Niocaill, *Annals of Ulster*, 152–3).

controversies, about the correct method for the calculation of Easter, which ultimately laid the ground for his later success. In the course of these vivid debates between AD 687 and 692 it would only have been natural if the Victorian reckoning was compared in detail with the Dionysiac one, and, as mentioned above, it must be presumed that some churches converted sooner than others, leading to conflicts between recent converts to the Dionysiac reckoning and southern Irish advocates of the Victorian one. The year AD 689 itself, falling right into this period, was of a very controversial nature, since the Victorian Easter table in the Sirmond MS (which, in all likelihood, represents an altered copy of the one used in southern Ireland) notes Easter Sunday on 18 April, a week later than in the other two reckonings, and on *luna* 22, which is uncanonical according to both the 84 (14) and the Dionysiac reckoning.[463] Consequently, the Cologne text appears to fit very well into the Irish context of the final two decades of the seventh century.

463 Oxford, Bodleian Library, Bodley 309, fol. 115r. Other (but continental) Victorian Easter tables list double-dates for this year, with the alternative date coinciding with the Dionysiac one and the *latercus*; cf. Krusch, *Studien* II, 33 and note 228 above. For the Dionysiac date see Wallis, *Bede*, 395; for the *latercus* Blackburn & Holford-Strevens, *Companion to the year*, 875.

RECEPTION AND INFLUENCE

As has been outlined earlier,[464] the Munich Computus was composed at a crucial time in the history of computistics: Three reckonings were known and used in the seventh century, none of which being satisfactorily explained in the computistical literature available at that time. Accordingly, seventh-century Insular computists faced the task of analyzing, comparing, and understanding the systems underlying these reckonings in order to decide which of these was to be considered as the theologically most justified and technically most accurate method. Once this task was accomplished, however, the study of computistics, and with this its literature, changed fundamentally. The focus, then, was solely on the one generally approved reckoning, which quite naturally led to an improvement of the modes of calculation within this one accepted system, while interest was lost in earlier and outdated reckonings and concepts.

From this perspective, it does not surprise that the reception and influence of the Munich Computus was rather limited. Together with the *Computus Einsidlensis* and *De ratione conputandi*, the Munich Computus laid the foundations of medieval computistics, since these Irish texbooks provided, for the first time, a thorough analysis of the Dionysiac reckoning, which became standard in the Latin West from this time to the Gregorian calendar reform. Not having any accurate model for certain calculations and concepts (like, e.g., the construction of the 19-year cycle), these Irish computists had to invent their own methods. As it is the case in the development of any science, the proofs established first are usually not the most suggestive, concise, and precise ones, so that those Irish methods were soon superseded. Moreover, these three Irish textbooks stood right at the end of the Insular Easter controversy of the seventh century and consequently still included technical details of earlier debates that lost their relevance and currency in the following decades and centuries. For these reasons, only very few of the theories and methods outlined in these Irish texts had a long-lasting influence, while the direct reception of these texts generally did not go beyond the eighth century (with few ninth-century exceptions). In the following, the influence of major, evidently seventh-century Irish computistical concepts found in the Munich Computus (and often also in the other two Irish computistical textbooks) are discussed first, followed by the direct reception of the Munich Computus by eighth and early ninth-century computists, analyzed text by text.

Starting with the concept of the 14 divisions of time, its reception and influence is almost as difficult to establish as its origin,[465] since the manuscript evidence has not been thoroughly analyzed to the present day, so that many important tracts dealing with this theory still remain hidden in the codices. Additionally, as Rabanus Maurus' example illustrates, the original Irish list of 14 divisions was liable to changes even if the total number of divisions, i.e. 14,

464 Cf. p. XLVII–LI above.

465 On the origin of this concept cf. the discussion p. CXLI–CXLIII above.

was kept: Rabanus omits the units 'week' and the rather vaguely defined *mundus*, but introduces *ostentum* as the second smallest unit and *pars* as the fourth smallest, placed between *momentum* and *minutum*.[466] Thus, certainly by Rabanus' time (AD 820) the original Irish list had received considerable alteration; in fact, the *ostenta* and *partes* were already known to Bede,[467] so that the reception of his work led to a diffusion of the original Irish concept. Before Bedan theory was, however, mixed with Irish concepts probably towards the end of the eighth century, the original 14 divisons of time listed by Irish computists enjoyed some popularity, as the Carolingian reworkings of *De divisionibus temporum* and numerous unpublished tracts still hidden in the manuscripts reveal.[468]

As mentioned in a different context,[469] late seventh- and early eighth-century Irish computist used specific modes of calculation. While the Dionysiac *argumenta* gained considerable popularity (in a rather slow process) towards the end of the seventh century, especially, it must be presumed, in Anglo-Saxon centres,[470] the Irish appear to have continued to use methods they had already applied to the 84 (14) and the Victorian reckoning before the Dionysiac *argumenta* became known in the *regiones Scottorum*. In these methods the chronological data of 1 January were taken as precondition (and thus revealing their earlier use for the 84 (14) and the Victorian reckoning[471]), from which the lunar and weekday data of any given Julian calendar date, particularly of the Easter full moon and of Easter Sunday, were then calculated with the help of certain calendrical rules. Interestingly enough, the lunar calculations outlined and used in various chapters of the Munich Computus did not find any imitators; they probably appeared to laborious to later computists, who preferred the more complex, but also more concise and effective Dionysiac *argumenta*, or their

466 *RM* 10; cf. also p. CXII above.

467 *DTR* 3.

468 For the Carolingian versions of *DDT* see notes 115 and 332 above. As for still unpublished texts, cf., e.g., note 319 and the five tracts in the following MSS: 1) Paris, Bibliothèque Nationale, Lat. 5543, fol. 126r; Vatican, Biblioteca Apostolica, Reg. Lat. 1038, fol. 126r; Vatican, Biblioteca Apostolica, Reg. Lat. 1260, fol. 117r; Strasbourg, Bibliothèque Nationale et Universitaire, 326, fol. 164v. 2) Karlsruhe, Landesbibliothek, 442, fol. 63v–64r; Geneva, Bibliothèque Publique et Universitaire, 50, fol. 139r–v; Leiden, Universiteitsbibliotheek, Scaliger 28, fol. 40r–v. 3) Paris, Bibliothèque Nationale, Lat. 2341, fol. 9r; Paris, Bibliothèque Nationale, Lat. 7418, fol. 89v–90r; Paris, Bibliothèque Nationale, Lat. 5543, fol. 125r–126r; Strasbourg, Bibliothèque Nationale et Universitaire, 326, fol. 164r–v; Vatican, Biblioteca Apostolica, Urbs Lat. 290, fol. 34v–35r. 4) Vatican, Biblioteca Apostolica, Reg. Lat. 1855, fol. 84r. 5) Karlsruhe, Landesbibliothek, 229, fol. 33r.

469 Cf. p. LXXII–LXXIII above.

470 Cf. especially Warntjes, 'Argumenta', p. 93–4.

471 Both the Victorian Easter table and that of the 84 (14) noted the chronolgical data (i.e. the weekday and the lunar age) of 1 January for every year. Cf. Krusch's edition of the Victorian table (*Studien* II, 27–52), and the reconstruction (Mc Carthy, 'Easter principles', 218–9), translation (Blackburn & Holford-Strevens, *Companion to the year*, 873–5), and facsimile of the Padua *latercus* (Warntjes, '84 (14)-year Easter reckoning', 80–2).

subsequent recensions.[472] On the other hand, some of the simple, but very suggestive rules for weekday calculations do not only feature in all three Irish computistical textbooks, but also in the Irish (as I would argue) *Dial. Langob.* (like the fact that the Calends of a month following a month of 31 days fall four – counted inclusively – or three – counted exclusively – weekdays later than the Calends of the preceding month), in *Quaest. Austr.* (like the fact that the Calends of January and of April have the same weekday in bissextile years, which falls on the day before the Calends of April in non-bissextile years), and in the Bobbio Computus (like the fact that the Calends of August fall a weekday later than the Calends of May);[473] the very useful calendrical tool of listing the Julian calendar dates with identical weekdays per class of months received an even wider reception; the first occurrence of this list is in Irish textbooks, but it can also be found as a separate table in numerous manuscripts.[474]

The most influential concept probably was the linear increase and decrease of daylight by 2⅔ *momenta* per day between the two solstices. This theory ultimately goes back to Anatolius (?)'s *De ratione paschali*, but it received considerable variations and additions during the seventh- and early eighth-century computistical discourse that took place in the *regiones Scottorum*.[475] Since this concept was completely disconnected from the calculation of Easter, it was unaffected by discussions of and changes in the system underlying this calculation.[476] This fact certainly explains the wide reception it received in subsequent centuries, which still has to be analyzed in detail. It figures prominently in the Irish influenced Bobbio Computus, the Irish *Dial. Langob.*, as well as the Frankish *Dial. Neustr.*, *Lect. comp.*, *Lib. ann.*, *Lib. comp.*, but also in various tracts on this question that still remain hidden in the manuscripts.[477]

The Munich computist's discussion of the bissextile day, as well as his more general lunar theory, was equally independent from the prevailing Easter

472 For the Munich computist's methods cf. especially the calculation of the Julian calendar date of the Easter full moon from the lunar age of 1 January outlined in MC 56.44–57, and the parallel calculation of the Julian calendar date (and weekday from the *feria* of 1 January) of *luna* 2 of the March lunation in order to establish the Julian calendar date of the *initium quadragesimae* in MC 58.40–74.

473 Cf. MC 29.2–23; *CE* p. 97–8; *DRC* 34–35; *Dial. Langob.* 12B, 16A,C; *Quaest. Austr.* 2.7; BC 105; and also p. CXXXVIII–CXL.

474 The first occurrence is in MC 29.24–35; *CE* p. 98; *DRC* 33; and it then found its way into many MSS, e.g., *BC* 2; Leiden, Universiteitsbibliotheek, Scaliger 28, fol. 40r; Paris, Bibliothèque Nationale, Nouvelle acquisition latine 1615, 10r; Vatican, Biblioteca Apostolica, Pal. Lat. 1447, fol. 23r; Vatican, Biblioteca Apostolica, Reg. Lat. 1260, fol. 15v; Vatican, Biblioteca Apostolica, Reg. Lat. 1855, fol. 47r.

475 *DRP* 13; MC 39; *CE* p. 99–103; *DRC* 50; *DDT* 12.

476 The only detail in this concept that was, in some way, connected to the general theory of the calculation of Easter was the date of the vernal equinox.

477 BC 30, 31, 114; *Dial. Neustr.* 28, 28a; *Dial. Langob.* 19; *Lect. comp.* 7.1 (citing *DDT*); *Lib. ann.* 54; *Lib. comp.* 3.14. For tracts on this question in manuscripts not studied by Borst for his monumental *Schriften*, see, e.g., Basel, Universitätsbibliothek, F III 15k, fol. 56v–57r; Oxford, Bodleian Library, Digby 63, fol. 33v–34r; Leiden, Universiteitsbibliotheek, Scaliger 28, fol. 23v. Cf. note 170.

reckonings. Several aspects outlined in these chapters of the Munich Computus appear to be of seventh-century Irish or Insular origin. Their reception, however, was less substantial than in the case of the theory of the increase and decrease of daylight. There are two particularly noteworthy examples from the Munich computist's discussion of the bissextile day. First, it is argued that the neglect of the bissextile day would result in a shift of the seasons within the Julian calendar, an observation that is not only noted in Irish textbooks, but also by Bede.[478] From Bede this theory found its way into *Lib. calc.* and Pacificus of Verona's *Computus*, while Irish sources influenced *Dial. Neustr.*, *Quaest. Austr.*, and *Comp. Col.*[479] Second, the bissextile increment per year (a quarter-day) is curiously explained mathematically as the remainder of three hours when dividing the 8760 hours of a year (consisting of 365 days) by the seven days of a week. The earliest datable occurrence of this explanation can be found in the *Computus Digbaeanus* of AD 675, which was presumably composed in an Anglo-Saxon centre in Ireland; it is attributed to a certain Theodore, who may be identified with Theodore of Tharsus (†690), the archbishop of Canterbury, in the *Computus Einsidlensis*.[480] Accordingly, an Anglo-Saxon origin of this explanation seems likely. Its earliest transmission, however, was based on the mentioned *Computus Digbaeanus*, the *Computus Einsidlensis*, and the Munich Computus, and then found its way into three texts heavily influenced by Irish computistical thought, namely the Bobbio Computus, the ps-Alcuin tract *De bissexto I*, and *Dial. Burg.*, and later also appeared in *Lib. comp.*[481] In lunar theory, one of the most noteworthy aspects of highly probable Irish origin is the observation that the kindling of the moon can take place above, on the same level as, and below the sun, applying the technical terms *superincensio*, *mediaincensio*, and *subincensio* for each of these phenomena respectively. This terminology first appears in the Munich Computus and *DRC*; from the former it found its way into *Comp. Col.*, from the latter into some glosses to *DTR*, and it also appears in the Irish influenced ps-Alcuin tract *De saltu lunae I*.[482]

Turning to epactal theory, the Munich computist explicitly states that the Irish had the habit of choosing 1 January as the *sedes epactarum* and therefore translated the Dionysiac epacts of 22 March to 1 January.[483] This practise is hardly surprising considering the fact that the two reckonings previously used in the *regiones Scottorum*, i.e. the Victorian reckoning and the 84 (14), both included a column of the epacts of 1 January in their Easter tables.[484] The op-

478 MC 41.54–61; *CE* p. 106–7; *DTR* 38.

479 PV §302; *Dial. Neustr.* 28a; *Quaest. Austr.* 2.9; *Comp. Col.* 3.1, 3.5.

480 Cf. p. CL–CLII above.

481 The relevant passage of the *Computus Digbaeanus* is published as part of the ps-Dionysiac *Argumentum XVI* (Krusch, *Studien* II, 80–1); this explanation can also be found in MC 41.92–106; *CE* p. 108; BC 39, 40; *De bissexto* I (*PL* 101, 994–5); *Dial. Burg.* 14A; *Lib. comp.* 1.3f.

482 MC 46.16–20; *DRC* 65; *Comp. Col.* 5.2A; *DTR* gloss (*PL* 90, 406); *De saltu lunae* I (*PL* 101, 988).

483 MC 49.14–15.

484 Cf. note 471 and also note 188 above.

tion of translating the epacts from one Julian calendar date to another is already mentioned in the *Computus Einsidlensis*, where the subsequent table contains both sets of Dionysiac epacts (i.e. the original one for 22 March and the one calculated for 1 January). Some Dionysiac Easter tables later also included both sets of epacts, as illustrated, e.g., by an Easter table starting from AD 798 as well as the one contained in St Dunstan's classbook, both being apparently influenced by Irish computistical thought.[485] Most remarkable in this respect, however, is the fact that in some computistical manuscripts these two sets of epacts constitute a separate table, apparently outlined for the purpose of translating the Dionysiac epacts from 22 March to 1 January without difficulty.[486]

Another, though less widely received feature of seventh- and early eighth-century Irish computistical thought was the designation in the Julian calendar year of an epactal period of eleven days (the difference between a lunar and a solar year). This question is discussed, to varying extents, in all three Irish computistical textbooks.[487] The reception of this concept, however, appears to have been rather limited, with the Angers glosses being based on the Munich computist's discussion, while *Comp. Col.*'s account of this question seems, interestingly enough, to be a synthesis of all three Irish textbooks.[488]

Even more characteristic of seventh-century Insular computistics are the comparisons of the Victorian with the Dionysiac system on various levels, which are, on the one hand, a direct result of attempts to understand the Dionysiac method, but which also orginated in a transitional period in which the two reckonings were tested for their practicability, accuracy, and theological foundation in order to determine the method to be followed. In the Irish context, such debates naturally took place in the southern rather than in the northern part of the *regiones Scottorum*, since the northern clergy followed the Victorian practice at no point in time. Discussions of the differences between these two systems must have been equally lively in Anglo-Saxon England in the decade before and the decades after the Synod of Whitby of AD 664; yet, no text recording the technical dimension of these debates survives from Anglo-Saxon England; Bede, for his part, only stands at the end of this process of comparing the two reckonings, ardently rejecting Victorian principles. From roughly the

485 *CE* p. 109. The Easter table starting from AD 798 can be found in: Vatican, Biblioteca Apostolica, Reg. Lat. 1260, fol. 112r–114r (note that this MS also contains *DRC*); Vatican, Biblioteca Apostolica, Pal. Lat. 1447, fol. 19v–22r; Vatican, Biblioteca Apostolica, Pal. Lat. 1448, fol. 13v–16r; see Appendix 8, which includes a facsimile. The Irish computistical influence in St Dunstan's classbook is impressively highlighted by the fact that the *only* computistica incorporated in this MS are the mentioned Easter table and two chapters of *DRC* (on the *luna abortiva*; cf. p. LXX–LXXI, CLXV–CLXVI): Oxford, Bodleian Library, Auct. F 4 32, fol. 21r, 22r–v.

486 St Gall, Stiftsbibliothek, 250, p. 80; London, British Library, Harley 3017, fol. 3r; Paris, Bibliothèque Nationale, Nouvelle acquisition latine 1615, fol. 152v (defective from the eighth entry in the column referring to 1 January); Paris, Bibliothèque Nationale, Lat. 7530, fol. 252v; Milan, Biblioteca Ambrosiana, H 150 inf, fol. 7v (as part of a larger table; published in BC *praefatio* (*PL* 129, 1282), but not in parallel columns).

487 MC 51; *CE* p. 110–111; *DRC* 68.

488 Angers, Bibliothèque Municipale, 477 (461), fol. 76v; *Comp. Col.* 4.2B–C.

last decade of the seventh century, the question whether the Dionysiac or the Victorian system should be followed became a major issue for the Frankish clergy due to contacts with Anglo-Saxon missionaries like Willibrord and Boniface, both strong supporters of the Dionysiac method.[489] When analyzing this question, those Frankish computists turned to Irish texts that outlined some comparative methods. After the gradual transition from the Victorian to the Dionysiac system had taken place in Francia – the last stronghold of the Victorian reckoning – by the end of the eighth century, the Victorian system finally lost its computistical (but not necessarily chronological) currency almost throughout western Europe.

This historical context explains the lively debates about the differences between the Victorian and the Dionysiac reckoning in seventh-century (southern) Ireland, and the reception of this analysis in Francia of the eighth century, but not beyond. Accordingly, the most comprehensive comparisons of the Dionysiac with the Victorian system can be found in the *Computus Einsidlensis* and the Munich Computus.[490] Among these comparative elements and methods, the Munich computist's justification of the difference between 'Greek' and 'Latin' tradition in the lunar age of Christ's passion and resurrection by means of a very peculiar alignement of the Victorian and the Dionysiac epacts received some reflection in *Comp. Col.* only.[491] Slightly more popular was the general idea of comparing the Julian calendar dates of the Easter full moons of the two reckonings; the Munich computist only recorded the result of this comparison (according to both the historical and the unhistorical method), without paralleling the 19 dates of each reckoning in detail;[492] such a correlation can later be found (for the historical method) as part of a larger (and to the present day not fully understood) table in Frankish texts and manuscripts, no doubt influenced by Irish computistical thought.[493] The most central aspect in the debates about

489 The Easter table followed by Willibrord, whose missionary work on the Frankish border to Frisia started in AD 690, was Dionysiac; cf. note 242 above; for Willibrord's Dionysiac training see especially my forthcoming article in the proceedings of the second conference on the science of computus; Boniface, a West-Saxon contemporary of Bede, probably knew at least the latter's early work on the reckoning of time, *De temporibus* of 703; towards the end of his life, ca. 746–754, Boniface asked in three letters addressed to the archbishop of York and the abbot of Bede's monastery Wearmouth and Jarrow explicitly for works of Bede, so that he may subsequently have received Bede's *De temporum ratione* of AD 725 (*Epistola ad Ecberthum archiepiscopum* (746–747) (*MGH Epp. sel.* 1, 158 (Nr. 75)); *Epistola ad Huetberhtum abbatem* (746–747) (*MGH Epp. sel.* 1, 159 (Nr. 76)); *Epistola ad Ecberthum archiepiscopum* (747–754) (*MGH Epp. sel.* 1, 207 (Nr. 91)); cf. Schüling, 'Handbibliothek des Bonifatius', 318–9); as a student of Willibrord's Boniface surely followed Dionysiac customs.

490 Cf. notes 184, 239, 252 above.

491 MC 52; *Comp. Col.* 5.4–5. Cf. p. CI–CII; CLIII–CLV, CLXXXI.

492 MC 55.46–52.

493 Cf. *Lib. comp.* 2.22, where four of the six columns record Julian calendar dates for Easter full moons (the remaining two list the Dionysiac epacts of 22 March and the year in the *cyclus lunaris*); these four columns are listed separately in Berlin, Staatsbibliothek, Phil-

the correct method of calculating Easter, however, were the lunar limits for Easter Sunday because of their theological implications. It does not surprise, therefore, to find the Victorian and the Dionysiac lunar limits (and, in most cases, also those of the *latercus*) discussed not only in all three Irish computistical textbooks, but also in the Frankish *Dial. Neustr.* and *Comp. Col.*, while lengthy (and often polemic) discussions of them can be found in Bede's *De temporum ratione* and Ceolfrith's letter to the Pictish king Nechtan of AD 710.[494]

Turning to purely Dionysiac theory, it will be remembered that the 19-year cycle underlying the Dionysiac reckoning was not explained in all its details by Dionysius himself, nor by Isidore, nor by any other pre-mid-seventh-century commentator on Easter computations. Accordingly, seventh-century computists had no authoritative instructions concerning the placement of the embolisms and of the *saltus lunae*. This led to the development of regional customs, since different computistical centres arrived at differing conclusions. Not even regional uniformity was achieved concerning the placement of the seven embolisms, which vary greatly in the known seventh- and eighth-century computistical texts, calendars, and still unedited tracts and tables;[495] only from the early-ninth century, when Bede's *De temporum ratione* had established itself as the standard computistical textbook, was uniformity in this question achieved by the general acceptance of Bedan placements. Yet, in the seventh century, the analysis of the lunations of the 19-year cycle led Irish computists to formulate a specific concept. Realizing that certain lunations fall completely between the Calends (i.e. the first day) of two successive Julian calendar months, they termed those lunations *lunae abortivae*.[496] The reason for this terminology is that such a lunation did, strictly speaking, not belong to any month, since lunations were commonly referred to according to the Julian calendar month in which they ended; consequently, the lunation ending on the Calends received the name of that Julian calendar month, while the subsequent lunation starting on the second day of that month and ending before the beginning of the following Julian calendar month had no reference month. Thus, such a lunation was simply called *luna abortiva*. This phenomenon is first described in a passage of the Sirmond manuscript datable to AD 658.[497] It is also referred to in the Munich Computus, without, however, the use of the terminus technicus itself.[498]

lipps 1831, fol. 128r (a facsimile of this table can be found in Appendix 3) and (according to Jones, *Bedae opera*, 24) Paris, Bibliothèque Nationale, Lat. 7296, fol. 111v.

494 MC 58.2–7; *CE* p. 119; *DRC* 99; *Comp. Col.* 5.3B, all four texts including the *latercus* data of these lunar limits, as does *HE* 5.21 (Ceolfrith's letter to Nechtan) and *DTR* 59; only the Dionysiac and the Victorian part of Bede's *DTR* account, then, is cited in *Lib. calc.* 93, and similarly *Dial. Neustr.* 13 also only refers to those two.

495 Cf. the commentary to MC 61.17–27 and note 111.

496 For the *luna abortiva* cf. p. LXX–LXXI above.

497 Oxford, Bodleian Library, Bodley 309, fol. 106v–107r; the relevant passage is transcribed in Ó Cróinín, *Cummian's letter*, 178–9; idem, 'Computistical works', 55; idem, 'Bede's Irish Computus', 206–7.

498 MC 59.44–47, 59.59–62.

The authoritative account of this concept was then composed by the author of *De ratione conputandi.*[499] His two chapters on the *luna abortiva*, namely a definition of the general concept and an outline of all *lunae abortivae* in the Dionysiac 19-year cycle, enjoyed some popularity, since they were not only copied into St Dunstan's classbook, but also reworked into a separate treatise on this technical feature at some stage in the ninth century.[500] Interestingly enough, the *luna abortiva* also features in a discussion of the first (or final) year of the Dionysiac 19-year cycle in an eleventh-century manuscript now in Zürich, which is generally based on seventh- and eighth-century Irish computistical thought.[501]

Concerning the *saltus*, which probably was the most discussed feature of the 19-year cycle, Irish computists, by ca. AD 700 at the latest, appear to have invented a division of its 24 hours by the 235 lunations of the 19-year cycle which proved to be very popular in subsequent centuries. The calculation itself is outlined in most detail in the two earliest Irish textbooks, the *Computus Einsidlensis* and the Munich Computus; only one later text, *Comp. Col.*, gives an equally detailed account of this division, which is directly based on the Munich text.[502] Early in the ninth century, Dicuil presents a more concise mode of this calculation, while employing the exact same terminology as his fellow countrymen more than a hundred years earlier.[503] This terminology (though without calculation) can also be found in *De ratione conputandi* and the Angers glosses,[504] whereas many later eighth- and ninth-century Frankish texts (and doubtlessly many more manuscripts) refer to the same results in different terms.[505]

Even more distinctive is the placement of the Dionysiac *saltus*. Until roughly the middle of the ninth century, two principal placements were followed, namely either at the end of the November or of the March lunation.[506] Since the *saltus* was implemented in the first year of the Dionysiac 19-year cycle (with the years starting on *luna* 15 of the Easter lunation), these placements coincided with 24 November and 22 March respecively. Now, Dicuil argues that this difference stems from regional preferences, attributing the November

499 *DRC* 70, 73.

500 For St Dunstan's classbook cf. Oxford, Bodleian Library, Auct. F 4 32, fol. 22r–v; the ninth century tract is edited in Appendix 6; note also that the definition of the term *luna abortiva* is formulated differently in the ninth-century manuscript Oxford, Bodleian Library, Digby 63, fol. 33v (cf. note 180 above).

501 Zürich, Zentralbibliothek, Car C 180, fol. 104v–106v; the whole manuscript is in desperate need of a thorough study of all its Irish computistical elements.

502 MC 62.14–61; *CE* p. 122–3 (both account in parallel on p. CXXXVI–CXXXVIII above); *Comp. Col.* 4.10B; for *Comp. Col.*'s dependency on the Munich Computus in this instance cf. p. CII above and p. CLXXXIII below.

503 *LDA* 4.3.2.

504 *DRC* 108 (cf. also p. CXCVII–CXCVIII below); Angers, Bibliothèque Municipale, 477 (461), fol. 46v.

505 BC 137; *De saltu lunae* I (*PL* 101, 986–8); *De saltu lunae* II (cited in *Lect. comp.* 7.4); *De saltu lunae* III–IV; *Lib. comp.* 4.7; *Lib. calc.* 48; RM 57.24–29; some glosses to *DTR* (e.g. Paris, Bibliothèque Nationale, Nouvelle acquisition latine 1632, fol. 13v).

506 For the different customs cf. the commentary under MC 62.98–111.

placement to Anglo-Saxon, the March one to Irish computists.[507] In fact, Dicuil's statement is fully supported by the surviving evidence: In the *Computus Einsidlensis* and *De ratione conputandi*, the March placement is the only one explicitly mentioned for the Dionysiac reckoning.[508] The Munich computist lists this placement among others, but his calculations reveal that 22 March certainly was the placement favoured by him.[509] Moreover, 22 March is implicit as the placement of the *saltus* in the Irish *Dial. Langob.*[510] Interestingly enough, this March placement is also mentioned in some, mostly Irish influenced eighth- and early ninth-century Frankish texts; yet, it appears not to be the one preferred by these computists.[511] Accordingly, the placement of the Dionysiac *saltus* on 22 March is a distinctive feature of Irish computistics.

This March placement of the Dionysiac *saltus* had more far-reaching consequences for calendrical calculations, since it fell within the Lenten period. Seventh-century Irish computists had developed some basic rules for the interdependency between the lunar age of Easter Sunday and that of the *initium quadragesimae*, i.e. the beginning of the Lenten fast. Since the Lenten fast began exactly six weeks (i.e. 42 days) before Easter Sunday, and since the lunation ending in this period, namely the March lunation, had 30 days, the lunar day difference between Easter Sunday and the *initium quadragesimae* was 12 days (42−30=12); e.g., if Easter Sunday fell on *luna* 15, the *initium* occurred on *luna* 3. If, however, the *saltus* is applied at the end of the March lunation, then the March lunation had only 29 days every 19th year, resulting in a lunar day difference between Easter Sunday and the *initium* of 13 instead of 12 days in those cases (42−29=13). Additionally, Irish computists used to place the lunar bissextile day on the same date as the intercalated Julian calendar day, 24 February, whithout, however, ever outlining this practice in detail. Accordingly, this extra Julian calendar day, and with it the extra lunar day, could also fall in the Lenten period. In these instances, the extra lunar day fell in the March lunation, increasing it from 30 to 31 days, which consequently led to a decrease of the lunar day difference between Easter Sunday and the *initium quadragesimae* by 1, from 12 to 11 lunar days (42−31=11). Precisely for the reason of avoiding this interference with the Lenten period, Bede not only placed the *saltus* outside of this six-week interval (in November rather than March), but also the lunar bissextile day, which he added to the end of the February lunation (increasing it

507 *LDA* 1.5.1–2, quoted p. LXXIV above.

508 *CE* p. 123–4; *DRC* 112; note that *DRC* differs formally from *CE* and the Munich Computus by placing the *saltus* on 21 March (ascribing *luna* 28 and 29 to that date, i.e. 'leaping' over *luna* 28 to *luna* 29) rather than 22 March (which would contain *luna* 29 and *luna* 30); this formal difference is due to the fact that the author of *DRC* had the idea of connecting the *saltus* with the vernal equinox as well as the first day of creation.

509 MC 62.68–72 and cf. the cross-references there.

510 *Dial. Langob.* 25.

511 *Dial. Neustr.* 25A; *De saltu lunae* I (*PL* 101, 985–6, 988–9); *Vers. Tur.* 13.31; *Comp. Col.* 4.10. Only in *De saltu lunae* III does 22 March appear to be the advocated placement of the Dionysiac *saltus*.

from 29 to 30 lunar days).[512] Consequently, with the reception of Bede's *De temporum ratione* the problem of the interference with the Lenten period of the lunar bissextile day and the *saltus* could have been removed from the computist's agenda. Yet, since the placements of the *saltus* and of the lunar bissextile day were questions of principles and of the correct interpretation of authoritative texts rather than purely of practicality, the seminal Irish discussions of this issue had some influence even after the reception of Bedan computistica, but the more so, of course, before it. The earliest and most detailed accounts can be found in the three Irish computistical textbooks. Note, however, that a change of perspective occurred between the *Computus Einsidlensis* and the Munich Computus on the one hand, and *De ratione conputandi* on the other, since the former approached this problem from Easter Sunday, the latter from the *initium quadragesimae*.[513] In the eighth century, similar discussions were included in the Irish *Dial. Langob.*, the Irish influenced Bobbio Computus, and the unfinished Fulda Computus of AD 789, as well as in still unedited passages and tracts.[514]

Turning to the reception of the Munich Computus in individual texts, it is the merit of Arno Borst to have uncovered the fact that some of the eighth-century Frankish texts on computistics are directly influenced by the Munich textbook. Borst highlights the importance of the Munich Computus for the study of computistics in general, and of eighth-century Frankish computistica in particular, by citing at the beginning of the introduction to his corpus the first few lines of the Munich text (dealing with the philosophical question about the nature of time). In the note to this quote, Borst then argues that at least three Frankish texts were directly influenced by the Munich Computus: *Dial. Burg.* of AD 727, *Dial. Neustr.* of AD 737, and *Quaest. Langob.*, which Borst later, in his introduction to this texts, dates to ca. AD 780.[515]

This picture is fundamentally broadenend and further specified by Borst in his introductions to the individual texts, which will be discussed here in the order in which they appear in Borst's *Schriften*. In line with his previous statement, he argues that the Munich Computus was one of the main sources for *Dial. Burg.* of AD 727,[516] while he attests a lesser influence on *Dial. Neustr.* of

512 For Bede's treatment of the *saltus* cf. note 111 above, for his treatment of the lunar bissextile day cf. notes 111, 390 above.

513 MC 58.8–32; *CE* p. 119–122; *DRC* 100–102. *DRC*'s account should be regarded as a revision of the two earlier discussions; for *DRC* being composed later that the other two textbooks cf. p. CXCI–CCI below.

514 *Dial. Langob.* 24B; BC 50; for the relevant passage in the unfinished Fulda computus of AD 789 (c. 31) see Basel, Universitätsbibliothek, F III 15k, fol. 44v–45r (cited in Paris, Bibliothèque Nationale, Nouvelle acquisition latine 1615, fol. 150r–v and, as part of a computus of AD 825/6, in Bern, Burgerbibliothek, 417, fol. 14v–15r); this problem is also discussed in Vatican, Biblioteca Apostolica, Reg. Lat. 1260, fol. 121r; cf. commentary to MC 58.8–32; cf. also note 183.

515 Borst, *Schriften*, 1. For the dating of *Quaest. Langob.* cf. Borst, *Schriften*, 509–10.

516 Borst, *Schriften*, 348.

AD 737.[517] A more prominent role is granted to the Munich Computus, and quite rightly so, concerning the composition of *Dial. Langob.*, even though this text does not appear in Borst's original list.[518]

As regards *Quaest. Austr.*, Borst believes that its author knew the Munich text from 'hearsay',[519] an interesting suggestion that quite neatly characterizes the relation between the two texts. In fact, only one passage of *Quaest. Austr.* may be proven to be directly based on the Munich Computus.[520] Still, the influence is clearly visible in the concepts transmitted: Certain themes, like the order of the months according to various traditions (Hebrew, Egyptian, Macedonian, and Roman), and more particularly the question why heathen terminology was adopted for the names of the months, are central to both text, while they are usually not included in contemporary computistical literature;[521] likewise, both computists apply the calculation of 4×365 (i.e. a bissextile interval of four years multiplied by the number of days per year), though in slightly different contexts;[522] and, in the discussions of the Victorian 19-year cycle, both authors point out that a certain lunation (the March lunation in the eleventh year) streches over three Julian calendar months (January, February, and March).[523] All of these aspects (except for the discussion of the order of months) are unique to these two texts, so that a certain influence of the Munich Computus on *Quaest. Austr.* (not necessarily direct, but equally likely through teaching) can hardly be doubted; more generally, *Quaest. Austr.* is heavily influenced by Irish computistical thought, far more than its editor admits.[524]

517 Borst, *Schriften*, 376 argues that the author of *Dial. Neustr.* used the Munich Computus only 'here and there'.

518 Borst, *Schriften*, 424.

519 Borst, *Schriften*, 474.

520 An Augustinian quote concerning the bissextile day shows the same variations (the addition of *Romani* and the omission of *enim*) in both the Munich Computus and *Quaest. Austr.*, but in no other text to my knowledge. It seems, however, more likely that both authors worked from a similar copy of Augustine's *De Trinitate*. Cf. Augustine, *De Trinitate* 4.4 (*CCSL* 50, p. 171; cited in *CE* p. 107; *DRC* 56.6–7; *DTR* 39.41–43 (in turn cited in *Lib. calc.* 42); *De bissexto* I (*PL* 101, 995); *Comp. Col.* 3.1C, which reads *poni* instead of *intercalari*): *quattuor enim quadrantes faciunt unum diem quem necesse est intercalari excurso quadriennio quod bissextum vocant ne temporum ordo turbetur.* MC 41.72–75: *Augustinus in epistula de resurrectione sic ait: Quadrantes IIII unum diem faciunt, quem necesse est inponi cursui quadriennio, quem Romani bissextum uocant, ne ordo temporum confusus esset. Quaest. Austr.* 2.9B: *Agustinus de bissexto ait: Quattuor quadrantes faciunt unum diem, quem necesse est intercalari ex cursu quadriennio, quod Romani bissextum vocant, ne temporum ordo turbetur.*

521 MC 24.19–31; *Quaest. Austr.* 1.1A,C, 1.2C, 2.1A,C, 2.2C.

522 MC 41.107–108; *Quaest. Austr.* 2.11B.

523 MC 50.78–80; *Quaest. Austr.* 2.9D.

524 In his introduction to *Quaest. Austr.*, Borst, *Schriften*, 462 only mentions the Bobbio Computus (*Comp. Graec.* in his notation) as having had an impact on this work, but none of the other Irish or Irish influenced texts, not even the Munich Computus; the Irish influence is particularly obvious in the weekday calculations (*Quaest. Austr.* 2.7; cf. *CE* p. 97; MC 29.2–12; *DRC* 35) and in the discussion of the consequences theoretically caused by a ne-

More controversial is Borst's inclusion of *Quaest. Langob.* in his original list of texts certainly influenced by the Munich Computus.[525] In the subsequent edition of *Quaest. Langob.*, Borst is more specific by stating that its author appears to have had a rather quick look through the Munich Computus, from which he copied very little.[526] When, however, *Quaest. Langob.* is analyzed in detail, not a single passage appears to be directly based on the Munich Computus.[527] Rather unconvincing is a random note in the introduction to *Comp. Col.*, suggesting that the Munich Computus left traces in *Dial. Neustr.*, *Quaest. Austr.*, and *Lect. comp.*[528] The first two texts have already been discussed in this context: *Dial. Neustr.* certainly relied on the Munich Computus, while some influence of the Munich Computus on *Quaest. Austr.* can be made out, but hardly a direct dependency. As for *Lect. comp.*, Borst says nothing about a possible dependency on the Munich Computus in his introduction to this text;[529] in the edition proper, the Munich Computus is mentioned only once as a primary source, and in this case not for a passage of the main text, but for an addition of AD 805. The passage in question is concerned with the placements of the embolisms, correlating the months in which they occur with the epacts of 1 January of the years in question. Such a correlation certainly derives from Irish tradition; *Lect. comp.*'s discussion, however, appears to be based on the *Computus Einsidlensis* (which, it must be noted, was not known to Borst) rather than the Munich text.[530] Accordingly, no passage of *Lect. comp.* nor any of its later additions derive from the Munich Computus.

Finally, the one text in the corpus of Frankish computistica edited by Borst that is most heavily influenced by the Munich Computus is *Comp. Col.* Borst

glect of the bissextile day (*Quaest. Austr.* 2.9D; cf. *CE* p. 106–7; MC 41.54–71; the entire section on the bissextile day is strongly influenced by Irish computistica).

525 One wonders, now, whether *Quaest. Langob.* is a typographical error for *Quaest. Austr.* or, more likely, *Dial. Langob.*?

526 Borst, *Schriften*, 509, 516.

527 Borst, *Schriften*, 518 indicates only one passage as being directly dependent on the Munich Computus, namely *Quaest. Langob.* 2A; this passage, however, may rather derive from *Dial. Langob.* 12A, since that text was frequently used by the author of *Quaest. Langob.* (cf., e.g., *Quaest. Langob.* 3C: *Sub idolo a Iunone sorrore vel coniunge Iovis fuisse testantur.* with *Dial. Langob.* 10A: *Iunius a Iunone, sorrore Iovis et coniunge.*; Borst, *Schriften*, 509 argues that *Dial. Langob.* was certainly known to the author of *Quaest. Langob.*, but in the subsequent edition he does not indicate a single passage of *Quaest. Langob.* as being based on *Dial. Langob.*).

528 Borst, *Schriften*, 885.

529 Borst, *Schriften*, 527–8.

530 A correlation between the placements of the embolisms and the epacts of 1 January of the respective years can be found in MC 61.17–27; *CE* p. 118; *Dial. Langob.* 23; and *Lect. comp.* 6.8a; all of these passages are quoted at length and can be compared in the commentary to MC 61.17–27, which reveals the textual closeness between *CE* and *Lect. comp.* in this instance. The *Computus Einsidlensis* was discovered in the monastery of Einsiedeln early in January 2006 (even though the publication date of the note announcing the discovery (Warntjes, 'Computus Einsidlensis') is 2005) and personally presented to Arno Borst in Konstanz only two days later; unfortunately, at this stage Borst had almost finished the corrections to the proofs of his monumental work.

believes that the author of *Comp. Col.* became aware of *Dial. Neustr.*'s and *Quaest. Austr.*'s dependency on the Munich text, and for that reason acquired a copy of it.[531] On the whole, however, Borst underestimates the Munich Computus' influence on *Comp. Col.*[532]

In the following, every text directly influenced by the Munich Computus is discussed separately, outlining this dependency in detail and analyzing its implications. The analysis starts with the texts of Borst's corpus, before turning to the neglected Angers glosses, the previously unknown Harvard fragment, while finally discussing the most controversial relationship in this context, namely the dependency of *De ratione conputandi* on the Munich Computus.

Dial. Burg.: *Dial. Burg.* is the earliest known computistical textbook written in Francia.[533] The final chapter explicitly refers to AD 727 as the year of composition, a fact that is confirmed by the initial year of the Easter table preceding this dating clause.[534] Based on the fact that *Dial. Burg.* is heavily influenced by Irish computistical thought, Borst very plausibly suggests that it was written in a Frankish centre with strong Irish connections, presumably in Luxeuil or any of its filiations.[535] Contrary to its Irish sources, however, *Dial. Burg.* promotes the Victorian rather than the Dionysiac reckoning. Among the Irish texts used by the author of *Dial. Burg.*, the Munich Computus certainly holds a prominent place, and direct influence can be proven by the following selection of the most striking textual parallels:

Munich Computus	***Dial. Burg.***
8.12–13: *emera in Greca a luce, emerat enim lux dicitur; dies Latine a diuidendo lucem a tenebris.*	11A: *Dies in graeca emira dicitur, sed in latino dies appellatur.*
9.33–34: *Hinc sanctus Hieronimus ait: Nox initium diei, non finis preteriti.*[536]	11C: *Et nox post diem, quod Hieronimus dicit: Nox initium diei sequentis, non finis praeteriti.*
11.37–40: *Quis primus dies factus est in mundi principio creaturarum? Primus dominicus.*	5A: *Quis primus dies fuit? Hoc est: Dominicus.*

531 Borst, *Schriften*, 885, 896.

532 Cf. the discussion of *Comp. Col.*'s dependency on the Munich Computus p. CII, CLXXIX–CLXXXIII.

533 *Dial. Burg.* has not received much attention in modern scholarship. It has been edited twice, by Krusch, *Studien* II, 53–7 and Borst, *Schriften*, 348–74. Unfortunately, Krusch's edition did not lead to an increased interest in this text, so that only sporadic references to it can be found in 20th-century computistical scholarship; cf. note 43. It can only be hoped that Borst's new edition will prove more successful.

534 *Dial. Burg.* 16–17.

535 Borst, *Schriften*, 218, 348–9. Krusch does not discuss the provenance of this text.

536 The only other computistical text known to me that cites this phrase from Jerome as the authority in the question of the order of day and night is *DDT* 8. Jerome's original phrase shows considerable variations: Jerome, *Commentarii in Ionam* 2.1 (*CCSL* 76, 394): *nam et in Genesi nox non praecedentis diei est sed sequentis, id est principium futuri, non finis praeteriti.*

24.2–11: *Primus mensis apud Hebreos Martius, quia in eo mundus factus est, et primum phase factum est, conceptio et passio et resurrectio Christi.* [...] *Apud Latinos uero primus mensis Ianuarius et caput anni pro natiuitate Christi in VIII Kalendas eius et circumcisione in Kalendas eius; et in VIII Kalendas eius tenebrę minuantur et lux augetur. Apud Grecos initium á Septimbrio, id est VII Kalendas Septimbris.*	4A–B: *Ideo Martius primus, quia in eo creatus est mundus, et prima Pascha facta est, et conceptio Christi ac passio eius et resurrectio eius fuit. Cur Ianuarius aput Latinos primus mensis fuit et caput anni legitur? Id est: Pro nativitate Christi in VIII Kalendas Ianuarii, et pro circoncisione in Kalendas ipsius, et quod in eo baptizatus est, et in ipso mense a magis adoratus est. Et in VIII Kalendas ipsius diminuuntur tenebrae et augetur lux.*
24.40–41: *Quot sunt principales dies de nominibus dierum mensium? Id est III: Kalende, Nonae et Idus.*	7B: *Quot dicunt principales partes in menses? Tres: Kalende, Nonas, Idus.*
46.2–5: *Luna dicitur quasi Lucina, ablata media syllaba. De qua Virgilius ait: casta faue Lucina. Luna autem per diriuationem a solis lumine nomen assumpsit. Deinde dicitur: Luminare maius et luminare minus.*	10B: *Luna a quadam dea nominatur, vel a lumine. Dicitur enim: Luminare maius et luminare minus.*

Accordingly, the Munich Computus was certainly read and discussed in some part of Francia less than ten years after its composition,[537] with the least controversial passages (i.e. those not immediately concerned with the methods of calculating Easter) leaving direct traces in *Dial. Burg.*, the earliest Frankish textbook on computistics.

Dial. Neustr.: Quite fortunately for modern scholars, the author of *Dial. Neustr.* mentions his *annus praesens*, referring to it as the 737th year after the incarnation of Christ, the year in which Theuderic, king of the Franks, died.[538] This passage is remarkable and revealing in various respects: First, it is the earliest known Frankish dating according to incarnation years; second, and more interesting for the present purpose, it appears to have been written in a computistical centre that showed at least some loyalty to the Frankish king (or rather his *major domus*, Charles Martell, who had installed him as a puppet king). Whether this latter fact necessarily points to Neustria as the provenance of this text, as Borst's title suggests, though very plausible as it is, must remain speculative (it should be remembered that the dating by regnal years was a long-standing custom in Francia, at least from the time of Gregory of Tours, so that this reference may not be indicative of any region).[539] Equally speculative is Borst's opinion

537 For the implications of this fact on the transmission of the Munich Computus cf. p. CIV above.

538 *Dial. Neustr.* 30C. For this dating clause cf. also p. LXXIV above.

539 For Borst's localisation of this text in Neustria cf. his *Schriften*, 375. Theuderic's death is not recorded in any chronicle or annal; in fact, it is only noted in *Dial. Neustr.*, while Theuderic himself is otherwise only mentioned in diplomata (cf. Breysig, *Jahrbücher*, 79–80); almost all of these diplomata signed by Theuderic belong to the Merovingian heartland, the region north-east of Paris right into modern-day Belgium (the exceptions are the

of a possible connection of this text with the circle of Boniface.[540] In that case, one would, in fact, expect a strong influence at least of Bede's *De temporibus* of AD 703 on *Dial. Neustr.*[541] Yet, not a single passage is based on this work (and obviously neither on his later *De temporum ratione*), so that only the use of AD reckoning would suggest a rather vague Bedan (or rather Anglo-Saxon) influence. The reception of this work, for its part, clearly points to the Cologne area as the provenance of *Dial. Neustr.*[542]

On the other hand, as has already been outline above, *Dial. Neustr.* depends heavily on Irish computistical thought, with certain concepts discussed in *Dial. Neustr.* being transmitted in only one pre-*Dial. Neustr.* text, the Munich Computus.[543] It must be kept in mind, however, that from the early period (seventh and early eighth centuries) more computistical texts appear to have been lost than that they would have survived, either because they soon lost their currency, or because the manuscripts transmitting them were lost or destroyed in subsequent centuries. Therefore, direct dependency can only in few cases be proven by the transmission of the same concept; the most significant indicator of direct dependency, then, is textual parallelism. In this respect, only very few passages (but then very) clearly reveal that the author of *Dial. Neustr.* read and used the Munich Computus for the composition of his text, particularly since the following textual parallels are not shared by any other known text:

Munich Computus	***Dial. Neustr.***
38.17–22: *Ab XI Kalendas Ianuarii in XII Kalendas Aprilis, in quo spatio sunt dies XC. Ab XI Kalendas Aprilis in XII Kalendas Iulii, ubi XCI dies apparent. Ab XI Kalendas Iulii in XII Kalendas Octimbris, in quo dies XCII sunt. Ab XI Kalendas Octimbris in XII Kalendas Ianuarii, ubi sunt dies XCII.*	29B: *Quantos dies <habet> ab XI Kalendas Ianuarii usque ad XII Kalendas Aprilis? Nonaginta. Et ab XI Kalendas Aprilis usque ad XII Kalendas Iulii quantos dies habet? Nonaginta unum. Et ab XI Kalendas Iulii usque ad XII Kalendas Octobris quot sunt dies? Nonaginta duo. Ab XI autem Kalendas Octobris usque ad XII Kalendas Ianuarii? Similiter nonaginta duo.*

diplomata for Le Mans; cf. *Diplomata Merovingica*, nr. 179–189 (*MGH DD Mer.* 1, 444–73)). The connection between Teuderic and the Merovingian heart-land, especially the north-eastern vicinity of Paris, is further accentuated by the facts that Theuderic was educated in Chelles and may have been buried in St Denis (cf. Ewig, *Merowinger*, 204–6). These facts would suggest a monastery in the Merovingian heart-land as the most likely candidate for the place of composition of this text; the reception of *Dial. Neustr.*, however, points slightly further east, to Cologne or its vicinities (cf. p. CI above). Consequently, the most likely place of composition of *Dial. Neustr.*, in my opinion, appears to be the eastern Merovingian heart-land close to Cologne (if not Cologne itself).

540 Borst, *Schriften*, 375. Cf. p. CIII–CIV above.

541 Cf. note 489.

542 Cf. note 539 and p. CIV above.

543 Cf. p. CIII above.

41.17: *Quis primus bissextum inuenit?*	27aB: *Quis primus invenit bissextum?*[544]
41.35–37: *Plenius est annus sine bissexto. Bissextus plenus est semetipse, ut mulier concipiens primo unum corpus, postea II corpora plena.*	28aC: *Cui simile est incrementum bissexti et ortus eius? Id: Mulieri concipienti in utero et parturienti.*

In summary, *Dial. Neustr.*, probably composed in the Cologne area in AD 737, heavily depends on seventh- and early eighth-century Irish computistical thought, while the textual parallels with the Munich Computus suggest that the author of *Dial. Neustr.* studied this text and copied (though hardly ever verbatim) many of its concepts, comparing and conflating those with discussions found in other (primarily Irish) texts.

Dial. Langob.: *Dial. Langob.* is even closer to the Irish computistical school of thought than the two Frankish texts just discussed. Several concepts and methods that are otherwise *only* attested in the three known Irish computistical textbooks of the late seventh and early eighth centuries (the *Computus Einsidlensis*, the Munich Computus, and *De ratione conputandi*) can also be found in *Dial. Langob.*: Particularly striking is *Dial. Langob.*'s treatment of weekday calculation, which includes a) the same categorisation of the Julian calendar months based on the number of nones occurring before and the number of calends occurring after the ides of a month,[545] and b) the same algorithms and calendrical observations as found in these three textbooks.[546]

It appears that this fact, among others, led its editor, Arno Borst, to the suggestion that *Dial. Langob.* originated in an Irish centre on the Continent, namely Bobbio; moreover, in absence of a dating clause, the historical context convinced Borst to date its composition to ca. AD 750.[547] Since this most interesting text has escaped the attention of historians before Borst introduced it to modern scholarship with his *editio princeps*, no second opinion on Borst's suggestion concerning the provenance and dating of *Dial. Langob.* exists.[548] A recent article has placed this textbook even stronger within Irish tradition, confirming Borst's supposition that the author was an Irishman.[549] But where and

544 Note, however, that the two texts differ in the answer to this question: The Munich computist argues that the bissextile day was an invention of the legendary Gignus, king of the Sabins, while *Dial. Neustr.* states (based on *DRC* 52, and therefore ultimately on Macrobius' *Saturnalia*) that its invention was due to Julius Caesar.

545 *Dial. Langob.* 12A; MC 28.2–5; *CE* p. 96; *DRC* 31.

546 *Dial. Langob.* 12B, 16A; MC 29.13–24; *CE* p. 98; *DRC* 34–35. Cf. p. CXXXVIII–CXL, CLX–CLXI above.

547 Borst, *Schriften*, 424–6.

548 *Dial. Langob.* was first mentioned in some detail in Valentin Rose's catalogue of Berlin manuscripts (Rose, *Handschriften-Verzeichnisse*, 284–5); however, Rose remained silent about provenance and date, and his catalogue description did not trigger further interest in or research of this text. Most strikingly, Krusch, 'Lehrbuch', 233–4 does not even mention *Dial. Langob.* when summarizing the content of the manuscript in which this text, as well as *Dial. Neustr.* (the topic of Krusch's article), are transmitted. Cf. Borst, *Schriften*, 423–4.

549 Bisagni & Warntjes, 'Latin and Old Irish', 7–8, 26–7 n. 86.

when did he compose this text? Central to this question is the evidence of the Munich Computus, as this text is securely dateable and the interdependency between the Munich text and *Dial. Langob.* is apparent not only in the use and transmission of identical methods and concepts, but also in identical phrasing. The textual parallels are outlined in the following:

Munich Computus	***Dial. Langob.***
1.2–3: *Interrogatio: Tempus, quid est? Responsio: Tempus est spatium tendens de principio usque in finem.*	1B: *Interrogatio: Quid est tempus? Responsio: Augustinus dicit: Tempus proprie dicitur spatium extendens ab initio usque ad finem.*
1.4: *Divisiones temporum quot sunt?*	3A: *Divisiones temporum quot sunt* [...] *?*
12.18–19: *Nomina mensuum IIII rebus assumpta sunt: A diis, a regibus, a numeris, a rebus.*	10A: *Interrogatio: Et menses anni, a quibus rebus nomina sumpserunt? Responsio: Quattuor sunt, a quibus nomina acceperunt, hoc est a diis, a rebus, a regibus, a numeris.*[550]
34.2–3: *Auctumnus ab augmento frugum dictus est. Vel Grece copiositas.*	6C: *Interrogatio: Quare autumnus dicitur? Responsio: Autumnus dicitur graece, copiositas frugum interpretatur. Vel latinum <est>, et de augmento frugum nomen accepit.*[551]
38.2–4: *Quater in IIII dies, id est solstitium hiemalem et aequinoctium uernale, solstitium aestiuale et aequinoctium auctumnale, <annus> diuiditur.*	17A–B: *Sed ex his diebus quattuor sunt principales, inter quos omnes dies anni dividuntur in quattuor partes.* [...] *Interrogatio: Quare et qua causa dividitur annus inter hos quattuor dies? Responsio: Quia in his quattuor diebus sunt duo aequinoctia et duo solstitia anni.*[552]
38.5–6: *Solstitium dicitur quasi solis statio, quia sole stante dies uel nox crescit.*	17C: *Solstitium dicitur quasi solis statio, quia sole stante dies vel nox incipiunt crescere.*[553]
38.7–8: *Aequinoctium uero dicitur, quia tunc dies et nox aequali horarum spatio consistunt.*	17B: *Aequinoctium dicitur, quia tam dies et nox aequali spatio horarum consistunt.*[554]
43.2–3: *Seculum secus colendo dictum, uel sex et colo, quia sex aetates mundi continet.*	2A, 3B: *saeculum tempus est, quod sex aetates in se copulat, id est custodiat.* [...] *Saeculum autem dicitur eo, quod copulat in se sex aetates mundi.*

550 In this passage, *Dial. Langob.* additionally relies on *CE* p. 93: *Quibus modis menses vocabula sumpserunt? Hoc est a quattuor: a diis et regibus, a rebus et numeris.*

551 Cf. also the parallel passage in *CE* p. 95: *Autumpnus ab autumento, hoc est ab augmento fructuum.* [...] *Aliter Greca sunt hęc nomina:* [...] *Autumpnus ergo copiositas fructuum* [...] *interpretatur.*

552 Note that the two equinoxes and solstices are also termed *principales dies* in MC 38.38–39.

553 Cf. Isidore's original citations: *Etym.* 5.34.1: *Solsticium dictum quasi solis statio, quod tunc sole stante crescant dies vel noctes. DNR* 8.2.7: *Solsticium autem dicitur quasi solis statio.*

554 Cf. Isidore's original statement: *Etym.* 5.34.1: *Aequinoctium appellatum quod tunc dies et nox horarum spatio aequali consistunt.*

53.14–15: *Annum autem solis Gignus rex Sabinorum in XII menses sub XII lunas ordinauit.*	13A: *Interrogatio: Quis primus divisit annum in menses? Responsio: Apud antiquos Latinos Rucingus rex Sabinorum primus divisit annum in decem menses.*

Additionally, the Munich Computus and *Dial. Langob.* are the only texts outlining the theories that the solar months were modelled on the lunar ones and that each month in the ten-month year ascribed to Romulus consisted of 36 days, so that they are clearly dependent in these concepts:

Munich Computus	***Dial. Langob.***
12.20–23: *Sciendum nobis est unde menses solis dicuntur. Id est a mensibus lunae, qui prius fuerunt ab incensione lunae usque ad extinctionem eius. Inde sub hos menses solis formati sunt.*	9B: *Interrogatio: Quare autem duodecim sunt menses, et plus aut minus non sunt? Responsio: Dicunt sapientes, quod ideo duodecim sunt propter convenientiam inter solem et lunam, quia sicut duodecim menses sunt in anno lunari, hoc est duodecim lunae, ita etiam et in anno solari duodecim menses oportuit conputari.*
12.28–31: *Romani uero auctore Romulo X mensium computabant.* [...] *Quoniam uero X, XXXVI dies et VI horae decima pars anni erat.*	13A: *Ita et Romolus Romae distinxit in decem, et unusquisque mensis triginta et sex diebus conputabatur.*

Now, did *Dial. Langob.* depend on the Munich Computus, or was the latter rather influenced by the former? Borst's main argument for dating *Dial. Langob.* later than the Munich text is *Dial. Langob.*'s final chapter, which deals with the indiction. Generally, the indiction played no central part in Irish computistics. The *Computus Einsidlensis* does not refer to the indiction at all, while it is mentioned only twice in the Munich Computus, and both instances can be shown to have been glosses originally that crept into the text at a later copying stage.[555] Consequently, the original version of the Munich Computus, like the *Computus Einsidlensis*, was not in the least concerned with the indiction, so that the only reference to this 15-year cycle in Irish computistical textbooks occurs in *De ratione conputandi*'s discussion of the columns of the Dionysiac Easter table.[556] In this respect, Borst is quite right in arguing that a text including a detailed discussion of the indiction would have been written in northern Italy rather than in Ireland. Yet, contrary to other captions occurring in this text, the heading *De indictione* is clearly separated from the end of the preceding passage in the only surviving manuscript witness;[557] moreover, even though the entire text is structured in dialogue (question and answer) form, this final paragraph on the indic-

555 MC 1.7, 1.14 and cf. notes there.

556 *DRC* 103.

557 Cf. the sole manuscript witness of this text, Berlin, Staatsbibliothek, Phillipps 1831, fol. 134r.

tion is not;[558] and finally, the text as a whole before the introduction of the indiction has a consistent structure, while a discussion of the indiction at the end appears more as a break in than a continuation of the narrative.[559] Accordingly, it seems reasonable to suggest that the final chapter on the indiction is a later addition.[560]

Concerning the rest (i.e. the original part) of *Dial. Langob.*, a few observations are important for the context of this text. The text begins with an introduction to the concept of time (c. 1–2), followed by a discussions of the various divisions of time (c. 3–4) and of the year with its seasons in particular (which were not listed among the 12 divisions of time) (c. 5–8), leading to a detailed analysis of the months and an outline of the Julian calendar (c. 9–16); in the context of the calendar, the division of a Julian calendar year into four parts and the increase and decrease of daylight are analyzed (c. 17–19); finally, the impact on calendrical and lunar calculations of the three artificially implemented technical devices, namely the bissextile day, the embolisms, and the *saltus lunae*, form the end of the text proper (c. 20–25). The discussion of the embolisms clearly reveals that the Dionysiac reckoning is the one followed. Neither alternative reckonings (and, in this respect, the absence of any discussion of the calculation of Easter proper is quite striking, leading to the conclusion that *Dial. Langob.* was not intended to present a full-scale computistical textbook[561]), nor alternative placements of the bissextile day and the *saltus* are discussed (the ones explicitly or implicitly favoured are the Irish ones, 24 February and 22 March respectively), contrary to the practice in other Irish computistical texts. From this perspective it appears that *Dial. Langob.* was written at a stage when discussions about alternative placements of technical devices and, more importantly, alternative methods of calculating Easter had not only ceased, but also appeared not even worth noting. Consequently, *Dial. Langob.* seems to have been composed later than the three Irish computistical textbooks, especially later than the datable Munich Computus, i.e. post-AD 719.

This impression is further confirmed by the two technical discussions of *Dial. Langob.* when compared to the Munich (and the Einsiedeln) computist's treatment of these questions: First, *Dial. Langob.* elegantly deals with the problem that the first year of the Dionysiac 19-year cycle consisted of only 353 (rather than 354) days due to the *saltus lunae* by arguing that the final day of the final year should, at the same time, be taken as the first day of the first year; the improvement compared to the Munich and the Einsiedeln texts lies in the fact that *Dial. Langob.* solves the problem by a double count (*geminatio*) of a Julian calendar day, rather than a division (*divisio*) of the day in question as done by the Einsiedeln computist, which is even, quite artificially, expressed in

558 *Dial. Langob.* 26.

559 For further details of this argument cf. notes 115, 324 above.

560 In this context, note especially that this chapter is edited as a separate treatise (*De indictione*) from a different manuscript (where it constitutes a commentary to Isidore's *Etymologiae*) by Arévalo (*PL* 82, 738–9). Cf. Borst, *Schriften*, 432, 459.

561 For this observation cf. p. CXI above.

terms of hours in the Munich Computus (though it is there already treated as a wrong approach).[562] Second, the impact of the *saltus lunae* and the lunar bissextile day on the lunar day difference between the beginning of Lent (*initium quadragesimae*) and Easter Sunday was one of the most prominent problems discussed by Irish computists in the late seventh, early eighth centuries. The three Irish textbooks on the subject discuss this question at great length; *Dial. Langob.*, on the other hand, only outlines the results of these analyses, clearly indicating that at the time of composition of *Dial. Langob.* the problem was well known and understood so that details could be neglected.[563]

Two aspects, however, appear to speak against this chronological order: First, *Dial. Langob.* lists only 12 divisions of time, not 14 as the Munich computist;[564] second, *Dial. Langob.* uses the Roman rather than the Greek dates for the equinoxes and solstices, and thus the, in terms of their application in the Latin West, apparently older custom.[565] This second feature can simply be explained by personal preference;[566] it should be remembered that *Dial. Langob.* avoids any discussions of the calculation of Easter, so that the Roman date of the spring equinox is mentioned only in the context of the division of a year, not in the more controversial Easter context. The first feature, on the other hand, can rather be explained by a preference in sources. Even though *Dial. Langob.* copied certain phrases from the Munich Computus, the main part of the passage in question is based on the *Computus Einsidlensis*.[567] Generally, Borst attests a strong influence of the Munich Computus on *Dial. Langob.*, a statement that is in need of some revision since the discovery of the *Computus Einsidlensis*; in fact, many passages indicated as deriving from the Munich Computus are more likely to be based on the *Computus Einsidlensis*.[568]

562 *Dial. Langob.* 20; *CE* p. 114; MC 59.79–90.

563 MC 58.8–32; *CE* p. 119–122; *DRC* 100–102; *Dial. Langob.* 24B–25; for this problem see also p. CLXVII–CLXVIII.

564 *Dial. Langob.* 3; MC 1.

565 The Latin equinoxes and solstices are outlined in *Dial. Langob.* 17, and they are applied in *Dial. Langob.* 19.

566 The reasoning for this preference of the Latin dates is given in *Dial. Langob.* 18, arguing that in the Greek custom only the spring equinox has a Biblical justification, namely the creation of the world, while in the Roman practise each individual solstice and equinox can be associated with a Biblical event. Note that the same preference can be found in *Quaest. Austr.* 2.8B–C.

567 For *Dial. Langob.* 3 cf. *CE* p. 85–6. Besides MC, Borst gives *DRC* as a source for this chapter of *Dial. Langob.*, but the passage indicated to derive from *DRC* is rather based on *CE*.

568 Borst, *Schriften*, 424. For the discovery of *CE* coming too late to be included in Borst's monumental work see note 530; one passage may suffice here as illustration that *Dial. Langob.* frequently cited *CE* rather than MC and *DRC* as argued in the edition: *Dial. Langob.* 12A: *Interrogatio: Quot sunt regulae, a quibus menses regulantur? Responsio: Sex sunt, id est duae ante Idus et quattuor post Idus. Duae regulae ante Idus haec sunt: Aut enim quartae Nonae sunt mensis ante Idus, ut Ianuarii, aut sextae Nonae fiunt ante Idus, ut est Martii. Quattuor vero regulae sunt mensium post Idus: Aut enim nonae decimae Kalendae fiunt post Idus, aut octavae decimae, vel septimae decimae, seu etiam sextae decimae fiunt.* certainly derives from *CE* p. 96: *Quot sunt regulę, quę superhabundant (!)*

In summary, then, *Dial. Langob.* appears to have used all three Irish computistical textbooks and therefore to have been composed later than those, particularly later than the datable Munich Computus (i.e. post-AD 719); its final chapter seems to have been a later addition (added, following Borst, in Bobbio ca. AD 750[569]), while the remaining part fits perfectly into the Irish context and may well have been written in the *regiones Scottorum* in the ca. AD 730s.

Comp. Col.: *Comp. Col.* certainly is the most curious of the texts edited among Borst's *Schriften*. In his introduction to this text, Borst highlights the importance of the sole codex transmitting *Comp. Col.*, namely Cologne, Dombibliothek, 83².[570] In the end, many of the texts incorporated in Borst's corpus can uniquely be found in this manuscript,[571] but according to its editor *Comp. Col.* holds a special place among these texts by presenting a summary of the entire codex, being composed in the same year as the codex itself, AD 805.[572] On a more general level, in the very late eighth and very early ninth century the Cologne school produced two outstanding computistical manuscripts within a few years, having the modern shelf-marks 83² and 103. The latter (Cologne, Dombibliothek, 103), presumably composed earlier than 83²,[573] is a col-

menses ante Idum et post Idum? Id est: duę ante Idum, IIII post Idum; aut enim quartas Nonas aut sextas ante Idum habent; post uero Idum aut XVIIII Kalendas, aut XVIII Kalendas, aut XVII Kalendas, aut XVI. rather than from *DRC* 31: *Sciendum nobis quot regulae sunt in mensibus ante Idus. Duae, id est aut enim menses quintanas Nonas habent aut septenas ante Idus. Post Idus vero quattuor. Sunt enim menses qui nonadecimas et alii qui octodecimas at alii qui septimdecimas et unus qui sexdecimas post Idus Kalendas habent.* or MC 28.3–6: *Quot regulis menses ante Idus mensurantur? Duobus: aut enim IIII Nonas, aut VI Nonas. Post, IIII regulis: aut XVIIII Kalendas, aut XVIII Kalendas, aut XVII, aut XVI Kalendas.* as argued in Borst's edition.

569 There cannot be any doubt that *Dial. Langob.* was well known in northern Italy by the early ninth century, since the only manuscript witness of this text (Berlin, Staatsbibliothek, Phillipps 1831, fol. 129r–134r) comes from Verona; moreover, Pacificus of Verona apparently used this text: Compare, e.g., the verbatim citations from *Dial. Langob.* 1B, 3 in §§ 176–7, 180–4 of Pacificus of Verona's *Computus.*

570 Borst, *Schriften*, 885. For descriptions of this manuscript see especially Jaffé & Wattenbach, *Coloniensis Codices*, 29–31; Krusch, *Studien* I, 195–205; Lowe, *Codices Latini Antiquiores* 8, 38 (No. 1154); Jones, *Script of Cologne*, 37–40; Jones, *Bedae pseudepigrapha*, 115; Bischoff, *Katalog* 1, 395; van Euw, 'Kompendium der Zeitrechung', 136–56; Borst, *Reichskalender*, 62–3; idem, *Schriften*, 236–8; Mc Carthy & Breen, *De ratione paschali*, 26; Englisch, *Zeiterfassung*, 11–6; Springsfeld, *Alkuins Einfluß*, 244–51.

571 These are, in chronological order, *Quaest. Austr.*, *Lect. comp.*, *Add. Col.*, *Comp. Col.*; cf. Borst, *Streit*, 82–3; idem, *Schriften*, 236–7. This most interesting codex is still not fully explored, as is illustrated by the fact that one of its texts, *De comparatione epactarum Dionysii et Victorii*, is discussed (p. CLII–CLVIII) and edited (Appendix 2) for the first time in the present book.

572 Borst, *Schriften*, 885. One of two dating clauses pointing to AD 805 for the composition of this codex is an integral part of *Comp. Col.*; see especially Krusch, *Studien* I, 198.

573 For this chronological relation between these two codices (103 being composed earlier than 83²) see especially Jones, *Script of Cologne*, 33, 40; van Euw, 'Beda Venerabilis', 129; idem, 'Kompendium der Zeitrechung', 136, 138; and furthermore Jones, *Bedae pseudepi-*

lection of Bedan computistica with few additions (including a calendar, a great Dionysiac Paschal cycle, and a copy of the *Computus Rhenanus* of AD 775);[574] the former appears to be designed as a supplement to this Bedan codex,[575] an anthology of texts presenting alternatives and / or additions to Bedan thought. In this respect, these two codices certainly have to be considered as one single great effort of the Cologne school to assemble in two codices all computistical material considered as important and fundamental at the turn from the eighth to the ninth century.

Certain texts that quite certainly formed part of the Cologne library at that time, like the Munich Computus, were apparently not considered worth being copied in full. Yet, certain ideas of these texts were felt important for the computistical discourse that took place in Francia during Charlemagne's reign. It was the purpose of *Comp. Col.*, then, to assemble all those ideas that would otherwise remain hidden in texts not included in these two codices, as well as to repeat essential ideas found in other tracts of 83^2 (which may, originally, not have been considered for incorporation into this codex). This purpose is particularly obvious in the inclusion of the so-called Cologne excerpts of the ancient and long outdated *Supputatio Romana* (the 84 (12)-year Easter reckoning followed in Rome in the fourth and fifth centuries).[576] From other texts, however, including the Munich Computus, the author of *Comp. Col.* copied ideas rather than full passages, expressing those concepts in his own, often hardly intelligible words and phrasings. In this respect, Irish computistical thought played a very prominent part in *Comp. Col.* in particular,[577] and, over a period of dec-

grapha, 115–6. The opposite is argued in Borst, *Reichskalender*, 62–5; idem, *Schriften*, 236–8.

574 Bede's early works, *De natura rerum* and *De temporibus*, can be found on fol. 23v–43v; his *Epistola ad Wicthedum* occurs on fol. 45r–52v, followed by his major work, *De temporum ratione*, on fol. 52v–184r. The calendar features on fol. 3r–8v, followed by the great Dionysiac Easter cycle on fol. 9r–22v; the *Computus Rhenanus* of AD 775 can be found towards the end of the codex, on fol. 184v–190v. For the MS cf. especially Jaffé & Wattenbach, *Coloniensis Codices*, 40–2; Jones, *Script of Cologne*, 32–3; van Euw, 'Beda Venerabilis', 129–35 (arguing for Cologne, ca. AD 795); Bischoff, *Katalog*, 397 (probably Cologne, saec. VIII./IX.); Jones, *Bedae pseudepigrapha*, 116 (Cologne, ca. AD 800); Springsfeld, *Alkuins Einfluß*, 82 (Cologne, AD 810–18); Borst, *Reichskalender*, 63–5; idem, *Schriften*, 238 (Borst argues in both accounts for Cologne, shortly after AD 810). For the *Computus Rhenanus* of AD 775, which is not identified as a separate text in any of the manuscript descriptions, cf. Warntjes, 'Argumenta', 75–6.

575 Cf. van Euw, 'Kompendium der Zeitrechung', 138.

576 *Comp. Col.* 6.2–3; these excerpts are discussed and edited in Krusch, *Studien* I, 39–42, 241–4.

577 When proof-reading Borst's *Schriften*, I realized that *Comp. Col.* was heavily influenced by Irish computistical thought. Since my comments on this text were too substantial to be included in the edition at proof-stage, Borst decided to mention in a note (*Schriften*, 890) that my interpretation of certain passages differed from his, adding that a future study of this text was to be expected. I may take the chance here to comment briefly only on the issues with immediate relevance to Irish computistics, hoping to publish a thorough commentary on this most interesting text in the near future: 2.3B deals with the division of the five days exceeding 360 in a Julian calendar year; these days are divided by the 90 days of

ades towards the end of the eighth century, in the Cologne school in general: The detailed division of the five solar days exceeding 360 in the course of a solar year by the 90 days of the quarter-years in which they occur,[578] as well as the designation of an eleven-day epactal period (the difference between a solar and a lunar year) within the Julian calendar year,[579] are both concepts found only in Irish texts and *Comp. Col.*; likewise, the discussion of the celestial location of the kindling of the moon relative to the sun is first found in Irish texts and then also features prominently in *Comp. Col.*[580]

Even though *Comp. Col.*'s direct dependency on the Munich text is perfectly obvious in terms of certain concepts, as these are almost exclusively found in these two texts (especially the comparison of the Victorian with the Dionysiac epacts in order to establish the *annus passionis* in the Dionysiac reckoning,[581] the comparison of the year of creation with the year of the Exodus

the quarter-year in which they occur; this idea is characteristically Irish (cf. *CE* p. 103–4; MC 40; *DRC* 50). Note 32: This passage deals with the concept of the annual bissextile increment of 3 hours (a day being reckoned as consisting of the 12 hours of daylight), which was supposedly due to a linear increase of 1⅓ *momenta* per day in the first quarter of the year, being reckoned as twilight when analyzing the linear increase and decrease of daylight; the use of the term *aurora* suggests that the source here was *Dial. Langob.* 19C. 4.1B: The ultimate source for the idea that the year of creation coincided with the second year of the Dionysiac *hendecas* (with the Easter full moon, the last day of this lunar year, falling on 25 March) is the Irish tract *De comparatione epactarum Dionysii et Victorii* (edition in Appendix 2 below, discussion p. CLII–CLVIII); cf. especially the sentence 1.11–12: *In secundo ogdoadis anno mundus coepit, quod XIIII luna primi mensis tunc occurente confirmatur*. 4.8A: This is possibly the most curious passage of the entire text; particularly interesting is the explicit mentioning of five *saltus* for the reckoning of Anatolius (?), though this reckoning had only one; it is tempting to assume, therefore, that these five dates referred to the *latercus* (mentioned right at the beginning of this list) rather than to Anatolius (?), particularly since the Munich computist lists five rather than the correct six *saltus* in the context of the *latercus* (MC 62.118–122); yet, the dates given in *Comp. Col.* do not agree with the lunar calendar of the *latercus*; still, it may be significant that the *saltus* are here placed in December, possibly pointing to a *latercus* practice of the full December lunation being reduced to a hollow one by the *saltus*. 4.10: The passage footnoted with note 74 deals with the division of 20 *momenta* by 235 lunations, which was an essential part of the division of the 24 hours of a *saltus* by the 235 lunations of a 19-year lunar cycle; this passage is directly based on MC 62.31–45. Note 89: The lunar limits for Easter Sunday of *luna* 14 to 20 are those of the *latercus*, and no other reckoning; is *Latini* here a scribal error for *laterci*? 5.5: One of the principal sources for this chapter is the text *De comparatione epactarum Dionysii et Victorii*, edited in Appendix 2.

578 *Comp. Col.* 2.3B; cf. *CE* p. 104; MC 40.10–20; *DRC* 50; see also the following page.

579 *Comp. Col.* 4.2B–C; cf. *CE* p. 110–1; MC 51; *DRC* 68; this concept is also found in the Angers glosses (Angers, Bibliothèque Municipale, 477 (461), fol. 76v), in this case directly based on MC.

580 *Comp. Col.* 5.2A; cf. MC 46.16–20; *DRC* 65; this question is also addressed in Irish influenced Bedan glosses (*PL* 90, 406) and *De saltu lunae* I (*PL* 101, 988); see also the following page and p. CLXII above.

581 MC 52; *Comp. Col.* 5.4–5. A different treatise in Cologne 83² (*De comparatione epactarum Dionysii et Victorii*, edited in Appendix 2 below and discussed on p. CLII–CLVIII above) is exclusively devoted to this dogmatic question; it appears to have been MC's source, and it may, in fact, have travelled to Cologne accompanying MC (cf. p. CI–CII above).

from Egypt and the year of Christ's resurrection,[582] as well as the discussion of the disruptions of the course of time recorded in the Bible,[583] but also the interest in the *latercus*[584]), a verbatim correspondence is less easy to detect. The most revealing passages are listed in the following:

Munich Computus	***Comp. Col.***
36.10–15: *Quis primus annus, in quo factus est mundus in mense Martio et die dominico? Ac in communi anno septimo ogdoadis, uel endecade, ut alii dicunt, id est in communi anno decimo, sine bissextu et saltu annus. Secundum Grecos mundus factus est XII Kalendas Aprilis in communi secundo ogdoadis.*	4.1A: *Embolismus primus annus mundi cum Latinis <est, post> communem secundum endecadi. Cum Graecis ogdoadus initium mundi est, communis secundus ogdoadi.*
40.2–3: *Sunt V dies, qui in tribus nonagintis superflui manent. Quos partiri in XCo, in quo sunt.*	2.3: *Item de sole dicendum, et de byssexto et quinque diebus superfluis in tribus nonagesimis, id est nonaginta uno, nonaginta duobus, nonaginta duobus. Quomodo ponendi sunt, si franguntur in suis nonaginta, in quibus dividuntur per annos artificaliter?*
41.80–88: *Qui quadrans hoc modo per totum annum diuiditur: In anno sunt XII menses XXXmi. V dies supersunt, qui habent horas CXX. Quas cum in decadas diuiseris, facient XII decies, ita ut unus mensis XXX dies et X horas habeat. Si item quadrantem, id est VI horas, diuiseris, inuenies XII semis, ita ut per omnem mensem, id est XXX dies et horas X, et XX momenta crescunt. Si minutius diuidas in omni die dimidium, et <bis> tertiam momenti partem repperies in hac preparatione.*	2.6A–B: *Item tertio intellectu hii quinque dies dividuntur per totum annum. Sic in anno sunt duodecim menses. Qui singuli duodecim menses secundum Aegyptios triginta dies possident. Quinque dies, qui superesse videntur, habent horas centum viginti. Quas cum in decadas diviseris, faciunt duodecim, ita ut unusquisque mensis triginta dies et decem horas habeat. Item si quadrantem, hoc est sex horas, dividi facias, invenies duodecim semisses, ita ut per omnem mensem, hoc est per triginta dies et decem horas, momenta viginti crescere videantur. Quodsi minutius dividas, in omni die sextam ferme momenti partem repperies.*
46.17–20: *Superincensio, id est super solem accenditur in breuitate diei; mediaincensio, id est communiter soli in aequalitate diei et noctis; subincensio, id est sub sole in longitudine diei.*	5.2A: *Superincensione super sole, in brevi? In diei media, comminus soli, aequa longitudine diei et noctis? Subincensione sub sole, in longiore die?*

582 MC 44.13–80; *Comp. Col.* 5.1–4. Note that the full moons of the first lunar month of these three years are also discussed in *Dial. Neustr.* 15.

583 MC 66; *Comp. Col.* 5.7, 6.1B; the first text to deal with this question is the southern Irish *De mirabilibus sacrae scripturae*, which was MC's source in this instance; for *De mirabilibus* and the dependency of the Munich text on this work cf. p. LXXVIII–LXXX above.

584 References to the *latercus* occur in *Comp. Col.* 4.8A (also mentioning St Patrick!) and 5.3B; cf. note 577 above.

62.17–19: *Quando dicis accensionem lunae, id est precederunt IIII momenta pro uelocitate accensionis lunę.*	4.9B: *Item interrogatur, si quattuor momenta semper in accensione praecedunt?*
62.41–42: *Tolle X quater de IIII LXtis, et remanent IIII L L L L, id est CC.*	4.10B (I cite the MS (Cologne, Dombibliothek, 83^2, fol. 50v) here rather than Borst's edition, since the notation of the numerals in the MS most clearly highlights the dependency of *Comp. Col.* on MC): *Tolle X de unoquoque LX, remanent IIII L L L L, id est CC.*
66.20–23: *Iterum ait: Sole et luna stantibus cum Iosue, uelociori cursu postea uicem more suo reddiderunt. Ita iuxta crucem tenebrae sole uelato cursum suum peragunt.*	5.7: *Non inmutat aliquid statio luminarium, quia aliqui dicunt: Duo luminaria in velociore cursu, postea autem morem reddiderunt et iterum concurre temporaliter.* [...] *Item tenebrae iuxta crucem non mutaverunt numerum, quia interim sol cursum suum currebat.*

Comp. Col., therefore, is an important witness to the fact that the Cologne school had, by the early ninth century at the latest, studied the Munich Computus, among other Irish computistica, with great interest. In fact, it appears that the Munich Computus was considered by the Cologne school as a central computistical text throughout the better part of the eighth century;[585] by the early ninth century, however, the Munich Computus had lost its currency as a textbook, particularly due to the outright acceptance of the Dionysiac reckoning by the Cologne school and the consequential preference for Bedan texts. Thus, only some of the more interesting concepts found in the Munich Computus were considered for the composition of *Comp. Col.*, while this textbook as a whole was not included in this Cologne codex designed to contain the most important non-Bedan computistical texts.

Angers glosses (Angers, Bibliothèque Municipale, 477 (461)): This manuscript was composed in the Breton monastery of Lendévennec in the late ninth century, presumably in AD 897, as a gloss on fol. 21r suggests.[586] Its principal texts are Bede's three scientific writings, his *De natura rerum*, *De temporibus*, and *De temporum ratione*, in this order.[587] All three texts are heavily glossed. Since some of these glosses are written in Old Breton, they have been of primary interest to Breton linguists, which led to a thorough analysis of their linguistic features, while their computistical content has almost completely been ne-

585 For the theory that the Munich Computus had reached Cologne by the mid AD 730s and for the state of the copy of this text kept in the Cologne library as deducible from the evidence of *Comp. Col.* cf. p. CI–CIV below.

586 For the provenance and date of this manuscript cf. especially Fleuriot, *Dictionnaire*, 9–11.

587 Fol. 10r–18v: *De natura rerum*; fol. 18v–22r: *De temporibus*; fol. 45–86: *De temporum ratione*. Note that Jones did not use (nor personally examine) this manuscript for his editions of Bedan computistica (cf. *CCSL* 123A, 174; *CCSL* 123B, 242–3).

glected.[588] In fact, Ó Cróinín is the only modern computist to have studied them in some detail, and it is his merit to have discovered that some of these glosses are closely related to Irish computistica;[589] a systematic analysis of this relation, however, has not been accomplished to the present day.

The glosses themselves can be divided into two sets, one (termed A, written in a dark black ink) corresponding to the late ninth-century main hand of the manuscript, the second (termed B, written in a brownish, at times hardly discernible ink) being added later, probably in the late tenth or early eleventh century.[590] Irish computistica occur exclusively among the A glosses, many of which are, in fact, direct quotations from the Munich Computus or *De ratione conputandi*.[591] It has to be noted, in this context, that the Angers glosses are the only known Bedan glosses that cite any of the Irish computistical textbooks directly (and many of these citations are not even shared by any other text, glosses or not). These facts, among others outlined above,[592] suggest that the glossing took place in Lendévennec or some other monastery in its vicinity, and did not travel any further in subsequent decades or centuries. Concerning the time of glossing, it appears very likely that these glosses were added to the Bedan texts immediately after their introduction into Brittany, i.e. some 150 years before the composition of the manuscript: With the introduction of more elegantly formulated and more practical texts in the course of the eighth and ninth centuries, not the least the Bedan texts themselves, the Irish textbooks lost their currency. Accordingly, it is hardly imaginable that a text like the Munich Computus, outdated by the ninth century, would be studied and memorized in such detail, as the Angers glossing suggests, by a ninth century computist. Thus, it appears that the Munich Computus and *De ratione conputandi* were known and studied in detail in Brittany before the introduction of Bedan computistica, and that their alternative definitions and theories (the sole basis for computistical teaching in the decades before the reception of Bede's works) were then added in the margins of Bede's texts immediately after these had become known and standard in that region. The Angers glosses, then, at least the ones transmitting Irish computistical theory, are likely to have been copied from earlier glossed copies of Bede's scientific works rather than to have been composed in the ninth century.

The direct citations from the Munich Computus, shared by no other text (if not stated otherwise), are the following:

588 For discussions and analyses of the Old Breton glosses found in this Angers MS see Fleuriot, *Dictionnaire*, 49–330; Lambert, 'Commentaires Celtique I', 119–39; idem, 'Commentaires Celtique II', 185–201; idem, 'Thirty', 29–43; note also the Anglo-Saxon gloss on fol. 96r (Ker, *Catalogue*, 578).

589 See especially the transcription of some of these glosses as comparative material to Irish computistica in Walsh & Ó Cróinín, *Cummian's letter*, 115–6, 118, 163.

590 Fleuriot, *Dictionnaire*, 8–9.

591 For direct quotation of *DRC* in the Angers glosses see note 307 above.

592 Cf. p. CV above.

Munich Computus	**Angers glosses**
6.2–7: *Hora nomen Latinum est ab ethimologia Greca deductum. Greci enim horan dicunt et umbra uel species interpretatur, a quo horologium et horam de horologio dicimus. Ora sine aspiratione circulus maris et uestimentorum dicitur. Inde poeta dicit: Pallium palliades ora circumdatum.*	fol. 46v (gloss to *DTR*): *Nomen latinum ab ethimologia greca dictum, id est horan, quod figura uel species interpretatur. Enim ora finis dicitur, ut finis temporis ora dicitur, ut fit horologium. Ora sine aspiratione finis maris uel uestimenta, id est circulus, ut pallii ora circulo datur.*
7.2: *Quadrans Latine, Grece autem dodras.*	fol. 49v (gloss to *DTR*): *Latine quadras, Grece dodras.*
7.5–6: *Quadrans autem secundum Ysidorum nomen numeri, quarta pars nummi.*	fol. 49v (gloss to *DTR*): *Nomen numeri, id est IIII pars numeri.*
8.2: *Dies a diuidendo lucem a tenebris dictus est.*	fol. 19r (gloss to *DT*): *<Dies> a diuidendo lucem a tenebris.*[593]
8.9–10: *Siue dictus est dies a diis, eo quod diis iocundus.*	fol. 19r (gloss to *DT*): *siue <dies> a diis, eo quod iocundus est eis.*[594]
11.2–3: *Feria a fando dicta est, uel a fiendo, dicente Domino: fiat lux et facta est lux.*	fol. 19r (gloss to *DT*): *feriae a fando dictae* [...] *uel feriae a fiendo dictae dicente domino: fiat lux.*[595]
11.9–11: *quę feriatus dies. Sabbatum dicitur, eo quod festiuitas in eo apud Hebreos agebatur.*	fol. 19r (gloss to *DT*): *Inde feriatum sabbatum dicitur, eo quod festiuitas cum Ebreis in eo agabatur.*
14.2–7: *Februarius <a febris> acris Lupercorum; ⌐ uel Lupercorum id est lignorum. ¬ Ut ait sanctus Hieronimus: Septem luperculis pauit inopem, quia omnia febricitant in eo, ut dicitur: nobiscum roborant, febricitant in eo. Vel Februarius a febribus, id est a lauacris, ut februm quodam habebant gentiles, id est lauacrum.*	fol. 19v (gloss to *DT*): *uel a febribus acris Lupercorum, id est lignorum, ut VII lupicinis satiauit (?) inopem. Vel februm Grece, Latine lauacrum, quia Luperci in eo se lauabant.*
31.11–12: *Alii annum dicunt ab innouatione, anath enim ab innouatione.*	fol. 19v (gloss to *DT*): *Alii annum dicunt ab innouatione, ana enim Grece innouatio dicitur.*
32: *Ver dicitur, eo quod uiret. Tunc enim post hiemem uestitur tellus herbis, et cuncta in florem rumpentur. Vel Grecum fructificatio et Latinę.*	fol. 20r (gloss to *DT*): *Ver, eo quod uiret. Tunc enim post hiemem uestitur (?) tellus in herbas cuncta rumpuntur. Vel uer Grece fructificatio.*[596]

593 *CE* p. 89 and *DRC* 22 give the same etymology.
594 *CE* p. 89 and *DRC* 22 give the same etymology.
595 *CE* p. 92 lists the same etymologies.
596 Cf. also *DRC* 42, 43.

33: *Aestas ab estu, id est a calore solis. Aestas quasi exusta, id est quasi usta et arida.*	fol. 20r (gloss to *DT*): *Aestas ab estu, id est a calore solis. Aestas quasi usta, id est exusta et arida.*[597]
34: *Auctumnus ab augmento frugum dictus est. Vel Grece copiositas. Siue auctumnus a tempestate uocatus quando folia arborum cadunt et omnia maturescunt.*	fol. 20r (gloss to *DT*): *Vel autumnus Grece copiositas. Vel ab agmento frugum. Autumnus a tempestate uenti uocatur, quando folia arborum cadunt et omnia maturescunt.*
37.3–5: *Gamse in Hebreo uel simpsia, elios in Greco, panath cum philosophis, foebe cum Syris, titan cum Chaldeis.*	fol. 13v (gloss to Bede's *DNR*): *Gamse in Ebreo uel simsia, elios in Greco, pinip apud philosophos, phebe cum Siris, titan cum Caldeis.*[598]
38.5–6: *Solstitium dicitur quasi solis statio, quia sole stante dies uel nox crescit.*	fol. 19v (gloss to *DT*): *Solstitium dicitur quasi solis statio, quia sole stante dies uel nox crescit.*[599]
41.17–20: *Quis primus bissextum inuenit? Grecorum Sabini, Gignus rex Sabinorum natura inuenit. Quid inuenit? Superfluum naturę lucis ingenitae. Per quod inuenitur? Per artem numerationis in signis ciuilibus.*	fol. 20r (gloss to *DT*): *Gingius rex Sabinorum natura inuenit primitus bissextum superfluum naturae lucis per artem numerationis et per signa ciuilia.*
41.21–24: *Apud calcentores asis uocatur, apud tractatores quadrans, cum Aegyptiis interkalatio, cum Grecis dies semedorus, id est XII horae bis, quia bissexti XXIIII horarum est.*	fol. 20r (gloss to *DT*): *Apud calculatores asis uocatur, apud tractatores quadras, apud Aegiptios calatio, apud Grecos dies simdoros, id est XII horae bissexti.*
41.32–37: *Bissextus duplex, in natura et in adnumeratione. In natura, id est in preparatione eius per momenta, id est momentum et tertia pars momenti et VIIII pars momenti. In adnumeratione, quando apparet dies. Plenius est annus sine bissexto. Bissextus plenus est semetipse, ut mulier concipiens primo unum corpus, postea II corpora plena.*	fol. 20r (gloss to *DT*): *Bissextus in duobus fit, in natura et in adnumeratione: in natura, quando per momenta preparatur, id est momentum et III pars momenti et IIII pars momenti; in adnumeratio, quando dies apparet. Numquid bissextus sine anno pleno et annus sine bissexto, utique altero, ut mulier concipiens prima unum corpus, et post conceptionem II corpora plena.*
42: *Ętas pro uno anno, pro VII annis et VI milibus dicitur. Et aetas quasi aeuitas, id est similitudo aeui. Nam aeuum aetas perpetua, cuius neque initium neque extremum noscitur. Quod Greci uocant aeuoas. Quod apud nos aliquando pro seculo, aliquando pro aeterno ponitur. Aetas aut hominis, ut infantia, aut mundi, ut aetas ab Adam usque ad Noe.*	fol. 10r (gloss to Bede's *DNR*): *Aetas pro uno anno vel pro VI annis et VI m accipitur. Aetas quasi aeuitas, id est similitudo aeui, nam euus aetas perpetua, cuius neque initium neque extremum noscitur. Quod Greci eonas uocant. Quod aliquando apud nos pro seculo, aliquando pro inifinito (!) ponitur. Aetas aut hominis aut mundi, ut prima aetas ab Adam usque Noe.*

597 The addition of *solis* is unique to MC and this Angers gloss; cf. the variations of this Isidorian quote listed in the edition (MC 33).

598 Cf. *DRC* 1; BC 116.

599 Note the variants, uniquely shared by MC and the Angers gloss, to the source, *Etym.* 5.34.1: *Solsticium dictum quasi solis statio, quod tunc sole stante crescant dies vel noctes.*

43.2–3: *Seculum secus colendo dictum, uel sex et colo, quia sex aetates mundi continet.*

fol. 10r (gloss to Bede's *DNR*): *Seculum sep\i/us colendo dictum, vel sex et colo, quia VI <aetates> mundi continet.*

51.3–12: *Et illo taliter ępactae decurrunt in XI Kalendas Aprilis: aepactas, id est adiectiones lunares, omnibus diebus mensis adsistimus, quae a IIII Idus Martii incipiunt et in XI Kalendas Aprilis finiri adsuescunt. Itaque uterque annus lunaris scilicet et solaris ab XI Kalendas Aprilis cursum incipit et in V Idus Martii parem gerīm extendit. In quo finito lunari anno XI dies de solari a IIII Idus Martii in XI Kalendas Aprilis supersunt. Quibus lunares totidem adieciuntur aetates, quae embolismum tertio anno prestant esse.*

fol. 76v (gloss to *DTR*): *Aepactę Grece adiectiones lunares, et plures in fine anni semper fiunt secundum solem, et incipiunt a IIII Idus Martii, licet naturaliter a II (recte V) Idus Martii, et finiunt in XI Kalendas Aprilis. Sciendum unde factę sunt ab initio mundi, quia facta est mundus in XI Kalendas Aprilis, et exinde incipit annus solis et lunae ab XI Kalendas Aprilis, sed in tribus primis diebus non luxit sol et luna. Ita et annus lunae V (?) Idus Martii finitur et restant XI dies de anno solaris a IIII Idus Martii usque in XI Kalendas Aprilis. Deinde adiciuntur (?) XI dies de anno lunaris ad suplementum incipienti harum XI dierum. Inde adiectiones lunares uero (?) necessitatę (?) sunt. Inde semper epactae fiunt in XI Kalendas Aprilis et hę (?) dies XI tertio anno praestant embolismum.*

The implications of this analysis are the following: Both the Munich Computus and *De ratione conputandi* appear to have been very closely studied in Brittany before the reception of Bede's computistical textbooks; in fact, for some decades in the eighth century, these two texts may even have been the standard computistical works in Breton schools. This observation further strengthens the argument that both texts, i.e. particularly the Munich Computus, were composed in southern Ireland,[600] which traditionally had close cultural links to Brittany. Moreover, it seems that the Munich Computus' route of transmission went through Britanny before continuing to monasteries further inland and further East.[601]

Harvard Fragment (Harvard, Houghton Library, MS Type 613, fol. 7r–v): The fragment in question is generally considered part of a set of binding fragments, which primarily contain passages of Isidore's *Etymologiae*. These fragments, 18 folios in total, had been found in the bindings of codices from St Emmeram in Regensburg (the monastery that produced the only surviving copy of the Munich Computus), but in modern times they were dispersed in three different libraries: Four folios survive in Munich, Bayerische Staatsbibliothek, Clm 29051 (b), another six in New York, G.A. Plimpton collection s. n., while the remaining 8 fo-

600 Cf. p. LXXVII–XCVI above.
601 Cf. p. CIV–CVI above.

lios used to be in Cheltenham, Phillipps collection, 20688.[602] 16 of these 18 folios are fragments of Isidore's *Etymologiae*; the last two folios of the eight formerly in Cheltenham (fol. 7 and 8), however, are of a different nature. The eight folios from Cheltenham were sold by Sotheby's on 30 November 1971, and were later donated to the Haughton Library of Harvard, where they received the shelf-mark MS Type 613. In Sotheby's catalogue, folios 7 and 8 are described as follows:[603] 'The last bifolium apparently contains portions of two different unidentified works, one dealing with the calendar (fol. 7; number of days in the months, etc.), the second of a theological nature (f. 8; quotations from St Jerome and St John, references to Gehenna).' It is this computistical folio 7 that concerns us here, which is here termed 'Harvard fragment' for easy referencing. Dáibhí Ó Cróinín, late in 2006, was the first to realize that part of the Harvard fragment deals with early medieval Irish computistics;[604] he passed this fragment on to me, and it transpired that a considerable part of this fragment shows otherwise unattested parallels to the Munich Computus. Some of these passages are slightly altered or amended by providing additional information; others, however, are verbatim copies of the Munich text. Most intriguingly, as has been just mentioned, the passages in question are unique to these two texts, not only in their phrasing, but also in content.

Munich Computus	**Harvard fragment**
49.14–18: *Unde Romani nec minus Scotti in Kalendis Ianuarii Grecorum obseruationes ępactas rimantur. Hinc ab VIII singulari in Kalendis Ianuarii ępactas incipimus: VIII, XX, I, XII, XXIII, IIII, XV, XXVI, VII, XVIII, XXVIIII, X, XXI, II, XIII, XXIIII, V, XVI, XXVII.*	fol. 7r: *<Apud> Grecos ępactę lunares in Kalendis Ianuarii: <VIII,> XX, I, XII, XXIII, IIII, XV, XXVI, VII, XVIII, <XXVIIII, X,> XXI, II, XIII, XXIIII, V, XVI, XXVII.*
50.67–76: *Apud latercum ępactae in Kalendis mensium: VIII in Kalendis Ianuarii; VIIII in Kalendis Februarii; VIII in Kalendis Martii; X Aprili; XI Maio; XII Iunio; XIII Iulio; XIIII Augusto; XV Septimbri; XVI Octimbri; XVII Nouimbri; XVIII Decimbri; XVIIII Ianuario.*	fol. 7r: *Aput latercum aepact<ae in Kalendis> Ianuarii (recte mensium[605]): VIII luna Ianuario; VIIII luna in Kalendis Februarii; VIIII (recte VIII) <luna in Kalendis Martii;> X luna in Kalendis Aprilis; XI luna in Kalendis Maii; XII luna in Kalendis <Iunii;> XIII luna in Kalendis Iulii; XIIII in Kalendis Agusti; XV luna in Kalendis <Septembris;> XVI luna in Kalendis Octembris; XVII luna in Kalendis Nouembris; XVIII luna <in Kalendis> Decembris; XVIIII luna in Kalendis Ianuarii.*

602 For this set of fragments see especially Lowe, *Codices Latini Antiquiores* 2, 8 (No. 144); Bischoff, *Schreibschulen*, 257.

603 Sotheby & Co, *Bibliotheca Phillippica* 6, 14–15.

604 Cf. p. XCIX–CI above.

605 The same mistake occurs in the MS containing the Munich Computus, which impressively reveals the direct dependency of the Harvard fragment on the Munich Computus; cf. *app. crit.* of the edition and p. C above.

Apud latercum lunae: Ianuario XXX luna; Februario XXVIIII luna; Martio XXVIIII luna; Aprili XXVIIII luna; Maio XXX luna; Iunio XXVIIII luna; Iulio XXX luna; Augusto XXX luna; Septimbri XXVIIII luna; Octimbri XXX luna; Nouimbri XXVIIII luna; Decimbri XXX luna.	*Quia aput latercum <Ianuarius XXXI> diem (recte dies) habet, lunae eius XXX; Februarius XXVIII dies habet, <lunae eius XXVIIII>; Martius XXXI diem (recte dies) habet, lunae eius XXVIIII; Aprilis XXX die<s habet, lunae eius> XXVIIII; Maius XXXI diem (recte dies) habet, lunae eius XXVIIII (recte XXX); Iunius <XXX dies habet, lunae eius> XXVIIII; Iulius XXXI habet, lunae eius XXX; Agustus XXXI di<es habet>, lunae eius XXVIIII (recte XXX); September XXX dies habet, lunae eius XX<VIIII; Octember> XXXI diem (recte dies) habet, lunae eius XXX; Nouember XXX dies h<abet, lunae eius> XXVIIII; December XXXI diem (recte dies) habet, lunae eius XXX.*
53.12–20: *Quid mensurauit annum lunȩ? Id est cursus lunae, id est XII lunae. Quis primus ordinauit annum lunae in XII lunas? Greci. Annum autem solis Gignus rex Sabinorum in XII menses sub XII lunas ordinauit.*	fol. 7v: *<Scien>dum, quid mensurat annum lunae. Id est lunae cur<sus, id est> XII lunae. Quis primus ordinauit annum in XII <lunas? Greci. An>num autem solis Gingius rex Sabinorum in XII menses <sub XII lun>as ordinauit.*
Ubi certus lunae cursus XVIIII annorum? Quicquid mensurauit numerum istum: XIIII luna primi mensis et plenitudo saltus; ut bissextus dominatur cursum solis, XXVIII annis usque bisȩxtus omnes septimanȩ dies circumierit.	*Sciendum est, ubi cursus certus <lunae XVII>II annorum. Quid mensurauit hunc numerum: lunȩ XIIII pri<mi mensis> et plenitudo saltus; ut bissextus dominatur cur<sum solis, XX>VIII annis usque bissextus omnes septimanȩ <dies circum>ierit.*
55.62–69: *Deinde XIIIIme lunȩ primi mensis hae sunt: Nonae Aprilis, VIII Kalendas Aprilis, Idus Aprilis, IIII Nonas Aprilis, XI Kalendas Aprilis, IIII Idus Aprilis, III Kalendas Aprilis, XIIII Kalendas Maii. Hucusque ogdoas. VII Idus Aprilis, VI Kalendas Aprilis, XVII Kalendas Maii, II Nonas Aprilis, VIIII Kalendas Aprilis, II Idus Aprilis, Kalendae Aprilis, XII Kalendas Aprilis, V Idus Aprilis, IIII Kalendas Aprilis, XV Kalendas Maii.*	fol. 7v: *Haec sunt XIIII lunae primi mensis: Nonae Aprilis, <VIII Kalendas Aprilis, I>dus Aprilis, IIII Nonas Aprilis, XI Kalendas Aprilis, IIII Idus Aprilis, III Kalendas Aprilis, XIIII Kalendas Maii. <Hucusque o>gdoas. VII Idus Aprilis, VI Kalendas Aprilis, XVII Kalendas Maii, II Nonas Aprilis, VIIII <Kalendas Aprilis, II Idus> Aprilis, Kalendae Aprilis, XII Kalendas Aprilis, V Idus Aprilis, IIII Kalendas Aprilis, XV Kalendas Maii. <Hucusque> endicas.*

Moreover, another, extremely fragmented passage at the end of the verso side of the Harvard fragment seems to be directly influenced by the Munich Computus. In this passage it is argued that in the Dionysiac reckoning, January and March have the same lunar age on the Calends as well as lunations of equal length, namely of 30 days, and that the same holds true for February and May, with the only difference being that their lunations are hollow rather than full. Now, a hollow May lunation presupposes the sequence of lunations uniquely attributed to the Dionysiac reckoning in the Munich Computus and the *Compu-*

tus Einsidlensis, and therefore it must be presumed that the passage of the Harvard fragment derives directly from the Munich text. The relevant passages are correlated in the following:

Munich Computus

50.3–39: *Per hunc ordinem epactarum, quę in Kalendis Ianuarii enumerauimus, per XII menses Grece menses et aetates in Kalendis mensium per XVIIII annos subtilis quisque intellegitur.*

VIII luna in Kalendis Ianuarii; VIIII in Kalendis Februarii; VIII luna in Kalendis Martii; saltu X in Kalendis Aprilis et Maii; transilit XI et XII in Kalendis Iunii; XIII in Kalendis Iulii; XIIII in Kalendis Augusti; transilit XV et XVI in Kalendis Septimbris et Octimbris; transilit XVII et est XVIII in Kalendis Nouimbris et Decimbris; transilit XVIIII, fit XX luna in Kalendis Ianuarii. Hucusque primus annus cum saltu.

Incipit II annus sine saltu: XX luna in Kalendis Ianuarii; XXI in Kalendis Februarii; XX in Kalendis Martii; XXI in Kalendis Aprilis et Maii; transilit XXII et XXIII in Iunio; XXIIII in Iulio; XXV in Augusto; transilit XXVI et fit XXVII in Septimbri et Octimbri; transilit XXVIII et est XXVIIII in Nouimbri et Decimbri; transilit XXX et est I in Kalendis Ianuarii. Hucusque secundus annus.

[...]

Qua auctoritate praefatas in Kalendis epactas mensium explora? Hoc est ex aetate lunae mensium. Extincta enim et accensa has praedictas explora epactas. Isto enim ordine lunae feruntur secundum Grecos: XXX luna in Ianuario; XXVIIII in Februario; XXX luna in Martio et Aprili; XXVIIII in Maio et Iunio; XXX in Iulio; XXVIIII Augusto; XXX in Septimbri; XXVIIII in Octimbri; XXX in Nouimbri; XXVIIII in Decimbri. Per has itaque lunae aetates quicumque obseruare uoluerit, ordinem euidenter repperiet et bissextum uigili sensu caueat, ne hanc obseruationem conturbet.

Harvard fragment

fol. 7v: *G<... men>sium apud Grecos: Ianuarius, <Martius unam> ętatem habent, lunas XXX. Februarius, Ma<ius unam ętatem> habent, nisi in anno saltus, lunas <XXVIIII.> Luna Maii XXVIIII; inde a Maio transilit* <...>

As for the provenance and date of the Harvard fragment, the facts alone that it used to be a binding fragment from St Emmeram in Regensburg and that it shows unique parallels to a text copied in that monastery (the Munich Computus) strongly suggest that the fragment itself was composed there; since the Harvard fragment preserves, in fact, better readings than the Regensburg copy of the Munich Computus, it appears that it was written earlier than this copy, presumably ca. AD 800, with its author working from the copy of the Munich Computus that had been brought to St Emmeram, the exemplar of the Regensburg copy.[606] Further studies of this fragment are needed, however, to be able to draw conclusions about the nature of the text from which this fragment originated, as well as this fragment's implications on computistical literature and teaching in St Emmeram ca. AD 800.[607]

De ratione conputandi: *De ratione conputandi* is one of three known Irish computistical textbooks from the post-Isidorian and pre-Bedan period, roughly to be dated to AD 650–750; the other two are the *Computus Einsidlensis* and the Munich Computus; a fourth text, *Dial. Langob.*, appears to be of Irish origin, but does not show a typical text-book structure, and should therefore rather be regarded as a treatise on specific questions of time reckoning.[608] Only one of these texts, the Munich Computus, contains a dating clause (for AD 719[609]) and can therefore be dated with certainty. Since all four texts are clearly related, the most promising method for establishing a chronological order among these appears to lie in the task of defining the relation to the only datable text of this group, the Munich Computus. This method has been applied to the *Computus Einsidlensis* and *Dial. Langob.* above, leading to the result that the *Computus Einsidlensis* was composed earlier than the Munich Computus, probably at some stage in the period AD 689–719, and that it was then, in AD 719, used for the composition of the latter;[610] *Dial. Langob.*, on the other hand, appears to be based on both the Munich and the Einsiedeln text, an observation that entails a post-AD 719 date of composition, presumably in the 730s.[611]

Concerning *De ratione conputandi*, its discoverer and editor, Dáibhí Ó Cróinín, has taken a different approach. In particular on the basis of a comparison between the sources used by Cummian in his famous letter to the Iona abbot Ségéne and those of *De ratione conputandi*, he argues that *De ratione conputandi* was probably written in the circle of Cummian, if not by Cummian himself, in ca. AD 650. Reassessing this argument and the more general context of *De ratione conputandi*, it has been argued above that geographical proximity between these two texts appears very plausible, while the similarity in sources

606 Cf. p. C–CI above.

607 An article on the Harvard fragment is currently being prepared by Dáibhí Ó Cróinín and myself.

608 For the Irish origin of *Dial. Langob.* cf. p. CLXXIV–CLXXIX above; for it not being a text-book in the strict sense cf. p. CIX above.

609 For the dating clause of the Munich Computus cf. p. LVIII–LXI above.

610 Cf. p. CXXXIII–CLII above and Bisagni & Warntjes, 'Early Old Irish material', 81–91.

611 Cf. p. CLXXIV–CLXXIX above.

does not necessarily suggest temporal closeness. This is to say that this text appears very likely to have been composed in the same (or a related) computistical centre as Cummian's letter, where both authors enjoyed a similar computistical training, used the same library, and therefore sometimes referred to the same sources; the context of *De ratione conputandi*, however, rather suggests it being composed several decades after AD 650.[612] This argument will here be further accentuated by the following comparison between *De ratione conputandi* and the Munich Computus.

First of all, the interdependency of the two texts is very obvious. This is, as will be seen below, more so the case in terms of the concepts discussed and methods applied than in verbatim correspondences. Nevertheless, the following examples of the most striking textual parallels do clearly reveal that the author of one of these two texts read the other:

Munich Computus	***De ratione conputandi***
1.4–8: *Diuisiones temporum quot sunt? Diuisiones temporum sunt XIIII: Atomos, momentum, minutum, punctum, hora, quadrans, dies, ebdoma, mensis, uicissitudo triformis, annus, seculum, aetas, mundus.*	13.1–3: *Sciendum nobis quot sunt diuisiones temporis. Quattuordecim, que sic nominantur: atomos, momentum, minutum, punctum, hora, quadras, dies, ebdoma, mensis, uicissitudo triformis, annus, aetas, seculum, mundus.*
2.2–7: *Atomos nomen Grecum est et interpretatur indiuisibile, Ysidoro dicente: Atomos philosophi dicunt quasdam in mundo partes minutissimas, ut uisui non pateant, nec sectionem recipiant. Huc illucque feruntur, sicut minutissimi pulueres, qui infusi per tenebras solis radiis uidentur.*	16.1–5: *Sciendum nobis quid sit atomos. Id est nomen Grecum, et indiuisum interpretatur. Quomodo difinitur? Isidorus dicit: Atomos philosophi dicunt quasdam in mundo minutissimas partes, ut uisui non pateant, nec sectionem recipiant. Huc illucque feruntur sicut minutissimi pulueres qui effusi per tenebras radiis solis uidentur.*[613]
20.3–6: *In hoc enim Cęsar Augustus in primum consulatum iniit et III triumphos de Aegyptis, de Macedonibus, de Parthis in urbem contulit finemque bellis ciuilibus inposuit.*	29.19–21: *Agustus autem ab Agusto Cesare nominatus est, quia in isto mense tres triumphos in urbem contraxit, id est de Antonio, uel de Parthis, Egyptiis, Macedonibus.*[614]
37: *Sol dictus est, eo quod solus luceat, uel a soliditate luminis. Gamse in Hebreo uel simpsia, elios in Greco, panath cum philosophis, foebe cum Syris, titan cum Chaldeis. Aut sol de soliditate luminis, ut Augustinus ait: Sol dicitur, quia solus per diem sui fulgoris ui*	1.1–7: *Sciendum nobis quomodo sol in principalibus linguis uocatur. Ita: Gamse uel Simsia apud Ebreos; Elios apud Grecos; Paniph apud philosophos; Phoebus apud Syros; Titan apud Caldeos; Sol apud Latinos. Sciendum nobis unde sol nomen accepit. Isidorus dicit:*

612 Cf. note 134 above.

613 Cf. the original Isidorian quote: *Etym.* 13.2.1: *Atomos philosophi vocant quasdam in mundo corporum partes tam minutissimas ut nec visui pateant nec τομήν, id est sectionem, recipiant; unde et ἄτομοι dicti sunt. Hi per inane totius mundi inrequietis motibus volitare et huc atque illuc ferri dicuntur, sicut tenuissimi pulveres qui infusi per fenestras radiis solis videntur.*

614 The specification of the three triumphs cannot be found in any other computistical text to my knowledge.

obtonsis cum luna stellis totum mundum inlustrat.	*Sol dictus est eo quod solus luceat,*[615] *uel soliditate luminis aut de solemnitate lucis. Unde Augustinus: Quia solus per diem sui fulgoris ui obtonsis cum luna stellis totum mundum inlustrat.*[616]
43: *Seculum secus colendo dictum, uel sex et colo, quia sex aetates mundi continet. Et pro infinito numero ponitur.*	60: *Sciendum nobis quid sit seculum. Nomen conpossitum, id est sex et copulum, uel cultus, eo quod sex mundi temporibus. Et seculum ad infinitatem dicitur.*
55.28–33: *Hieronimus ait: Cum Dominus, uerus agnus, et uerum pascha progreditur, in mundum aliqua permanere uolens custodiuit, alia non seruare cupiens mutauit. Nam XIIII luna umbralem relinquens et figuralem preuidens ueritatem per se expressam in dominicum diem cum luna transiuit.*[617]	98.22–26: *Item Hieronimus dicit: Cum dominus, uerus agnus, uerum pascha progreditur, in mundum aliqua permanere uolens custodiuit, aliqua non obseruare cupiens motauit. Nam XIIII lunam umbralem relinquens, et figuralem praeuidens ueritatem per sé expressam, in dominicam diem cum luna transiuit.*
59.28–35: *In communibus et embolismis ordo euentus est: per decennouenalem ciclum tertio embolismi occurrunt, nisi in ultimo anno ogdoadis et endecadis, hoc ordine: Communis, communis, embolismus, communis, communis, embolismus, communis, embolismus, ogdoas; communis, communis, embolismus, communis, communis, embolismus, communis, communis, embolismus, communis, embolismus, endecas.*	80: *Sciendum nobis quomodo lunares anni in decinnouenali ciclo currunt. Ita etiam, id est duo commones simul et tertio anno embolesmus, excepto fine ogduadis et endicadis, ubi precedit unus commonis embolesmus propter rationem paschalem. Nisi enim hii duo embolesmi fuissent anni, XII Kl Aprelis pascha inueniretur, quod non congrum fuit. Sic itaque currunt: Commonis, commonis, embolesmus, commonis, commonis, embolesmus, commonis, embolesmus. Hucusque ogduas. Item endicas: Commonis, commonis, embolesmus, commonis, commonis, embolesmus, commonis, commonis, embolesmus, commonis, embolesmus. Hucusque endicas.*
59.36–37: *Ogdoas Grece Latinae VIII, hoc est anni. Endecas Grece Latinę XI. Ena enim Grece Latine unum, deca X.*	81.1–3: *Sciendum nobis quod Greca sunt ogduas et endicas. Ogduas enim Grece octo interpretatur. Enna enim Grece unum Latine; deca uero Grece decim Latine.*

615 Cf. the sources for this phrase: Cassiodorus, *Expositio psalmorum, Expositio in psalmum 103*, §22 (*CCSL* 98, 935): *Sol enim dictus est, quod solus ita lucet, ut ex eo dies fiat. Etym.* 3.71.1: *Sol appellatus eo quod solus appareat, obscuritatis fulgore suo cunctis sideribus.*

616 Cf. the source for this sentence, which is not Augustine, but rather Virgilius Maro's *Epitomae* 11.49–50: *Sol dicitur, quia solus per diem sui fulgoris ui totum orbem obtunsis una stellis cum luna inlustrat.*

617 Cf. the ultimate sources for this passage, *De sollemnitatibus* 2 (Walker, *Sancti Columbani opera*, 200–1), which is quoted at length in the edition below.

62.2–9: *Saltus non libris hoc nomen inueniri quidam ferunt, sed qualitatem ei inponi tradunt. Tamen Augustinus ait: Saltus quasi saliendo quoddam signum, quia compotus transilit unum diem, ut ab VIII in Kalendis Ianuarii in XX et a XVIII in XXX in XI Kalendas Aprilis, transilit enim XXVIIII luna in XXX. Aut saltus dicitur non saliendo, sed geminando dicitur duas aetates uno die, ut VIII luna <Kalendae> Ianuarii et Martii.*	111: *Sciendum nobis cur saltus dicitur. Augustinus ostendit dicens: Saltus, quasi quoddam saliendi signum, quia conpotus transilit unum diem. Quamuis non hic saltus sed geminatio uideatur, et in numeri libris hoc nomen non habetur, sed tractatores pro qualitate dant ei.*
62.9–10: *Saltus euentus famosus in luna, cui nulla uis est secundum solem.*	107.7–8: *Item Augustinus dicit: Saltus est euentus famosus in luna, ut bissextus in sole habetur.*
62.17–19: *Quomodo fit? Quando dicis accensionem lunae, id est precederunt IIII momenta pro uelocitate accensionis lunę.*	109.1–3: *Sciendum nobis quomodo saltus exploratur in naturali lunae cursu. Hoc ita est: Quando dicis accensionem lunae precesserunt IIII momenta <secundum> opinionem nostram.*
62.112–117: *Nomina enim naturalia ei augmentum lunare, incrementum lunare; haec nomina, cum preparatur, ei conueniunt. Una luna, cum apparet. Adiectio lunae, cum addit aetatem, ut XXX facit in XI Kalendas Aprilis. Mutatio lunę, cum transilit aetatem, id est XXVIIII, in XXX.*	110: *Sciendum nobis quot nomina saltus habet. Sex videlicet, id est naturalia duo, id est augmentum lunare et incrementum lunare; et haec duo nomina dum preparatur sibi conueniunt. Quattuor uero in adnumeratione: luna una cum apparet, adiectio lunae cum addit aetatem, ut XXX facit in XI Kl Aprelis, motatio lune cum transilit XXVIIII in XXX. Et sic saltus dicitur, quo nomine nos utimur.*
63.28–29: *Ysidorus ait: Quicumque ante XV luna pascha celebrari iubet, transgreditur mandatum.*	98.21–2: *Et Isidorus dicit: Quicumque ante XV lunam pascha celebrari iubet, transgreditur mandatum domini.*[618]

Turning from verbatim correspondences to analogies in ideas, it has to be kept in mind that Irish schools of the late seventh and early eighth centuries appear to have had similar curriculas of computistics; accordingly, the same concepts and methods were taught in these schools, so that certain themes are recurring in all Irish computistical texts of that period, as well as later texts directly influenced by this Irish school of thought. Yet, there are some ideas that only occur in the Munich Computus and *De ratione conputandi* (as well as in direct copies of these two texts), a fact that impressively highlights the interdependency of these two textbooks. These ideas are, in particular: the assumption that an *annus magnus* (defined as the period taken by all planets to return to their initial position in the zodiac) consisted of 30 years (i.e. the time taken by Saturn, which is the longest among the moveable celestial bodies); the question

618 This citation cannot be found in any of Isidore's works, nor in any other computistical text to my knowledge; cf. p. CXXVIII.

of which parameters determine the length of the 19-year cycle; and the discussion of the origin of the term Pascha. Since the comparison of these passages also reveals considerable similarities in phrasing, they are listed in the following:

Munich Computus	***De ratione conputandi***
31.23–25: *Annus magnus omnibus planetis recurrentibus in eodem loco, qui fit post solstitiales annos plurimos, id est XXX.*	46.7–9: *Sciendum nobis quid sit annus magnus. Qui XXX annis finitur, et a stella Saturni moderatur, ut Isidorus dicit: Stella Saturni XXX annis fertur cursum suum implere.*
53.16–20: *Ubi certus lunae cursus XVIIII annorum? Quicquid mensurauit numerum istum: XIIII luna primi mensis et plenitudo saltus; ut bissextus dominatur cursum solis, XXVIII annis usque bisęxtus omnes septimanę dies circumierit.*	105: *Sciendum nobis quid constituit hunc numerum XVIIII annorum in lunari et in decinnouenali ciclo. Saltus uidelicet. Sicut bissextus XXVIII annorum ciclum solis mensurauit, omnes septimane dies in hoc spatio circumiens, sic saltus cursum lunaris XVIIII praebuit. Nisi enim esset saltus in finé, XX anno non esset reuersio. Cum uero saltus <in> XVIIII reuersio in epactis et XIIII luna primi mensis inuenitur, et in fine XVIIII annorum integra preparatio saltus inuenitur.*
54.9–13: *Phase in Hebreo, quia Hebrei non habent p; <Petrus et Paulus> Fetrus et Faulus dicuntur. Sed Grecus posuit pro tanta lenitate uocis suae et extremam syllabam per chi pro eadem tanta lenitate scripsit. Pascha uox uulgi. Inde commune est apud Grecos et Latinos.*	84.13–17: *Sed hoc nomen, quod fasse dicitur apud Ebreos, leuius sonatur, quod Ebrei 'p' litteram non habent, sed 'fhe' huius littere nobiscum uicem tenet 'p'. Et quaecumque nomina apud nos per 'p' scribuntur, apud Ebreos per 'fhe', ut Falistina, Faulus, Fasse, quod Greci ita dicunt: Palestina, Paulus, Pascha.*

In summary, then, there can hardly be any doubt that the Munich Computus and *De ratione conputandi* are directly related in the sense that they do not only derive from the same computistical background (if not the same computistical school), but that the author of one of them worked from the other. These two texts, however, should not be considered in isolation, especially since the *Computus Einsidlensis* shows equally close similarities to both texts, as has been noted in the discussion of the relation between the Einsiedeln and the Munich text above.[619] One passage in particular highlights the interdependency of the three Irish computistical textbooks, since it contains an obvious technical mistake shared by all three witnesses:[620] The lunar limits of the *initium quadragesimae* applicable to the *latercus* are incorrectly given as *luna* 2 to 8, based on a method of calculation that works for the Victorian and Dionysiac lunar calendar, but not for that of the *latercus*. This transmission of a common error

619 Cf. p. CXXXIII–CLII above.

620 Cf. the full discussion p. CXLVIII–CL above.

does not only clearly reveal the close interdependency of these three texts, but the differences in phrasing additionally suggest that this mistake originated with the Einsiedeln computist. Consequently, the *Computus Einsidlensis* has to be regarded as the oldest of these three Irish computistical textbooks and therefore needs to be taken into consideration when analyzing the chronological development and order of the theories and concepts outlined in the Munich Computus and *De ratione conputandi*. In fact, already the general observation that the Munich Computus shares far more features than *De ratione conputandi* with the *Computus Einsidlensis* (the oldest of the three texts) suggests that the Munich textbook is older than *De ratione conputandi*, with *De ratione conputandi* marking a later stage in the development of Irish computistics.

Before analyzing in detail the question of how certain concepts may have developed in order to determine the chronological order of the texts transmitting these concepts, a few more words on methodology are necessary. It seems possible and appropriate to formulate general rules applicable to texts of technical content (especially to those originating in the same school, in the same intellectual milieu), which have already been applied rather silently in the analyses above, but which become absolutely crucial in the following discussion: 1) If a certain technical concept is described in two dependent texts, the more sophisticated and more precise description generally tends to be the more recent one, since the way mathematical (and other scientific) proofs generally develop is by finding any working solution first, which is then rendered more concise and precise in subsequent reworkings of this proof.[621] 2) On the other hand, certain ideas, concepts, and algorithms first formulated in an older text may have become common knowledge since, so that a more recent text would, in these cases, tend to abandon detailed explanations. 3) Moreover, a more recent text would tend to reject or silently abandon outdated theories and methods that proved false or unworkable after they had been first formulated in an older text.

On the basis of these rules the relation between the Munich Computus and *De ratione conputandi* is analyzed in the following under constant cross-reference to the *Computus Einsidlensis*, since not only do most methods and concepts discussed here also occur in the Einsiedeln text, but, in fact, this textbook will prove to stand at the beginning of their development.

Concerning the first aspect, the development of a mathematical proof and concept from a loosely formulated, sometimes experimental origin to a well-defined and accurately explained later stage, one passage in particular quite clearly reveal that *De ratione conputandi* was compiled later than the Munich Computus. First, the Einsiedeln and the Munich computists struggled to find a proof for the cyclic character of the 19-year cycle; both relate to the same earlier attempts of proving the *ogdoas* (the first eight years of the 19-year cycle) and the *hendecas* (the final eleven years) each to be true cyclic, before continuing to outline a lengthy and rather simplistic, but nevertheless correct description of the 19-year cycle and its cyclic character by analyzing the relation to the Julian calendar year of the initial and final days of each lunar year of this cy-

621 For this see also p. CLIX, CLXXVII–CLXXVIII above.

cle.[622] This treatment of the question stands in sharp contrast to *De ratione conputandi*'s account, which ignores earlier unsuccessful attempts, as well as the successful, but rather simplistic approach of the Einsiedeln and the Munich computists. In a very concise (only 12 lines in the printed edition compared to four and a half manuscript pages in the *Computus Einsidlensis*), mathematically precise and for its time very sophisticated way, the author of *De ratione conputandi* calculates the total difference in days between 19 common lunar and 19 solar years (209 solar days), while continuing to prove that this number equals the number of lunar days provided by the seven embolisms minus the *saltus lunae* (7×30–1=209).[623] If either the Einsiedeln or the Munich computist had known this calculation, they would certainly not have applied their clumsy methods, nor would they have related to misguided earlier attempts of proving the cyclic character of the 19-year cycle. In this instance, the difference between *De ratione conputandi* on the on hand, and the *Computus Einsidlensis* and Munich Computus on the other, can only be explained by the Einsiedeln and the Munich Computus standing at the beginning, *De ratione conputandi* at a later stage of a scientific development.

The second aspect mentioned above, the neglect of further explanations after a concept had become common theory, is well illustrated by the division of the 24 hours of the *saltus lunae* by the 235 lunations of the 19-year lunar cycle as found in the three Irish computistical textbooks. Such a division, which includes unusually complex fractions (like 1/564), certainly was not a trivial matter in a society that would generally not use fractions smaller than a quarter. Accordingly, the first computists to have presented the result of this division certainly had to outline a detailed calculation if they wanted their audience to trust this result. However, once the result became generally accepted and common knowledge, it could be referred to without any further explanation. And this is precisely what appears to have happened concerning the division in question. While the *Computus Einsidlensis* and the Munich Computus provide very detailed accounts of the calculation, the author of *De ratione conputandi* contents himself with the mentioning of the bare result, followed by the simple statement that whoever wants to verify the result will find the exact same numbers (*quicumque hoc probauerit, sic inveniet*);[624] essentially, this final phrase means that the calculation is considered known and rather trivial and not worth or in need of a lengthy proof.

The third aspect, i.e. the explicit rejection or silent abandonment of outdated theories and methods, is probably the most significant of the rules outlined above concerning the chronological order of scientific texts. In respect of the chronological order of the three Irish computistical textbooks in particular, one example of this third category is the placement of the solar bissextile day.

622 *CE* p. 114–118; MC 60; cf. p. CXXXIV–CXXXVI, CXLIV–CXLV above, where this passage is analyzed in the context of MC's dependency on *CE*.

623 *DRC* 79.

624 *CE* p. 122–3; MC 62.14–61 (these passages are discussed and printed in parallel p. CXXXVI–CXXXVIII above); *DRC* 108; cf. also p. CLXV above.

As has been discussed above in the analysis of the relation between the *Computus Einsidlensis* and the Munich Computus, both texts mention three placements, namely 24 February, 2 March, and 21 March. All three placements are seriously considered in the *Computus Einsidlensis*, while the Munich Computus rejects the last, 21 March.[625] In this context it is extremely noteworthy that *De ratione conputandi* refers only to the first two placements, 24 February and 2 March, but does not even mention the one placement rejected by the Munich computist, 21 March. Of the two placements discussed in *De ratione conputandi*, its author strongly and vehemently favours the first, a statement of preference not found in the other two Irish texts.[626] It appears, therefore, that the *Computus Einsidlensis* represents the earliest known medieval discussion of the placement of the solar bissextile day, taking three different places into serious consideration; this account was studied by the Munich computist some time later, rejecting the last of these three placements, which then led the author of *De ratione conputandi* to silently pass over the placement rejected in the earlier Munich Computus, and to advocate the Macrobian placement, 24 February, more vehemently than any previous Irish text.

Another good example of this third category is the treatment of the Victorian reckoning. Especially the detailed comparisons between the Dionysiac and the Victorian reckoning as found in the Einsiedeln and the Munich Computus[627] do not feature in *De ratione conputandi*. This seems to indicate that *De ratione conputandi* was written at a later stage when the Dionysiac reckoning was undisputed and so well understood that such comparisons with an older system were considered outdated and unnecessary. Moreover, in *De ratione conputandi* the discussion of features of the Victorian reckoning is limited to the mentioning of the lunar limits of Easter Sunday and of the *initium quadragesimae*, as well as the placement of the Victorian *saltus*.[628] *De ratione conputandi*'s account of the Victorian *saltus* is particularly interesting for the present purpose. It is argued that the Victorian *saltus* is to be placed on 31 December (reckoning the lunar ages 14 and 15 on this date) rather than on 17 November, as others would falsely do (*licet alii falso in XV Kl Decimbris ponunt*). Now, both the *Computus Einsidlensis* and the Munich Computus give only one option for the Victorian *saltus*, namely the one rejected by the author of *De ratione conputandi*,[629] and it consequently appears that the *alii* referred to in *De ratione conputandi* are to be identified with the Einsiedeln and the Munich computists.

In yet another passage the author of *De ratione conputandi* appears likewise to refer to the Einsiedeln and the Munich Computus. The latter two textbooks are the only known texts attributing a peculiar sequence of lunations to the Dionysiac reckoning, and even applying it for their calculations. Generally, sev-

625 Cf. p. CXLIII–CXLIV above.

626 *DRC* 35, 52, 57.

627 See especially *CE* p. 109–111; MC 49.16–32; 50. For these comparisons of the two reckonings cf. also notes 184, 239, 252 above.

628 *DRC* 98.7–12, 99.4–6, 112.4–6.

629 *CE* p. 123–4; MC 62.87–95.

enth-century computists were completely uninformed about what kind of sequence of lunations formed the basis of the Dionysiac lunar year, since neither Dionysius himself, nor Isidore, nor any other pre-mid-seventh-century text deals with this feature. Accordingly, as is the case for many other details, seventh-century computists had to reconstruct a working sequence of lunations. The only lunar data they could use as orientation were the epacts (of 22 March) and the lunar information for Easter. From this data (especially that of the 9th and the 17th year of the Donysiac 19-year cycle, in which the epact of 22 March fell in the March lunation) it was obvious that the March lunation consisted of 30 days. Three further years, the 8th, 11th, and 19th of the 19-year cycle, additionally suggested that the lunation ending in April, commonly regarded as the April lunation, also consisted of 30 days. Accordingly, both the Einsiedeln and the Munich computists believed that the Dionysiac sequence of lunations contained full lunations (i.e. of 30 days) for the successive months of March and April.[630] However, the *Computus Einsidlensis* and the Munich Computus stand alone with this observation. In every other computistical text, the Dionysiac sequence of lunations is conventionally regarded as alternating between full and hollow lunations, which means that the March lunation was full, the April lunation, on the other hand, hollow (i.e. of 29 days); in the three years that appear to disagree with this theory, i.e. the 8th, 11th, and 19th of the 19-year cycle, the full lunation ending in April was declared as an embolism (or being the consequence of a shift of the March lunation to April due to an earlier embolism). Now, the author of *De ratione conputandi* does not only list the alternating sequence rather than the peculiar one used in the other two Irish computistical textbooks, but he also explicitly condemns the latter practice:[631]

> Sciendum nobis quomodo menses lunares mensibus solis apud Grecos et Latinos secundum hanc regulam iuguntur. Ita etiam, id est XXX luna Ianuario, XXVIIII Februario, XXX Martio, XXVIIII Apreli (etsi alii dicunt lunam Aprelis XXXmam esse semper apud Grecos, quod non sinit natura, nisi in tribus annis tantum, id est quando XXVI uel XXVII uel XXVIIII in Kalendis Ianuarii fit. Nisi enim in his tribus annis XXXma fuerit luna Aprelis, XIIII luna primi mensis, quae contra Maium regulatur, non erit in naturalibus mensis diebus); XXX Maio, XXVIIII Iunio, XXX Iulio, XXVIIII Agusto, XXX Septimbri, XXVIIII Octimbri, XXX Nouimbri, XXVIIII Decimbri.

630 This sequence of lunations is explicitly given in MC 50.29–39; the Dionysiac sequence of lunations outlined in *CE* p. 111 is the alternating one, but the discussion of three Dionysiac lunar years on p. 109–110 clearly reveals that the Einsiedeln computist used the peculiar sequence outlined in MC; consequently, the Dionysiac sequence in *CE* p. 111 was apparently corrected by a later copyist to agree with his own customs (a hypothesis strengthend by the fact that the Dionysiac sequence is followed by the alternating sequence attributed to the 'Latins', i.e. Victorius; if the Dionysiac sequence was originally also regarded as alternating, there would have been no need to explicitly list it twice, a mentioning of both the 'Greeks' and the 'Latins' in the heading would have done the trick). Cf. note 254 and also the use of this sequence of lunations by the author of the Harvard fragment p. CXC–CXCI above.

631 *DRC* 72.

Consequently, this passage reveals that the author of *De ratione conputandi* knew the practice outlined in the *Computus Einsidlensis* and the Munich Computus (in all likelihood directly from these texts), plainly rejected it, and by substituting it with a different theory set the standard for the following centuries.

In summary, then, the chronological order of the three known Irish computistical textbooks proposed here is that the *Computus Einsidlensis* is the oldest of the three, followed by the Munich Computus, while *De ratione conputandi* appears to be the most advanced and sophisticated, and therefore the most recent one. Since the Munich Computus in datable to AD 719, this order suggests that the *Computus Einsidlensis* was composed pre-AD 719, *De ratione conputandi*, on the other hand, post-AD 719. It has been argued above that the *Computus Einsidlensis* was, in all likelihood, written between AD 689 and 719.[632] But can a more specific date also be assigned to *De ratione conputandi*? Borst attests a strong influence of *De ratione conputandi* on *Dial. Burg.*, composed in AD 727.[633] In fact, the following textual parallels in particular suggest that the two texts are dependent:

De ratione conputandi	***Dial. Burg.***
22.2–4: *Sciendum nobis quis primitus diem nominauit. Deus, ut in Genesi dicitur: Appellauit lucem diem et tenebras uocauit noctem.*	5B: *Quis primus diem ac noctem vocavit? Id est: Deus, ut dicitur: Apellavitque diem lucem et tenebras vocavit noctem.*[634]
27.2: *ebdoma in Greco; septimana in Latino.*	5B: *Ebdomada graece dicitur, in latino septimana.*
29.3–5: *Sciendum nobis quot modis menses uocabula sumpserunt. Quattuor: ex rebus, ex regibus, ex numeris, ex diis.*	6A: *Nomina mensium quatuor modis vocantur. Id est: ex diis et ex rebus et ex regibus et ex numeris.*

Yet, there are no indications within these two texts that could solve the problem of their chronological order. On the other hand, the fact that *Dial. Burg.* generally depends heavily on Irish computistical thought[635] renders it likely that *De ratione conputandi* was among the corpus of Irish computistical texts available to the author of *Dial. Burg.* Accordingly, *De ratione conputandi* seems to have been composed between AD 719 and 727.

The chronological order of the three Irish computistical textbooks proposed here (the *Computus Einsidlensis* being the oldest, followed by the Munich Computus, while *De ratione conputandi* appears to be the latest) certainly has some general implications on the development of knowledge and intellectual training of late seventh- and early eighth-century Irish scholars in general, but also on the specific use of the vernacular for didactic purposes in particular.

632 Cf. p. CXLI–CLII above.

633 Borst, *Schriften*, 348. The date is obvious from the Victorian Easter table and the following dating clause in *Dial. Burg.* 16–17.

634 Cf. the Vulgate source, Gen 1:5: *appellavitque lucem diem et tenebras noctem.*

635 Cf., e.g., *Dial. Burg.*'s dependency on the Munich Computus discussed p. CLXXI–CLXXII above.

Among the three texts in question, most instances of Old Irish occur in the *Computus Einsidlensis*,[636] significantly less in the Munich Computus,[637] and not a single instance can be found in *De ratione conputandi*.[638] Consequently, a linear decrease in the use of Old Irish can be observed in the Irish computistical textbooks of the late seventh and early eighth centuries. The reason for this decrease may lie in the simultaneous increase in the knowledge of computistical matters (and maybe also of the Latin language). According to the outlined chronology, the *Computus Einsidlensis* is the earliest known comprehensive computistical textbook, not only in the *regiones Scottorum*, but in the Latin West; not being able to rely on a model for the difficult task of composing a comprehensive textbook on computistics, the Einsiedeln computist felt the need to illustrate or explain certain ideas and concepts in his own and his audience's native tongue, primarily to provide for a better understanding of the passages in question. In subsequent decades, however, computistical theory became more familiar (while knowledge in the second language, Latin, presumably also increased further). For these reasons, students at a later stage appear to have found it easier to understand (at that stage well-known) concepts conveyed in Latin, while their teachers had less difficulties in expressing ideas in Latin that had become more common theory.

636 *CE* p. 90, 93, 97, 123; for a discussion of these instances cf. Bisagni & Warntjes, 'Earliest Old Irish material', 91–105.

637 For the occurrence of Old Irish in the Munich Computus cf. p. XIX, LXXV above and especially Bisagni & Warntjes, 'Latin and Old Irish', 8–33.

638 Cf. Bisagni & Warntjes, 'Latin and Old Irish', 8 note 25.

THE EDITION

This edition is designed to meet the special character and requirements of scientific texts written in what some scholars tend to designate as 'barbaric' Latin. Quite naturally, a scientific text is not usually intended to be a literary masterpiece. Its purpose is to be as concise and precise as possible, and it is this style, combined with the fact that some authors, like the Munich computist, wrote in their second language, which renders the Latin sometimes difficult to understand and to appear 'barbaric'. On the other hand, it is precisely the scientific content that provides for a good understanding of difficult passages, since the concepts outlined are usually not unique, but also appear in other texts, either in the author's sources, or in contemporary treatises on the same subject. This fact already highlights the importance of the analysis of sources and comparable texts for the edition of a scientific work and the special requirements that such an edition has to meet: This analysis cannot be restricted to concordances in wording, but the scope has to be extended to concordances in more general scientific definitions, methods, and concepts to be able to reconstruct the meaning of difficult passages. At the same time, such detailed analyses also serve a different purpose in that they provide the context for the edited text, its place in the transmission of certain scientific concepts and ideas. Accordingly, detailed apparatus of sources and of comparable texts are of paramount importance to any edition of a scientific text. In the case of the Munich Computus this is even more important, since it survives in one manuscript only, with the text being highly corrupted by various copying stages, so that sources and comparable texts are essential for the reconstruction of certain passages in some instances, to supply better (and the original) readings in others. But even after trying to restore the original text, and despite full references of sources and comparable texts, the content may still remain obscure to the non-specialist. Therefore, a translation is provided facing the Latin text, and, for an easy overview, brief summaries of each passage are given in the left margin of the Latin text. In the bottom half of the pages four apparatus are arranged in the following order, floating from the Latin text page to the translation page: 1) *apparatus criticus*, 2) *apparatus fontium*, 3) *apparatus comparationis*, 4) *apparatus commentationium modernarum*. The editorial principles in each of these sections are outlined in the following. This is followed by a detailed description of the sole manuscript in which the Munich Computus survives, to provide the reader with the more general context of the transmission of this text.

EDITORIAL PRINCIPLES

The Latin text: The Munich Computus, as preserved in the sole manuscript witness, is highly corrupted by various copying stages.[639] It appears that the copyists had little difficulty with simple definitions and citations from well-known authors, but they were at a loss with passages describing technical concepts, which they clearly did not understand. This is especially obvious in the punctuation of the text, which is well executed in straight-forward passages, but often contrary to the meaning in more difficult discussions of calendrical methods. It appears, therefore, that the copyists did not slavishly copy the text word for word, but they read and clearly understood the more intelligible passages. Accordingly, the content more than the script posed obstacles for the copyists trained in reading, writing, and copying, but not in computistics, though in few instances Insular abbreviations appear to have been wrongly interpreted, too. Yet, there are two more principal reasons for the corrupted state of the text as transmitted in the Regensburg copy: 1) Due to the fortunate fact that a passage was mistakenly repeated it can be shown that at least the Regensburg copyist of the text was not particularly careful and probably worked in haste, since the two versions of the passage are by no means identical, as the following comparison of the transcript shows (the differences are highlighted in bold):[640]

*Sec**undu-** latinos · ab VIII k- ian- · in VIII k- Apr- annus diuiditur **et kl-** (recte rl-) ·*
*Sc**d-m** latinos ab VIII k- ian- in VIII k- Apr- annus diuiditur ·*

*Sciendu- **tam-** quid **g**reci in hac difinitione distant **ab hoc** · Quia greci*
*Sciendu- **ut** quid **c**reci in hac difinitione distant · **Id · e-** · quia greci*

uiciniores sunt ortui qua- latini · hinc solare- cursum facilius inuestigant ·
*uiciniores sunt ortui **solis** qua- latini · hinc solare- cursum facilius inuestigant*

2) The text appears to have been sporadically glossed at some stage of its transmission, and some of these glosses then crept into the main body of the text at a later copying stage. Additionally, it should be noted that the Munich Computus, as it survives, provides neither a table of contents nor chapter numbers. Headings are given for few passages, but more often a change of topic is not indicated in the manuscript.

Consequently, the editor faces a number of difficulties, and to provide the reader with an intelligible and structured text, the following emendations have been implemented: The text is divided into chapters according to content, and headings are provided where necessary (and a table of contents based on these

639 For a reconstruction of these copying stages cf. p. XCVII–CVI above.

640 Fol. 18v; MC 38.26–30.

headings is provided as part of the general table of contents of this book on p. VII–IX). These chapters and their lines (in the right margin) are numbered throughout, and accordingly all references are to chapter and line numbers. In previous studies of the Munich Computus, however, the referencing quite naturally referred to folio numbers. As a guide for these studies, a vertical slash (|) occurs in the text at every change of folio, and the number of the folio that starts from that point is listed in the right margin of the text. Since the punctuation of the text as it survives is often misleading and contrary to the original argument, it has been ignored for this edition. The punctuation applied here is designed as a help for the reader, to facilitate an easy grasp of the text by highlighting the structure of the argument. The parts of the text that appear to have been copied directly from sources are indicated in italic. Non-Latin, i.e. Old Irish and Old English words are underlined and edited according to the common principles applied for vernacular texts. Glosses, as far as they can be detected, are given in rectangular dashes (e.g. ⌐ *indictio* ¬ ; this is modelled on the symbol ‘ ¬ ’ for logical negation). Wherever a word is clearly missing, it is added in angle brackets. Note, however, that this policy is reduced to an absolute minimum in order to remain faithful to the text. Modern editors tend to excessively add words to clarify matters for the modern reader. In computistical texts, and especially in the Munich Computus, however, the style is concise, sometimes stenographic. Very simple examples are: *a XII XXIIII* would normally be emended to *a XII <in> XXIIII*; likewise *XI kl-* would always be specified, e.g. as *XI Kalendas <Aprilis>*, even though the context already provides the specification. Such emendations are helpful to the modern reader, but the added words were quite certainly silently understood by the medieval computist and therefore not necessarily part of the original text. In the present edition, the translation serves as a help for the modern reader and should provide for a full understanding, so that excessive emendation contrary to the style of the original author are, in this respect, not essential and can be avoided. Moreover, the Latin itself is not rendered into Classical Latin. The grammatical principles of the author are kept, so that, e.g., accusative and ablative are sometimes interchangeable (e.g., *in luna* was as correct to the Munich computist as *in lunam*), as are the gender of some nouns (e.g. *dies* appear interchangeably in masculine and feminine). The same applies to orthography, where e for i and vice versa, a for o and vice versa, b for v and vice versa, t for d and vice versa, single consonants instead of double consonants, etc. are frequent; e can stand for ae, while e-caudata (ę) is used both for ae and e. Here, corrections are only applied if the change in orthography changed the meaning of the word in any way, e.g. if the plural *menses* appears instead of the singular *mensis*. Numbers appear inconsistently either as Roman numerals, or they are (though less frequently) spelled out, and here again the edition remains faithful to the manuscript witness. Roman numerals are used for all kinds of numbers, i.e. for cardinals, ordinals, distributives, as well as multiplicatives. If spelled out, numbers are corrected to the right kind wherever necessary. Note, however, that ordinals (as well as the names of the Julian calendar months) appear to have been used in a special adjectival sense in some instances,[641] so that this form is accordingly kept in the edition. In the manuscript, *Kalendae*, *Nonae*, and *Idus* are almost exclusively written in abbreviated forms as *k-*, *kl-*, or *kal-*, *n-*, *no-*, or *non-*, and *id-* respectively, without any further indication of the case. The same applies to the Julian calendar months and other technical terminology. In most instances, it must be presumed that the medieval computist was not concerned with the case of these nouns, and some modern editors follow medieval practice by simply letting these abbreviations speak for themselves. A grammatical understanding of the passage can, however, only be achieved by expansion of the abbreviations to their inflected form, and accordingly this is done throughout.

Translation: The translation is not designed to be an independent piece of literature, but rather as a tool for the understanding of the Latin text, to which it is accordingly placed in parallel.[642] Passages of the Latin text that are grammatically straight-forward, as well as those in which the Latin grammatical structure has no equivalent in the English language, are translated rather liberally. Technical passages, however, in which the argument is not immediately apparent to the modern non-computist, are translated closer to the Latin text, and additional comments necessary for a full understanding of the argument are provided in round brackets (while in general every word that cannot be found in the Latin is given in round brackets). It is hoped that this method will prove particularly helpful for scholars of Hiberno-Latin without profound knowledge in computistics. Glosses, wherever they occur in the Latin text, are indicated by the same means in the translation. In only two instances, the meaning of the Latin remains obscure, so that cruces († ... †) replace the translation of these phrases.

641 Cf. Bisagni & Warntjes, 'Latin and Old Irish', 26–7.

642 Computistical texts are very often only marginally commentated, even though a detailed commentary would be essential for the understanding of the text, especially for the interested non-specialist. The most precise commentary of any text, and the most thorough interpretation, is a translation. Concerning computistical literature, only few texts have been translated. Van Wijk, 'Comput', 3 describes the situation in 1951 in the following words: 'S'il est vrai, comme on le pretend souvent, que le meilleur commentaire d'un texte consiste en une traduction, le comput est bien un texte mal commenté. En effet, je ne connais que trois traductions de computs, dont une n'est encore sortie que de ma plume.' The need for translations as tools for a thorough analysis and better understanding of difficult computistical texts has been felt more pressing in recent years. This is especially stressed by Strobel, *Texte*, V and Springsfeld, *Alkuins Einfluß*, 322, 329. When printed editions already existed, the translations are often not accompanied by the Latin text, as in the case of Strobel's book and the recent translation of the most influential medieval computistical work, Bede's *De temporum ratione* by Wallis (*Bede*). In all other cases, the translations are placed parallel to the Latin texts by the editors to provide for a direct understanding of the Latin text. Cf. the editions with parallel translations in van Wijk, 'Comput', 10–80; Walsh & Ó Cróinín, *Cummian's letter*, 56–97; Springsfeld, *Alkuins Einfluß*, 322–75; in Mc Carthy & Breen, *De ratione paschali*, the translation (p. 63–70) immediately follows the Latin text (p. 45–53) and a facsimile (p. 54–62).

apparatus criticus (*app. crit.*): All variants and noteworthy features of the MS are listed in the *apparatus criticus*. If corrections are based on the evidence of a source, or a comparable text, or the study of a previous scholar, this is noted in round brackets, and can easily be checked against the information given in the following three apparatus respectively. In this respect, 'corr. according to x' indicates that the correction in question is suggested by x, whereas 'corr. from x' indicates that the correction in question is explicitly taken from x, which contains the exact same, but uncorrupted phrase as the Munich Computus. But note that corruptions of the Munich Computus, which can also be found in certain manuscripts of the source, and / or in texts with close parallels to the Munich Computus (like the *Computus Einsidlensis* and *De ratione conputandi*), appear in most cases to have been part of the original text, and as such remain uncorrected (as long as they make sense in the context). If reasons for the occurrence of a certain corruption in the text are apparent (e.g. a misinterpretation of an Insular abbreviation by a continental copyist), then these are noted. Additionally, explanations for the assumption that a certain part of the text was originally a gloss are also outlined in this apparatus. Concerning abbreviations, 'add.' equals 'added (in)', 'om.' equals 'omitted (in)', while M stands for the sole manuscript witness of the Munich Computus (Munich, Bayerische Staatsbibliothek, Clm 14456, fol. 8r–46r), M* for the manuscript reading before correction by the same scribe; MC denotes the original text and its author. All other abbreviations can be easily checked against the list of abbreviations given at the beginning of the present book.

apparatus fontium (*app. font.*): All sources used by the Munich computist are listed in the *apparatus fontium*. The source analysis of this text is of special interest, since the Munich Computus stands at the beginning of the development of a literary genre, the computistical textbook.[643] Accordingly, the Munich text could not draw on comprehensive treatises of computistical matters (save for the *Computus Einsidlensis*), but its author had to assemble material from a variety of texts, namely from all texts available in the library of his monastery that dealt, in some way or another, with computistical matters. Therefore, the source analysis of the Munich Computus provides an insight into the scientific section of the library of one particular monastery in the *regiones Scottorum* in AD 719. Additionally, the negative results of this analysis are equally interesting, in that they highlight the problems faced by Irish computists of the late seventh and early eighth centuries due to a lack of information on certain matters. In this respect, concepts for which no apparent source (or only the *Computus Einsidlensis*, but no earlier text) could be located, appear, in most instances, to have originated in late seventh-century Irish computistical circles.

If the wording reveals that the Munich computist drew directly on certain sources, these are classified as such in this apparatus, and the textual parallels are highlighted in italics in the Latin text. Variants of the sources shared by the Munich Computus, as far as they can be determined by means of the editions

643 Cf. p. LII–LVI above.

available, are also noted. If, however, the Munich text shows no textual parallels, but parallels in content, in the concepts discussed, to any evidently pre-Munich text, the possibility exists that the Munich computist was aware of this text, and had consulted it at some point prior to the composition of his text; these cases are also discussed in this apparatus. The same applies to pre-Munich texts that clearly were not the Munich computist's source, but which contain the same information as his source and which lay within the range of texts available to him. Other computistical tracts contemporary or near contemporary to the Munich Computus that quote the same sources, directly or through an intermediary, are listed for each source to establish the popularity and commonness of a citation in the compustical milieu between the late seventh and the early ninth century; by this method, the place of the Munich Computus in the reception of specific passages of certain authors can be analyzed, resulting in various degrees of reception, from a unique occurrence in the Munich Computus to popularity only among Irish computistical text, to popularity only among Insular computistical texts, to popularity throughout western Europe. All source references are spelled out, since the Munich computist hardly ever quotes a source verbatim; especially in cases in which he reproduces the argument of his source in his own words without any textual parallel, the full text is absolutely essential for an understanding of the relationship between the Munich text and its source. Whenever it is felt that this relationship needs further comment, this is provided. Equally, whenever the source provides the key for an understanding of the Munich text, a discussion of the passage in question appears in this apparatus.

Most sources, especially those of the Church fathers, exist in printed editions. Seventh century computistical material, however, is only inadequately edited, sometimes not even identified. Since these seventh century tracts or texts are transmitted in manuscripts of the eighth or later centuries and frequently do not contain dating clauses, their identification often is a difficult task. Yet, because of the great number of unedited computistical tracts and texts from the early medieval period, any editor of eighth to tenth century computistical texts certainly must consult at least the most important and relevant manuscripts of that time if he wants his source analysis and apparatus to stand the test of time and if he wants to do the author of the text to be edited some justice. I have studied some 150 of the more important codices in my search for sources used by the Munich computist, but made only few – though some very lucky – discoveries. The text that shows most parallels to the Munich Computus is the *Computus Einsidlensis*, which has escaped the attention of modern scholars and which I have identified in Einsiedeln, Stiftsbibliothek, 321 (647), p. 82–125; the case for this text being older than the Munich Computus and therefore one of its sources is outlined above.[644] Additionally, a previously unnoticed tract in Cologne, Dombibliothek, 83², fol. 176v–178r shows very close and unique parallels to one chapter of the Munich Computus (c. 52); again, the reasoning for the assumption that this text was one of the Munich computist's sources rather than

644 Cf. p. CXXXIII–CLII.

vice versa is given above;[645] it is edited, for the first time, under the title *De comparatione epactarum Dionysii et Victorii* in Appendix 2.

The *Computus Einsidlensis* is referred to as *CE* in this apparatus, and the referencing is to MS pages. The Cologne tract is referred to by the short title *De comparatione*, following the title given to this tract in its *editio princeps* in the appendix; the referencing is to chapter.line number of this edition. Abbreviations of frequently cited sources can be checked against the list of abbreviations at the beginning of the pesent book. All other source references are self-evident. The referencing system varies, depending on the editions cited; when these editions are consulted, the referencing principles will be immediately apparent. For the editions see the list of abbreviations and the bibliography.

apparatus comparationis (*app. comp.*): This apparatus of comparable texts is of special importance to the present edition for three principal reasons: 1) In some cases these comparable texts provide the data that make an understanding and reconstruction of the Munich text possible in the first place. 2) The place of the Munich Computus in the history and transmission of computistical concepts becomes apparent. In this respect, the source analysis establishes the medieval origin of the computistical ideas formulated by the Munich computist, and places this text either within or at the beginning of a line of transmission. The apparatus of comparable texts, then, highlights the importance and reception of these concepts in the computistical discourse from the early eighth to the early ninth century. 3) More specifically, this analysis of comparable texts singles out the treatises that were directly influenced by the Munich Computus, and therefore is of primary importance for the study of the reception of the Munich Computus itself.[646]

The general question for the compilation of this apparatus obviously was: Which texts should it contain? Many important treatises still remain hidden in the manuscripts, and it was impossible to thoroughly analyze all of the hundreds of pre-AD 1200 computistical manuscripts that appear to be of relevance. Relief came with Arno Borst's *Schriften zur Komputistik im Frankenreich von 721 bis 818*, which he so generously allowed me to study (and proof-read) before publication. This publication tremendously enlarged (but certainly did not complete) the corpus of published computistical texts from the early eighth to the early ninth century, to a degree that it can be regarded as representative for (Frankish) computistical thought of this period (though, it should be noted, still unedited texts like the unfinished Fulda Computus of 789 will shed different light on certain aspects). Accordingly, the basis of this apparatus is formed by a thorough analysis of all printed computistical texts from the early eighth to the early ninth century. Computistical tracts later than ca. AD 830 are sometimes considered (whenever they provide an interesting perspective on the concept in question), but they are not systematically analyzed. Generally, the computistics

645 Cf. p. CLII–CLVIII above.

646 The main results of this analysis are outlined in the chapter on the reception of the Munich Computus on p. CLIX–CCI above.

of the period between the early ninth century and the advent of Arabic science in the Latin West in the very late tenth and eleventh centuries is very inadequately covered by only random editions of texts that can be attributed by name to certain authors; the more interesting anonymous texts have not yet received the attention they deserve.

Anyway, as in the case of the Munich computist's sources, the existing corpus of printed computistical texts still proved to be insufficient to do justice to the Munich Computus. Ó Cróinín drew attention to the fact that some glosses to Bede's computistical texts in Angers, Bibliothèque Municipale, 477 (461), are very similar to the Munich text.[647] In fact, in some instances the Angers glosses together with the Munich text are the only surviving witnesses of certain ideas. This is even more so the case for a computistical fragment constituting fol. 7r–v in Harvard, Houghton Library, MS Type 613, which Dáibhí Ó Cróinín discovered only some months before the completion of the present dissertation. In the edition, the Angers MS will be referred to as Angers 477, the fragment transmitted in the Harvard MS as Harvard fragment. In addition to these, I only randomly and unsystematically refer to manuscript witnesses whenever it appears appropriate.

Whenever the translation and the source analysis do not provide enough information for a full understanding of a certain passage, further explanations are given in this apparatus. Likewise, the context of the Munich Computus is very briefly discussed in this apparatus for every single passage of the text on the basis of the comparable data here provided. All references are spelled out in full to facilitate a thorough and detailed comparison of these texts with the Munich Computus, as well as a full comprehension of the basis of the corrections and reconstructions applied to the Munich text. As for the abbreviations used, these can be checked against the list of abbreviations at the beginning of the present book. Again, the referencing system varies, depending on the editions cited; when these editions are consulted, the referencing principles will be immediately apparent. For the editions see the list of abbreviations and the bibliography.

apparatus commentationium modernarum (*app. comm.*): It is one of the primary tasks of an *editio princeps* to acknowledge and include all studies prior to the publication of a definitive text. Many passages of the Munich Computus have been commented on over the past 130 years, especially the ones either dealing with the *latercus* or incorporating Old Irish words. Some of these passages were additionally transcribed, corrected, and sometimes even translated. For certain passages, these studies are therefore to be considered as precursors of the present edition, and as such they deserve to be noted at the appropriate places. Accordingly, every passage that has been previously discussed is marked and the bibliographical references are provided in this apparatus. Concerning notation, 'cf. x' indicates that a discussion of the marked passage can be

647 Cf. the comparison of Munich and Angers passages in Walsh & Ó Cróinín, *Cummian's letter*, 115–6, 163.

found in x. If the passage in question is additionally transcribed and / or translated in any of the studies, this is explicitly stated. Note that passages were transcribed and / or translated for the sole purpose of forming the basis of discussion, so that, e.g., 'transcribed in x' naturally implies that not only a transcription, but also a discussion of that passage can be found in x. Whenever corrections made by previous scholars agree with the corrections of the present edition, these are noted in the *apparatus criticus*. Finally, cross-references to discussions of the marked passages in the present book are also provided in this apparatus.

THE MANUSCRIPT (MUNICH, BAYERISCHE STAATSBIBLIOTHEK, CLM 14456)

A comprehensive catalogue of computistical manuscripts is one of the main desiderata in computistical studies. The closest to a full scale catalogue of computistical manuscripts are Borst's manuscript descriptions in his corpus of Frankish computistical texts of AD 721–818.[648] Yet, as the title of his work implies, only manuscripts incorporating Frankish material of the mentioned period are considered, so that many important early manuscripts are not listed.[649] Moreover, despite providing a welcome overview of the more important texts to be found in each of these manuscripts, Borst's manuscript descriptions are by no means comprehensive for any codex. For future research, a folio by folio description of every single computistical manuscript will be needed. This has been done for some computistical manuscripts in library catalogues, too often, unfortunately, by non-specialists. Accordingly, many tracts are either wrongly or insufficiently described, giving the impression that these tracts are well-identified and even well-known, whereas quite often they are not. Since no single comprehensive description of the manuscript containing the Munich Computus exists, this is provided in the following after listing all previous catalogue descriptions of it.

Catalogue descriptions: Docen, 'Notizen', 515–9; Piper, *Kalendarium*, 77; Halm et al., *Catalogus*, 175 (Nr. 1438); Bischoff, *Schreibschulen*, 195–6 (Nr. 16); Leonardi, 'Codici II', 95 (Nr. 124); Bierbrauer, 'Regensburger

648 Borst, *Schriften*, 205–317. Usefull is also Jones' list in *Bedae pseudepigrapha*, 111–40.

649 Borst, *Schriften*, 205 is aware of the provisional character of his collection, calling it 'vorläufige Übersicht komputistischer Schriften vom 8. bis zum 12. Jahrhundert'; he also sets the limits of his list: 'Meine Liste beschränkt sich auf diejenigen komputistischen Handschriften, die im Frankenreich während des Jahrhunderts von 721 bis 818 geschrieben oder nach Vorlagen aus dieser Zeit in diesem Raum kopiert wurden'. Borst's limits led to the omission of highly important computistical manuscripts from that period, e.g. the two oldest computistical manuscripts (besides, obviously, the Zeitz fragments and the Gotha Victorian table), Bern, Burgerbibliothek, 645 and London, British Library, Cotton Caligula A XV, as well as the late eighth-century St Gall, Stiftsbibliothek, 225 (for which see now especially Springsfeld, 'Beschreibung'), the ninth-century Padua, Biblioteca Antoniana, I 27 (for which see now Ó Cróinín, 'Dionysius Exiguus', 262–4 and especially idem, 'Continuity', 341–4), Leiden, Universiteitsbibliotheek, Scaliger 28, Munich, Bayerische Staatsbibliothek, Clm 14725, St Gall, Stiftsbibliothek, 251, the tenth-century Cologne, Dombibliothek, 102, and the eleventh-century Zürich, Zentralbibliothek, Carolinus C 180. Even more noteworthy is the omission of one of the most important Frankish computistical MS of the period studied by Borst, Basel, Universitätsbibliothek, F III 15k; this MS was not included, since it does not transmit any of the texts edited by Borst; yet, it contains one of the most important Frankish computistical texts of the period, an unfinished Computus of AD 789, which probably is to be linked to Charlemagne's *Admonitio generalis* of the same year; it appears that the provenance of this text is Fulda, since Rabanus Maurus evidently used it for the composition of his *De computo*.

Buchmalerei', 20 (Nr. 4); eadem, *Handschriften*, 61 (Nr. 111); Borst, *Reichskalender*, 75 (Siglum: a2, *Kal. Rat. I*); idem, *Schriften*, 261–2 (MS My). Additionally, this MS is listed in the following catalogues, but with almost no information concerning its content or physical features: Holder-Egger, 'Handschriften', 560; Swarzenski, *Denkmäler*, 20; Zinner, *Verzeichnis*, 374, 537–8; Bischoff, *Katalog 2*, 257.

Codicological and palaeographical description: This vellum codex, written in Carolingian minuscule, has 86 folios of 240×160 mm. It consists of five parts, which vary in their *Schriftspiegel* and number of lines: 1) single ternion (fol. 1–6); 2) single folio (fol. 7); 3) nine quaternions with the two final folios cut (fol. 8–77); 4) single bifolium (fol. 78–79); 5) single quaternion with the final folio cut (fol. 80–86). These five parts are clearly separated by bold duplicated lines in the description of the content of this MS below. Hair-side is facing hair-side, and consequently flesh-side is facing flesh-side. For the different hands see Bischoff's study.[650] Initials and headings, as far as they are highlighted by a different colour, appear in red; colours used in the diagrams are red, brown, and yellow.

Date: Five items in this manuscript give indications of dates: 1) The chronological data in the ps-Dionysiac *Argumentum XIV* of folios 46v–47v (*concurrentes* 5, 6, and 7 for consecutive years) refer either to AD 589–592 or, more likely, to AD 684–686; 2) the Munich Computus of folios 8r–46r can be dated to AD 719; 3) the Easter table of folios 64r–65v extends from AD 779 to 873; 4) the note on Louis the Pious of folio 1v refers to either AD 817 or 832; 5) the annals of folios 80v–86r break off at AD 823. Accordingly, the first two items are copied from older exemplars, while the Easter table narrows the date of the MS down to a 95-year period. More conclusive are items 4 and 5; especially the fact that the annals break off at AD 823 appears to indicate that the MS was written in that very year; this would allow a precision of the dating clause given in item 4, which consequently appears to refer to AD 817; Bischoff argues on palaeographical grounds that the first and the last part of the manuscript, incorporating the dating clauses for 817 and 823 respectively, were written by a slightly later hand than the core middle part of the MS.[651]

Provenance: The annals of folios 80v–86r clearly reveal a close interest in the monastery of St Emmeram in Regensburg.[652]

650 Bischoff, *Schreibschulen*, 195–6.
651 Bischoff, *Schreibschulen*, 196.
652 Cf. Pertz, *MGH SS* 1, 92–3.

Munich, Bayerische Staatsbibliothek, Clm 1445

Folios	Content	Facsimile	Edition	References to these folios
1r–v	Note on a donation by Louis the Pious; indiction 10 is given, which occurred only in AD 817 and 832 during his reign; 1r is erased, but parts of the lost text were recovered and reconstructed by Bischoff.		Docen, 'Notizen', 516; Swarzenski, *Denkmäler*, 20; Bischoff, *Schreibschulen*, 196.	Docen, 'Notizen', 515–6; Swarzenski, *Denkmäler*, 20; Bischoff, *Schreibschulen*, 196, 217; Bierbrauer, *Handschriften*, 61.
1v–2r	*De decem nominibus, quibus apud Hebreos Deus vocatur* (Isidore, *Etymologiae* 7.1.2–17 in a truncated version)		Not from this MS in Lindsay, *Etymologiarum sive originum libri*.	Halm et al., *Catalogus*, 175.
2r–3v	*De angelis* (Isidore, *Etymologiae* 7.5 in a truncated version)		Not from this MS in Lindsay, *Etymologiarum sive originum libri*.	Halm et al., *Catalogus*, 175.
4r–6r	Glossary		Graff, *Sprachschatz* 1, lx–i (only the Latin–Old High German glosses); Steinmeier & Sievers, *Glossen* 4, 326–7 (only the glosses from the grammatical tract *Clementis Ars*).	Docen, 'Notizen', 516; Halm et al., *Catalogus*, 175; Manitius, *Geschichte*, 458; Bischoff, *Schreibschulen*, 196; Bierbrauer, *Handschriften*, 61.
6v	blank			
7r	List of termini quadragesimae (i.e. the Julian calendar dates of luna 2 of the lunation preceding the Easter lunation) and the corresponding regulars for calculating the weekday of the respective terminus (by addition of these regulars to the concurrentes of the respective year).		The list of *termini quadragesimae* is edited from this MS as c. 1.12 of *Lect. comp.* (cited in *Lib. comp.* 2.21) in Borst, *Schriften*, 556–7.	Halm et al., *Catalogus*, 175; Borst, *Buch der Naturgeschichte*, 168; idem, *Schriften*, 261.
7r	Etymology of *tempora*, which consists of two citations from Isidore, *Etymologiae* (5.35.1 and 5.29.1).			

Folios	Content	Facsimile	Edition	References to these folios
7v	blank			
8r–46r	Munich Computus	Partially in Warntjes, '84 (14)-year Easter reckoning', 83–5.	In the present volume.	Cf. the present study.
46v–47v	ps-Dionysius, *Argumentum XIV* (with *concurrentes* 5, 6, and 7 in the examples).		Not from this MS in Krusch, *Studien* II, 78–9 (with the canonical *concurrentes* 4, 5, and 6).	Mac Carthy, *Annals of Ulster* 4, lxvii; Borst, *Reichskalender*, 75; idem, *Schriften*, 261.
48r–53v	Calendar (a2 in Borst's edition)		Not a full edition, but valuable information about the details of this calendar can be found in the commentary of Borst's edition of the *Reichskalender* (Borst, *Reichskalender*, 392–1644) under the siglum *Kal. Rat. I.* Cf. also Borst, *Schriften*, 1103.	Sanftl, *Catalogus* III, 1729; Docen, 'Notizen', 517; Piper, *Kalendarium*, 77; Halm et al., *Catalogus*, 175; Zinner, *Verzeichnis*, 374, 537; Lechner, *Kirchenfeste*, 4; Bischoff, 'Ostertagtexte', 193; Freise, 'Grundformen', 449; Walsh & Ó Cróinín, *Cummian's letter*, 147; Borst, *Kalenderreform*, 138, 305–6, 465; idem, *Reichkalender*, 75.
54r–63v	Table of lunar letters for every year of the Dionysiac 19-year cycle.			Sanftl, *Catalogus* III, 1729; Halm et al., *Catalogus*, 175; Borst, *Kalenderreform*, 486.
63v	Text on the pronunciation of the letters of the alphabet from book 3 (*De arte grammatica*) of Martianus Capella's *De nuptiis Philologiae et Mercurii.*		Not from this MS in Willis, *Martianus Capella*, 68–9.	Halm et al., *Catalogus*, 175; Leonardi, 'Codici I', 468; idem, 'Codici II', 95; Borst, *Kalenderreform*, 514.

64r–65v	Dionysiac Easter table AD 779–873, which consists of the same columns as the original Dionysiac Easter table (ed. Krusch, *Studien* II, 69–74), additionally indicating final years of *ogdoades* and *hendecades*.	Cf. Borst, *Schriften*, 1106–8 (which actually does not edit this table; to me the table in M appears not to be an excerpt from *Lib. comp.* 1.4, but older).	Sanftl, *Catalogus* III, 1729; Docen, 'Notizen', 517; Piper, *Kalendarium*, 94; Halm et al., *Catalogus*, 175; Zinner, *Verzeichnis*, 374, 537; Freise, 'Grundformen', 449; Borst, *Reichskalender*, 75.
65v–66v	ps-Dionysius, *Argumentum XVI*; the second paragraph (Krusch, *Studien* II, p. 80, l. 26 – p. 81, l. 6) is given before the first (Krusch, *Studien* II, p. 80, ll. 18–25).	Not from this MS in Krusch, *Studien* II, 80–1.	
66r	A table listing the length of moonlight per lunar day from the first to the fifteenth moon; the data appear to be corrupt.		
66v	The Anatolian (?) table (cf. *DRP* 10) for counting the number of days from 1 January to the Nones, Ides, and Calends of each month.	Edited as *Lect. comp.* 1.11 in Borst, *Schriften*, 555–6.	Borst, *Schriften*, 261; Jones, *CCSL* 123B, 351.
66v	The weekday regulars for calculating the weekday of the Calends of each month by addition to the *concurrentes* of the respective year (starting from October with regular 1, i.e. Oct. 1, Nov. 4, Dec. 6, Jan. 2, Feb. 5, Mar. 5, …, Sept. 7).	Edited as *Lect. comp.* 1.3 in Borst, *Schriften*, 546–7 (correct, but difficult to reconstruct).	Borst, *Schriften*, 261.
66v	The lunar regulars for calculating the lunar age of the Calends of each month by addition to the epact of 22 March of the respective year (starting from September with regular 5, i.e. Sept. 5, …, Aug. 13).	Edited as *Lect. comp.* 1.4 in Borst, *Schriften*, 547–8.	Borst, *Schriften*, 261.

Folios	Content	Facsimile	Edition	References to these folios
66v–67r	Rule for the calculation of the weekday on the Calends of any given month of any given year by means of the weekday regulars and the *concurrentes* of the year in question.			
67r	Text on the Julian calendar limits for the Easter new and full moons.		Edited as *Lib. ann.* 33–35 (cited in *Lib. comp.* 4.18A–C) in Borst, *Schriften*, 729–31, 1216–7.	Borst, *Buch der Naturgeschichte*, 168; idem, *Reichskalender*, 75; idem, *Schriften*, 261.
67v	Table of the lunar age of the Calends of each month for the 19 years of the *cyclus decemnovenalis*.		Edited as *Lib. comp.* 4.1 in Borst, *Schriften*, 1198–1200.	Borst, *Buch der Naturgeschichte*, 168; idem, *Reichskalender*, 75; idem, *Schriften*, 261.
68r–69r	Four diagrams on the three divisions of philosophy: theory, practice (two diagrams), logic.			Swarzenski, *Denkmäler*, 11; Bierbrauer, *Handschriften*, 61.
69v	Diagram on the seven sages.			Swarzenski, *Denkmäler*, 11; Bischoff, 'Das griechische Element', 263; Bierbrauer, *Handschriften*, 61.
70r (top half)	Rota displaying the translation of the Egyptian calendar year into Julian calendar terms; it starts from 28 September, and the 365 days are divided into 12 equal months of 30 days and 5 remaining days; based on Isidore, De natura rerum 4.7.		From a different MS in Fontaine, *Traité*, 190*bis*.	Bierbrauer, *Handschriften*, 61; Obrist, 'La représentation', 3–12.
70r (bottom half)	*Horologium* in semi-circle with the depiction of a man holding a gnomon.	Obrist, 'Astronomical sundial', 111; Wiesenbach, 'Wilhelm von Hirsau', 128.		Bierbrauer, 'Regensburger Buchmalerei', 20; eadem, Handschriften, 61; Obrist, 'Astronomical sundial', 80; eadem, 'La représentation', 3–12; Wiesenbach, 'Wilhelm von Hirsau', 128.

70v (top half)	Rota on the daily course of the sun in the sky and consequently on the hours of daylight, with the three examples referring to the equinoxes (in the middle), the summer and winter solstices.	Obrist, 'Astronomical sundial', 114 (the facsimile is printed upside down).	Bierbrauer, *Handschriften*, 61; Obrist, 'Astronomical sundial', 97; eadem, 'La représentation', 3–12.
70v (bottom half)	Rota on the phases of the moon.		Bierbrauer, *Handschriften*, 61; Obrist, 'La représentation', 3–12.
71r (top half)	Rota on the Dionysiac 19-year cycle, listing the Julian calendar date of the Easter new moon, the cycle number of the *cyclus lunaris*, the 28 *concurrentes* of the solar cycle, the epact of 22 March, the regular for calculating the Julian calendar date of the Easter full moon, the Julian calendar date of the Easter new moon, the date of the Easter full moon in successive counts of days, the corresponding Julian calendar date, notation on the Easter full moon (*luna* 14), multiples of 20 up to 100, sequences of common and embolismic years; a depiction of a man is placed in the centre.		Bierbrauer, *Handschriften*, 61; Obrist, 'La représentation', 3–12.
71r (right centre)	Number of days, *puncti*, *minuta*, and *momenta* in a week.		
71r (bottom left)	Rota on the division of a day in hours, *puncta*, *momenta*, and *atomos*; a depiction of a hand is placed in the centre.		Swarzenski, *Denkmäler*, 11; Bierbrauer, *Handschriften*, 61; Obrist, 'La représentation', 3–12.
71r (bottom right)	Rota on the division of a week into seven days of 24 hours, 96 *puncta*, 240 *minuta*, and 960 *momenta* each; depiction of a man in the centre.		Bierbrauer, *Handschriften*, 61; Obrist, 'La représentation', 3–12.

Folios	Content	Facsimile	Edition	References to these folios
71v (top half)	Rota on the Dionysiac 19-year cycle, noting the Easter new and full moons; additionally, the length of moonlight per lunar day from the first to the fifteenth moon are listed.			Bierbrauer, *Handschriften*, 61; Obrist, 'La représentation', 3–12.
71v (bottom left)	Rota on Dionysiac 19-year cycle, listing the epacts of 1 January and singling out the year of the *saltus*.			Bierbrauer, *Handschriften*, 61; Obrist, 'La représentation', 3–12.
71v (bottom right)	Rota on the twelve winds (cf. Isidore, *De natura rerum* 37).		The Isidorian rota is ed. in Fontaine, *De natura rerum*, 296*bis*.	Bierbrauer, *Handschriften*, 61; Obrist, 'La représentation', 3–12.
72r (top half)	Rota on the course of the sun through the zodiac signs.			Bierbrauer, *Handschriften*, 61; Obrist, 'La représentation', 3–12.
72r (bottom half)	Rota on the relation between solar and lunar months; note that the sequence of lunations is given as alternating, but starting with a hollow month in January.			Bierbrauer, *Handschriften*, 61; Obrist, 'La représentation', 3–12.
72v (top half)	Rota on the course of the sun through the zodiac signs; depiction of a man symbolizing the sun in the centre.	Obrist, 'La représentation', 10.		Bierbrauer, *Handschriften*, 61; Obrist, 'La représentation', 3–12.
72v (bottom half)	Rota on the direction of moonrise for each moon of a lunation; depiction of a man symbolizing the moon in the centre.	Obrist, 'La représentation', 10.		Bierbrauer, *Handschriften*, 61; Obrist, 'La représentation', 3–12.

73r	Rota on the relation between zodiac signs and Egyptian, Macedonian, Julian calendar, and Hebrew months; the zodiac signs are illustrated; depiction of a man symbolizing the sun in the centre; two-headed, serpent-like animal at the bottom.	Bierbrauer, 'Regensburger Buchmalerei', 16; Obrist, 'La représentation', 10; Kühnel, *The end of time*, 346; Graff, 'Thirteenth figure', 321.	Graff, 'Thirteenth figure', 333–4.	Swarzenski, *Denkmäler*, 11; Bischoff, *Schreibschulen*, 196; Bierbrauer, 'Regensburger Buchmalerei', 20; eadem, *Handschriften*, 61; Obrist, 'La représentation', 3–12; Kühnel, *The end of time*, 166–7; Graff, 'Thirteenth figure', 321–34.
73v (top half)	Rota on the five climate zones of the earth (cf. Isidore, *De natura rerum* 10).		The rota ed. by Fontaine in his edition of *De natura rerum* (p. 210*bis*) has the same content, but the layout of the diagram differs considerably.	Bierbrauer, *Handschriften*, 61; Obrist, 'La représentation', 3–12.
73v (bottom half)	Calculation of the *longitudo adae* and *altitudo turris*; distance between the earth and the moon, sun, and stars.			
74r (top half)	*Ratio septizodi* with illustrations of the seven planets.	Saxl, 'Beiträge', Abb. 19; Mütherich & Dachs, *Regensburger Buchmalerei*, Tafel 84; Kühnel, *The end of time*, 331.		Swarzenski, *Denkmäler*, 11; Saxl, 'Beiträge', 166; Bierbrauer, 'Regensburger Buchmalerei', 20; eadem, *Handschriften*, 61; Kühnel, *The end of time*, 152.
74r (bottom half)	Rectangle as source of the four rivers of paradise.	Saxl, 'Beiträge', Abb. 19; Mütherich et al., *Regensburger Buchmalerei*, Tafel 84; Kühnel, *The end of time*, 331.		Bierbrauer, *Handschriften*, 61; Kühnel, *The end of time*, 151–2.
74v (top half)	Table of the concordance between Julian calendar months and zodiac signs.			Bierbrauer, *Handschriften*, 61.

Folios	Content	Facsimile	Edition	References to these folios
74v (bottom half)	Drawing of the arch.		From a different MS in slightly altered form in Mac Carthy, *Codex Palatino-Vaticanus*, 14 and *PL* 90, 1179.	Bierbrauer, 'Regensburger Buchmalerei', 20; eadem, *Handschriften*, 61.
75r–v	*Romana computatio* (tract on finger-counting).		Jones, *CCSL* 123C, 669–72; not from this MS in Jones, *Bedae pseudepigrapha*, 106–8; repr. and transl. in Williams & Williams, 'Finger Numbers', 605–6.	Sanftl, *Catalogus* III, 1729; Halm et al., *Catalogus*, 175.
75v	Tract on the impact of the weekday of 1 January on the seasons, harvest, and social life.			
76r	blank			
76v	Lists of the *termini quadragesimae* (i.e. the Julian calendar dates of *luna* 2 of the lunation preceding the Easter lunation), the Easter full moons, and the *termini rogationis* (i.e. the Julian calendar dates of *luna* 20 of the lunation following the Easter lunation).		Edited as *Lect. comp.* 1.12–14 in Borst, *Schriften*, 556–60.	Borst, *Schriften*, 261; Halm et al., *Catalogus*, 175.
76v	Table of the *concurrentes* of the 28-year solar cycle.		Edited as *Lect. comp.* 1.1B in Borst, *Schriften*, 545.	Borst, *Schriften*, 261.
77r	Correlation of epacts and corresponding lunar letters on 1 January. Bottom half is left blank.			
77v	Discussion of the number of years from the creation to the incarnation and the passion, and of the consequences on the chronology of these years.		Edited as *Lib. comp.* 1.7C–E in Borst, *Schriften*, 1122–3.	Halm et al., *Catalogus*, 175; Borst, *Reichskalender*, 75; idem, *Schriften*, 261.

78r–79v	'Der Regensburger Protestbrief von 809' (*Epist. Rat.*).		Borst, *Schriften*, 1021–33.	Sanftl, *Catalogus* III, 1729; Docen, 'Notizen', 517; Halm et al., *Catalogus*, 175; Bischoff, *Schreibschulen*, 196; Boschen, *Annales Prumienses*, 21; Borst, 'Alkuin', 70; idem, *Buch der Naturgeschichte*, 154; idem, *Schriften*, 261, 1021–6.
80r	blank			
80v–86r	*Annales Sancti Emmerammi Rastiponensis maiores*, breaking off at AD 823.	fol. 81v–82r (with transcription): Chroust, *Monumenta palaeographica 1,1* (2. Lieferung, Tafel 1).	Mabillon, *Veterum analectorum* 4, 476–7; Sanftl, *Catalogus* II, 934–6, IV, 443–4; Pertz, *MGH SS* 1, 91–93.	Docen, 'Anzeige', 428; idem, 'Notizen', 515, 517–8; Holder-Egger, 'Handschriften', 560; Halm et al., *Catalogus*, 175; Chroust, *Monumenta palaeographica 1,1* (2. Lieferung, Tafel 1); Mac Carthy, *Annals of Ulster* 4, lxvii; Bischoff, *Schreibschulen*, 196, 217; idem, *Kalligraphie*, 12; idem, *Mittelalterliche Bibliothekskataloge* 4,1, 102; Leonardi, 'Codici II', 95; Freise, 'Grundformen', 449; Borst, *Reichskalender*, 75; idem, *Schriften*, 261 (wrong connection to the Easter table of this MS).

THE MUNICH COMPUTUS: TEXT & TRANSLATION

Text:

| Compotus sancti Augustini, sancti Hieronimi, sancti Ysidori, sancti Dyonisii, sancti Quirilli Greciae, et ceterorum. fol. 8r

<I. DE DIVISIONIBUS TEMPORUM>

Definition of time

Interrogatio: *Tempus, quid est?* Responsio: *Tempus* est spatium tendens de principio *usque* in *finem.*

I,2 Interrogatio…est] This philosophical question about the nature of time may ultimately have been motivated by Augustine, *Confessiones* 11.14 (*CCSL* 27, p. 202): *Quid est enim tempus?* The answer derives from CE p. 95: *Hinc inquiritur, quantum valet tempus magnitudine et paruitate? Hoc ita soluendum est: Magnitum, ut dicitur unum tempus ab initio usque ad finem.*

I,2–3 Interrogatio…finem] MC's passage on the nature of time is directly copied in DIAL. LANGOB. 1B (cited in PV §§176–177): *Interrogatio: Quid est tempus? Responsio: Augustinus dicit: Tempus proprie dicitur spatium extendens ab initio usque ad finem.* In a short tract on the divisions of time found in Paris, Bibliothèque Nationale, Lat. 5543, fol. 126r; Vatican, Biblioteca Apostolica, Reg. Lat. 1038, fol. 126r; Vatican, Biblioteca Apostolica, Reg. Lat. 1260, fol. 117r; Strasbourg, Bibliothèque Nationale et Universitaire, 326, fol. 164v the same phrase is used not to define *tempus*, but *seculum*: *seculum autem est totum spatium ab initio mundi usque ad finem.* DRC 43.1–2 preferred CE's to MC's account on the nature of time (cf. *app. font.*): *Sciendum nobis qua magnitudine et qua breuitate tempus dicitur. Magnitudine, quando sex mundi aetas et minae temporis dicuntur.* Interestingly enough, the *Anonymus ad Cuimnanum* 18 (*CCSL* 133D, p. 119) introduced his discussion of the tenses of a verb with the more general philosophical question about the nature of time, which he felt unable to answer: *Sed ante omnia requiri non inconueniens est, quid est tempus ... In tempore loquimur, et nescimus, quid est tempus.*

3–5 Compotus…ceterorum] Transcribed in Krusch, *Studien* I, p. 10; idem, 'Einführung', p. 162–3; Mommsen, *MGH AA* 9, p. 34; Halm et al., *Catalogus*, p. 175; Schwartz, 'Ostertafeln', p. 89; Cordoliani, 'Traités', p. 59; Jones, *Bedae pseudepigrapha*, p. 67; Stevens, *CCCM* 44, p. 170 (in a wrong context); Mc Carthy, 'Easter principles', p. 223; Machielsen, *CCSL Clavis Patristica* 3A, p. 188; transcribed and translated in Mac Carthy, *Annals of Ulster* 4, lxvii; partially transcribed in Docen, 'Annales Ratisponensis', p. 516; only the first three words in Esposito, '*De mirabilibus*', p. 200. **I,2–7** Interrogatio…mundus] Translated into German in Borst, *Schriften*, p. 1; only the first question transcribed in Halm et al., *Catalogus*, p. 175. **2–3** Tempus[2]…finem] Transcribed in Borst, *Schriften*, p. 360; cf. p. CXXX–CXXXI, CLXXV of the introduction.

Translation:

Computus of St Augustine, St Jerome, St Isidore, St Dionysius, St Cyril of Greece, and others.

1. ABOUT THE DIVISIONS OF TIME

Question: What is time? Answer: Time is the interval extending from the beginning to the end.

14 divisions of time

Diuisiones temporum quot sunt? Diuisiones temporum sunt XIIII: *Atom*os, *moment*um, *minutum*, *punctum*, *hora*, *quadran*s, *die*s, ebdoma, *mens*is, uicissitudo triformis, *annu*s, *secul*um, *aeta*s, mundus, ⌐ indictio ¬ .

I,7 uicissitudo triformis] *om. M**. **8** indictio] *The indiction is out of sequence here, and no separate chapter is dedicated to it later in the text, contrary to all the other units of time mentioned in this list; in fact, the indiction only occurs in the following passage, but nowhere else in this computus; therefore I believe that it originally was a gloss that crept into the text at a later copying stage (likewise Borst), which is confirmed by the fact that the inclusion of the indiction here would make 15 rather than the postulated 14 divisions of time.*

4–16 Diuisiones[1]...DXXXII] The concept of the 14 divisions of time and their relation goes back to an Irish tract on the topic, *De divisionibus temporum*; a number of Carolingian recensions of this texts survive (one printed in *PL* 90, col. 653–664 (cf. *app. comp.*, where it is listed as DDT), for the MSS of the others cf. Jones, *Bedae pseudepigrapha*, p. 48–50; Graff, 'Recension of two Sirmond texts', p. 118–9), the Irish archetype, however, is lost. This now lost Irish text probably was the source for MC here. The concept itself appears to be an expansion of Isidore's list in ETYM. 5.29.1 (which was used by Bede with the omission of *lustris*: DT 1.2–3, DTR 2.3–4 (cited in RM 9.9–10): *Tempora autem momentis, horis, diebus, mensibus, annis, lustris, saeculis, aetatibus dividuntur*. CE p. 85, 86, 92–3 does not refer to the concept of the 14 divisions of time when assembling all the information about the divisions of time available to him, including Isidore's list: *Quid minimum sit numeri, quod meruit nominationem specialem? Hoc est momentum. Alii athomum minorem esse contendit, quod falsum non est, nam indiuisum interpretatur. Quomodo crescunt numeri in nominationibus specialibus? A momento in minutum, a minuto* (corr. from *minutu*) *in punctum, a puncto in horam, ab hora in quadrantem, a quadrante in diem, a die in septimanam, a septimana in mensem, a mense in tempus, a tempore in annum, ab anno uero in circulum XIX annorum secundum lunae cursum ratione saltus. Si secundum solis cursum in XXVIII annos ratione bissexti. Communi itaque ordine solis scilicet et lunę, hoc est in die mensis et in die septimanę et in etate lunę, in circulum annorum DXXXII numerorum series crescit. Et hoc modo in eundem circulum peracto cursum quo graditur circulus solis et lunę reuoluitur: Isidorus numerorum incrementa ita diuidit, dicens: Tempora autem momentis, horis, diebus, mensibus, annis, lustris, sęculis, aetatibus diuiduntur. ... Itaque annus IIII temporibus; tempus III mensibus; menses IIII septimanis; septimana VII diebus; dies IIII quadrantibus; quadrans III horis; hora IIII punctis; punctum II minutis et dimidio; minutum IIII momentis. ... momenta in minutis; minuta in punctis; puncta in horis; et reliqua usque annos et ciclos*. Accordingly, it seems that CE was composed before and set the ground for the Irish *De divisionibus temporum* now lost, and that MC worked from both texts.

4–8 Diuisiones[1]...indictio] MC's initial question is copied in DIAL. LANGOB. 3A (cited in PV §180): *Divisiones temporum quot sunt ... ?*, though this text continues to argue for 12 rather than 14 divisions of time, omitting *atomos* and *mundus*, in this apparently following CE. The same 14 divisons as in MC are found in DRC 13.1–3: *Sciendum nobis quot sunt diuisiones temporis. Quattuordecim, que sic nominantur: atomos, momentum, minutum, punctum, hora, quadras, dies, ebdoma, mensis, uicissitudo triformis, annus, aetas, seculum, mundus*. DDT 1A2–7 (with the later addition of *cyclus*): *Discipulus. Divisiones temporis quot sunt? Magister. Quatuordecim. D. Quae? M. Atomus, momentum, minutum, punctus, hora, quadrans, dies, hebdomada, mensis, vicissitudo triformis, annus, cyclus, aetas, saeculum, mundus*. They can also be found in the following (and doubtlessly in many other) unpublished tracts, often in the same wording as in DDT (but without the later addition of *cyclus*) and all of them clearly related: 1) Paris, Bibliothèque Nationale, Lat. 5543, fol. 126r; Vatican, Biblioteca Apostolica, Reg. Lat. 1038, fol. 126r; Vatican, Biblioteca Apostolica, Reg. Lat. 1260, fol. 117r; Strasbourg, Bibliothèque Nationale et Universitaire, 326, fol. 164v. 2) Karlsruhe, Landesbibliothek, 442, fol. 63v–64r; Geneva, Bibliothèque Publique et Universitaire, 50, fol. 139r–v; Leiden, Universiteitsbibliotheek, Scaliger 28, fol. 40r–v. 3) Paris, Bibliothèque Nationale, Lat. 2341, fol. 9r;

How many divisions of time exist? There are 14 divisions of time: the *atomos*, the *momentum*, the *minutum*, the *punctum*, the hour, the quarter-day, the day, the week, the month, the season, the year, the *saeculum* (millennium), the *aetas* (age), the *mundus*, ⌐ the indiction ¬ .

Paris, Bibliothèque Nationale, Lat. 7418 , fol. 89v–90r; Paris, Bibliothèque Nationale, Lat. 5543, fol. 125r–126r; Strasbourg, Bibliothèque Nationale et Universitaire, 326, fol. 164r–v; Vatican, Biblioteca Apostolica, Urbs Lat. 290, fol. 34v–35r. 4) Vatican, Biblioteca Apostolica, Reg. Lat. 1855, fol. 84r. 5) Karlsruhe, Landesbibliothek, 229, fol. 33r. A very interesting unpublished prologue to Bede's DT is related to these texts, listing 13 instead of 14 divisions of time, omitting *mundus*; it can be found in Vatican, Biblioteca Apostolica, Pal. Lat. 1448, fol. 1v–2r; Vatican, Biblioteca Apostolica, Pal. Lat. 1447, fol. 3r. Most of these MSS continue to discuss the divisions of time individually, but since they are usually close to DDT, they have not been collated with MC in the following; a thorough study of all texts dealing with these 14 divisions of time is desperately needed. Note that the order of *saeculum* and *aetas* is changed compared to MC in all of the texts mentioned above; later, MC himself discusses *aetas* before *saeculum* (c. 42–43). RM 10.34–39 also refers to 14 divisions of time, though at his time, in AD 820, the list has been considerably modified, abandoning the week and *mundus*, while introducing the *ostenta* and *partes* from DTR 3. A different number of divisons can be found in various unpublished tracts; 16 are given in BC 89D13–A3, which are, however, nothing but the 15 listed in DDT, i.e. the 14 divisions with the addition of *cyclus*, since the season is listed twice under two different terms (*tempus* and *vicissitudo triformis*); this chapter of BC is closely related to MC in certain features (cf. c. 2, ll. 14–18 below).

4–8 Diuisiones[1]...indictio] Cf. Borst, *Schriften*, p. 434 and p. CXLI–CXLIII, CLIX–CLX, CLXXV, CLXXVIII, CXCII–CXCIII of the introduction.

The relation between these units of time

Atomos XVI in momentum; *momen*ta *IIII* in *minutum*; *duo minut*a *et med*ium in *punctum*; *puncti IIII* in *hor*am; *hor*e *III* in *quadran*tem; *quadrant*es *IIII* in *di*em; *di*es *VII* in ebdomadem; ebdomadae *IIII* in *mens*em; *mens*es *III* in *tempus*; *tempor*a *IIII* in *annu*m; anni mille in seculum; ⌐ per tér quinos annos indictio ¬ ; *an*ni *XVIIII* in *cursum lunae*; in *cursum solis XXVIII*; in *solis et lunae* adunatum *curs*um *DXXXII*.

---

9 XVI] XV *M (corr. from* BC, *cf. app. comp. c. 2, ll. 14–18).* **13** anni] annos *M, Borst.* **14** per…indictio] *I suspect that this phrase was a gloss originally, connected to the earlier interpolated reference of the indiction above.* | anni] annos *M, Borst.* | XVIIII] XVIII *M**.

---

14 per…indictio] This gloss may be based on PROLOGUS DIONYSII (Krusch, *Studien* II, p. 64): *indictiones per annos XV solita revolutione decurrant.*

9–16 momenta…DXXXII] This passage on the relation between the units of time clearly influenced the second part (4B) of DIAL. LANGOB. 3B, 4B (only the first part cited in PV §183): *A momento in minutum, de minuto in punctum, de puncto in horam, de hora in quadrantem, de quadrante in diem, de die in septimanam, de septimana in mensem, de mense in annum, de anno in cyclum, de ciclo in aetatem, de aetate in saeculum decurrunt. ... Quattuor momenta in minutum. Duo minuta et dimidium in punctum. Quattuor puncti in horam. Tres horae vel sex in quadrantem. Quattuor quadrantes in diem. Septem dies in ebdomadam. Quattuor ebdomadas integras in mensem. Duodecim menses in annum. Anni in cyclos. Cicli in saeculum.* The relation outlined in DRC 14–15 appears to reflect a later stage in Irish computistics: *Sciendum nobis quomodo diuisiones temporis crescunt. Ita: a momento in minutum, minuto in punctum, a puncto in horam, ab horam in quadrantem, a quadrante in diem, a die in ebdomadam, ab ebdomoda in mensem, a mense in uicissitudinem triformis, a uicissitudine triformis in annum, ab anno in ciclum XXVIII annorum secundum solem; secundum lunam XVIIII annorum; DXXXII secundum solem et lunam, quando fit integra reuersio. Sciendum nobis quomodo maiores diuisiones a minoribus suplentur. Ita: duo momenta et duae tertiae partes momenti in minutum. Tria minuta et duo momenta in punctum; quattuor puncti in horam; sex horae in quadrantem; quattuor quadrantes in diem; VII dies in septimanam; IIII septimanae in mensem; tres menses in uicissitudinem trifomis; quattuor uicissitudines triformis in annum exiunt, id est ab annis suplentur cicli.* The relation given in DDT 1A10–D14 is mostly based on Irish thought, but also shows Bedan (five *puncta* constituting an hour in lunar reckoning; cf. DTR 3.11–12) and Carolingian influence: *et sic crescunt de minoribus numeri ad majora; hoc est ab atomis in momentum, a momento in minutum, a minuto in punctum, a puncto in horam, a hora in quadrantem, a quadrante in diem; a die in hebdomadam, ab hebdomada in mensem, a mense in tempus, a tempore in annum, ab anno in cyclum, a cyclo in aetatem, ab aetate in saeculum, a saeculo in mundo. ... Quingenti sexaginta quatuor atomi unum momentum efficiunt. Quatuor momenta unum minutum faciunt. Decem minuta, unum punctum. Quinque puncti in luna horam faciunt. Sex horae quadrantem complent. Quatuor quadrantes unum diem complere videntur. Septem dies hebdomadam faciunt. Quatuor septimanae unum mensem faciunt. ... Tres autem menses unum tempus efficiunt. ... haec quatuor tempora unum annum efficiunt. Quatuor anni bissextilem cyclum faciunt. ... Item quindecim anni cyclum indictionum complent, et novemdecim anni cyclum lunarem implere videntur. ... Cyclus vero solaris, viginti octo annis cursum suum finire perhibetur. Cyclus magnus, in quo concordia inter solem et lunam usque dum in se revertantur post quingentos et triginta duos annos, tertio anno incipiente, in se revertitur.* The same relation as in DDT can be found in many of the MSS listed in the previous note; the difference to MC in the number of *atomos* constituting a *momentum* (564) is based on the smallest fraction (1/12×1/47=1/564 of a *momentum*) in the division of the 24 hours of a *saltus* by the 235 lunations of a 19-year lunar cycle (cf. c. 62, ll. 14–61), and may have been first introduced as the numerical value of *atomos* in Carolingian computistics. Note that in the basic units of time (and most importantly in the numerical value of *minutum* and *punctum*) Bede

There are 15 *atomos* in a *momentum*; four *momenta* in a *minutum*; two and a half *minuta* in a *punctum*; four *puncta* in an hour; three hours in a quarter-day; four quarter-days in a day; seven days in a week; four weeks in a month; three months in a season; four seasons in a year; 1000 years in a *saeculum*; ⌜ an indiction goes through three times five years ⌝ ; 19 years in the course of the moon; 28 in the course of the sun; 532 in the combined course of the sun and the moon.

---

agrees with MC (cf. DT 1.7–8; DTR 3.10–11). For the 532-year Easter cycle being a combination of the 19-year lunar and the 28-year solar cycle cf. c. 65, ll. 34–38. For the solar cycle only DIAL. BURG. 13B: *cursus solis per viginti acto annos collegitur*. For the indiction cycle DIAL. LANGOB. 26C: *indictio, quia per ter quinque annos ... exigebatur*. DT 13.4–5, DTR 48.2–5 (cited in LIB. CALC. 88, PV §210, RM 66.2–10). For the 19-year lunar cycle cf. also c. 53, ll. 16–18; c. 60; for the solar cycle cf. c. 53, ll. 18–20; c. 65.

9–10 momenta…punctum] Transcribed in Borst, *Schriften*, p. 417; cf. p. CXLII of the introduction. **10** puncti…horam] Cf. p. CXLII of the introduction. **13** anni…seculum] Transcribed in Borst, *Schriften*, p. 435; cf. also Borst, *Schriften*, p. 374. **14** per…indictio] Transcribed in Borst, *Schriften*, p. 370; cf. p. CLXXVI of the introduction. **14–15** anni…lunae[1]] Transcribed in Borst, *Schriften*, p. 365. **15** in[1]…XXVIII] Transcribed in Borst, *Schriften*, p. 364.

<II. DE ATOMO>

Etymology and definition of atomos

Atomos nomen Grecum est et interpretatur indiuisibile, Ysidoro dicente: *Atomos philosophi* dicu*nt quasdam in mundo partes minutissimas, ut uisui non pateant, nec sectionem recipiant. Huc illuc*que *fer*untur, *sicut* minut*issimi pulueres, qui infusi per* tene*bras solis radiis uidentur*. Inde atomos indiuisibile interpretatur.

Classific.

Atomos autem in IIII partibus manet, | id est *in corpore* et *in tempore*, *in numero* atque in sole. fol. 8v

Atomos in corpore

In corpore, *ut* si *lapi*dem *diuida*s *in partes*, qu*as* parti*ris in grana,* ut *arene*, quas *rursum diuid*is *in puluerem*, *qu*em *diuidi non po*tes. Inde illum puluerem atomos nominas.

II,4 uisui] usui *M (as* BC*; corr. from* ETYM., DRC, DDT, RM*)*. **11** quas] quae *M.* | diuidis] diuides *M (corr. from* ETYM., RM*)*.

II,2–7 Atomos…interpretatur] The entire chapter on the *atomos* appears to be based on the Irish *De divisionibus temporum*. Since this text is now lost, I can only refer to the ultimate sources here. As it is the case in most of the following passages, the source for the definition of the *atomos* is Isidore: ETYM. 13.2.1: *Atomos philosophi vocant quasdam in mundo corporum partes tam minutissimas ut nec visui pateant nec τομήν, id est sectionem, recipiant; unde et ἄτομοι dicti sunt. Hi per inane totius mundi inrequietis motibus volitare et huc atque illuc ferri dicuntur, sicut tenuissimi pulveres qui infusi per fenestras radiis solis videntur*. MC's main source, CE, could not be consulted for this smallest unit of time, since its only reference to the *atomos* is CE p. 85: *Alii athomum minorem esse contendunt, quod falsum non est, nam indiuisum interpretatur.* **8–9** Atomos…sole] Three of the four categories of *atomos* listed in MC here ultimately come from Isidore: ETYM. 13.2.2: *Sunt autem atomi aut in corpore, aut in tempore, aut in numero*. The fourth, *in sole*, is an interpretation of ETYM. 13.2.1 (cf. c. 2, ll. 2–7, 29–33), and was first added to the list, I believe, in the Irish *De divisionibus temporum* now lost. MC itself appears to represent a transition stage from four to five categories of *atomos*: It copied the list of four categories from *De divisionibus temporum*, but in the subsequent discussion added a further category, the *atomos in oratione*. All subsequent texts, then, followed MC's example, recording five categories from the outset. **10–13** In…nominas] The definition of *atomos in corpore* is taken from ETYM. 13.2.2: *In corpore, ut lapis. Dividis eum in partes et partes ipsas dividis in grana, veluti sunt harenae; rursumque ipsas harenae grana divide in minutissimum pulverem, donec, si possis, pervenias ad aliquam minuttiam, quae iam non sit quae dividi vel secari possit. Haec est atomus in corporibus.*

II,2–7 Atomos…interpretatur] The definition of *atomos* being the Greek term for indivisible, which is not explicitly expressed in, but ultimately derives from ETYM. (cf. *app. font.*), can be found in almost all Insular computistical texts of the late seventh, early eighth centuries. Cf. DRC 16.1–2: *Sciendum nobis quid sit atomos. Id est nomen Grecum, et indiuisum interpretatur.* DDT 2A11–12: *Atomos Graecum nomen est et compositum, et interpretatur indivisibilis, aut insecabilis.* BC 89A3–5: *Atomus nomen Graecum est, quod interpretatur indivisibilis, nam atomon apud Graecos indivisum, vel indivisibile interpretatur.* DTR 3.23–25: *atomum graece, hoc est indivisibile siue insectibile, nominant.* Likewise, the Isidorian quote given by MC was a very popular one; in fact, wherever it appears in contemporary computistical texts, the readings are closer to MC than to the Isidore's original. Cf. DRC 16.2–5: *Isidorus dicit: Atomos philosophi dicunt quasdam in mundo minutissimas partes, ut uisui non pateant, nec sectionem recipiant. Huc illucque feruntur sicut minutissimi pulueres qui effusi per tenebras radiis solis uidentur.* DDT 2B4–8: *Isidorus diffinivit, dicens: Atomos philosophi dicunt quasdam in mundo partes minutissimas, ut visui facile non pateant, nec sectionem recipiant: huc illucque feruntur, sicut tenuissimi pulveres, qui infusi per fenestras radiis solis fugantur.* BC 89A8–12: *Sicut ait Isidorus:*

2. ABOUT THE *ATOMOS*

Atomos is a Greek term and is to be translated as indivisible, as Isidore says: The philosophers call certain smallest parts in the cosmos *atomos*, as these can neither be seen, nor can they be divided further. They are carried hither and thither, like the smallest particles of dust, which appear to have been poured through the shadows into the rays of the sun. Therefore, *atomos* is to be translated as indivisible.

Atomos appear, however, in four domains, i.e. in an object and in time, in a number and in the sun.

(*Atomos* appear) in an object, as if you divide a stone into pieces, which you divide into grains, just like sand, which you divide then further into dust, which (in turn) you cannot divide any further. Therefore, you term this dust *atomos*.

---

authomus (!) *philosophi dicunt quasdam partes in mundo minutissimas ut usui* (! as M) *non pateant nec sentione* (!) *recipiant. Huc illucque feruntur, sicut minutissimi pulueres, qui effusi per tenebras solis radiis uidenter* (!). RM 11.3–8: *Atomos philosophi uocant quasdam in mundo minutissimas partes corporum, ita ut nec uisui facile pateant nec sectionem recipiant. Vnde et atomi dicti sunt, nam tomus graece diuisio dicitur, atomus uero indiuisio. Denique huc illucque uoliant atque feruntur sicut tenuissimi pulueres, qui infusi per fenestras radiis solis fugantur*. For the allegory of *atomos* being represented by particles of dust seen in particular sunlight see also c. 2, ll. 29–33 below. **8–9** Atomos…sole] All post-MC texts that discuss the classification of *atomos* list five rather than four categories, a development initiated by MC (cf. *app. font.*): DDT 2B9–13: *D. Quot sunt genera atomorum? M. Quinque. D. Quae? M. Atomus in corpore, atomus in sole, atomus in oratione, atomus in numero, atomus in tempore*. BC 89A12–14: *Athomorum generi* (!) *V sunt, id est athomus in re, athomus in corpore, athomus in oratione, athomus in numero, athomus in tempore*. RM 11.12–14: *Quinque ergo species sunt atomorum, id est atomus in corpore, atomus in sole, atomus in oratione, atomus in numero, atomus in tempore*. **10–13** In…nominas] The three texts (other than MC) discussing the classification of *atomos* (cf. previous note) give similar definitions of *atomos in corpore*, all ultimately based on ETYM., but considerably different in wording: DDT 2B14–17: *D. Atomus in corpore, quomodo est? M. Quidquid minimum in corporis, quod secari aut dividi non potest, atomus dicitur, veluti sunt minutissima grana arenarum*. BC 89A15–B3: *In corpore ueluti cum partiris quamlibet partem lapidis mille particulas. Millesima pars, quae partiri uel dividi non potest, athomus in corpore dicitur*. RM 11.16–19: *Atomus quoque in corpore est, cum corpus aliquod in partes diuidis, partes in alias partes, et hoc totiens donec ad tales minutias peruenias, quae ob sui paruitatem ullomodo diuidi non possint*.

II,2–7 Atomos…uidentur] Cf. p. CXCII–CXCIII of the introduction.

Atomos in tempore

In tempore uero atomos fit, ut *annum diuidis in menses*, *menses* uero in septimanas, septimanas *in dies*, *dies* autem *in horas*, hore uero in momenta, momentum in *quandam stillam* paruissimam. Inde haec stilla *diuidi non potest* et *atomos* dicitur.

Atomos in numero

*In numer*o autem atomos nominatur, *ut puta VIII diuid*is *in IIII*, *IIII* uero diuidis *in II*, *duo in unum*. *Un*um, quod diuidi non potest, et *atomos* dicitur.

Atomos in oratione

Et littera <atomos> in oratione pronuntiatur, Donato grammatico dicente: *Littera est pars minima uocis articulate*. Id ipso dicente *litteram partem esse minimam*, eo *quod diuidi non potest*. Iterum Donato dicente: Sententia soluitur in oratione, *oratio soluitur in syllab*is, *sillaba soluitur in litter*is. *Littera non habet* in *quo soluatur*, et *atomos dicitur*.

---

14 diuidis] diuides *M (corr. from* Etym., Ddt*)*. **20** diuidis[1]] diuides *M (corr. from* Etym., Ddt*)*. diuidis[2]] diuides *M (corr. from* Etym., Ddt*)*. **22** atomos] *om. M (add. from* Ddt*, according to* Mc*)*. oratione] corpore *M (corr. from* Ddt*)*. **27** in[2]] a *M (corr. from* Ddt*)*.

14–18 In...dicitur] The definition of *atomos in tempore* is also taken from Etym. 13.2.3: *In tempore vero sic intellegitur atomus. Annum, verbi gratia, dividis in menses, menses in dies, dies in horas; adhuc partes horarum admittunt divisionem, quousque venias ad tantum temporis punctum et quandam momenti stillam, ut per nullam morulam produci possit; et ideo iam dividi non potest. Haec est atomus temporis*. **19–21** In...dicitur] The definition of *atomos in numero* is again taken from Etym. 13.2.4: *In numeris, ut puta octo dividuntur in quattuor, rursus quattuor in duo, deinde duo in unum. Vnus autem atomus est, quia insecabilis est*. **22–28** Et...dicitur] The *atomos in oratione* is the only category of *atomos* that does not feature in the classification of *atomos* above, and it may therefore have been subsequently added by Mc. The incentive for its inclusion may again have come from Etym. 13.2.4: *Sic et littera: nam orationem dividis in verba, verba in syllabas, syllabam in litteras. Littera, pars minima, atomus est, nec dividi potest*. Yet, Mc preferred to quote the grammarians directly rather than through Isidore. The first quote in this passage attributed to Donatus is taken verbatim from Donatus, *Ars maior* 1.2 (Holtz, *Donat*, p. 603). The other two citations, even though also ascribed to Donatus, were rather taken from Sergius (?), *De littera, de syllaba, de pedibus, de accentibus, de destinctione commentarius* (Keil, *Grammatici Latini* 4, p. 475): *ideo dixit partem minimam esse litteram vocis articulatae, quod, cum omnis oratio solvatur in verba, verba denuo solvantur in syllabas, rursum syllabae solvantur in litteras, littera sola non habet quo solvatur. ideo a philosophis atomos dicitur*. Priscian, *Institutiones grammaticae* (Keil, *Grammatici Latini* 2, p. 6): *Litera est pars minima vocis compositae, hoc est quae constat compositione literarum, minima autem, quantum ad totam comprehensionem vocis literatae – ad hanc enim etiam productae vocales brevissimae partes inveniuntur – vel quod omnium est brevissimum eorum, quae dividi possunt, id quod dividi non potest*. It may well be that Mc cited these from memory, which could explain the wrong attributions.

14–18 In...dicitur] The discussions of the *atomos in tempore* are interesting for the numerical value ascribe to *atomos*; in post-Mc times, an *atomos* was generally regarded as constituting 1/564 of a *momentum*, a value taken from the division of the 24 hours of a *saltus lunae* by the 235 lunations of a 19-year lunar cycle (cf. c. 62, ll. 14–61 and c. 1, ll. 9–16); this value is given in the following discussions of *atomos in tempore*: Ddt 2C14–19: *D. Quomodo est atomus in tempore? M. Sic est: Momentum dividis in duodecim partes, unamquamque partem de duodecim partibus momenti dividis in quadraginta septem partes, quadragesima septima pars, quingentesima sexagesima quarta pars momenti. Sic est atomus in tempore*. Rm 11.26–33: *Denique atomus in tempore de quo incipiebamus*

In time, however, an *atomos* emerges when you divide a year into months, months into weeks, weeks into days, days into hours, hours into *momenta*, (and finally) a *momentum* into certain smallest drops. These drops, then, cannot be divided further and are (therefore) called *atomos*.

(What) is termed '*atomos* in a number', however, (is,) e.g., when you divide eight into four, four into two, two into one. One, which cannot be divided further, is then called *atomos*.

And a letter is declared to be an *atomos* in a sentence, as the grammarian Donatus says: A letter is the smallest part of any utterance. The same person says that a letter is the smallest part, because it cannot be divided. Furthermore, Donatus says: A passage is dissolved into sentences, a sentence is dissolved into syllables, a syllable is dissolved into letters. A letter has nothing into which it could be dissolved further, and (consequently) it is called *atomos*.

---

disputare taliter constat: cum maiora spatia temporis sicuti est dies uel hora per punctos uel etiam ceteras minores partes diuidens, ad talem particulam peruenias, quae ob sui pusillitatem nullam habeat moram, talem quae ullomodo diuidi possit, sicuti uelocissimus ictus est oculi, ipsam scias esse atomus. Est ergo atomus CCC et LXXVI pars unius ostenti. Note that this value is explicitly rejected in DRC 16.7–8, where the belief is expressed that no numerical value should be ascribed to *atomos*. MC himself, however, reflects a different, earlier tradition; in c. 1, l. 9, *M* has 15 *atomos* constituting a *momentum*; a similar number, namely 16 *atomos*, is given in the discussion of *atomos in tempore* in BC 89B11–C2, clearly reflecting the same tradition as MC: *Athomus in tempore, sicuti soluitur saeculum in aetatis, aetas in anno, annos* (!) *in tempore, tempus in mensaes* (!), *mensis in ebdomadas, ebdomadas in dies, dies in quadrantes, quadrans in horas, hora in punctis* (!), *punctis* (!) *in minuta, minutum in momentum, momentum in athomos XVI. XVIa pars, quae dividi non potest, athomus dicitur.* I do not presently know the origin of this value, but numerically BC's 16 appear more likely than *M*'s 15, and therefore I have emended MC accordingly above. **19–21** In...dicitur] The same example for explaining the *atomos in numero* as outlined in ETYM. and MC (8/2=4; 4/2=2; 2/2=1) is also given in DDT 2C10–14: *D. Quomodo est atomus in numero? M. Sic autem est: Octo dividis in bis quaternos, quatuor dividis in bis binos, duos dividis in bis, unus est: unum dividere non potes, propter hoc atomus dicitur.* RM 11.23–26: *Atomus in numero est unum; cum ergo numerum aliquem disperties, uerbi gratia, octo si diuidias in bis quattuor, item quattuor in bis bino, item binos in singulos, sola unitas indiuisibilis permanet.* A different, numerically less appropriate example can be found BC 89B7–10: *In numero sicuti soluitur millias in mille, mille in centum, centum in XX, XX in X, X in V, V in II, II in unum. Unum uero, quod diuidi uel solui non potest, athomus in numero dicitur.* **22–28** Et...dicitur] All texts discussing the *atomos in oratione* in the context of the categories of *atomos* base their account on Sergius (?); DDT even refers to him by name (in contrast to MC's ascription of this quote to Donatus). Cf. DDT 2C2–9: *Atomus in oratione, ut est littera. Dividis enim partem orationis quamlibet in syllabas. Syllabas dividis in litteram. Litteram dividere non potes. Inde Sergius Graecus dicit: Omnis oratio solvitur in verba, verba denuo solvuntur in syllabas, syllabae solvuntur rursum in litteras; sola litteras non habet in quo solvatur. Ideo a philosophis atomus dicitur.* BC 89B3–7: *Athomus in oratione, sicuti oratio in versos in partes in psillabas, psillabam in litteras. Litteras autem dividi vel solvi non potest. Ideo athomus a Latinis philosophis dicta est.* RM 11.20–23: *Atomus in oratione est minima portio, ut est littera; cum enim partem quamlibet orationis diuidis in sillabas, sillabam denuo in litteras, sola littera non habet quo soluatur.* Bede, in his very brief treatment of *atomos*, generally treats *atomos* as a literary feature: DTR 3.23–29: *Minimum autem omnium, et quod nulla ratione dividi queat tempus, atomum graece, hoc est indivisibile sive insectibile, nominant; quod ob sui pusillitatem grammaticis potius quam calculatoribus visibile est. Quibus cum uersum per uerba, uerba per pedes, pedes per syllabas, syllabas per tempora diuidant, et longae quidem duo tempora unum breui tribuant, ultra quid diuidant non habentibus, hanc atomum nuncupari complacuit.*

Atomos in sole

Atomos autem in sole nominamus, ut sicut stillicidia ignita, quae ante auroram solis prorumpunt, et inspici in illam non uidemus nec discerni. Inde philosophi, id est Pitagoras et Plato et Aristotiles, atomos nuncupauerunt eo, quod discernere in illam non potuerunt.

<III.> DE MOMENTO

Def. and etym. of *momentum*

| *Momentum est minimum atque angustissimum tempus, a moto siderum dictum. Est enim extremitas horę in breuibus interuallis, cum sibi aliquid cedit atque succedit.* Momentum autem a mouendo dictum. fol. 9r

Classific. of *momenta*

Nam *momentorum* diuersa *sunt genera*: *Momentum* autem *oculi* naturaliter dicimus, dicente sancto *Augustin*o: *In momento oculi* facta sunt *omn*ia. Sed *a moto solis momenta* numeri pensa*mus*: XL momenta in *hora*. Fisici philosophi dicunt: *Momentum* aereus *lectus solis* in caelo.

III,3 dictum] dictus *M (corr. from* ETYM., *etc.)*. **8** moto] motu *Borst (contra M)*.

29–33 Atomos...potuerunt] This passage appears to be ultimately based on ETYM. 13.2.1 (cited in c. 2, ll. 3–7; cf. *app. comp.*). **III,2–5** Momentum...dictum] Isidore's etymology and definition of *momentum* cited here was extremely popular in seventh- and eighth-century computistical literature: ETYM. 5.29.1–2 (cited in CE p. 86, DRC 17.5–7, DDT 3D11–15, BC 35C14–D3, 96B5–8, DT 1.3–5; partially in DTR 3.21–23; as a gloss to that passage of DTR in ANGERS 477, fol. 46v): *Momentum est minimum atque angustissimum tempus, a motu siederum dictum. Est enim extremitas horae in brevibus intervallis, cum aliquid sibi cedit atque succedit.* I could not find a source for nor an imitator of the etymology *momentum a movendo.* **6–10** Nam...caelo] Isidore could not be consulted for the classification of *momenta*, since he does not deal with this question. Again, MC's classification of *momenta* into natural and numerical here may derive from the now lost *De divisionibus temporum*. MC's other principal source, CE p. 85–86, does not give an explicit number of *genera momentorum*, but continues to list six: *Sed momentorum genera non sunt pauca: Aliud enim est momentum solis, momentum lunę, momentum aquę, momentum rotę, momentum ligni, momentum oculi quod diuidi non potest.* The *momentum oculi* may ultimately derive from AMBROSE, *Exameron* 4.4 (*CSEL* 32,1, p. 121; based on 1 COR 15:51–52, which is cited in the context of *atomos* in DTR 3.37–38): *non esse dubium quod tempus in atomo et in momento oculi sit adducor ut credam, quando omnes in atomo, in momento oculi resuscitamur, ut apostolus protestatur dicens: ecce mysterium dico. omnes quidem resurgemus, non omnes autem inmutabimur in atomo, in momento oculi, in nouissima tuba, et mortui resurgent incorrupti et nos inmutabimur.* In this context, cf. CE's citation of Augustine: CE p. 85–86 (citing AUGUSTINE, *Epistulae, Epistula 205*, §14 (*CSEL* 57, p. 335)): *momentum oculi, quod diuidi non potest; ut illud Augustini in momento oculi, hoc est in summa celeritate.* The connection of *momentum* with the motion of the sun may ultimately stem from AUGUSTINE, *Confessiones* 11.23 (*CCSL* 27, p. 209): *quoniam ergo dies expletur motu solis atque circuitu ab oriente usque orientem, quaero, utrum motus ipse sit dies an mora ipsa, quanta peragitur, an utrumque. Si enim primum dies esset, dies ergo esset, etiamsi tanto spatio temporis sol cursum illum peregisset, quantum est horae unius.* It is also referred to in CE p. 86, as well as the *lectus solis* analogy: *A motu igitur solis momentum priuatim dictum credimus ... Item dictus momentum est certus lectus solis.*

29–31 Atomos...discerni] The parallel passages in DDT, BC, and RM do more clearly reveal that the concept of *atomos in sole* is based on ETYM. Cf. DDT 2C1–2: *Atomus in sole, illi tenuissimi pulveres quos diximus radiis solis fugari.* BC 89A14–15: *In re sicuti praedicti pulveres.* RM 11.19–20: *Atomus in sole est ille tenuissimus puluis, quem diximus radiis solis fugari.* **III,6–10** Nam...caelo] A two-fold

We term something '*atomos* in the sun', however, like a glowing haze, which breaks forth before sunrise, and which we cannot really make out (clearly enough to be able) to examine or distinguish it at dawn. Hence the philosophers, namely Pythagoras, Plato, and Aristoteles, named the *atomos* from that which they could not distinguish at dawn.

3. ABOUT THE *MOMENTUM*

The *momentum* is the smallest and narrowest unit of time, so called from the motion of the stars. It is, in fact, the end of an hour in short intervals, when something departs and succeeds at the same time. The *momentum* is so called from moving (*mouendo*).

For there are different types of *momenta*: We speak naturally about a blink of the eye (*momentum oculi*), as St Augustine says: Everything is done in a blink of the eye. But we also consider, from the motion of the sun, *momenta* as belonging to a number: There are 40 *momenta* in an hour. The natural philosophers say: *Momentum* is a couch of copper belonging to the sun in the sky.

---

division of the *genera momentorum* similar to the one outlined in MC can be found in DRC 17.7–11: *Sciendum nobis quot sunt genera momenti. Duo. Primum naturale, de quo dicitur: Omnia in momento oculi fecisti. Secundum artificale, ut fuit momentum liquoris apud philosophos, ut scirent quot momenta habet artificialis dies, id est CCCCLXXX. Naturalis uero dies <D>CCCCLX.* A variation of DRC's account occurs in the Angers glosses: ANGERS 477, fol. 46v (gloss to DTR): *Multa momentorum genera sunt, tamen duo principalia: spetiale et naturale* (corr. from *generale*). *Spetiale, si quod hunc mare* (?) *in plenitudine sua. Naturale occuli momentum, de quo dicitur: omnia in momento occuli facta sunt.* DIAL. NEUSTR. 29aB lists three categories of *momenta* (the first two being closely related to the ones found in MC), *momentum oculi*, *momentum solis*, and *momentum maris*, without further explanation. DDT 3D16–A13 discusses four *genera momenti*, two of which showing parallels to MC: *D. Quot sunt genera momenti? M. Quatuor. ... Augustinus dicit: Unum momentum est certissimus lectus solis in caelo, et per quadraginta vices jam hora est. ... Tertium momentum est indivisibile, ut Paulus dicit: Omnia facta sunt in momento, id est, in summa celeritate.* BC 96B10–A2 refers to six *genera momentorum* but continues to list more than that, including *momentum in sole* and *momentum in oculo*: *Momentorum genera VI sunt. ... Momentum in sole, id est de eo quod sol hora orbis sui superiore ostendit, quo usque inferiora horam ibidem posuerit. Id est dum se totum momenta erit mora temporis, qui ibidem esses* (!) *videtur, certus lectis solis est in caelo. ... Momentum in oculo dicitur, de quo philosophi dicunt, quia facta sunt omnia in monumento* (!), *uel ictu oculi*. Additionally, the *motu solis* etymology can also be found in DIAL. LANGOB. 3B: *Momentum dicitur a motu solis*, the *lectus solis* allegory in RM 13.2–3: *D. Quid est momentum? M. Certus lectus solis in caelo, hoc per quadraginta uices ita emensum, horam iam reddit integram*. For the division of an hour into 40 *momenta* see also c. 1, ll. 9–16 above, and especially c. 6, l. 7; c. 7, ll. 12–13 below.

III,6–9 Momentum…hora] Cf. Borst, *Schriften*, p. 418, 439, 482. **8–10** Sed…caelo] Partially transcribed in Borst, *Schriften*, p. 434. **9** XL…hora] Transcribed in Borst, *Schriften*, p. 457.

<IIII.> DE MINUTO

Etymology and definition of *minutum*

Minutum a *minuendo* dictum, eo quod *IIII momenta* in lucem crescentem de *tenebr*is minuat *ab* XI *Kalendas Ianuarii in* XII *Kalendas Aprilis*. Ab *hoc* autem *minutum non augmentum dicitur*, quia *maiori* minorem auget, siue quia pręcedit minutio augmentum. Ita autem definitur secundum Ysidorum dicentem: Minutum est uelut minus momentum, quia minus numerat quod maius implet.

<V.> DE PUNCTO

Etym. and def. of *punctum*

Punctum uero *a pungendo* aciem oculorum dictum. Quod *ita* diuisit *Ysidorus*: *Punctus a pungendo, eo quod quibusdam punctionibus certe designationis in horologiis*

IIII,8 quod] qui *M (corr. from* DRC, RM*)*. **V,4** quibusdam] quibus *M (corr. from* CE, DDT, DRC, RM*)*.

IIII,2–8 Minutum...implet] The numerical value of *minutum* is generally based on Anatolius (?)'s theory of the increase and decrease of daylight (cf. c. 39); yet, in this theory various constants existed, namely the increase and decrease of daylight of 2⅔ *momenta* per day, the bissextile increase of 1⅓ per day, and their sum, i.e. 4 *momenta* per day. All three options for the numerical value of *minutum* are outlined in CE p. 86 (as is the etymology *minutum a minuendo*): *Quo sensu intelligitur illud nomen, quod est minutum, id est minuendo lucis scilicet et tenebrarum. Quot sunt diuisiones minuti? Duos: prima est duo momenta et duae tertiae partes momenti, altera IIII. Vel momentum unum et tertia pars momenti momenta (recte minutum?).* Of these, CE appears to have favoured the *minutum* to consist of four *momenta*, since this is the only numerical value given to the *minutum* in CE's discussion of the relation of time (p. 86; cf. *app. font.* to c. 1, ll. 4–16). MC, then, followed CE's example, while additionally justifying his choice by cross-reference to Anatolius (?)'s theory: DRP 13.214–216: *Per XV dies et horae dimidium sole ascendente per singula momenta, id est per IIII in una die, et ab VIII Kal. Ianuarii in VIII Kal. Aprilis hora deminuitur*. The following further explanation for the term *minutum*, deriving from decrease / decreasing, comes directly from CE p. 86: *Cui diuisioni utrum maiori an minori ingenitum est. Hoc nomen, quod est minutum, maiori scilicet. Quaeritur ut quid minutum dicitur et non augmentum. Nam quid \d/e nocte minuit, ad diem addit*. Finally, the citation attributed to Isidore does not occur in any of Isidore's works, nor in CE; yet, it was quite popular among Irish computists (cf. *app. comp.*) and the Isidore in question may have been an Irish teacher rather than the bishop of Seville, or this citation may have circulated among the Irish under Isidore's name in lists of exerpts. **V,2–6** Punctum...pungebant] Most of this passage appears to have been based on the now lost *De divisionibus temporum*. Isidore gives no etymology for *punctum* (cf., however, ETYM. 18.6.6: *Pugio a pungendo et transfigendo vocatus. Est enim gladius parvus et bis acutus lateri adhaerens.*); the citation ascribed to Isidore here, however, was very popular in Irish computistics, where it always appears under Isidore's name (cf. *app. comp.* and CE below); again, an Irish teacher rather than Isidore of Seville may have been referred to here, or more likely it circulated under Isidore's name in lists of excerpts. CE p. 87–8 has an account very similar to MC's, and certainly was consulted for the composition of this passage here (especially for MC's last sentence; note also that the citation attributed to Isidore occurs verbatim in CE): *Punctum a pungendo ethimologiam accepit. ... Cum geometricis uero lapis vel stipis segregans milia passuum punctum dictum est. ... Isidorus ita ut alibi definiuit: Punctus a pungendo dictus est, eo quod quibusdam punctionibus certe designationis in horologi<i>s sit designatus*.

IIII,2–4 Minutum...Aprilis] BC 33D9–13 outlines the same numerical value for *minutum* (four *momenta*) as well as the exact same explanation for it as MC: *Momentum Grecum est. Anatholius Grecus episcopus dixit: Per XV dies sole ascendente per singula momenta, id est per IIIIor in una die ab XI Kalendas Ianuaris usque in XI Kalendas Aprilis hora diminuitur et dimidia.* DRC 18.2–5, on the other

4. ABOUT THE *MINUTUM*

The *minutum* is so called from decreasing (*minuendo*), since it decreases the shadows by four *momenta* per day during the increase of daylight between 22 December and 21 March. From this it is called decrease (*minutum*) (and) not increase, because it increases less rather than more, or because the decrease (*minutio*) precedes the increase. It is defined in the following way by Isidore, who says: The *minutum* is just like a minor *momentum*, because that, which fills up more, counts less.

5. ABOUT THE *PUNCTUM*

The *punctum* is so called from penetrating (*pungendo*) the pupil of the eye. Isidore distinguished this (term) in the following way: The *punctum* (is so called) from puncturing (*pungendo*), because (this measure of time) should have been designated on horologues by some little holes (*punctionibus*) of a certain design.

---

hand, explicitly rejects MC's interpretation, arguing for 2⅔ *momenta* constituting a *minutum*: *De tenebris ad praesentiam lucis die VIII Kl. Ian., id est duo momenta et duae tertiae partes momenti. Et deinde transfertur ad id tempus, quamuis alii dicunt minutum quattuor momentos habere per ascensionem solis in primo XC, quod non est uerum.* DDT 4A16–B1, then, like most later texts, only mentions 2⅔ *momenta* as the value of a *minutum*; to avoid ambiguity, the Julian calendar interval referred to as justification for this choice is one that was considered as containing a linear increase of daylight (of 2⅔ of a *momentum* per day), but no bissextile increase (cf. c. 39): *Isidorus dicit: Minutum dictum a minuendo, quia duo momenta et duae partes tertii momenti minuuntur de tenebris in unoquoque die, ab undecimo Kalendarum Aprilis, usque in duodecimum Kalendarum Julii.* **6–8** Ita…implet] This quote is also attributed to Isidore in DDT 4B3–5: *Isidorus diffinivit dicens: Minutum dicitur velut minus momentum, quia minus numerat, sed majus implet.* DRC 18.6–7: *Isidorus dicit: Minutum dicitur uelut minus momentum, quia minus numerat quod maius implet.* It occurs without ascription in BC 96A4–6: *Iterum minutum uelut minus momentum, quia minus numerat id, quod maius implet.* RM 15.4–6: *D. Vnde dicitur minutum? M. A minore interuallo quasi minus momentum, quia minus numerat quod maius implet.* **V,2** Punctum…dictum] The etymology *punctum a pugendo aciem oculorum* was quite popular in the Irish or Irish influenced eighth-century computistical discourse. It was copied directly from MC into the Angers glosses: ANGERS 477, fol. 46v (gloss to DTR): *Punctus a pungendo aciem oculorum.* Moreover, it can be found in BC 96A6–7: *Punctum pugendo aciem oculorum dictum.* DRC 19.2–3: *Super pugendum aciem oculorum, quo sol oculos pungit, addito puncto super diem.* **3–5** Quod…designatus] This quote is unanimoulsy ascribed to Isidore in all Irish or Irish influenced texts; besides MC and CE (cf. *app. font.*), cf. DDT 5B9–11: *Isidorus ostendit, cum dixit: Punctus a pungendo dictus est, eo quod quibusdam punctionibus certae designationis in horologiis designantur.* DRC 18.6–8: *Isidorus dicit: Punctus a pungendo dictus, eo quod quibusdam conpunctionibus certe designationis in horologiis sit designatus.* It is also cited by Rabanus Maurus, though without attribution: RM 16.5–7: *Punctus quippe a pungendo dictus est, eo quod quibusdam punctionibus certe designationibus in horologiis designatur.*

IIII,2–4 Minutum…Aprilis] Cf. p. CXXVII, CXLII of the introduction. **6–8** Ita…implet] Cf. p. CXXVII of the introduction. **V,3–5** Quod…designatus] Cf. p. CXXVII of the introduction.

sit designatus. Quod primo genitum super lignum cum ferro, quo *geumetrici* terram pungebant.

Division of *punctum*

Et alii dicunt inde transferri ad nomen numeri: Duo minuti et medium in punctum ueniunt.

<VI.> DE HORA

Etymology and definition of hour (*hora*)

| *Hora nomen Latinum est* ab ethimologia *Greca* deductum. *Grec*i enim *hora*n dicunt et *umbra uel species interpretatur*, a quo horologium et *hora*m de *horologio* dicimus. *Ora* sine aspiratione *circulus maris* et *uestimentorum* dicitur. Inde poeta dicit: Pallium palliades ora circumdatum. fol. 9v

Div. of hour

*Quattuor punct*is *hora* impletur, hoc est *XL moment*is.

5 Quod] quem *M.*

7–8 Et…ueniunt] In its discussion of *punctum*, CE p. 86 only mentions that a *punctum* consists of 10 *momenta*, not referring to *minuta*; in the earlier detailed outline of the relation of the units of time (CE p. 85; cf. *app. font.* c. 1, ll. 9–16), however, the Einsiedeln computist explicitly states that 2½ *minuta* constitute a *punctum*, and this may be the source here. **VI,2–7** Hora…circumdatum] The fact that Latin *hora* derives from a Greek term ultimately comes from ETYM. 5.29.2 (cited in CE p. 88, DT 1.5–6, DRC 20.7–8): *Hora Graecum nomen est, et tamen Latinum sonat*. The translation of Greek *horan* as shadow or image appears to stem from CE p. 88: *Hora Grece finis uel umbra uel series interpretatur ... nam hora ab horologio artificiose, horologium autem ab hora naturaliter regulatur. Ora* meaning edge or boundary is taken either directly from ETYM. 5.29.2 (cited in DT 1.6–7, DRC 20.8–9): *Hora enim finis est temporis, sicut et ora sunt finis maris, fluviorum, vestimentorum.* or through CE p. 88 (almost verbatim BC 96A14–B2): *Hora enim finis est temporis, sicut et horę sunt fines maris, fluuiorum, uestimentorum; sed hora finium per o, hora dierum per h scribendum est*. The connection between *circulus* and *ora* may derive from JEROME, *Commetarii in Hiezechielem* 4.16.13b (*CCSL* 75, p. 177): *circulos in ora pendentes*. The poet's quote is unique to MC. **8** Quattuor…momentis] CE p. 88 also outlines the numerical value of an hour in its discussion of this unit of time, and therefore appears to have been the source here: *In hora itaque, si quis diligenter inuestigauerit, XL momenta et X minuta IIIIque puncta inueniet.*

5–6 Quod…pungebant] The reference to the *geometrici* in the context of *punctum* may have had its origin in CE (cf. *app. font.*), it was then taken up by MC, and from there found its way into ANGERS 477, fol. 46v (gloss to DTR): *A geometricis terrę primo dictus est punctus.* **7–8** Et…ueniunt] The only other texts referring to the numerical value of *punctum* in their discussion of this unit of time are DDT 4B4–5: *Duo minuta et dimidium, hoc est decem momenta, unum punctum faciunt*. RM 16.7–8: *Punctus autem habet minuta duo et dimidium*. Usually, the numerical value of *punctum* is mentioned in the relation of the units of time (cf. c. 1, ll. 9–16 above); note that the value of *punctum* directly depends on the value of *minutum*: If a *minutum* is reckoned as 4 *momenta*, then a *punctum* consists of 2½ *minuta*; this is MC's practice. If a *minutum* is reckoned as 2⅔ *momenta*, then a *punctum* consists of 3 *minuta* and 2 *momenta* (cf., e.g., DRC in *app. comp.* to c. 1, ll. 9–16). The reason for this dependency of the value of *punctum* on that of *minutum* is that the relation of a *momentum* and that of a *punctum* to an hour were generally accepted as 40 *momenta* as well as 4 *puncta* constituting an hour (cf. c. 6, l. 8), and therefore a *punctum* always consisted of 10 *momenta*. **VI,2–7** Hora…circumdatum] A gloss in ANGERS 477, fol. 46v (gloss to DTR) seems to derive directly from MC: *Nomen latinum ab ethimologia greca dictum, id est horan, quod figura uel species interpretatur. Enim ora finis dicitur, ut finis temporis ora dicitur, ut fit horologium. Ora sine aspiratione finis maris uel uestimenta, id est circulus, ut pallii ora circulo datur*. Similar accounts can be found in DRC 20.1–6: *Horá enim Grece umbra interpretatur. Sciendum super quod fundamentum primitus dictus sit. Id est super umbram quae obumbrabat sol lineas*

This was at first done on wood with an iron instrument, with which geometers (usually) punctured the ground.

Others say that (this term) was then transferred into the name of a number: two and a half *minuta* occur in a *punctum*.

6. ABOUT THE HOUR

Hour (*hora*) is a Latin term taken from a Greek etymology. In fact, the Greeks call it *horan*, (which) translates to shadow or image, from which we derive *horologium*, and *hora* (in turn) from *horologium*. *Ora* without aspiration means edge (*circulus*) of the sea and of garments. Thence the poet says: The cloak of Pallas that is surrounded by a decorative edge (*ora*).

An hour is completed by four *puncta*, i.e. 40 *momenta*.

---

horologii. Aliter nomen Grecum et Latinum hora, id est finis. Philosophi tabulam horologii in duodecim partes diuidebant ad significandas duodecim partes diei, et unaquaque párs hora dicebatur. DDT 6C4–D4: *D. Hora, cujus linguae est, et quomodo diffinitur? M. Hora Graecum nomen est et interpretatur finis vel tempus. D. Horologium, quid est? M. Hora, finis vel tempus; locus, series, vel umbra interpretatur. Inde horologium, umbra vel series, finiumque temporum, quia in horologio tempora per umbras mensurantur. ... Oras maris et fluviorum et vestimentorum dicimus, id est, extremitates sive terminos. Hora autem finis temporis est. D. Quomodo discernitur in orthographia inter oras maris et fluviorum et vestimentorum, et inter horam temporis?. M. Oras maris et fluviorum et vestimentorum quando dicimus, sine aspiratione profertur; hora autem, hoc est finis temporis, cum aspiratione.* BC 96B4–6 (and cf. *app. font.*): *Hoc nomen Grecum est, quod Latini dicere species uel pulchritudo.* RM 17.9–18: *D. Hora unde nomen accepit? M. De horologio, sicut horologium de hora nomen sumpsit, horae quoque nomen ex greca origine descendens, interpretatur series uel umbra siue etiam finis. Inde horas maris et fluuiorum et uestimentorum dicimus extremitates siue terminos. D. Numquid eodem modo scribitur hora temporis et hora ceterarum quas predixisti rerum? M. Aspiratione quoque sola discernuntur, quia hora temporis cum aspiratione ceterorum autem sine aspiratione scribuntur.* **8** Quattuor…momentis] Within the discussion of this unit of time, the relation of an hour to the smaller divisions of time is also given in DDT 5B16–17: *Quattuor puncti sunt quadraginta momenta, et unam horam faciunt.* RM 17.19–23: *D. Hora quas subdiuisiones habet? M. Habet quoque punctos quattuor in solari computatione, ... Habet et minuta decem, ... momenta XL ...* . For the relation of these units of time cf. also c. 1, ll. 9–16 above, and c. 7, ll. 12–13 below.

VI,2–7 Hora…circumdatum] Cf. p. CLXXXV of the introduction. **8** Quattuor…momentis] Transcribed in Borst, *Schriften*, p. 457; cf. p. CXLII of the introduction.

<VII.> DE QUADRANTE

Definition, etymology of and names for quarter-day (*quadrans*)

*Quadran*s Latine, Grece autem dodras, *Hebra*icę qu*odra*s. Quadrans quartam partem diei tenet. Si diei abusiue, tres horis, si uero diei naturalis, horis sex conficitur. *Quadrans* autem secundum Ysidorum nomen numeri, *quarta pars nummi.* Inde Latine transfertur, et quarta pars diei dicitur. Unde in Euangelio ait: *Non* ueni*es inde* usque ad *nouissimum quadrantem*. Quattuor enim deo reddimus: Cogitatio, uerbum, opus, perfectio. Unde quadrans in initio mundi primum est finis lucis angelicę antequam oreretur sol in VIIII Kalendas Aprilis.

VII,4 tres] *corr. to* tribus *Borst*. **6** nummi] numbi *M (corr. from* JEROME, CE*)*.

VII,2–11 Quadrans…Aprilis] The Hebrew term for *quadrans* is copied from ETYM. 16.25.17: *Quadrantem Hebraei similiter codrantem vocant*. The connection between *quadrans* and *dodrans* may ultimately derive from a wrong equation of these terms where they appear in the same sentence (note that *dodrans*=¾ is the complement of *quadrans*=¼). Cf. BOETHIUS, *Liber de diffinitione* (*PL* 64, col. 906C; cited in CASSIODORUS, *Institutiones* 2.3.14 (Bürsgens, *Cassiodor* 2, p. 370)): *ut si quaeratur quid sit quadrans, respondeatur cui dodrans deest, ut sit assis*. PRISCIAN, *Institutiones grammaticae* (Keil, *Grammatici Latini* 2, p. 319): *'hic quadrans huius quadrantis', 'dodrans dodrantis'*. More likely, however, is the possibility that a misinterpretation of CE p. 89 lies behind this confusion; in the passage in question, a possible Greek origin of *quadras* or *quadrans* is analyzed by comparison with the Greek term for *integritas*, namely *doras* or *dodras*, gen. *dodrantis*: *Quadrans, quo sensu hoc nomen intellegitur? Aut dubium quo\d/ III significationes dicitur, hoc est: Quadras, ut dicitur gigas, gigantis; et quadrans participium a uerbo quadro, et nomen est quadrans; Grece autem integritas doras uel dodras, dodrantis*. Furthermore, the definition of *quadrans* as the fourth part of a day (for the *dies naturalis* and *dies abusive* in this respect cf. c. 8, ll. 7–9, 14–20) and its origin as a monetary unit is also taken from CE p. 89: *Quomodo definitur, sic soluitur hoc modo: Quadrans quarta pars diei est. Super quod generatur hoc nomen, quod est quadrans? Non dubium qui ponderi primo generatum est, ut dicitur quadrans quarta pars nummi. Haec autem omnia nomina translatiue tempori usurpata sunt*. The ultimate source for the fact that *quadrans* constituted a monetary unit is, here, not Isidore (as MC suggests), but JEROME, *Expositio in Evangelium secundum Matthaeum* (*PL* 30, col. 546C): *Quadrans est quarta pars nummi*. Cf., however, ETYM. 16.25.17 (and *app. comp.*): *et vocatur quadrans quod unciae quartam partem adpendeat*. The Gospel-reference appears to be a (considerable) variation of MAT 5:26 (cited in BC 97C1–2, introduced with the phrase *Ut in Euangelio legitur*): *non exies inde donec reddas novissimum quadrantem*. I could not locate a source for the final two sentences of this passage.

VII,2–5 Quadrans…conficitur] Cf. ANGERS 477, fol. 49v (gloss to DTR), which is copied directly from MC: *Latine quadras, Grece dodras*. The mistaken belief that *quadrans* and *dodra(n)s* refer to the same thing in different languages (rather than being complementary; cf. *app. font.*) can exclusively be found in Irish or Irish infuenced text: DRC 21.2–4: *Isidorus dicit: Quadras quartam unciae habet partem, quam Ebrei quasi quodrantem, Greci dodrantem, Latini quadrantem uocant*. DDT 7D11–12: *quem Hebraei quadrantem, Graeci vero dodrantem, Latini quadrantem vocant*. BC 97C10–13: *Quadrans dicitur, eo quod quartam partem diei suspendit, quem Ebrei quadrantem, Greci dodrantem, Latini quadrantem uocant*. The origin of this mistake seems to lie in MC's (or rather the now lost *De divisionibus temporum*'s?) misinterpretation of CE (cf. *app. font.*). Bede in the early eighth, and Helperic in the early tenth century do not make the same mistake: DTR 4.36–37 (cited in RM 8.35–36, PV §41): *Si in quattuor, quarta pars quadrantis nomen; residuae tres dodrantis accipiunt*. HELPERIC, *De computo* 1 (*PL* 137, col. 21C): *Si quadrans quid sit dilucide nosse desideras, scito quia quidquid in quatuor aequas partes divideris, unaquaeque earum quadrans, tres autem reliquae nominantur*

7. ABOUT THE QUARTER-DAY

(A quarter-day is called) *quadrans* in Latin, *dodras* in Greek, *quodras* in Hebrew. A quarter-day comprises the fourth part of a day. If it belongs to an artificial day (*dies abusive*), it is completed by three hours, if it belongs to a natural day, it is completed by six hours. According to Isidore, *quadrans* is the term for a specific numerical value, (namely) the fourth part of a coin. Thence it is transferred into Latin, (where) it denotes the fourth part of a day. Whence it says in the Gospel: You will not come from there before the final quarter of a day (*quadrans*). In fact, we offer God four things: meditation, the word, labour, and perfection. Therefore, the quarter-day at the beginning of the world constituted the end of the angelic light before the sun rose on 24 March.

---

dodrans. Cf. also RM 18.6–7. The distinction between an artificial and a natural quarter-day of three and six hours respectively (or at least the terminology) can also be found in DRC 21.6–8: *Sciendum nobis quod inuenitur apud Grecos dodras sex horarum et dodras trium punctorum uel horarum, ut sit apud Latinos quadras sex horarum atque trium horarum.* DDT 7D15–16: *Sic etiam et quadrans in tempore sex horarum, quarta pars est diei naturalis.* For the *dies naturalis* and *dies abusive* cf. c. 8, ll. 7–9, 14–20. For the quarter-day of six hours cf. c. 41, ll. 77–79. **5–7** Quadrans…dicitur] The statement that *quadrans* is a numerical term is copied directly from MC into the Angers glosses: ANGERS 477, fol. 49v (gloss to DTR): *Nomen numeri, id est IIII pars numeri.* In most contemporary computistical texts, Isidore's rather than Jerome's definition is given: DRC 21.2–3: *Isidorus dicit: Quadras quartam unciae habet partem.* DDT 7D9–10: *Quadrans dicitur a quarta parte unciae.* BC 97B9–10: *Quadrans dicitur eo quod quarta parte untiae appendit.* DRC, however, also quotes Jerome's definition, though from a different text (JEROME, *Commentarii in Matthaeum* 1.598–599; *CCSL* 77, p. 30): *Quadras genus nummi est.* **7–9** Unde…perfectio] This definition of *quadrans* referring to any one of the four human actions is uniquely quoted in ANGERS 477, fol. 49v (gloss to DTR): *Inde in Euangelio legitur: usque ad nouissimum quadrantem, id est usque qua (?) reddemus cogitationem (?) rectam, uerbum bonum, opus perfectum, perfectionem perfectam.* Related and more explicit is BC 97B12–C3: *Item in scripturis diuinis quarta pars actus humani quadrans dicitur. Actus enim hominis quattuor modis probatur consistere siue in bonum siue in malum. Idem opere, consensu, uerbo et cogitationem, ut in Euangelio legitur: non exies inde donec reddat nouissimum quadrantem, id est ultimum culpae cogitationem.* **9–11** Unde…Aprilis] For the question of the first quarter-day, i.e. the quarter-day at the beginning of the world, see c. 41, ll. 38–42; c. 44, ll. 24–26; for the angelic light at the creation of the world cf. notes to c. 8, ll. 3–4 and the eighth-century Irish reference Bible: PP I §91: *Fiat lux, id est lux angelica. Ista lux, utrum temporalis an intemporalis? temporalis in angelis, intemporalis in deo.*

VII,2–3 Quadrans…quodras] Cf. p. CXLVII–CXLVIII, CLXXXV of the introduction. **3–5** Quadrans…conficitur] Transcribed in Borst, *Schriften*, p. 364. **5–6** Quadrans…nummi] Cf. p. CLXXXV of the introduction. **9–11** Unde…Aprilis] Cf. Borst, *Schriften*, p. 516.

Division of quarter-day

Horę autem III in quadrantem, puncti uero XII, et CXX momenta.

<VIII.> DE DIEI NOMINE

Etymology and definition of day (*dies*)

Dies a *diuidendo lucem a tenebris* dictus est, dicente Ysidoro, uel deficiendo dictus est. *Dies super* opus *angelo*rum *primo creat*us. Inde translatiue pręsentia solis dicitur, ut in Genesi nuntiatur: | *Lucem diem appellauit.* fol. 10r Inde dicitur: *Dies est solis orientis pręsentia, quousque ad occasum ueniat. Dies gemine appellari solet: proprie a solis ortu donec rursu*m *oriatur; abusiue a solis ortu usque dum ueniat <ad> occasum* solis. Siue *dict*us est *dies a diis, eo quod diis iocundus.*

---

13 momenta] momentis *M*. **VIII,2** tenebris] tenebras *M**. **4** pręsentia] pręsentiae *M*. **9** ad] *om. M (add. from* ETYM., CE, DRC, DDT, BC*).*

12–13 Horę…momenta] MC's main source, CE, does not refer to subdivisions of a quarter-day in its discussion of this unit of time (p. 89); likewise, the lost *De divisionibus temporum* appears not to have mentioned them (cf. *app. comp.*). **VIII,2–10** Dies…iocundus] The etymology of day deriving from dividing light and shadow is ultimately based on GEN 1:4–5 (cited from *divisit* in RM 19.5–6): *et vidit Deus lucem quod esset bona et divisit lucem ac tenebras appellavitque lucem diem et tenebras noctem.* MC, however, appears to have taken this etymology from CE p. 89: *Aliter dies diuidendo lucem a tenebris*. It does not occur in Isidore's works, as MC claims, but cf. EPIT. 11.146–147: *sed nos dicimus 'diem' nominari, quia <a> tenebris disiungat ac diuidat*. The theory that the day was created as part of the work of the angles also derives from CE p. 90: *Quaeritur, super quod prima creatione nominatus est nomen, quod est dies? Hoc est super angelos, ut dicitur: fiat lux, et quando feci cęlum et terram et reliqua. Nam lux angelica prima in mundo facta est*. The citation from Genesis is GEN 1:5 (cited in DRC 22.3, DIAL. BURG. 5B, RM 19.5–6): *appellavitque lucem diem et tenebras noctem*. The characterisation of the day as the presence of the sun and the related two definitions of the day (the period of daylight and the period from sunrise to sunrise) are Isidore's: DNR 1.1.1–4 (*ortu* instead of *exortu* in MSS CLV, *usque dum* instead of *usquequo* in MS V): *Dies est solis orientis praesentia, quousque ad occasum perueniat. Dies gemine appellari solet: proprie a solis exortu, donec rursus oriatur, abusiue a solis ortu usquequo ueniat ad occasum*. Cf. ETYM. 5.30.1 (cited in DDT 8C13–15, DT 2.1, DTR 5.6, in turn cited in PV §51): *Dies est praesentia solis, sive sol supra terras, sicut nox sol sub terris*. MC, however, again appears to cite at least part of this through CE p. 89–90, or a copy of DNR also used by CE: *Dies est solis orientis presentia. ... ut Isidorus dicit: Dies gemine appellari solet, proprie a solis ortu donec rursum oriatur; abusiue a solis ortu usque dum ad occasum ueniat*. The etymology of day (*dies*) from gods (*diis*) ultimately seems to be a combination of ETYM. 5.30.5 (cited in DDT 8A4–5, PV §55): *Dies dicti a diis*. and EPIT. 11.145–146: *'Dies' nominatur a quibusdam, quia diis iucundus est*. which can also be found in CE p. 89: *Aliter dies a diis, eo quod iocundum erat diis ad sacrificandum utique in die idolis*.

12–13 Horę…momenta] Generally, a quarter-day is referred to as consisting of the six hours of a *dies naturalis* rather than the three hours of a *dies abusive* given here, the latter being a specifically Irish concept that enjoyed some popularity in the seventh and eighth centuries, but hardly beyond (cf. especially DIAL. BURG. 12A, as well as the bissextile increase in the first quarter of a year c. 39, ll. 10–12 and the pseudo-mathematical explanation of the bissextile increment per year c. 41, ll. 92–106), particularly since Bede (DTR 39.10–17, cited in LIB. CALC. 42A, PV §304) vehemently rejected this concept. Further divisions of this unit of time are not explicitly mentioned in other texts discussing the quarter-day; for these division cf. also c. 1, ll. 9–16; c. 6, l. 8 above. **VIII,2–3** Dies…est] MC is here cited directly in ANGERS 477, fol. 19r (gloss to DT): *<Dies> a diuidendo lucem a tenebris*. The *dies a diuidendo* etymology was a very popular one in Irish or Irish influenced texts. Cf. DDT 8A2–3: *D. Dies*

There are three hours in a quarter-day, which equal 12 *puncta* or 120 *momenta.*

8. ABOUT THE NOMINATION OF THE DAY

The day (*dies*) was so called from separating (*diuidendo*) light from shadows, as Isidore argues, or from being abandoned (*defiendo*). The day was first created during the work of the angels. Thence it is methaphorically called 'the presence of the sun', as it is announced in Genesis: He called the light day. Accordingly it is said: The day is the presence of the rising sun until it arrives at its setting. A day is usually defined in two ways: properly, from one sunrise to the next; artificially, from sunrise to sunset. Or the day (*dies*) was so called from the gods (*diis*), because it pleases the gods.

---

unde nomen accepit? M. A divisione, eo quod dividat lucem a tenebris. DRC 22.7: *dies diuidendo lucem a tenebris dicta.* BC 70A15–B1: *Isidorus dicit: ..., quae ideo dies dicta est a diuidendo, ad (!) iacendo, lucem a tenebras.* PP I §94: *inde dicitur dies diuidendo lucem a tenebris.* DIAL. BURG. 11A: *Dies dicitur ab eo, quod dividat inter lucem et tenebras.* Bede's account is based on EPIT. (cf. *app. font.*): DTR 5.2–3 (cited in PV §50, RM 19.2–4): *nomen inde sumens quod tenebras disiungat ac diuidat.* Likewise DDT 8A5–6. **3–4** Dies…creatus] For the day being the work of the angels see also the Irish reference Bible PP I §94. **6–9** Dies…solis] This Isidorian citation occurs with minor variation in the following two related accounts: DRC 23.1–2, 24.1–3: *Isidorus dicit: Dies est solis orientis praesentia quousque ad occasum perueniat. ... Isidorus dicit: Dies gemine appellari solet: proprie a solis ortu, donec rursus oriatur; abussiue a solis ortu usque quo ad occasum perueniat.* BC 70A12–15: *Dies solis orientis praesentia quousque ad ortum perueniat. Isidorus dicit: Dies gemine appellari solet propriae ab ortu sole (!) donec rursus oriatur; abusivae uero ab ortu solis usque ad occasu perueniat.* Only Isidore's definition of a proper and an artificial day is found in DDT 8D1–4: *Isidorus diffinivit, dicens: Dies gemine appellari solet; proprie, ab ortu solis, donec rursus oriatur; abusive, dies ab ortu solis, quousque ad occasum perveniat.* BC 69C2–5: *Dies \est/ solis orientis praesentia quousque ad occasum perueniat. Dies autem geminatur (!) appellari solet: propriae a solis ortu usquequo ueniat ad occasu.* A variation in RM 19.10–11, 26–27: *Naturalis quippe ac dies legitimus est ab ortu solis donec rursus oriatur ... Vulgaris siue artificalis dies est ab ortu solis usque ad occasum.* Cf. also DTR 5.5–8 (cited in PV §51). **9–10** Siue…iocundus] MC is directly cited in ANGERS 477, fol. 19r (gloss to DT): *siue <dies> a diis, eo quod iocundus est eis.* A very close reading to MC can also be found in DRC 22.6–7: *Ita dies ... dicta, uel eo quod iucundus sit diis.* For the etymology *dies a diis* cf. also DDT 8A4–5.

---

12 Horę…quadrantem] Transcribed in Borst, *Schriften*, p. 364. **VIII,2–3** Dies…Ysidoro] Cf. p. CXXVIII, CLXXXV of the introduction. **7–10** Dies…iocundus] Transcribed in Ó Cróinín, 'Virgilius Maro Grammaticus', p. 199; cf. p. CLXXXV of the introduction.

Names for day

Ema in Hebreorum lingua a dominatione, quia emat dominatio dicitur; emera in Greca a luce, emerat enim lux dicitur; dies Latine a diuidendo lucem a tenebris.

Classific. and division of the day

Dies legitimus XXIIII *horarum usque dum dies et nox spatia sui cursus ab oriente* in occidentem *sol*is *uolubilitate concludat.* Augustino dicente: Inde *diei spatia II sunt, interdianum et nocturnum.* Sed *dies uocatu*r *a parte melio*ri, quia *nox nihil est nisi absentia solis. Dies abusiue unu*m *spatium* habet, *ab oriente sole usque* in *occidentem.*

The three parts of the day

*Partes diei abusiu*e *III sunt: mane, meridies, su*b*prima.*

---

11 a] de *M, Ó Cróinín (corr. according to* MC*).*

11–13 Ema…tenebris] It seems that MC's source for the names for day in the three sacred languages was the lost *De divisionibus temporum* (cf. *app. comp.*). **14–20** Dies…occidentem] This whole passage is primarily based on Isidore. The definition of the legitimate and the artificial day derives from ETYM. 5.30.1: *Dies legitimus viginti quattuor horarum, usque dum dies et nox spatia sui cursus ab oriente usque ad alium orientalem solem caeli volubilitate concludat. Abusive autem dies unus est spatium ab oriente sole usque ad occidentem.* Note, however, that MC's account of the *dies legitimus* varies considerably from Isidore's original; in fact, the substitution of *orientalem* with *occidentem* leads to the exact opposite meaning, namely that the *dies legitimus* is the same as the *dies abusiue*, extending from sunrise to sunsets, which contradicts the first part of the sentence that argues that the *dies legitimus* consists of 24 hours, comprising both day and night. Since the same mistake occurs in DDT (cf. *app. comp.*), I am inclined to think that it was already part of the now lost *De divisionibus temporum*, from where it was copied into MC as well as DDT; accordingly, MC did not correct this quote according to ETYM. here. The division of the day into two intervals, daytime and nighttime, does not occur in Augustine's works, as MC claims, but is taken directly from DNR 1.1.4–5: *Spatia diei duo sunt, interdianum et nocturnum; et est dies horarum XXIIII, spatium horarum XII.* ETYM. 5.30.2: *Sunt autem diei spatia duo, interdianum atque nocturnum; et est dies quidem horarum viginti quattuor, spatium autem horarum duodecim.* In this instance, it is certainly not cited through CE p. 90: *Dies proprie duas diuisiones habet, interdianum et nocturnum.* Likewise, the argument that the term day (*dies*) is taken from daytime (*dies*) is copied from ETYM. 5.30.3: *Vocatus autem dies a parte meliore.* The justification for this statement, i.e. that the night is nothing but the absence of sunlight, appears to be a combination of CASSIODORUS (?), *Expositio in cantica canticorum* 4 (*PL* 70, col. 1075A): *nox nihil est aliud nisi umbra terrae.* and DNR 2.1.1–3 (cited in BC 69A3–5; the first sentence only in DRC 25.1–2): *Nox est solis absentia, quamdiu ab occasu rursus ad exortum recurrat. Noctem autem fieri umbra terrarum.* Cf. also CASSIODORUS, *Expositio psalmorum, Expositio in psalmum 104*, §16 (*CCSL* 98, p. 947)): *nox appelatur solis absentia.* AUGUSTINE, *De Genesi contra Manichaeos* 1.14 (*PL* 34, col. 183): *noctem vero nobis fieri solis absentia.* CE p. 92: *Nox est absentia lucis.* **21–22** Partes…subprima] The division of the artificial day, or daytime, into three parts, morning, noon, and afternoon, is directly taken from DNR 1.2.6 (almost all MSS read *abusiue*; MSS DL read *subpr-*): *Partes abusiui diei tres sunt: mane, meridies et suprema.* Cf. ETYM. 5.30.13: *Partes diei tres sunt: mane, meridies et suprema.* CE p. 90: *Abusiue III diuisiones habet: mane, meridies, supprima.*

11–13 Ema…tenebris] The names for day in the three sacred languages is a popular theme in Irish or Irish-influenced texts of the eighth century: DRC 22.2: *Sciendum nobis quomodo dies in tribus linguis principalibus uocatur. Ita: ella in Ebreo, emero in Greco, dies in Latino.* DDT 8A8–9: *Ema in Hebraeo, hemera in Graeco, dies in Latino.* BC 66A5–6: *ema cum Ebreis, apud Grecos emera, poenenes* (! confusion with *poenes* earlier in this passage) *Latinos dies.* The Frankish DIAL. BURG. 11A omits the

(The day is called) *ema* in the tongue of the Hebrews from rule, because rule is called *emat*; (it is called) *emera* in Greek from light, because light is called *emerat*; (and it is called) day (*dies*) in Latin from separating (*diuidendo*) light from shadows.

The legitimate day consists of 24 hours, (and it is) as long as day and night confine the period of their course from sunrise to sunset according to the circular motion of the sun. Augustine says: There are two intervals of a day, daytime and nighttime. The day is so called from the better one of these, because the night is nothing but the absence of the sun. The artificial day has only one interval, from sunrise to sunset.

Three parts belong to the artificial day: morning, noon, afternoon.

Hebrew name: *Dies in graeca emira dicitur, sed in latino dies appellatur.* **14–16** Dies…concludat] The same mistake as in MC, namely the definition of the *dies legitimus* as stretching from sunrise to sunset, contradicting the fact that it consists of both daytime and nighttime, of 24 hours, can be found in DDT 8B7–10: *Isidorus diffinivit, dicens: Dies legitimus est viginti quatuor horarum usque dies et nox spatia sui cursus ab oriente in occidentem, solis sui volubilitate concludat.* More correct is RM 19.10–11: *Naturalis quippe ac dies legitimus est ab ortu solis donec rursus oriatur, et hic dies XXIIII horas habet.* **16–17** Augustino…nocturnum] Isidore's division of a day into daytime and nighttime was extremely popular among Irish computists, which also influenced Rabanus Maurus. Cf. the related DRC 24.3–4: *Sciendum nobis quot sunt spatia naturalis diei. Id est duo: interdianum et nocturnum.* DDT 8B19–C3: *D. Quot sunt divisiones diei naturalis? M. Duae, id est, dies et nox. Inde Isidorus dicit: Spatia diei naturalis duo sunt: interdianum scilicet et nocturnum.* RM 18.12–14: *D. Quot sunt diuisiones diei naturalis? M. Principalis diuisio istius diei duo habet spatia, id est interdianum et nocturnum.* It can also be found in BC 69C5–6, though, interestingly enough, in relation to the artificial day of 12 hours (with this revealing Irish influence): *Spatia die* (!) *duo sunt, interdiurnum et nocturnum; et est dies horarum XII.* **17–20** Sed…occidentem] The definition of the night as the absence of sunlight can also be found in DDT 11B19–C2 (cited in RM 19.2–4): *D. Nox quid est? M. Solis absentia, terrarum umbra conditi, donec ab occasu redeat ad exortum.* DT 3.1: *Nox est solis absentia terrarum umbra conditi.* DTR 7.3–5: *Est autem nox solis absentia terrarum umbra conditi donec ab occasu redeat ad exortum.* Rabanus Maurus uses a different teminology for the artificial day: RM 19.26–28: *Vulgaris siue artificalis dies est ab ortu solis usque ad occasum omne tempus lucis super terram complectens, sicut e contrario nox est absentia lucis super terram.* **21–22** Partes…subprima] Cf. BC 69C6–7: *Partes abusiuae diei III sunt: mane, meridiae et subprema.* BC 70C6–7: *Partes diei tres sunt: mane, meridies, subprema.* DRC 23.2-4: *Sciendum nobis quot sunt diuisiones diei artificalis. Tres: mane et meridies et suprima.* DDT 8D5–8: *D. Quot sunt spatia artificialis diei? M. Tria. D. Quae? M. Mane, meridies et suprema.* PV §63: *Partes diei tres sunt: mane, meridies, supprema.* RM 20.2–5: *D. Quas subdiuisiones dies uulgaris siue artificalis habet? M. Tres. D. Quas? M. Mane, meridiem, et supremum.*

11–13 Ema…tenebris] Transcribed in Ó Cróinín, 'A seventh-century Irish computus', p. 107; cf. p. CLXXI of the introduction. **14–16** Dies…concludat] Transcribed in Borst, *Schriften*, p. 457, the first part only ibidem, p. 418. **16–17** Augustino…nocturnum] Cf. p. CXXVI of the introduction.

Etym. and def. of *mane*

Mane enim *a mano dictum. Manum enim antiqui bonum dicebant. Quid enim melius luce?* Vel *mane lux matutina et plena.*

Def. and etym. of *meridies*

Meridies, quod est *medidies,* id *est medius dies.* Meridies secundum Ysidorum *purus dies* dicitur. *Merum enim* Grece *purum dicitur* Latine. *In toto enim die nihil clarior* est *meridie, quando* e *medio cęlo sol rutilat et pari claritate* | totum *orbem inlustrat.* fol. 10v

Def. and etym. of *subprima*

Subprima, cum *sol* inclinatur. Inde *dicitur* subprima, *quę* sicut *superest ultima pars diei. Serum uocatur á clausis seris quando uenit nox.*

---

26 medidies] medii dies *M (corr. from* ETYM., CE; *but cf.* PV*).* **28** die] *a lacuna follows in M.*

23–25 Mane…plena] The etymology *mane a mano* ultimately derives from Isidore's very popular quote ETYM. 5.30.14 (cited in DDT 8D17–18, CE p. 90, BC 70C10–12, RM 20.9–11, PV §64): *Et dictum mane a mano; manum enim antiqui bonum dicebant. Quid enim melius luce?* It seems, however, that MC did not cite Isidore directly here, but rather through the now lost *De divisionibus temporum*, since the following statement, a variation of ETYM. 5.30.14 (cited in CE p. 90, PV §64): *Mane lux matura et plena, nec iam crepusculum*, appears verbatim in DDT (cf. *app. comp.*). **26–30** Meridies…inlustrat] The definition and etymology of noon stems from ETYM. 5.30.15 (cited in CE p. 90, PV §65 which reads *mediidies*): *Meridies dicta quasi medidies, hoc est medius dies; vel quia tunc purior dies est. Merum enim purum dicitur. In toto enim die nihil clarius meridie, quando sol de medio caelo rutilat et omnem orbem pari claritate inlustrat.* Note that the same misreading as in MC occurs in PV. **31–33** Subprima…nox] MC's definition and etymology of afternoon ultimately derives from ETYM. 5.30.16–17 (only the first part cited in CE p. 90, PV §66): *Suprema est postrema pars diei, quando sol cursum suum in occasum vertit: dicta quod superest ad partem ultimam diei. Serum vocatum a clausis seris, quando iam nox venit, ut unusquisque somno tutior sit.* Note, however, that MC also shares some features with DDT (cf. *app. comp.*) and may therefore have cited the lost *De divisionibus temporum* here.

23–25 Mane…plena] For other texts citing the Isidorian etymology *mane a mano* cf. *app. font.* The last sentence of this passage appears verbatim in DDT 8D16–17: *Mane lux matutina et plena*, which suggests that this variation of Isidore's statement already featured in the now lost *De divisionibus temporum*, the source for MC and DDT here. **26–30** Meridies…inlustrat] Isidore's definition of noon appears also in the Irish based BC 70C13–D2: *Meridies dicta est, quasi medies, hoc est medies dies; tunc enim purior micat lux, quando sol medio caelo rutilat, et totum orbem pari claritatem inlustrat.* DDT 8A(657)5–8: *D. Meridies quomodo diffinitur? M. Isidorus diffinit, dicens: Meridies dicitur, quasi purus dies. In toto enim die nihil purius meridie. Antiqui enim merum purum dicebant.* RM 20.17–20 is directly dependent on DDT: *D. Vnde meridies nomen traxit? M. Meridies dicitur quasi medius dies, siue meridies dicitur quasi purus dies. In toto enim die nihil purius est meridie; antiqui enim merum purum dicebant.* **31–33** Subprima…nox] Again, Isidore's definition of afternoon appears in variation also in the Irish based BC 70D2–3: *Subp-ma dicta est pars die nouissima, ut quia superest ultima.* DDT 8A(657)9–11: *D. Supremum quare dicitur? M. Quasi supprimendo. Tunc enim sol inclinatur ad occasum.* A version of the latter, then, influenced RM 20.24–26: *D. Supremum quare dicitur? M. A suppremendo; tunc enim sol quasi cursum supprimens inclinatur ad occasum.*

27–28 Merum…Latine] Transcribed in Bischoff, 'Das griechische Element', p. 250.

Morning (*mane*) was so called from *manus*. In fact, the ancients called something good *manus*. And indeed, what is better than light? Or morning (*mane*) is (defined as) the full light of the early morning (*matutina*).

Noon, that is midday, i.e. middle-day. According to Isidore, noon (*meridies*) means the pure day (*purus dies*). In fact, *merum* in Greek means pure (*purus*) in Latin. (And) indeed, nothing in the course of an entire day appears brighter than noon, when the sun shines (right) from the middle of the sky and illuminates the whole world with equal brightness.

Afternoon (is the time of the day) when the sun turns. Accordingly, (that part of the day) is called afternoon (*subprima*), which remains as the final part of the day (*superest ultima*). The late hour (*serum*) is so called from deadbolts (*seris*), which get closed when night falls.

Different customs for the beginning of the day

Dies apud *Aegyptios ab occasu solis inchoat*, lunam enim sequuntur. Apud *Persas* et *Chald*eos *ab ortu solis* usque ad ortum, quia secundum solem numerant. H*ebrei a sexta hora usque* in *sextam horam*, *luna*rem enim cursum inuestigant. Quos *Athenienses* sequi properant. *Romani uero media nocte diem* incipiunt, prouidentes Christum incarnatum *media nocte* in uterum uirginis et *surg*entem ac *uenturum iudicare uiuos et mortuos*. Inde sanctus Hieronimus ait: Media nocte factus est mundus et in media nocte iterum destruitur.

---

35 sequuntur] sequitur *M*. **39–40** diem…nocte] *om. Ó Cróinín*. **43** destruitur] destruetur *Ó Cróinín*.

34–39 Dies…incipiunt] The original impetus for the discussion of the various traditions concerning the beginning of the day comes from two Isidorian passages: ETYM. 5.30.4: *Dies secundum Aegyptios inchoat ab occasu solis: secundum Persas ab ortu solis: secundum Athenienses a sexta hora diei; secundum Romanos a media nocte*. DNR 2.7–11 (cited in BC 69C7–13, LIB. ANN. 22): *Initia diei alii a solis ortu putant, alii ab occasu, alii a media nocte. Nam Chaldaei a sole exorto diei initium faciunt, totum id spatium unum diem appellantes. Aeyptii autem ex initio noctis sequentis diei originem trahunt. Romani autem a medio noctis oriri diem uolunt, et in medio noctis finiri*. All the information given in MC can be found in these Isidorian passages, except for the Hebrew custom; this is additionally also mentioned in CE p. 90–1, which does not refer to the Athenians: *Hic inquiritur unde habet dies exordium secundum Chaldeos et Ebreos et Ęgyptios et Romanos. Chaldei autem a solis ortu donec rursum oriatur diem exordiunt ... Ebrei autem a sexta hora usque ad sextam horam diem arbitrantur secundum lunam conputantes ... Aegyptii a uespere usque ad uesperum diem iudicant ... Romani uero a media nocte usque ad mediam noctem diem dicunt*. Yet, MC's source rather appears to have been the now lost *De divisionibus temporum*, since MC, in this passage, shares more features with DDT (cf. *app. comp.*) than with any other text. The custom ascribed to the Hebrews and Athenians is the one followed by MC for the lunar day (cf. especially c. 44, ll. 18–80; c. 48, ll. 2–4), which gained special weight through DRP 5.97–102 (the second part cited in DRC 66.2–5): *quia nesciunt XIII et XIIII, XIIII et XV, XV et XVI, XVI et XVII, XVII et XVIII, XVIII et XVIIII, XIX et XX, XX et XXI luna posse probatissime in uno die inuenire. Omnis namque dies in lunae conputatione non eodem numero quo mane initiatur ad uesperum finitur, quia dies qui mane in luna, id est usque ad sextam et dimidium horae, XIII adnumeretur, idem ad uesperum XIIII inuenitur*. **38–43** Romani…destruitur] The fact that Christ rose in the middle of the night may derive from PSALM 118:62: *Media nocte surgebam ad confitendum tibi super iudicia iustificationis tuae*. The phrase *uenturum iudicare uiuos et mortuos* occurs verbatim in *Fides Romanorum* 7 (*CCSL* 69, p. 268). Cf. 1 PETER 4:5: *qui reddent rationem ei qui paratus est iudicare vivos et mortuos*. The quote ascribed to Jerome cannot be found in any of the saints works, but was also known and attributed to Jerome in contemporary Irish texts (cf. *app. comp.*).

34–39 Dies…incipiunt] Concerning the different traditions for the beginning of the day, the account closest to MC is DDT 8(657)A17–B9, leading to the supposition that MC worked from the now lost *De divisionibus temporum* in this passage: *D. Quomodo dies habet initium apud Hebraeos, et Chaldaeos, et Aegyptios, et Romanos, et Persas? M. Dies secundum Hebraeos et Athenienses a sexta hora diei incipit, quia Hebraei secundum lunam numerant; et sic computant quasi media die aetas lunae commutetur aut accendatur. Secundum Chaldaeos et Persas, ab ortu solis incipit dies, quia Chaldaei primum solem adorabant. Secundum Aegyptios dies incipit ab occasu solis, quando Vesper stella oritur, quae dicitur alio nomine Lucifer, et illam stellam Aegyptii primum adorabant. Secundum Romanos et Persas, dies naturalis est a media nocte usque ad mediam noctem*. DRC 26.1–5 adds the Greeks as following the same practice as the Egyptians, but omits the Hebrews: *Sciendum nobis quot principia naturalis sunt diei. Quattuor. Caldei ex ortu solis dicunt initium diei, quia solem colebant. Ebrei uero meridie initium sumunt, quia <secundum> lunam numerant. In meridie enim aetas lunae motari putantur. Greci uero et*

Among the Egyptians the day begins with sunset, as they follow the moon. Among the Persians and the Chaldeans (the day stretches) from sunrise to sunrise, as they reckon according to the sun. The Hebrews (reckon the day) from the sixth hour (i.e. noon) to the sixth hour, for they investigate the lunar course. The Athenians hasten to follow these. The Romans, however, begin a day in the middle of the night, forseeing Christ to be conceived in the womb of the Virgin in the middle of the night and his rising and his (future) coming to judge the living and the dead. Thence St Jerome says: The world was created in the middle of the night, and in the middle of the night it will again be destroyed.

Aegypti ex initio noctis initium diei putant. Romani uero ex medio noctis incipere diem et finiri uolunt. The Irish reference Bible (PP I §130) gives the same four beginnings of the day as MC, but attributes the custom of starting the day with sunrise to the Greeks rather than the Persians and Chaldeans: *In dies – id est, a sexta hora usque in sexta, ut Ebrei dicunt, uel ab ortu in ortum solis, ut Greci dicunt, uel ab initio noctis usque initium sequentis, ut Aegypti dicunt, uel a media noctis usque medium sequentis noctis, ut Romani.* BC 70B3–C2 ascribes customs to the Hebrews and Chaldeans that differ from the ones mentioned in MC, while omitting the Persians and adding the Greeks to the Egyptian tradition (like DRC): *Inicia dierum IIIIor sunt: Secundum Ebreos ab ortu solis dies; ... secundum Chaldeos a media die; secundum Grecos et Aegyptios ab ortu stellarum; secundum Romanos a media nocte; ab ortu solis Ebrei computant. ... Chaldei ideo a media die incipiunt, quia cursum lunae obseruant, quae assiduae circa media die accenditur et extinguitur. Aegyptii uero, quia astrologi sunt, et circum siderum obseruant, a uespera et ab ortu stellarum incipiunt, quarum cursos et tempora obseruant. Romani a media nocte incipiunt.* In DT 2.3–6, Bede does not refer to the Hebrews, but includes the Umbrians instead: *Hunc chaldei et persae inter duos solis exortus, aegyptii inter duos occasus, romani a medio noctis in medium, umbri et athenienses a meridie computant ad meridiem. Moyses autem a mane ad mane unum diem appellat.* In DTR 5.90–94 (cited in PV §53, RM 23.7–12), he repeats his previous account, but then also includes the Hebrews as starting the day with the morning rather than MC's noon: *Quem hebraei, chaldaei, et persae sequentes iuxta primae conditiones ordinem diei cursum a mane ad mane deducunt, umbrarum videlicet tempus luci supponentes. At contra aegyptii ab occasu ad occasum. Porro romani a medio noctis in medium. Umbri et athenienses a meridie ad meridiem dies suos computare maluerunt.* **38–43** Romani...destruitur] The justification of the Roman practice (starting the day at midnight) attributed to Jerome, namely that the world was created and will also be destroyed at that time of the day, was well known among contemporary Irish computists. Cf. DRC 26.5–7: *Romani uero ex medio noctis incipere diem et finiri uolunt sequentes Hieronimum dicentem: Media nocte factus est mundus, et in media nocte iterum distruetur.* DDT 8(657)B7–11: *Secundum Romanos et Persas, dies naturalis est a media nocte usque ad mediam noctem, propter illam auctoritatem Hieronymi, qua dixit: Quia in media nocte factus est mundus, et in media nocte iterum destruetur.* BC 70C1–4: *Romani a media nocte incipiunt, sequentes illud quod scriptum est: media nocte factus est mundus et media nocte iterum distruetur.* This quote appears in a different context also in the Irish REG. LAT. 49 (Wilmart, 'Catéchèses celtiques', p. 105): *Ideo in horis nocturnis, idest in media nocte, natus est ac resurrexit, quia in media nocte factus est mundus et in media nocte distruetur.* and the Irish influenced DE BISSEXTO I 995C13–14 (the correction is mine): *de quo Andromachus historiographus Judaeorum aliis pervigile scribens illum quadrantem mediatum noctis XI, quia in media nocte factu<s> est mundus, et in media nocte iterum destruetur.*

38–43 Romani...destruitur] Partially transcribed in Ó Cróinín, *Cummian's letter*, p. 134; cf. also p. XXII of the introduction. **41–43** Inde...destruitur] Cf. Springsfeld, *Alkuins Einfluß*, p. 205 and p. CXXXII of the introduction.

<VIIII.> DE NOCTE

Etym. of night (*nox*)

*Nox á nocendo dic*itur. Nox super opus angelorum malorum primo creata est. Inde tenebris translatius dicitur.

Reasons for its existence

Sciendum est quur Deus noctem fecit. Id est pro *animalibus* aliis *non fer*entibus *sol*em, *et* ad *temper*antiam *labor*is human*i. Noctem fieri* philosophi dicunt *quia longo itinere lassatur sol*; *cum ad ultimum cęli spatium peruen*erit, *languescit* et a*fflat ignes suos;* uel *eadem ui, qua sub terris cogitur, sup*ra *terra*m extend*it lumen et sic umbra terrae facit noctem.*

The 7 intervals of the night

| *Noctis* autem spatia *VII sunt* secundum Ysidorum, *id est crepusculum, uespert*inum, *conticinium, intempestum, gallicinium, matutinum, diluculum.* fol. 11r

Evening twilight

Crepusculum est dubia lux. Creperum enim *dubium dicimus.*

VIIII,6 quia] quam *M (corr. from* ETYM.*)*. **7** itinere] itineris *M (corr. from* ETYM.*)*.

VIIII,2–3 Nox[1]…dicitur] The etymology *nox a nocendo* is Isidore's: ETYM. 5.31.1 (cited in PV §78): *Nox a nocendo dicta, eo quod oculis noceat*. DNR 2.1.6: *Nox autem a nocendo dicta, quod oculis noceat*. Cf. EPIT. 11.147–8 (cited in DIAL. BURG. 11B): *Nox dicitur ab eo, quod humanis noceat negotiis*. CE p. 92 (cited in DRC 25.3–4): *Hinc dicta quod humanis negotiis uel oculorum obtutibus noceat*. The statement that the night was the work of the evil angels corresponds to c. 8, ll. 3–4, where the day is said to be the work of the angels. **4–10** Id…noctem] The various reasons for the existence of the night ultimately derive from ETYM. 5.31.1–3 (MSS C^2K read *pervenerit*; only the first part cited in PV §§78–9): *et ut quibusdam animantibus, quae lucem solis ferre non possunt, ad sufficientiam temparetur. ... et ut operis diurni laborem noctis requies temperet. Noctem autem fieri, aut quia longo itinere lassatur sol, et cum ad ultimum caeli spatium pervenit, elanguescit ac tabefactus efflat suos ignes; aut quia eadem vi sub terras cogitur qua super terras pertulit lumen, et sic umbra terrae noctem facit*. The references to the animals and human labour, however, rather seem to have been copied from the lost *De divisionibus temporum*, as they appear almost verbatim in DDT (cf. *app. comp.*). **11–13** Noctis…diluculum] The list of the seven intervals of the night given by MC ultimately is a synthesis of two Isidorian passages: DNR 2.2.7–8 (MS B reads *vespertinum*; cited in DRC 25.5–7 with the notable difference of *intempestiuum* – cf. BC in *app. comp.*): *Noctis partes sunt septem: crepusculum uesperum conticinium intempestum gallicinium crepusculum matutinum*. ETYM. 5.31.4 (cited in PV §81): *Noctis partes septem sunt, id est vesper, crepusculum, conticinium, intempestum, gallicinium, matutinum, diluculum*. The same list as in MC can be found and may originate in CE p. 91: *Nox VII partes habet: crepusculum, uesperum, conticinium, intempestum, gallicinium, matutinum, diluculum*. Note, however, that MC's subsequent discussion of each interval does not follow this outline, but rather DNR 2.2.8–2.3.16. **14–15** Crepusculum…dicimus] MC's etymology of evening twilight (*crepusculum*) is essentially taken from ETYM. 5.31.7 (cited in CE p. 91, PV §83, DT 3.5–6, DTR 7.58–59, in turn cited in RM 22.8–9): *Crepusculum est dubia lux. Nam creperum dubium dicitur, hoc est inter lucem et tenebras*. Additionally, MC appears also to have read DNR 2.2.8–10 (cited in DRC 25.10–11): *Crepusculum dicitur, id est creperum, quod dubium dicimus, hoc est inter lucem et tenebras*.

VIIII,2–3 Nox[1]…dicitur] The Irish reference Bible (PP I §99) is the only text to my knowledge that connects the night with the work of the angels, or, in this case, rather the fall of the angels (it also contains the *nox a nocendo* etymology): *Augustinus inquit: Noctem, quid hic nomine noctis dicitur? Lapsus anglorum, nocens luci; inde transfertur ad praesentiam tenebrarum corporalium, quia nox nocendo oculis dicitur*. The *nox a nocendo* etymology is a regular feature, in various forms, in the

9. ABOUT THE NIGHT

The night (*nox*) is so called from hurting (*nocendo*). The night was first created during the work of the evil angels. Thence it is more metaphorically so called from shadows.

It has to be known why God made the night. I.e., for other animals which do not bear the sun, and for setting bounds to human labour. The philosophers argue that the night happens because the sun is exhausted by its long journey; when it arrives at the final part of the sky, it becomes faint and breathes out its flames; or light stretches over the earth's surface with the same strength, with which the sun is driven below the earth's surface, and thus the shadow of the earth creates the night.

There are, however, seven intervals of the night according to Isidore, namely evening twilight, early evening, late evening, cavernous night, pre-dawn, morning twilight (dawn), daybreak.

Evening twilight (*crepusculum*) is dubious light. In fact, we call *creperus* dubious.

Insular computistical context. Cf. the related DDT 11C3–4 (cited in RM 21.5–6): *D. Unde dicta est nox? M. Quod noceat aspectus, vel negotiis humanis*. DTR 7.2: *Nox dicta quod noceat aspectibus vel negotiis humanis*. and BC 71D11–12, based on CE: *Nox dicta est a nocendo, eo quod obtutibus nocet humanis*. **4–10** Sciendum…noctem] The first two reasons given here for the existence of the night, namely that it was created for animals shy of the sun as well as for humans to rest from labour, were of some attraction to Irish computists and exegets, as well as their Frankish successors. Cf. PP I §129: *Cur non diem tantum fecit, cum in eo laboratur tantum? Id est, propter piaetatem in homines, ne ultra modum laborarent, uel propter animalia que solem ferre non possunt, ut consulentur*. and especially DDT 11C7–10 (cited in RM 21.8–11), which is very close to MC, leading to the supposition that both are based on the same exemplar, the lost *De divisionibus temporum*: *D. Ob quam causam facta est nox? Pro temperantia humani laboris, ut corpora requiem haberent, et ut animalibus quibusdam solem non ferentibus, victum quaesitandi daretur occasio*. On the other hand, the remaining reasons for the existence of the night copied from Isidore by MC do not occur in any other computistical text. **11–13** Noctis…diluculum] Bede gives the exact same list as CE and MC in both of his textbooks, probably working from an Irish exemplar: DT 3.2–9: *Nox .... Cuius partes sunt septem: crepusculum ...; uesperum ...; conticinium ...; intempestum ...; gallicinium ...; matutinum ...; diluculum ....* DTR 7.56–58 (cited in RM 22.2–6): *Noctes sane partes sunt septem: crepusculum, uesperum, conticinium, intempestum, gallicinium, matutinum, diluculum*. A correction of the list found in DNR is given in BC 71A3–5: *Partes noctis sex sunt: uespere, crepusculum, conticinum, intempestiuum, gallicinium, matutinum*. **14–15** Crepusculum…dicimus] Isidore's etymology of *crepusculum* was generally accepted and hardly ever altered (cf. *app. font.*). A notable extended version can, however, be found in BC 71B2–11: *Crepusculum dictum est nocturni temporis spacium a creppero* (!), *id est conmixtione tenebrarum et luminis, qui a Greci diafodi nominant, id est dubrantia inter excessum diei et introitu noctis aut exitum noctis et introitum diei; et ob hoc dubium, ...* Interestingly enough, Isidore's etymology also appears in Irish glossaries; cf. SANAS CORMAIC Y 325 (Meyer, *Sanas Cormaic*, p. 28).

VIIII,2–10 Nox[1]…noctem] Cf. Borst, *Schriften*, p. 363.

Early evening

Vesperum a *stella oriente cui hoc nomen est.* Unde *haec stella a* maioribus Romam appete*ntibus* uesperum uocata est, quae noctis initium *trans Hesperiam oriebatur.* Itaque I*speria ab Espero rege* dicta, *quę nunc Italia ab Italo duce* dicitur.

Late evening

Conticinium, quando omnes silent. Conticiscere enim silere est.

Cavernous night

Intempestum, quando nihil agi potest. Tempus enim *per se non intellegitur, nisi per actos humanos.*

Pre-dawn

Gallicinium propter gallos lucis prenuntios.

Morning twilight (dawn)

Crepusculum inter ex*cessum noctis et diei* principi*um.*

---

18 Hesperiam] uesperam *M (corr. from* CE*).* **21** Conticiscere] contitiscere *M**.

16–20 Vesperum…dicitur] The etymology of early evening (*uesperum*) deriving from a star of the same name (i.e. the first sentence of this passage) is taken directly from DNR 2.2.10–11: *Vesperum stella oriente cui hoc nomen est.* Cf. ETYM. 5.31.5 (cited in CE p. 91, PV §82): *Vesperum ab stella occidentali vocatum, quae solem occidum sequitur et tenebras sequentes praecedit.* The rest of the passage, explaining the origin of the name of the star *uesperum*, comes from CE p. 91: *Vesperum quidam cum adspiratione hesperum dicunt a prouintia Hesperea, que ab Hespero Romanorum duce nomen sortita est. Hesperus ab Hesperea ciuitate, quam construxit Hespera filia Romuli. Tale uero nomen quod est Hesperus Macedonii appellauerunt, eo quod eis a Macedonia ad Ęgyptum nauigantibus haec stella oriebatur trans Hesperiam, quae nunc Italia ab Italo siculorum rege nominata est.* Ultimately, the connection between the star *vesperum* and the king Hesperus derives from ETYM. 3.71.19: *Vesperus stella <est> occidentalis, quam cognominatam perhibent ab Hespero Hispaniae rege.* Later in ETYM. (14.4.19), Isidore connects *Hispania* with *Italia*: *Italia autem et Hispania idcirco Hesperiae dictae quod Graeci Hespero stella navigent et in Italia et in Hispania.* Isidore's own source appears to have been SAT. 1.3.15: *deinde vespera, quod a Graecis tractum est. illi enim ἑσπέραν a stella Hespero dicunt: unde et Hesperia Italia, quod occasui subiecta sit, nominatur.* **21–22** Conticinium…est] The etymology for late evening (*conticinium*) is directly taken from DNR 2.3.12–13 (cited in DRC 25.27–28; the first sentence only in DT 3.6–7): *Conticinium quando omnes silent: conticescere enim silere est.* Cf. ETYM. 5.31.8 (cited in CE p. 91, PV §84): *Conticinium est quando omnes silent. Conticescere enim silere est.* **23–24** Intempestum…humanos] The etymology for cavernous night (*intempestum*) is an abbreviated version of ETYM. 5.31.9 (cited in CE p. 91, PV §85; the first sentence only in BC 70B14–C1; the first half of the first sentence only in DT 3.7): *Intempestum est medium et inactuosum noctis tempus, quando agi nihil potest et omnia sopore quieta sunt. Nam tempus per se non intellegitur, nisi per actus humanos.* Cf. DNR 2.3.13: *Intempesta, id est inportuna, quando agi nihil potest et omnia quieta sunt.* **25** Gallicinium…prenuntios] Again, MC cites ETYM. 5.31.11 (cited in CE p. 91, PV §87): *Gallicinium propter gallos lucis praenuntios dictum.* Cf. DNR 2.3.14–15 (cited in DRC 25.28–29): *Gallicinium autem dictum est propter gallos lucis praenuntios.* **26–27** Crepusculum…principium] The etymology for dawn (*crepusculum*) is based on DNR 2.3.15 (cited in DRC 25.33–34): *Crepusculum matutinum inter abscessum noctis et diei aduentum.* Note that MC here diverts from his original list (c. 9, ll. 11–13), which has *matutinum* rather than *crepusculum* at this point. Had MC been faithful to his list, he would have quoted ETYM. 5.31.12 (cited in CE p. 91, PV §88; the first sentence only in DT 3.8–9, DTR 7.64–65, in turn cited in RM 22.15): *Matutinum est inter abscessum tenebrarum et aurorae adventum; et dictum matutinum quod hoc tempus inchoante mane sit.* Like DNR, MC ends the discussion here, and therefore does not turn to the last interval of the night, daybreak (*diluculum*); had MC continued, its discussion of *diluculum* would have been based on ETYM. 5.31.13–14 (cited in CE p. 91–2, PV §§89–90; the first two sentences only DT 3.9–10, DTR 7.65–66, in

The early evening (*uesperum*) (is so called) from the rising star that bears this name. This star, which rose above (the country) *Hesperia* at the beginning of the night, was called *uesperum* by the ancestors travelling to Rome. *Isperia*, which is now called *Italia* from the lord *Italus*, was so called from the king *Esperus*.

Late evening (*conticinium*) (is the part of the night) when everything is silent. In fact, *conticiscere* means 'to be silent'.

Cavernous night (*intempestum*) (is the part of the night) when nothing can be done. In fact, time itself is not perceived unless through human actions.

Pre-dawn (*gallicinium*) (is so called) because of the cocks (*galli*) announcing (the coming of) daylight (at this time).

Morning twilight (*crepusculum*) (is the time) between the withdrawal of the night and the beginning of the day.

---

turn cited in RM 22.16–17; the first sentence only in DRC 25.35): *Diluculum quasi iam incipiens parva diei lux. Haec et aurora, quae solem praecedit. Est autem aurora diei clarescentis exordium et primus splendor aeris, qui Graece ἠῶς dicitur; quam nos per derivationem auroram vocamus, quasi eororam. Vnde est illud: et laetus Eoos, Eurus equis. et Eosque acies.*

16–20 Vesperum...dicitur] In its etymology for *vesperum*, DRC 25.13–15 seems to solely rely on ETYM. (cf. *app. font.*): *Sciendum nobis unde uesperum nominatur. A uespere sidere, quod ab Hisperia nominatum est, quae a Hespero duce nominata est.* On the other hand, BC 71A5–10 appears to be more directly influenced by the Irish textbooks: *Vesperum dictum est ab stella occidentale, quae uesper uocatur, et ipsa stella ab Hesperia regione nomen adsumpsit, mutata u pro h, uesper dicitur. Hesperi (!) autem dicta est pridiem Spania ab Hespero regionis filio, qui hanc regionem bellicis uirtutibus subiugauit.* Bede also relates the essentials of this etymology: DT 3.6: *vesperum apparente stella huius nominis.* DTR 7.59–60 (cited in RM 22.9–10): *Vesperum apparente stella eiusdem nominis.* **21–22** Conticinium...est] Variations of Isidore's etymology for *conticinium* can be found in BC 70B11–13: *Conticinium, hoc est silentium; conticiscere enim silere est. Et inde dictum est, quia in ipso tempore omnes taceunt, et silent, et primi somnus gratia adgrauantur.* DTR 7.62 (cited in RM 22.11–12): *Conticinium, quando omnia conticescunt, id est silent.* **23–24** Intempestum...humanos] The etymology for *intempestum* found in DRC 25.28 seems to be a conflation of ETYM. and DNR (cf. *app. font.*): *Intempestiuum, quando agi nihil potest et ita diffinitur, ut Isidorus dicit: Intempestiuum, id est inactuosum, quando agi nihil potest et omnia quieta sunt.* Bede gives a variation of ETYM. (which he cites almost verbatim in DT) in DTR 7.62–63: *Intempestum, media nox, quando omnibus sopore quietis nihil operandi tempus est.* RM 22.12–14 is a conflation of DTR and DT: *Intempestum media nox, id est quasi inactuosum, quando omnibus sopore quietis nihil operandi tempus.* **25** Gallicinium...prenuntios] Variations of Isidore's etymology for *gallicinium* can be found in BC 71C1–3: *Gallicinium dictum est ..., propter gallos uidelicet lucis praenuncius, ...* DT 3.8: *Gallicinium, quando gallus resonat.* DTR 7.64 (cited in RM 22.14): *Gallicinium, quando galli cantum leuant.* **26–27** Crepusculum...principium] As most other texts analyzing the intervals of the night, MC should have discussed *matutinum* here (cf. *app. font.*). Only BC 71C6–8 gives an etymology different to Isidore's widely accepted one: *Matutinum dictum est noctis tempus a maturitate lucis, accendente die, recedente aurora.* Contrary to MC, most other texts analyzing the intervals of the night continue to discuss the final one, daybreak (*diluculum*) (cf. *app. font.*; note especially that DRC also continues to discuss *diluculum*, even though it had not been part of its original list, which was copied from DNR). Only BC breaks off at the same point as MC, but BC at least remains faithful to its original outline.

16–20 Vesperum...dicitur] Cf. p. CXL of the introduction.

The order of day and night

In prima mundi creatione lux inciperat et nox sequębatur, quia omnia a bonitate incipiebant, *ad significandum hominis lapsum*. In secunda uero per *Christ*um *nox prece*debat *diem*, quia a bonitate incurrunt omnia, *ne dies obscuretur in tenebris, sed nox lucescat in diem*. Hinc sanctus Hieronimus ait: *Nox* initium *diei, non finis preteriti*.

---

32 in^{2}] *om. M**. **33** diei] <sequentis> *add. Borst from* DIAL. BURG.

28–34 In…preteriti] Ultimately, the principal sources for this passage describing the order of day and night is DNR 1.3.12–16 (cited to *clarescere* in DRC 24.5–9, DDT 8B(657)15–C1; DRC adds *facit* after *clarescere*, DDT *fecit*): *Dies in principio operum Dei a lumine habebat exordium, ad significandum hominis lapsum; nunc autem a tenebris ad lucem, ut non dies obscuretur in noctem, sed nox lucescat in diem, sicut scriptum est de tenebris lumen clarescere, quia a delictorum tenebris liberatus homo ad lucem fidei scientiaeque peruenit*. In this instance, however, MC certainly also consulted CE p. 90: *Utrum dies primo an nox facta est? Dies utique ante aduentum Christi praecessit noctem, ut Isidorus dicit: Ad significandum hominis lapsum. Post aduentum uero Christi nox praecessit diem, ut Isidorus dicit: Ne dies obscuretur in tenebris sed nox luceat in diem*. The Jerome quote comes from JEROME, *Commentarii in Ionam* 2.1 (*CCSL* 76, p. 394): *nam et in Genesi nox non praecedentis diei est sed sequentis, id est principium futuri, non finis praeteriti*, but may have been cited through the now lost *De divisionibus temporum* (cf. *app. comp.*).

28–34 In…preteriti] The question about the order of day and night was central to early medieval computistics and was commonly answered in Isidore's words (cf. *app. font.*). DDT 8(656)A13–18, 8B(657)12–15 (followed by the citation from DNR given above; cited in RM 22.20–22) gives the most comprehensive account; note the textual parallels in the Jerome quote, which suggests that both DDT and MC cited it through the same exemplar, the now lost *De divisionibus temporum*: *D. Si nox ante diem aut post diem? M. Diei ascribitur, ut Hieronymus revelavit: Nox est enim initium diei, non finis praeteriti, quia ante crucem Christi dies praecedebat noctem, nunc autem post resurrectionem Domini nox praecedit diem. Dies, sine dubio, ab initio mundi usque ad resurrectionem Christi praecedebat noctem; a resurrectione autem Christi usque ad judicium nox praecedit diem*. Especially the Jerome quote reveals that the Frankish DIAL. BURG. 11C is based on Irish sources, if not directly dependent on MC: *Sed quid primum fuit, dies an nox? Dies primum fuit, quia Deus primum condidit lucem. Et nox post diem, quod Hieronimus dicit: Nox initium diei sequentis, non finis praeteriti*. The unanimously accepted theory that day preceded night from the creation of the world, but that this situation was reversed by Christ's crucifixion, received a wide reception through Bede's DTR 5.94–97 (cited in PV §54; up to *vesperam* in LIB. CALC. 64D; up to *vesperam* with additions from DTR 5.104–105 in RM 22.22–28): *Divina autem auctoritas, quae in genesi dies a mane usque ad mane computandos esse decrevit, eadem in evangelio totius diei tempus a vespera inchoari et consummari sanxit in vesperam; quia qui in exordio mundi prius lucem vocavit diem, deinde tenebras noctem, ipse in fine saeculorum primo noctem gloria suae resurrectionis illustravit, et sic diem se discipulis ostendendo consecravit*. It also appears in glosses to DTR (*CCSL* 123B, p. 288): *Antea enim nox sequebatur diem, id est a mane ad mane computabatur, nunc autem sequitur dies noctem, id est computatur a uespera ad uesperam, qui antea post gaudia paradisi secuta est mors et tenebrae, nunc autem post tenebras mortis et peccati sequitur caeleste gaudium*. The popularity of this theory is best illustrated by the fact that it was also discussed outside of the immediate computistical context. Cf. PS-BEDE, *De sex dierum creatione* (*PL* 93, col. 221D): *Et in principio, si nox sequebatur diem, aut dies noctem? id est, nox diem sequebatur. Et modo post adventum Salvatoris dies sequitur noctem, quod designavit hominem lapsum, quando transgressus fuit, et nox eum secuta fuit, id est, peccatum*.

28–29 In…incipiebant] Transcribed in Borst, *Schriften*, p. 363. **33–34** Hinc…preteriti] Transcribed in Borst, *Schriften*, p. 363; cf. p. CLXXI of the introduction.

At the first creation of the world, light came before night, because everything started from goodness to foreshow the failing of men. In the second (creation of the world), however, (which happened) through Christ, night preceded day, because everything turned away from goodness, so that the day does not hide in the shadows, but rather the night shines during the day. Hence St Jerome says: The night is the beginning of a day, not the end of the preceding one.

<X.> DE EBDOMADA

Def. and etym. of week (*ebdoma*)

Ebdoma est *apud Grecos et Romanos VII dierum cursus. Apud Hebreos autem VII anni sunt. Declarat hoc Danihel de LXX ebdomatibus.* | *Ebta enim Grece VII* dicitur. Latinę *hanc septimanam vocamus. Ebdoma VII feriis constat.* fol. 11v

<XI.> DE FERIIS

Etym. of *feria*

Feria a fando dicta est, uel *a fiendo*, *dicente Domino: Fiat lux et facta est lux.*

X,3 cursus] *follows* peragitur *in M. It seems that the author worked with and conflated two sources, namely the now lost De divisionibus temporum and the ultimate source,* DNR, *which resulted in this sentence having one verb too many. So for the sake of readable Latin, I have deleted* peragitur *from the text here.* | hoc] autem *M (corr. from* DNR, CE, DE HEBDOMADIBUS, BC*; it seem that the continental copyist mistook the Irish abbr. for* hoc *for the Irish abbr. for* autem*; but cf. MSS AS of* DNR, *which reads* autem hoc, *and this may well have been* MC*'s original reading).*

X,2–6 Ebdoma...constat] The fact that the first sentence in *M* contains two verbs (cf. *app. crit.*) reveals that MC worked from and conflated two sources, apparently the now lost *De divisionibus temporum* (cf. the textual parallels to DDT in *app. comp.*) and DNR 3.1.1–3 (MSS AS read *autem hoc* instead of simply *hoc*, almost all MSS *Danihel*, MSS CLASB[2] *ebdomadibus*, MS V *ebdomatibus*; cited in DE HEBDOMADIBUS, BC 56C9–11 and 57B1–3 both with the notable variant *ebdomadibus*; the first two sentences only in DRC 27.3–4; the last two in CE p. 92): *Hebdomada apud Graecos et Romanos septem dierum cursu peragitur. Apud Hebraeos autem septem anni sunt. Declarat hoc Daniel de septuaginta hebdomadis.* Cf. also ETYM. 5.32.1 (cited in PV §91): *Hebdomada dicta a numero septem dierum, quorum repetitione et menses et anni et saecula peragunt.* The etymologies for *ebdomada* and *septimana* ultimately stem from ETYM. 5.32.1 (MS T reads *Grece septem dicitur*; cited in CE p. 92, PV §91): *ἑπτὰ enim Graeci septem dicunt. Hanc nos septimanam vocamus, quasi septem luces.* MC, however, may again have taken these etymologies from the lost *De divisionibus temporum* (cf. the textual parallels to DE HEBDOMADIBUS in *app. comp.*). The final statement that a week consists of seven weekdays, then, is again ultimately based on Isidore's DNR 3.1.3–4 (cited in BC 56C11–12, where *VII* is omitted): *Hebdomada autem septem feriis constat*, but CE p. 92 is closer to MC and may therefore have been the direct source here: *Ebdoma VII feriis constat apud Grecos et Latinos.* **XI,2–3** Feria...lux[2]] The ultimate source for the *feria a fando* etymology is DNR 3.1.4 (cited in DRC 27.10–11 omitting *quasi faria*, BC 56C12–14, ANONYMUS AD CUIMNANUM 7.52–53 (*CCSL* 133D, p. 57)): *Feria quoque a fando dicta quasi faria, eo quod in creatione mundi per singulos dies dixit Deus fiat.* Cf. ETYM. 5.30.12 (cited in DDT 10A11–13, PV §62 with *appellatę* instead of *nuncupatae*): *A fando autem feriae nuncupatae sunt, quod sit in eis nobis tempus dictionis, id est in divino vel humano officio fari.* MC, however, cited this etymology through CE p. 92, where he also found the second etymology, *feria a fiendo*, and the reference to the scripture: *Feria a fando dicta est, quasi faria, hoc est opera VII dierum fando. Aliter feria a fiendo dicta est, eo quod dixit Deus: Fiat lux et reliqua.* The Bible citation is GEN 1:3.

X,2–6 Ebdoma...constat] MC's first sentence of this passage shows parallels to DDT 13A14–15 (cited in RM 24.2–3): *Hebdomada est septem dierum cursus*, which suggests that MC also used the lost *De divisionibus temporum* in this instance. The same holds true for the etymologies of *ebdoma* and *septimana*, in which MC shares close readings with DE HEBDOMADIBUS: *ebta autem grece septem dicitur latine. Item Isidorus dicit, hanc septemanam vocamus nos quoniam septem lucis habet, id est septem dies.* These etymologies, or at least the statement that *septimana* is the Latin term for Greek *ebdoma*, both meaning the same thing (week), was basic knowledge among eighth-century computists. Cf. DDT 13A15–16 (cited in RM 24.5–6): *Hebdomada ..., et a septenario numero nomen accepit; nam hebdomada Graece, septimana dicitur Latine.* DRC 27.2: *ebdoma in Greco; septimana in Latino.* DIAL.

10. ABOUT THE WEEK

The week is (defined as) the course of seven days among the Greeks and the Romans. Among the Hebrews, however, seven years constitute (a week). Daniel explains this (in his tract) about the 70 weeks. In fact, seven (*septem*) is called *ebta* in Greek. (Consequently,) we call a week (*ebdoma*) *septimana* in Latin. A week consists of seven weekdays.

11. ABOUT WEEKDAYS

The weekday (*feria*) was so called from speaking (*fando*), or from being done (*fiendo*), as the Lord said: Let there be light, and there was light.

BURG. 5B: *Ebdomada graece dicitur, in latino septimana.* BC 57A12–13: *Ebdomada dicta est a Greca appellatione quasi ebdomada a numero VII dierum dicta.* DTR 8.1 (cited in LIB. CALC. 67): *Hebdomada graece a septenario numero nomen accepit.* Gloss to DTR (*CCSL* 123B, p. 299): *Ebda enim grece, VII latine.* The fact that the week consisted of seven days is also regularly mentioned. Cf. DRC 27.12–13: *Isidorus dicit: Ebdoma septem feriis constat.* DT 4.2: *Hebdomada septem diebus constat.* DDT 13A1–17 (cited in RM 24.7): *Hebdomada ..., id est septem dies in se complectatur.* **XI,2–3** Feria...lux[2]] A direct copy of MC's account of the etymology of *feria* can be found in ANGERS 477, fol. 19r (gloss to DT): *feriae a fando dictae ... uel feriae a fiendo dictae dicente domino: Fiat lux.* The only other text that gives both of MC's etymologies is DDT 10D13–14,A3–6: *Feria enim a fando nomen accepit. Et dicitur feria quasi faria, id est nomen quod est faria. ... Vel si feria a fiendo dicitur, ut alii aestimant; fit enim verbum activum, cujus passivo fieri potest saltem feria nominari. In eo enim die primitus dictum est: Fiat lux.* All other texts only refer to the *feria a fando* etymology; of these DRC 27.10 (followed by the citation from DNR given above): *Feria enim a fando nomen accepit.* is related to DDT, while DE HEBDOMADIBUS seems somehow related to MC and CE: *Quia feria a fando dicitur quia in illo die dominico dixit Deus, Fiat lux.* which (or a different version of it) in turn influenced RM 27.10–12: *D. Unde dicitur feria? M. A fando sicilet, et ideo dominicus dies in quo dixit Deus: Fiat lux.* This etymology also appears frequently in glosses to DTR (cf., e.g., Munich, Bayerische Staatsbibliothek, Clm 18156, fol. 12v) and in various Irish commentaries to Donatus (e.g. by Muretach and Sedulius Scottus; cf. *CCCM* 40, p. 90; *CCCM* 40B, p. 133).

X,2–6 Ebdoma...constat] Cf. Borst, *Schriften*, p. 357, 418. **4–5** Ebta...vocamus] Cf. Ó Cróinín, 'A seventh-century Irish Computus', p. 107. **XI,2–3** Feria...lux[2]] Cf. Borst, *Schriften*, p. 517 and p. CXL, CLXXXV of the introduction.

The *feria* term.

Quia tota facta in eis dicebat: Et hoc nomen omnibus diebus ebdomatis communis est. Quod reuincit Y*sidorus* *dicen*do: *Prima feria* dicta *est*, eo quod *primus* dies *a feria*; *secunda feria*, eo quod *secundus* dies *a feria est*, *et sic* soluitur. Sunt enim communes et feriati dies. Hii sunt communes dies ebdomatis excepto *sabbato*, *quę feriatus* dies. Sabbatum dicitur, eo quod festiuitas in eo apud Hebreos agebatur.

Mom/week

In ebdoma sunt VĪDCCXX momenta.

The Hebrew *sabbati* term.

Cum *Hebr*eis et Grecis *dies prima una sabbati*, *secunda* dies *secunda sabbati*. Item *tertia sabbati* usque ad *sabbatum dicitur* sic.

XI,8–11 Hii…agebatur] *Note that M and* ANGERS 447 *rather suggest the following arrangement of these two sentences (but the punctuation in M is evidently often faulty):* Hii sunt communes dies ebdomatis excepto sabbato. Quę feriatus dies sabbatum dicitur, eo quod festiuitas in eo apud Hebreos agebatur. **10** Sabbatum] sabbato *M.* **15** sic] *follows* eo quod festiuitas *M, which is a mistaken repetition from c. 11, l. 10; the copyist abruptly stopped transcribing this phrase when realizing his mistake.*

4–11 Quia…agebatur] The *feria* terminology derives from DNR 3.1.6–10 (MS A reads *sabbato, quae est feriatus*; cited in BC 56C14–D4:): *Inde dies solis prima feria nuncupatur, quia primus est a feria. Item lunis proinde secunda feria quia secundus est a feria, id est a sabbato, qui est feriatus. Sic et ceteri dies tali ex numero sumpserunt uocabula.* Cf. also DNR 1.4.59–62 (cited in BC 69C15–D3): *Feriati dies, in quibus res diuina fit et abstinere homines a litibus oportet. ... Atri dies sunt qui et communes uocantur.* MC may also have read CE's p. 92 account of this terminology: *Isidorus paulo post subiungit dicens: Prima feria eo quod prima a feria est ... dicendo uero prima feria, hoc est prima a feria.* For the meaning of *sabbatum* cf. ETYM. 5.30.10 (cited in PV §60): *Sabbatum autem ex Hebraeo in Latinum requies interpretatur, eo quod Deus in eo requievisset ab omnibus operibus suis.* **13–15** Cum…sic] The Hebrew terminology for the weekdays derives from ETYM. 5.30.9–11 (cited in PV §59 with the same omissions as given in the following): *Apud Hebraeos autem dies prima una sabbati dicitur ... Secunda sabbati secunda feria ... Tertia sabbati tertia feria ... Quarta sabbati quarta feria ... Quinta sabbati quinta feria est ... Sexta sabbati sexta feria dicitur ... Sabbatum.* Isidore's source for at least the beginning of this passage in ETYM. probably was AUGUSTINE, *Enarrationes in psalmos, Enarratio in psalmum 93, Sermo 3* (*CCSL* 39, p. 1302), which was also known to MC (cf. c. 43, ll. 13–81): *Una sabbati, dies dominicus est; secunda sabbati, secunda feria ... tertia sabbati, tertia feria ... Quarta ergo sabbatorum, quarta feria.* The origin of the differing usages of *prima* and *una sabbati* (cf. *app. comp.*) is nicely explained by AUGUSTINE, *Epistula 36* (*CCSL* 31, p. 150): *Vna enim sabbati tunc appellabatur dies qui nunc dominicus appellatur, quod in euangeliis apertius inuenitur. Nam dies resurrectionis domini 'prima sabbati' a Matthaeo, a ceteris autem tribus 'una sabbati' dicitur.* Cf. also CE p. 92: ... *Ebraicam rationem secutus quia dicebatur prima sabbati, secunda sabbati, tertia sabbati et reliqua.*

4–8 Quia…soluitur] The fact that the *feria* terminology applies to all days of the week is also mentioned in DDT 10A13–14: *Omnes autem hebdomadae dies, feriae dicunt*, which earlier (DDT 10A7–9) shows some minor textual parallels to MC: *Item quaeritur, dum dicitur: Prima feria, quo sensu hic utuntur, id est, prima dies a feria, ut Isidorus dicit.* Lists of the seven *feriae* were quite popular, which are either introduced or given in full in DRC 27.8, DE HEBDOMADIBUS, DTR 8.58–63 (cited in LIB. CALC. 67, PV §§101–2, RM 27.5–9), QUAEST. LANGOB. 1C. **9–11** quę…agebatur] This sentence is directly quoted in ANGERS 477, fol. 19r (gloss to DT): *Inde feriatum sabbatum dicitur, eo quod festiuitas cum Ebreis in eo agabatur.* Isidore's definition of *sabbatum* (cf. *app. font.*) appears in variation in DE HEBDOMADIBUS: *Proptera autem a sabbato dies septimane nominantur apud hebreos quia sabbatum principalis dies erat apud illos et in eo requies erat secundum preceptum domini.*

(This is so,) because he said (the following) after everything had been done in these (days): This name (i.e. *feria*) is common to every day of the week. Isidore confirms this by saying: The first *feria* was so called, since (it is) the first day after the *feria*; the second *feria*, since it is the second day after the *feria*, and thus it continues. There are, in fact, common days and holidays. These (weekdays) are common, except for Saturday (*sabbatum*), which (is) a holiday. Saturday (*sabbatum*) is so called, since a feast was celebrated on that day among the Hebrews.

There are 6720 *momenta* in a week.

Among the Hebrews and the Greeks the first day (is called) 'one from the Sabbath', the second day 'the second from the Sabbath'. Likewise 'the third from the Sabbath' is so called, (as are the following days) up to the Sabbath.

QUAEST. LANGOB. 1B: *Sabbatum dicitur Hebraea lingua et interpretatur apud Latinos requies*. DDT 10B6–7 simply states that *sabbatum* stands for *feria* in Hebrew: *Sabbatum vero apud Hebraeos dicitur feria*. **12** In…momenta] The number of *momenta* per week is calculated according to the relation of the units of time outlined in c. 1, ll. 9–16 and especially c. 6, ll. 7: A week has seven days, a day 24 hours, an hour 40 *momenta*: 7×24×40=6720. The only other text that calculates the number of *momenta* in a week is DIAL. NEUSTR. 29aC: *Ebdoma quoque ... sex milia septingenta viginti momenta*. **13–15** Cum…sic] The Hebrew weekday terminology was an integral part of many computistical texts dealing with the week; only in MC, however, is this terminology also ascribed to the Greeks. Cf. DRC 27.8–9: *Ordine, ut est prima feria, secunda feria, ut fuit apud ueteres prima sabbati, secunda sabbati, id est prima dies a sabbato*. BC 57C1–5: *Nomina dierum ebdomadis secundum Aebreos duos modis dicuntur, id est ordinem et requiem. Hoc modo prima sabbati, IIa, IIIa, IIIIa, Va, VIa sabbati, sabbatum*. DT 4.7–8: *imitatus hebraeos qui prima sabbati, secundam sabbati, et sic caeteras a numero nominant*. QUAEST. LANGOB. 1B: *Secundo modo dicitur prima sabbati, secunda sabbati, tertia sabbati, quarta sabbati, quinta sabbati, sexta sabbati sive sabbatum*. Bede's account in DTR 8.19–22 (cited in LIB. CALC. 67, RM 26.4–7): *Quae a populo Dei hebdomada ita computabatur antiquitus: prima sabbati vel una sabbati sive sabbatorum, secunda sabbati, tertia sabbati, quarta sabbati, quinta sabbati, sexta sabbati, septima sabbati vel sabbatum*. is directly based on DE HEBDOMADIBUS: *Sic dicunt: prima sabbati vel una sabbati, secunda sabbati, tertia sabbati, quarta sabbati, quinta sabbati, sexta sabbati, sabbatum*.

9–11 quę…agebatur] Cf. p. CLXXXV of the introduction. **13–15** Cum…sic] Cf. Borst, *Schriften*, p. 516.

The Roman planetary weekdays

Romani uero *nomina die*rum *diis* et *sideribus* con*secrauerunt. Primum enim diem a sole, princi*pe *siderum,* nominant. *Secundum á luna, quae soli splendore proxima est. Tertium a stella Martis, quae Vesper uocatur. Quartum a stella Mercurii, quam candidum circulum* nomina*nt. Quintum a stella Iouis, quam* Fe*tontem* uoca*nt. Sextum a stella* | *Veneris, qua*e *Lucifer* dicitur, *quae plus lucis habet inter omnia sidera*, sicut Ysidorus ait: Sicut *sol et luna, ita et haec umbram facit. Septimum a stella Saturni, quae* in *s*eptim*o caelo XXX annis fertur* im*plere cursum suum.* fol. 12r

Charact. of the planets

Ab *his VII s*ideribus putaba*nt gentiles* Romani *aliquid s*e *habere*: *A sole* enim *spiritum, a luna* uero *corpus, a Marte* audacia*m*, *a Mercurio* autem *ingenium et linguam, a Ioue temperantiam, a Vener*i *uoluptatem, a Saturno humorem* uel *tarditatem.*

---

16 dierum] *follows* á *M, which stems from the Isidorian quote (cf. app. font.), but which makes no grammatical sense here.* **18** soli splendore] solis splendoris *M (corr. from* ETYM. *(note, however, that MS B reads* solis*),* PV*).* **24** umbram facit] umbra in facię *M (corr. from* DNR*).*

16–31 Romani…tarditatem] The principal source for the Roman planetary weekdays and the characteristics of these planets is ETYM. 5.30.5–8 (MS B reads *solis* instead of *soli*; the dett. MSS read *septimo* instead of *sexto*; cited in PV §§55–58; only the first part up to *cursum suum* in DE HEBDOMADIBUS): *Dies dicti a diis, quorum nomina Romani quibusdam sideribus sacraverunt. Primum enim diem a Sole appellaverunt, qui princeps est omnium siderum, sicut et idem dies caput est cunctorum dierum. Secundum a Luna, quae Soli et splendore et magnitudine proxima est, et ex eo mutuat lumen. Tertium ab stella Martis, quae Vesper vocatur. Quartum ab stella Mercurii, quam quidam candidum circulum dicunt. Quintum ab stella Iovis, quam Phaetontem aiunt. Sextum a Veneris stella, quam Luciferum asserunt, quae inter omnia sidera plus lucis habet. Septimus ab stella Saturni, quae sexto caelo locata triginta annis fertur explere cursum suum. Proinde autem ex his septem stellis nomina dierum gentiles dederunt, eo quod per eosdem aliquid sibi effici existimarent, dicentes habere a Sole spiritum, a Luna corpus, a Mercurio ingenium et linguam, a Venere voluptatem, a Marte sanguinem, a Iove temperantiam, a Saturno humorem*. Some information is added from an almost identical passage in DNR 3.2–4 (MSS E^{2}AS read *septimo caelo*; cited in BC 56D4–A8): *Apud Romanos autem hii dies a planetis, id est erraticis stellis, nomina accepterunt. Primum enim diem a sole uocatum, ... Septimum ab stella Saturni, quae sexto caelo locata triginta annis fertur explere cursum suum. Proinde autem gentilis ex his septem stellis nomina dierum dederunt, eo quod per eosdem aliquid sibi effici extimarent, dicentes habere ex aere ignem, ex sole spiritum, ex luna corpus, ex Mercurio linguam et sapientiam, ex Venere uoluptatem, ex Marte feruorem, ex Iove temperantiam, ex Saturno tarditatem.* The sentence explicitly ascribed to Isidore is taken from DNR 23.1.8–9, 26.10.66–67: *quemadmodum sol et luna, ita et haec umbram facit*. These two passages of DNR are identical, but the latter seems more likely to have been the source, since it is placed in a general discussion of all the stars, and this would have been the place to look for additional information about Venus, or rather Lucifer. This citation was inserted to explain why Venus was the brightest star, and may originally have been a gloss. I could not locate any source explicitly connecting *audacia* to *Mars*. MC's other main source, CE p. 92, only lists the Roman planetary weekdays: *et ut mutaret uocabula gentium quibus dies uocabant, hoc est dies solis, dies lunę, dies Martis, dies Mercorii, dies Iouis, dies Veneris, dies Saturni.*

16–26 Romani…suum] The Roman planetary weekdays were an integral part also of early Frankish computistics. Cf. DIAL. BURG. 5A: *Nam gentiles de idolis nomina illorum ad dierum nomina posuerunt:*

The Romans dedicated the names of the days (of the week) to the gods and the stars. In fact, they call the first day after the sun, the foremost among the stars; the second after the moon, which is next to the sun in terms of brightness; the third after the star Mars, which is called the evening-star (*Vesper*); the fourth after the star Mercury, which they call the bright circle; the fifth after the star Jupiter, which they call *Phaeton*; the sixth after the star Venus, which is called the morning-star (*Lucifer*), (and) which is the brightest among all stars, as Isidore says: Like the sun and the moon, this (star) also creates shadows; the seventh after the star Saturn, which is believed to complete its course through the seventh sky in 30 years.

The heathen Romans believed that they possessed certain things from these seven stars: The breath from the sun, the body from the moon, bravery from Mars, talent and eloquence from Mercury, moderation from Jupiter, pleasure from Venus, mood or slowness from Saturn.

Dominica dies solis nomen accepit. Sic postea dies lunae et dies Martis et dies Mercurii et dies Iovis et dies Veneris et dies Saturni vocabulum acceperunt. QUAEST. LANGOB. 1D: *Illi dicitur dies solis et dies lunae et dies Martis vel Mercurii et Iovis vel Veneris ac Saturni.* Despite their pagan origin, they were used throughout the middle ages, and besides Isidore's works (cf. *app. font.*), DTR 8.25–29 (cited in LIB. CALC. 67, RM 26.14–19) became the main source for them: *Verum gentiles cum observationem a populo israhel hebdomadis ediscerent, mox hanc in laudem suorum deflexere deorum; primam videlicet diem soli, secundam lunae, tertiam Marti, quartam Mercurio, quintam Iovi, sextam Veneri, septimam Saturno dicantes.* **27–31** Ab…tarditatem] The characteristics of the Roman planets obviously were not of immediate relevance to the early medieval computist. Nevertheless, they are at least hinted at as an explanation for the nomination of the planets in DE HEBDOMADIBUS, ponting to Isidore as the main source: *Inde Isidorus dicit, ex his septem stellis nomina dierum gentiles dederunt eo quod per easdem gradus aliquid sibi effici aestimabant.* Bede, then, took up this hint and worked from Isidore's works for his own discussions. In DT 4.3–6, he conflates the accounts of DNR and ETYM. (cf. *app. font.*): *His nomen gentilitas a planetis indidit, habere se credens a sole spiritum, a luna corpus, a marte sanguinem, a mercurio ingenium et linguam, a iove temperantiam, a venere voluptatem, a saturno tarditatem.* In DTR 8.31–33 (cited in LIB. CALC. 67, RM 26.20–23), however, he solely relies on DNR: *Existimabant enim se habere a sole spiritum, a luna corpus, a marte fervorem, a mercurio sapientiam et verbum, a iove temperantiam, a venere voluptatem, a saturno tarditatem.* Note that MC as well as Bede rearranged Isidore's discussions by bringing the order of planets in line with the weekdays. A different account of the characteristics of the Roman planets, or rather of the gods after whom these planets were named, can be found in BC 57C7–15: *Secundum Latinis a diis gentilium et reliquiae nominantur. Habuerunt enim duos deos caelestes et quinque terrenos. Martem patrem Romoli, Mercurium eloquentiam* (?) *et ingenii deum* (?), *Iovem uirtutes et humor eius idolum, Venere libidinis concupiscentiae et omnis turpitudinis inuentricem, Saturnum patrem Iouis frugum et uentorum moderatorem.*

16–31 Romani…tarditatem] Cf. Borst, *Schriften*, p. 356, 517.

Pope Sylvester changed the Roman custom

Siluester uero sanctus episcopus urbis Romę, ne haberent dies uocabula secundum consuetudinem gentilium, commutauit; et primum diem prima feria, secundam uero diem secundam feriam nominauit, sic et reliqui usque ad sabbatum.

32–36 Siluester…sabbatum] The origin of the notion that pope Sylvester had changed the pagan planetary weekdays to the *feria* terminology is uncertain. MC may have relied on the lost *De divisionibus temporum* (cf. *app. comp.*) or CE p. 92: *Haec autem nomina dierum (i.e. feriae) Siluester episcopus Romae indidit.*

32–36 Siluester…sabbatum] The belief that pope Sylvester had introduced the *feria* terminology was certainly common among seventh-century Irish computists, since the following statements ultimately derive from the now lost *De divisionibus temporum*: DDT 10A9–10: *Secundum Sylvestrum papam, prima feria dicitur quasi prima dies.* DE HEBDOMADIBUS: *Silvester sanctus apostolicus sic docuit et praedicavit Christianis ut non nominarent dies septimane ritu gentilium sed Christiana observatione nominarent, et sic nominavit prima feria sicut diximus, et cetera.* This Irish school of thought then influenced Bede: DT 4.6–8: *Sed sanctus Silvester ferias appellari constituit, primum diem dominicum nuncupans imitatus hebraeos qui primam sabbati, secundam sabbati, et sic caeteras a numero nominant.* DTR 8.54–55 (cited in LIB. CALC. 67, PV §100, RM 27.4–5): *Ferias vero habere clerum primus papa Silvester edocuit.* For the list of *feriae* cf. c. 11, ll. 5–8 above.

32–36 Siluester…sabbatum] Cf. Borst, *Schriften*, p. 368, 517.

Silvester, the holy bishop of the city of Rome, changed this Roman denomination, so that the weekdays would not have names according to heathen custom; and he called the first day the first *feria*, the second day the second *feria*, and so on up to the Sabbath.

The first day of creation

Quis primus dies factus est *in mundi principio* creaturarum? *Primus dominicus, XII Kalendas Aprilis* secundum Grecos, *VIII Kalendas Aprilis secundum Romanos.*

---

39 VIII] VII *M**.

37–40 Quis…Romanos] Mc's question about the first day of creation is motivated by two sources: De ordinatione feriarum paschalium (*PL* 90, col. 607C–608B) argues that the weekday of this first day of creation was a Sunday and that it fell on the spring equinox, 25 March, which is the tradition ascribed to the Romans in Mc: *Tunc pariter omnes dixerunt episcopi, nisi prius quomodo mundus fuerit a principio investigatus, nihil potest de observantia Paschae salubriter ordinari. Dixerunt ergo episcopi: Quem credimus factum fuisse in mundo primum, nisi Dominicum diem? ... Ergo cum novissimum diem signet sabbatum, quis potest esse primus, nisi Dominicus dies? ... Theophilus episcopus dixit: In quo loco caput mundi esse creditis? In principio temporis an medio temporis, an in medio tempore, aut in fine? Episcopi responderunt: In aequinoctio octavo Kalendarum Aprilium ... Nunc ergo investigamus quomodo in principio factus fuerit mundus, id est, die Dominico.* Martin of Braga, *De pascha* 4–6 (Barlow, *Opera omnia*, p. 272–3; cited in Tractatus Adthanasi 4–5 (Krusch, *Studien* I, p. 331–2)) also refers to the first day of creation as a Sunday, but he assigns a different date to it; in his opinion, the equinox of 25 March was the fourth rather than the first day of creation, since the sun and the moon were created on that fourth day; consequently, the first day occurred three days earlier, on 22 March: *Definiamus igitur quis primus mensis in mensibus anni, deinde quis primus eius dies. ... Inchoasse mundum veris tempore Genesis docet. ... In quo germinare omnia videmus, atque ita in eo esse principium mundi non dubitamus. ... VIII autem Kal. Apr. aequalis est nox et dies, sicut factum mundi initium Genesis docet. ... Ita in quo aequalitatem noctis et diei invenimus, in eo initium mundi constitutum intellegamus. Sed non sine causa maiores nostri super VIII Kal. Apr. tres dies addiderunt, ut primum diem mundi invenirent. XI enim Kal. Apr. primum mensem mundi et diem maiores nostri existimaverunt, quia antequam sol in principatum mundi conderetur triduum ante praecesserat. Refert enim Genesis quarta die facta luminaria solis et luna. ... Diem autem Dominicam primam diem esse mundi dubitare non potest.* The Greek date given in Mc, 21 March, derives from the fact that this date was regarded as the spring equinox in the Dionysiac reckoning; Mc's source for this Greek date in the context of the creation may have been Disputatio Morini (Pitra, *Spicilegium Solesme*, p. 30–1), which also gives the Roman date: *VIII kalendas aprilis, ut in aequinoctio renovatio mundi novissima ... in equinoctio fuit mundi constructio ... et in eodem equinoctio post XII kalendas aprilis ... equinoctium quod est VIII kalendas aprilis ... XII kalendas aprilis quod est equinoctium vernale.* For the Roman tradition cf. also Drp 2.36–39 (cited in c. 55, ll. 6–11): *Est ergo in primo anno initium primi mensis ... secundum Romanos uero Martii mensis XXV die, id est VIII Kal Aprilis*. Prologus Victorii 7 (Krusch, *Studien* II, p. 23): *octavo Kalendarum aprilium, quo mundus traditur institutus*.

37–40 Quis…Romanos] An account of this question similar to Mc can be found in Bc 82D13–A14 (the first part being cited in Bc 148A(1359)9–14, where the second part reads slightly later (B(1359)7–8): *quod primum XII Kalendas Aprilis in principio diei Dominicae incipit*): *Interrogatio de mundi principio, quomodo factus est initio creaturarum Dei? In quo die? Et in <quibus> diebus septimanae? Et in quo anno? Et in quo tempore? Si in die anni? In nocte? Et in qua hora? interrogamus. In aequinoctio autem et in die Dominico ... Latini namque aequinoctium VIII Kalendas Aprilis facere uolunt, ... ut aequinoctium ab XI Kalendas Aprilis computetur.* Generally, like the two accounts of Bc just referred to, eighth-century computists varied between 21 and 22 March as the Greek date for the first day of creation; the confusion is neatly illustrated by Dial. Neustr., which states 25A, 30B: *Secundum autem Diunisium in XI Kalendas Aprilis, quia inicium est mundi. ... Initium vero mundi in vere, in mense Martio, in VIII Kalendas Aprilis; in XI Kalendas Aprilis, ut alii; in dominico die fuit factum.* while in c. 27a it is argued: *Alii in XII Kalendas Aprilis, quia initium mundi est.* 21 March, the date mentioned in Mc, is also given in Drc 68.24–5; note that the Roman date is not mentioned in this context, indicating

Which day was the first to be created at the beginning of the creation of the world? The first (day was) a Sunday, 21 March according to the Greeks, 25 March according to the Romans.

that 25 March was rejected as a possible date for the first day of creation when DRC was compiled: *cum factus esset mundus in XII Kl Aprelis*. Most other Irish or Irish influenced texts preferred Martin of Braga's date, 22 March: DIAL. LANGOB. 18B: *Graecus naturam sequitur, quia in XI Kalendas Aprilis factus est mundus, qui est primus dies saeculi secundum solem*. PP I §64 (MS V reads *XII*): *quia mundus incipit in mense Martio, in XI Kl Aprilis*. PP II §14 (MS S reads *XII*): *Interrogatio: Quo tempore mundus factus est, uel in quali mense? Responsio: In mense Martio in XI Kl. Aprilis*. COMP. COL. 2.3A (the MS reads *XI*): *Prius ituerit debet initium mundi XII Kalendas Aprilis*. For the Greek date for the beginning of the world see also c. 36, ll. 10–15; c. 44, ll. 18–19 below. The early Frankish DIAL. BURG. 4A, 5 regarded the Roman date of 25 March as the first day of creation, apparently following Victorius: *VIII Kalendas Aprilis factus est mundus. ... Quis primus dies fuit? Hoc est: Dominicus. ... Quo tempore dies apellatus est, aut quo tempore factus est mundus? Id est: VIII Kalendas Aprilis*. Bede had a different opinion on this matter, and his point of view became generally accepted from the ninth century onwards at the latest: In DTR 6.2–5 (MS B reads *XIo*; cited in PV §72, where both MSS LB actually read *XImo*) he first introduced the common opinion of his time, namely that either 21 or 25 March were regarded as the first day of creation, before outlining his own theory (DTR 6.65–69; cited in PV §76; up to *notandum* in LIB. CALC. 66D), namely that the Greek equinox, 21 March, was to be considered as the fourth day of creation (the day when the heaveny bodies were created), leading to 18 March as the first day of creation: *Quo sane in loco primus saecli dies sit, nonulli quaerentes octavo kalendarum aprilium, alii duodecim kalendarum supradictum, die magis adnotandum putarunt, uno utrique, hoc est aequinoctio, argumento nitentes. ... Nunc admonere contenti XII kalendarum earundem, primum saeculi diem esse notandum, cuius ad indicium primatus ibi reor antiquos initium zodiaci circuli voluisse praefigere*. Bede's theory is one of the Carolingian additions to the otherwise largely Irish based DDT 9D4–11; the success of Bede's theory is most impressively illustrated by the calendar tradition (see Borst, *Reichskalender*, p. 687).

37–40 Quis...Romanos] Transcribed in Borst, *Schriften*, p. 356, 420; cf. also p. LXV, CLXXI of the introduction.

<XII.> DE MENSIBUS

Etymology and definition of month (*mensis*)

Mensis a mensura dicitur, quia unus*quisque eorum mensuratur.* Augustinus ait: *Veteres menses dic*ebant a *mensa communi*, quia terram *suis fructibus* quasi *quibusdam dapibus reple*bit. C*himales* uero *menses* esse *nega*bant*, sed spatia dic*ebant, quia nullum dapibus dant modum. Hucusque Augustinus. Aliter *menses* | ab ethimologia *Grec*a mene, id est *lun*a. Inde Ysidorus dicit: *Mensis est circuitus luminis lunaris* et *reintegratio*; *siue noua ad nouam. Unde apud Hebreos menses legitimi non* a *solis* cursu, *sed* a *lunę cursu* dicti sunt. Sed *Aegypti primi uelociorem cursum lunae, ne error computationis eius uelocitate accederit,* a *solis cursu* numerau*erunt, qu*ia *tardior solis motus facili*or *poterat* ad*prehendi.*

fol. 12v

XII,2 mensura] mensara *M**. | quia] qui *M (corr. from* EPIT.*; also Borst).* **6** sed] secundum *M (corr. from* EPIT.*; I suspect that the Irish abbr. for* sed *was misinterpreted as* secundum *by the continental copyist).* **11** cursu[1]] cursui *M**.

XII,2–14 Mensis…adprehendi] At the beginning of this passage, MC relies heavily on the enigmatic Virgilius Maro. The first etymology (*mensis a mensura*) as well as the second one (*menses a mensa communi*) ascribed to Augustine are taken from EPIT. 11.155–160 (MS O reads *unusquisque*; MS N adds *id est terram* after *communem*): *Mensis a mensura dicitur, quia quisque eorum mensuratur; sed veteres alio dixere modo, quia menses ideo dicti sunt, eo quod mensam communem omnibus suis fructibus uelut quibusdam dapidus repleant, unde et himales menses dici negarunt, sed spatia.* Note that in another instance (c. 37, ll. 5–7), a quote from Virgilius Maro is also attributed to Augustine, leading to the supposition that his *Epitomae* may have circulated under Augustine's name in Irish computistical circles. For the first etymology cf. also CE p. 92–3: *Mensis a mensurando dicitur, hoc est tempora et annos et ciclos, vel mensurando momenta in minutus, minuta in punctis, puncta in horis, et reliqua usque annos et ciclos.* The rest of this passage, then, is based on Isidore: The third etymology (*menses* deriving from the Greek *mene*) comes from ETYM. 5.33.1 (cited in CE p. 93, DRC 28.2–5, PV §103; less closely in COMP. COL. 2.1A): *Mensis nomen est Graecum de lunae nomine tractum. Luna enim μήνη Graeco sermone vocatur.* The definition of the month constituting the period between two new moons is based on DNR 4.1.1–2 (most MSS read *sive nova*; cited in CE p. 93, DRC 28.6, BC 59B10–11, all three read *sive nova*, DDT 14C11–13, DTR 11.50–51, RM 28.12–13, these three read *sit* and *redintegratio de nova*, COMP. COL. 2.1A, which reads *siue de luna nova usque*): *Mensis est luminis lunaris circuitus ac redintegratio: siue a noua ad nouam.* Finally, the statement that the Hebrews reckoned their months according to the moon's course, while the Egyptians then applied the concept of the month to solar reckoning, is taken from ETYM. 5.33.1–2 (cited in CE p. 93; only up to *enumerantur* in DRC 28.4–5; only up to *novam* in DIAL. NEUSTR. 14, cited in turn in COMP. COL. 2.1A; from *Aegyptii* in PV §104, DDT 14B10–14, which reads *quia* instead of *quoniam*): *unde apud Hebraeos menses legitimi non ex solis circulo, sed ex lunae cursu enumerantur, quod est de nova ad novam. Aegyptii autem primi propter lunae velociorem cursum, et ne error conputationis eius velocitate accideret, ex solis cursu diem mensis adinvenerunt; quoniam tardior solis motus facilius poterat conprehendi.*

XII,2–3 Mensis…mensuratur] Virgilius Maro's etymology *mensis a mensura* was very popular in early eighth century computistics. Closest to the original is DTR 11.2 (cited in LIB. CALC. 69): *Menses dicti a mensura, qua quisque eorum mensuratur.* It occurs in variation in the Irish or Irish influenced BC 58D7–8: *Mensis a mensura quadam nomen accepit, quoniam omnis mensis secundum lunae cursum XXVIIII diebus et semis mensuratur.* DIAL. BURG. 6A: *Mensis a mensura dicitur, eo quod mensuratur.* DIAL. LANGOB. 11A: *Mensis dicitur a mensurando, eo quod mensuram in se trigenos dies <habent>*

12. ABOUT THE MONTHS

The month (*mensis*) is so called from measuring (*mensura*), since every single one of them is measured. Augustine says: The ancients named the month (*menses*) after a common table (*mensa communi*), since (a month) fills up the earth with its fruits, as if (this happens especially) for certain feasts. They denied, however, that the winter-months are months (in this sense, and consequently) they rather called them intervals, because they provide nothing for the feasts that would be worth noting. Thus far Augustine. According to a different interpretation, the (term) month (*menses*) (derives) from the Greek word *mene*, which means moon. Hence Isidore says: A month is the cycle of moonlight and its renewal; in other words, from one new moon to the next. For that reason, the legitimate months among the Hebrews are said not (to accord) with the course of the sun, but with the course of the moon. Because of the swift course of the moon, the Egyptians were the first to reckon the months according to the course of the sun, so that no computational error should occur due to the moon's swiftness, as the slower motion of the sun could more easily be observed.

---

vel etiam dies tregenos singulos. **3–7** Veteres...modum] As could have been expected, Virgilius Maro's second etymology (*menses a mensa communi*) was less popular; in fact, the only text quoting it in the eighth or early ninth century other than MC, COMP. COL. 2.1A, plainly rejects it to be a valid etymology for the reason that it would not assign the term *mensis* to the three winter months: *Alii mensem putant esse dictum eo quod mensam terrae fructiferam facit, et tres menses in hieme non menses dicuntur, sed spatia, eo quod non fructificant. Sed hoc falsum est.* **7–8** Aliter...luna] The most popular etymology certainly was Isidore's statement that the term *mensis* derives from Greek *mene* (cf. *app. font.*). It occurs in variation in DRC 28.1–2 (followed by the Isidorian citation given above): *Nomen diriuatiuum a nomine Greco, quia mene apud Grecos luna interpretatur.* and DDT 14C6–7, which, in this passage, certainly transmits the account of the now lost *De divisionibus temporum*: *Mensis dicitur a luna, quae Graeco sermone mene dicitur; nam et apud eos menses vocantur menes.* On this, then, relied Bede as well as Rabanus Maurus. Cf. DTR 11.2–4 (cited in LIB. CALC. 69): *Sed melius a luna quae graeco sermone mene vocatur; nam et apud eos menses vocatur menes.* RM 28.4–6: *D. Unde dicitur mensis? M. A luna quae greco sermone moene dicitur nam et apud eos menses uocatur moenes.* **8–14** Inde...adprehendi] Besides the texts listed in *app. font.*, Isidore's account of the development of the month from a lunar to a solar concept occurs also, in considerable variation, in BC 58D9–A1: *Mensis primitus in lunari cursu reperti sunt, et ab Hebreis haec obseruatio primitus abstracta est, apud enim illos primum Kalende et neominiae dicta sunt. Menes enim apud Aebreos luna, uel neomenia dicitur. Inde neomenias nouae lunae observationis caelebrant. Deinde apud Latinos mensis solares reperti sunt.* DTR 11.6–7,10–11,38–40,70–4 (the first two phrases cited in LIB. CALC. 69): *Antiqui enim menses suos non a solis sed a lunae cursu computare solebant; ... a qua semper hebraei, quibus credita sunt eloquentia Dei, antiquo patrum more menses observare non cessant ... Verum haec atcumque acta vel comutata fuerint, claret tamen hebraeos ad lunae cursum suos menses observare consuesse. ... Denique aegyptii, qui primi propter ociorem lunae discursum, ne videlicet error calculandi eius velocitate gigneretur, ad solis cursum cuius motus tardior facilius poterat comprehendi, suos menses putare coeperunt.* Cf. also c. 12, ll. 20–23 below.

XII,2–7 Mensis...Augustinus] Transcribed in Ó Cróinín, 'A seventh-century Irish Computus', p. 110; idem, 'Virgilius Maro Grammaticus', p. 199; the first sentence only transcribed in Borst, *Schriften*, p. 357, 441. Cf. Borst, *Schriften*, p. 893 and p. CXXVI, CXXIX of the introduction. **8–10** Inde...nouam] Cf. Borst, *Schriften*, p. 893.

The names of Julian calendar months

Hii *sunt menses* Latinorum: *Ianuarius, Februarius, Martius, Aprilis, Maius, Iunius, Iulius, Augustus, Septimber, Octimber, Nouimber, Decimber.*

*Nomina mens*uum *IIII* rebus as*sump*ta s*unt*: *A diis*, a *regibus*, a *numeris*, *a rebus*.

The naming and origin of the solar months

Sciendum nobis est unde menses solis dicuntur. Id est a mensibus lunae, qui prius fuerunt ab incensione lunae usque ad extinctionem eius. Inde sub hos menses solis formati sunt.

---

18 mensuum] mensium *Borst.*

15–19 Hii…rebus] The immediate source for this list of Julian calendar months as well as for the classification of the origins of their names is CE p. 93: *Quibus modis menses uocabula sumpserunt? Hoc est a quattuor: a diis et regibus, a rebus et numeris. Quot menses sunt anni? Hoc est XII: Ianuarius, Februarius, Martius, Aprilis, Maius, Iunius, Iulius, Augustus, Septimber, Octimber, Nouimber, Decimber*. CE, for its part, may have relied on ETYM. 5.33.3–11 for a list of the Julian calendar months (if a written source was in fact needed). A calendar would appear to be another possibility (the only extant calendars that could have been known to CE and MC are Polemius Silvius' (Mommsen, *Corpus inscriptionum latinarum* 1, p. 335–57) and Willibrord's (Paris, Bibliothèque Nationale, Lat. 10837, fol. 34r–40r)). The classification of the origins of the names may ultimately have been stimulated by DNR 4.2.7–8 (cited in BC 59C1–3), which appears to have also been consulted by MC: *Antiqui autem gentiles mensibus nomina quaedam ex diis suis, quaedam ex causis, quaedam uero ex numero inposuerunt.* **20–23** Sciendum…sunt] The theory outlined here that the solar months are based on the lunar ones ultimately derives from ETYM. 5.33.1–2, quoted c. 12, ll. 10–14. The more immediate source for MC probably was CE p. 95: *Nominationes mensium cuius sunt proprie, utrum solis vel lunę? Hoc soluendum est, specialiter lunę sunt. Sed sol rapuit eas quasi uiolenter nominationes, nec in commune habent eas*.

15–17 Hii…Decimber] The only other computi that list the twelve months in one sentence are the early eighth-century DRC 29.1–3, DIAL. NEUSTR. 2B, DIAL. LANGOB. 9A, and abridged in QUAEST. AUSTR. 2.1A. Accordingly, in the second half of the eighth century at the lastest, presumably due to an increased production of calendars, the Julian calendar months were regarded as common knowledge, so that no need was felt to explicitly list them separately. **18–19** Nomina…rebus] The classification of the origins of the names of the Julian calendar months first appears in CE, presumably stimulated by DNR (cf. *app. font.*), so that this concept presumably originated in seventh-century Ireland. This is confirmed by the fact that it only appears in Irish or Irish influenced texts. DRC 29.3–5 directly relies on CE: *Sciendum nobis quot modis menses uocabula sumpserunt. Quattuor: ex rebus, ex regibus, ex numeris, ex diis*. DIAL. LANGOB. 10A appears to depend mainly on MC, but also on CE: *Interrogatio: Et menses anni, a quibus rebus nomina sumpserunt? Responsio: Quattuor sunt, a quibus nomina acceperunt, hoc est a diis, a rebus, a regibus, a numeris*. See also BC 49D12–13 (cited in QUAEST. LANGOB. 3A): *Quibus modis dicuntur mensi? IIIIor. Quomodo? Sub idolis, sub rebus, sub regibus, sub numeris*. which influenced RM 32.30–1: *Quattuor ergo modis menses nominantur, hoc est sub idolis, sub rebus, sub regibus, sub numeris*. Cf. also the undoutedly Irish tract *De mundi principio* in BC 148(1360)B14–C1 and St Gall, Stiftsbibliothek, 251, p. 14: *Quibus modis nomina sortiti sunt menses? Apud Romanos IIIIor dicuntur: Idolis et rebus et regibus et numeris*. All of these proceed to explicitly classify each month in these categories. This is not the case in DIAL. BURG. 6A (citing DRC): *Nomina mensium quatuor modis vocantur. Id est: ex diis et ex rebus et ex regibus et ex numeris.* **20–23** Sciendum…sunt] The theory that the solar months are modelled on the lunar ones is also implicit in the following Irish or Irish influenced texts that hardly reveal their Isidorian exemplar: BC 60C9–11: *Sed Greci et hii supradicti et omnes hii XII menses habent, et sub mensae eorum dies XXX semper praeter Aegyptios et concordant omnes*. DIAL. LANGOB. 9B: *Interrogatio: Quare autem*

These are the months of the Latins: January, February, March, April, May, June, July, August, September, October, November, and December.

The names of the months are borrowed from four things: from gods, from kings, from numbers, and from objects.

We have to know whence the solar months took their names. From the lunar months, which formerly extended from the kindling of the moon to its extinction. Hence the solar months were formed by following their structure.

---

duodecim sunt menses, et plus aut minus non sunt? Responsio: Dicunt sapientes, quod ideo duodecim sunt propter convenientiam inter solem et lunam, quia sicut duodecim menses sunt in anno lunari, hoc est duodecim lunae, ita etiam et in anno solari duodecim menses oportuit conputari. For the solar months being modelled on the lunar months see also c. 53, ll. 2–4, 12–15.

15–17 Hii…Decimber] Cf. Borst, *Schriften*, p. 439 and p. CXLI of the introduction. **18–19** Nomina…rebus] Transcribed in Borst, *Schriften*, p. 357; cf. also Borst, *Schriften*, p. 440 and p. CXLI, CLXXV of the introduction. **20–23** Sciendum…sunt] Transcribed in Borst, *Schriften*, 440; cf. p. CLXXVI of the introduction.

The history of the Julian calendar

Antiquus annus tribus mensibus extiterat, dicente Macrobio: *Arcades annum suum tribus mensibus explicabant.* Aliquando *VI* menses fuerunt, quia auctumnus in aestate, vernum in hieme computabant. Veteres pro similitudine caloris et frigoris faciebant. *Romani* uero *auctore Romulo X mens*ium *computaba*nt. | Quoniam VI fol. 13r
menses, LXI dies sexta pars anni erat. Quoniam uero X, XXXVI dies et VI horae decima pars anni erat. *A Romulo haec fuit annua dimensio* in *X mensi*bus *dierum uero CCCtorum IIII; quattuor ex his trigen*i *singul*i, *VI uero trigen*i tantum era*nt. Cum* hic *numerus neque solis cursui neque lunae conuenera*t. *Solo ingenio magistro uel Grecorum obseruatione* L *dies addidit. CCCLIIII die*bus constat iste *annus, quibus XII lunę cursus confic*itur. Et de

34 cursui] cursus *M (corr. from* SAT.*).* **36** L] uel *M. It seems that the copyist confused the numeral* l *with the Irish abbr.* ł *for* vel.

24–39 Antiquus…Februarius] The principal source for the history of the Julian calendar was a part of Macrobius' *Saturnalia* circulating under the title DISPUTATIO CHORI ET PRAETEXTATI in early medieval computistical circles. MC refers to his source right before the first quote, which is taken from SAT. 1.12.2 (cited, with some variation, in DRC 28.17): *Arcades annum suum tribus mensibus explicabant, Arcanes sex*. In a six months year, four months can be ascribed to the four seasons, while the remaining two quite naturally have to cover the end of one and the beginning of another season. MC's argument, then, is that one of these two months covers a part of summer and autumn, the other a part of winter and spring. The analogy (or rather antagonism) of warmth and cold (a cold and a warm season being combined in each of these two months) may be based on DNR 7.4.41–45: *Annus. Humidum calium uer: oriens. Calida sicca aestas: meridies. Siccus frigidus autumnus: occidens. Frigida humida hiemps: septentrio*. MC then returns to his main source when arguing that Romulus invented a year of 10 months: SAT. 1.12.3: *non igitur mirum in hac varietate Romanos quoque olim auctore Romulo annum suum decem habuisse mensibus ordinatum*. In this instance, DCH is evidently closer to *M* than SAT.: *Romani, auctore Romulo, X habuisse mensibus ordinatum*. This citation may have been conflated with DNR 4.4.29–30 (cited in BC 59D11–13): *Itaque apud antiquos Latinos decem mensibus cursus anni conputabatur*. The following calculations are MC's own and they are not in line with the following statements about the length of Romulus' year, since 6×61=366 days and 10×36½=365 days, if a day is reckoned as consisting of 12 hours. The details about Romulus' year of ten months are taken from SAT. 1.12.38–39 (cited in considerable variation in DDT 27(663)B2–7, RM 32.4–7; up to *dies*, with additions from SAT. 1.12.3 given below, in DTR 12.36–40, in turn cited in LIB. CALC. 70): *Haec fuit a Romulo annua ordinata dimensio qui, sicut supra iam diximus, annum decem mensium, dierum vero quattuor et trecentorum habendum esse constituit, mensesque ita disposuit ut quattuor ex his tricenos singulos, sex vero tricenos haberent dies. sed cum is numerus neque solis cursui neque lunae rationibus conveniret*. Cf. SAT. 1.12.3 (cited in considerable variation in DRC 28.18–23): *non igitur mirum in hac varietate Romanos quoque olim auctore Romulo annum suum decem habuisse mensibus ordinatum, qui annus incipiebat a Martio et conficiebatur diebus trecentis quattuor, ut sex quidem menses, id est Aprilis Iunius Sextilis September November December, tricenum essent dierum, quattuor vero, Martius Maius Quintilis October, tricenis et singulis expedirentur*. Numa Pompilius' alteration of Romulus' construction is quoted from SAT. 1.13.1–3 (the latter half partially cited in DTR 12.50–58, in turn cited in LIB. CALC. 70 and the first part only in RM 32.10–15): *Sed secutus Numa, quantum sub caelo rudi et saeculo adhuc impolito solo ingenio magistro conprehendere potuit, vel quia Graecorum observatione forsan instructus est, quinquaginta dies addidit, ut in trecentos quinquaginta et quattuor dies, quibus duodecim lunae cursus confici credidit, annus extenderetur. atque his quinquaginta diebus a se additis*

The ancient year developed from an original three months, as Macrobius says: The Arcadians laid out their year in three months. At some point in time, a year consisted of six months, since they reckoned autumn (as belonging) to summer (and) spring to winter. The ancients did this according to the analogy of warmth and cold. The Romans, however, reckoned a year (as consisting) of 10 months, according to Romulus' authority. At the time when (a year had) six months, 61 days constituted the sixth part of a year. At the time when (a year had) ten months, 36 days and 6 hours constituted the tenth part of a year. According to Romulus, however, the extent of a year was 304 days in 10 months; four of these (months) consisted of 31 days; six, however, of only 30 days, though this total number (of days) agreed neither with the course of the sun, nor (with that) of the moon. He (i.e. Numa) (then) added 50 days, following his own genius or the

---

adiecit alios sex retractos illis sex mensibus qui triginta habebant dies, id est de singulis singulos, factosque quinquaginta et sex dies in duos novos menses pari ratione divisit, ac de duobus priorem Ianuarium nuncupavit primumque anni esse voluit, tamquam bicipitis dei mensem, respicientem ac prospicientem transacti anni finem futurique principia; secundum dicavit Februo deo, qui lustrationum potens creditur. Cf. also c. 53, ll. 12–15.

24–31 Antiquus…erat] Bc 58A1–9 also mentions all three types of years discussed in Mc, namely of three, six, and ten months, and assigns a number of days per month to the first two: *Archades enim annum suum tribus mensibus numerabant, id est in nomine centum XXti et unum diebus conputabant. Archanenses uero sex, id est singulos menses sexaginta et duobus diebus computabant. ... Romani uero auctore Romolo annum suum X mensibus computabant. Incipiebant a Martio mense, perficiebatur a Decembrio mense, dies CCC et unum* (*recte quattuor*). The early developments of the calendar are also outlined in De Anno, adding an Egyptian year of four months: *licet, ut auctores plurimi prodiderunt, apud Aegytios quattuor, apud Arcades tribus, apud Acarnanes sex mensibus computatus fuisse feretatur. Annus primum decem mensium fuit qui trecentos et quattuor dies habet*. Except for the texts citing Sat., or rather DCh, directly (cf. *app. font.*), all other texts were not concerned with the early stages of this development, only with the ten-month year ascribed to Romulus. Among these, Dial. Langob. 13A is the most interesting, since it assigns 36 days to each month of this year, in this resembling and being possibly influenced by Mc: *Ita et Romolus Romae distinxit in decem, et unusquisque mensis triginta et sex diebus conputabatur*. Cf. also Bc 59D11–13: *Apud antiquos itaque Latinos X mensibus cursus anni conputabatur*. Dt 6.2–3: *Romani, auctore Romulo, decem mensibus annum, diebus vero CCCIIII, agebant*. Rm 32.2–4: *D. Menses autem Romanorum quibus auctoribus sunt ordinati? M. Primus apud eos Romulus decem menses ordinauit.* **31–35** A…conuenerat] For other texts citing this section of Sat. cf. *app. font.* **35–37** Solo…conficitur] A variation of Sat.'s account of Numa introducing an extra 50 days to Romulus' year of 304 days occurs in Ddt 27(663)B9–12, undoubtedly taken from the original *De divisionibus temporum*: *Sed secutus Numa, quasi hunc omnem ablaturus errorem, in trecentos quinquaginta quatuor dies, quibus duodecim lunae cursus confici credidit, annum extendit*. This was then copied by Bede: Dtr 12.48–50 (cited in Lib. calc. 70, Rm 32.8–10): *Sed secutus Numa quinquaginta dies addidit ut in trecentos quinquaginta et quattuor dies, quibus duodecim lunae cursus confici credidit, annus extenderetur.* **37–40** Et…Februarium] The fact that Numa Pompilius added two months to Romulus' calendar, namely January and February, was quite popular in Insular texts. Dt 6.12–14 derives from Sat.: *His Numa duos menses adiciens, ianuariium a Iano, februarium a Februo deo lustrationeum nominando*. Since this fact is found in late antique and early medieval chronicles, it also found its way into the Annals of Tigernach (Stokes, *Annals of Tigernach*, p. 9), citing Isidore's chronicle: *Romanorum secundus Numa Pampilius* (*sic MS*) ... *duosque menses, Ianuarium et Februarium, X mensibus anni adiecit*. Dial. Langob. 13A is rather based on Jerome's chronicle: *Sed Numma Pamphilius rex Romanorum de his superfluis sex diebus addidit duos*

24–40 Antiquus…Februarium] Cf. Borst, *Schriften*, p. 443, 479 and p. CLXXVI of the introduction.

diebus additis duos nouos menses conficit, *Ianuarius* et *Februarius*. ⌐ *Nummus Pampilius* uero post Romulum *duos* nouos *menses adiecit, Ianuarium et Februarium.* ¬

<XIII.> DE IANUARIO

Etym. of January

Ianuarium in honore *Iani nominauit*. Ianus rex *gentilis*, *Latini* genere, *primus Latinorum rex* fuit: *Ianus, Saturnus, Picus, Faunus, Latinus*. Autem cuius *tempore Aeneas capta Troia Italiam uenit*. Et in eo mense Ianus concrematus est Romę. Vel dicente Ysidoro: *Ianuarius dictus* est, *eo quod ianua anni sit*.

39–40 Nummus…Februarium] *Since this sentence basically repeats the previous statement, it seems to me that it was a gloss originally, commenting on the previous sentence by citing a different source; the reason for its addition may have been the fact that* MC *does not mention Numa by name, so that the gloss was intended to provide just that information.* **XIII,2** Ianuarium] <Numa> *add. before* Ianuarium *Borst*. **3** Latini] Latinus *M, Borst*.

39–40 Nummus…Februarium] This gloss, which crept into the main text at some later copying stage (cf. *app. crit.*), stems from ISIDORE, *Chronica* §§152–153 (*MGH AA* 11, p. 444): *Numa Pompilius ... duosque menses Ianuarium et Februarium decem mensibus anni adiecit*. Cf. JEROME, *Chronicon* (Helm, *Chronik des Hieronymus*, p. 91b; cited in CASSIODORUS, *Chronica* §§78–79 (*MGH AA* 11, p. 122)): *Numa Pompilius duos menses anno addidit, Ianuarium et Februarium*. SAT. 1.12.34: *a Numa Ianuarium ac Februarium retinuit nomen*. DNR 4.4.30–31 (cited in BC 59D13–14, KAL. F4): *Sed Ianuarium Romani, Februarium Numa Pompilius addidit, atque in duodecim mensibus annum distinxit*. **XIII,2–7** Ianuarium…sit] The etymology of January deriving from the god Janus also appears in CE p. 93: *Vel Ianuarius ab Iano rege Epirotarum*, but MC seems to have rather relied directly on DNR 4.4.26 (cited in BC 59D7–8, KAL. F4,6–7, G1–2): *Porro Ianuarium ex nomine Iani uocauerunt*. ETYM. 5.33.3 (cited in PV §105): *Ianuarius mensis a Iano dictus, cuius fuit a gentilibus consecratus*. Isidore's source presumably was AUGUSTINE, *Contra Faustum* 18.5 (*CSEL* 25, p. 494): *Porro Januarius a Jano appellatus est*. The list of the early Latin kings stems from chronicle tradition: JEROME, *Chronicon* (Helm, *Chronik des Hieronymus*, p. 62b; cited from *Latinis* in BEDE, *Chronica* §72 (*MGH AA* 13, p. 261), from *post III annum* in the ANNALS OF INISFALLEN I §73 (Mac Airt, *Annals of Inisfallen*, p. 10)): *Primus rex Latinorum post captam Troiam. Latinis, qui postea Romani nuncupati sunt, post III annum captiuitatis Troiae siue, ut quidam uolunt, post annum VIII regnauit Aeneas ann. III. Ante Aeneam Ianus Saturnus Picus Faunus Latinus in Italia regnarunt ann. circiter CL*. The information that Aeneas came from Troy to Italy may also be based on this passage from Jerome's chronicle, but the wording suggests that it is rather taken from AUGUSTINE, *De civitate Dei* 18.19 (*CCSL* 48, p. 610): *Eo tempore post captam Troiam atque deletam Aeneas cum uiginti nauibus, quibus portabantur reliquiae Troianorum, in Italiam uenit, regnante ibi Latino*. ISIDORE, *Chronica* §§95–96 (*MGH AA* 11, p. 438): *Labdon ann. VIII. Cuius anno tertio Troia capta est et Aeneas Italiam venit*. The fact that Janus was cremated in January seems to derive from the now lost *De divisionius temporum* (since this information is also mentioned in DDT; cf. *app. comp.*) or CE p. 93: *Ipse (i.e. Ianus) quoque in linteaminibus incensis peri\i/t*. Finally, the etymology of January deriving from the entrance (*ianua*) of a year ultimately stems from DNR 4.4.27 (cited in BC 59D9, KAL. F4,6–7, e4): *Ianuarius appellatur eo quod ianua sit anni atque principium*. Cf. ETYM. 5.33.3 (cited in PV §105, KAL. H5, f2–3, a17): *Ianuarius mensis ... dictus ...; vel quia limes et ianua sit anni*. MC, however, cites Isidore through CE p. 93: *Ianuarius dicitur, eo quod ianua anni sit*.

---

menses Ianuarium et Februarium. Cf. also KAL. H1–6, especially H3–5, which read (with few variations), quoting Isidore: *Ianuarium et Februarium menses Numa Pompilius anno adiecit*.

observations of the Greeks. That year corresponds to 354 days, in which the course of 12 lunations is completed. And he created two new months out of these added days, (namely) January and February. ⌐ After Romulus, Nummus Pampilius introduced two new months, (namely) January and February. ¬

13. ABOUT JANUARY

He (i.e. Numa) named January in honour of Janus. Janus (was) a pagan king, from the house of Latinus, (who) was the first king of the *Latini*. (The kinglist reads:) Janus, Saturnus, Picus, Faunus, Latinus. At the time of this (king) (i.e. Janus), Aeneas came to Italy after the fall of Troy. And in this month Janus was cremated in Rome. Or, as Isidore says: January (*Ianuarius*) was so called because it is the entrance (*ianua*) of the year.

XIII,2–6 Ianuarium…Romę] Isidore's etymology *Ianuarius a Iano* was a very popular one. DRC 29.5 may be based directly on Isidore: *Ex deo, ut est Ianuarius, id est ab Iano deo.* Likewise DT 6.12–13: *Ianuarium a Iano ... nominando.* DIAL. BURG. 6B, however, seems to be influenced by MC: *Quod est Ianuarius? De Iano, qui primus princeps paganorum in Roma erat.* DDT 15D6–11 (cited in KAL. F6, d2) is similar to CE (cf. *app. font.*), so that CE and the now lost *De divisionibus temporum* may have been related: *Januarius ... ex idolo, hoc est ex Jano bifronte rege Epirotarum.* Directly based on DDT or the lost *De divisionibus temporum* are: BC 49D13–A2: *Ianuarius ... sub idolo dicitur, eo quod a Iano pagano accepit nomen et bifrons fuit.* RM 32.33–35: *Ianuarius ... ex idolo ergo, hoc est ex iano bifronte simulacre uidelicet Iani Epirotarum regis.* From a similar background derive DIAL. LANGOB. 10A: *Ianuarius a Iano, idulo Romanorum.* QUAEST. LANGOB. 3B: *Ianuarius quibus modis dicitur? ... Sub idolo a Iano accepit nomen.* See also KAL. H 1–6. The following information about Janus are, in the computistical context, unique to MC, except for the fact that he was cremated in January, which is also related in DDT 15A10 (cited in KAL. F6, d2): *Janus vere igne consumptus est.* RM 32.35–36: *qui (i.e. Ianus) argumento excogitato semetipsum igni tradidit.* **6–7** Ianuarius…sit] Even more popular was Isidore's derivation of January from *ianua*, but again most citations are not directly based on Isidore's works. Most importantly for the the present study, DRC 29.5–6 has the exact same wording as MC, and consequently seems to have quoted that text: *Isidorus dicit: Ianuarius dictus est, eo quod ianua sit anni.* DDT 15A20–B1 (cited in RM 32.40–41, KAL. F6, d2) is probably based on the now lost *De divisionibus temporum*, which in turn presumably relied directly on DNR: *Item Janus (recte Januarius) ex re dicitur, eo quod sit janua anni, hoc est principium.* Somehow related to DDT are: BC 49B13–A4: *Ianuarius sub re dicitur, quia sicut homo ingreditur per ostium in domo, ita anni ingrediuntur per istum Ianuarium.* QUAEST. LANGOB. 3B: *Ianuarius quibus modis dicitur? ... sub re verbi gratia, eo quod ianua sit anni.* DIAL. BURG. 6A is a direct citation from DNR: *Vel Ianuarius eo, quod ianua sit anni.* Bede, in DTR 12.56 (cited in LIB. CALC. 70, KAL. H1–4,6), preferred ETYM.: *Quidam autumant ianuarium nuncupatum ex eo quod limes et ianua sit anni.* DIAL. LANGOB. 10B seems to be a conflation of DNR and ETYM.: *Ianuarius eo, quod sit ianua et limes anni.* Cf. also KAL. G1–2: ... *vel specialiter eo, quod ianua sit anni.* KAL. h9: *Ianuarius appellatur, eo quod sit ianua vel introitus, limes et fines.*

XIII,2–6 Ianuarium…Romę] Partially transcribed in Borst, *Schriften*, p. 358. **2–XXIII,3** Ianuarium…uocantur] Cf. Borst, *Schriften*, p. 440–1. **6–7** Vel…sit] Cf. Borst, *Schriften*, p. 358.

<XIIII.> DE FEBRUARIO

Etym. of February *Februarius <a febris> acris Lupercorum*; ⌐ uel Lupercorum id est lignorum. ¬ Ut ait sanctus Hieronimus: Septem luperculis pauit inopem, quia omnia febricitant in eo, | ut dicitur: Nobiscum roborant, febricitant in eo. Vel fol. 13v *Februarius a febribus*, id est a lauacris, ut *februm quodam habebant gentiles, id est lauacrum*. Vel *Februarius a Februo deo, qui lustrationum potens*.

XIIII,2 a febris] *om. M (corr. from* CE*; it may rather have read* a febribus*, as* ANGERS 477, DRC, DDT, *but I suspect that the etymology here differed from the one below).* **2–3** uel…lignorum] *I suspect that this phrase was originally a gloss, intended as an explanation of* Lupercorum *(but note that this phrase is also copied in* ANGERS 477*).* **4** quia] qui *M (this might be a wrong expansion of the Irish abbr. for* quia *by the continental copyist).* | febricitant] febricant *M.* **8** Februo] feruo *M (corr. from* SAT., CE, DRC*).* | qui…potens] inlustrationem potentem *M (corr. from* SAT., DRC, CE, DDT*)*

XIIII,2–8 Februarius…potens] The first etymology for February has been reconstructed from MC's immediate source, CE p. 93: *Februarius dictus est a febris acris Lupercorum. Hoc enim tempore acra febris fit, quo parturiunt lupę et corui.* This etymology is a misreading ('heavy fever' (*febris acris*) instead of '*februa*, the holy ceremonies' (*februis, sacris*)) of DNR 4.4.28 (MSS VSB read *ac-* instead of *sacris*; cited in BC 59D10–11, KAL. F4, f2, both KAL. also with the reading *acris*): *Februarium autem a februis, sacris Lupercorum, appellauerunt.* or Isidore's source: AUGUSTINE, *Contra Faustum* 18.5 (*CSEL* 25, p. 494): *Februarius a Februis sacris Lupercorum.* The following explanation of the *Luperci* and the citation ascribed to Jerome must remain obscure. The etymology *Februarius a febribus* appears to be a curious conflation of CE p. 93–4: *Februarius dictus est a febris acris Lupercorum. Hoc enim tempore acra febris fit, quo parturiunt lupę et corui. ... Aliter Februarius a febribus sacris, quia hoc tempore Luperci stagnum quoddam natantes transfigurant se in lupos. Hoc stagno transuadato lupi efficiuntur et nisi humanam carnem comederint prius quam mensis finiatur. Per annos IIIIor lupi fiunt. Aliter Februaribus a febro; februm enim quoddam habebant gentiles, id est lauacrum, quo se gentiles lauabant.* The underlying theme that February is a purifying month may ultimately derive from AUGUSTINE, *De civitate Dei* 7.7 (*CCSL* 47, p. 191–2): *Ideo Terminalia eodem mense Februario celebrari dicunt, cum fit sacrum purgatorium, quod uocant Februm, unde mensis nomen accepit.* Cf. POLEMIUS SILVIUS, *Fasti* (Mommsen, *Corpus inscriptionum latinarum* 1, p. 337): *Februarius ... dictus a febro verbo, quod purgamentum veteres nominabant.* Finally, the etymology of February deriving from the Roman god Februus may either be a direct quote from SAT. 1.13.3: *secundum dicavit Februo deo, qui lustrationum potens creditur.* or through CE p. 94: *Aliter Februarius a Februo deo, qui potens lustrationum creditur.* Cf. also ETYM. 5.33.4 (cited in PV §106; the second part only in KAL. F4,6): *Februarius nuncupatur a Februo, id est Plutone, cui eo mense sacrificabatur. ... Ergo Februarius a Februo, id est Plutone, non a febre, id est aegritudine nominatus.*

XIIII,2–7 Februarius…lauacrum] The Angers glosses are directly based on MC, also transmitting the explanation of *Lupercorum*, and the obscure citation ascribed to Jerome in considerable variation: ANGERS 477, fol. 19v (gloss to DT): *uel a febribus acris Lupercorum, id est lignorum, ut VII lupicinis satiauit* (?) *inopem. Vel februm Grece, Latine lauacrum, quia Luperci in eo se lauabant.* For the explanation of *Lupercorum* cf. also a gloss to MS E of DNR: *lup. id est nomen arboris illa folia sit hominis in ipso*. The misreading of Isidore's etymology as 'heavy fever' sparked Irish seventh-century imagination. DDT 16B16–C7 (cited in KAL. F6, d2, e4, c11) has the most wonderful explanation of this etymology, undoubtedly copied from the now lost *De divisionibus temporum*, since it shares important readings at the end with MC and CE: *Sive Februarius dictus est a re, hoc est a febribus acris Lupercorum. Luperci enim et Lurcones, duae gentes sunt in Oriente, qui post immunditias totius anni lavabant corpora sua in quodam lacu qui inter illos est, et febricitabant, et sicut tradidit gentilitas tunc mutabant figuras. Et quia in Februario mense hoc faciebant, et lavabant se, et deinde febricitabant,*

14. ABOUT FEBRUARY

February (*Februarius*) (is so called) from the heavy fever (*febris*) of the *Luperci*. ⌐ 'Of the *Luperci*' means 'of the woods'. ¬ As St Jerome says: † ... †, because everything has fever in this (month), as it is said: They become strong again through us, when they have fever in this (month). Or February (*Februarius*) (is so called) from *febribus*, i.e. from baths, since the heathens had a certain *februm*, i.e. a bath. Or February (is so called) from the God Februus, who was powerful in purifications by sacrifice.

Februarius, a lavatione vel a lavacro dictus est. Hinc scriptum est: Februm quoddam habebant gentiles, hoc est lavacrum. Cf. also BC 49A4–10 (cited in ANGERS 477, fol. 19v (gloss to DT)): *Februarius ... sub idolo erat generatio, quod appellabantur Luperci uel Lurconis, quia omnia inmunditia, quod per totum annum faciebant, uel lauabant eorum corpora, nisi tantum in istum februarium, et sic febrizizabant in aqua frigida.* which directly influenced QUAEST. LANGOB. 3B: *Februarius quibus modis dicitur? ... Sub idolis a Lupercis vel a Lurconis febrizandis appellatus est.* Related, but more concise etymologies can be found in DRC 29.8 (the MSS reading is *acris*): *uel ex re, id est a febribus, sacris lupercorum.* DIAL. BURG. 6B (cited in DIAL. LANGOB. 10B – or vice versa? – or both sharing the same source?): *Febroarius eo, quod febricitat mundos.* KAL. G1–2: *vel ut Macrovius dicit a febribus sacris Lupercorum et Lurconum, qui post inmunditias totius anni lavabantur in quodam stagno et febricitabant; vel a febro, id est lavacro.* See also KAL. F6, h9. **7–8** Vel...potens] The etymology of February deriving from the Roman god Februus also was very popular. DRC 29.6–7 is related to MC, CE, and the now lost *De divisionibus temporum*: *Februarius uero a Februo deo nominatus est, qui lustrationum potens esse credebatur.* DDT 16B12–14 (cited in RM 32.45–46, KAL. d2, e4, c11), 27(663)B20–C1, for its part, probably is a direct copy of the lost *De divisionibus temporum*, being ultimately based on both SAT. and ETYM.: *Numa Pompilius secundum mensem Februarium nominavit, et dedicavit Februo deo, qui alio nomine Pluton dicitur, qui lustrationum potens esse apud gentiles credebatur. ... Secundum dicavit Februo deo, qui lustrationum potens credebatur.* This influenced DTR 12.56–57 (cited in KAL. F4,6–7, H1–6, a19, f2,3, h9): *Secundum dicavit Februo, id est Plutoni, qui lustrationum potens credebatur.* Cf. also DT 6.13: *februarium a Februo deo lustrationum nominando.* DIAL LANGOB. 10A: *Februarius a Februo Plutone, idulo Lupercorum.*

XIIII,2–8 Februarius...potens] Cf. Borst, *Schriften*, p. 358; for the Jerome citation see p. CXXXII, CLXXXV of the introduction.

<XV.> DE MARTIO

Etym. of March

*Romul*us *a Martio incipiebat annu*m *genitori*s *su*i *Marti*s. *Huius etiam prim*o *die ignem nouum uest*i *aris accend*unt sacerdotes, *ut incipien*s *an*nus hoc *ign*e consecraretur. Siue *Martius* aut a mare, *qu*ia *eo tempore cuncta animal*ia *agantur ad marem*, hoc est ad cretum.

<XVI.> DE APRELIO MENSE

Etym. of April

Aprelis autem, *cum aspiratione* h *quasi a*ffre*lem, a spuma, quam Greci a*ff*ron uocant, unde Venus orta creditur. Romul*us *h*u*nc mensem* in honore *Vener*is *matr*is *Aeneae nomina*uit. *Sed antiqu*i negabant, quia *nullus dies festus Veneri* in *hoc mense* aga*tur*.

XVI,2 quasi] si *M (corr. from* SAT., *etc.)*. | affrelem] affredem *M (corr. following* SAT., *etc.)*. **3** spuma] spumo *M (corr. from* SAT., *etc.)*. **5** quia] qui *M (this might be a wrong expansion of the Irish abbr. for* quia *by the continental copyist)*.

XV,2–6 Romulus…cretum] The first sentence of this passage is a curious conflation of SAT. 1.12.3: ... *auctore Romulo ..., qui annus incipiebat a Martio.* and SAT. 1.12.5 (cited in DT 6.3): *Haec fuit Romuli ordinatio, qui primum anni mensem genitori suo Marti dicavit.* which results in a change of meanings from March being named after Mars to 'the year of Mars'. For March deriving from Mars cf. also ETYM. 5.33.5 (cited in DDT 17D5–6, in turn cited in KAL. F6; PV §107, KAL. F4,6, G1–2): *Martius appellatus propter Martem Romanae gentis auctorem.* DNR 4.2.9–11 (cited in BC 59C5–7, KAL. F4,6–7): ... *incipientes a Martio ... Hunc autem Martium propter honorem Romuli sic appellauerunt quia eum Martis filium esse crediderunt.* CE p. 94: *Martius a Marte, Romuli patre, dicitur. Pater enim Romuli Mars erat.* That the beginning of a year was celebrated by lighting a fire on 1 March is copied from SAT. 1.12.6: *huius etiam prima die ignem novum Vestae aris accendebant, ut incipiente anno cura denuo servandi novati ignis inciperet.* The etymology *Martius a mare* derives from ETYM. 5.33.5 (cited in PV §107, KAL. F4,6): *Martius appellatus ..., vel quod eo tempore cuncta animantia agantur ad marem et ad concumbendi voluptatem.* **XVI,2–6** Aprelis…agatur] The etymology of April deriving from the Greek *aphron* is ultimately based on SAT. 1.12.8 (MS T reads *aphron*, all other MSS *afron*; cited in DRC 29.12–14, KAL. G1–2; from *a spuma* in ANGERS 477, fol. 19v (gloss to DT), all with the latinized Greek term *aphron* and *Venus orta* (KAL. G1–2 *Venus creditur orta*); the first half only in DTR 12.15, in turn cited in LIB. CALC. 70): *Secundum mensem nominavit Aprilem, ut quidam putant cum adspiratione quasi Aphrilem, a spuma quam Graeci ἀφρὸν vocant, unde orta Venus creditur.* In this case, however, MC appears to have either cited CE p. 94 or a version of DCH close to CE's readings: *Vel Aprilis cum adspiratione quasi aphrilis, a spuma, quam Greci aphron uocant, unde Venus orta creditur.* This etymology then led Macrobius to the theory that Romulus named April in honour of Venus, which is also cited by MC here, but hardly ever in other computistical texts: SAT. 1.12.8: *et hanc Romuli asserunt fuisse rationem, ut primum quidem mensem a patre suo Marte, secundum ab Aeneae matre Venere nominaret.* Cf. ETYM. 5.33.7 (cited in PV §109, KAL. F4,6–7 and less closely in DIAL. LANGOB. 10A): *Aprilis pro Venere dicitur, quasi Aphrodis; Graece enim Ἀφροδίτη Venus dicitur.* The rejection of this theory also derives from SAT. 1.12.12: *sed Cingus in eo libro quem De fastis reliquit ait imperite quosdam opinari Aprilem mensem antiquos a Venere dixisse, cum nullus dies festus nullumque sacrificium insigne Veneri per hunc mensem a maioribus institutum sit, sed ne in carminibus quidem Saliorum Veneris ulla ut ceterorum caelestium laus celebretur.* Cf. DNR 4.2.11–12 (cited in BC 59C7, KAL. F4,6–7): *Aprilem uero nullo deorum suorum nomine, sed de re propria quasi aperilem nominauerunt, eo quod tunc plurimum germinis aperiatur.*

15. ABOUT MARCH

Romulus started the year of his father Mars from March. On the first day of this (month) the priests light the new fire on the altars of Vestus, so that the beginning of the year is consecrated by means of this fire. Or March (*Martius*) is so called from the sea (*mare*), because at this time all animals are driven to the sea, i.e. to the risen (sea level).

16. ABOUT THE MONTH OF APRIL

April, with the aspirate h (sounding) like *affrelem*, (is so called) from the foam of the sea, which the Greeks call *affron*, (and) from which Venus is supposed to have been born. Romulus named this month in honour of Venus, the mother of Aeneas. But the ancients disagreed (with this etymology), since no festival was dedicated to Venus in this month.

XV,2–6 Romulus…cretum] The etymology *Martius a Marte* can be found in many computistical texts of the eighth and early ninth centuries; similar are CE (cf. *app. font.*) and DDT 17C14–18 (cited in KAL. F6, c11, d2, e4), with CE presumably somehow related to the lost *De divisionibus temporum*: *Martius ... Sub idolo, hoc est a Marte, patre Romuli nominatus est. Inde Macrobius dixit: Romulus primum anni mensem Martium constituit et nominavit, et genitori suo Marti dedicavit.* Influenced by these texts are the early eighth-century DIAL. BURG. 6B: *Mars nomen idoli.* DIAL. LANGOB. 10A: *Martius a Marte, patre Romoli.* BC 49A11–12 (cited in QUAEST. LANGOB. 3B): *Martius sub idolo accepit nomen et a Marso pagano.* DRC 29.8–9: *Martius a Marte idulo nominatus est, qui pater Romuli esse credebatur.* Interestingly enough, the latter (or a related text) appears then to have been cited by RM 32.51–53 in the early ninth century: *D. Martius ergo unde dictus est? M. A marte idolo, qui pater ROMVLI et romanae gentis auctor esse estimabatur.* Cf. also DTR 12.10–11 (cited in LIB. CALC. 70): *Primum mensem Marti, cuius se filium credi voluit, dicavit* (*i.e. Romulus*). Note that the lighting of a fire on 1 March to mark the beginning of a new year copied from SAT. in MC appears not to feature in any other computistical text. The *Martius a mare* etymology derived from Isidore became through DDT 17D5–10 (cited in RM 32.53–55, KAL. F6, c11, d2, e4, G1–2) popular in the calendar tradition: *Item Martius a re nomen accepit, hoc est a maribus; nam et eo tempore cuncta animantia terrae mares desiderant, ad concumbendi voluptatem.* Besides these texts, it is found in the Irish DIAL. LANGOB. 10B: *Martius a mare eo, quod isto mense mares agantur ad greges.* **XVI,2–6** Aprelis…agatur] An influential variation of Macrobius' account of the etymology of April (cf. *app. font.*) can be found in DDT 18D15–A1 (cited in RM 32.57–60, KAL. F6, c11, d2, e4), which was probably already part of the now lost *De divisionibus temporum*: *Sub idolo, a Venere matre Aeneae. Aphron enim Graece, spuma interpretatur, unde Venus orta creditur, quae Aphrilis, vel Aphronis, apud Graecos nominatur.* The most popular etymology for April, however, was not mentioned by MC, namely Isidore's (cf. *app. font.* and ETYM. 5.33.7), which is cited in or influenced CE p. 94 (cited in DRC 29.9–10), DDT 18A1–3 (cited in KAL. F6, c11, d2, e4), BC 49A13–14 (cited in QUAEST. LANGOB. 3C), DIAL. BURG. 6B, DTR 12.15–18 (cited in LIB. CALC. 70A, RM 32.60–64, KAL. G1–2, H1–6), PV §109.

XV,2–6 Romulus…cretum] Cf. Borst, *Schriften*, p. 358. **XVI,2–6** Aprelis…agatur] Cf. Borst, *Schriften*, p. 358.

<XVII.> DE MAIO

Etym. of May

Maius a maioribus, quia *Romulus populum diuisit, dicit*ur; *ut altera pars armis altera consilio rém publicam tueretur*, *Nobilior in fastis* narrat. Aliter *a*pud *Tusculanos deus* Maris, *qui est Iuppiter*. In huius honore *hic mens*is *Maius uocatur, magnitudine scilicet ac maiestate*. | Gign*us* fol. 14r
uero hunc *mensem a Maia uxore Ulcani dicit*, *q*uia *flamen* U*lculan*is *huic deae Kalendis Maii rem diuinam facit. Sed* S*iso uxorem Ulcani Maistam dicit.*

XVII,2 quia] qui *M (this might be a wrong expansion of the Irish abbr. for* quia *by the continental copyist).* | diuisit] diuisi *M (corr. from* SAT., DDT, DRC*).* **3** publicam] publica *M (corr. from* SAT., DDT, DRC*).* **4** Tusculanos] Tusculanus *M (corr. from* DDT*).* **5** hic mensis] hoc mense *M.* **6** ac] a *M (corr. from* SAT.*).* **7** Ulcani] *corr. to* Vulcani *Borst.* **7–8** flamen Ulculanis] fla-mȩ Ulculane *M (corr. from* SAT., *according to* MC*).* **8** Maii] maps *M (corr. from* DDT*; four dots arranged as a rhombus appear in the margin here, which may indicate that this word or phrase was in need of correction).* | rem diuinam] rediuiuam *M (corr. from* SAT.*).*

XVII,2–9 Maius…dicit] The etymology *Maius a maioribus* is primarily based on SAT. 1.12.16: *Maium Romulus tertium posuit, de cuius nomine inter auctores lata dissensio est. nam Fulvius Nobilior in fastis, quos in aede Herculis Musarum posuit, Romulum dicit postquam populum in maiores ioniores que divisit, ut altera pars consilio altera armis rem publicam tueretur, in honorem utriusque partis hunc Maium, sequentem Iunium mensem vocasse.* MC seems also to have consulted ETYM. 5.33.8 (cited in PV §110, KAL. G1–2): *Maius dictus ... vel a maioribus natu, qui erant principes reipublicae.* or CE p. 94: *Maius, Iunius a maioribus et iunioribus populi dictis, quos Romulus ad consiliandum et ad bellandum regni constituit.* Cf. POLEMIUS SILVIUS, *Fasti* (Mommsen, *Corpus inscriptionum latinarum* 1, p. 343): *Maius ... dictus a maioribus.* DNR 4.3.15–16 (the first part cited in BC 59C10–11, KAL. F4,6): *... Maium ... siue propter maiores. ... Alii autem, sicut Maium pro maioribus.* The etymology of May deriving from a Tusculan god is taken from SAT. 1.12.17: *sunt qui hunc mensem ad nostros fastos a Tusculanis transisse commemorent, apud quos nunc quoque vocatur deus Maius, qui est Iuppiter, a magnitudine scilicet ac maiestate dictus.* The etymology of May deriving from Maia, the wife of the Roman god Vulcanus, as well as its rejection stem from SAT. 1.12.18 (MS T reads *Cignus*; cited with some variations in DDT 19B5–10, in turn cited in KAL. F6, c11, d2; the third sentence is not cited here, but in DTR 12.21–23, in turn cited in LIB. CALC. 70, KAL. F4,6–7, H1–6, a17, f2,3, h9): *Cingius mensem nominatum putat a Maia quam Vulcani dicit uxorem, argumentoque utitur quod flamen Vulcanalis kalendis Maiis huic deae rem divinam facit. sed Piso uxorem Vulcani Maiestam, non Maiam dicit vocari. contendunt alii Maiam Mercurii matrem mensi nomen dedisse, hinc maxime probantes quod mense mercatores omnes Maiae pariter Mercurioque sacrificant.* Cf. ETYM. 5.33.8 (cited in CE p. 94, DRC 29.14, DT 6.4, DIAL. BURG. 6B, DIAL. LANGOB. 10A, PV §110, RM 32.67, BC 49A15, in turn cited in QUAEST. LANGOB. 3C): *Maius dictus a Maia matre Mercurii.* DNR 4.3.15–16 (cited in BC 59C10–11, KAL. F4,6): *Inde mensem Maium pro Maia Mercurii matre quam deam colunt.*

XVII,2–9 Maius…dicit] Macrobius' *Maius a maioribus* etymology can also be found in the related DDT 19B12–C1 (cited in KAL. F6, c11, d2): *Sicut Maius a re dicitur, hoc est, a majoribus Romanorum. Hinc Macrobius dicit: Romulus Maium mensem tertium esse posuit. Postquam enim populum Romanum divisit in duas partes, hoc est majores minoresque constituit, ut altera pars consilio, hoc est majores, altera vero, hoc est juniores, rempublicam tueretur; deinde, in honorem utriusque partis hunc Maium mensem, sequentemque Junium, vocavit.* DRC 29.14–18: *Maius ... ex maioribus populi. Iunius uero ex iunioribus populi, ut Macrobius dicit: Romulus vero posteaquam populum in maiores minorsque diuisit, ut altera párs consilio, altera armis rém publicam tueretur, in honorem utriusque partis hunc Maium, sequéntem uero Iunium uocauit.* RM 32.68–69: *a sub re, hoc est a maioribus Romanorum, quem Romulo ita uocari placuit.* Bede gives an abridged version of this Macrobian citation: DTR 12.19–21

17. ABOUT MAY

May (*Maius*) is so called from the elders (*maioribus*), because Romulus divided the people; Nobilior tells (us) in (his) *Fasti* that one part (of the male population) was supposed to support the republic by means of arms, the other through (their) council. According to a different interpretation, a god Maris (existed) among the Tusculans, who is (the equivalent to) Jupiter. In his honour this month is called May (*Maius*) from greatness (*magnitudine*) and, of course, from majesty (*maiestate*). Gignus, however, named this month after Maia, the wife of Vulcanus, because the priest of Vulcanus performed a sacrifice for this goddess on 1 May. On the other hand, Siso calls the wife of Vulcanus Maista.

(cited in LIB. CALC. 70, KAL. F7, H1–6, a11, f2,3, h9; the latter part only in KAL. G1–2): *Maium tertium, quartum iunium posuit, in honorem videlicet maiorum ac iuniorum in quos divisit populum, ut altera pars armis altera consilio rem publicam tueretur.* Cf. also DIAL. LANGOB. 10B: *Maius a maioribus natu, quos eo tempore Romolus constituit ad consilium.* BC 49A15–B1 (cited in QUAEST. LANGOB. 3C): *Maius ... a maiorum sapientia Romanorum.* The theory that May derives from a Tusculan god is less frequently cited; it appears to have been part of the now lost *De divisionibus temporum* and from there found its way into the related DDT 19B1–5 (cited in KAL. F6, c11, d2): *Sub idolo Maius dictus est, a Maio deo apud paganos, qui dictus est Jovis. Inde Macrobius dicit: Apud Tusculanos deus Maius vocabatur, qui est Jupiter, a magnitudine scilicet et majestate dictus.* RM 32.66–7: *hoc est sub idolo maio, quem iouem esse uoluerunt, quasi maiestate principium.* The Macrobian etymology of May deriving from the name of the wife of the Roman god Vulcanus can, besides MC, only be found in DDT and DTR, and texts citing these two; more popular was Isidore's statement taken from Macrobius that May derived, in fact, from the name of Mercury's mother (cf. *app. font.*).

XVII,2–9 Maius...dicit] Cf. Krusch, *Studien* II, p. 58; Borst, *Schriften*, p. 358–9. **7** a...Ulcani] Transcribed in Borst, *Schriften*, p. 358.

<XVIII.> DE MENSE IUNIO

Etym. of June

Iunius a iunioribus. Gignus ait *quod Iunonius* prius *apud Latinos uocatus* est. Inde *post*ea *detritis* aliis *litteris Iunius* uocatur. *Aedes munitae Iunoni*o deo *Kalendis Iunii dedicat*e. Vel *Iunium mensem a Iunio Bruto* a *prim*o *consulę Rome* dicunt *nominatum*, *quod* in *Kalendis Iunii pulso Tarquinio in cęlio monte sacrum deae carneę fecerit*.

XVIII,2 iunioribus] iunoribus *M (corr. from* CE, DDT, DRC, BC*; Mac Carthy corrects* iuinoribus *to* iuinioribus*)*. **4** Aedes] inedes *M (corr. from* SAT., DTR*;* in edes *Mac Carthy)*. **5** Iunium mensem] Iunio mense *M (corr. from* SAT., DDT*)*. **6** nominatum quod] nominatu-q; *M (corr. from* SAT.*; I suspect that the continental copyist mistook the Irish abbr. for* quod *for* q;*)*.

XVIII,2–8 Iunius…fecerit] The etymology *Iunius a iunioribus* ultimately derives from SAT. 1.12.16 cited in the previous note. MC probably also consulted CE p. 94 (cf. previous note) and DNR 4.3.18 (cited in BC 59C14, KAL. F4,6): *ita pro iunioribus Iunium uocari dixerunt*. Cf. also ETYM. 5.33.8–9 (cited in PV §§110–111): *Nam hunc mensem maioribus, sequentem vero minoribus Romani consecraverunt. Vnde et Iunius dicitur. Antea enim populus in centurias seniorum et iuniorum divisus erat.* and especially POLEMIUS SILVIUS, *Fasti* (Mommsen, *Corpus inscriptionum latinarum* 1, p. 345): *Iunius ... dictus a iunioribus*. The two following etymologies (July being formerly called *Iunonius*, which was then changed to *Iunius*; July being named after Iunius Brutus) are directly taken from SAT. 1.12.30–31 (note especially, however, MC's confusion when changing *Monetae* to *munitae* and attributing it to *aedes* rather than *Iunoni*): *Iunius Maium sequitur, aut ex parte populi, ut supra diximus, nominatus, aut ut Cingius arbitratur quod Iunionius apud Latinos ante vocitatus, diuque apud Aricinios Praenestinosque hac appellatione in fastos relatus sit adeo ut, sicut Nisus in commentariis fastorum dicit, apud maiores quoque nostros haec appellatio mensis diu manserit, sed post detritis quibusdam litteris ex Iunonio Iunius dictus sit. nam et aedes Iunoni Monetae kalend. Iuniis dedicata est. non nulli putaverunt Iunium mensem a Iunio Bruto qui primus Romae consul factus est nominatum, quod hoc mense, id est kalendis Iuniis, pulso Tarquinio sacrum Carnae deae in Caelio monte voti reus fecerit.* Cf. also CE p. 94: *Iunius a Iunone, qui et Iunonius, sed detritis paucis litteris Iunius dicitur.*

XVIII,2–8 Iunius…fecerit] The *Iunius a iunioribus* was a very popular etymology for June in eighth- and early ninth-century computistical texts, particularly those of Irish origin or relying on Irish thought. Cf. DDT 20C16–D1 (cited in KAL. F6, d2, c11; the first sentence only in RM 32.71–72): *Sive Junius a re dicitur, hoc est a junioribus Romanorum, qui armiis defendebant rempublicam. Inde Macrobius dicit: Junius Maium sequitur, ex parte populi nominatus.* BC 49B1–2 (cited in QUAEST. LANGOB. 3C): *Iunius a iunioribus sapientia Romanorum.* DIAL. LANGOB. 10B: *Iunius a parte populi, id est iunoribus, quos constituit, ut rem publicam contuerentur armis.* See also previous note and *app. font.* Macrobius' theory that June was originally called *Iunonius* was also widespread. Bede remains very close to the original: DTR 12.23–27 (cited in LIB. CALC. 70, KAL. F4,6, a17, e4, f2,3, h9; only the beginning in KAL. F7; partially in KAL. H1–6, G1–2): *Iunius mensis aut ex parte populi, ut diximus, nominatur aut, ut Cingius arbitratur, iunonius apud latinos ante vocatus est; et haec appellatio mensis apud maiores diu permansit, sed, post detritis quibusdam literis, ex iunonio iunius dictus est; nam et aedes Iunoni Monetae kalendis iuniis dedicatae sunt.* Macrobius had further specified this *Iuno* (*Iunonius* being the adjectival form) with the attribution *Moneta*; contrary to MC, who clearly misunderstood this attribution, other seventh- and eighth-century computists appear to have been more knowledgeable in Roman mythology, since this goddess (rather than a god, as MC would have it) was then correctly identified as Juno, the daughter of Saturn, and the wife and sister of Jupiter: DDT 20C11–14 (cited in KAL. F6, d2, c11): *Sub idolo, hoc est a Junone filia Saturni, quae apud Gentiles quasi dea adorabatur, et a suo nomine Junium mensem vocaverunt. Inde Macrobius dicit: Junius apud Latinos, antea Junonius, ab Junone vocatus est. Postea, detritis quibusdam litteris, hoc est n et o, ex Junonio, Junius*

18. ABOUT THE MONTH OF JUNE

June (*Iunius*) (is so called) from the younger (*iuniores*). Gignus says that (this month) was formerly called Iunonius among the Latins. Later, then, after various letters had be suppressed, it was called June (*Iunius*). The fortified temples were dedicated to the god Iunonius on 1 June. Or they called the month of June after Junius Brutus, i.e. after the first consul of Rome; this was done, because (Junius Brutus) performed the worship of the goddess Carnea on the mountain Caelius on 1 June after Tarquinius had been expelled.

---

dictus est, nam et templum Junoni in Kalendis Junii dicatum est. DIAL. LANGOB. 10A: *Iunius a Iunone, sorrore Iovis et coniunge.* QUAEST. LANGOB. 3C: *Sub idolo a Iunone sorrore vel coniunge Iovis fuisse testantur.* RM 32.71: *D. Iunius unde nomen habet? M. A iunione filia saturni sorore iouis.* Without further specification: DT 6.4–5: *quartum Iunoni sacravit.* DIAL. BURG. 6B: *Iunius nomen est idoli, id est Aginum.* QUAEST. AUSTR. 2.2B: *Iunius de Iunio.* Finally, Macrobius' connection of June with Junius Brutus can, besides MC, only be found in DDT 20C14–16 (cited in KAL. F6) and probably was part of the lost *De divisionibus temporum* (and subsequent versions of this text still to be edited): *Alii putaverunt Junium mensem ab Junio Bruto nominatum, qui primus in urbe Roma consul factus est.*

XVIII,1–5 De…dedicate] Transcribed in Mac Carthy, *Annals of Ulster* 4, p. clxxix; cf. Borst, *Schriften*, p. 359.

<XVIIII.> DE IULIO

Names for and etym. of July

Iulius Quintilis a Martio nominatus. *Postea in honore Iulii Cesaris dictatoris a Marco Antonio Marci filio consul*is *Iulius appellatus est, qu*ia in quinto *Idus Quintil*is il*lius procreatus* est.

<XX.> DE AUGUSTO

Names for and etym. of August

*Augustus Sextilis uoca*tus est *donec* in *hono*re *Augusti consult*um *daretur.* In *hoc* enim *Cęsar Augustus* in *primum consulatum iniit et III triumphos* de *Aegypt*is, de

---

XVIIII,5 illius] filius *M (corr. from* CE; *it may well have read* Iulius *as in* SAT.*).* **XX,4** III] *a superscript* s *was later added above the second* I *of* III.

XVIIII,2–5 Iulius…est] The fact that July was formerly called 'the fifth (month from March)' as well as the etymology of July from Julius Caesar are directly taken from SAT. 1.12.34 (or rather DCH, which reads *Marco Antonio Marci*; note that MC attributes the consulship to Mark Antony's father, rather than to Mark Antony himself as SAT. and all other texts; only the second part cited in DTR 12.29–31, in turn cited in cited in LIB. CALC. 70A, KAL. F5, G1–2, H1–6, a17, e4, f2,3, h9): *Sequitur Iulius qui, cum secundum Romuli ordinationem, Martio anni tenente principium, Quintilis a numero vocaretur. ... Sed postea in honorem Iulii Caesaris dictatoris legem ferente M. Antonio M. filio consule Iulius appellatus est, quod hoc mense a. d. quartum Idus Quintiles Iulius procreatus sit.* Mark Antony's names were spelled out not only in DCH, but also in CE p. 94: *Iulius in honore Iulii Cesaris a Marco Antonio Marci filio consule Iulius appellatus est, quod hoc mense ad quartas Idus Quintilis illius procreatus sit. Et hic mensis Quintilis primo uocatus est.* Cf. also POLEMIUS SILVIUS, *Fasti* (Mommsen, *Corpus inscriptionum latinarum* 1, p. 347): *Iulius ... dictus a Iulio Caesare, cum Quintilis antea diceretur.* ETYM. 5.33.10 (cited in PV §112; only the second part partially in KAL. F5, H1–6): *Iulius vero et Augustus de honoribus hominum, Iulii et Augusti Caesarum, nuncupati sunt. Nam prius Quintilis et Sextilis vocabantur: Quintilis, quia quintus erat a Martio, quem principem anni testantur esse Romani; Sextilis similiter, quod sextus.* DNR 4.3.19–21 (cited in BC 59C15–D4, KAL. F4,6–7; only the first sentence in DRC 29.18–19): *Item Iulium a Iulio Caesare, Augustum uero ab Octauiano Augusto uocauerunt. Nam prius Iulius Quintilis et Augustus Sextilis uocabatur, sed eorum nomina a Caesaribus Iulio siue Augusto sunt commutata.* **XX,2–6** Augustus…inposuit] Like in the case of July, MC's account of August is directly based on SAT. 1.12.35: *Augustus deinde est qui Sextilis antea vocabatur donec honori Augusti daretur ex senatus consulto cuius verba subieci: cum imperator Caesar Augustus mense Sextili et primum consulatum inierit et triumphos tres in urbem intulerit et ex Ianiculo legiones deductae secutaeque sint eius auspicia ac fidem, sed et Aegyptus hoc mense in potestatem populi Romani redacta sit finisque hoc mense bellis ciuilibus impositus sit atque ob has causas hic mensis huic imperio felicissimus sit ac fuerit, placere senatui ut hic mensis Augustus appelletur.* MC may also have consulted CE p. 94: *Augustus deinde est, qui Sextilis antea uocabatur, donec Augusto daretur honor ex senatus consultu. Nam hoc mense primum consulatum iniit et III triumphos in urbem contulit et ex Ianiculo legiones deduxit et Aegyptus hoc mense in potestatem populi Romani redacta est finisque hoc mense bellis inpositus est atque ab has causas placuit senatui, ut hic mensis Augustus apellaretur.* Cf. POLEMIUS SILVIUS, *Fasti* (Mommsen, *Corpus inscriptionum latinarum* 1, p. 349): *Augustus ... prius Sextilis; dictus an Octaviano Augusto.* and previous note.

XVIIII,2–5 Iulius…est] Both facts, namely that July was formerly called *Quintilis* and that its later name derived from Julius Caesar, were common knowledge among computists and often repeated. Cf. DDT 21D6–12 (cited in Kal. F6, c11, d2; with the last sentence in variation in RM 32.73–77): *D. Julius unde nomen accepit? M. A quondam imperatore, hoc est a Julio Caesare, quia in quarto Idus Julii mensis creatus est Caius Julius Caesar, qui primus arripuit imperium apud Romanos. Deinde propter suum honorem de suo nomine Julius mensis dictus. Nam antea Quintilis vocabatur.* DT 6.5–6: *Quintili*

19. ABOUT JULY

July was (first) named the fifth month (*Quintilis*) from March. Later it was named July (*Iulius*) by Mark Antony, the son of the consul Mark, in honour of the dictator Julius Caesar, because the latter was born on the eleventh day of the fifth month (*Quintilis*).

20. ABOUT AUGUST

August was called the sixth month (*Sextilis*) until a decree was passed in honour of Augustus. In fact, in this (month) Caesar Augustus entered his first

---

et sextili, qui nunc a nativitate Iulii Caesaris et Augusti triumpho nuncupantur. DTR 12.27–29 (followed by a direct quote from SAT. (cf. *app. font.*); cited in LIB. CALC. 70A, KAL. F4,6–7, G1–2, a17, e4, f2,3, h9): *Iulius mensis nomen quintilis quod a numero sumpserat, etiam post praepositos martio duos menses servavit*. DIAL. LANGOB. 10A: *A regibus, ut Iulius ab Iulio Caesar, qui in eo mense natus est, et antea Quintilis dicebatur*. QUAEST. LANGOB. 3D: *Iulius dicitur sub idolo, et a regibus Caesaris accepit nomen vel pro dignitate, vel <quod> quintus erat a mense Martio*. Interestingly enough, QUAEST. AUSTR. 2.2B applies the theory that July and August were formerly named according to the number of months from March also to June: *Prius igitur Iunius et ceteri duo sequentes quartus et quintus et sextus vocabatur. Nunc autem Iunius de Iunio, Iulius de Iulio Caesare, Agustus de caesare Agusto haec sortiti sunt vocabula*. The following two early texts content themselves with the present terminology: DIAL. BURG. 6B: *Iulius et Agustus ex nominibus regum sunt*. BC 49B2–3: *Iulius Caesar et Agustus ad dignitatem acceperunt nomen*. **XX,2–6** Augustus…inposuit] The specification of the three triumphs appears to have originated in MC and was then copied in DRC 29.19–21: *Agustus autem ab Agusto Cesare nominatus est, quia in isto mense tres triumphos in urbem contraxit, id est de Antonio, uel de Parthis, Egyptiis, Macedonibus*. It cannot be found in any other text. The most detailed and infuential account of the names for and etymology of August presumably was that of the now lost *De divisionibus temporum*; at least this is what DDT 22A7–16 (cited in KAL. F6) suggests: *D. Augustus unde nomen accepit? M. Ab Octaviano Caesare Augusto imperatore, qui in Kalendis mensis Augusti vicit Antonium, de regno contendentem, et Cleopatram reginam Aegypti, uxorem Antonii, et auxilia et exercitum Antonio afferentem. Hos duos vicit Caesar Augustus, in una die Kalendarum Augusti, et postea Caesar principatum habuit totius mundi. Sic Augustus a nomine Augusti dictus, qui eo mense primum consulatum initii, et tres triumphos urbi intulit. Deinde Augustus a populo Romano appellatus est, et in suum honorem mensis Augustus vocatus est, ut sua memoria esset in tempore. Nam ille mensis antea apud Romanos Sextilis dictus est, eo quod sextus mensis sit a Martio. Deinde dicit Macrobius, Augustus antea Sextilis vocabatur, donec honori Augusti ex senatus consulto Augustus dictus est*. It influenced DTR 12.31–32 (cited in LIB. CALC. 70; from *donec* in KAL. F 4,6, H1–6, f2,3, h9, a17): *Augustus mensis sextilis antea vocabatur; donec honor Augusti daretur ex senatus consultu, eo quod ipse die primo huius mensis Antonium et Cleopatram superaverit et imperium populi romani firmaverit*. DIAL. LANGOB. 10A: *Augustus ab Octaviano Augusto nominatus est, qui prius Sextilis vocabatur*. KAL. G1–2: *Augustus autem prius Sextilis vocatus est, sed ex senatu consultu a Caesare Augusto Augustus appellatus est, eo quod primo die mensis huius Antonium et Cleopatram idem Augustus superaverit et imperium dilataverit bellisque civilibus finem inposuerit et quia hoc tempore eius fausta primum floruerint. Inde placuit senatui, ut et ipse Caesar Augustus et mensis ex illo appellaretur*. and especially RM 32.78–83: *D. Augustus nomen unde accepit? M. Ab Octaviano Caesare Augusto, qui in kalendis istius mensis Antonium et Cleopatram uicit et imperium populi Romani firmauit. Cum autem Augusti nomen ille inde assumpserit, ex senatus consultu mensem ob memoriam sui nomen hoc habere uoluit*. Cf. also the previous note and QUAEST. AUSTR. 2.2B: *Agustus de caesare Agusto*. QUAEST. LANGOB. 3D: *Et Augustus similiter accepit vocabulum, vel eo quod sextus esset a Martio*.

XVIIII,2–5 Postea…est] Cf. Borst, *Schriften*, p. 359. **XX,3–6** In…inposuit] Cf. p. CXCIII of the introduction.

Macedonibus, de Parthis *in urbem* co*ntulit fin*em*que bellis ciuilibus* in*pos*uit.

<XXI.> DE SEPTIMBRIO

Names for and etym. of September

Septimber á conpositione *dicit*ur: *sept*e*m* et *imber* et *uér*. *Qu*i postea *Germanic*us á Diocletiano, qui uicit Germaniam in eo mense.

| <XXII.> DE OCTIMBRIO fol. 14v

Names for and etym. of October

Octimber a conpositione uocatur. Postea *Domitianus suo nomine* uocauit, *sed* senatus nominat *tiranic*e, *appellationis* abdicauit, et sua *pr*opri*a nomina* inposuit.

<XXIII.> DE NOUIMBRIO ET DECIMBRIO

Etym. of Nov/ Dec

*Nouimber et Decimber a numero imbrium a Martio uoca*ntur.

XXII,3 tiranice] iránice *M (corr. from* SAT.*).*

XXI,2–4 Septimber…mense] The etymology of September being a composition of the Latin terms for seven, rain, and spring ultimately derives from DNR 4.4.23–24 (cited in BC 59D4–7 (the *PL* edition reads *acris* at the end, but the MS has the correct *ueris*); the first sentence only with *autumni* instead of *veris* in KAL. F4,6, a14, f2,3): *Iam September eo quod septimus sit a Martio qui est principium ueris. Simili quoque ordine October et Nouember et December ex numero imbrium atque ueris acceperunt uocabulum.* Yet, MC appears to have rather worked from CE p. 94 in this instance: *Septimber a numero dictus, eo quod septimus sit a Martio, imber, vere: sept enim corrupte dicitur septem Latine, im autem imber, ber uero uer exponitur, hoc est septimus imber vere. Ita Octimber, Nouimber, Decimber hoc modo explanatur.* Cf. ETYM. 5.33.11 (cited in PV §113): *September nomen habet a numero et imbre, quia septimus est a Martio et imbres habet. Sic et October, November atque December ex numero et imbribus acceperunt vocabula.* POLEMIUS SILVIUS, *Fasti* (Mommsen, *Corpus inscriptionum latinarum* 1, p. 351; this etymology is repeated for October, November, and December): *September ... dictus a numero.* The fact that September used to be called Germanicus derives from SAT. 1.12.36: *Mensis September principalem sui retinet appellationem: quem Germanici appellatione ... Domitianus invaserat.* **XXII,2–4** Octimber…inposuit] For the etymology of October cf. previous note. The fact that October used to be called Domitianus is copied from SAT. 1.12.36–37: *Octobrem vero suo nomine Domitianus invaserat. sed ubi infaustum vocabulum ex omni aere vel saxo placuit eradi, menses quoque usurpatione tyrannicae appellationis exuti sunt: cautio postea principium ceterorum, diri ominis infausta vitantium, mensibus a Septembri usque ad Decembrem prisca nomina reservavit.* **XXIII,2–3** Nouimber…uocantur] The etymology of November and December derives from DNR 4.4.23–25, ETYM. 5.33.11, quoted in c. 21.

XXI,1–XXIII,3 De…uocantur] Concerning the etymology of September, October, November, and December, all three elements mentioned in MC, of which each of these four names of Julian calendar months is supposed to be a composition of, are mentioned only in DTR 12.34–36 (cited in LIB. CALC. 70, KAL. F7, G1–2, e4, g1): *September mensis, october, november, et december principalem sui retinent appellationem, significantes nomine quoti sint a verno mense, id est martio, vel quod imbres in eis immineant.* Most other texts do not refer to the element *ver*, only to the number of months from March and *imber*: DDT 23B1–2 (cited in KAL. F6, c11, d2), 24B11–12 (cited in KAL. F6, c11), 25B19–C1 (cited in KAL. F6, c11), 26C5–6 (cited in KAL. F6, c11): *September a numero nomen accepit, eo quod septimus imber sit a Martio. ... October dicitur, eo quod octavus sit imber a Martio. ... November autem*

consulate, brought three triumphs to the city, (namely) over the Egyptians, over the Macedonians, and over the Parthians, and brought the civil war to an end.

21. ABOUT SEPTEMBER

September is so called from the composition (of the following words): seven (*septem*), rain (*imber*), and spring (*uer*). Later, it (was named) Germanicus by Diocletian, who conquered Germania in this month.

22. ABOUT OCTOBER

October is so called by composition (as the previous month). Later, Domitian named this month according to his own name, but the senate called this despotic, renounced these namings (i.e. Domitianus for October, Germanicus for September), and reimposed their appropriate names.

23. ABOUT NOVEMBER AND DECEMBER

November (*Nouimber*) and December (*Decimber*) are so called from the number (of months) of rain (*imbrium*) (counted) from March.

---

propterea dicitur, eo quod nonus imber sit a Martio. ... December a numero nomen accepit, eo quod a mense Martio decimus sit imber. BC 49B3–5 (cited in QUAEST. LANGOB. 3D, omitting *imber pluuiae*): *September, October, Nouember et December a numerum acceperunt nomina: September, quia septimus est a mense Martio, et October, eo quod octauus, et Nouember, eo quod nonus, et December, eo quod decimus et imber pluuiae est.* DIAL. LANGOB. 10A: *A numero, ut Septimber, id est septimus imber a Martio; Octimber octavus imber; Noimber nonus imber; December decimus.* RM 32.84–89: *D. September unde vocatur? M. A numero, eo quod sit septimus imber a martio. D. Octimbrem ergo nouembrem et decembrem unde iudicas esse dictos? M. a numero quippe similiter et ab imbre, quia his temporibus imbres ternae imminent.* KAL. H1–6 (and very similarly KAL. h9): *September quasi septimus imber appellatur, eo quod septimus a Martio.* Some texts refer only to the numerical element: DT 6.6–7 (following the discussion of July and August): *et caeteris sequentibus a numero nomen imposuit.* DRC 29.21–23: *September uero et Octimber et Nouember nominibus priscis permanserunt, quia omnes menses primitus <ex> nominibus numeri appellabantur.* Macrobius' statement that September as well a October once had a different name is not considered by any text other than MC; note, however, the fact that DIAL. BURG. 6B quite generally argues that October, November, and December were, among other aspects, named after kings, which may be a direct reflection of MC's account: *September, eo quod septimus a Marcio sit. October, November, December aput gentiles ex nominibus regum ac rebus vel numero apellantur.*

XXI,2 Septimber…imber] Transcribed in Borst, *Schriften*, p. 359.

<XXIIII. CONTINUATIO: DE MENSIBUS>

Different traditions concerning the first month of a year

Primus mensis apud Hebreos Martius, quia in eo mundus factus est, et primum phase factum est, conceptio et passio et resurrectio Christi. *Apud Orientales Octimber primus mensis, post collectionem fruguum, quando decimae offerebantur in templo*. Apud Latinos uero primus mensis Ianuarius et caput anni pro natiuitate Christi in VIII Kalendas eius et circumcisione in Kalendas eius; et in VIII Kalendas eius tenebrę minuantur et lux augetur. Apud Grecos initium á Septimbrio, id est VII Kalendas Septimbris.

XXIIII,2–11 Primus…Septimbris] There is no single source for this account of the different traditions concerning the first month. Mc had to assemble the information from a variety of texts: The fact that the Hebrew year begins in March may have been taken from Dnr 6.2.10–12 (cited in Bc 59B4–6, Drc 45.2, Dt 9.3–4, Dtr 36.38–39, in turn cited in Pv §145): *Principium anni ... alii ab aequinoctio uerno, ut Hebraei*. All the information concerning the Biblical events in March given here (except for the conception) seem ultimately to derive from Prologus Victorii 9 (Krusch, *Studien* II, p. 24–25), which is too long to be quoted here; for the conception (as well as the passion) see ps-Dionysius, *Argumentum XV* (Krusch, *Studien* II, p. 80): *conceptus est Christus dominus noster, in die dominica VIII Kl. Aprl ... passum VIII Kl. Aprl.* For the term *phase* cf. Lev 23:5: *mense primo quartadecima die mensis ad vesperum phase Domini est*. The Oriental praxis of starting the year in October was copied from Jerome, *Commentarii in Hiezechielem* 1.1.3a (*CCSL* 75, p. 6): *apud orientales enim populos, post collectionem frugum et torcularia quando decimae deferebantur in templum, October erat primus mensis*. The Latin beginning of the year in January may derive from Sat. 1.13.3 (cited in Dtr 12.53–54, in turn cited in Lib. calc. 70, Rm 32.14–15): *ac de duobus priorem Ianuarium nuncupavit primumque anni esse voluit*. Dnr 4.4.27 (cited in c. 13, ll. 6–7, Bc 59D9, Kal. F4,6–7, e4), 6.2.10–11 (cited in Bc 59B4–5, Dtr 36.38–39, in turn cited in Pv §145): *Ianuarius appellatur eo quod ianua sit anni atque principium. ... Principium autem anni alii a bruma putant, ut populi Romani*. For the Biblical events falling in January (or rather on Julian calendar dates that refer to January) see: Christ's birth in ps-Dionysius, *Argumentum XV* (Krusch, *Studien* II, p. 80): *natus est in III feria, VIII Kal. Ian. Christus dominus noster*. Prologus Cyrilli (Krusch, *Studien* I, p. 342–3): *VIII kalendarum ianuarium in natale domini nostri Jesu Christi fit aliud solestium*. His circumcision in Lev 12:3: *et die octavo circumcidetur infantulus*. Luke 2:21: *et postquam consummati sunt dies octo ut circumcideretur vocatum est nomen eius Iesus quod vocatum est ab angelo priusquam in utero conciperetur*. The information about the beginning of the increase of daylight is found in Drp 13.215–216: *et ab VIII Kal. Ianuarii in VIII Kal. Aprilis hora deminuitur*. Finally, the Greek beginning of the year in September may derive from Dnr 6.2.10–13, 6.7.52–53 (cited in Drc 45.3, 39.4–5, Bc 59B4–10, which reads *VII Kalendas Septembris*; the first sentence only in Dtr 36.38–39, in turn cited in Pv §145): *Principium autem anni ... alii a solstitio, ut Graeci; alii ab autumno, ut Aegyptii. ... Annus Aegyptiorum sine bissexto habet initium quarto kalendarum septembrium.*

XXIIII,2–11 Primus…Septimbris] Generally, in the discussions of the Julian calendar months found in other computistical texts, there is hardly ever a section on the first month of the year according to different traditions; such discussions usually appear in the context of the year (cf. c. 36, ll. 2–5), or, with a different perspective, in the context of Easter calculations (cf. c. 55, ll. 2–12). The only similar account was written only eight years later than and directly influenced by Mc: Dial. Burg. 4A–B: *Ideo Martius primus, quia in eo creatus est mundus, et prima Pascha facta est, et conceptio Christi ac passio eius et resurrectio eius fuit. Cur Ianuarius aput Latinos primus mensis fuit et caput anni legitur? Id est: Pro nativitate Christi in VIII Kalendas Ianuarii, et pro circoncisione in Kalendas ipsius, et quod in eo baptizatus est, et in ipso mense a magis adoratus est. Et in VIII Kalendas ipsius diminuuntur tenebrae et*

24. CONTINUATION: ABOUT THE MONTHS

The first month among the Hebrews (is) March, since the world was created and the first Pasch took place in this (month), (as did) the conception, the passion, and the resurrection of Christ. October is the first month among the Orientals, (which happened to fall) after the collection of fruits, when the tithes were brought to the temple. Among the Latins, however, the first month and the beginning of the year (is) January, according to the birth of Christ on 25 December and (his) circumcision on 1 January; and from 25 December the shadows diminish and light increases. Among the Greeks, (the year) starts from September, i.e. (more precisely) on 26 August.

---

augetur lux. Nevertheless, within the discussions of indiviual months, statements about them being the first month in certain traditions can frequently be found: For the practice ascribe to the Hebrews cf. especially DE ANNO: *cuius initium ... Martio cum Iudaeis habetur*. and c. 55, ll. 6–12. The tradition of October being the first month is mentioned in DDT 24B13–14: *Et hic mensis* (i.e. October) *in principio anni ponitur apud Orientales*. DTR 11.20–22 (cited in LIB. CALC. 69): *septimus theseri octobri, quem propter collectionem frugum et celeberrimas in ipso festivitates novum annum appellant*. Moreover, cf. c. 24, ll. 19–26. Some of the information given by MC can be found elsewhere in this text: for the creation of the world cf. c. 7, ll. 9–11; c. 11, ll. 37–40; c. 24, ll. 20–21; c. 36, ll. 10–15; c. 44, ll. 18–19; for Christ's birth and conception c. 38, ll. 32–35; for his passion and resurrection c. 44, ll. 55–80; for the increase of daylight c. 38, ll. 32–33; c. 39, ll. 5–10.

XXIIII,2–4 Primus…Christi] Transcribed in Borst, *Schriften*, p. 355. **2–11** Primus…Septimbris] Cf. p. CLXXII of the introduction. **6–10** Apud…augetur] Transcribed in Borst, *Schriften*, p. 355.

The developm. of the length of the months

Omnes menses Grecorum XXXmi sunt et lunas XXXmas habent, et *V dies* superfluos habent, *quos interkalares appellant*, et sub hos omnes menses diuiduntur. Inde *Februarium XXX dierum* faciunt, quia *tollunt* II *dies* de *septem*, et alios *V dies* diuidunt per horas et per momenta. *Romulus* uero *ingenio acri XXXI alios* XXX.

XXIIII,16 septem] septe *M.* **17** XXXI] XXXII *M (corr. from* SAT., DDT, DTR*; it may well have read* XXXImi *or* XXXIi *originally).*

12–18 Omnes...XXX] The theory that a (lunar) year originally consisted of 12 months of equal length (namely 30 days) is taken from PROLOGUS DIONYSII (Krusch, *Studien* II, p. 65): *Sed nec hoc praetereundum esse putavimus, quod nimis errant, qui lunam peragere cursum sui circuli XXX dierum spatiis aestimantes, XII lunares menses in CCCLX diebus adnumerant quibus etiam V dies adiciunt, quos intercalares appellavit antiquitas, ut solare annum adimplere videantur.* This kind of year was usually attributed to the Egyptians: DNR 1.5.67–69 (cited in DRC 68.10–12, BC 69D6–8, LIB. ANN. 23), 4.7 (cited, in some form, in COMP. COL. 2.4): *Intercalares dies sunt quinque, qui iuxta Aegyptios supersunt XII mensibus et incipiunt a nono kalendas septembris et quinto kalendarum memoratarum finiuntur. ... Apud Aegytios autem principia mensuum ante kalendas quattuor aut quinque dies pronuntiantur, iuxta quod formula subiecta declarat: Ianuarius, VI kl. ianuarias, diebus XXX; Februarius, VII kl. Februarias, diebus XXX; Martius, V kl. Martias, diebus XXX; Aprilis, VI kl. aprilis, diebus XXX; Maius, VI kl. Maias, diebus XXX; Iunius, VII kl. iunias, diebus XXX; Iulius, VII kl. iulias, diebus XXX; Augustus, VIII kl. augustas, diebus XXX; September, IIII kl. septembres, diebus XXX; October, IIII kl. octobres, diebus XXX; Nouember, V kl. nouembres, diebus XXX; December, V kl. decembres, diebus XXX. Dehinc reuerteris ad quartum kalendas septembris talique ratione conplentur dies CCCLX duodecim mensuum Aegyptiorum. Quinque dies supersunt, quos epagomenus uel intercalares siue addictios uocauerunt, de quibus superius memoratum est.* COMPUTUS COTTONIANUS fol. 79r (cited in BC 15B6–8, which reads *XIII*, in turn cited in LIB. ANN. 58): *Aegypti faciunt in omni anno XII menses, in unumquisque mensem XXX dies habet (!), super sunt quinque dies.* For the *dies interkalares* cf. also AUGUSTINE, *De civitate Dei* 15.12.19 (*CCSL* 48, p. 469): *Propter quinque dies enim reliquos, quibus solaris annus impletur, ..., quod dies Romani intercalares uocabant.* CE p. 103: *V utique dies non conputauimus per XC dies quadrifarie diuisas, quas interkalares vel superfluas alii dicunt. Sed saltim superfluę dicuntur, naturales tamen sunt.* The concept of turning the Julian calendar year into a year of 12 months of 30 days each by subtracting a day each from the seven months of 31 days, adding two of those days to February, with the remaining five constituting the *dies interkalares*, comes directly from PROLOGUS CYRILLI, *Addendum* (Krusch, *Studien* I, p. 343): *XII sunt menses in anno et VII ex ipsis mensibus sunt, qui habent supra XXX singulos dies, id est ianuarius, martius, maius, iulius, agustus, october, december. Et ideo tolle decembri et ianuario singulos dies et adde eos februario, qui habet dies XXVIII, et fiunt dies XXX. Reliquos vero dies, qui in capite quinque mensibus super trecesimum numerum adiacent, tolle eos, et aequantur omnes menses, ut habeant dies XXX.* Finally, the statement that Romulus invented a year consisting of months of either 31 or 30 days is obviously copied from SAT. 1.15.5–6 (MSS NZFR'A' read *triginta* instead of *undetriginta*; cited in DDT 27(662)C10–D3, DTR 12.3–10 which reads *XXX* instead of *undetriginta*, in turn cited in LIB. CALC. 70): *Romulus cum ingenio acri quidem sed agresti statum proprii ordinaret imperii, initium cuiusque mensis ex illo sumebat die quo novam lunam contigisset videri. ... sic factum est ut alii triginta et unum, alii undetriginta sortirentur dies.*

12–18 Omnes...XXX] Based on Isidore (cf. *app. font.*), the year of 12 months of equal length, namely 30 days, with five intercalary days being placed at the end of August, is unanimously attributed to the Eygptians in eighth- and early ninth-century computistical texts. Cf. DRC 68.5–9: *Ita etiam apud Aegyptios, qui in omnibus lunaribus mensibus XXX dies numerantes, quinque dies superfluos de cursu solis habent ... Et hae quinque dies, quos epactas uel interkalares uocant.* BC 60C9–11: *Sed Greci et hii supradicti et omnes hii XII menses habent, et sub mensae eorum dies XXX semper praeter Aegyptios et*

All Greek months consist of 30 solar and lunar days, and (when taken together) they have 5 superfluous days, which they call *interkalares*, and which are divided among all of these months. Accordingly, they turn February into a month of 30 days by taking two days from the seven (months that consist of 31 days in the Julian calendar), and they divide the remaining five days into hours and *momenta*. Romulus, however, with subtle genius, (introduced months of) 31 (and) others (of) 30 (days).

---

concordant omnes. DT 5.8–11: *Aegyptii ... XXX dies suos menses putantes; residuos quinque dies intercalares appellant*. DTR 11.70–82 (cited in LIB. CALC. 69C, RM 30.4–5,13–4): *Aegyptii ... tricenis hos produntur includere diebus ... residuos quinque dies epagomenas vel intercalares sive additos vocant*. COMP. COL. 2.4: *Item in alio intellectu stant menses Egyptiorum triginta diebus, interkalares cum eis. ... De quo numero sequuntur quinque dies interkalares, qui sic nominantur: VIIII Kalendas Septembris, ..., V Kalendas*. The intercalary days only: DIAL. NEUSTR. 3: *secundum Egypcios ... Dies interkalares, qui sic numerantur: VIIII Kalendas Septembris, ... V Kalendas Septembris*. PP I §131: *Annus solaris ... apud Aegyptios a IIII Kl. Septembris incipit et finit in X Kl. Septembris, et intercalares dies V finiuntur in V Kl. Septembris*. DE BISSEXTO I 998A11–14: *Illi autem V dies qui supersunt, intercalares dies apud antiquos erant, sicut modo apud Aegyptios sunt a VIIII Kal. Septembris usque in IIII Kalendas Septembris*. LIB. COMP. 2.16A (cited in LIB. CALC. 13): *Hos quinque dies Aegyptii, qui mensem tricenis tantum diebus concludunt, quasi residuos post completos duodecim menses in fine anni sui intercalare consuerunt*. Lists of the Julian calendar dates of the beginning of each Egyptian month of 30 days, starting with *IIII Kalendas Septembris*, are given in DRC 37, DTR 11.74–79 (cited in LIB. CALC. 69C, partially in RM 30.7–12 where only the first date is mentioned), DIAL. NEUSTR. 3A, COMP. COL. 2.4A. For the five intercalary days cf. also c. 45, ll. 2–6; for the division of these five days by the 12 months of a year referred to here cf. c. 41, ll. 80–86; for a different division c. 38, ll. 23–25; c. 40; for a month consisting of 30 days cf. c. 46, ll. 11–12.

12–15 Omnes...diuiduntur] Transcribed in Borst, *Schriften*, p. 417; cf. also Borst, *Schriften*, p. 899 and p. XXII of the introduction.

The order of months

Sciendum est, quis ordo in positione mensium. Id est: ordo naturalis, | incipere a Martio, in quo mundus factus est; ordo frugiferalis apud Aegyptios, incipere á Septimbrio, ubi fructus condiuntur; ordo uoluptarius apud Macedones, incipere a Nouimbrio, hoc est a cheme, ubi initium laetitiae et uoluptatis post laborem; ordo principatus apud Romanos, incipiebat ab Ianuario deo maximo. fol. 15r

Why was a heathen calendar adopted?

Sciendum est, quur nominata sunt nomina mensium, quae apud gentiles erant. Ut facile suaderent gentiles nomina naturalia ád fidem. Ideo prolixior est Ianuarius quam Februarius. Sic per omnes menses hic intellectus crescit.

---

20 naturalis] natura *M (corr. by Borst).* **25** Ianuario] Ionario *M (corr. by Borst).*

19–26 Sciendum…maximo] No pre-MC text contains a similar passage. Note that the various orders of months outlined here conflict, at times, with the different traditions of what was supposed to constitute the first month of a year discussed in c. 24, ll. 2–11: The beginning of a year in September is there ascribed to the Greeks rather than to the Egyptians, and a November beginning is not recorded at all. Cf. also the discussion of the beginning of a year in c. 36, ll. 2–5. **27–31** Sciendum…crescit] I could not locate any pre-MC text that may have motivated this question about the reasons for adopting heathen names for the months that formed the basis of what was to become a Christian calendar.

19–26 Sciendum…maximo] MC's question about the order of months was directly taken up by QUAEST. AUSTR. 1.1A, 2C, 2.1A, 2C: *Menses singulorum annorum in solari cursu licet interrogare: Quot sunt, et qui sunt, et quis eorum ordo est? ... Nunc, iuxta Latinorum duodecim mensium interrogata ratione, animo occurrit iuxta Hebraeos et Aegyptios et Macedones quaerere de mensibus: Quem ordinem secundum singulos teneant et que vocabula apud eos habent? Et quo mense iuxta singulos annos currere incipiet, ut nobis primus mensis, ille scilicet solis, aput Hebraeos et Aegypcios et Macedones et Latinos lucide tescat? ... Quorum ordo et nomina haec sunt iuxta Latinos: Ianuarius, Februarius, sic usque ad Decembrem. ... Quid vero iuxta Hebraeos et Aegyptios et Macedones menses vocentur, dicendum est, quis sit ordo, quis in his linguis primus mensis solaris sit iuxta singularum linguarum proprietatem. Nam iuxta fidem et lunae rationem idem est mensis, sed alibi <aliter> scriptus*. The *ordo naturalis* and the *ordo principatus* seem to have influenced DIAL. LANGOB. 18B: *Interrogatio: Cur ergo Graeci sic faciunt, et cur Latini aliter? Graecus naturam sequitur, quia in XI Kalendas Aprilis factus est mundus, qui est primus dies saeculi secundum solem. Ideo ab XI volunt ordinari rationem temporis per aequinoctium et solstitium. Latinus vero secutus est auctoritatem et principatum, quia in VIII Kalendas Ianuarii lux renovatur, quasi initium lucis sic dicitur*. Rather than discussing the order of months according to different traditions, eighth-century computists preferred to list the names of the months followed by other people. In agreement with MC, all lists of Egyptian months start with the equivalent to September: BC 151B14–C3: *De mensibus Aegyptiorum. Thot September ... Mensuri Agustus*. DRC 36.1–4: *Sciendum nobis quomodo Egyptii menses uocant. Ita: Thod, Pothi, Athir, ..., id est September, Octuber, Nouember, et reliqua, usque Agustum*. DTR 11.74–79 (cited in LIB. CALC. 69C, only the list of names in RM 30.7–12): *aegyptii ..., quorum primus mensis thoth, IIII kl. septembris; ... duodecimus mesore, VIII kl. augustarum die sumit exordium*. Yet, the situation is more complicated concerning the Macedonian months. These months are listed without reference to their Roman equivalents in DRC 40.1–2: *Sciendum nobis quomodo Macidoniorum menses nominatur. Ita: Dios, Appollonios, Admios, Feritios, Districs, ...* . and with attribution to the Egyptians rather than the Macedonians in BC 58B4–8: *Nomina mensuum secundum Aegyptios ita dicuntur: Dius, Apollonius, et Dyneus, Filisteus, Distreus, ...* . The November beginning mentioned in MC finds support in BC 151B8–13: *De Mensibus Macaedonum. Nouember Apolus ... October* (follows blank space). The

It has to be known, what kind of order concerning the placing of the months (exists). (There are the following orders): the 'natural order', beginning with March, in which the world was created; the 'fruit-bearing order', according to the Egyptians, beginning with September, when the fruits are preserved; the 'pleasant order', according to the Macedonians, beginning with November, i.e. with (the Macedonian month) *cheme*, when delight and pleasure starts after (months of) labour; the 'principal order', according to the Romans, which begins with January, (named after) the greatest god (i.e. Janus).

It has to be known, why the names of months which originated among the heathens were adopted. So that the natural names would easily persuade the heathens of the (true) faith. For that reason January is longer than February, and accordingly the (general) understanding (of the Christian calendar) increases by (examining) all of the (heathen) months.

---

following two texts, however, rather suggest December as the first month of the Macedonians: DDT 25B14–26C7: *De mense Novembri ... Dios apud Macedones ... De mense Decembri ... Apileos apud Macedones ... Et hic mensis in principio anni ponitur apud Macedones*. DTR 14.13–17 (cited in LIB. CALC. 72, RM 31.12–17): *Vocatur autem apud eos* (i.e. the Greeks) *ipse december, apileos; ...; november, dios*. In none of these accounts is any of the Macedonian months called *cheme* or anything similar. For the creation of the world occurring in March cf. c. 11, ll. 37–40; c. 24, ll. 2–3; c. 36, ll. 10–15; c. 44, ll. 18–19. **27–31** Sciendum…crescit] The question about the reasons for adopting the heathen Roman names for the Julian calendar months is also addressed in QUAEST. AUSTR. 1.1C, 2.1C, and may well have been directly influced by MC; the answer is not as in MC (namely to facilitate an easier conversion of the heathens to the Christian faith), but rather points to the need of keeping customs to preserve continuity: *Nam fortasse quis dicat, duodecim mensium vocabula et vocabulorum causas et numerum et caput numeri et ordinem cursus de veteri gentilium constare consuetudine. Licet interrogare quidem: Quod ecclesia catholica et fides Christi superveniens haec omnia silentio praeterierit et nihil horum, quasi placeret, mutare curaverit? ... Sed non est nunc temporis disserere, quod ecclesia reticuit mutare nomina de vocabulis idolorum haerentia et de vana gentilium superstitione conposita. Haec fortasse causa est, ne in cunctis consuetudine contrairet, in his dumtaxat, in quibus posset salva fides fieri. Durum est enim veterem consuetudinem, tempore longo linguae adhaerentem, parvo mutare tempore.*

19–21 Sciendum…est] Transcribed in Borst, *Schriften*, p. 449. **19–26** Sciendum…maximo] Cf. Borst, *Schriften*, p. 480 and p. CLXIX of the introduction. **22–23** apud…Nouimbrio] Cf. Borst, *Schriften*, p. 412. **24–26** ordo…maximo] Transcribed in Borst, *Schriften*, p. 449. **27–31** Sciendum…crescit] Cf. p. CLXIX of the introduction.

Why do individual names not exist for lunar months?

Sciendum est, quur menses lunę non habent nomina specialia ut menses solis. Quia menses lunę in una ré adfiguntur, non ut solis mensis a diis, a rebus, et reliqua. Menses lunę cum Hebreis uno nomine appellari solent: teuelat, hoc est XXVIIII et semis. Sic et cum Aegyptiis unum nomen habent: tebet, id est XXXmi. Apud Grecos uero duo nomina possedebant, id est tebat, XXVIIII, et tebat, XXXmi.

Three principal days

*Quo*t *sunt principales dies* de nominibus dierum *mens*ium? Id est *III*: *Kalende*, *Nonae* et *Idus*.

32–39 Sciendum…XXXmi] Mc here reflects on an earlier passage (c. 12, ll. 18–23), in which it is argued that the concept of solar months is based on an earlier lunar construction, and that the names of the Julian calendar months are borrowed from four things, the names of gods, kings, numbers, and objects. The theory that Hebrew lunar months consist of 29½ days (a theory which is applied c. 38, ll. 5–6) is probably based on Prologus Cyrilli 6, 8, *Addendum* (Krusch, *Studien* I, p. 341–3): *XX autem et novem semis dies lunam habere manifestum est. ... Luna autem, quae ab his dicitur tertia hodiequae a Judaeis et Grecis dicitur prima, et merito videtur ad vesperam, quia iam et semisse habere de XXX, et que eque et prima. ... Et unus mensis lunaris XXVIIII et semis dies lucet.* or Computus Cottonianus fol. 78v–79r (cited in Bc 15B6–8, in turn cited in Lib. ann. 58): *IIII differentiae lunae: ... apud Ebraeos XXVIIII et semis*. For Egyptian lunar months consisting of 30 days see c. 24, ll. 12–14; c. 46, ll. 11–12; for lunar months alternating between 29 and 30 days see c. 50, ll. 29–39, 61–66. The terminology for lunar months given in this passage here appears to derive from the equivalents to January in the respective languages: cf. Jerome, *Commentarii in Hiezechielem* 9.29.1–3a (*CCSL* 75, p. 402): *qui hebraice appellatur 'tebeth' et apud Aegyptios 'tybi', apud Romanos 'ianuarius' dicitur.* Polemius Silvius, *Fasti* (Mommsen, *Corpus inscriptionum latinarum* 1, p. 335): *Ianuarius ... vocatur ... apud Aegyptios Tibi.* The Egyptian *tybi* is the equivalent to December in Pliny, *Naturalis historia* 6.106 (Brodersen, *Plinius* 6, p. 80): *mense Aegyptio Tybi incipiente, nostro Decembri.* The Hebrew *tebet* is the equivalent to December in Polemius Silvius, *Fasti* (Mommsen, *Corpus inscriptionum latinarum* 1, p. 357): *December ... vocatur apud Hebraeos Tebet.* **40–41** Quot…Idus] This introduction to the following discussion of the Calends, Nones, and Ides is directly copied from Ce p. 95: *Quod sunt principales dies, quibus menses conputantur? II uel III, id est Kalendę, Nonę, Idus.*

32–39 Sciendum…XXXmi] Even though this passage has no parallel in any other computistical text, the details mentioned can frequently be found elsewhere: For Hebrew lunar months consisting of 29½ days (a theory which is applied c. 38, ll. 5–6) cf. Bc 61D14–A1 (cited in Quaest. Austr. 2.4A, in turn cited in Comp. Col. 6.4): *Unde apud Ebreos omnes menses XXVIIII et semis semper habent aetates.* Bc 62D5–7: *Diximus namque unoquoque mense luna habere dies XXVIIII et dimidium.* Dtr 36.7–12 (cited from *accipitur* to *prolixior* in Lib. ann. 1A1–2, in turn cited in Lib. comp. 4.4, Lib. calc. 39, Rm 36.11–14, Pv §§128–129): *Sed lunaris annus quadrifarie accipitur: primus est namque cum luna XXVII diebus et VIII horis zodiacum percurrens ad id signum ex quo egressa est revertitur; secundus duobus diebus et quattuor horis proloxior, qui consuete mensis appellatur cum solem, a quo digressa est, XXVIIII diebus et XII horis exactis iam defecta*; for Egyptian lunar months consisting of 30 days see c. 24, ll. 12–14; c. 46, ll. 11–12; for lunar months alternating between 29 and 30 days see c. 50, ll. 29–39, 61–66. As mentioned in *app. font.*, the terminology given here for a lunar month derives from the equivalents to January (or, in some cases, December) in the respective languages: The Hebrew January is called *Tibiath* in Drc 41.2, *Thabeth* in Ddt 15D4, *Tereth* in Bc 151B6, *Tebet* in Dtr 11.23 (cited in Lib. calc. 69A, as *Tebeth* in Rm 29.12); the Hebrew December is called *Tebet* in De Anno, *Thebet* in Quaest. Langob. 4A; the Egyptian January is called *Tebethi* in Drc 36.2, *Tibi* in Ddt 15D4, Dtr 11.76 (cited in Lib. calc. 69C, as *Thubi* in Rm 30.10), *Tybi* in De Anno, *Tubi* in Bc 151C1. **40–41** Quot…Idus] In this passage, Mc appears to have directly influenced the two early eighth-century Irish texts Drc 30.1–2: *Sciendum nobis quot sunt dies uel stipites contra quos*

It has to be known, why lunar months do not have individual names like solar months. Because lunar months are labelled according to one thing only, not according to gods, objects, etc. like solar months. Lunar months are customarily called by one name according to the Hebrews: *Tevelat*, which means 29½. Likewise (lunar months) have only one (general) name according to the Egyptians: *Tebet*, which means 30. Among the Greeks, however, they possessed two names, namely *Tebat*, (which means) 29, and *Tebat*, (which means) 30.

How many principal days are there concerning the names of the days of the months? Three, namely the Calends, Nones, and Ides.

---

regulantur menses. Tres, id est Kalendae, Nonae, Idus. DIAL. LANGOB. 14B: *Interrogatio: Unde nomina acceperunt dies mensium? Responsio: Omnes dies omnium mensium a tribus diebus nominantur et ordinantur. Interrogatio: Qui sunt hi tres? Responsio: Kalendae, Nonae et Idus*. and also the early eighth-century Frankish text DIAL. BURG. 7B: *Quot dicunt principales partes in menses? Tres: Kalende, Nonas, Idus*. Cf. also QUAEST. AUSTR. 1.5A: *Nunc festinat animus cognoscere de tribus adsuetis singulorum mensium, verbi gratia: ... Kalendae ... Nonas ... Idos vel Idis ....*

40–41 Quot...Idus] Transcribed in Borst, *Schriften*, p. 360; cf. p. CXLI, CLXXII of the introduction.

<XXV.> DE KALENDIS

History and etym. of Calends

Kalendae singularem numerum non habent, quia apud antiquos tres dies Kalendares appellabantur. | fol. 15v *Kalendas, Nonas et Idus Romani propter dies festos inposuerunt. In his enim diebus conueniebant sacrificio celebrato a rege et pontifice. Kalendę a calando dictae*, uel *colendo*, siue *uocando*, quia *calo Greci uoco*. Ut Macrouius ait: *Minor pontifex*, cui *noue* accensio *lunę* apparere, ⌐ primum erat officium ¬ *numerum, qui* in *nonas* restabat, k*alando prodebat*.

XXV,2 non] *om. M**. **3–4** Kalendas] Kalendae *M (corr. from* ETYM., BC*)*. **6** calando] calendo *M (corr. from* SAT., *etc.; cf. c. 25, l. 10, where M also reads* kalendo *instead of the Macrobian* calando*)*. **7** uoco] uoca *M (corr. from* SAT., *etc.)*. **9** primum…officium] *This seems to have been a gloss originally to explain that the described act was the first duty of the minor priest; its later inclusion breaks the structure of the sentence.* **10** nonas] nouas *M (corr. from* SAT., *etc.)*. | kalando] kalendo *M (corr. according to* SAT., *etc.)*.

XXV,2–10 Kalendae…prodebat] The grammatical fact that the term Calends (like Nones and Ides) only exists in plural form is probably derived from POMPEIUS (MAURUS), *Commentum artis Donati* (Keil, *Grammatici Latini* 5, p. 195): *et multa nomina tantum numeri pluralis <sunt>, kalendae nundiae feriae*. MC's account of the introduction of the Calends (as well as Nones and Ides) by the Romans as their festive days is a synthesis of ETYM. 5.33.12 (cited in BC 20B6–9, which reads congregabuntur): *Kalendas autem, Nonas et Idus propter festos dies Romani instituerunt; vel propter officia magistratuum. In his enim diebus conveniebatur in urbibus.* and SAT. 1.15.10 (cited in DDT 27D(662)13–15, DRC 30.17–18, DTR 13.4–5, in turn cited in RM 33.7–9, PV §119, LIB. CALC. 71; from *calata* only in DT 6.8): *itaque sacrificio a rege et minore pontifice celebrato idem pontifex calata, id est vocata, in Capitolium plebe*. This latter Macrobian passage also supplied the etymology *a vocando*, in combination with SAT. 1.15.11 (cited in DDT 27A(663)2–4, DTR 13.8–9, in turn cited in PV §120, LIB. CALC. 71; KAL. F1,6, a17; the first part only in DRC 30.21–22): *verbum autem καλῶ Graecum est, id est voco, et hunc diem, qui ex his diebus qui calarentur primus esset, placuit kalendas vocari*. For this etymology MC may also have consulted CE p. 95: *Aliter Kalendę a uerbo Greco, quod est kalo, id est uoco. Inde Kalendę uocationes interpretantur.* and cf. POLEMIUS SILVIUS, *Fasti* (Mommsen, *Corpus inscriptionum latinarum* 1, p. 335): *Kalendae dictae ἀπὸ τοῦ καλεῖν hoc est vocando, quia tum in rostris Romae ad contionem populus vocabatur*. The etymology *a colendo* is directly taken from DNR 4.6.41 (MSS PCS read *Kalendae ... dictae*; cited in BC 59A10–11, which also reads *Kalendae ... dictae*): *Kalendas autem a colendo dictas*. ETYM. 5.33.13 (cited in BC 20B9–10): *Quidam autem Kalendas a colendo appellari existimant*. The etymology *a calando* and the detailed explanation of it in the final sentence stem from SAT. 1.15.12 (cited in DDT 27A(663)7–12, DTR 13.11–15, in turn cited in RM 33.16–20, PV §121, LIB. CALC. 71): *ideo autem minor pontifex numerum dierum qui ad nonas superesset calando prodebat, quod post novam lunam oportebat nonarum die populares qui in agris essent confluere in urbem, accepturos causas feriarum a rege sacrorum sciturosque quid esset eo mense faciendum*. A similar explanation based on SAT. can be found in CE p. 95–6.

XXV,2–10 Kalendae…prodebat] The grammactical fact of *Kalendae* only existing in plural form was common knowledge, and therefore not stressed in any computistical text other than MC. The Isidorian explanation for the origin of Calends, Nones, and Ides cited in MC was not very popular in computistical literature, whereas the Macrobian passage aditionally consulted by MC was, especially since it was quoted by and received a wide reception through Bede (cf. *app. font.*). Concerning the three etymologies given by MC, many eighth and early ninth-century computists used the same sources (cf. *app. font.*). All three etymologies can also be found in various versions in DRC 30.2–4, 30.23–24: *Sciendum nobis quid sint Kalendae. Id est, nomen deriuatiuum, a uerbo Greco quod est kalo, quod*

25. ABOUT THE CALENDS

The term 'Calends' does not have a singular form, since three days were called *Kalendares* among the ancients. The Romans introduced Calends, Nones, and Ides on account of their festivals. In fact, they assembled on these days, after a sacrifice had been offered by the king and the high-priest. The Calends (*Kalendae*) were so called from calling (*calando*), or from honouring (*colendo*), or from calling (*uocando*), since *calo* in Greek means 'to call' (*uoco*) in Latin. As Macrobius says: The minor priest, to whom the kindling of the new moon appeared, announced the number of nones remaining by calling (*kalando*); ⌐ this was his first and foremost duty ¬ .

---

interpretatur uoco. Kalendae uocationes dicuntur ... Aliter Kalendae a colendo nominatae sunt, ut Isidorus dicit: Kalendae a colendo dictae sunt. BC 19A12–15: *Kalendae Grecus sermo est, sed ad colendum dicitur, in Latinum vocationes dicuntur. Apud Grecos de calo Kalendae, apud Latinos derivantur de voco vocationes.* BC 20B9–11: *Quidam autem Kalendas a colendo appellari exaestimant, vel a calo, id est voco.* DIAL. LANGOB. 15A: *Kalendae graeco vocabulo, latine vocationes dicuntur. Kalo enim verbum graecum est, quod significat voco, et de quo verbo graeco nomen graecum deducitur, id est Kalendae, quod significat vocationes. ... Item Kalendae a colendo dictum.* QUAEST. LANGOB. 2B: *Interrogatio: A quo dicitur Kalendas? Responsio: A colendo dicitur, quia apud veteres semper ipsas colebantur. Interrogatio: Cuius sermo est Kalendas? Responsio: Graecum est. <Interrogatio:> Quod interpretatur in Latino? Responsio: Praevocationis. Interrogatio: Unde derivatur Kalendas? Responsio: De calo. Interrogatio: Unde derivatur vocationes? Responsio: De voco; de calo Kalendas, de voco vocationes.* Only the *a colendo* etymology is given in DIAL. BURG. 7A: *Sed Kalendas a colendo dictis dicitur.* Cf. also Muretach's and Sedulius Scottus' commentaries to Donatus (*CCCM* 40, p. 89; *CCCM* 40B, p. 133). The Macrobian explanation for the *a calando* etymology appears in rather lengthy variations in BC 20B11–15, DRC 30.4–22, DIAL. LANGOB. 15A.

XXV,1–XXVII,12 De…immolabat] Cf. Borst, *Schriften*, p. 360, 445.

<XXVI.> DE <NOMINE> NONARUM

Etym. and history of Nones

*Non*ae autem *dict*ae *quasi nouae obseruationis initium, uel quod ad Idus no*nus *die*s; aut *Nonę a nundinis* ⌐ *nundinae enim sunt publice conuentiones siue mercimonia* ¬ *uoca*ntur. Omnes primos menses IIII Nonae fuerint. In hoc die scientes rustici conuentum occidebant duces. Inde commutauerunt duces alios in VI Nonas.

XXVI,1 nomine] *om. M (add. according to* MC*).* **3** ad Idus] auidus *M (corr. from* SAT.*, etc.; it may rather have read* ab Idus *originally, as in* DIAL. LANGOB.*, which seems more logical in connection with the ordinal* nonus*; but one would then rather expect the ablative* Idibus*, as in* DRC*, or* Ido*, as later in* MC*, c. 27, l. 7).* **4–5** nundinae…mercimonia] *This seems to have been a gloss originally to explain the term* nundinis*; its later inclusion breaks the structure of the sentence.* **4** enim] autem *M (corr. from* ETYM., BC, DIAL. LANGOB.*; I suspect that the copyist mistook the Irish abbr. for* enim *for the Irish abbr. for* autem*).* | conuentiones] inuentiones *M (corr. from* ETYM., BC, DIAL. LANGOB.*).*

XXVI,2–7 Nonae…Nonas] The first two etymologies, i.e. the Nones being so called from new observations or from the number nine, are directly copied from SAT. 1.15.13 (cited in DDT 27A(663)12–13, DRC 30.26–27, DTR 13.15–16, in turn cited in LIB. CALC. 71, RM 30.20–22, PV §121, KAL. F1,6, a17): *unde quidam hinc nonas aestimant dictas quasi novae initium observationis, vel quod ab eo die semper ad idus novem dies putentur*. MC may also have consulted CE p. 96: *Nonae a numero dicitur, eo quod nonus dies sit ante Idus, vel Nonae a noua obseruatione*. Cf. POLEMIUS SILVIUS, *Fasti* (Mommsen, *Corpus inscriptionum latinarum* 1, p. 335): *Nonae dictae ideo, quia nonus dies eas discernit ab idibus*. The etymology *nonae a nundinis* and the glossed explanation for *nundinae* are direct quotes from ETYM. 5.33.14 (cited in BC 20B15–C3, DIAL. LANGOB. 15C): *Nonae a nundinis vocatae. Nundinae enim sunt publicae conventiones sive mercimonia*. Cf. DNR 4.6.43 (cited in CE p. 96, BC 69A13, DRC 30.25): *nonas a nundinis*. Differently SAT. 1.13.29: *nonae a nundinis segregarentur*. I could not locate any source for MC's curious explanation of the fact that some months have four, while others have six nones. A different explanation, given in SAT. 1.15.5–7, was certainly known to MC, but apparently rejected.

XXVI,2–7 Nonae…Nonas] MC's account of the Macrobian etymologies for the term Nones, deriving from 'new observations' or the number nine, appears to be directly copied in DIAL. LANGOB. 15C: *Sive Nonae dicuntur, quia nonus dies est ab Idus, vel Nonae quasi novae initium conversationis*. Especially the etymology from the number nine, which was generally regarded as the most convincing explanation for the term Nones, enjoyed wide popularity: DRC 30.24–25: *Sciendum nobis unde Nonae nominatae sunt. ..., uel eo quod nonus dies retro sit ab Idibus*. DT 6.9–11: *Nonas autem quod nono ante idus die confluerent in urbem scituri quid esset eo mense faciendum*. QUAEST. LANGOB. 2C: *Apud gentiles et pro hoc dicitur Nonas, eo quod nonus sit ante Idus*. The Isidorian etymlogy *Nonae a nundinis* is also frequently cited, either verbatim (cf. *app. font.*) or in variation: BC 19B1–3: *Nonae a nundinis deriuantur apud Grecos. Apud Latinos mercatum, uel mercimonia dicuntur*. QUAEST. LANGOB. 2C: *Interrogatio: Cuius sermo est Nonas? Responsio: Graecum est. Nonas a nundinis dicitur et interpretatur Nonas apud Latinus mercatus vel mercimonia vel dies epularum*. Cf. also Muretach's and Sedulius Scottus' commentary to Donatus (*CCCM* 40, p. 89–90; *CCCM* 40B, p. 133). Interestingly enough, an explanation similar to MC's of the fact that some months have four, others six nones, is outlined in BC 20C3–9 (with thieves rather than peasants trying to rob and kill merchants rather than lords): *Quare antiqui IIII Nonas, vel VI Nonas diuisissent? IIII Nonas, uel VI Nonas ideo diuiserunt, ut non scirent latrones mercatum illorum, quando esse debeat, qui abscondebant se per diuersa latibola in secretis locis iuxta uiam, et uenientes ad mercatum occidebant, et spoliabant.*

26. ABOUT THE DENOMINATION OF NONES

The Nones (*Nonae*) were so termed as though (they were) the beginning of a new observation (*nouae obseruationis initium*), or because (they constitute) the ninth (*nonus*) day before the Ides. Or the Nones (*Nonae*) are so called from *nundinae*; ⌐ *Nundinae* are public conventions or goods ¬ . Originally, there happened to be four nones in all months. (Then) some peasants, being informed about the assembly, killed certain lords on that day. As a result of this, the lords changed (some) other (months) to six nones.

<XXVII.> DE NOMINE IDUARUM

Etym. and history of Ides

*Idu*s *nomen sumptum a Tuscis, apud quos is dies* <*Itis*> *uocatur* et *interpretatur Iouis fiducia*, quia *Iouem auctorem lucis acce*perunt, quem *Romani Dispitrem* uoc*ant*, id est *diei patrem, cuius* et *splendor* <*luna*m> *inlustra*t. Inde *luna in pleni*um in *medio mense*, id est in Ido, lucet. Inde *a uidendo uidus* dicitur, et *u littera detracta idus* nominatur. | *Iduare enim Ethrusca lingua* fol. 16r *diuidere est, idua diuisa est. Omnes Idus Ioui* et *Kalendas Iunoni tributas pontificalis auctoritas adfirmat.* Inde *kalend*arum die *Rome uxor regis Iunoni,* rex uero Ioui, *agnum immola*bat.

XXVII,3 Itis] *om. M (add. from* SAT.*).* | interpretatur] interpetratur *M (corr. according to* SAT.*).* **5** lunam] *om. M (add. according to* SAT.*).* **11** regis] regi *M**.

XXVII,2–12 Idus...immolabat] The entire section on the term Ides is a variation of SAT. 1.15.14–19 (only the middle part, from *nobis vero* (or slightly thereafter) to *a viro divisa* (or slightly before), cited in CE p. 96, DDT 27(663)A15–18, DRC 30.29–31, DT 6.11–12, DIAL. LANGOB. 15B, DTR 13.17–19, in turn cited in LIB. CALC. 71, RM 33.23–25, PV §122, KAL. F1,6, a17): *Iduum porro nomen a Tuscis, apud quos is dies Itis vocatur, sumptum est. Item autem illi interpretantur Iovis fiduciam. nam cum Iovem accipiamus lucis auctorem, unde et Lucetium Salii in carminibus canunt et Cretenses Δία τὴν ἡμέπαν vocant, ipsi quoque Romani Diespitrem appellant ut diei patrem, iure hic dies Iovis fiducia vocatur, cuius lux non finitur cum solis occasu, sed splendorem diei et nocte continuat inlustrante luna, quod semper in plenilunio id est medio mense fieri solet. ... alii putant idus, quod ea die plena luna videatur, a vivendo vidus appellatas, mox litteram v detractam, sicut contra, quod Graeci ἰδεῖν dicunt, nos v littera addita videre dicimus. ... nobis illa ratio nominis vero proprior aestimatur, ut idus vocemus diem qui dividit mensem. iudare enim Etrusca lingua dividere est, unde vidua quasi valde idua id est valde divisa, aut vidua id est a viro divisa. ut autem idus omnes Iovi, ita omnes kalendas Iunoni tributas et Varronis et pontificalis adfirmat auctoritas. ... Romae quoque kalendis omnibus, praetor quod pontifex minor in curia Calabra rem divinam Iunoni facit, etiam regina sacrorum, id est Regis uxor, porcam vel agnam in regia Iunoni immolat.* For the Ides constituting the middle of a month cf. also POLEMIUS SILVIUS, *Fasti* (Mommsen, *Corpus inscriptionum latinarum* 1, p. 335): *Idus dictae ἀπὸ τοῦ εἰδεῖν, a videndo, quia priusquam annus hic qui est fuerat, mense medio luna completa, quae incipiebat kalendis, de qua menses dictos accepimus, videbatur.* CE p. 96: *Vel Idus ergo medium mensis.*

XXVII,2–12 Idus...immolabat] The middle part of the Macrobian section cited here, namely the etymology of Ides deriving from the Etruscan verb for 'to divide', was often quoted in eighth- and early ninth-century computistical literature (cf. *app. font.*). But also some of the other features mentioned by Macrobius here made their way into various texts: The etymologies *Idus a fide Iovis* and *a videndo* were not very popular, but they are nevertheless found in DRC 30.31–32: *Aliter Idus a fide Iouis uocatae sunt, uel uidendo, quia ea die plena luna uidetur.* On the other hand, the fact that the Ides constitute the middle of a month was a very popular observation. Cf. DIAL. BURG. 7A: *Idus graece dicitur, in latine medio mense vel divisio.* QUAEST. AUSTR. 2.5A: *Idus autem vel Idis iuxta qualitatem medium mensis significat.* BC 19B3–4: *Idus Grecus sermo est, in Latinum dicitur cor mensuum, uel medietas mensuum.* which ultimately appears to have influenced the following accounts: BC 20C9–14: *Qua causa dicti sunt Idus cor mensuum uel medietas mensuum? Idus ... uel propter medium menses, quia iduare Greci medium dicitur.* DIAL. LANGOB. 15B: *Interrogatio: Quid est Idus? Responsio: Idus nomen graecum est et medietas interpretatur, quia in medio stat inter duas Kalendas, id est <diem Kalendarum mensis ipsius et> diem Kalendarum mensis sequentis.* QUAEST. LANGOB. 2D: *Interrogatio: Idus cuius sermo est? Responsio: Graecum est et dicitur cor mensium vel medietas mensium, quia aliqui menses in Idus*

27. ABOUT THE DENOMINATION OF IDES

The Ides received (their) name from the Tuscans, among whom this day was called *Itis* and interpreted as the trust in Jupiter, since they regarded Jupiter as the author of light, whom the Romans call *Dispiter*, which means 'the father of the day', and his splendour also illuminates the moon. Therefore the moon shines in its fullness in the middle of the month, i.e. on the Ides. In fact, *uidus* is so called from observing (*uidendo*), and after removing the initial *u*, it is called Ides (*Idus*). *Iduare* means 'to divide' in the Etruscan language, (and accordingly) *idua* means 'divided'. The pontifical authority affirms that all Ides are dedicated to Jupiter, and all Calends to Juno. Therefore, the wife of the king of Rome offered a lamb to Juno on the day of the Calends, the king, however, to Jupiter.

---

dies quindecim habent. Cf. also Sedulius Scottus' commentary to Donatus (*CCCM* 40B, p. 133). The curious misspelling in CE p. 96 (*ove* instead of *Iove*): *Vel Idus ab oue iduli*. combined with Macrobius' statement that certain animals were offered to Jupiter on the Ides, led to DIAL. LANGOB. 15B: *Alii dicunt Idus ab ove, quae offerebatur idolo in illa die, id est Iovi. Idis enim ovis interpretatur, et de illa ove dies nominabatur*.

<XXVIII. *DE REGULIS MENSIUM*>

Patterns for the days of a month

*Quot regul*is *menses ante Idu*s mensurantur? *Du*obus: *aut enim IIII Nonas, aut VI* Nonas. *Post*, IIII regulis: *aut XVIIII Kalendas, aut XVIII Kalendas, aut XVII, aut XVI* Kalendas.

All possible days of the month

Haec sunt nomina dierum mensium solis: Kalendae, VI Nonas, V Nonas, IIII Nonas, III Nonas, II Nonas, Nonae, VIII Idus, VII Idus, VI Idus, V Idus, IIII Idus, III Idus, II Idus, Idus, XVIIII Kalendas, XVIII Kalendas, XVII Kalendas, XVI Kalendas, XV Kalendas, XIIII Kalendas, XIII Kalendas, XII Kalendas, XI Kalendas, X Kalendas, VIIII Kalendas, VIII Kalendas, VII Kalendas, VI Kalendas, V Kalendas, IIII Kalendas, III Kalendas, II Kalendas, Kalendae.

XXVIII,1 DE…MENSIUM] CE p. 96 gives this heading to the passage that served MC as model in the following discussion. **2–5** Quot…Kalendas] The patterns described here obviously have the Ides as their point of reference: If the Ides fall on the 13th day of a month, this month has four nones, if they fall on the 15th day, it has six nones; likewise, the number of calends following the Ides depends on the placement of the Ides in, as well as the total number of days of the month in question. These patterns may ultimately derive from similar statements in SAT. 1.14.11, 1.15.7 (cited in DDT 27(662)D8–9): *licet ab incremento non iam a. d. sextum decimum kalendas sed a. d. septimum decimum, si unus, aut a. d. octavum decimum, si duo sunt additi, diceretur. ... hinc aliis quintus a kalendis dies, aliis septimus nonas facit*. MC, however, directly quotes CE p. 96 here: *Quot sunt regulę, quę superhabundant (!) menses ante Idum et post Idum? Id est: duę ante Idum, IIII post Idum; aut enim quartas Nonas aut sextas ante Idum habent; post uero Idum aut XVIIII Kalendas, aut XVIII Kalendas, aut XVII Kalendas, aut XVI*. **6–13** Haec…Kalendae] MC's model, CE p. 96–97, does not list all days found in Julian calendar months, but rather outlines only one example, the month of January. Any calendar could also have served as source here; pre-MC calenders that may have been known to its author are the ones of Polemius Silvius (Mommsen, *Corpus inscriptionum latinarum* 1, p. 335–57) and Willibrord (Paris, Bibliothèque Nationale, Lat. 10837, fol. 34r–40r).

XXVIII,2–5 Quot…Kalendas] The description of the patterns regulating the structure of the Julian calendar months are a very prominent feature of late seventh-, early eighth-century Irish computistics. These patterns ultimately derive from Macrobius, but they appear to have been formulated in a condensed version in CE (cf. *app. font.*), from which they were copied into DIAL. LANGOB. 12A: *Interrogatio: Quot sunt regulae, a quibus menses regulantur? Responsio: Sex sunt, id est duae ante Idus et quattuor post Idus. Duae regulae ante Idus haec sunt: Aut enim quartae Nonae sunt mensis ante Idus, ut Ianuarii, aut sextae Nonae fiunt ante Idus, ut est Martii. Quattuor vero regulae sunt mensium post Idus: Aut enim nonae decimae Kalendae fiunt post Idus, aut octavae decimae, vel septimae decimae, seu etiam sextae decimae fiunt*. DRC 31.1–4, on the other hand, worked not only from CE, but also consulted SAT.: *Sciendum nobis quot regulae sunt in mensibus ante Idus. Duae, id est aut enim menses quintanas Nonas habent aut septenas ante Idus. Post Idus vero quattuor. Sunt enim menses qui nonadecimas et alii qui octodecimas at alii qui septimdecimas et unus qui sexdecimas post Idus Kalendas habent*. In eighth-century Frankish computistics, these patterns were applied rather than copied: DIAL. NEUSTR. 7 lists the names of the days of the Julian calendar months according to this pattern: *Scire debemus dies mensis: Quomodo nominatur? Kalendae, IIII Nonas, VI Nonas, VIII Idus, VII Idus, VI Idus et reliqua, XVIIII Kalendas, XVIII Kalendas, XVII Kalendas, XVI Kalendas post Idus*. copied in QUAEST. LANGOB. 2A: *Interrogatio: Quibus modis dicuntur dies mensis? Responsio: Quattuor. Quomodo? Kalendas, sex Nonas aut quattuor, octo Idus et novemdecim aut octodecim aut septendecim aut sedecim Kalendas post Idus*. For each of these patterns, QUAEST. AUSTR. 2.5B, 6, lists the Julian calendar months that show these characteristics: *Nunc ordo est scire, in quibus quattuor Nonae <sint>.*

28. ABOUT THE RULES OF THE MONTHS

According to how many patterns are the months structured before the Ides? According to two: (they have) either four or six nones. After (the Ides, they are structured) according to four patterns: (they have) either 19, or 18, or 17, or 16 calends.

These are the names of the days of the solar months: Calends, 6th nones, 5th nones, 4th nones, 3rd nones, 2nd nones, Nones, 8th ides, 7th ides, 6th ides, 5th ides, 4th ides, 3rd ides, 2nd ides, Ides, 19th calends, 18th calends, 17th calends, 16th calends, 15th calends, 14th calends, 13th calends, 12th calends, 11th calends, 10th calends, 9th calends, 8th calends, 7th calends, 6th calends, 5th calends, 4th calends, 3rd calends, 2nd calends, Calends.

... Et hii sunt quattuor, in quibus sex Nonae <sint>: ... Et hii sunt tres, in quibus novemdecim Kalendae ab Ido <sint>: ... Et hii sunt quattuor, in quibus octodecim Kalendae ab Ido <sint>: ... Et hii sunt, in quibus septendecim Kalendae ab Ido <sint>: ... Et hic est unus, in quo sedecim Kalendae sint: ... Ordo est nunc dicere singulorum dierum nomina. Primus dies dicitur Kalendae. Deinde, si Martius vel Maius vel Iulius vel Octuber: VI Nonas, V Nonas; si vero praeteribis hos duos dies, et dices: IIII Nonas, III Nonas, II Nonas, Nonas. Dein <in> singulis mensibus cursu simili conputabis, ut ad Idus: VIII Idus ... Idus. Acsi denique, si in Ianuario vel Agusto vel Decembre: XVIIII Kalendas; si in Aprile, Iunio, Septembre, Novembre: XVIII Kalendas; si vero in Martio, Maio, Iulio, Octubre: XVII Kalendas; si in Februario: XVI Kalendas. Dehinc per singulos aequalite numerabis: XV Kalendas ... Kalendas. **6–13** Haec...Kalendae] DIAL. LANGOB. 14A, based on CE (cf. *app. font.*) lists, in abridged form, the days of January as an example, followed by a more general account in DIAL. LANGOB. 14B; DRC 32 lists abridged the days of the month for one representative of each class of months (for these classes see the following passage); for DIAL. NEUSTR., QUAEST. AUSTR., and QUAEST. LANGOB. see the previous note.

XXVIII,2–5 Quot...Kalendas] Cf. p. CLXXIV, CLXXIX of the introduction. **2–22** Quot...Idus] Cf. Ó Cróinín, *Cummian's letter*, p. 143–4; Borst, *Schriften*, p. 388, 442, 485, 518.

Classific. of the months |

Qui *sunt menses* una regula concordes? *Ianuarius, Augustus* et *Decimber*: tres trigeni singuli, trī quarti nōnaich, trī noī decem *Kalendas post Idus*; *Aprilis, Iunius, Septimber, Nouimber*: quattuor *tri*cesimi, IIII quater *Nonas habent*, quaterque octecim *Kalendas post Idus*; *Martius, Maius, Iulius, Octimber*: <IIII *trigeni singuli*,> IIII *sex Nonas*, IIIIque septecim *Kalendas post Idus*. *Februarius* singularis a ceteris discors: *XXVIII dies*, IIII *Nonas*, *XVI Kalendas post Idus*.

XXVIII,16 nōnaich] noinaic *M (for this reconstruction cf. Bisagni & Warntjes).* **19** IIII[1]...singuli] *om. M (add. according to* MC, CE, DRC, DIAL. LANGOB.*).*

14–22 Qui...Idus] The same groups of months, though in a different context (i.e. the development of the Roman calendar), can already be found in SAT. 1.14.7 (cited in DDT 27(664)A9–19; cf. also DRC 28.43–56): *dies autem decem, quos ab eo additos diximus, hac ordinatione distribuit. in Ianuarium et Sextilem et Decembrem binos dies inseruit, in Aprilem autem Iunium Septembrem Novembrem singulos; sed neque mensi Februario addidit diem, ne deum inferum religio immutaretur, et Martio Maio Quintili Octobri servavit pristinum statum, quod satis pleno erant numero, id est dierum singulorum tricenorumque*. The earliest occurrence of the classification as outlined in MC appears to be CE p. 96: *Menses conregulares hi sunt: Ianuarius, Augustus, Decimber dies trigenos singulos habent, et quartas Nonas, et XVIIII Kalendas post Idus. Aprilis, Iunius, Septimber, Nouimber dies trigenos, quartas Nonas, XVIII Kalendas post Idus habent. Martius, Maius, Iulius, Octimber dies trigenos singulos, sextas Nonas, XVII Kalendas post Idus habent. Februarius solus dies XXVIII, quartas Nonas, XVI Kalendas post Idus habet*. Note the difference between CE and MC in the use of the vernacular; it seems that MC applied the vernacular here because this classification was presumably taught in Old Irish in the classroom (for the Old Irish forms cf. the article referred to in *app. comm.*).

14–22 Qui...Idus] This classification of the months was extremely popular in early medieval computistics; note, however, that the order of classes as oulined in CE and MC is unique to these two texts; in all subsequent texts, the class of April, June, September, and November is mentioned after the class of March, May, July, and October. DRC 31.5–11 appears to be a synthesis of CE, MC, and SAT.: *Sciendum nobis qui menses conregulares sunt secundum hanc ordinationem. Hii sunt, id est Ianuarius, Agustus, Decimber, tres triceni singuli sunt, et quintanas Nonas habent, et XVIIII Kl illis post Idus rediunt. Martius uero et Maius et Iulius et Octimber quattuor triceni singuli sunt et septenas habent Nonas, et XVII Kl post Idus habent. Aprelis uero et Iunius et Septimber et Nouimber quattuor triceni sunt et quintanas Nonas habent, et XVIII Kl illis post Idus rediunt*. DIAL. LANGOB. 12B seems to have relied on both CE and MC: *Interrogatio: Quot sunt menses, qui concordant sub his regulis supradictis, et unius eiusdem regulae fiunt?* <*Responsio:*> *Ianuarius, Augustus, December, hi tres menses unam regulam servant: Trigenos singulos dies habent, quattuor Nonas ante Idus et novemdecim post Idus. Martius, Maius, Iulius, Octimber, hi quattuor menses pariles sunt: Trigeni singuli sunt, sex Nonae fiunt ante Idus, septendecim post Idus. Aprilis, Iunius, September, Noimber, hi quattuor sub una regula sunt: Trigeni, quattuor Nonae ante Idus, octodecim post Idus. Februarius solus sedit. Cuius dies solares fiunt viginti octo, quattuor Nonas ante Idus et sedecim Kalendas post Idus habere videtur*. The version of this classification that became standard in the eighth to tenth centuries is outlined in BC 1(1281)C1–7 (cited in PV §5, LECT. COMP. 1.9, in turn cited in LIB. COMP. 2.14 and numerous calendars (cf. Borst, *Reichskalender*, p. 403–4) and MSS; the occurrence of this passage in London, British Library, Cotton Caligula A XV, fol. 66r, makes it possible to date it to pre-AD 743): *Ianuarius, Augustus, December IIII Nonas habent, XVIIII post Idus, et dies XXXI. Marcius, Maius, Iulius, et October VI Nonas habent, XVII post Idus, et dies XXXI. Aprelis, Iunius, September, et Nouember IIII Nonas habent, XVIII post Idus, et dies XXX. Mensis Februarius IIII Nonas habet, XVI post Idus, et dies XXVIII. Omnes mensi (!) VIII Idus habent*. Note that by this classification the entire structure of the Julian calendar is covered. DIAL.

Which months follow the same pattern? January, August, (and) December, (since these) three (months have) 31 days, four nones, (and) 19 calends after the Ides. April, June, September, (and) November, (since these) four (months have) 30 days, four nones, and 18 calends after the Ides. March, May, July, (and) October, (since these) four (months have) 31 days, six nones, and 17 calends after the Ides. February alone does not agree with any of the other (months, since it has) 28 days, four nones, (and) 16 calends after the Ides.

NEUSTR. 5 defines this classification through the number of nones and calends only, QUAEST. AUSTR. 2.5B through the number of calends only.

14–22 Qui…Idus] Facsimile, transcribed and translated in Bisagni & Warntjes, 'Latin and Old Irish', 21–2; cf. also p. XIX, LXXV of the introduction.

<XXVIIII. DE CALCULATIONE FERIARUM>

Weekday calculation for Easter

Quomodo mensium initia ebdomatis diebus intelleguntur? Dies in *quo Kalendae Ianuarii fit, quarto* sequenti lumine *Kalendae Februarii et Martii et VIII et Idus et XI et IIII Kalendas Aprilis*, n*isi fuerit bissextus*. | *Si* fol. 16v
autem contigerit, *quinto die post Kalendas* <*Ianuarii*> *VIII Idus Martii et* relicum erit; ita et *Kalendae Martii, si in* VI Kalendas eius *bissextus* esse *ful*serit. *Dies* in *quo Kalendae Ianuarii* innotescit, *priori die Kalendae Aprilis* esse fulgebit, n*isi bissextus* sit. Nam *si* ipse euen*erit*, uno *eodem*que *die Kalendae Ianuarii et Aprilis* copulari apparebit.

XXVIIII,6 Ianuarii] *om. M (add. from* CE*)*. **12** apparebit] kl- *add. M.*

XXVIIII,2–12 Quomodo…apparebit] This passage outlines MC's principal method for calculating the weekday of Julian calendar dates in the Easter period from the weekday data of 1 January; this method is then later applied for calculating the weekday of the Easter full moon (c. 56, ll. 15–43) and of *luna* 2 of the March lunation that regulates the beginning of Lent (c. 58, ll. 35–80); it is directly taken from CE p. 97, 106 (followed by a detailed proof of these calendrical rules): *Quocumque die Kalendae Ianuarii fuerint, quarto die postea Kalendae Februarii erint et Kalendae Martii et VIII Idus Martii et Idus Martii et XI Kalendas Aprilis et IIII Kalendas Aprilis, si non fuerit bissextus. Si autem fuerit bissextus et si in Februario: Quocumque die Kalendae Ianuarii fuerint, IIII die Kalendae Februarii post Kalendas Ianuarii erint, et quinto die Kalendae Martii et VIII Idus Matii et Idus Martii et XI Kalendas Aprilis et IIII Kalendas Aprilis fiet. Si uero in Martio bissextus: Quocumque die Kalendae Ianuarii fuerint, quarto die Kalendae Februarii et Kalendae Martii post Kalendas Ianuarii erint, VIII Idus autem Martii et Idus Martii et XI Kalendas Aprilis et IIII Kalendas Aprilis quinto die erit post Kalendas Ianuarii. Item Kalendae Ianuarii quocumque die fuerint, pridie Kalendae Aprilis erit, si non fuerit bissextus. Si autem fuerit bissextus, eodem die erit Kalendae Ianuarii et Kalendae Aprilis. ... Nam diximus quolibet die fuerit Kalendas Ianuarii eodem die Kalendas Aprilis, si bissextus fuerit. Si autem non fuerit bissextus, pridie Kalendae Aprilis erit Kalendis Ianuarii.* Note that both CE and MC in this passage seriously consider the option that the bissextile day may be placed on 2 March rather than 24 February.

XXVIIII,2–12 Quomodo…apparebit] This method of weekday calculation from the weekday data of 1 January is a characteristic of late seventh-, early eighth-century Irish computistics, since it is found in all three Irish computistical textbooks of that period, including DRC 35.1–7 (based on CE): *Sciendum nobis utrum sunt Kalendae contra quas priuatim regularentur Kalendae aliae mensium. Sunt utique, id est Kalendae Ianuarii, quia quacumque die Kalendae eius fuerint, Kalendae Februarii et Martii IIII die erunt, si non fuerit bissextus. Si uero fuerit bissextus, transiliunt Kalendae Martii de illo die super diem sequentem. Quocumque die Kalendae Ianuarii fuerint, in die antecedente Kalendae Aprelis erunt, si non fuerint bissextus. Et si fuerit, in eodem die fiunt.* Since the composition of the Dionysiac *argumenta* in AD 525, but especially since the late seventh century, numerous other, more mathematical and sophisticated methods for the calculation of the weekday of any Julian calendar date in general, of the Calends of the respective months in particular, were invented and applied, so that the rather simplistic Irish methods, which certainly had been invented at a time when the *latercus* was the predominant Easter reckoning, did not find many imitators in later centuries. A noteworthy exception is the Frankish QUAEST. AUSTR. 2.7B: *Kalendae Ianuarii dominico die, ... Kalendas Februarii quarta feria, eadem quinquies revoluta, quia bissextus interponitur, II Kalendas Martii est. Denique Kalendas Martii quinta feria. ... Quocumque die anno bissextus interseritur, Kalendae Ianuarii et Aprilis eundem diem ebdomadis observabunt. Sin vero byssextus non fuerit, quo die ebdomadis Ianuarii Kalendas ingredi perspexeris, in Kalendas Aprilis pridie illius observabis. Verbi gratia, ut si Kalendas Ianuarii dominico*

29. ABOUT WEEKDAY-CALCULATIONS

How are the beginnings of the months related concerning their weekdays? (Take) the (week)day of the Calends of January (i.e. 1 January); the Calends of February and March (i.e. 1 February and 1 March), the 8th (ides of March, i.e. 8 March), the Ides (of March, i.e. 15 March), the 11th (calends of April, i.e. 22 March), and the 4th calends of April (i.e. 29 March) (will be) four weekdays (*lumine*) later (reckoned inclusively), if the bissextile day does not occur (in that year). If, however, (the bissextile day) should occur (in this year), the 8th ides of March (i.e. 8 March) and so on will fall on the fifth (week)day after the Calends of January (i.e. 1 January) (reckoned inclusively); and likewise the Calends of March (i.e. 1 March), if the bissextile day occurs on the 6th calends of that (month) (i.e. 24 February). (Take) the (week)day of the Calends of January (i.e. 1 January); the Calends of April (i.e. 1 April) will fall on the previous (week)day, if the bissextile day does not occur. For if it does, the Calends of January and April (i.e. 1 January and 1 April) will appear to be united by one and the same (week)day.

---

venerit et non sit bissextus, Kalendas Aprilis sabbatum erit. Kalendae quoque Februarii et Martii, si non fuerit bissextus, eundem diem observabunt. Si vero <bissextus> fuerit, die, quo Kalendas Februarii provenerit, Kalendas Martii in sequentem observabit.

XXVIIII,2–35 Quomodo…Martii] Cf. Walsh & Ó Cróinín, *Cummian's letter*, 145; Borst, *Schriften*, p. 442, 445; p. LX, LXXIII, CXXXIX, CXLIV, CLXI, CLXIX of the introduction.

General rules for the weekdays of the Calends

Omnis *mensis XXXI*: *dies* in quo <*Kalendae eius*> esse contigerit, *quarto die Kalendae sequentis mensis erit*. In *triginta* uero *mense*, *III*o *die* <*Kalendae* sequentis *mensis*> inueniri ardebit.

Dies in quo fit Kalendae Ianuarii, sequenti die Kalendae Maii, nisi fuerit bissextus. Si enim bissextus fit, tertio die erit. Sequenti uero lumine in quo Kalendae Maii uiget, Augustus initium habet; in postero die Nouimbrio initium fulget; cui secundo die Kalendae Februarii conuenit. *Dies* in *quo Kalendae Maii fi*eri innotescit, in *eo VIII Kalendas Ianuarii* sequentis et *Kalendae* eius indubitanter euenire florebit.

---

13 Kalendae eius] *om. M (add. from* CE, DRC; *it may rather have read* Kalendae huius, *which can also be found in* CE *and* DRC*)*. **14** quarto] quar *M (corr. from* CE*)*. **15** IIIo] III *M**. | Kalendae...mensis] *om. M (add. according to* MC, CE, DRC; *it may rather have read* Kalendae alterius mensis *as in* CE*)*. **17** Dies] die *M (corr. according to* MC*)*. **18** Maii] Mar- *M (corr. according to* DRC*)*. **20** Augustus] Augusti *M*. **21** fulget] fulgit *M*. **22** quo] fit *add. M (om. according to* MC*)*. **22–23** in^{2}...VIII] VIII in eo *M (corr. from* CE, DRC*)*. **23** sequentis] sequens *M*.

13–16 Omnis...ardebit] This general rule that the weekday of the Calends of a month falls four weekdays later (reckoned inclusively) compared to the weekday of the Calends of the previous month if that month consists of 31 day, and three weekdays later in case of a month of 30 days, is again directly copied from CE p. 98: *Item mensis XXXI dies habens IIIIto die post Kalendas eius sequentis mensis Kalendae erit. Mensis autem XXX dies habens III die post Kalendas huius alterius mensis Kalendae erit. Mensis XXVIII dies habens eodem die Kalendae eius et sequentis mensis erit.* **17–24** Dies...florebit] The observation of the linear increase by one of the weekday data of 1 January, 1 May, 1 August, 1 November, and 1 February of the following year is unique to MC. Yet, its ultimate source becomes apparent when this statement is put into context: 1 May, 1 August, 1 November, and 1 February mark, in Irish computistical theory, the beginning of summer, autumn, winter, and spring; CE p. 97–8 has already argued that the beginnings of the seasons fall on the four weekdays following the weekday of 1 January: *Item Kalendae Ianuarii quolibet die fuerint, IIII subsequentibus diebus printipia (!) IIII temporum erunt, si non fuerit bissextus. Si autem fuerit bissextus, IIII temporum initia uno intercidente (corr. from intercidendte) die IIII diebus subsequentibus post Kalendae Ianuarii erunt*. To be precise, what is meant here is that if, e.g., January is a Sunday, the four following weekdays, Monday to Thursday, correspond to the weekday at the beginnings of the seasons on 1 May, 1 August, 1 November, and 1 February respectively (cf. BC 105: *Si vis scire, qua die septimanae intrant quatuor tempora, hoc est ver, aestas, autumnus, et hyems, sume principium a Kalendis Maii, quacumque enim feria fuerint Kalendae Maii, sequenti die septimanae intrant Kalendae Augusti, hoc est principium autumni, tertia die Kalendae <Nouimbris, hoc est principium> hyemis, quarta die principium veris, hoc est Kalendae Februarii, nec bissextus mutat hanc rationem*. DIAL. LANGOB. 16C (I give the correct MS readings): *A prima enim die Ianuarii monstrantur quattuor tempora anni, quibus diebus septimanae inchoantur, ut est verbi gratia Kalendas dominico die, aestas secunda feria, autumnus tertia eria, hiemps quarta feria, deinde inchoatur ver*. see also QUAEST. AUSTR. 2.7B); this passage from CE quite certainly motivated MC's statement here. The observation mentioned subsequently in MC that 1 May, 25 December, and 1 January fall on the same weekday, is then again directly copied from CE p. 98: *Item Kalendae Maii quolibet die fuerit eodem die VIII Kalendas Ianuarii et Kalendae Ianuarii erit, si bissextus fuerit an non.*

13–16 Omnis...ardebit] The observation outlined in this passage is, again, a characteristic of late seventh-, early eighth-century Irish computistics. It is not only found in the third Irish computistical

(The following holds true for) every month (of) 31 (days): (Take) the (week)day of the Calends of that (month); the Calends of the following month will be four (week)days later (reckoned inclusively). In (the case of) months with 30 (days), however, the Calends of the following month will occur three (week)days later (reckoned inclusively).

(Take) the (week)day of the Calends of January (i.e. 1 January); the Calends of May (i.e. 1 May) (will fall) on the following (week)day, if the bissextile day does not occur (in that year). If the bissextile day occurs, (the Calends of May, i.e. 1 May) will be three (week)days (later reckoned inclusively). August, however, has its beginning (i.e. 1 August) on the weekday (*lumine*) following the weekday of the Calends of May (i.e. 1 May); the beginning of November falls on the (week)day following (that of 1 August); it agrees (with this) that the Calends of (the following) February (i.e. 1 February) (fall) on the (week)day following (that of 1 November). (Take) the (week)day of the Calends of May (i.e. 1 May); the 8th calends of the following January (i.e. 25 December) and the Calends of January (i.e. 1 January) will undoubtedly fall on that very same (weekday).

---

textbook, DRC 34: *Sciendum nobis quomodo Kalendae mensis sequentis contra Kalendas mensis praecedentis regulantur. Hoc modo: Quisquis mensis triceni singuli est, quocumque die Kalendae huius fuerint, IIII die Kalendae mensis sequentis erunt. Uerbi gratia, Kalendae Ianuarii in dominico, IIII die Kalendae Februarii. Quicumque uero mensis tricenum dierum est, quocumque die Kalendae eius fuerint, III die Kalendae mensis sequentis erunt. Uerbi gratia, Kalendae Aprelis in dominico, III feria Kalendae Maii. Februario uero, quocumque die Kalendae eius fuerint, in eodem die Kalendae mensis sequentis erunt, si non fuerit bissextus.* but also in DIAL. LANGOB. 12B: *Item observa hanc rationem omnis mensis, quacumque die inchoatur. Si fuerit trigenus singularis, post tres dies intrabis in alterum <mensem>, si vero trigenus, post duos dies.* **17–24** Dies…florebit] DRC 35.7–10 describes the relation between 1 January and 1 May, influenced by both CE and MC: *Et quocumque die Kalendae Ianuarii fuerint, crastino die Maio; et quocumque die Kalendae Maii fuerint, in eodem die VIII Kl Ianuarii et Kalendae Ianuarii erunt. Et sic regulantur Kalendae mensium contra Kalendas Ianuarii et Maii.* Interestingly enough, BC 105D12–14 mentions the fact that 1 August falls on the weekday following that of 1 May, obviously influenced by Irish computistics, if not directly by MC: *quacumque enim feria fuerint Kalendae Maii, sequenti die septimane intrant Kalendae Agusti.* DIAL. LANGOB. 16A lists the months with identical weekdays on the Calends and illustrates the relation between these. QUAEST. AUSTR. 2.7A lists the weekdays of the Calends of each month for a bissextile year, starting with Sunday on the Calends of January; in 2.7B, the case of a non-bissextile year is discussed, and the months with identical weekdays on the Calends are referred to. Like CE, BC 105 and LANGOB. 16C (cf. *app. font.*) describe the linear increase of weekdays at the beginnings of the four seasons.

Days of the month with identical weekday data, listed per class of months

Omnes *menses trige*simi *singul*i, *IIII Non*ę: *uno die* inueniuntur *Kalendae* eorum *et VI Idus et XVIII Kalendas et XI Kalendas et IIII Kalendas*. Cuncti uero *menses XXX*mi, *quatt*uor Nonae: in *eodem die* ebdomatis fiunt *Kalend*ę eorum *et VI Idus et XVII Kalendas et X Kalendas et III Kalendas*. Omnes *menses XXXI, sex Nona*e: *eodem die* ebdomatis *Kalendae* eorum *et VIII Idus et Idus et XI Kalendas et IIII Kalendas* repperiuntur. *Februarius XXVIII*, IIII Nonae: in *eodem die Kalendae eius et VI Idus et XV Kalendas et VIII Kalendas* | *et Kalendae mensis sequentis*, (fol. 17r) id est mensis Martii.

<XXX.> DE TEMPORE

Etymology of season (*tempus*)

*Tempus temperament*um dicitur, eo *quod* designationem IIII tempora anni *temperant*, et *unumquodque* ex illis IIII temporibus alterius introitu *tempera*tur. Sed quidam in anno IIII tempora disputare student, ut mox operari uidebimur.

---

32 repperiuntur] repperitur *M (corr. according to* MC*)*. **XXX,3** tempora] *corr. to* temporum *Borst*. temperant] temperat *M (corr. from* ETYM., DIAL. LANGOB., *c. 31, l. 30)*. **4** unumquodque] unumquo *M (corr. from* EPIT., DTR*; also Borst)*.

25–35 Omnes…Martii] These lists of days of the month with identical weekday data, given for each of the four classes of months outlined in c. 28, ll. 14–22, is again directly copied from CE p. 98: *De epdomadibus regulandis in diebus mensis. Quicumque mensis trigenos singulos dies habuerit et IIIItas Nonas, uno eodemque die Kalendae ipsius et VI Idus et XVIII Kalendas et XI Kalendas et IIII Kalendas sequentis mensis erit, ueluti Ianuarius, Augustus, et December. Itemque quicumque mensis trigenos possiderit dies et IIIItas, uno eodemque die Kalendae ipsius et VI Idus et XVII Kalendas et X Kalendas et III Kalendas mensis sequentis erit, ueluti Aprilis, Iunius, Septimber, et Nouimber. Itemque quicumque mensis trigenos singulos dies habuerit et VItas Nonas, eodemque die ipsius Kalendae et VIII Idus et Idus et XI Kalendas et IIII Kalendas sequentis mensis erit, ueluti Martius, Maius, Iulius, et Octimber. Item mensis dies XXVIII habens, eodem die Kalendae eius et VI Idus et XV Kalendas et VIII Kalendas et Kalendae sequentis mensis erit, ueluti Februarius.* **XXX,2–6** Tempus…uidebimur] The first etymology of the season (*tempus*) being so called from the combination (*temperamentum*) of two out of four characteristics, namely humidity, drought, heat, and cold, derives from ETYM. 5.35.1–2 (cited directly in c. 31, ll. 28–30, BC 28B12–C2): *Dicta sunt autem tempora a communionis temperamento, quod invicem se humore, siccitate, calore et frigore temperent. Haec et curricula dicuntur, quia non stant, sed currunt. Constat autem post factum, mundum ex qualitate cursus solis tempora in ternos menses fuisse divisa*. Cf. also the more specific account in DNR 7.4.39–45: *Vnde etiam sunt tempora a communionis temperamento dicta, cuius communionis haec est figura: Annus. Humidum calidum uer: oriens. Calida sicca aestas: meridies. Siccus frigidus autumnus: occidens. Frigida humida hiemps: septentrio*. The second etymology of each season being limited (*temperatur*) by the beginning of the following one clearly stems from EPIT. 11.151 (cited in variation in DTR 2.2–5, in turn cited in RM 9.8–11; the first part only in DIAL. BURG. 8A): *Tempus a temperamento nomen accipit, eo quod diebus mensibus annis frigore aestateque temperatur, siue quod unumquoque spatium separatim temperatum est*. At the end of this passage MC refers to the discussions in c. 31, ll. 28–30; c. 32–35. For the term *tempus* interpreted as time rather than season cf. c. 1, ll. 2–3; interestingly enough, CE p. 94–5, when discussing the seasons, defines the term *tempus* in the context of its underlying units of time: *Tempus a temperando dictum est, hoc est momentis, minutis, punctis, horis, quadrantibus, diebus, septimanis, mensibus*.

(The following holds true for) all months of 31 (days and) 4 nones: The Calends of these, the 6th ides, the 18th calends, the 11th calends, and the 4th calends can be found on the same (week)day. (The following holds true for) all months of 30 (days and) 4 nones: The Calends of these, the 6th ides, the 17th calends, the 10th calends, and the 3rd calends fall on the same weekday. (The following holds true for) all months (of) 31 (days and) 6 nones: The Calends of these, the 8th ides, the Ides, the 11th calends, and the 4th calends can be found on the same weekday. (The following holds true for) February, (having) 28 (days and) 4 nones: The Calends of this (month), the 6th ides, the 15th calends, the 8th calends, and the Calends of the following month, i.e. March, (fall) on the same (week)day.

30. ABOUT THE SEASON

A season (*tempus*) is called a combination (*temperamentum*), because the four seasons of a year combine (*temperant*) (their individual) characteristics, and every one of these four seasons is kept within limits (*temperatur*) by the beginning of another (season). But certain people wish to discuss the four seasons in (the context of) a year, as we will be pleased to do soon.

25–35 Omnes…Martii] These lists of days of the month with identical weekday data per class of months was central in late seventh-, early eighth-century Irish computistics, and also quite popular in later centuries, since they were frequently copied in MSS as a tool for calendrical calculations. Cf. DRC 33: *Sciendum nobis quot sunt conspectus in mensibus. Quinque. Omnes enim menses excepto Februario quinque conspectus habent. Sciendum nobis quomodo intellegitur. Sic, id est pro Ianuario terno rithmo, quia quacumque die Kalendae eorum fuerint, in eadem VI Idus et XVIII Kl et XI Kl et IIII Kl erunt. Martio autem et conregularibus eius, quacumque die Kalendae eorum fuerint, in eo die VIII Idus et Idus et XI Kl et IIII Kl erunt. Apreli uero et conregularibus eius, quacumque die Kalendae eorum fuerint, in eodem VI Idus et XVII Kl et X Kl et III Kl erunt. Februario uero, quacumque die eius Kalendae fuerint, in illa VI Idus et XV Kl et VIII Kl et Kalendae mensis sequentis inuenitur.* BC 2: *In Ianuario, Augusto, et Decimbri Kalendis et VI Idus, et XVIII Kalendas, et XI Kalendas, et IIII Kalendas, unus dies est. In Marcio, Maio, Iulio, et Octobri Kalendis, et VIII Idus, et Idus, et XI Kalendas, et IIII Kalendas unus dies est. In Aprili, Iunio, Septembri, et Nouembri Kalendis, et VI Idus, et XVII Kalendas, et X Kalendas, et III Kalendas unus dies est. In Februario vero Kalendis, et VI Idus, et XV Kalendas, et VIII Kalendas unus dies est.* **XXX,2–5** Tempus…temperatur] The etymology *tempus a temperamento*, based on Isidore, was most popular, in numerous variations, in early eighth-century computistics. It occurs in two Irish texts of this period, DRC 42.1–4: *Sciendum nobis unde nominatum est. A temperamento, ut Isidorus dicit: Tempus a temperamento quodam siderum nomen accepit, quia diebus ac noctibus, mensibus et annis, frigore aestateque conponitur*. and DIAL. LANGOB. 2B (which is based on EPIT.): *Alii sic interpretantur tempus: A temperamento nomen accepit, quia quattuor anni et frigorem et calorem inter se invicem temperant, sive in ascensu solis, sive in discensu*. It is also used by Bede and received some reception through his work: DTR 35.6–9 (cited in LIB. CALC. 83, PV §314, RM 34.8–10): *A quo temperamento videtur temporibus inditum nomen; vel certe quia quadam suae similitudine qualitatis ad invicem contemperata volvuntur, tempora recte vocantur*. Virgilius Maro's alternative etymology also received some attention by Bede (cf. *app. font.*).

25–30 Omnes…Kalendas] Transcribed and translated in Bisagni & Warntjes, 'Latin and Old Irish', p. 25–6. **XXX,2–5** Tempus…temperatur] Transcribed (and corrected differently) in Borst, *Schriften*, p. 360; translated differently into German in Borst, *Schriften*, p. 7. Only the beginning transcribed in Ó Cróinín, 'Virgilius Maro Grammaticus', p. 199.

<XXXI.> DE ANNO

Etymology of year (*annus*)

*Annus dic*tus est *quasi circulus. Unde et anuli diminut*i *dicti sunt*, qui sunt *menses*. *Annus* enim *ad id signum*, a *quo* annus incipit, *peractis CCCLXV diebus* regredi innotescit. Augustinus ait: *Annus solis anfractus peractis CCCLXV diebus ad eadem loca siderum redit.* Hucusque Augustinus. Inde Macrouius ait: *<Ateius Capito> annum a circuitu temporis putat dictum, quia ueteres annum pro circum ponere solebant, ut Cato in Originibus Oratorum*: *annum terminum, id est circa terminum, et ambire dicitur*, hoc est *pro circumire. Alii annum dicunt* ab innouatione, anath enim *ab innouatione*; *renouatur enim semper.*

Classific.

Annorum tria genera sunt:

XXXI,5 solis] per *add. M (corr. from* ETYM., *etc.).* **7–8** Ateius Capito] *om. M (add. from* SAT.*;* DRC *and* DCH *read* Ataius Capito, CE Taius Capito*).* **9** pro] per *M (corr. from* SAT.*, etc.; it appears that the continental copyist mistook the Irish abbr. for* pro *for that of* per*).* **11** pro] per *M (corr. from* SAT.*, etc.; it appears that the continental copyist mistook the Irish abbr. for* pro *for that of* per*).*

XXXI,2–13 Annus…semper] The first etymology of *annus* (or rather *anus* with the diminutive *anuli*) being another term for circle is primarily based on DNR 6.2.9–10 (cited in DRC 44.5–7, where the sole MS B reads *anuli diminute menses dicti sunt*): *Annum autem quasi anum dici quidam putant, id est circulum. Vnde et anuli dicti sunt diminutiue*. The additional information that the diminutive form is to be interpretetd as month is taken directly from CE p. 95: *Annus dicitur quasi anus. Unde anuli diminutiue dicti sunt, id est menses*. Cf. ETYM. 5.36.1 (cited in PV §124; the second sentence only in DIAL. LANGOB. 5A): *Annus autem dictus quia mensibus in se recurrentibus volvitur. Vnde et anulus [dicitur], quasi annuus, id est circulus, quod in se redeat*. The first of the following explanations for the association of a year with a circle, namely that a year (or rather the sun) returns after 365 days to that zodiac sign from which it set out originally, is a combination of SAT. 1.14.4: *ita solis annus hoc dierum numero colligendus est quem peragit dum ad id signum se denuo vertit ex quo digressus est*. and ETYM. 5.36.1 (cited in PV §124, DTR 36.18–19 (adding a quarter-day), in turn cited in LIB. ANN. 1C8, in turn cited in LIB. COMP. 3.4, LIB. CALC. 39, PV §133, RM 36.19–20; from *peractis* only in DIAL. LANGOB. 5A): *Annus est solis anfractus, cum peractis trecentis sexaginta quinque diebus ad eadem loca siderum redit*. It seems that Isidore's source was AUGUSTINE, *De Genesi ad litteram* 2.14 (*CSEL* 28,1, p. 55; cited in DRC 55.13–17): *annos uero uel istos usitatos solis anfractus, non cum ad orientem, quod cotidie facit, sed cum ad eadem loca siderum redit – quod non facit nisi peractis trecentis sexaginta quinque diebus et sex horis, id est quadrante totius diei, quae pars quater ducta cogit interponi unum diem, quod Romani bissextum uocant, ut ad eumdem circuitum redeatur*. This may explain the attribution of this citation to Augustine in MC. Note that the length of a solar year is also given as 365 days in c. 45; c. 46, ll. 11–12; but as 365¼ days in c. 31, ll. 15–16; c. 41, ll. 29–31. The Macrobian citation is from SAT. 1.14.5 (almost all MSS read *oratorum*; cited in CE p. 95, DRC 44.2–5, DTR 36.2–5, in turn cited in RM 35.6–9; all four read *oratorum*, CE reads *id est circa* and *an-* instead of *an* (in both cases), MSS BKMuPS of DTR read *annum* instead of *an* in both cases, which is then copied in RM): *hinc et Ateius Capito annum a circuitu temporis putat dictum, quia veteres an pro circum ponere solebant, ut Cato in Originibus: arator an terminum, id est circum terminum, et ambire dicitur pro circumire*. Note that MC is closer to CE's quote of this Macrobian passage than to the original; accordingly, MC may have cited Macrobius through CE, or (and more likely) MC and CE worked from a similar, if not the same, copy of SAT. (or rather DCH). The alternative etymology of *annus* deriving from the Greek term for renewal is taken from ETYM. 5.36.2: *Alii annum dicunt ἀπὸ τοῦ ἀνανεοῦσθαι, id est ab innovatione; renovatur enim semper*. Cf. EPIT. 11.155 (MSS AN read *innouat*;

31. ABOUT THE YEAR

The year (*annus*) was so called as though (it was) a circle (since *anus* means circle). From this (term, i.e. *anus*,) the 'little years' (*anuli*), which are the months, took their name as the diminutive form. Indeed, a year is known to return to that (zodiac) sign from which it set out (originally) after 365 days. Augustine says: A revolving solar year returns to the same place among the stars after 365 days. Thus far Augustine. Hence Macrobius says: Ateius Capito believes that a year was so called from the circular course of time, because the ancients used to employ (the term) *annus* (as a substitute) for (the term) 'around' (*circum*), as Cato (says) in his *Origines Oratorum*: '*annum terminum*, i.e. *circum terminum*', and *ambire* is used (as a substitute) for *circumire*. Others term a year such from a renewal, since (the Greek) *anath* (means) 'from a renewal'; in fact, (a year) is always renewed.

There are three types of years:

---

cited in DIAL. BURG. 9A, which also reads *innovat*): *Annus dicitur ab eo, quod cuncta annouat.* **14** Annorum…sunt] Even though MC continues to list eight different types of years, he argues for only three in this passage here; the statement itself is a direct quote from ETYM. 5.36.3: *Tria sunt autem genera annorum.* which reveals that the basis for the classification of a year was Isidore's account, to which information from other texts, particularly from DNR 6, was subsequently added. Note especially that such a list of types of years cannot be found in CE.

XXXI,2–13 Annus…semper] The etymology of year being a different term for circle was extremely popular in Irish and Irish influenced texts. Cf. ANGERS 477, fol. 19v (gloss to DT): *Annum dicitur quasi anus, hoc est circulus, quod in se redeat, et menses diminutiue quasi anula dicuntur.* DIAL. BURG. 9A: *Et similatur rotae currus vel anulo, quasi nec fines habent, semper in se volvuntur.* DIAL. NEUSTR. 2A: *Unde dicitur annus? Ut multi volunt, ab anulo, quasi ab: in se redeat. Unde impletur annus ex mensibus.* PP I §131: *Annus dictus est uelut anus, id est circulus.* RM 35.2–3: *D. Quid est annus? M. Solis circuitus ac reditus per duodecim mensis.* The fact that a year, or rather the sun, returns to that sign in the zodiac from which it set out originally, receive a wide reception through Bede's quote of ETYM. (cf. *app. font.*) and is also noted in DT 10.2–4 (cited in DE BISSEXTO I 994B7–10, RM 54.3–5): *Bissextus ex quadrantis ratione per quadriennium conficitur, dum sol ad id signum ex quo egressus est non in CCCLXV diebus sed quarta diei parte superadiecta revertitur.* The Macrobian explanation for a year representing a circular course of time was very popular among late seventh- and early eighth-century Insular computist (cf. *app. font.*). The alternative etymology of a year deriving from the Greek term for renewal can also be found in ANGERS 477, fol. 19v (gloss to DT): *Alii annum dicunt ab innouatione, ana enim Grece innouatio dicitur.* DTR 36.1 (cited in PV §126, RM 35.5): *Annus vel ab innovando cuncta quae naturali ordine transierant.* DIAL. LANGOB. 5B: *Alii sic dicunt: Ab innovatione dicitur eo, quod in eo iterum fructus innovantur, vel eo, quod rursum innovat in suum circulum redire, vel quod innovat in se eadem tempora.* **14** Annorum…sunt] Bede mentions four types of years in DT 9, twelve in DTR 36.5–27, so that MC's eight, at least from the numerical total, fit neatly between these two texts. The Irish DRC 46.1–4 lists thirteen types of years, including all of MC's eight: *Sciendum nobis quot sunt genera annorum. Tredecim. Id est annus solis, annus lunaris, annus breuis, annus magnus, annus naturalis, annus commonis, annus embolesmus, annus byssextilis, annus solstitialis, annus equinoctialis, annus iubilius, annus acceptabilis, annus ciuilis.* This fact suggests that DRC represents a later stage in the composition of the classification of a year, and thus lends support to the theory that DRC was composed later than MC. Note that the version of *De divisionibus temporum* transmitted in the

---

XXXI,2–3 Annus…menses] Transcribed in Borst, *Schriften*, p. 361. **2–13** Annus…semper] Cf. Borst, *Schriften*, p. 436. **5–7** Augustinus…Augustinus] Cf. p. CXXVI, CXXIX of the introduction. **11–12** Alii…innouatione[1]] Transcribed in Borst, *Schriften*, p. 361; cf. p. CLXXXV of the introduction.

Annus certus *solis diebus CCCLXV et quadrante*, id est quarta pars diei.

Annus ciuilis, quem ciuiles senes enumerant.

Annus naturalis, in quo *luna sol*em contig*it*.

Annus solstitialis ab *VIII Kalendas Ianuarii* usque in *VIII Kalendas Iulii.*

| *Annus bissextilis, in quo unus die*s superaddit*ur*. fol. 17v

Duo *ann*i *lun*ae, *communis* et *embolismus*.

15–16 Annus…diei] The information about the *annus solis* was probably taken from SAT. 1.14.3–4: *post hoc imitatus Aegyptios, solos divinarum rerum omnium conscios, ad numerum solis, qui diebus trecentis sexaginta quinque et quadrante cursum conficit, annum dirigere contendit. ... solis annus ...* . Cf. PROLOGUS CYRILLI, *Addendum* (Krusch, *Studien* I, p. 343): *Nam et sol in anno lucet dies CCCLXV et quadrans*. ETYM. 5.36.1, DNR 6.4.24–25, cited in c. 31, ll. 5–6, reckon the *annus solis* as consisting of only 365 days, omitting the bissextile increment. For the solar year of 365¼ days cf. c. 41, ll. 29–31; for only 365 days c. 31, ll. 3–6; c. 45; c. 46, ll. 11–12. **17** Annus…enumerant] The term *annus civilis* is probably taken from DNR 6.2.13–15: *Ciuilis annus est qui in unius astri recursu per menses XII terminatur*. or DNR 6.4.24–25, cited in c. 31, ll. 5–6; the explanation of this term, however, appears to be MC's own invention. **18** Annus…contigit] The information about the natural year, defined as the year of a solar eclipse, stems from DNR 6.3.16–19 (cited in DRC 46.13–15, BC 67A13–B1): *Annus naturalis est cum se soli luna supponit, ut inter orbem solis et oculos nostros media facta tenebras totius orbis efficiat, quod dicitur eclipsis.* **19–20** Annus…Iulii] The definition of the solstitial year as the period from one solstice to the next derives ultimately from ETYM. 5.36.3, 5.34.2 (except for the first sentence cited in BC 30B8–C8): *aut solstitialis, qui duodecim continet menses. ... Duo sunt autem solstitia: unum aestivum, VIII Kal. Iul., ...; aliud hiemale, VIII Kal. Ian. ... quia annus olim in duas tantum partes dividebatur, hoc est in aestivum et hiemale solstitium.* DNR 6.4.23–24 (cited in BC 67B6–8), 8.1.1–3: *Solstitialis annus est cum sol expleto per omnia signa circuitu, in id unde principium cursus sui sumpsit recurrit. ... Solstitia duo sunt: primum hibernale VIII kl. ianuarias, ...; alterum aestiuum, VIII kl. iulias*. Note, however, that MC's solstitial year extends from winter to summer solstice and therefore contains six months, whereas Isidore's contains all twelve months, extending from one solstice to the same (i.e. from winter solstice to winter solstice, or from summer solstice to summer solstice). For these dates of the solstices cf. c. 38, ll. 26–38. **21** Annus…superadditur] The bissextile year of 366 days is copied from DNR 6.4.29–31: *Annus bissextilis est in quo unius diei per quadriennium ex quadrantis ratione summa colligitur.* **22** Duo…embolismus] The fact that two different lunar years exist was obviously common computistical knowledge in the early eighth century; MC may have relied directly on DNR 4.26–29: *Annus lunaris communis est, id est qui per duodecim lunares menses decurrit, id est diebus CCCLIIII. Annus embolismus est qui lunas tredecim et dies CCCLXXXIIII habere monstratur*. For common and embolismic lunar years see also c. 59; c. 60, ll. 3–7; 61.

---

ninth-century Fulda MS Basel, Universitätsbibliothek, F III 15k (the relevant passage on fol. 31r) argues for twelve divisions of a year but continues to list only eleven, omitting the *annus soltitialis* and *equinoctialis* when compared DRC: *D. Genera annorum quot sunt? M. Duodecim. Dicitur enim annus ciuilis, annus naturalis, annus iobeleus, annus bissextilis, annus lunaris, annus quod dicitur aera, annus solaris, annis (recte annus) communis, annus embolismus, annus breuis, annus magnus*. At the end of the eighth century, then, LIB. ANN. 1 (cited in LIB. COMP. 3.4, LIB. CALC. 39, PV §§127–144, RM 36.2–40) gives a list of 17 types of years, which is headed as follows in RM 36.2–3: *D. Quot species sunt annorum? M. XVII.* **15–16** Annus…diei] The solar year obviously was the most common definition of a year, so that it does not only occur in the various lists of types of years. Bede relied on Isidore's defintion, but added a quarter-day to Isidore's total of only 365 days: DT 9.2–3: *Annus solaris vel civilis est dum sol CCCLXV diebus et quadrante zodiacum peragit.* DTR 36.18–19 (cited in LIB. ANN. 1C8, in turn cited in LIB. COMP. 3.4, LIB. CALC. 39, PV §133, RM 36.19–21): *Item solis est annus cum*

The fixed solar year with 365 and a quarter-day, i.e. a fourth part of a day.

The civil year, which the old citizens (*ciuiles*) reckon.
The natural year, in which the moon takes hold of the sun.
The solstitial year, (which extends) from 25 December to 24 June.

The bissextile year, in which one (extra) day is added.
(There are) two lunar years, a common and an embolismic (year).

---

ad eadem loca siderum redit, peractis CCCLXV diebus et sex horis, id est quadrante totius dies. Like MC, DRC 46.4–6, in this context, also refers to a quarter-day without any further specification concerning the number of hours: *Sciendum nobis quid sit annus solis. Id est cum solis cursus finitur, peractis CCCLXV diebus et quadrante*. DIAL. BURG. 12A specifies this quarter-day as consising of three hours (i.e. a quarter of daytime): *Secundum solis cursum ratione annus solaris per menses duodecim trecentos sexaginta quinque dies habet et horas tres*. while the six hours mentioned by Bede in DTR were obviously more common: LECT. COMP. 9.8a: *annus trecentos sexaginta quinque dies et sex horas*. KAL. A1: *Annus solaris habet dies trecentos sexaginta quinque et sex horas*. But note that if a distinction is made between bissextile and non-bissextile years, a non-bissextile solar year would consist of 365 days only: QUAEST. AUSTR. 2.8A: *Solaris autem annus binas longitudines habet: Si non habeat bissextum, trecenti sexaginta quinque dies*. **17** Annus…enumerant] In DNR (which is the basis for DT cited in the previous note), Isidore argued that a solar and a civil year are synonymous. This is taken up by DRC 46.33–34, which shows some resemblence to MC: *Aliter annus ciuilis dicitur, eo quod ex eo omnes ciues fuguntur, et idem est hic annus et annus solis*. **18** Annus…contigit] The few texts that refer to the natural year cite DNR's definition (cf. *app. font.*). **19–20** Annus…Iulii] The same definition of the solstitial year can be found in DRC 46.20 which may have been based on MC: *Annus solstitialis, a solsticio usque ad alterum*. Besides these two Irish texbooks, the *annus solstitialis* is hardly ever mentioned. The Irish DIAL. LANGOB. 17D, however, divides a year into two halfs on the basis of the solstices: *Ergo duo solstitia diuidunt annum inter se. Sex menses sunt cum illo, et sex sunt cum altero solstitio*. This account, as well as MC, may have influenced the Frankish QUAEST. AUSTR. 2.8C: *Si vero in duas partes annus dividitur, dimidia pars eius ab VIII Kalendas Ianuarii usque in VIII Kalendas Iulii, dies habens centum octoginta unum, pars altera ab VIII Kalendas Iulii usque <in VIII Kalendas> Ianuarii, dies habens centum octoginta quattuor*. **21** Annus…superadditur] The bissextile year of 366 days is regularly mentioned in eighth- and early ninth-century computistical literature: DRC 46.19–20: *Annus bissextilis quando fit bissextus*. DTR 36.21–24: *Quartus solaris gyri annus bissextilis est, caeteris tribus uno die prolixior*. QUAEST. AUSTR. 2.8A: *Solaris autem annus binas longitudines habet: Si non habeat bissextum, trecenti sexaginta quinque dies; si vero habeat, trecenti sexaginta sex sunt dies*. LIB. ANN. 1C9 (cited in LIB. COMP. 3.4, LIB. CALC. 39, PV §134, RM 36.21–22): *Nonus est item solaris, dum quarto anno bissextus inseritur, et fiunt dies trecenti sexaginta sex*. KAL. A1: *annus bissextilis trecentos sexaginta sex dies*. **22** Duo…embolismus] The two different types of lunar years appear in almost every discussion of a year: DRC 46.18–19: *Annus autem commonis et embolesmus in luna numerantur*. DT 9.4–6: *Annus lunaris communis XII lunis, id est diebus CCCLIIII; embolismus tredecim lunis et diebus trecentis octuaginta quattuor implentur*. DTR 36.12–15 (cited in LIB. ANN. 1A3–4, in turn cited LIB. COMP. 3.4, LIB. CALC. 39, PV §§130–131, RM 36.14–17): *Tertius, qui XII mensibus huiusmodi, id est diebus CCCLIIII, expletur et vocatur communis eo quod duo saepissime tales pariter currant. Quartus, qui embolismus graece dicitur, id est superaugmentum, et habet XIII menses, id est dies CCCLXXXIIII*. QUAEST. AUSTR. 2.8A: *Similiter et lunaris annus duas longitudines habere probatur: Si communis sit, trecenti quinquaginta quatuor; si embulesmus, trecenti octoginta quatuor*.

15–16 Annus…diei] Transcribed in Borst, *Schriften*, p. 363; cf. Borst, *Schriften*, p. 447. **21** Annus…superadditur] Cf. Borst, *Schriften*, p. 489. **22** Duo…embolismus] Cf. Borst, *Schriften*, p. 364.

Annus magnus omnibus planetis recurrentibus in eodem loco, *qui fit post solstitiales annos plurimos*, id est *XXX*.

Half- and quarteryear

*Ann*us est et dualis et quaternalis, in quo enim *IIII tempora sunt: uér, aestas*, *auctumnus, et hiemps*.

Etymology of seasons (*tempora*)

Haec *dicta sunt tempora a communionis temperamento, quod inuicem se h*o*more, siccitate, calore, et frigore temper*ant.

---

24 solstitiales annos] solstitialis annus *M (corr. from* ETYM.*)*. **29** quod] quo *M (corr. from* ETYM., BC*)*.

23–25 Annus…XXX] The definition of the *annus magnus* as the period required by all movable celestial bodies to return to their initial position in the zodiac is copied from ETYM. 5.36.3, 3.66.1–2: *aut magnus, omnibus planetis in eundem locum recurrentibus, qui fit post annos solstitiales plurimos. ... Numerus circularis stellarum, per quod cognosci dicitur in quanto tempore circulum suum unaquaeque stella percurrat, sive per longitudinem, sive per latitudinem. Nam luna totannis fertur explere circulum suum, Mercurius annis XX, Lucifer annis IX, Sol annis XIX, Vesper [annis] XV, Phaeton annis XII, Saturnus [annis] XXX. Quibus peractis ad reversionem circuli sui isdem signis et partibus revertuntur*. Isidore mentions an exact length of the *annus magnus* only in DNR 6.3.19–21 (cited in DRC 46.9–12, BC 67B2–5, omitting *undeuicesimo anno*), 23.4, arguing that the ancients believed it to contain 19 years: *Annus magnus dicitur quando omnia sidera, certis temporibus numerisque conpletis, ad suum locum uel ordinem reuertuntur. Quem annum antiqui undeuicesimo anno finiri uel adinpleri dixerunt. ... Anni autem singularum stellarum hi sunt qui in sphaera subiecta continentur. Quibus peractis ad reuersionem circuli sui hisdem signis et partibus ueniunt. Nam luna octo annis fertur explere circulum suum; Mercurius annis XX; Lucifer annis VIIII; sol annis XVIIII; Vesper XV; Phaeton XII; Saturnus XXX*. Isidore's accounts must have appeared rather confused and outdated to a late seventh-, early eighth-century computist (especially the definitions of the lunar course of 8, the solar course of 19 years); nevertheless, MC obviously based his own theory of an *annus magnus* consisting of 30 years on Isidore, regarding the longest period taken by any of the planets to travel through the zodiac, namely Saturn's 30 years, as the appropriate interval (the solstitial years referred to here certainly are those in Isidore's sense, covering 12 months; cf. c. 31, ll. 19–20); the definition of an *annus magnus*, however, would rather have suggested the least common multiple of the various periods ascribed to these planets as the length of that type of year. **26–27** Annus…hiemps] The terminology employed by MC here (*annus dualis* and *annus quaternalis*) cannot be found in pre-MC texts. The division of a year into four seasons, however, was obviously common knowledge. Textually closest to MC is CE p. 95: *Tempora autem anni IIII sunt: ver, aestas, autumpnus, hiemps*. and ETYM. 5.35.1 (cited in BC 28B11–12): *Tempora anni quattuor sunt: ver, aestas, autumnus et hiems*. but similar stements can be found in numerous pre-MC texts: DNR 7.2.11–12 (cited in DRC 42.10–11): *Iuxta Latinos autem unius anni quattuor tempora adscribuntur: hiemis, ueris, aestatis atque autumni*. PROLOGUS CYRILLI, *Addendum* (Krusch, *Studien* I, p. 342): *Annus habet dies CCCLXV, mensis XII, tempora IIIIor: ver, aestas, authumnum et hiems*. DE ORDINATIONE FERIARUM PASCHALIUM (*PL* 90, col. 607D): *Quatuor enim mundi tempora accipiuntur: ver, aestas, autumnus, hiems*. For MC suggesting 1 February, 1 May, 1 August, and 1 November respectively as the beginnings of the seasons cf. c. 38, ll. 41–43, c. 45; for the division of a year into two parts see also c. 31, ll. 19–20. **28–30** Haec…temperant] The etymology *tempora a temperamento* is a direct quote from ETYM. 5.35.1 (cited in BC 28B12–14): *Dicta sunt autem tempora a communionis temperamento, quod invicem se humore, siccitate, calore et frigore temperent*. The overlapping nature of the characteristics ascribed to the seasons is outlined in DNR 7.4.39–45: *Vnde etiam sunt tempora a communionis temperamento dicta, cuius communionis haec est figura: Annus. Humidum calidum uer: oriens. Calida sicca aestas: meridies. Siccus frigidus autumnus: occidens. Frigida humida hiemps: septentrio*. Cf. also c. 30, ll. 2–5.

The great year (defining the period of time it takes for) all planets (to) return to their initial place (in the zodiac), which happens after numerous solstitial years, namely 30.

A year is either two-parted or four-parted (i.e. it can be divided into two as well as four parts), (since) there are four seasons in a year: spring, summer, autumn, and winter.

These are called seasons (*tempora*) from the combination (*temperamentum*) of (their) common characteristics, since they overlap (*temperant*) in turns according to humidity, drought, heat, and cold.

KAL. A1: *annus lunaris et communis dies trecentos quinquaginta quattuor, annus lunaris embolimalis trecentos octoginta quattuor.* **23–25** Annus…XXX] The main question concerning the *annus magnus* was its length. Only MC and DRC 46.7–9 (followed by a quote from DNR 6.3.19–21; cf. *app. font.*) argue for 30 years, a fact that clearly demonstrates that the two texts are closely related: *Sciendum nobis quid sit annus magnus. Qui XXX annis finitur, et a stella Saturni moderatur, ut Isidorus dicit: Stella Saturni XXX annis fertur cursum suum implere.* Bede employs the authority of Josephus in arguing for a length of 600 years: DT 9.6–8: *Annus magnus est dum omnia sidera certis cursibus exactis ad locum suum revertuntur, quem sexcentis annis solaribus Iosephus dicit impleri.* DTR 36.25–38: *Annus magnus est cum omnia simul errantia sidera ad sua quaeque loca quae simul habuere recurrunt. De quo Iosephus in primo antiquitatum libro, cum longevitatem primorum hominum describet, ita meminit: … nisi sexcentis viverent annis – per tot enim annorum curricula magnus annus impletur.* The influential late eighth-century Frankish LIB. ANN. 1F17 (cited in LIB. COMP. 3.4, LIB. CALC. 39, PV §144, RM 36.36–40) ascribes 532 years to the *annus magnus*, obviously referring to the length of a paschal cycle: *Septimus decimus annus est, qui vocatur magnus annus planetarum, dum omnia sidera ad locum revertuntur, ubi primum statua fuerunt, et quingentis triginta duobus annis impletur.* **26–27** Annus…hiemps] MC's terminology (*annus dualis* and *annus quaternalis*) is not repeated by any post-MC author. On the other hand, the fact that a year comprises four seasons was common knowledge and is frequently recorded: BC 27A15–B1 (cited in LIB. COMP. 1.3a): *Tempora mensuum quod sunt? IIIIor: uer, aestas, authumnus et hiems.* DRC 43.7–8: *Aliter uero alii intellegunt haec nomina quatuor temporum, id est uer, estas, autumnus, hiems.* DTR 35.2 (cited in LIB. CALC. 83, PV §314): *Tempora sunt anni quattuor.* DIAL. BURG. 9B: *Quattuor mundi tempora in anno accipiuntur. Hoc est: Ver, aestas, autumnus et hiems.* DIAL. NEUSTR. 30A: *Dividitur in quattuor tempora, id est ver, aestas, autumnus, hiems.* DIAL. LANGOB. 5B: *In circulo anni quattuor sunt tempora et menses duodecim. Quattuor tempora sunt ver, aestas, autumnus, hiems.* RM 34.4–5: *Vicissitudines ergo quas alii tempora nominant quattuor sunt: hiems, uer, aestas, autumnus.* **28–30** Haec…temperant] For the etymologies of *tempus* in post-MC texts cf. c. 30, ll. 2–5.

23–25 Annus…XXX] Cf. p. CXCV of the introduction. **26–30** Annus…temperant] Cf. Borst, *Schriften*, p. 361. **26–XXXV,5** Annus…dicitur] Cf. Borst, *Schriften*, p. 420, 436.

<XXXII.> DE VERNO

Etym. of spring (*ver*)

*Ver dic*itur, eo *quod uiret. Tunc enim post hiemem uestitur tellus herbis, et cuncta in florem rumpentur*. Vel *Grec*um *fructificatio* et Latinę.

<XXXIII.> DE AESTATE

Etym. of summer (*aestas*)

Aestas ab estu, id est a calore solis. *Aestas quasi exusta, id est quasi usta et arida.*

<XXXIIII.> DE AUCTUMNO

Etym. of autumn (*autumnus*)

Auctumnus ab augmento frugum dictus est. Vel *Grec*e *copiositas*. Siue *auctumnus a tempestate uocatus quando folia arborum cadunt et omnia maturescunt.*

XXXII,2 uestitur] auertitur *M (corr. from* ETYM., BC*; but it may rather have read* vertitur *as* DRC *and MS A of* ETYM.*).* **XXXIII,2** id…a] ita *M (corr. from* ANGERS 477, ETYM., BC, DRC*).* **3** usta] exuta *M (corr. according to* ANGERS 477, ETYM., BC*).*

XXXII,1–3 Ver…Latinę] The etymology of *ver* deriving from the verb *virere* is again a direct quote from ETYM. 5.35.3 (MS A reads *vertitur*; cited in BC 28C8–11): *Ver autem dictum quod viret. Tunc enim post hiemem vestitur tellus herbus, et in florem cuncta rumpuntur*. Cf. EPIT. 11.164–165 (cited in DTR 35.63, in turn cited in LIB. CALC. 83, PV §318, RM 34.24–25): *uer ab eo, quod in eo cuncta uernent, id est uirescant*. CE p. 95: *Ver a uiriditate nomen accepit*. The argument that *ver* is a Greek term is copied from CE p. 95: *Aliter Greca sunt hęc nomina: Ver ergo fructificatio ... interpretatur*. **XXXIII,2–3** Aestas[1]…arida] As in the previous two passages, MC relies on ETYM. 5.35.4 (cited in BC 28C11–12; the first sentence only in DRC 43.10–11, which reads *estu*) for his etymology of summer: *Aestas dicitur ab aestu, id est a calore; et aestas quasi usta, id est exusta et arida*. Cf. CE p. 95: *Aestas ab aestu. ... Aliter Greca sunt hęc nomina: ... Aestas ergo calor ... interpretatur*. **XXXIIII,2–4** Auctumnus…maturescunt] The first two etymologies, i.e. *auctumnus ab augmento* and autumn being the Greek term for abundancy, derive from CE p. 95: *Autumpnus ab autumento, hoc est ab augmento fructuum. ... Aliter Greca sunt hęc nomina: ... Autumpnus ergo copiositas fructuum ... interpretatur*. Cf. EPIT. 11.162–163 (cited in DRC 43.12–13, DIAL. BURG. 8B, which reads *signatione* instead of *autumnatione*, DTR 35.64–65, in turn cited in LIB. CALC. 83, PV §318, RM 34.28–29): *Autumnus uocatur de autumnatione fructuum, qui in eo colliguntur*. The third explanation for the term autumn is then again a direct quote from ETYM. 5.35.6 (cited in BC 28C13–14, DRC 43.11–12): *Autumnus a tempestate vocatus quando et folia arborum cadunt et omnia maturescunt*.

XXXII,1–3 Ver…Latinę] MC's discussion of spring is directly quoted in ANGERS 477, fol. 20r (gloss to DT): *Ver, eo quod uiret. Tunc enim post hiemem uestitur* (?) *tellus in herbas cuncta rumpuntur. Vel uer Grece fructificatio*. Irish texts of the first half of the eighth century give very similar accounts of spring: DRC 42.14–15, 43.9–10 is based on CE and ETYM.: *Sciendum nobis quod dicunt alii Greca esse haec nomina temporum, et ita interpretatur, id est uer fructificatio. ... Uer dicitur a uiriditate, ut Isidorus dicit: Uer dictum est, eo quod uiret. Tunc enim post hiemem uertitur herbis et in florem cuncta uertuntur*. DIAL. LANGOB. 6A draws primarily on CE: *Interrogatio: Quare ver dicitur? Responsio: Ver dicitur a fructificatione vel a viriditate. Ver enim graece, fructificatio latine interpretatur. Vel latinum est, et a viriditate nomen accepit, quia in eo virescunt omnia germina terrae*. The early Frankish DIAL. BURG. 8B shows similarities to DIAL. LANGOB.: *ver eo quod omnia virescunt*. These Irish explanations were superseded towards the end of the eighth century by Bede's discussion (DTR 35.63–66), not only in the case of spring, but of the other seasons as well; Bede's etymologies of the seasons are primarily based on Virgilius Maro. **XXXIII,2–3** Aestas[1]…arida] Again, ANGERS 477, fol. 20r (gloss to DT) shares notable readings with MC: *Aestas ab estu, id est a calore solis. Aestas quasi usta, id est exusta et*

32. ABOUT SPRING

Spring (*uer*) is so called since it is green (*uiret*). Indeed, at that time (of the year), after winter, the earth is covered with grass, and everything will break forth in blossom. Or (it is) a Greek term, (which means) 'a bearing of fruit' in Latin.

33. ABOUT SUMMER

Summer (*aestas*) (is so called) from heat (*estu*), i.e. from the warmth of the sun. (And) summer (appears) as if dried out, i.e. as if burned and dry.

34. ABOUT AUTUMN

Autumn (*auctumnus*) was so called from the growth (*augmento*) of fruits. Or (it means) abundancy in Greek. Or autumn was so called from the period of time when the leaves of the trees fall to the ground and everything becomes ripe.

---

arida. DRC 42.14–15 (which later cites ETYM. 5.35.4 quoted in *app. font.*) rather relies on CE: *Sciendum nobis quod dicunt alii Greca esse haec nomina temporum, et ita interpretatur, id est ... estas calor*. A similar account is also given in the other early eighth-century Irish text, DIAL. LANGOB. 6B: *Interrogatio: Quare aestas dicitur? Responsio: Aestas dicitur graece calor, vel siccitas interpretatur. Vel latinum est, et ab aestuante sole aestas nominatur*. Bede's discussion, then, proved to be very influential in the late eighth, early ninth centuries: DTR 35.63–64 (cited in LIB. CALC. 83, PV §318, RM 34.26–27): *aestas ab aestu, qui in ea maturandis fructibus datur*. **XXXIIII,2–4** Auctumnus...maturescunt] The Angers glosses seem to be based on MC in this instance: ANGERS 477, fol. 20r (gloss to DT): *Autumnus a tempestate uenti uocatur, quando folia arborum cadunt et omnia maturescunt. Vel autumnus Grece copiositas. Vel ab agmento frugum*. Generally, the interpretation of autumn being a Greek term for abundancy is a specific feature of Irish computistical literature, first formulate in CE, then repeated in MC, DRC 42.14–15 (which later cites EPIT. 11.162–163 quoted in *app. font.*): *Sciendum nobis quod dicunt alii Greca esse haec nomina temporum, et ita interpretatur, id est ... autumpnus copiositas*. DIAL. LANGOB. 6C: *Interrogatio: Quare autumnus dicitur? Responsio: Autumnus dicitur graece, copiositas frugum interpretatur. Vel latinum <est>, et de augmento frugum nomen accepit*. The most popular etymology is the one outlined by Virgilius Maro, which received a wide reception through Bede's works (cf. *app. font.*).

XXXII,1–3 Ver...Latinę] Cf. p. CLXXXVI of the introduction. **1–XXXV,4** Ver...uescendi] Partially transcribed in Borst, *Schriften*, p. 361; cf. Borst, *Schriften*, p. 437. **XXXIII,2–3** Aestas[1]...arida] Cf. p. CLXXXVI of the introduction. **XXXIIII,2–4** Auctumnus...maturescunt] Cf. p. CLXXV, CLXXXVI of the introduction.

<XXXV.> DE HIEME

Etym. of winter (*hiems*)

*Hiem*ps dicitur, *quia breuiori* soluitur *circulo.* I*nde hoc tempus bruma dicitur, quasi* baxin, *id est breuis. Vel a cibo, qu*ia *maior* est *appetitus tunc uescendi.* A*edacitas enim Grec*is brumon dicit*ur.*

<XXXVI. CONTINUATIO: DE ANNO>

The beginning of a solar and lunar year

Annus initium habet a Kalendis Ianuarii cum Latinis. Incipit cum Grecis ab XI Kalendas Aprilis. Alterum initium anni ab VIII Kalendas Aprilis, | a VII Kalendas Aprilis cum Aegyptiis, a quo pascha numeretur. fol. 18r

Initium lunaris anni dubium esse fertur.

XXXV,2–5 Hiemps…dicitur] The entire discussion of winter is again based on ETYM. 5.35.6 (MS B reads *baxin*; cited in BC 28C14–D5, which reads *braxim*): *Hiemem ratio hemisphaerii nuncupavit, quia tunc breviori sol volvitur circulo. Vnde et hoc tempus bruma dicitur, quasi βραχύς, id est brevis; vel a cibo, quod maior sit tunc vescendi appetitus. Edacitas enim Graece βρῶμα appellatur.* **XXXVI,2–5** Annus…numeretur] All of the different beginnings of a solar year listed here have their origin in the various traditions of Easter calculations; the earliest Easter table used in the *regiones Scottorum*, the *latercus*, noted the weekday and lunar data for 1 January, as did the Victorian table, giving the impression that this date was the beginning of the year. However, the January beginning of a year may also have been derived from one of the following texts: SAT. 1.13.3 (cited in DTR 12.53–54, in turn cited in LIB. CALC. 70, RM 32.14–15): *ac de duobus priorem Ianuarium nuncupavit primumque anni esse voluit.* MACROBIUS, *Commentarii in somnium Scipionis* 2.11.13 (Willis, *Commentarii in somnium Scipionis*, p. 129): *igitur sicut annum solis non solum a kalendis Ianuariis usque ad easdem vocamus.* CAESARIUS OF ARLES, *Sermones*, *Sermo 192*, c. 1 (*CCSL* 104, p. 780): *Et quia aput illos ianuariae kalendae unum annum inplere, alterum incipere dicebantur, istum Ianum quasi in principio ac termino posuerunt; ut unum annum finire, alterum incipere crederetur.* DNR 4.4.27 (cited in c. 13, ll. 6–7, BC 59D9, KAL. F4,6–7, e4), 4.6.43–44: *Ianuarius appellatur eo quod ianua sit anni atque principium. ... Menses autem omnes apud Latinos ex kalendis sumunt principium.* For January being the first month of the year cf. also c. 12, ll. 6–7; c. 24, ll. 6–10; for the 'principal order of months' starting with January see c. 24, ll. 24–26. The following beginnings of the year, 22 March and 25, or rather 26 March, are based on the 'Greek' and 'Roman' vernal equinoxes respectively; note, however, that in both cases the day after the equinox, the earliest possible date for Easter Sunday, rather than the equinox itself is mentioned. The general information that these equinoxes are considered as the beginning of the year may derive from one of the following texts: DNR 6.2.10–12 (cited in BC 59B4–6, DT 9.3–4, DTR 36.38–39, in turn cited in PV §145): *Principium autem anni alii a bruma putant, ut populi Romani; alii ab aequinoctio uerno, ut Hebraei; alii a solisticio, ut Graeci; alii ab autumno, ut Aegyptii.* DRP 2.36–39 (cited in c. 55, ll. 6–11): *Est ergo in primo anno initium primi mensis, quod est X et VIIII annorum circuli principium, secundum Aegyptios quidem mensis Famenoth XXVI die, iuxta Macedones uero Dystri mensis XXII, secundum Romanos uero Martii mensis XXV die, id est VIII Kal. Aprilis.* DISPUTATIO MORINI (Pitra, *Spicilegium Solesmense*, p. 31): *equinoctium quod est VIII kalendas Aprilis ... XII kalendas aprilis quod est equinoctium vernale, hoc est initium anni solis.* For *XI Kalendas Aprilis* being the beginning of the year see also c. 51, ll. 6–8; for March being the first month of the year see c. 24, ll. 2–4; for the 'natural order of months' beginning with March see c. 24, ll. 20–21; for the different dates of the spring equinox see c. 38, ll. 8–11, 26–2, 34–36; for these dates being the first day of creation see c. 11, ll. 37–40. The final sentence of this passage here is particularly noteworthy, since the argument that Easter Sunday is reckoned from 26 March can only derive from the *latercus* paschal tradition. **6** Initium…fertur] Even though MC argues here that the beginning of the lunar year is dubious, he lists three possibilities in c. 53, ll. 5–11, namely *luna* 1, 14, or 15 of the Easter lunation.

35. ABOUT WINTER

Winter is so called because it is completed by a shorter (*breuiori*) circle. Accordingly, this season is called *bruma*, like *baxin*, which means short (*breuis*). Or it is so called from food, since the appetite for eating is greater at that time. In fact, gluttony is called *brumon* among the Greeks.

36. CONTINUATION: ABOUT THE YEAR

The year (has) its beginning on 1 January according to the Latins. It begins on 22 March according to the Greeks. Another beginning of the year (is) on 25 March, (or rather) on 26 March according to the Egyptians, (and) Easter should be reckoned from this date.

The beginning of a lunar year is said to be dubious.

XXXV,2–5 Hiemps…dicitur] Again, the Angers glosses share notable readings with MC: ANGERS 477, fol. 20r (gloss to DT): *Hiems dicitur a breuitate, quia tunc breuiore sol uoluitur circulo. Unde hoc tempus bruma dicitur quasi braxin, id est breuis. Vel ab edacitate, quia tunc maior sit uescendi apetitus. Edacitas dicitur brumon Grece*. Readings close to MC (especially *solvitur* instead of ETYM.'s *sol volvitur*) can also be found in PS-ALCUIN, *Disputatio puerorum* 6 (*PL* 101, col. 1119B): *Hiems est ratio emisferii nuncupata. Inter. Quomodo? Resp. Quia tunc breviori solvitur circulo. Unde et hoc tempus bruma dicitur, quasi braxin, id est brevis: vel cibo, quod major sit tunc vescendi appetitus. Inter. Qua ratione? Resp. Edacitas enim Graece bruma appellatur*. All other texts do not cite ETYM. as MC, but rather EPIT., or they give a completely different etymology for this season. **XXXVI,2–5** Annus…numeretur] The principal discussion of the beginning of the year was Isidore's in DNR (cf. *app. font.*). This is also the basis of DRC 45, which, however, changed Isidore's statement concerning the beginning in January, in this showing parallels to MC: *Sciendum nobis quot sunt principia anni. Quattuor: Ab VIII Kl Ianuarii, uel a Kl Ianuarii apud Latinos; ab equinoctio uernali apud Ebreos; a solstitio aestiuo apud Grecos; ab equinoctio autumnali apud Aegyptios*. Two of the beginnings listed in the Irish reference-Bible agree with MC: PP I §131: *Annus solaris apud Latinos a Kl. Ianuarii incipit; apud Aegyptios a IIII Kl. Septembris incipit ... Ebreorum anni incipiunt ab XI Kl. Aprilis, quia in eo die mundus factus est. Greci a solstitio estuo incipiunt anni*. Interestingly enough, QUAEST. AUSTR. 2.10A explicitly connects the different traditions for the beginning of the year with the various Easter reckonings, and in this may have been influenced by MC: *Sed iuxta Victorium et Latinos initium in Kalendas Ianuarii ponamus, quamvis novimus non hinc incipere iuxta Aegiptios initium anni, sed ab XI Kalendas Aprilis*. In most eighth- and early ninth-century computistical texts that do not cite Isidore directly, the vernal equinox is, for various reasons, considered as the beginning of the year: DIAL. BURG. 9B: *Et incipit annus solis ab equinoctio vernali, hoc est XII Kalendas Aprilis*. COMP. COL. 3.3B: *Deinde principium anni XII Kalendas Aprilis <est>*. Gloss to DTR (*CCSL* 123B, p. 294): *Naturaliter enim in XII kl. Aprilis est intium anni*. DE BISSEXTO I 997A4–B14: *Principium ergo anni secundum Graecos XI Kal. Aprilis est. ... quia proptera Latini initium anni faciunt ab VIIII (recte VIII) Kal. Aprilis, eo quod illa die luminaria facta sunt sol et luna*. LDA 3.4.1–2: *Dum solaris annus in vernali aequinoctio hoc in vigesimo primo die mensis Martii naturaliter incipit. ... Annus etiam dominicae incarnationis, si in dies illius incarnationis inceperit, in vigesimo quinto semper die mensis Martii inchoabit. ... Cur primus dies anni solis in duodecimis Kalendis Aprelis naturaliter fieri a Graecis et Aegyptiis*. **6** Initium…fertur] For the beginning of the lunar year in post-MC texts see c. 53, ll. 5–11.

XXXVI,2–5 Annus…numeretur] Cf. p. XXII of the introduction.

Weeks/*mom. per year*

LII dies dominicos annus habet et tot ebdomades habet. Momenta uero in anno solis sunt <XXX> muriades et V muriades CCCC simplicia.

The place of the year of creation in the *cyclus decemnov.*

Quis primus annus, in quo factus est mundus in mense Martio et die dominico? Ac in communi anno septimo ogdoadis, uel endecade, ut alii dicunt, id est in communi anno decimo, sine bissextu et saltu anno. Secundum Grecos *mundus* factus est XII Kalendas Aprilis *in* communi *secundo ogdoadis*.

XXXVI,7 dies] habet *add. M.* **8** XXX] *om. M (add. according to* MC*).* **13** bissextu] *corr. to* bissexto *Borst.*

7–9 LII...simplicia] The knowlegde of the term *myriades*, meaning 10 000, may have been transmitted through AMBROSE, *De excidio urbis Hierosolymitanae* 5.49 (*PL* 15, col. 2302D): *decies centena millia (CX myriades, ut plerique asserunt)*, or JEROME, *Commentarii in Danielem* 2.7.10b (*CCSL* 75A, p. 846): *Ista sunt milia, et istae myri-dyades de quibus et in Psalmis legimus: Currus Dei decem milibus multiplex, milia laetantium; Dominus in eis*. If, however, DE COMPUTO DIALOGUS is accepted to be a pre-MC Irish text, it is more likely that MC was aware of this Greek term through this text (*PL* 90, col. 651), or its Irish exemplar (if the *PL* text is a Carolingian recension): *Mille decies multiplicatus, myriadam facit*. The fact that a year has 52 Sundays and weeks could have easily been calculated by MC himself (365/7=52 1/7); the calculation of the 350 400 *momenta* of a year was certainly more demanding (365×24×40=350 400). But since both facts are not mentioned in any evidently pre-MC text, it must be presumed that MC executed these calculations himself. For the underlying numerical relations cf. c. 1, ll. 9–13; c. 11, l. 12. **10–15** Quis...ogdoadis] For the world being created in March, on 21 March, a Sunday, see c. 11, ll. 37–40; c. 24, ll. 2–3; c. 44, ll. 18–19. The theory that the year of creation was the second year of the Dionysiac *ogdoas*, obviously the opinion favoured here, may have been based on DE COMPARATIONE EPACTARUM 1.11–12 (see Appendix 2): *In secundo ogdoadis anno mundus coepit, quod XIIII luna primi mensis tunc occurente confirmat*. The argument is that in this second year of the Dionysiac *ogdoas* the full moon after the vernal equinox falls on 25 March, i.e. on the date assigned to the creation of the moon (as a full moon) at the beginning of the world (cf. c. 44, ll. 27–34). The second option mentioned by MC, the world being created in the tenth year (of the *cyclus decemnovenalis*, i.e. the second year of the *hendecas*), appears to be a projection of the Victorian *annus passionis* on the year of the creation of the world: Victorius started his Easter table with the *annus passionis*, which corresponds to the second year of the *hendecas* when compared to the Dionysiac Easter table; in c. 44, it is argued that the year of creation and the *annus passionis* have identical chronological data, so that MC thought it legitimate to assume an identical placement within the 19-year cycle of the Victorian *annus passionis* and the year of creation (note that this theory is confirmed by the information of COMP. COL.; cf. *app. comp.*). On the other hand, Victorius himself argues in his prologue (Krusch, *Studien* II, 24) that the creation of the world occurred 5228 years before the resurrection; since Victorius placed the resurrection in AD 28, AM 5199=2 BC, which is the final year of a *cyclus decemnovennalis*; the creation, then, took place in −5199mod19=−12mod19=7mod19, i.e. in the seventh year of the *cyclus decemnovenalis*, and therefore in the seventh year of the *ogdoas*; this seems to be the argument behind the first option given by MC for the year of creation.

7–9 LII...simplicia] The late eighth-century Frankish LIB. ANN. gives twice a detailed account of the various subdivisions of a year, including the number of weeks and *momenta*; the number of *momenta*, however, is wrongly calculated in both instances, illustrating the diffiulties that early medieval computists had with the calculation of great numbers (correct would have been 350 640 *momenta*, since LIB. ANN. reckoned a year as consisting of 365¼ days rather than MC's 365): LIB. ANN. 60 (cited in LIB. COMP. 3.2A, LIB. CALC. 40): *Annus solaris habet ... ebdomades quinquaginta duas, ... momenta trecenta quinquaginta novem milia et trecenta quadraginta*. LIB. ANN. 64 (cited in KAL. A1–2,4–5,7–9, C6,8,

A year has 52 Sundays, and the same number of weeks. However, there are 350 400 single *momenta* in a solar year.

What kind of a year (is) the first year, in which the world was created in the month of March, on a Sunday? (The world was created) in a common year, the seventh of the *ogdoas*, or in the *hendecas*, as others argue, namely in a common year, the tenth (of the *cyclus decemnovenalis*), a year without bissextile day and *saltus*. According to the Greeks, the world was created on 21 March, in a common year, the second year of the *odgoas*.

---

a14,19, c3,6, f9, g3): *Expliciunt quattuor tempora anni, duodecim menses, ebdomadae quinquaginta duae et unus dies, ... momenta trecenta quinquaginta milia sescenta sexaginta*. Cf. also KAL. E3 (52 weeks, 350 650 *momenta*). Only early eighth-century texts note the number of weeks within a year: DRC 20.14–15: *septimanae uero quinquaginta duae et unus dies*. DIAL. NEUSTR. 29aB: *Annus enim solaris decet habere ... ebdomadas quinquaginta duas*. QUAEST. AUSTR. 2.8A: *De ebdomadibus sermo est uniuscuiusque solaris anni. Ebdomades quinquaginta due <sunt>, sed ultima diebus octo, non ut caeterae septem diebus terminatur*. Cf. also KAL. a24. Later texts are not concerned with the number of weeks per year (which certainly was or became common knowledge in the eighth century), but some list, among other data, the number of *momenta* per year; only the following texts give the correct total (for a year including the bissextie increment): LECT. COMP. 9.3C: *Et sic per duodecim menses uniuscuiusque <anni> computabis in annum solarem: Momenta trecenta quinquaginta milia sescenta quadraginta*. RM 36.41–46: *D. Annus ergo solaris ... Quot momenta? M. CCCL milia DCXL*. Also KAL. A9, g13 (350 640 *momenta*). Most other texts arrive at different results, and not a single texts gives MC's correct total for a year of 365 days: BC 92D1: *Momenta habet CCCL milia XXXX*. BC 95A16–17: *CCCLII CCCCLXXX. Tot momenta sunt in anno. In bissextili autem anno sunt momenta CCCLII DXL*. QUAEST. LANGOB. 5C: *et fiunt in totius anni circulo momenta milia centum septuaginta <tres> trecenta viginti*. For the term *myridiades* cf. DRC 9.15 (which reads *muriades* after corr. in the sole MS B): *Decim uero milia murias dicuntur*. **10–15** Quis...ogdoadis] The few early medieval texts that deal with the problem of the placement of the year of creation within the *cyclus decemnovalis* agree with MC in the preference for the second year of the *ogdoas*. Two of the three options mentioned by MC are outlined in COMP. COL. 4.1A, 7A, with the second year of the *ogdoas* being clearly favoured; the information probably derives directly from MC: *Embolismus primus annus mundi cum Latinis <est, post> communem secundum endecadi. Cum Graecis ogdoadus initium mundi est, communis secundus ogdoadi. ... Cur praecedit ogdoadus? Quia sic fuit initium mundi. Anno secundo ogdoadi initium mundi <fuit>*. BC 82D13–A5 (cited in the undoutedly Irish tract *De principio mundi* transmitted in BC 148A(1359)9–15 and St Gall, Stiftsbibliothek, 251, p. 13–14) refers only to the second year of the *ogdoas* as the year of creation, sharing important readings with MC: *Interrogatio de mundi principio, quomodo factus est initio creaturarum Dei? In quo die? Et in <quibus> diebus septimanae? Et in quo anno? Et in quo tempore? Si in die anni? In nocte? Et in qua hora? interrogamus. In aequinoctio autem et in die Dominico, et in communi, in anno secundo, et in ogdoade, et in prima hora creatus est.*

7–9 LII...simplicia] Cf. Borst, *Schriften*, p. 417, 511. **10–15** Quis...ogdoadis] Transcribed in Borst, *Schriften*, p. 909 (with different punctuation); cf. p. CLXXX of the introduction.

<XXXVII.> DE SOLE

Etymology of and names for sun (*sol*)

*Sol dic*tus est, *eo quod solus luceat*, uel a soliditate luminis. Gamse in Hebreo uel simpsia, elios in Greco, panath cum philosophis, foebe cum Syris, titan cum Chaldeis. Aut sol de soliditate luminis, ut Augustinus ait: *Sol dicitur, quia solus per diem sui fulgoris ui obtonsis cum luna stellis totum* mundu*m inlustrat.*

<XXXVIII. *DE DIVISIONE ANNI*>

Division of the year in four parts

Quater in IIII dies, id est solstitium hiemale et aequinoctium uernale, solstitium aestiuale et aequinoctium auctumnale, <annus> diuiditur.

XXXVII,3 simpsia] simpsé *or* simpsea *Ó Cróinín (both contra M).* **4** foebe] Foebus *or* Foebis *Ó Cróinín (both contra M).* **6** ui] uim *M (corr. from* EPIT., DRC*).* **7** cum luna] luna cum *M (corr. from* EPIT., DRC*).* **XXXVIII,1** De…anni] *I have employed* CE*'s heading here (cf. app. font.).* **2** hiemale] hiemalem *M.* **4** annus] *om. M (add. according to* MC, CE*).*

XXXVII,2–7 Sol…inlustrat] The etymology explaining that the sun (*sol*) is so called from 'shining alone' (*solus luceat*) seems to ultimately derive from CASSIODORUS, *Expositio psalmorum, Expositio in psalmum 103*, §22 (*CCSL* 98, p. 935): *Sol enim dictus est, quod solus ita lucet, ut ex eo dies fiat.* A similar statement can be found in ETYM. 3.71.1 (cited in PV §438): *Sol appellatus eo quod solus appareat, obscuritatis fulgore suo cunctis sideribus.* MC, however, appears to have relied on CE p. 95 here: *Sol dicitur, eo quod solus luceat obscuratis fulgore suo cunctis syderibus.* The second mentioning of this etymology at the end of this passage, where it is attributed to Augustine, is a direct quote from EPIT. 11.49–50: *Sol dicitur, quia solus per diem sui fulgoris ui totum orbem obtunsis una stellis cum luna inlustrat.* As an interpretation of this statement, MC appears to have invented the etymology of the sun (*sol*) deriving from the strength (*soliditas*) of its light. Likewise, the names of the sun in the various languages cannot be found in any pre-MC text. **XXXVIII,1** De…anni] The heading given to the parallel passages to MC c. 39–40 is *De diuisione anni, que magis naturam possidit* in CE p. 98. **2–4** Quater…diuiditur] MC seems to have taken the terminology for the solstices and equinoxes from ETYM. 5.34.2–3 (cited in BC 30B7–C3): *Duo sunt autem solstitia: unum aestivum ... aliud hiemale. ... Item duo sunt aequinoctia, unum vernale et aliud autumnale.* DNR 8.1: *Solstitia duo sunt: primum hibernale ... alterum aestiuum. ... His contraria duo aequinoctia sunt: unum uernale ... alterum autumnale*; note, however, the difference in the term for the summer solstice. The fact that a year is divided into four parts by these four dates derives from CE p. 98: *Annus igitur IIII diuisiones habet, hoc est II equinoctia et II solstitia.*

XXXVII,2–7 Sol…inlustrat] Etymologies in general, as well as the question of terminology in different languages, were very popular in Irish intellectual (not only computistical) circles of the seventh and eighth centuries, so that it does not surprise to find parallel passages to MC's account of the sun almost exclusively in Irish and Irish influenced texts of that period. The discussion of the sun in DRC 1.1–7 seems to be directly based on MC (with the addition of the etymology *sol a solemnitate* from EPIT. 11.50–52): *Sciendum nobis quomodo sol in principalibus linguis uocatur. Ita: Gamse uel Simsia apud Ebreos; Elios apud Grecos; Paniph apud philosophos; Phoebus apud Syros; Titan apud Caldeos; Sol apud Latinos. Isidorus dicit: Sol dictus est eo quod solus luceat, uel soliditate luminis aut de solemnitate lucis. Unde Augustinus: Quia solus per diem sui fulgoris ui obtonsis cum luna stellis totum mundum inlustrat.* The Angers glosses certainly copied directly from MC the names of the sun in the various languages: ANGERS 477, fol. 13v (gloss to Bede's DNR): *Gamse in Ebreo uel simsia, elios in Greco, panip apud philosophos, phebe cum Siris, titan cum Caldeis.* Clearly dependent on MC is the

37. ABOUT THE SUN

The sun (*sol*) is so called, because it is used to shine alone (*solus*), or from the strength (*soliditas*) of its light. (The sun is called) *gamse* or *simpsia* in Hebrew, *elios* in Greek, *panath* among the philosophers, *foebe* among the Syrians, and *titan* among the Chaldeans. Or the sun (*sol*) (is so called) from the strength (*soliditas*) of its light, as Augustine says: The sun (*sol*) is so called, because it alone (*solus*) illuminates the entire world throughout a day with the strength (*ui*) of its light, while the stars, including the moon, are blurred.

38. ABOUT THE DIVISION OF THE YEAR

The year is divided four times by four days, namely by the winter solstice, the vernal equinox, the summer solstice, and the autumnal equinox.

---

short tract constituting chapter 116 in BC (citing ETYM. at the end, cf. *app. font.*): *De uariis nominibus solis per uarias linguas. Quam se in Aebreo luna simplicia, aelios in Greco, phanath cum philosophis, tithan cum Chaldeis, sol cum Latinis de soliditate luminis, uel eo quod solus appareat obscuratis sideribus radiis eius potentibus*. The Irish reference Bible also gives etymologies for the sun that must derive from Irish computistical textbooks: PP I §126: *Isidorus: Sol dicitur quod solus lucet, uel soliditate naturae*. The late Merovingian DIAL. BURG. 10A, for its part, relies on Virgilius Maro (cf. *app. font.*): *Sol de solempnitate dicitur, quia solus per diem lucet*. Rabanus Maurus only refers to the etymology of the sun being so called from 'shining alone', but in his account shares some readings with the Irish texts: RM 37.11–13: *D. Vnde dicitur sol? M. Quod solus inter omnia sidera luceat, oriente enim sole cetera sidera mox radios suos abscondunt.* **XXXVIII,2–4** Quater...diuiditur] The division of a year into four parts between the solstices and equinoxes was particularly popular in late seventh- and early eighth-century Irish or Irish influenced texts: DRC 47.1–2: *Sciendum nobis quot sunt dies contra quos regulatur annus. Quatuor. Id est duo solstitia et duo equinoctia*. DIAL. NEUSTR. 29A asks the question: *Ubi sunt quattuor anguli in anno?*, followed by the dates of the equinoxes and solstices. DIAL. LANGOB. 17A–B employs, in this context, the same terminology as can be found later (c. 38, ll. 39–40) in MC (namely *principales dies* for the solstices and equinoxes): *Sed ex his diebus quattuor sunt principales, inter quos omnes dies anni dividuntur in quattuor partes. ... Interrogatio: Quare et qua causa dividitur annus inter hos quattuor dies? Responsio: Quia in his quattuor diebus sunt duo aequinoctia et duo solstitia anni*. Note that this division of a year was the fundamental basis for the discussion of the increase and decrease of daylight discussed in c. 39, so that it can primarily be found in texts that continue to deal with that problem.

XXXVII,2–5 Sol...Chaldeis] Transcribed in Ó Cróinín, 'A seventh-century Irish Computus', p. 105; idem, *Cummian's letter*, p. 115; the first sentence only transcribed in Borst, *Schriften*, p. 362; cf. p. CLXXXVI of the introduction. **6–7** Sol...inlustrat] Transcribed in Ó Cróinín, 'Virgilius Maro Grammaticus', p. 200; cf. p. CXXVI, CXXIX of the introduction. **XXXVIII,2–8** Quater...consistunt] Cf. p. CLXXV of the introduction. **2–43** Quater...aequinoctia] Cf. Borst, *Schriften*, p. 416, 448, 903.

Etym. of solstice

*Solstitium dicitur quasi solis statio, qu*ia *sole stante dies uel no*x *cresc*it.

Etym. of equinox

Ysidoro dicente: *Aequinoctium* uero dicitur, *qu*ia *tunc dies | et nox aequali horarum spatio consistunt.* fol. 18v

Division of the year based on the 'Greek' dates

Ita *duo solstitia et II ęquinoctia*, quae *annu*m *habe*n*t*, *Grec*is censiri innotescunt: XII Kalendas Ianuarii, XII Kalendas Aprilis, XII Kalendas Iulii, XII Kalendas Octimbris. Sicque <spatia> computari debent: *ab XI Kalendas Ianuarii in XII Kalendas Aprilis, ab XI Kalendas Aprilis in XII Kalendas Iulii, ab XI Kalendas Iulii in XII Kalendas Octimbris, ab XI Kalendas Octimbris in XII Kalendas Ianuarii.*

---

12 spatia] *om. M (add. according to* MC*).* **13** Ianuarii] *om. M**.

5–6 Solstitium…crescit] The etymology for solstice is copied from ETYM. 5.34.1 (cited in BC 30B7–8): *Solsticium dictum quasi solis statio, quod tunc sole stante crescant dies vel noctes.* MC may also have read DNR 8.2.7: *Solsticium autem dicitur quasi solis statio.* **7–8** Aequinoctium…consistunt] Likewise, the etymology for equinox is copied from ETYM. 5.34.1 (cited with few variations in BC 30C3–5): *Aequinoctium appellatum quod tunc dies et nox horarum spatio aequali consistunt.* Cf. DNR 8.2.7–9: *aequinoctium uero quod tunc dies et nox in aequitatem horarum duodenarum reuertantur, coaequatis spatiis suis.* **9–16** Ita…Ianuarii] The only texts available to MC, in which Julian calendar dates are explicitly given for the two solstices as well as the two equinoxes, were ETYM. 5.34.2–3 and DNR 8.1; there, the 'Roman' dates are given, i.e. the *VIII Kalendas* of the respective months; on the basis of these dates, Anatolius (?) (DRP 13.215–233) had divided the year in four parts (cf. c. 38, ll. 26–27). Only one of the 'Greek' dates given in MC was explicitly mention in pre-ca. AD 650 texts, namely *XII Kalendas Aprilis* for the spring equinox: EPISTOLA PROTERII c. 8 (Krusch, *Studien* I, p. 277; cited in EW 1.14–15): *XII kl aprl, aequinoctium esse cognoscitur. ... in aequinoctio, ..., qui est XII kl aprl.* PROLOGUS DIONYSII (Krusch, *Studien* II, p. 65): *quae quia cum solis cursu non aequaliter volvitur, tantorum dierum spatiis occursum vernalis aequinoctii consequatur, qui a XII kl. apr. die, cunctorum orientalium <spatii> sententiis et maxime Aegyptiorum, qui calculationis prae omnibus gnari sunt, specialiter adnotantur.* DISPUTATIO MORINI (Pitra, *Spicilegium Solesmense*, p. 31): *XII kalendas aprilis, quod est equinoctium vernale.* It seems, therefore, that seventh-century Irish computists calculated the 'Greek' Julian calendar dates of the equinoxes and solstices from the information of Isidore's 'Roman' dates and the one 'Greek' date found in the above mentioned sources by invariably adding four days to Isidore's dates; this method is outlined by Bede (cf. *app. comp.*). The 'Greek' dates are already used by CE, from which MC copied the four intervals: CE p. 98: *Annus igitur IIII diuisiones habet, hoc est II equinoctia et II solstitia, id est ab XI Kalendas Ianuarii usque in XII Kalendas Aprilis, ab XI Kalendas Aprilis usque in XII Kalendas Iulii, ab XI Kalendas Iulii usque in XII Kalendas Octobris, ab XI Kalendas Octobris usque in XII Kalendas Ianuarii secundum Grecos.*

5–6 Solstitium…crescit] If an etymology of solstice is given in eighth-century computistical texts, this is always, at least ultimately, based on Isidore: cf. *app. font.* and DRC 47.3: *Sciendum nobis quid sit solstitium. Id est solis statio, ut Isidorus dicit: Solsticium quasi solis satio.* The Angers glosses and the Irish DIAL. LANGOB., however, cite Isidore through MC: ANGERS 477, fol. 19v (gloss to DT) is a verbatim copy of MC, while DIAL. LANGOB. 17C reads: *Solstitium dicitur quasi solis statio, quia sole stante dies vel nox incipiunt crescere.* **7–8** Aequinoctium…consistunt] As in the case of the etymology of solstice, DIAL. LANGOB. 17B copied the etymology of equinox directly from MC: *Aequinoctium dicitur, quia tam dies et nox aequali spatio horarum consistunt.* Likewise ANGERS 477, fol. 19v (gloss to DT): *Aequinoctium dictum, eo quod tunc dies et nox aequali horarum spatia consistunt.* DRC 47.4–5, 7–8 shows more variation to MC as well as the Isidorian original: *Isidorus dicit: Equinoctium dicitur quando dies et nox equis spatiis diuiduntur. ... Et ideo equinoctium dicitur*

The solstice (*solstitium*) is so called as if (it were) the standing still of the sun (*solis statio*), since the day or the night grows while the sun stands still.

Isidore says: The equinox (*aequinoctium*) is so called because at that time day and night (*nox*) comprise an equal (*aequali*) number of hours.

The following two solstices and two equinoxes, which occur (in the course of) a year, are known to be estimated among the Greeks: 21 December, 21 March, 20 June, (and) 20 September. And the intervals have to be reckoned in the following way: from 22 December to 21 March, from 22 March to 20 June, from 21 June to 20 September, and from 21 September to 21 December.

---

quia equalis est nox diei in mensura horarum. **9–16** Ita…Ianuarii] Particularly the question of the date of the spring equinox was vital for computists, since the Easter full moon was not supposed to fall before that date. The 'Greek' dates for the equinoxes and solstices outlined in MC are also referred to in DRC 47.8–11: *Sciendum nobis in quibus diebus mensis duo solstitia et duo equinoctia inueniuntur. Ita: In XII Kl Ianuarii et in XII Kl Iulii duo solstitia inueniuntur apud Grecos. In XII Kl Aprelis et in XII Kl Octimbris duo equinoctia inueniuntur.* Bede gives an interesting account of how these 'Greek' dates were to be calculated from the older 'Roman' dates (at least as far as western tradition is concerned): DTR 30.48–52, 74–77 (cited in LIB. CALC. 82B,D, followed by the explicit dates (cf. below); PV §§311, 313, RM 53.9–15, 42–47): *Verum quia, sicut in ratione paschali didicimus, aequinoctium vernale duodecim kalendarum aprilium die, cunctorum orientalium sententiis et maxime aegyptiorum quos calculandi esse peritissimos constat, specialiter adnotatur, caeteros quoque tres temporum articulos putamus aliquanto priusquam vulgaria scripta continent esse notandos. ... hoc aequinoctium duodecim kl. aprilium die veraciter adscribendum, sicut non solum auctoritate paterna sed et horologica consideratione docemur. Sed et caetera tria temporum huiusmodi confinia simili ratione aliquot diebus ante octavum kl. sequentium esse notanda.* LIB. CALC. 82F adds to the just cited Bedan account the explicit dates for the solstices and equinoxes according to the Greeks: *Ita ut diligenti industria multis artibus a quibusdam fertur esse repertum, solstitium aestivale XII Kalendas Iulii esse atque aequinoctium autumnale XII Kalendas Octobris, solstitium quoque brumale XII Kalendas Ianuarii esse aut XIII.* Interestingly enough, DIAL. NEUSTR. 29A and DIAL. LANGOB. 18A list these four dates a day later, on the *XI Kalendas* of the respective months; in BC 29A13–14 (cited in BC 45B7–8, in turn cited in LIB. COMP. 1.3b) and 30C5–6 only the vernal equinox is corrected to *XI Kalendas*, for the others the 'Roman' dates are given. The intervals between these 'Greek' dates for the solstices and equinoxes as first defined in CE and then copied into MC cannot be found in any other text, impressively illustrating the interdependency between these two Irish textbooks.

5–6 Solstitium…crescit] Cf. p. CLXXXVI of the introduction.

The lengths of these four intervals

Ab XI Kalendas Ianuarii in XII Kalendas Aprilis, in quo spatio sunt dies XC. Ab XI Kalendas Aprilis in XII Kalendas Iulii, ubi XCI dies apparent. Ab XI Kalendas Iulii in XII Kalendas Octimbris, in quo dies XCII sunt. Ab XI Kalendas Octimbris in XII Kalendas Ianuarii, ubi sunt dies XCII.

90-day intervals, superfluous days

Quos V superfluos dies in XC IIII partiri tandem apparebimus. In his namque incremento dierum et diminutione in superficie non indigebimus.

The 'Latin' division

Secundum Latinos *ab VIII Kalendas Ianuarii in VIII Kalendas Aprilis* annus diuiditur et reliqua.

---

25 diminutione] diminutio *M.* **27** reliqua] kl *M (an obvious misreading of* rl, *the abbr. for reliqua).*

17–25 Ab…indigebimus] The ultimate model for calculating the lengths of the four intervals of a year may have been DNR 7.5 (cited in variation in BC 27), in which Isidore calculated the lengths of the four seasons; these differ from the lengths given here: spring 91 days, summer 91 days, autumn 93 days, winter 90 days (91, 91, 92, 91 in BC 27). The lengths of these intervals are not explicitly calculated in CE, so that it must be presumed that the calculation has been executed by MC himself; CE p. 98 rather treats these intervals as being of equal length, consisting of 90 days each: *Hi numeri quaterni XC noscuntur*. As MC argues in the final sentence of this passage here, the theory of a linear increase and decrease of daylight (c. 39) depended on intervals of equal length of 90 days; this, however, leaves a remainder of five days in a 365-day year (4×90+5=365). This apparent problem was solved in a curious way by late seventh- and early eighth-century Irish computists: the argument was that if these five days were divided by the 90 days of the interval in which they occur and then distributed equally among these 90 days, they could not be counted as full days, and therefore would not have any impact on the theory of the linear increase and decrease of daylight; for the details of this division of the five 'superfluous' days cf. c. 40; for a different division cf. c. 24, ll. 12–17; c. 41, ll. 80–88. **26–27** Secundum…reliqua] The principal source for the divison of a year between the 'Roman' dates of the solstices and equinoxes, and especially the explicit reference to the intervals between them as outlined here in MC, is Anatolius (?)'s DRP 13.215–233: *ab VIII Kal. Ianuarii in VIII Kal. Aprilis ... ab VIII Kal. Aprilis usque in VIII Kal. Iulii ... ab VIII Kal. Iulii in VIII Kal. Octobris ... ab VIII Kal. Octobris usque in VIII Kal. Ianuarii*. Cf. also especially ETYM. 5.34.2–3 (cited in BC 30B8–C6 – with the notable difference of *XI k Ap* for the vernal equinox): *Duo sunt autem solstitia: unum aestivum, VIII Kal. Iul. ... aliud hiemale, VIII Kal. Ian. ... Sunt autem haec aequinoctia die VIII Kal. Apr. et VIII Kal. Oct.* DNR 8.1: *Solstitio duo sunt: primum hibernale VIII kl. ianuaris, ... alterum aestiuum, VIII kl iulias. ... His contraria duo aequinoctia sunt: unum uernale VIII kl. aprilis ... alterum autumnale, VIII kl. octobres*. MC's immediate source, however, seems to have been CE p. 98, 99: *Secundum autem Romanos ab VIII Kalendas Ianuarii usque in VIII Kalendas Aprilis, ab VIII Kalendas Aprilis usque in VIII Kalendas Iulii, ab VIII Kalendas Iulii usque in VIII Kalendas Octobris, ab VIII Kalendas Octobris usque in VIII Kalendas Ianuarii ... Utriumque autem tali conputatione differunt, quod Greci diem, quo incipiunt et quo finiunt in XC conpotu numerant, hoc est ab XI usque in XII, ut dicitur, et dominabitur a mari usque ad mare. Romani diem, quo incipiunt, enumerant in XC presenti et non conputant diem, quo finiunt, nam ab VIII Kalendas usque in VIII Kalendas XC dies enumerant*. Note that these are the dates of the equinoxes and solstices followed by Columbanus, *Regula monachorum* 7 (Walker, *Sancti Columbani opera*, p. 130–2). For these 'Latin' dates see also c. 38, ll. 31–38; c. 47; for the equivalent 'Greek' intervals c. 38, ll. 12–22.

17–25 Ab…indigebimus] MC's calculation of the lengths of the intervals between the 'Greek' solstices and equinoxes is directly copied in the Frankish DIAL. NEUSTR. 29B: *Quantos dies <habet> ab XI Kalendas Ianuarii usque ad XII Kalendas Aprilis? Nonaginta. Et ab XI Kalendas Aprilis usque ad XII*

From 22 December to 21 March, in this interval there are 90 days. From 22 March to 20 June 91 days appear. From 21 June to 20 September there are 92 days. From 21 September to 21 December there are 92 days.

We will demonstrate at length that the five remaining days (i.e. the days that exceed 90 days in the respective intervals) are divided among the four (intervals of) 90 (days). Accordingly, we will leave aside(, in the following chapter,) the marginal increase and decrease of daylight on these days.

According to the Latins, the year is divided from 25 December to 25 March, etc.

Kalendas Iulii quantos dies habet? Nonaginta unum. Et ab XI Kalendas Iulii usque ad XII Kalendas Octobris quot sunt dies? Nonaginta duo. Ab XI autem Kalendas Octobris usque ad XII Kalendas Ianuarii? Similiter nonaginta duo. The Irish DIAL. LANGOB. 19B gives the same lengths as MC, the implicit dates are, however, the 'Roman' rather than the 'Greek' ones: *Cum omnes dies anni in quattuor partes dividuntur, cur non aequa fit divisio in his quattuor, quia in prima divisione nonaginta dies tantum inveniuntur, in secunda divisione nonaginta unus dies, in tertia divisione nonaginta et duo, et <in> quarta divisione nonaginta duo?* QUAEST. AUSTR. 2.8B (cited in the following note), then, outlines the same lengths as MC explicitly between the 'Roman' rather than the 'Greek' dates (and also including a quarter-day in the first interval); for a different, seasonal division of the year into intervals of 90, 91, 92, 92 days, cf. c. 38, ll. 41–43, c. 45; BC 31 and RM 53.48–61 divide the year into equal intervals of 91 days, among which the remaining day is then evenly distributed (and in RM also the quarter-day; cf. also RM 34.58–62). The fact that a year is divided into four 90-day intervals for calculation purposes, particularly for the theory of the linear increase and decrease of daylight, is also referred to in the Irish DRC 50.1–8: *Anatolius uero naturalem diuisionem ostendit, id est diuidit annum in quatuor nonaginta. De quatuor uero diebus confinia facit et circa que annum diuidit. ... Ordinauit autem Anatolius annum solis ... in quatuor nonaginta aequalia.* DIAL. LANGOB. 17A, 19A: *Sed ex his diebus quattuor sunt principales, inter quos omnes dies anni dividuntur in quattuor partes, id est in quattuor nonaginta. ... Quare ergo arithmetici dividunt in quattuor partes, id est quattuor nonaginta?* as well as the Irish influenced COMP. COL. 4.6B: *Qui ab eodem Anatolio per totum nonaginta dierum tempus aequis partibus dividuntur.* **26–27** Secundum...reliqua] A discussion of these intervals between the 'Roman' dates of the solstices and equinoxes can also be found in the Irish DRC 50.6–41, which appears to be based on DRP: *Anatolius ... incipit prima pars apud se ab ortu solis in VIII Kl Ianuarii usque <quo> oriatur in VIII Kl Aprelis. ... Incipit uero secunda pars apud se ab <ortu> solis in VIII Kl Aprelis usque <quo> oriatur in VIII Kl Iulii. ... Incipit autem tertia pars apud se ab ortu solis in VIII Kl Iulii usque quo oriatur in VIII Kl Octimbris. ... Incipit item quarta pars apud se ab ortu solis in VIII Kl Octimbris usque quo <oriatur in VIII Kl Ianuarii>.* and in QUAEST. AUSTR. 2.8B, which may well have been influenced by MC, particularly considering the explicit reference to the lengths of these intervals (for which cf. the previous note): *De partione anni in quattuor quadrantes nunc disserimus. Primus quadrans est ab VIII Kalendas Ianuarii usque in VIII Kalendas Aprilis, nonaginta dies habens et quadrantem. Secundus quadrans ab VIII Kalendas Aprilis usque in VIII Kalendas Iulii, nonaginta dies et unum habens. Tertius quadrans ab VIII Kalendas Iulii usque in VIII Kalendas Octubris, nonaginta et duos dies habens. Quartus quadrans ab VIII Kalendas Octubris usque in VIII Kalendas Ianuarii, nonaginta et duos dies habens.* For the 'Roman' dates of the solstices and equinoxes (but not the intervals between them) cf. also the following note and DRC 47.11–13, DT 7.2–3 (cited in RM 53.6–8), DIAL. NEUSTR. 29A.

17–22 Ab...XCII] Cf. Borst, *Schriften*, p. 451 and p. CIII, CLXXIII of the introduction. **17–27** Ab...reliqua] Cf. Borst, *Schriften*, p. 490. **26–30** Secundum...inuestigant] Cf. p. CCIII of the introduction.

The reasons for the difference between the 'Greek' and the 'Latin' division |

Sciendum tamen *qu*id *Greci* in hac difinitione distant ab hoc. <Id est> *quia Greci* uiciniores sunt ortui <solis> quam Latini, hinc *sol*arem cursum facilius inuestigant. Aut in honore factorum, quę in his operari, euenit: | Inter haec fol. 19r dierum loca Latini annum disputant: *In* VIII *Kalendas Ianuarii est* enim et *natiuitas Christ*i et augmentum diei; *in* VIII *Kalendas Aprilis* incarnatio *eius uirginis in utero* et *aequinoctium*; *in* VIII *Kalendas Iulii natiuitas Iohannis Baptistae*, eo quod plene in ea dies; *in* VIII *Kalendas Octimbris* autem *est conceptio Iohannis* Baptistae et *aequinoctium*.

---

29 Id est] *om. M (add. from M)*. | solis] *om. M (add. from M)*. **30** inuestigant] *follows* Secundum Latinos ab VIII Kalendas Ianuarii in VIII Kalendas Aprilis annus diuiditur. Sciendum ut quid Creci in hac difinitione distant. Id est quia Greci uiciniores sunt ortui solis quam Latini, hinc solarem cursum facilius inuestigant. *by mistaken repetition M (forming the basis for the previous two corrections).* **37** autem] hoc *M (it appears that the Irish abbr. for* autem *was mistaken for the abbr. for* hoc *by the continental copyist).*

31–38 Inter…aequinoctium] The question about the reasons for the difference between the 'Greek' and the 'Roman' tradition concerning the dates of the equinoxes and the solstices is first addressed in CE p. 98–9, which was the immediate source for MC: *Hic inquiritur, quare conpotant Greci ab XI Kalendas Ianuarii primum XC dies, Romani autem ab VIII Kalendas Ianuarii. Ideo Greci conputare incipiunt ab XI Kalendas Ianuarii, quia ab ea die sol incipit conscendendo lucem augere. Latini autem ab VIII Kalendas Ianuarii conputare incipiunt, ... et per eo quod sollempnitates domini et Iohannis IIII VIII Kalendas tales habuerint, hoc est natiuitas domini in VIII Kalendas Ianuarii, crux eius in VIII Kalendas Aprilis, natiuitas Iohannis Baptistę in VIII Kalendas Iulii, conceptio eius et adsumptio in VIII Kalendas Octobris, in quo differtur Greci et Romani in conputatione XC dierum.* Concerning some of the theological justifactions of the 'Roman' dates and their references to the equinoxes and solstices, it appears that MC additionally relied on PS-DIONYSIUS, *Argumentum XV* (Krusch, *Studien* II, p. 79–80; partially cited in EPIST. RAT. 10): *Equinoctium primum est in VIII Kl. Aprl. ... Eodem die Gabriel nuntiavit sancte Mariae, dicens: Spiritus sanctus supervenient in te et virtus altissimi obumbrabit te. Propterea quod nascetur ex te vocabitur filius dei. ... Solstitium secundum est VIII Kl. Iulii, quando etiam natus est sanctus Johannes Baptista. ... Equinoctium secundum est VIII Kl. Octb., in qua die conceptus est Johannes Baptista. ... natus est in III feria, VIII Kal. Ian. Christus dominus noster.* and MAXIMUS OF TOURS (?), *Sermones, Sermo 60* (*PL* 57, col. 651B, 653A): *Sancti praecursoris Domini Joannis Baptistae nativitatem, quam hodie celebrat sancta universalis Ecclesia, fratres carissimi, multis modis antea sancti prophetae Domini praenuntiaverant. ... In mense autem septimo aequinoctio autumnali conceptio est S. Joannis in utero matris suae. In mense autem primo alterius anni in aequinoctio vernali conceptio est Domini Salvatoris in utero Virginis matris.* Cf. also PROLOGUS CYRILLI, *Addendum* (Krusch, *Studien* I, p. 342–3): *Nono kalendarum aprilium unum aequinoctium, et VIII kalendarum iuliarum in natale sancti Johannis fit unum solestitium et VIII kalendarum ianuarium in natale domini nostri Jesu Christi fit aliud solestium.* Note that MC here, as well as in c. 24, ll. 2–10, connects the theory of the linear increase of daylight with the 'Roman' rather than the 'Greek' dates as in c. 39. For these 'Latin' dates cf. also c. 38, ll. 26–27; c. 47; for dates referring to Christ's life cf. especially c. 24, ll. 2–10.

28–38 Sciendum…aequinoctium] The question about the reasons for the difference between the 'Greek' and the 'Roman' dates of the solstices and equinoxes is, besides CE and MC, explicitly addressed only in DIAL. LANGOB. 18, which suggests that this question was well debated in the late seventh-, early eighth-century Irish computistical mileu; even though DIAL. LANGOB. gives an explanation different from MC's for the 'Greek' dates, connecting the 'Greek' date of the vernal equinox with the first day of

It has to be known, however, (for) what (reasons) the Greeks differ in the definition (of the division of the year) from the just mentioned (Latin one). Because the Greeks are closer to sunrise than the Latins, (and) so they can more easily investigate the course of the sun. Or (this difference) occurred in honour of events that happened on these days, (since) the Latins discuss the year on the basis of these dates (for the following reasons): 25 December marks the birth of Christ and the increase of daylight; 25 March (marks) his incarnation in the womb of the virgin and the equinox; 24 June (marks) the birth of John the Baptist, because (daylight) is full on that date; 24 September marks the conception of John the Baptist and the equinox.

---

the creation of the world rather than with the course of sun, it still appears that this account is based on MC: *Item querendum est, cur Graeci et Latini dissentiunt in solstitiis et aequinoctiis? Graeci enim in XI Kalendas faciunt, Latini vero aequinoctia et solstitia anni per VIII Kalendas faciunt. ... Interrogatio: Cur ergo Graeci sic faciunt, et cur Latini aliter? Responsio: Haec est ratio. Graecus naturam sequitur, quia in XI Kalendas Aprilis factus est mundus, qui est primus dies saeculi secundum solem. Ideo ab XI volunt ordinari rationem temporis per aequinoctium et solstitium. Latinus vero secutus est auctoritatem et principatum, quia in VIII Kalendas Ianuarii lux renovatur, quasi initium lucis sic dicitur. Ab hodernio die nox minuitur, dies crescit, concutiuntur tenebrae, lumen augetur. In eodem die natus est Christus, filius Dei, in carne. In VIII Kalendas Aprilis Christus conceptus est et Christus passus est, et in illa die facta sunt sol et luna quarta die in principio, in figura Christi <et> ecclesiae. VIII Kalendas Iulii natus est Iohannes, praecursor Christi; in VIII Kalendas Octimbris conceptus est Iohannes, praecursor Domini, et passus est eodem die, sicut et Christus eodem die conceptus est et passus fuit. Et quia hi dies quattuor consecrati sunt per Christi et praecursorem suum, ideoque primatum tenent apud Latino et Christianos*. Bede, in a lengthy chapter about the equinoxes and solstices, additionally argues that the 'Greek' dates are confirmed by observations of the horologue (DTR 30.74–77; cf. c. 38, ll. 9–12). The theological justification for the 'Roman' dates can also be found in BC 29 (cited in BC 45B1–C5, in turn cited in LIB. COMP. 1.3b): *Solestitia duo sunt. Primum hiemnale VIII Kalendas Ianuarii ... est ipsa die in Bethleem nativitas domini nostri Iesu Christi secundum carnem. Alterum aestivale VIII Kalendas Iulii ... est ipsa die in provintia Palestina civitate natale sancti Johannis Baptiste ... . His econtrario duo equinoctia sunt, unum uernale secundum Latinos VIII Kalendas Aprilis ... est conceptio sanctae Dei genetricis Mariae, quando salutata est ab angelo. Alterum authumnale VIII Kalendas Octobris ... fuit in Macheronta castello conceptio sancti Iohannis Baptistae*. DTR 30.2–4 (cited in LIB. CALC. 82A, PV §310), 40–44 (cited in LIB. CALC. 82B): *De aequinoctiis, quod octavo kl. aprilium et octavo kl. octobrium, et de solstitiis, quod octavo kl. iuliarum et de octavo kl. ianuarium die sint notanda. ... VIII kl. apriles in aequinoctio verno dominum conceptum et passum, eundem in solstitio brumali VIII kl. ianuaris natum; item beatum praecursorem et baptistam domini VIII kl. octobres in aequinoctio autumnali conceptum, et in aestivo solstitio VIII kl. iulias natum*. The validity of these dates was still lively debated in the early tenth century: cf. HELPERIC, *De computo* 30 (*PL* 137, col. 40C): *De ipso autem aequinoctio varia non tantum veterum, verum etiam modernorum exstat opinio, aliis XII Kal. April., aliis VII (recte VIII) Kal. asserentibus*.

The relation of equinoxes and solstices to the seasons

Et hii sunt IIII principales dies in anno: duo aequinoctia et duo solstitia, in medio temporum sita secundum naturam, quamuis sint plures dies antecedentes his iuxta mensium concursum, quia prius menses inuenti sunt ordines, ita et postea solstitia et aequinoctia.

---

42 his] se *M.*

39–43 Et…aequinoctia] MC took the information that the solstices and equinoxes were supposed to fall right in the middle of the respective seasons from DRP 14 (cited in DRC 48.2–9, DTR 35.48–55, in turn cited in RM 34.48–57): *Hoc autem non ignores, quod ista IIII quae praediximus temporum confinia, licet mensium sequentium Kalendis approximatur, unumquodque tamen medium temporis, id est uerni et aestatis, autumni et hiemis, teneat. Et non exinde temporum principia inchoantur unde mensium Kalendae initiantur, sed ita unumquodque tempus inchoandum est ut a prima die ueris tempus aequinoctium diuidat et aestatis VIII Kal. Iulii et autumni VIII Kal. Octobris et hiemis VIII Kal. Ianuarii similiter diuidat.* and MARTIN OF BRAGA, *De Pascha* 5 (Barlow, *Opera omnia*, p. 273; cited in TRACTATUS ADTHANASI 4 (Krusch, *Studien* I, p. 332)): *Sed cum tres menses vernum tempus habeat, horum trium medius est qui initium mundo dedit, nec solum mensis medius, sed etiam dies mensuum medii. Ex V enim Id. Febr. (veris est inchoatio) in V Id. Mart. unus est mensis. Ex V Id. autem Mart. in VIII Kal. Apr. quindecim dies sunt, id est, medietas mensis. Ita unus et dimidius mensis ex hoc antecedit et unus et dimidius mensis subsequitur*. In Irish tradition, however, the seasons began on 1 February, 1 May, 1 August, and 1 November: cf. CE cited in c. 29, ll. 17–24, and especially Columbanus, *Regula monachorum* 7 (Walker, *Sancti Columbani opera*, 130): *initium hiemis, id est Kalendas Novembris*. That MC also followed this rule is obvious from c. 45. Accordingly, the solstices and equinoxes occurred roughly after ⅔ of a season had passed according to Irish tradition, and this is what MC means by the final statement of this passage here that there are more days of a season before than after these dates.

39–43 Et…aequinoctia] The question whether or not the equinoxes and solstices mark the middle of the seasons appears to have only been addressed in the Irish computistical milieu of the early eighth century. The first option, i.e. the 'Roman' dates of the solstices and equinoxes marking the middle of the seasons, is based on Anatolius (?) and especially on Martin of Braga (cf. *app. font.*), with the seasons beginning on *V Idus* of February, May, August, and November respectively (this placement of the seasons within the Julian calendar is outlined in detail in BC 32 and other MSS containing this tract listed by Walsh & Ó Cróinín, *Cummian's letter*, p. 158, as well as a gloss to DTR (*CCSL* 123B, p. 394)). Irish computists, then, had their doubts about this theory, since, in their opinion, the seasons started on 1 February, 1 May, 1 August, and 1 November respectively, as suggested in Columbanus' writings and explicitly outlined in BC 105 (cf. also c. 45); according to this placement of the seasons within the Julian calendar, the equinoxes and solstices do not fall in the middle of the months. MC was aware of both options and treated them equally, terming the first *secundum naturam*, the second *iuxta mensium concursum*. The most detailed account of this question, which is strongly influenced by MC, is given in DRC. After citing Anatolius (?) in DRC 48 (cf. *app. font.*), DRC 49.1–6 continues: *Sciendum nobis quomodo hoc fieri potest, dum praecedentes dies haec confinia quam sequentes plures esse uidimus. Uerbi gratia, solstitium hiemale in VIII Kl Ianuarii, et LIII dies praecedunt, XXX uero VIIII sequuntur. Haec est causa huius rei, quod gentiles Kalendas mensium prima lunae nascentis ostensione possuerunt, et non secundum naturam. Racionabilius enim erat, ut solstitium quod in VIII Kl Ianuarii fit tempus hiemis equaliter diuideret*. DIAL. LANGOB. 8 only discusses the beginnings of the seasons, but in this relied on the just cited passage from DRC. Similarly, Bede touches the question of the placement of the solstices and equinoxes within the seasons only accidently by citing Anatolius (?) when discussing the various traditions concerning the beginnings of the seasons (cf. *app. font.*).

These are the four principal days within a year: the two equinoxes and the two solstices, which are situated in the middle of the seasons according to nature, though there are more days (of a season) preceding these (dates) according to the alignment of the months, because the sequence of the (Julian calendar) months was invented earlier, and accordingly the equinoxes and solstices (were implemented into this structure) later.

<XXXVIIII. DE AUGMENTO DIEI ET NOCTIS>

The increase of daylight in the first quarter of a year

Hinc animus fuit sapientibus ut mutarentur menses. Quattuor igitur XC in XXX duodecies diuidunt, et unumquodque XXX in XV duo diminutius partiri eloborant. Itaque *ab XI Kalendas Ianuarii in XII Kalendas Aprilis per XV dies* hora *et dimidium horae sole ascendente per singula momenta, id est per IIII* momenta *in una*quaque *die.* VI horę pręsentiae solis adunantur, et III in nube noscuntur. Duo enim *momenta et tertia par*s *momenti* in augmentum diei naturaliter addicunt. *Momentum* uero *et*

XXXVIIII,3 duodecies] dudecies *M*. **8** adunantur] adunatur *M*.

XXXVIIII,2–17 Hinc…Finit] This chapter deals with the theory of the increase and decrease of daylight throughout a year: It is generally argued that the summer solstice consists of 18 hours of daylight and 6 hours of darkness, a relation that is reversed at the time of the winter solstice, while the equinoxes have daytime and nighttime of equal length. Accordingly, daylight increases by six hours from the winter solstice to the spring equinox as well as from the spring equinox to the summer solstice, while it decreases by the same ratio from the summer solstice to the autumnal equinox as well as from the autumnal equinox to the winter solstice. For the sake of easy calculation, the quarter-years were reckoned as consisting of 90 days, so that the increase and decrease of six hours of daylight per quarter-year was distributed equally among these 90 days, leading to an increase or decrease of 2⅔ *momenta* per day. Ultimately, this theory comes from Anatolius (?)'s DRP 13; MC, however, relies more directly on the very detailed description of this theory in CE. Concerning the first quarter of a year, from the winter solstice to the vernal equinox, Anatolius (?) mentioned a daily increase of 4 rather than 2⅔ *momenta* (note that Anatolius (?) refers to the 'Roman' rather than the 'Greek' dates of the equinoxes and solstices): DRP 13.214–216 (cited in DRC 50.19–21, cf. *app. comp.*; BC 33D10–12 with variations, most notably the dates *XI Kalendas Ianuarii* and *XI Kalendas Aprilis*): *Per XV dies et horae dimidium sole ascendente per singula momenta, id est per IIII in una die, et ab VIII Kal. Ianuarii in VIII Kal. Aprilis hora deminuitur.* Irish computists of the late seventh and early eighth centuries understood these 4 *momenta* to be the sum of the increase of daylight of 2⅔ *momenta* per day and a bissextile increment of 1⅓ *momenta* per day, the latter adding up to 3 hours in the course of these 90 days (1⅓×90=120 *momenta*=3 hours). This, however, left these computists with the problem of including this bissextile increment in the general theory; CE set the standard by dividing the 12 hours of nighttime at the vernal equinox into 9 hours of darkness and 3 hours of 'pale light', to be understood as 1½ hours each for dusk and dawn (cf. also the following note). In MC's terminology, these three hours constituted a 'nocturnal cloud' within the 12 nocturnal hours. Besides Anatolius (?)'s brief statement cited above, MC's immediate source here is CE p. 99, which outlines the calculations underlying MC's account in detail: *Hic interrogandum est, quomodo crescit solis lumen ab XI Kalendas Ianuarii usque in XII Kalendas Aprilis. Hoc est per II momenta et tertia partem bis momenti. Cuius probatio hęc est: II momenta in unoquoque die crescentia per XV dies XXX momenta faciunt. Tertia autem pars bis momenti coniunctis ternis singulis diebus per XV dies X facit momenta. Additis his X ad XXX XL faciunt momenta. Hinc VI horę ab XI Kalendas Ianuarii usque in XII Kalendas Aprilis crescunt, duę enim horę per menses singulos augentur. Est quoddam augmentum, quod quadrantem efficit, id est momentum uniuscuiusque diei et tertia pars momenti ante ortum solis et post occasum. Cuius probatio hęc est: Per XV dies, hoc est momentum unius cuiusque diei, XV momenta sunt. Tertia pars momenti per XV dies V facit momenta. Coniunctis ternis singulis diebus inde horę III per III menses efficiuntur ab XI Kalendas Ianuarii usque in XII Kalendas Aprilis, quę lucis pallide sunt. Inde hęc dies horas XV horas habet, nox uero VIIII horas.*

XXXVIIII,2–17 Hinc…Finit] The theory of the increase and decrease of daylight first received considerable attention in the seventh- and early eighth-century Irish context. The terminology applied in DIAL. LANGOB. 19C reveals that it directly relied on MC, though the 'Roman' dates for the equinoxes

39. ABOUT THE INCREASE OF DAYTIME AND NIGHTTIME

Hence, there was a feeling among the wise men that the months should be arranged differently (after the introduction of the solstices and equinoxes). So they divided the four 90(-day intervals of a year) into 12 times 30, and they further decreed that each of (these) 30 be divided further into two (parts of) 15 (days). Accordingly, from 22 December to 21 March, an hour and a half (emerges) through the course of 15 days, as the sun ascends through single *momenta*, i.e., by four *momenta* in every single day. (At the end of this period, i.e. on 21 March,) six hours belonging to the presence of the sun are summed up, and three hours are known to be in a cloud. In fact, two *momenta* and (two) thirds of a *momentum*

---

and solstices originally employed by Anatolius (?) are here prefered to the 'Greek' ones: *In VIII Kalendas Ianuarii, tunc primitus dies accipit augmentum lucis, et per quattuor momenta omni die augetur dies usque ad aequinoctium vernale, hoc est VIII Kalendas Aprilis. Et sic hoc ordine crescit dies, id est per duo momenta et duas tertias momenti in praesentia solis, et unum momentum et unam tertiam partem momenti in aurora ante solem et post solem. Et sic in praesentia solis per augmenta cottidiana sex horas accipit dies in primo nonaginta super id, quod ante habebat, et ita duodecim horae fiunt in die in aequinoctio, in VIII Kalendarum Aprilis, quod est aequinoctium vernale.* The 'Roman' dates are also favoured in DRC 50.13–21, which gives the most detailled explanation for the bissextile increment: *Et in primo nonaginta crescentis lucis frangit quadrantem preparationis bissexti ideo, quia in primo nonaginta, id est in sexto Kl Martii in Februario, Iulius Cesar diem qui de eo efficitur possuit. Et unusquisque dies in nonaginta momentum et terciam partem momenti de quadrantis fractione habet, id est dimedium horae in XV diebus. Et quadras trium horarum hic frangiur et hora de tenebris ad praesentiam lucis a XV diebus et dimedio horae quadrantis crescit, Anatolio dicente: Per XV dies et hore dimedium, sole ascendente per singula momenta, id est per quatuor momenta in unaquaque die, ab VIII Kl Ianuarii usque in VIII Kl Aprelis hora diminuitur.* This theory is not touched by Bede and its wide popularity in subsequent centuries is clear proof for the fact that early medieval computistics cannot solely be understood by reference to Bede's works, but different, most notably Irish traditions have also to be taken into account. The late Merovingian DIAL. NEUSTR. 28A, 28aA appears to be directly influenced by CE in this instance: *Ab XII Kalendas Ianuarii usque XII Kalendas Iulii sol duobus momentis et duabus partibus momenti ab ortu solis usque ad occasum adcrescit. Id est: Per quindecim dies hora facitur. Hoc est: Sex horae crescunt ab XI Kalendas Ianuarii usque in XII Kalendas Aprilis. ... Ad materiam autem bissexti: De XI Kalendas Ianuarii usque XII Kalendas Aprilis momentum et tertia pars momenti adcrescunt.* This theory can also be found in numerous early and high medieval MSS and it will only be possible to adequately analyze the reception of this theory when all tracts still hidden in these MSS will have been edited. Usually, only the results, not the calculations are recorded in these later tracts: For this first quarter, only the increase of daylight between the 'Roman' dates in BC 30B9–12; only the increase of daylight between the 'Greek' dates in BC 31C15–A3, 114B4–C10; only the increase of daylight without specification of the dates in DDT 12D8–18 (cited in LECT. COMP. 7.1B), LIB. COMP. 3.14; the increase and decrease of daylight without specification of any period in LIB. ANN. 54; only the bissextile increment between the 'Latin' dates COMP. COL. 3.2C, 4.6 (apparently related to DIAL. LANGOB.): *Item sic scies, quod alii dicunt, Anatolium in illa parte anni, quae in caeteris brevior esse videtur, calculandi conpendio supputasse quadrantem, eumque per nonaginta crescentem momento in unoquoque die et tertia parte momenti. ... Aliter: <Sol> simpliciter stabit prima aurore initio sex horarum, aurore novissime in fine sex horarum, quod non difficile ostendit res. Et labuntur cum nonaginta in preteritum, et remanent semper tres horae invicem quadrantis, ut dixi, usquequo in quarto anno integer dies efficiatur. ... Huius rei nobis Anatolius dedit exemplum, qui ab VIII Kalendas Ianuarii usque in VII Kalendas Aprilis quadrantem crescere dixit.*

XXXVIIII,2–17 Hinc…Finit] Cf. Borst, *Schriften*, p. 901. **2–55** Hinc…innotescerit] Cf. p. LXVI–LXVII, CIII, CXLV–CXLVI, CLXI of the introduction. **2–XL,20** Hinc…prȩponas] Cf. Borst, *Schriften*, p. 413.

tertia pars momenti ad preparationem bissexti, ut alii dicunt. Et *cresc*ente *per XV sex*ies | *hor*a solis in presentiam, addita *hor*a *per* XXX dies ter, *XV hora*e *in XII Kalendas Aprilis* cum luce sunt, et *VIIII* in *noct*e remanent. Et ideo *III hore* quasi in nube diximus. Equiperari noctis horas diurnas prestant. Hoc ordine aequinoctium disputant. Finit. fol. 19v

21 March

De XII Kalendas Aprilis incipit. Cum quo *VI horae* in *occasum solis* innotescunt, ab *occasu* igitur usque ad *ortum solis XII hor*e sunt in nocte, et *ab ortu usque ad* horam sextam diei *VI hore* sunt. Hinc parilis horarum diei et noctis numerus in XII Kalendas Aprilis ita manet.

The increase of daylight in the second quarter of a year

Hinc inde *ab XI Kalendas Aprilis in XII Kalendas Iulii duo momenta et ter*tia *par*s *momenti*, interueniente *nub*e, quae in hac parte *cresc*ere desinit, in XII Kalendas Aprilis esse expleta desinit. Et *in XII Kalendas Iulii XVIII hor*arum magnitudinem *in XII* horis efficiunt cum sole, VI uero in nocte. Sed *III* in *tenebr*is, *III* uero in *nub*e periti esse adfirmant.

---

13 horae] horas *M*. **15** quasi] quas *M*. **16** diurnas] diurnis *M*. **17–18** Finit…quo] *In red ink like an explicit and incipit in M*. **26** XVIII] VIIII *M (corr. from* DRP, CE*)*. **27** magnitudinem] efficiunt *add. M (copying mistake from the* efficiunt *that occurs three words later)*.

18–22 De…manet] This discussion of the length of day and night on the vernal equinox is principally based on CE p. 99–100, but note that, contrary to MC here, CE divides the 12 hours of nighttime into 9 hours of night and 1½ hours of dusk and dawn respectively (cf. also the previous note): *Sed per cunctari debet: Quomodo nominatur hęc dies? id est aequinoctium uernale. Cuius mensuratio? Hęc est a sexta hora usque ad sextam horam, id est VI horę a meridie usque ad occasum solis, hora et semis post occasum, VIIII horę tenebrarum, hora et semis ante ortum solis, VI horae ab ortu usque ad meridiem. Coniunge VI horas ante occasum et VI horas post ortum solis, XII horas fecisti. VIIII uero horas tenebrarum et horam et dimi<di>am post occasum et horam et dimidiam ante ortum coniunge, que XII sunt. Hinc equinoctium dicitur haec dies*. **23–29** Hinc…adfirmant] The increase of daylight in the period between the vernal equinox and the summer solstice, as well as the number of hours of daytime and nighttime at the summer solstice, are ultimately based on DRP 13.227–229 (cited in DRC 50.28–30), 234–238: *Et ab VIII Kal. Aprilis usque in VIII Kal. Iulii per XV dies et II horas hora deminuitur, per duo momenta et dimidium et VI partem momenti per singulos dies sol ascendente. ... Eousque namque dies et nox deminuitur, ut XII horae quae in uernali aequinoctio in principio domini dispensatione constituae sunt, in VIII Kal. Iulii nocte deminuta, sole ascendente per singulos quos supra diximus gradus, XVIII in XII longiore spatio reperiantur adiunctae*. MC relied on Anatolius (?) particularly for the idea that the 18 hours of daylight are supposed to be considered within the frame of 12 hours of daytime; MC's principal source, however, was CE p. 100, 101 here, particularly for the information that the 'cloud' constituting the bissextile increment would not increase further in this quarter: *Item dies crescit duobus momentis et ternis bis partibus momenti ab XI Kalendas Aprilis usque in XII Kalendas Iulii. Nubs autem pallida non crescit, saltim mobili cursu in eo tempore fit. ... Hinc accedit, ut XVIII horę in XII Kalendas Iulii die repperiri uideantur, nocte uero III horę tenebrarum et III horę nubis fiant.*

contribute naturally to the growth of daylight. One *momentum* and a third part of a *momentum*, however, (are assigned) to the preparation of the bissextile day, as some say. On 21 March, 15 hours belong to the light and nine remain with the night, since one hour of daylight grows in the course of 15 days, (which happens) six times (in this 90-day interval), (and) one hour per 30 days is added to this, (which happens) three times (in this interval). And for that reason we have argued that three hours (appear) as if in a cloud. (These three hours) ensure that the hours of the night are equal to the ones belonging to the day. In this way they discuss the equinox. End of this discussion.

Here begins (the discussion) about 21 March. As six hours are known (to have passed) at sunset on this (day), there are twelve hours in the night from sunset to sunrise, followed by six hours from sunrise to midday. Consequently, the number of hours of day and night thus remain equal on 21 March.

Then, from 22 March to 20 June, two *momenta* and (two) thirds of a *momentum* (constitute the increase of daylight per day), while the cloud intervenes, which ceases to grow in this period, since it ceases to increase on 21 March. And (this linear increase) leads to a length of 18 hours of daylight in (a space of) twelve hours on 20 June; six hours, however, (remain) in the night. But the learned affirm that (only) three (of these nocturnal hours) are in the shadows, the (remaining) three, however, in the cloud.

18–22 De…manet] The characteristic reckoning of a day (and the vernal equinox in particular) from midday to midday as outlined in CE and MC can also be found in COMP. COL. 3.4A, which appears to be primarily basd on CE: *Ideo dicitur aequinoctium XII Kalendas, quia duodecim horae in die, duodecim in nocte <sunt>, sex horae a medio diei heri usque ad vesperum et dimidium quadrantis horae et dimidium horae, sex horae hodie ab ortu solis usque ad sextam horam diei et dimidium quadrantis horae et dimidium horae. Nox: In media noctis hora et dimidium quadrantis heri et dimidium hodie iuguntur, et efficiunt tres horas, et iunguntur nocti. Et fiunt duodecim horae in die et duodecim horae in nocte.* **23–29** Hinc…adfirmant] The Frankish DIAL. NEUSTR. 28B, 28aA is directly influenced by the two Irish computistical textbooks in its discussion of the increase of daylight in the second quarter of a year: *Et ab XI Kalendas Aprilis usque ad XII Kalendas Iulii altere sex horae crescunt. ... Ab XI Kalendas Aprilis usque XII Kalendas Iulii materia bissexti non crescit.* The increase of daylight between these 'Greek' dates is also recorded in BC 31C15–A14, 114C10–D10; the increase of daylight between the 'Latin' dates in DIAL. LANGOB. 19C; only the increase of daylight without specification of the dates in DDT 12A1–5 (cited in LECT. COMP. 7.1C), LIB. CALC. 3.14. Concerning the summer solstice, the 18 hours of daylight and 6 nocturnal hours are given for the 'Greek' date in BC 114D10–A1, for the 'Latin' date in DIAL. LANGOB. 19C, for the summer solstice without specification of a date in DDT 12A11–12 (cited in LECT. COMP. 7.1C); cf. also LIB. COMP. 3.14B.

The decrease of daylight in the third quarter of a year

*Ab XI Kalendas Iulii in XII Kalendas Octimbris duo momen*ta *et tertia par*s *momenti* confinibus lucis *noc*tem uicissim supplent et *nocturna*m *nub*em, quam alii hic crescere ęstimant. Et subtilius ceteri disputantes certa obiectione uetant. Et dierum quadrantem in duo XC | fol. 20r integre egredi noctemque illo adlabi puplice uetant, in quem II momenta et duę tertiae partes momenti, quae noctem augent, labi noscunt; cuius nubes consumptis per XV <sexies> III horis; illud, quod in XI Kalendas Iulii presentiam solis minuit, noctem addit. Hinc dies et nox in XII Kalendas Octimbris auctis per ipsum spatium VI horis, haeque solis in absentia in XII Kalendas Iulii fiebant, pari numero consistunt.

The decrease of daylight in the fourth quarter of a year

Ab XI *Kalendas Oct*imbris *in* XII *Kalendas Ianuarii* idem *numer*us *momentorum* de luce *minu*tus in nubem incedit. *Quae* qualis mensura tenebras auget? Et *moment*um *et terti*a *par*s *momenti*. *Nub*em semetipsum, quae in primo XC cum luce creuit, demere innotescit. *Quę* in hoc spatio ab XI Kalendas Octimbris in XII Kalendas Ianuarii minuta in tenebris apparere in XII Kalendas Ianuarii ratio non permittat. Hinc *XVIII* horarum magnitudo *in XII* horis *in nocte* XII *Kalendas Ianuarii* esse probatur, et *VI in XII diuisę* in pręsentia solis morari feruntur, quae III essent tantum, si noctualis quadrans, quem respuimus, creuisset, et diurnalis in confinibus lucis resedisse innotescerit.

35 adlabi] adiubi *M.* **37** labi] laui *M.* **38** sexies] *om. M (add. according to* MC*).* **41** haeque] hisque *M.* **44** numerus] numeris *M.* **52** feruntur] fertur *M.*

30–42 Ab…consistunt] Again the ultimate source for the decrease of daylight in the third quater of a year was DRP 13.229–231 (cited in DRC 50.37–39): *et ab VIII Kal. Iulii in VIII Kal. Octobris similiter per XV dies, et per IIII horas hora deminuitur descendente sole in unaquaque die per eundem numerum momentorum*. MC's immediate source for this passage is CE p. 100, 101: *Ab XI Kalendas Iulii usque in XII Kalendas Octimbris nox crescit, id est duobis momentis et ternis bis partibus momenti. Nubs autem nocturna momento uno et tertia parte crescit. Inde XV horę sunt cum III horas nubis in nocte XII Kalendas Octimbris, in die autem IX horę sunt. Hinc queritur quomodo hęc dies equinoctium est. ... Hinc XII horę diei et XII noctis sunt, quę ęquinoctium conplent. ... Item nox crescit ab XI Kalendas Iulii usque in XII Kalendas Octimbris II momentis et tertia parte bis momenti ad nocte pertinentibus, momento autem et tertia parte momenti ad nubem*. Note that MC exlicitly rejects CE's statement that a bissextile increment of 1⅓ *momenta* per day is to be reckoned in this quarter adding up to a 'cloud' of three hours; it appears from MC that CE's account of the increase and decrease of daylight was well debated in Irish computistical circles of the early eighth century, and that MC than recorded the results of these debates. **43–55** Ab…innotescerit] In this passage on the decrease of daylight in the fourth quarter and the length of day and night on the winter solstice, it is particularly obvious that MC had consulted the ultimate source for this question, DRP 13.231–233 (cited in DRC 50.45–47), 238–240: *Et spatium quod superest ab VIII Kal. Octobris usque in VIII Kal. Ianuarii simili numero dierum horarum ac momentorum finitur. ... Et iterum XII horae, quae in autumnali aequinoctio solis descensu repleantur, in VIII Kal. Ianuarii VI horae in XII diuisae inueniuntur disiunctae, nocte XVIII in XII diuisas*

On the other hand, from 21 June to 20 September two *momenta* and (two) thirds of a *momentum* from the light of dawn replenish the night as well as the nocturnal cloud, which some hold to grow in this period. The rest, who reckon more accurately, oppose this view with a certain objection. They publicly object (to the idea) that a quarter-day becomes complete in two (intervals of) 90 (days) and that (such an additional increment) falls to the night in this (period), in which two *momenta* and two thirds of a *momentum* are known to pass as the nocturnal increment (per day); the cloud of this (quarter-day had already been established) by three hours that had elapsed through six times 15 days; that (part), which diminishes daylight from 21 June onwards, adds (only) to the night (and not to the cloud also). Therefore, day and night consist of an equal number (of hours) on 20 September, after six hours had grown during this period, and these emerged in the absence of the sun from 20 June onwards.

From 21 September to 21 December the same number of *momenta*, which is reduced from daylight, enters into the cloud. What kind of measure (is) this, (which) increases the shadows? A *momentum* and a third part of a *momentum*. It is known that the cloud, which grew with daylight in the first (interval of) 90 (days), subtracts from itself the same number. The reckoning does not allow this, which is reduced (from daylight in the period) from 21 September to 21 December, to appear in the shadows on 21 December. Thus the length of 18 hours is proven to be in (a space of) twelve hours in the night on 21 December, and six hours distributed in (a space of) twelve (hours) are said to remain in the presence of the sun, which would only be three, if the nocturnal quarter-day had grown, which we

tenente. All additional information, and especially the fact (here rejected) that the 'cloud' is reduced in this quarter, derives from CE p. 102: *Ab XI Kalendas Octimbris usque in XII Kalendas Ianuarii nox crescit II momentis et duabus partibus momenti, id est ternis duabus. Quę sex horas adiciunt super IX, quę sunt in nocte XII Kalendas Octimbris. Quę omnes XV horas faciunt. III autem horę derunt (recte desunt), quas conpleri facies ex tribus horis nubis, quę in XII Kalendas Octimbris sunt. Quę nubs cotidie momento et III parte momenti minuitur et ad noctem augetur et in XII Kalendas finitur et ab XI Kalendas Ianuarii coepit incrementum. Hinc XVIII horę sunt nocte in XII Kalendas Ianuarii, die autem sex sunt, quę nube non habet.*

30–42 Ab...consistunt] Since DIAL. NEUSTR. 28B, 28aA records a bissextile increment for this third quarter of a year, it is obvious that its account in based on CE: *XII Kalendas Iulii ... Et ab illo loco sic sol decrescit. ... Et ab XI Kalendas Iulii usque XII Kalendas Octobris materia bissexti in nocte crescit.* Only the decrease of daylight between the 'Greek' dates is mentioned in BC 31A15–B15, 114A1–15; only the decrease of daylight from the 'Latin' date in BC 30B12–C1, DIAL. LANGOB. 19C; only the decrease of daylight without specification of the dates in DDT 12C13–D3, LECT. COMP. 7.1a, LIB. COMP. 3.14. **43–55** Ab...innotescerit] Interestingly enough, DIAL. NEUSTR. 28aA argues that no bissextile increase is supposed to be reckoned in this quarter of a year: *Ab XI autem Kalendas Octobris usque ad XII Kalendas Ianuarii bissextus non crescit.* Only the decrease of daylight and the length of day and night on the 'Latin' date of the winter solstice is mentioned in DIAL. LANGOB. 19A,C: *In solstitio hyemale nox habet horas octodecim, in die vero sex horas tantum . ... Et ab illa die nox crescit, dies minuitur per duo nonaginta usque ad VIII Kalendas Ianuaris, id est solstitium hyemale. Et sicut dies subtrahebat de nocte, sic et nox eadem regula et eadem mensura subtrahit tempora de die, usquedum fiunt in nocte decem <et> octo horae et sex in die tantum.* Only the decrease of daylight between the 'Greek' dates in BC 31A15–C11, also the length of day and night on the winter solstice BC 114A1–B3.

<XL. DE V DIEBUS SUPERFLUIS>

The problem of the five remaining days

Sunt *V dies*, qui in tribus *nonag*intis *superflu*i manent. Quos partiri in XCo, in quo sunt, | facimus, ne quidam fol. 20v disputandi non subtilis quidem quo tempore non indiget. Et *moment*is *XVI* in his super annum crescentibus, quolibet tempore utrum sudarent. Quę peractis *LX annis XVI momenta* in unoquoque *anno* enumerata *XXIIII hora*s efficerent. Quod quis sollicitus in <horas> XVI momentis *per V annos* diuisis facile cognoscet.

XL,1 De…superfluis] *cf.* DIAL. NEUSTR. **4** tempore] tempus *M.* **6** Quę] que- *M.* | LX] XL *M (corr. according to* CE*).* **7** momenta] momentum *M.* | horas] horarum *M.* **8** sollicitus] solcitus *M.* | horas] *om. M.* | momentis] momentum *M.*

XL,2–9 Sunt…cognoscet] The previous theory of the linear increase and decrease of daylight is based on a division of the year into four intervals of equal length, namely 90 days; however, the second quarter, between the vernal equinox and the summer solstice, constists of 91 days, the third and the fourth of 92 days. This chapter here, then, deals with the question of how these five remaining days are to be treated. It is argued that they are to be divided by and distributed equally among the 90 days of the quarter in which they occur; the division itself is discussed in the following paragraph. The logic of this division seems to be that once these five days are splitted up into 90 parts, they were not to be considered as full days anymore, and therefore would have had no impact on the theory of increase and decrease of daylight. The present paragraph, then, tries to explain the problem that would occur if these five days would be included in that theory. Generally, this whole chapter is difficult to understand, and a reconstruction, at least in sense and the mathematics involved, is only possible through the more detailed and more intelligible account of MC's direct source, CE. In this paragraph here, MC worked from CE p. 103: *Hinc interrogatur utrum aliquod in conputatione nonagesima superfluum obliti sumus. V utique dies non conputauimus per XC dies quadrifarie diuisas, quas interkalares vel superfluas alii dicunt. Sed saltim superflueę dicuntur, naturales tamen sunt. Igitur hos V dies comminui intra XC quater dies debet hac de causa, ne quidam calculatores postea aduenientes audacia elati bissexti diem horum V dierum incremento conpotarent, id est postquam praescierunt quosdam bissextum naturaliter posuisse IIII annis peractis quinto incipiente anno, hi de nouo gloriantes diem bissexti de his V diebus fecissent. Sed quaeritur: quantum sit tempus, quo bissextus esset super his V diebus, et quomodo incrementum eius esset? Hoc ita soluendum est: Dies, qui est primo XCmo crescit II minutis (recte momentis) et ternis bine partibus momenti. IIdus uero dies in II XCmo II momentis et ternis bis partibus momentis utrique crescunt. Haec coniungendo momenta VIII fecisti. In tertio XCmo II dies superflui II minutis crescunt, hoc est IIII momentis utrique crescit, quę VIII faciunt momenta. Hęc iungendo XVI fecisti momenta. Hinc LX annis hoc incremento bissextum perficies hoc modo: Annos V copula quibus momenta XVI per singulos annos crescunt; quinquies X et quinquies V et unum quinquies. Haec autem iugens L et XXX faciunt, quibus inpleuisti XL bis momentorum, hoc est duas horas. Adde V alios annos, quibus duas horas perficies, quascum coniunxeris IIII horas inplebis per annos X. Hinc ab annis XX VIII horę sunt, et ab XX annis VIII similiter sunt horae. Hinc diem horarum XXIIII per annos LX fecisti.* One of the five 'superfluous' days occurs in the second quarter (from the vernal equinox to the summer solstice), and two each in the third and fourth quarters. The ratio of the increase or decrease of daylight is by $2\frac{2}{3}$ *momenta* per day. CE applies this ratio to three of the five 'superfluous' day, namely to the one day from the second quarter of a year, and to the two days of either the third or the fourth quarter. This leads to a sum of $3\times2\frac{2}{3}=8$ *momenta* for these three days. Concerning the remaining two 'superfluous' days, CE does not only assign the ratio of the increase and decrease of daylight of $2\frac{2}{3}$ *momenta* to these, but additionally an increase of $1\frac{1}{3}$ *momenta*, adding up to a ratio of 4 *momenta* per day. Accordingly,

reject, and the diurnal (quarter-day) would have been known to have remained in the light of dawn.

40. ABOUT THE FIVE SUPERFLUOUS DAYS

There are five days, which remain superfluous in three (of these four periods of) 90 (days). We carry out the division of these in (that quarter of) 90 (days) in which they occur, so that any knowledgeable person does not need this time for discussing (this matter). And since 16 *momenta* grow in these (five days) every year, they would have a great impact in any given period of time. These 16 *momenta*, which are calculated in (the course of) a single year, would amount to 24 hours after 60 years. A careful person will easily verify this by dividing into hours (the sum, to which) 16 *momenta* (per year add up) every five years (i.e. 16×5=80 *momenta*=2 hours).

---

the sum of these two days is 2×4=8 *momenta*, and combined with the 8 *momenta* of the other three 'superfluous' days, this leads to a total of 16 *momenta* for all five days. On this basis, the argument is that if these five days were included in the theory of the increase and decrease of daylight, they would lead to an error of 16 *momenta* per year, thus of 80 *momenta*=2 hours in 5 years, of 24 hours in 60 years. This is the mathematical logic behind this passage in CE and MC. Two aspects, however, remain unclear: 1) The first of these five 'superfluous' days falls into a period of an increase of daylight, the remaining four in a period of a decrease of daylight; this difference is not reflected in the calculation. 2) It is not quite clear which of the final two quarters was supposed to provide the additional increment of 1⅓ *momenta* per day; I suspect that it was the third quarter, since CE explicitly refers to this additional increase in the third quarter when discussing the theory of the increase and decrease of daylight (cf. *app. font.* to c. 39, ll. 30–42); note in this respect that MC rejects the formation of a *nubs* by this additional increase at that point; in MC's version of this theory, an additional increase of 1⅓ *momenta* only occurs in the first quarter of a year, i.e. the one quarter of a year that has exactly 90, and therefore no 'superfluous' days; consequently, the calculation executed in this passage here fits only into CE's theory, not in MC's; it appears, therefore, that MC changed CE's theory of the increase and decrease of daylight, but did not realize the impact of this change on the theory of the five 'superfluous' days. For these five superfluous days and their distribution over 12 months see c. 24, ll. 12–15; for their division within the quarters of a year and their omission in the calculation of the increase and decrease of daylight see c. 38, ll. 23–25.

XL,2–9 Sunt…cognoscet] The late Merovingian DIAL. NEUSTR. 29aA certainly referred to CE when stating that some (Irish computists) argue for dividing the five 'superfluous' days in their respective quarter of the year, so that they would not contribute to the bissextile increment: *De quinque autem diebus superfluis: Alii eos in suis temporibus frangere volunt, ut caveatur, ne crescat materia bissexti.* The Carolingian COMP. COL. 2.3, then, cited DIAL. NEUSTR. directly, but may also have consulted MC in this question: *Item de sole dicendum, et de byssexto et quinque diebus superfluis in tribus nonagesimis, id est nonaginta uno, nonaginta duobus, nonaginta duobus. Quomodo ponendi sunt, si franguntur in suis nonaginta, in quibus dividuntur per annos artificaliter?... Si franguntur superflui dies quinque, ut caveatur, ne crescat materia byssexti.* The curious calculation of the error of 16 *momenta* that would arise in the theory of the increase and decrease of daylight if these five days were not distributed equally among the 90 days of their quarter-year is unique to CE and MC. This division of the five 'superfluous' days and their omission in the calculation of the increase and decrease of daylight is already referred to in c. 38, ll. 23–25; for a different division of these days cf. c. 24, ll. 12–15; c. 41, ll. 80–88; for these five superfluous days cf. also c. 45, ll. 2–6.

XL,2–20 Sunt…pręponas] Cf. Borst, *Schriften*, p. 896–7 and p. CIII, CXLV–CXLVI, CLXXX–CLXXXI of the introduction.

The division of these days

Ideo hos dies tali ordine diuidimus: Primi diei duas in horas XV ponentes, ut quando incipias XV ab XI Kalendas Aprilis. *Verbi gratia*: Prima feria; peractis *XV die*bus et *II hor*is in *secunda*m *feria*m; finem reddas in *tertia feria*; impletis *XV diebus* et *VI hor*is <in *IIII feria*;> in *V feria* pensatis *XV die*bus et *VIII hor*is; in *VI feria* exactis *XV* diebus et *X hor*is; in VII feria transactis *XV diebus* et *XII hor*is. Hinc VII feriae <XII> horas in preteritos sexies XV dies, hoc <est> nisi dispersas. Cum uero dies partiaris in sequentibus nonagintis, *XV* diebus *IIII hor*is sexies diuidi pręponas.

---

14 in[1]...feria[1]] *om. M (add. according to* CE*).* **17** XII] *om. M (add. according to* CE, MC*).* **18** est] *om. M.* | nisi] niri *M.* | dispersas] dispersimos *M.*

10–20 Ideo...pręponas] In this paragraph, MC procedes to a discussion of the division of the five 'superfluous' days. A day, here, is reckoned as consisting of the 12 hours of daytime. If these 12 hours are divided by 90 days, the result is 5⅓ *momenta* per day (12 hours=480 *momenta*; 480/90=5⅓ *momenta*); cf. especially DRC in *app. comp*. Accordingly, this division leads to two hours per 15 days (5⅓×15=80 *momenta*= 2 hours). Thus, if the one 'superfluous' day of the second quarter (from the vernal equinox to the summer solstice) is divided by the six 15-day intervals of this 90-day period, the result is 2 hours per 15 days; in the third and fourth quarter, however, there are two rather than one 'superfluous' day, leading to four hours per 15 days. Anatolius (?) had explicitly recorded these surpluses per 15-day interval in his theory of the increase and decrease of daylight, and his account certainly is the ultimate source for the divison of the five 'superfluous days': DRP 13.227–233 (cited in DRC 50.28–47): *Et ab VIII Kal. Aprilis usque in VIII Kal. Iulii per XV dies et II horas hora deminuitur ... et ab VIII Kal. Iulii in VIII Kal. Octobris similiter per XV dies, et per IIII horas hora deminuitur ... ab VIII Kal. Octobris usque in VIII Kal. Ianuarii simili numero dierum horarum ac momentorum finitur.* Once again, however, MC's direct source is CE; in fact, MC's account (corrupted by various copying stages), and especially the references to the weekdays in the example, become intelligible only through CE; note that the division of the two 'superfluous' days in each of the final two quarters is only hinted at in MC, while all the details are outline in CE p. 104: *Verbi gratia dominicus dies unus est, secundus dominicus octauus est, III dominicus XVmus est. Trahe II horas de II feria et adde eas ad XV dies, quos enumerasti. Item dinumera alios dies XV hoc modo: secunda feria unus dies est; item alia secunda feria octauus dies est; tertio secunda feria XVmus dies est. De tertia feria sequente IIII horas adde ad XV dies praetermissos. Item alios dies XV conputa ab hac III feria, quę est dies unus; secunda uice III feria octauus dies est; et alia III feria XVmus dies est. Adice sex horas a IIII feria his XV diebus. Item IIIIta feria dies primus est; secunda octauus dies est; tertio XV dies est. His diebus, quos numerasti, adde VIII horas a Vta feria. Item Vta feria dies in numero primus est; secundo octauus est; tertio XVmus est. His adde X horas a sexta feria. Item sexta feria dies primus est; iterum octauus est; tertio XVmus est. His XV diebus adde XII horas de sabbato, cuius fractione hoc modo peracta est et dies XII Kalendas Iulii est in diebus mensis. Item in secundo XCmo II dies sunt, qui hoc modo franguntur: dies dominicus primus est; secundo octauus est; tertio XVmus est. His adde IIII horas a secunda feria. Item secunda feria unus dies est; secundo octauus; tertio XVmus est. His adpone VIII horas a III feria. Item III feria unus dies est; secundo octauus est; tertio XVmus est. His XV diebus adde XII horas a IIIIta feria, quam plene comminuisti vice sabbati, et non a IIIIta feria conputabis, sed a Vta hoc modo: Vta feria dies unus est; secundo octauus est; tertio XVmus est. His adde IIII horas* (a section seems to be missing here) *a sabbato. Item sabbatum dies primus est; secundo octauus est; tertio XVmus est. His adde XII horas a dominico die, quem plene comminuisti. Hi II dies in diebus ebdomadis dies sabbati et dies dominicus uocantur. In diebus autem mensis XIII Kalendas Octimbris et XII Kalendas Octimbris dicuntur. Hoc autem modo in XCmo ultimo bini dies franguntur.*

Therefore, (to avoid this,) we divide these days in the following way: 15 (days) produce two hours of the first (of these five) day(s), as you should begin with the 15 (days) from 22 March. For example: (Set out from a) Sunday; after 15 days and two hours (you will arrive at a) Monday; you should turn to a Tuesday at the end of a month (i.e. after 30 days and four hours); to a Wednesday after (another) 15 days and six hours; to a Thursday after (another) 15 days and eight hours; to a Friday after (another) 15 days and ten hours; to a Saturday after another 15 days and twelve hours. Hence, (you will get) twelve hours of a Saturday in (the course of) the preceding six times 15 days, i.e. were (these hours) not divided up. In case you should want to divide (the superfluous) days in the following (two periods of) 90 (days), you have to take into account that they are divided into four hours (per each of) the six times 15 days (as two rather than one day are divided in these quarter-years).

10–20 Ideo…pręponas] DRC 50.22–44 does not only give the most detailed and precise account of the division of the five 'superfluous' days, but even connects them to specific Julian calendar dates: *Incipit uero secunda pars apud se ab <ortu> solis in VIII Kl Aprelis usque <quo> oriatur in VIII Kl Iulii; et dies in his nonaginta diebus frangitur, et specialiter ille dies qui secundum ordinationem Romuli in Maio supra XXX crescit hic frangitur, et V momenta et terciam partem momenti de fractione diei unusquisque dies habet, id est duae horae in XV diebus ... Incipit autem tertia pars apud se ab ortu solis in VIII Kl Iulii usque quo oriatur in VIII Kl Octimbris. Et duo dies hic frangit. Specialiter autem frangit hós dies qui in Iulio et Agusto supra XXX sunt, et unusquisque dies in XC, decim momenta, id est punctum, et duas tercias partes momenti de fractione duorum dierum habet, id est quatuor horae in XV diebus ... Incipit item quarta pars apud se ab ortu solis in VIII Kl Octimbris usque quo <oriatur in VIII Kl Ianuarii; et in his> XC diebus duos dies frangit, quos in Decimbri et in Octimbri supra XXX uidemus. Et unusquisque dies in XC, decim momenta, id est punctum, et duas tercias partes momenti de fractione duorum dierum, id est quatuor horae in XV diebus.* COMP. COL. 2.3B, for its part, relied directly on MC: *Sic haec fractio est: Postquam enumeraveris quindecim dies, addes de quindecim diebus, qui restant, duas horas, et iterum addes de quindecim diebus quattuor horas, et iterum addes de quindecim diebus sex horas, et iterum addes de quindecim diebus octo horas, et iterum addes de quindecim diebus decem horas, et iterum de quindecim diebus addes duodecim. Et sic egisti totum diem. Sic facies in caeteris diebus nonaginta uno, nonaginta duobus, nonaginta duobus. Si incipies a dominico die, sic consummabis in die Saturni. Haec observatione facilis est semper.*

<XLI.> DE BISSEXTO

Definition and etymology of bissextile day (*bissextus*)

Bissextus euentus quidam famosus in sole, quem Greci inuenerunt. *Bissextus bis sex*. Inde Ysidorus ait: *Bissextus quasi bis sexies ductus assem facit, quod unus dies*, id est *XII horarum, quia tenebrae nihil sunt*. Hinc Augustinus ait: Bissextus famosus solaris euentus, | fol. 21r cui nulla uis est secundum lunam. Siue bis sex Kalendas, si in Februario *bissextu*m ponas. Inde Macrouius ait: *Ante ultimos V dies Februarii mensis bissextu*s inuenitur. Si autem in *Marti*o *bissextu*s, <*bis sex*> *Nonas*. Inde Ysidorus dicit: *Bissextus autem a VI Nonas Martii usque ad diem* II *Kalendas Ianuarii cursu*i *lunae* opp*onitur*. Haec

---

XLI,10 bis sex] *om. M (add. from* CE*)*.

XLI,2–16 Bissextus…facit] The fact that the bissextile day is a famous solar event, mentioned in the first sentence of this passage and slightly later attributed to Augustine, cannot be found in any of Augustine's works or, in fact, any other pre-MC text; note that it is ascribed to Isidore in DRC (cf. *app. comp.*), and that the same phrase is later used by MC to characterize the *saltus* (c. 62, ll. 9–10). The statement attributed to Isidore that the bissextile day takes its name from twice six, i.e. the 12 hours of daytime, is a conflation of the following three source: ETYM. 6.17.26 (cited in DE BISSEXTO I 994C5–7, BC 140(1354)B11–13): *Dictus autem bissextus quia bis sexies ductus assem facit, quod est unus dies.* CE p. 105: *Bissextus autem quasi bis sexies ductus, non quia dies XII horarum sit, quę bis sex dicuntur, licet alii dicunt.* AUGUSTINE, *De Genesi contra Manichaeos* 1.4 (*PL* 34, col. 177): *Sicut autem silentium nihil est, sic et tenebrae nihil sunt.* The theory that the bissextile day derives its name either from *bis sex Kalendas* or *bis sex Nonas* is based on CE p. 105: *Bissextus autem quasi bis sexies ductus, ... sed quod nomen accipit de nomine diei mensis aut a sex Kalendas Martii, aut a sex Nonas Martii, ut diceretur bis sex Kalendas Martii uel bis sex Nonas. Verbi gratia: VI Kalendas Martii hodie, VI Kalendas Martii cras, aut VI Nonas hodie, et VI Nonas Marti cras.* For the etymology *bis sex Kalendas* cf. also PS-DIONYSIUS, *Argumentum XVI* (Krusch, *Studien* II, p. 80; cited in BC 42B11–12, DIAL. NEUSTR. 27, QUAEST. AUSTR. 2.9A, in turn cited in LIB. ANN. 5, in turn cited in LIB. COMP. 3.10B): *Et ideo bissextus dicitur, quia bis VI Kl. Mar. habet Februarius.* The Macrobian and Isidorian citations given as the authoritative basis for these etymologies are quoted from SAT. 1.14.6 (cited in DE BISSEXTO I 997D7–998A5, DTR 12.86–91, in turn cited RM 32.19–25, partially in LIB. CALC. 70B; partially in DRC 52.4,10–13): *Iulius igitur Caesar ... statuit ut quarto quoque anno sacerdotes ... unum intercalarent diem, eo scilicet mense ac loco quo etiam apud veteres mensis intercalabatur id est ante quinque ultimos Februarii mensis dies, idque bissextum censuit nominandum.* and DNR 6.7.50–52 (MS B reads *cursui lunae opponitur*; cited in DRC 57.2–3, DE BISSEXTO I 997C12–15, both also reading *cursui lunae opponitur*): *Bissextus autem a sexto nonas martias usque ad diem pridie kalendarum ianuarium in lunae cursu adponitur.* Cf. ETYM. 6.17.27 (cited in BC 140(1354)B15–C2): *A VI autem Non. Mart. usque in diem prid. Kal. Ian., in lunae cursu bissextus adponitur atque inde detrahitur.* Finally, the statement attributed to Augustine that a bissextile day is produced in the course of four years is rather Isidore's: ETYM. 6.17.25 (MSS BK read *annum*; cited in BC 140(1354)B8–11, DE BISSEXTO I 994C2–7, both reading *bissextilem annum*): *Bissextus est per annos quattuor unus dies adiectus. Crescit enim per singulos annos quarta pars assis. At ubi quarto anno assem conpleverit, bissextum unum facit.*

XLI,3–7 Bissextus…lunam] The theory that a bissextile day takes its name from *bis sex horae*, and that it is only a solar, not a lunar phenomenon, is copied from MC into DRC 52.5, 57.25–28: *Bis sex horae, quia tenebrae nihil sunt nisi absentia lucis. ... Sciendum nobis si aetas lunae diei bissexti deputatur. Non, et si habet luna aetatem non ei adscribitur, ne motet epactas in Kalendis mensium, ut Isidorus dicit. Bissextus est effectus quidam solaris, cui nulla uis est cum tenebris.* Bede refers directly to these

41. ABOUT THE BISSEXTILE DAY

The bissextile day (*bissextus*) is a certain famous feature (of the) solar (year) invented by the Greeks. *Bissextus* (derives from) 'twice six' (*bis sex*). Therefore Isidore says: The bissextile day (*bissextus*), when regarded as twice six (*bis sexies*), forms a whole, namely one day (consisting) of 12 hours, since the shadows are nothing. Hence Augustine says: The bissextile day is a famous solar feature, which has no effect on the moon. Or (this term denotes) twice the sixth (*bis sex*) calends, if you place the bissextile day in February. Therefore Macrobius says: The bissextile day is to be found before the last five days of the month of February. If, however, the bissextile day (is placed) in March, (then the term denotes) twice the sixth (*bis sex*) nones. Therefore Isidore says: The bissextile day, however, is opposed by the course of the moon from 2 March to 31

Irish computists when condemning the reckoning of a bissextile day as consisting of the 12 hours of daytime only: DTR 39.10–12 (cited in LIB. CALC. 42, PV §304): *Nec omnino putamus eorum suscipiendam esse sententiam qui, quasi nocti nihil tribuentes, tres tantum horas per annum bissexto accrescere confirmant*. For the question whether or not a lunar age should be assigned to the *bissextus* see also DIAL. NEUSTR. 28aD, QUAEST. AUSTR. 2.9B, COMP. COL. 3.7. Note that Irish computists of the late seventh, early eighth centuries did apply a lunar bissextile day, which they placed on the same day as the additional Julian calendar date (cf. especially c. 58, ll. 8–32), even though this practice in nowhere explicitly described (note that the common practice, used by Bede and everyone following him, was to place the extra lunar bissextile day at the end of the February lunation). Besides CE and MC, the etymology of *bissextus* deriving from *bis sex* (or *sextus*) *Nonas* can only be found in ANGERS 477, fol. 20r (gloss to DT), which is certainly based on these Irish texts: *Bissextus quasi bis sextus Kalendas Martii uel bis sextus Nonas*. All other post-MC computists, then, took 24 February as the only valid placement of the solar bissextile day (cf. c. 41, ll. 43–49), and the etymology of *bissextus* deriving from *bis sex* (or *sextus*) *Kalendas* enjoyed wide popularity. DRC 52.6–10 is again directly based on MC, which is revealed by the fact that it repeats MC's argument that the first of the two 24 Februaries is to be regarded as the intercalated day: *uel bis sextus. Sciendum nobis cur sextus. Id est, quia sextus dies est a Kl Martii. Cur dicitur bis? Id est, quia sextus bís inuenitur dies a Kl Martii, quando fit bissextus, id est sextus dies bissexti, et sextus dies iterum, VI Kl naturalis*. DIAL. LANGOB. 24A may be based on CE: *Verbi gratia, si hodie VI Kalendas Martii, et crastino VI Kalendas Martii. Propterea dicitur bis, quia bis ibi legitur*. Quite certainly influenced by Irish computistical thought in this respect are DE BISSEXTO I 993D2–994A1 (cf. p. XXII of the introduction): *Bissextus ergo propterea dicitur propter bis sextas Kalendas nominatas; sic enim dicitur sexto Kalendas hodie et sexto Kalendas cras*. and QUAEST. AUSTR. 2.9B: *Ab aliquibus ergo vocatur bissextus, ab aliquibus bissex, hoc est VI Kalendas bis*. This etymology is also applied by Bede: DT 10.9–10: *romani VI kal. martiarum, unde et nomen accepit*. DTR 38.26–27 (cited in RM 54.12–13, PV §301, LIB. COMP. 3.6, LIB. CALC. 41): *romani autem sexto kalendarum martiarum die, unde et bissextum vocant*. Cf. also LECT. COMP. 7.2D: *quod est integer dies, qui appellatur in VI Kalendas Martii et pro hoc dicitur bissextus*. and COMP. COL. 3.5. Most commonly, however, this etymology is cited from PS-DIONYSIUS, *Argumentum XVI* (cf. *app. font.*). For the popularity of MC's direct sources for this passage among other eighth century computists cf. *app. font.*; in the last citation of this passage, MC is particularly close to DRC 53.3–7, though both texts differ in the ascription of the quote: *Isidorus dicit: Bissextus per quatuor annos unus dies adiectus. Crescit enim per singulos annos quarta pars assis. Ubi uero anno assem conpleuerit, bissextilem annum facit*.

XLI,2–3 Bissextus…inuenerunt] Cf. Borst, *Schriften*, p. 412. **2–12** Bissextus…opponitur] Cf. p. CXLIV of the introduction. **4–5** Bissextus…sunt] Transcribed in Borst, *Schriften*, p. 411. **5–7** Hinc…lunam] Transcribed in Walsh & Ó Cróinín, *Cummian's letter*, p. 169, from *bissextus* in Borst, *Schriften*, p. 415; cf. also Borst, *Schriften*, p. 906 and p. CXXIX of the introduction. **10–12** Inde…opponitur] Cf. Borst, *Schriften*, p. 411.

difinitio diei nomini pertinet. Augustinus ait: *Bissextus unus dies per* IIII *annos adiectus. Crescit enim per singulos annos quarta pars assis. Ubi* uero *assem compleuerit, bissext*ilem an*num facit.*

The origin of the bissextile day

Quis primus bissextum inuenit? Grecorum Sabini, Gignus rex Sabinorum natura inuenit. Quid inuenit? Superfluum naturę lucis ingenitae. Per quod inuenitur? Per artem numerationis in signis ciuilibus.

The names for bissextile day

Apud calcentores asis uocatur, apud tractatores quadrans, cum Aegyptiis interkalatio, cum Grecis dies semedorus, id est XII horae bis, quia bissexti XXIIII horarum est.

---

18 Sabinorum] Sauinorum *M.* **19** inuenitur] bissextus *add. Ó Cróinín.* **21** calcentores] *corr. to* calcenteros *Ó Cróinín.* | asis] ayis *M (corr. from* ANGERS 477*); aigis Ó Cróinín (contra M).* tractatores] tractores *Ó Cróinín (contra M).* **22** interkalatio] intotkalatio *M (corr. according to* ANGERS 477, DRC*).*

17–20 Quis…ciuilibus] The theory that Gignus, the king of the Sabines, invented the bissextile day is in line with MC's statement in c. 53, ll. 12–15 that Gignus modelled the solar months (and with this the entire solar year) on the lunar ones. It appears that this like many of the following passages in this chapter are MC's own compositions, since no apparent source can be found for them. The idea that the Greeks invented the bissextile day may ultimately derive from SAT. 1.13.9–10 (cited in COMP. COL. 2.2A, up to *intercalarent* in DTR 12.74–80): *nam et Graeci cum animadverterent temere se trecentis quinquaginta quattuor diebus ordinasse annum – quoniam appareret de solis cursu, qui trecentis sexaginta quinque diebus et quadrante zodiacum conficit, deesse anno suo undecim dies et quadrantem – intercalares stata ratione commenti sunt, ita ut octavo quoque anno nonaginta dies, ex quibus tres menses tricenum dierum composuerunt, intercalarent. id Graeci fecerunt, quoniam operosum erat atque difficile omnibus annis undecim dies et quadrantem intercalare.* The Sabines, however, were, of course, an Italic rather than a Greek tribe. Interestingly enough, Macrobius was quite explicit that it was Julius Caesar who introduced the bissextile day, relying on Egyptian observations: SAT. 1.14.3 (cited in COMP. COL. 2.2C), 1.14.6 (cited in DTR 12.86–91, in turn cited LIB. CALC. 70B, RM 32.19–25; partially in DE BISSEXTO I 997D7–998A5, COMP. COL. 3.1A): *ergo C. Caesar exordium novae ordinationis initurus dies omnes qui adhuc confusionem poterant facere consumpsit, eaque re factum est ut annus confusionis ultimus in quadringentos quadraginta tres dies protenderetur. post hoc imitatus Aegyptios, solos divinarum rerum omnium conscios, ad numerum solis, qui diebus trecentis sexaginta quinque et quadrante cursum conficit, annum dirigere contendit. ... Iulius igitur Caesar decem dies observationi veteri super adiecit, ut annum trecenti sexaginta quinque dies, quibus sol zodiacum lustrat, efficerent, et ne quadrans deesset, statuit ut quarto quoque anno sacerdotes, qui curabant mensibus ac diebus, unum intercalarent diem.* **21–24** Apud…est] There is no apparent source for this list of names for the bissextile day.

17–20 Quis…ciuilibus] This passage is directly cited in ANGERS 477, fol. 20r (gloss to DT): *Gingius rex Sabinorum natura inuenit primitus bissextum superfluum naturae lucis per artem numerationis et per signa ciuilia.* DRC 52.1–2 uses the same terminology as MC: *Quot nomina bissextus habet? Quinque. ... superfluum lucis apud Latinos.* The same question about the origin of the bissextile day is asked verbatim to MC in DIAL. NEUSTR.; the answer, however, is different, namely that it was an invention of Julius Caesar, and appears to be taken from DRC (ultimately from SAT.; cf. *app. font.*): DIAL. NEUSTR. 27aB (cited in COMP. COL. 3.1A): *Quis primus invenit bissextum? Iulius Caesar. Et ipse nominavit primitus his causis, ut integer servatur cursus anni.* DRC 52.3: *Sciendum quis primitus bissextum nominavit. Iulius, ut dicitur. Idque bissextum censuit nominandum.* Julius Ceasar is also

December. This definition refers to the name of the day. Augustine says: The bissextile day (constitutes) one day which is added in the course of four years. The fourth part of the whole grows, in fact, in the course of a single year. When it has completed a whole, it forms a bissextile year.

Who (was) the first (person) to discover the bissextile day? The Sabine tribe among the Greeks, i.e. Gignus, king of the Sabines, discovered it in nature. What did he discover? An overflow of light inherent in nature. By what means was it discovered? Through the art of counting through the official constellations (i.e. by observing/counting the days of the course of the sun through the zodiac).

(The bissextile day) is called 'a whole' (*asis*) (i.e. the full day that is intercalated every fourth year) among the *calcentores*, 'a quarter-day' (i.e. the annual bissextile increment) among the *tractatores*, *interkalatio* among the Egyptians, and *dies semedorus*, i.e. twice twelve hours, among the Greeks, because it consists of the 24 hours of a bissextile day.

---

referred to as the inventor of the bissextile day in VERS. TUR. 3.6; cf. also the texts citing Macrobius listed in *app. font.* **21–24** Apud…est] MC's list of different names for the bissextile day is copied in ANGERS 477, fol. 20r (gloss to DT): *Apud calculatores asis uocatur, apud tractatores quadras, apud Aegiptios calatio, apud Grecos dies simdoros, id est XII horae bissexti.* DRC 52.1–3 shows considerable variation: *Quot nomina bissextus habet? Quinque. Id est interkalares dies apud Aegyptios; <as> apud alios, superfluum lucis apud Latinos, aeque nundinarum diem et bissextum.*

13–16 Augustinus…facit] Cf. Walsh & Ó Cróinín, *Cummian's letter*, p. 165 and p. CXXVI of the introduction. **17–18** Quis…inuenit[1]] Transcribed in Walsh & Ó Cróinín, *Cummian's letter*, p. 163. **17–20** Quis…ciuilibus] Cf. Borst, *Schriften*, p. 412, 443, 899 and p. CLXXIV, CLXXXVI of the introduction. **19** Superfluum…ingenitae] Transcribed in Walsh & Ó Cróinín, *Cummian's letter*, p. 163. **19–24** Per[1]…est] Transcribed in Walsh & Ó Cróinín, *Cummian's letter*, p. 163, only to *semedorus* p. 63 (*calcenteros*, *tractores*), only to *ciuilibus* p. 156; only to *uocatur* in Ó Cróinín, 'Hiberno-Latin *calcenterus*', p. 56. Cf. Walsh & Ó Cróinín, *Cummian's letter*, p. 211 and p. CLXXXVI of the introduction.

Placement of the bissextile quarter-day

Ubi ponantur hii quadrantes donec dies de his fit? Id est initium anni et annum initium mutat. Tamen non facile, *qui*a ortus *sol*is *anni* solaris *orbem in confusa uarietate* admutat, ut Augustinus dicit.

Bissextus as intercalated day

Bissextus non nomen numeri et diei ideo fit, quia non implent CCCLXV dies cursum solis, nisi fuisset | quadrans cum his. fol. 21v

---

30 implent] implet *M (corr. from* BC*)*.

25–28 Ubi…dicit] If a year is reckoned as consisting of 365¼ days, then every year begins six hours later than the previous year. MC here argues that the quater-day exceeding 365 days should be placed at the beginning of a year, leading to a variation of six hours in the beginning of a year when compared to the beginning of the previous year. This theory is based on CE p. 107–8: *Ipse autem Augustinus diem bissexti per IIII annos comminuit, uerbi gratia Kalendas Ianuaris dominicus et finitur dominico die annus primus diebus CCCLXV. Adde VI horas usque ad dimidium noctis secundae ferię, finiturque III feria conpletis diebus CCCLXV post sex in nocte III ferię. Adde sex horas usque ad ortum solis tertię ferię et finitur IIIIta feria eadem hora diebus conputatis CCCLXV. Adde VI horas a solis ortu usque ad dimidium diei. Annus IIII incipit a medio diei quartę ferię finiturque eadem hora quinta feria. Adde VI horas usque ad uesperam. Hinc comminuisti diem, id et quintam feriam, uerbi gratia ita donec Vtus annus incipiat a VIta feria et non a quinta.* The quote ascribed to Augustine appears to stem from MISSALE MIXTUM SECUNDUM REGULAM B. ISIDORI DICTUM MOZARABES, *In dominica ante epiphaniam domini* (*PL* 85, col. 225C): *quibus Sol certa cursus sui dimensione anni orbem in confusa varietate distingueret.* **29–31** Bissextus…his] When etymologizing and defining the bissextile day at the beginning of this chapter (c. 41, ll. 2–16), MC argued that *bissextus* could either be interpreted as a number (*bis sex* – 'twice six', i.e. 12), or as the doubling of a Julian calendar day (*bis sextus* – 'twice the sixth' calends or nones). In this passage here, MC vehemently argues for the second interpretation.

25–28 Ubi…dicit] The theory that the beginning of a year differed by six hours from year to year due to the quarter-day exceeding 365 days was a widespread theory in the eighth century. Cf. BC 44D7–11: *Primo anno incipit quadrans a solis ortu usque in mediam noctis. Incipit quadrans anno secundo a dimidiam noctis usque a solis ortum. Quadrans tertio anno usque ad sextam horam. Quadrans quarto anno a sexta hora usque a solis occasum.* DT 10.4–6 (cited in RM 54.6–8, DTR 38.20–23, in turn cited in LIB. COMP. 3.6, LIB. CALC. 41, PV §301): *Verbi gratia, si nunc aequinoctialem caeli locum <sol> oriens intraverit in hunc, anno sequente meridie, tertio vespere, quarto medio noctis, quinto rursum in exortu recurrens.* DE BISSEXTO I 995D6–996A2 (based on DTR): *unde fit ut si, verbi gratia, nocte aequinoctiali coeli locum mane oriens sol intraverit in principio anni, in hunc locum aequinoctialem, anno sequenti non mane oriente sole illum annum terminat, sed sequenti meridie, et sic secundus annus finem vespere habebit; in tertio autem anno media nocte, in quarto anno iterum in exordio diei sol in loco eodem atque in eodem signo apparebit.* Cf. also EW 3, and especially c. 41, ll. 89–91 below. For the citation ascribed to Augustine cf. COMP. COL. 3.2C: *Hoc semper in omnibus annis curabis numerare, ut quarto anno tempora inter aequinoctia et solistitia certis quidem dimensionibus computes et in singulis anni partibus nonaginta dies et viginti <momenta> naturaliter videas, quibus sol in ascensione anni orbem in confusa varietate distinguit.* **29–31** Bissextus…his] MC's general conclusion here that *bissextus* is the name of a day rather than a number is not repeated in any other text, at least not in such a direct way. The fact, however, that the quarter-day exceeding 365 days was integral for the proper length of a year is often stressed, with BC 44D3–6 (cited in LECT. COMP. 7.2A) even expressing this fact in the same words as MC: *Quare fit bisextus ... id est quia non implent CCCL\X/V dies solarem annum, nisi fuisset quadrans.* Cf. also DT 10.2–4 (cited in DE BISSEXTO I 994B7–10, RM 54.3–5): *Bissextus ex quadrantis ratione per quadriennium conficitur, dum sol ad id signum ex quo egressus est non in CCCLXV diebus, sed quarta diei parte superadiecta revertitur.* DTR 38.18–20 (cited in LIB. COMP. 3.6, LIB. CALC. 42, PV §300): *Haec est ratio quia sol annuum caeli ambitum ... non in CCCLXV diebus sed superadditis sex horis, adimplere cognoscitur.* QUAEST. AUSTR.

Where are these (bissextile) quarter-days (*quadrantes*) placed until the time when a day is produced by them? At the beginning of a year, and (every) year (this) beginning varies. (This is,) however, not easy, since sunrise changes the cycle of the solar year by a confusing variation, as Augustine says.

Bissextus (is) not the name of a number, and therefore happens to be that of a day, since 365 days do not complete the course of the sun, if the (bissextile) quarter-day (is) not (added) to these.

2.9D: *De eo autem, quod interrogavimus de bissexto, quare praeponendus sit, haec causa est. Trecenti sexaginta quinque dies ad rationem solaris anni non sufficiunt.* For a solar year consisting of 365¼ days cf. especially c. 31, ll. 15–16.

26–28 Tamen…dicit] Cf. Borst, *Schriften*, p. 901.

Classific. of bissextile day

*Bissext*us duplex, in *natura* et in adnumeratione. In natura, id est in preparatione eius per *momenta*, id est momentum et *tertia par*s *momenti* et VIIII pars momenti. In adnumeratione, *quando* apparet *dies*. Plenius est annus sine bissexto. Bissextus plenus est semetipse, ut mulier concipiens primo unum corpus, postea II corpora plena.

---

32 adnumeratione] numeratione *M (corr. according to* Mc, Angers 477*)*.

32–37 Bissextus...plena] In Ce p. 105–6 there is a discussion whether the bissextile day is naturally produced by the increment, called *lux pallida*, of 1⅓ *momenta* in the first quarter of every year (from 22 December to 21 March); the answer is negative for the reason that the bissextile day is implemented either on 24 February or 2 March in the fourth year, which is 26 or 21 days respectively short of the end of the period in which the bissextile increment is produced, i.e. 21 March; accordingly, on 24 February or 2 March of the fourth year a full bissextile day is not yet produced: *Item querendum est, numquid dies bissexti de luce pallida secundum naturam factus est. Dies enim bissexti et lux pallida naturalis est, utique haec dies bissexti de hac luce effici non potest, quia res incorporalis reseruari per quadriennium dificile est, et IIIIto anno in unum coniugi die, ueluti res corporalis patitur. Et hoc uerum esse firmamus, quia non de luce pallida fit hęc dies bissexti quando ponimus illum supra VI Kalendas Martii uel supra VI Nonas Martii. Ab altero enim XXVI dies, ab altero XXI dies desunt, hoc est momenta XXVI uel XXI cum tertia parte momenti in unoquoque die usque ad XII Kalendas Aprilis*. This passage served as the basis for Mc's classification of the bissextile day into two categories here, *in natura* and *in adnumeratione*. Mc defines the *bissextus in natura* as the bissextile increment in the first quarter of a year, which is outlined in detail in c. 39, ll. 10–12; interestingly enough, Mc here lists the daily increment in this first quarter as 1⅓ 1/9 *momenta*, whereas it consisted only of 1⅓ *momenta* in the earlier passage. The reason for this difference lies in Ce's passage quoted above: Mc regards 24 February here as the correct placement of the bissextile day; this is 26 days before the end of the period in which the bissextile day is commonly produced every year (22 December to 21 March); hence, the bissextile day is produced in 90 days in three of four years, in only 90−26 = 64 days in the fourth year, leading to a total of 3×90+64=334 days; for easier calculation, this total seems to have been reduced to 333; now, if an increment of 1 ⅓ 1/9 momenta is assigned to each of these 333 days, the total after 333 days is 333×1⅓ 1/9=481 *momenta*, i.e. almost accurately the 480 *momenta* of a 12-hours day. The *bissextus in adnumeratione* is defined as the summation of this increment to a full day (of 12 hours) in the course of four years; the idea behind the last two sentences of this passage is that as long as this bissextile day grows, it is considered as an integral part of a year, the year thus being longer than 365 days; once the extra day is produced after four years, it is an entity in its own right, disconnected from the entity of the year, which in that case consists of only 365 days.

32–37 Bissextus...plena] This passage is directly copied, in different order and variations in wording, in the Angers glosses: Angers 477, fol. 20r (gloss to Dt): *Numquid bissextus sine anno pleno et annus sine bissexto, utique altero, ut mulier concipiens prima unum corpus, et post conceptionem II corpora plena. Bissextus in duobus fit, in natura et in adnumeratione: in natura, quando per momenta preparatur, id est momentum et III pars momenti et IIII pars momenti; in adnumeratio, quando dies apparet*. This classification of the bissextile day did not receive a wider reception; the allegorical comparison of the bissextile day with a pregnant woman, however, apparently first formulated by Mc, became fairly popular in eighth-century Frankish computistics: Dial. Neustr. 28aC (cited in Comp. Col. 3.1B): *Cui simile est incrementum bissexti et ortus eius? Id: Mulieri concipienti in utero et parturienti*. Quaest. Austr. 2.9C: *Autem rationem numeri byssextus quarto anno et saltus nono decimo anno ponitur, ut filius de ventre matris, nunc usque invisus, partu monstratur, et nucleus de nuce illa, huc usque circumdatus, enucleando foras producitur*.

32–37 Bissextus...plena] Cf. Borst, *Schriften*, p. 413–4, 482 and p. CIII, CLXXIV, CLXXXVI of the introduction. **34** momentum...momenti[2]] Transcribed in Borst, *Schriften*, p. 413.

The bissextile day (is) twofold, according to nature and according to summation. According to nature means according to the preparation of the bissextile day by *momenta*, namely one *momentum* and a third of a *momentum* and a ninth of a *momentum*. According to summation (is the case), when a day appears (i.e. when the bissextile increment sums up to a whole day every four years). A year is fuller without a bissextile day, (since) a bissextile day is full in itself (i.e. it is an entity in its own right), just as a pregnant woman (represents) one body at first, (but) two full (and distinct) bodies later.

36–37 Bissextus…plena] Transcribed in Borst, *Schriften*, p. 414; cf. Borst, *Schriften*, p. 900.

The first bissextile quarter-day

Quo quadrante *quęritur* bissextus initiatur. Ut alii *quadrante di*urnali. Et non est facile eum in uernali tempore *pon*i, nisi a *quadrante noct*urno appareat incipi. *Hinc noct*urnum precedere bissextum propterea dicunt. Diurnum enim preesse *mundu*m ferunt.

---

39 diurnali] diurnalis *M.*

38–42 Quo…ferunt] A bissextile day consisted of four quarter-days, each of these being produced by a single year, adding up to a full day in the course of four years. Since this bissextile day quite naturally consisted of day and night, two of these quarter-days had to be diurnal, the other two nocturnal. It appears that the question about the quality of the first of these four quarter-days, i.e. whether it was diurnal or nocturnal, was lively debated among late seventh- and early eighth-century Irish computists. The longest account of this question can be found in CE p. 105, ultimately arguing for a nocturnal quarter-day; this is MC's source here: *Quaeritur, quem primum quadrantem ponemus in hoc die, id est utrum diei an noctis quadrantem? Cui quesitioni cogimur redire ad initium mundi, qui factus est in XII Kalendas Aprilis secundum Grecos, VI horis restantibus at dimidia de XII Kalendas Aprilis. Igitur si ante sol esset, diei quadrans crescere a XI Kalendas Ianuaris usque ad XII Kalendas Aprilis. Hinc si conputaueris IIII quadrantem noctis, uidebis quartum quadrantem noctis deesse, ueluti quadrans diei et quadrans noctis in primo anno, quadrans diei et quadrans noctis in II anno, quadrans diei, quadrans noctis in III anno, quadrans diei in IIII anno. Si autem quadrantem noctis computaueris, transibis XI Kalendas Aprilis indecenter. Etsi iterum diei quadrantem addideris, quintum quadrantem diei adponemus. Hinc quadrantem noctis primum ponamus et ad locum quo factus est mundus curiosius redeamus, hoc est ad XII Kalendas Aprilis, in qua VI horę et semis restabant. Etsi ante hunc diem sol fuisset, ad mensurandum finis esset bissexti in plenitudinem. Enim mundus factus est sicut homo in perfectione factus est XXXma aetate et luna in plenilunio. Hinc primo quadrans noctis naturaliter factus est. Etsi IIII quadrantes noctis et diei conputaueris, IIIIto anno congrue uidebis IIII quadrantes noctis et diei conuenire, qui diem bissexti conplent quadrante diei terminante diem ordinaliter.*

38–42 Quo…ferunt] Again, the Angers glosses are in this question directly based on MC: ANGERS 477, fol. 20r (gloss to DT): *Notandum quid prius factum quadrans diurnalis an nocturnalis, id est quadrans nocturnalis, quia quadrans diurnalis primo anno precessit mundum, uel quadrans diurnalis prior, quia omnia integra facta sunt.* DE BISSEXTO I 996A2–4, 996D12–997A4 also states that the first *quadrans* should be regarded as a nocturnal one: *Sic ergo fuit in principio mundi ille quadrans a media nocte aequinoctiali usque ad mane ... Scire nos oportet, secundum Graecos in aequinoctio vernali, hoc est, in XII Kal. Aprilis, sed in fine illius diei positus dicitur. Sed quomodo fuit ab initio, quando dicunt, quod in illo quadrante noctis, quando primum mundus factus est, propterea ille quadrans, de quo nascitur bissextilis dies, in illo loco naturaliter in fine anni poni debet.* COMP. COL. 3.3 appears to be based on CE (cf. *app. font.*): *De primo quadrante: Si quadrans nocti vel die factus est? Prius intueri debet initium mundi XII* (MS *XI*) *Kalendas Aprilis. Id: Propter dimidiam partem diei, ut alii dicunt, a mane usque ad sextam horam cum dimidio quadrante horae et dimidium horae, quod verum est. Ille sex horae iunguntur anno praeterito cum quadrante, cum esset annus primus. Deinde principium anni XII* (MS *XI*) *Kalendas Aprilis <est>, et inde incipiunt per quattuor annos quattuor quadrantes noctis et quattuor quadrantes diei, et faciunt diem in XII Kalendas Aprilis et dicitur sic: XII hodie, XII cras. Et sic ponitur byssextus secundum naturam et secundum nomen numeri, ut alii, et sic semper praecedit noctis quadrantes.* For the quarter-day at the creation of the world see c. 7, ll. 9–11; c. 44, ll. 24–26; for the more specific concept that the first of the four bissextile quarter-days occurs between sunset and midnight cf. c. 41, ll. 25–28, 89–91.

38–42 Quo…ferunt] Cf. Borst, *Schriften*, p. 902 and p. XXII of the introduction.

It is asked, by what kind of quarter-day the bissextile day is initiated. As some (say,) by a diurnal quarter-day. But the bissextile day (or rather its increment) cannot (readily) be placed in springtime, if it does not appear to be initiated by a nocturnal quarter-day. For this reason, they argue that a nocturnal (quarter-day) precedes the bissextile day. In fact, they believe that a diurnal (quarter-day) preceded (the creation of) the world.

Different placements of the bissextile day

Quem in VI Kalendas Martii quidam ponunt, ut Macrouius ait, eo quod interim *ueteres* k*alaba*nt. Et primus dies bissextus, quod retro exegitur. Quidam uero in VI Nonas esse statuunt, ut Ysidorus dicit. Ceteri in XII Kalendas Aprilis fieri dicunt, quod rationi conuenire, quod si de quadrantibus dies prepari innotescerit. Sed ne primi mensis initium turbet, illa ratio remouetur.

Three time-changing factors

Quot numeri ordinem mutant et temporum consuetudinem? Id est tria: Bissextus, embolismus, saltus.

---

47 quod[1]] quot *M*. **49** remouetur] remouet *M*. **50** mutant] mutat *M*.

43–49 Quem…remouetur] MC lists three possible placements of the bissextile day within the Julian calendar here, namely 24 February, 2 March, and 21 March. The first, 24 February, is based on Macrobius, as MC himself remarks: SAT. 1.14.6 (cited in DE BISSEXTO I 997D7–998A5, DTR 12.86–91, in turn cited in RM 32.19–25, partially in LIB. CALC. 70B; partially in DRC 52.11–13, c. 41, ll. 8–9): *unum intercalarent diem, eo scilicet mense ac loco quo etiam apud veteres mensis intercalabatur id est ante quinque ultimos Februarii mensis dies, idque bissextum censuit nominandum*. Based on this citation, MC outlines the theory that the first of the two *VI Kalendas* is to be regarded as the intercalated day; interestingly enough, the opposite opinion is given in PS-DIONYSIUS, *Argumentum XVI* (Krusch, *Studien* II, p. 80; cited in BC 42B6–11, QUAEST. AUSTR. 2.9A): *Ergo in IIII annis terne hore, que sunt XII, diem faciunt I, qui addatur Februario, cum VI Kl. Mar. habuerit, ut in crastino sic habeat. Verbi causa, si hodie VI Kl. Mar. additur ille dies in IIII anno expleto, nihilomnius et in crastino VI Kl. Mar. habeatur*. The second possible placement, 2 March, is ultimately an interpretation of Isidore's statements: DNR 6.7.50–52 (cited c. 41, ll. 11–12): *Bissextus autem a sexto nonas martias usque ad diem pridie kalendarum ianuarium in lunae cursu adponitur*. ETYM. 6.17.27 (cited in BC 140(1354)B15–C2): *A VI autem Non. Mart. usque in diem prid. Kal. Ian., in lunae cursu bissextus adponitur atque inde detrahitur*. Isidore's account is not perfectly intelligible, but it is apparent that he talks about the lunar rather than the solar bissextile day; note that if the lunar bissextile day is placed at the end of the March lunation, the latest possible date in the Dionysiac 19-year cycle would be 2 March. The third placement of the solar bissextile day listed by MC here is 21 March, the date considered as the beginning of the world; the idea behind this placement is that if the bissextile day grows gradually, it would have started to grow on the first day of creation, and would be completed after exactly four years, i.e. on 21 March; MC, however, rejects this placement, since it could interfere with the reckoning of the first month (i.e. in the case that a bisextile year coincides with an Easter full moon on 21 March). All three placements are also related to in CE p. 105–6, where they are even used for calculations; note that CE treats all three placements equally, without any obvious preference. **50–53** Quot…saltus] As neither the course of the sun through the zodiac nor the phases of moon could be reckoned by the use of integral numbers only, three technical devices had to be implemented in order to construct a mathematically working system, namely the bissextile day in the Julian calendar, and the embolisms and the *saltus* in the 19-year lunar cycle, and this is the context into which MC places the bissextile day here. There is no obvious source for this list.

43–49 Quem…remouetur] Discussions of the possible placements of the bissextile day, including all three option mentioned here, can be found in the following texts, all of which being heavily influenced by Irish computistical thought: ANGERS 477, fol. 20r (gloss to DT): *Sciendum ubi est locus bissexti inuenire: In Februario, in VI Kalendas Martii, hoc est VI Kalendas hodie, VI Kalendas cras. ... Bissextus quasi bis sextus Kalendas Martii, uel bis sextus Nonas, uel bis sextus qui et* (?) *Kalendas Aprilis*. DIAL. NEUSTR. 27a: *De locis bissexti: Secundum Romanos, ut diximus, in VI Kalendas Martii. Secundum*

Some place this (bissextile day) on 24 February, as Macrobius relates, since the ancients had inserted (it there) in the meantime. And the first day (of the two 24 Februaries) is the bissextile (i.e., the intercalated) day, since it was implemented backwards (i.e. reckoned from the Calends of March). Some, however, are of the opinion that (the bissextile day) falls on 2 March, as Isidore states. The rest argues that (the bissextile day) occurs on 21 March, whenever it is appropriate for the computation, namely when a day is known to be produced by the quarter-days. But this (latter) method of placing the bissextile day on 21 March) is abolished, lest (the bissextile day) disturbs the beginning of the first month.

How many (chronological features) change the order and customary reckoning of time? Three, namely the bissextile day, the embolism, and the *saltus*.

Isidorum in VI Nonas Martii. Alii in XII Kalendas Aprilis, quia initium mundi est. DE BISSEXTO I 997C4–D7: *Interrogandum est ergo, iste quadrans, de quo nascitur, in quot locis anni positus dicitur? in V locis, secundum Graecos in aequinoctio, hoc est, in XII Kal. Aprilis; ... secundum Isidorum bissextus dies dicitur in VI Nonas Martii ... secundum Romanos autem, et secundum auctoritatem Julii Caesaris bissextus in VI Kal. Martii ponitur, inde in Macrobius legitur: ....* These three placements are also given in COMP. COL. 3.3B, 5A, 8A. The first two placements only in DRC 57.1–2 (followed by the Isidorian quotation cited c. 41, ll. 11–12): *Sciendum nobis in co loco primum diém bissexti possuerunt. In Kl Martii et in Nonis eius, ut Isidorus dicit.* For the first two placements cf. also the discussion of the etymology of *bissextus* in c. 41, ll. 6–13. The Irish computistical textbooks in particular, from CE through MC to DRC, reveal that the placement of the bissextile day was a well-debated question in the seventh- and early eighth-century Irish computistical milieu and that the Macrobian placement on 24 February that became standard in the middle ages was not, *a priori*, considered as the only option. By the time of CE, all three placements were seriously considered, while MC then excluded 21 March, and finally DRC, though still listing the Isidorian 2 March, only calculated with 24 February (particularly interesting, in this respect, is chapter 57 of DRC). In Bede's opinion there was no alternative to 24 February, though stating that in the Egyptian calendar the bissextile day would occur on 29 August: DTR 38.25–27 (cited in RM 54.11–13, PV §301, LIB. COMP. 3.6, LIB. CALC. 41): *Quem aegyptii anno suo solemniter expleto, id est quarto kalendarum septembrium, romani autem sexto kalendarum martiarum die, unde et bissextum vocant, intercalare consuerunt.* Consequently, HELPERIC, *De computo* 1 (*PL* 137, col. 22D–23A), in the early tenth century, knows only the Macrobian placement. MC's argument that the first of the two sixth calends is to be regarded as the intercalated day can also be found in DRC 52.7–10 (followed by the Macrobian quotation cited above), 57.18–21, which is quite certainly directly influenced by MC: *Cur dicitur bis? Id est, quia sextus bis inuenitur dies a Kl Martii, quando fit bissextus, id est sextus dies bissexti, et sextus dies iterum, VI Kl naturalis, ut Macrobius ... Sciendum nobis utrum primus án secundus dies dies byssexti est, quando dicis sextus diés hodie, sextus cras. Utique primus, ut Macrobius dicit: Reliquas autem Februarii mensis dies, qui erant quinque, post interkalationem subiungebant.* **50–53** Quot…saltus] Even though the bissextile day is often compared to the *saltus lunae* (cf., e.g., c. 62, ll. 123–130), all three artificially implemented technical devices are usually not listed.

43–44 Quem…kalabant] Transcribed in Borst, *Schriften*, p. 411, up to *ait* p. 491. **43–49** Quem…remouetur] Cf. p. XXII, CIII, CXLIV, CXCVIII of the introduction. **45–46** Quidam…dicit] Cf. Borst, *Schriften*, p. 907. **46–47** Ceteri…dicunt] Transcribed in Borst, *Schriften*, p. 411, 491. Cf. also Borst, *Schriften*, p. 902.

Unum in anno solis, duo in annum lunae, | embolismus et saltus. fol. 22r

How a wrong application of the bissextile day changes the order of the seasons

Sciendum est, *quid esset, si* non numeraretur *bissextus* et *naturaliter esset*, et *si* remaneretur et *non esset naturaliter*. *Temp*ora *mutaret*, nam si non numeraretur, itaque peractis *CCCLX annis menses III* in alios III insinuaret. *Si* uero remaneretur *et non esset*, sic nominationem *III mens*ium in *III alios* mutaret, isdem peractis *CCCLX annis*. Ita aut hiems nomine ueris uocaretur.

---

54 est] *om. Borst (contra M)*. **57** III] *om. Borst (contra M)*. **60** hiems] hiemi *M (corr. by Borst)*.

54–61 Sciendum…uocaretur] The general importance of the bissextile day becomes obvious by analyzing the following question: What would happen, if the bissextile day is not applied? Less obvious is the second question asked by MC here: What would happen, if the bissextile day is applied, but if there was no correspondence to this application in nature? In both instances, the answer is that the seasons would change in the course of time. In the first instance (the bissextile not being applied), 90 calendar days (i.e. the length of a season) would be missing after 360 years (4×90=360), so that what used to be denoted as spring previously, would then be denominated as summer in the calendar, i.e. spring would occur under the name of summer. The second instance (the application of the bissextile not having a correspondence in nature) would lead to the opposite effect, so that what used to be denoted as spring previously, would then be denominated as winter in the calendar, i.e. spring would occur under the name of winter. Ultimately, the impetus for these questions presumably was AUGUSTINE, *De Trinitate* 4.4, cited further below (c. 41, ll. 73–75). MC's direct source, however, was CE p. 106–7: *Item quaeritur quid esset, si bissextus non esset? Utique nihil fastu dii esset, nisi naturaliter esset. Secundo quaeritur, si non conputaremus bissextum, quid esset? Dum est, hoc est tempus in aliut rediendo (!) retrorsum mutaretur. ... De cursis vero annis CCCLX III menses retro conputares. Item queritur quid esset, si bissextum conputaremus et non esset? Hoc est tempora in tempora in ante precedissent. ... CCCLX annis completis III menses super alios III adpones. ... CCCLXmis annis perfectis III menses super alios III nominares et congrua essent post tergum tempora.* For more details on the consequence of the neglect of the application of the bissextile day cf. the following passage.

54–61 Sciendum…uocaretur] Only the first and more obvious of the two questions analyzed in CE and MC, the consequences of the neglect of the application of the bissextile day, is discussed in other eighth-century computistical texts. In these discussions, the perspective and the clear distinction between a calendrical and a factual season apparently posed some difficulties. Bede is very precise by arguing that spring time would occur under the name of summer, i.e. factual spring in calendrical summer: DTR 38.27–32 (cited in LIB. CALC. 41, PV §302): *Quod si qui calculatorum facere negligens CCC solum ac LXV diebus omnes se annos agere debere putauerit, magnum sibi mox inueniet annui circuitus occurrisse dispendium ita ut, post aliquot annorum uertentium curricula, aestiuis mensibus uernum tempus, uernis brumale, brumalibus autumnale, autumnalibus austiuum se offendisse peruersus computator horrescat*. DIAL. NEUSTR. 28aC argues that the beginning of spring would occur in summer, a statement that should be interpreted as the beginning of factual spring occurring in calendrical summer: *Quid esset, nisi fuisset bissextus? Id: Ruissent initia temporum in alia tempora, initium veris in aestatem, initium aestatis in autumnum, initium autumni in hiemem et reliqua*. COMP. COL. 3.1C, 3.5A–B, then, working from DIAL. NEUSTR. and probably CE, got confused by these accounts: DIAL. NEUSTR. argued that the beginning of spring would move forward into summer, while CE stated that three months would move backwards if the bissextile day was not applied; in the end, COMP. COL. did not realize that both texts mean the same thing, DIAL. LANGOB. arguing for factual spring occurring in calendrical summer, CE for calendrical summer moving backwards into what used to be the months of

One (of these appears) in the solar year, (the other) two in the lunar year, (namely) the embolism and the *saltus*.

It has to be known, what would happen (on the one hand), if the bissextile day was not included in the reckoning of time, but nevertheless occurred naturally, and (on the other hand), if the bissextile day remained (part of the reckoning of time), but did not occur naturally. (In the first case) it would change the seasons, since if it was not included in the reckoning of time, then three months would correspond to three different months after 360 years. If, however, it remained (part of the reckoning of time), but did not occur naturally, then it would change the denomination of three months into three different months after the same (total of) 360 years. Accordingly, winter would be called by the name of spring (if the bissextile day was not applied).

---

spring: *Quid esset, si non fuisset byssextus? Id: Ruissent initia temporum in alia tempora, initium veris, initium aestatis, initium autumni. Alii dicunt, nisi fuisset byssextus, retro initia temporum venissent, quod prohibet dies interiectus. ... Alii dicunt ideo: Hic dies byssexti adiecitur, ne veniant retro initia temporum in alia tempora, dum praeterierunt quadrantes annorum. ... Hoc semper faciendum est, ne inveniatur initium hiemis in autumno retro et initium veris in hieme retro, vel ne ruant tempora in alia tempora, ut aestas in autumnum retro*. QUAEST. AUSTR. 2.9D appears to have preferred the backward movement of the calendrical seasons as outlined in the Irish texts: *Nam si quis parvipendat quarto anno bissextum interserere, solaris anni ratio incipiet vacillare. ... Et sic omnia tempora per multos annos confusa apparebunt, ut autumnus in aestatem et aestas in ver et ver in hiemem promoveatur*. In the early tenth century, HELPERIC, *De computo* 1 (*PL* 137, col. 23A), discusses this question on the basis of the solstices and equinoxes, the calendrical spring equinox occurring on the factual winter solstice, etc.: *Qui dies si negligatur, eveniet post aliquot annos ut hieme aestivi, et contra aestate hiberni menses occurrant. Nam per CCCLXIV annos tantum calculatio regradabit, ut in Kalendis aequinoctiorum solstitia, et contra in solstitiorum eveniant aequinoctia, id est, cum XII Kal. April. pronuntiaveris, et debeat esse aequinoctium vernale, appareat solstitium brumale, quod est XII Kal. Januarii. Cumque pronuntiaveris XII Kal. Julii, quod est solstitium aestivum, occurrat aequinoctium vernale, quod esse debet XII Kal. April. Sicque in caeteris eveniet anni diebus. Ad hunc evitandum errorem bissextilis dies quarto semper anno intercalatur*.

54–61 Sciendum…uocaretur] Partially transcribed in Borst, *Schriften*, p. 414. Cf. also Borst, *Schriften*, p. 900 and p. CLXII, CLXIX–CLXX of the introduction.

Why was the Julian calendar date, but not the weekday doubled?

Sciendum est, quid diebus ebdomatis bissextus non geminatur, sicut in diebus mensis quare duplicatur. Ideo quia difficile fuit septimanum numerum ebdomatis dierum in principio mundi consecratum geminari aut mutari, quem Deus benedixit. Quid prestaret, si in diebus mensis geminatus non esset? Nominationem scilicet *mens*ium confunderet, ut uerbi gratia: *Si* esset *bissextus* et *non* geminatus, ita mutaret *Kalendas Ianuarii* in *II Kalendas*, nominares Kalendas, et IIII Nonas initium esset. Accederet inde amplius error esset in nominationibus mensium.

Augustine's argument on the bissextile day

Augustinus in epistula de resurrectione sic ait: *Quadrantes IIII unum diem faciunt*, *quem necesse est in*poni *cursu*i *quadriennio*, *qu*em Romani *bissextum uocant*, *ne ordo temporum* confusus esset. Ita Augustinus quadrantes IIII annis IIII coniungit, ut annus singularis quadrantem singularem sumat. Qui *quadrans VI horas habet.* | *Totus enim dies cum sua nocte,* id est *XXIIII* fol. 22v
*hora*rum, *quarum pars quarta VI horę*.

---

70 Accederet] accede *M*. **73** Quadrantes] quadrante *M (corr. from* AUGUSTINE, *etc.)*.

62–71 Sciendum…mensium] There is no apparent source for the question why only the Julian calendar date was doubled, and not the weekday also. The following proof of the necessity for the doubling of the Julian calendar date is directly connected to the previous passage, and is likewise based on CE p. 106–7: *quando quadriennio completo sine bissextus, II Kalendas Ianuaris hodie dicas, et Kalendas Ianuarius cras. Si non conputaremus bissextus, super II Kalendas esset Kalendas Ianuaris IIIIto anno, super III Kalendas VIII anno, super IIII Kalendas XII anno et ita conpletis CXX annis mensem reductum in alium mensem conputabis.* **72–79** Augustinus…horę] After having outlined in some detail the consequences that the neglect of the bissextile day would have had on the reckoning of time in general, and on the seasons (*tempora*) in particular, MC here points to the source that had ultimately triggered this discussion, namely AUGUSTINE, *De Trinitate* 4.4 (*CCSL* 50, p. 171; cited in CE p. 107 (under the heading *De bissexto Augustini*), DRC 56.6–7, DTR 39.41–43, in turn cited in LIB. CALC. 42; DE BISSEXTO I 995B9–14, QUAEST. AUSTR. 2.9B, which adds *Romani* and omits *enim*, COMP. COL. 3.1C, which reads *poni* instead of *intercalari*), where it is argued that the bissextile day is applied *ne temporum ordo turbet*: *quattuor enim quadrantes faciunt unum diem quem necesse est intercalari excurso quadriennio quod bissextum vocant ne temporum ordo turbetur*. This is followed by another citation from the same text, which, it will be noted, does not occur in CE: AUGUSTINE, *De Trinitate* 4.4 (*CCSL* 50, p. 171–2; cited in DTR 39.47–50, DE BISSEXTO I 995B14–C2, COMP. COL. 3.1C): *ipse autem quadrans sex horas habet; totus enim dies, id est cum sua nocte, viginti quattuor horae sunt quarum pars quarta qui est quadrans diei sex horae inueniuntur*.

62–71 Sciendum…mensium] MC's question why the weekday was not doubled, answered by the fact that the week was God's creation and therefore immutable, is repeated in DRC 57.21–25: *Sciendum nobis cur non per diem septimane bissextus exploratur. Uerbi gratia, dominicus dies hodie, dominicus cras. Hoc modo soluitur, quia septenum quem deus creauit numerum nemo ausus est augere uel minuere; uel quia sempiterna esset turbatio, si per diem septimanae bissextus exploraretur*. Nothing similar can be found in other eighth-century computistical texts. QUAEST. AUSTR. 2.9D analysis of the impact that a neglect of the bissextile day would have on the Julian calendar, resulting in a shift of the Julian calendar dates by one day in four years, two days in eight years, etc., is similar to MC here, since it is like MC based on CE: *Nam si quis parvipendat quarto anno bissextum interserere, solaris anni*

It has to be known, why the bissextile day is not doubled in terms of weekdays, (and) likewise why it is duplicated in terms of the days of the month. For the reason that it was difficult to double or change the number of weekdays, namely seven, which was consecrated at the beginning of the world and blessed by the Lord. What would happen, if (the bissextile day) would not be doubled in terms of the days of the month? (This) would obviously disturb the denomination of the months, as, e.g.: If the bissextile day occurred but (the day of the month was) not doubled, then 1 January would change to 31 December, which you would call 1 January, and 2 January would be the beginning (of the month). It would happen, therefore, that the error in the denomination of the months would become greater.

Augustine argues as follows in his letter about the resurrection: Four quarter-days add up to one day, which the Romans call a bissextile day, and it is necessary that this is implemented in the course of four years, so that the order of time would not be confused. Accordingly, Augustine combines four quarter-days in four years, since every single year takes a single quarter-day. This quarter-day has six hours. In fact, a day (is only) complete with its night, i.e. (it consists) of 24 hours, of which the fourth part (is) six hours.

---

ratio incipiet vacillare. Verbi gratia primo bissexto, id est quarto anno, uno die futurus annus ante aequinoctium incipiet; secundo bissexto, id est nono anno, duobus diebus; tertio bissexto, id est duodecim anno, tribus diebus; quarto byssexto, id est sexto decimo anno, quatuor diebus; quinto bissexto, id est vicesimo anno, quinque diebus. Et sic omnia tempora per multos annos confusa apparebunt, ut autumnus in aestatem et aestas in ver et ver in hiemem promoveatur. **72–79** Augustinus…horę] The first Augustinian citation was very popular among eighth-century computists, and QUAEST. AUSTR. and COMP. COL. in particular share some variant readings with MC (cf. *app. font.*). An evidently direct copy from MC is ANGERS 477, fol. 20r (gloss to DT): *Augustinus ait: IIII quadrantes unum diem faciunt, quem necesse est quadriennio cursui inponi, quem Romani bissextum dicunt, ne ordo temporum turbaretur, nisi enim numeraretur ordo temporum confusus esset.* The fact related to by Augstine that the neglect of the bissextile day would lead to a disturbance of the reckoning of time can also be found in BC 44D12–13: *Turbaretur ordo temporum, si non fuisset quadrans*. DIAL. NEUSTR. 28aB: *Hoc semper faciendum est, ne turbetur ordo temporum*. Besides the texts listed in *app. font.* citing Augustine directly, the Augustinian argument that the bissextile day consists of 24 hours, i.e. of daytime and nightime, can also be found in DTR 38.15–16 (cited in BC 44D13–A2, LIB. COMP. 3.6, LIB. CALC. 41, PV §299): *et ideo quarta pars diei qui XXIIII horis cum sua nocte completur, id est horae sex, quadrans consuete vocatur*. QUAEST. AUSTR. 2.9B: *Agustinus vero diem cum nocte sua denumerat, et ideo iuxta illum quadrans sex horae sunt, id est, dies cum nocte sua.* For the length of the quarter-day, the annual bissextile increment, cf. c. 7, ll. 3–5.

62–71 Sciendum…mensium] Cf. Borst, *Schriften*, p. 904 and p. CLXIX–CLXX of the introduction. **66–68** Quid…confunderet] Transcribed in Borst, *Schriften*, p. 906. **72–79** Augustinus…horę] Cf. Borst, *Schriften*, p. 492, 900 and p. CLXIX of the introduction.

Division of quarter-day and calculation of 12 months of equal length

Qui quadrans hoc modo per totum annum diuiditur: In anno sunt XII menses XXXmi. V dies supersunt, qui habent horas CXX. Quas cum in decadas diuiseris, facient XII decies, ita ut unus mensis XXX dies et X horas habeat. Si item quadrantem, id est VI horas, diuiseris, inuenies XII semis, ita ut per omnem mensem, id est XXX dies et horas X, et XX momenta crescunt. Si minutius diuidas in omni die dimidium, et <bis> tertiam momenti partem repperies in hac preparatione.

Quarterday at beginning of year

Primus annus incipit a uespere in VIII Kalendas Iulii et in media nocte in VIII Kalendas eiusdem Iulii finitur, et reliqua.

---

81 XXXmi] XXXI *M (corr. according to* MC*; it may well have been a misreading of* XXXi*).* **87** bis] *om. M (add. according to* DE BISSEXTO, LDA*; but cf.* DIAL. LANGOB.*).*

80–88 Qui…preparatione] This division of the five days exceeding 360 in the course of a year, as well as of the annual bissextile increment of a quarter-day, by the 12 months of a year leads to the following results: If equal length was to be achieved for every month of a year, a month would have to consist of 30 days, 10 hours, and 20 *momenta*; the 10 hours are the result of the division of five days of 24 hours by 12 months (5×24/12=10 hours), while the 20 *momenta* represent the bissextile increment per month (a quarter-day=6 hours=240 *momenta*; 240/12=20 *momenta*), which can be further divided into an increment of ⅔ *momenta* per day (20/30=⅔). There is no apparent source for this passage, but note that in PS-DIONYSIUS, *Argumentum XVI* (Krusch, *Studien* II, p. 80; cited in BC 42, DIAL. NEUSTR. 27, QUAEST. AUSTR. 2.9A) a quarter-day of three hours is divided by 12 months. **89–91** Primus…reliqua] The more general theory that the first year of the four-year bissextile cycle started at sunset and ended at midnight, the second year at midnight and ended at sunset, etc., resulting in the fact that a six-hours difference occurred between the beginnings of two successive years, derives from CE p. 107–8: *Ipse autem Augustinus diem bissexti per IIII annos comminuit, uerbi gratia Kalendas Ianuaris dominicus et finitur dominico die annus primus diebus CCCLXV. Adde VI horas usque ad dimidium noctis secundae ferię, finiturque III feria conpletis diebus CCCLXV post sex in nocte III ferię. Adde sex horas usque ad ortum solis tertię ferię et finitur IIIIta feria eadem hora diebus conputatis CCCLXV. Adde VI horas a solis ortu usque ad dimidium diei. Annus IIII incipit a medio diei quartę ferię finiturque eadem hora quinta feria. Adde VI horas usque ad uesperam. Hinc comminuisti diem, id et quintam feriam, uerbi gratia ita donec Vtus annus incipiat a VIta feria et non a quinta.*

80–88 Qui…preparatione] Similar calulations can be found in many computistical texts. COMP. COL. 2.6A–B is directly based on MC: *Item tertio intellectu hii quinque dies dividuntur per totum annum. Sic in anno sunt duodecim menses. Qui singuli duodecim menses secundum Aegyptios triginta dies possident. Quinque dies, qui superesse videntur, habent horas centum viginti. Quas cum in decadas diviseris, faciunt duodecim, ita ut unusquisque mensis triginta dies et decem horas habeat. Item si quadrantem, hoc est sex horas, dividi facias, invenies duodecim semisses, ita ut per omnem mensem, hoc est per triginta dies et decem horas, momenta viginti crescere videantur. Quodsi minutius dividas, in omni die sextam ferme momenti partem repperies*. The following Irish or Irish influenced texts record the same results: DE BISSEXTO I 998B5–C10: ... *in unoquoque die crescunt duae partes momenti ... et ita in XXX diebus et in X horis crescunt XX momenta ...* . DE BISSEXTO II 998C13–999B12: ... *et ita in unoquoque die duae partes momenti, in XXX diebus et X horis XX momenta ...* . LDA 4.1.2: ... *unus quidem XXX dierum atque decem horae, ac de preparatione bissexti XX momenta penitus remaneant. ... Sic unus dies ac duae tertiae partes dimidiae horae habere duas partes tertias unius bissextilis momenti intelliguntur*. For the increment per month (or zodiac sign) of the five superfluous days and the quarter-day see DRC 55, DDT 14C13–16 (cited in RM 28.13–16), DTR 11.51–58 (cited in PV §§115–116), LECT. COMP. 9.3C, 8a, RM 28.17–18; for the increment of the quarter-day per month and

This quarter-day is distributed (equally) through the entire year in the following way: There are 12 months of 30 days in a year. Five days remain, which have 120 hours. If you divide these into decades, they will produce twelve times ten, so that one month has 30 days and 10 hours. If you divide the quarter-day, i.e. six hours, in the same way, you will find twelve half (hours), so that (an additional) 20 *momenta* (which equals one half hour) grow per month, i.e. (per) 30 days and ten hours. If you divide this half-hour (per month into) smaller (units) per day, you will find two thirds of a *momentum* in this preparation (of the quarter-day).

The first year begins in the evening of 25 June and it is terminated at midnight of 25 June (of the following Julian calendar year), and so on.

---

day cf. BC 36, LECT. COMP. 7.2B (in 7.2C the increment of each of the five superfluous days per month is discussed); for the increment of the quarter-day per month only cf. BC 41 (only 3 hours), LIB. ANN. 5 (6 hours; cited in LIB. COMP. 3.10B); for the increment of the quarter-day (of 3 hours) per day only see DIAL. LANGOB. 24A: *Et iste quadrans non uno loco tantum crescit, sed sparsim per totum annum, et de materia diei bisexti tertia pars momenti crescit omni die*; for the increment of the five superfluous days per month cf. BC 37; for the division of the quarter-day (of 3 hours) by the 360 days of a year cf. c. 60, ll. 34–37; for a different division of the five superfluous days cf. c. 38, ll. 23–25, c. 40; the division of the five superfluous days by the 12 months of a year is already referred to in c. 24, ll. 12–15. **89–91** Primus...reliqua] Cf. c. 41, ll. 25–28.

80–88 Qui...preparatione] Cf. p. XXII, CLXXXII of the introduction. **87** dimidium...partem] Transcribed in Borst, *Schriften*, p. 899.

A 'Greek' mathematic explanation of the bissextile increment

Aliter Grecorum bissextus preparari artificiose intellegitur. Greci autem *ann*i *hora*s rimari sollicitant. Quas per *VII* diuidunt in figuram VII dierum per quos mundi circulus inuoluitur. *Quod* itaque extra VII sentiunt *super*esse, ex eo diem *bissexti f*aciunt. Hoc modo horas anni numerabis: Nam X horę per X dies duabus horis relictis sunt uniuscuiusque diei, CXX efficiunt. Ac deinde *C C C die*bus *$\overline{III}$* numerantur et duabus *hori*s relictis *DC*. *LX die*bus | *DCCXX hora*e rimantur. Et de *V die*bus *LX hor*ę fol. 23r sunt. Hic est totus numerus, quem diximus, *$\overline{IIII}$CCCLXXX*. Quibus parem *horarum no*ctium *numerum* coniungas, *V$\overline{III}$DCCLX* utrumque inuenire scias. Inde per *VII diui*sas *III hor*as *super*esse noscas. Quibus per quadriennium XII horas effici intellegas. Quas *per*acto IIII annorum curso in *diem bissexti* unari putant.

---

97 duabus] duabus *corr. to* duabas *M.* **98** relictis] relicti *M (corr. according to* MC*).* **99** $\overline{III}$] III *M.* **101** diximus] dixmus *M.* **103** V$\overline{III}$DCCLX] V$\overline{III}$CCLX *M (corr. from* CE, *Argumentum XVI).* **103–104** diuisas…horas] diuisę VI horę *M.* **105** curso] cursum *M.*

92–106 Aliter…putant] This passage provides a curious mathematical explanation for the bissextile day: The argument is that if the 8760 hours of a year, not including the annual bissextile increment itself (365×24=8760), are divided by seven, the remainder is three hours (8760mod7=3); these, then, are supposed to constitute the annual bissextile increment, a quarter-day, with a day being reckoned as consisting of 12 hours. The earliest datable occurrence of this proof is in the *Computus Digbaeanus* of AD 675 as PS-DIONYSIUS, *Argumentum XVI* (Krusch, *Studien* II, p. 80–1; cited in BC 43B14–C13): *Sex dies fecit deus mundum, septimo requievit. Ut ergo plenius intellegatur, conputa qantas horas habet unus dies, et divides illas in VII partes et quantas remanent, exinde fit bissextus. Primo conputa dies CCC, quomodo horas habent, decies tricenti, sunt $\overline{III}$. Iterum facis: bis tricenti, sexcenti, fiunt in tricentis diebus horae $\overline{III}$ DC. Iterum facies: decies sexageni DC, et bis sexageni CXX. Fiunt ergo in sexagenis diebus horae DCCXX. Iterum facis: decies quini L, et bis quini X. Ecce habes in quinque diebus horas LX. Fiunt simul integro anno in diebus CCCLXV horae $\overline{IIII}$ CCCLXXX, et alias tantas in nocte, fiunt simul dierum et noctium totius anni horae $\overline{VIII}$ DCCLX. Divide illas in VII partes. Primum facis: septies milleni VII, remanent $\overline{I}$ DCCLX. Item facis: septies ducenteni, fiunt $\overline{I}$ CCCC, remanent CCCLX. Item facis: septies quinquageni, fiunt CCL, remanent X. Item facis: septies as, VII, remanent III. Iste tres horae faciunt in III (recte IIII) annis diem.* MC's immediate source, however, appears to have been CE p. 108, where this theory is ascribed to a certain Theodore (of Tharsus?): *De bissexto, quem secundum Theodorum (!) formamus hoc modo: ut in numero horarum diei per totum annum septenum ponamus numerum et quodcumque superfuerit septenum numerum, hoc fiat materia bissexti. Cuius probatione hęc est: Annus diebus CCCLXV conputatur. C uero dies horas habent $\overline{I}$ CC; ita et CCC dies habent horas $\overline{III}$ DC; LXta uero dies habent horas DCCXX; V autem dies horas LXta habent; que simul faciunt horas $\overline{IIII}$ CCCLXXX. Ita et nox eundem numerum horarum per totum annum habet. Hinc conpotus est horarum totius anni $\overline{VIII}$ DCCLX. Hoc septeno numero diuide VII milia et DCC, remanent $\overline{I}$ et LXta. Remitte itidem DCC, remanent CCLX. Partire iterum hunc numerum in quinquaginta, hoc est septies quinquaginta, remanent X. Deduc VII ex hoc numerum, remanent III horę, quę septenum numerum supersunt, et quę per quadriennium diem bissexti licet abusiuum faciunt et non diem proprium.*

92–106 Aliter…putant] As outlined in *app. font.*, this mathematical explanation of the bissextile day may have originated in the second half of the seventh century, possibly in Theodore of Tharsus' Canterbury school, whence it found its way into the *Computus Digbaeanus* of AD 675, presumably composed in an Anglo-Saxon monastery in Ireland; at least by the late seventh, early eighth centuries

The bissextile day of the Greeks is known to be skilfully prepared in a different manner. The Greeks require the hours of a year to be calculated. They divide these by seven, according to the nature of the seven days by which the circle of the world revolves. Then, they obtain the bissextile day from that part, which they make out to remain above seven (i.e. the remainder of this division by seven). You will count the hours of a year in the following way: As ten hours, with two hours remaining, belong to every single day, these result in 120 hours per ten days (12×10=120 hours). Consequently, 3000 (hours) are counted in three times 100 days, (plus) 600 hours from the two remaining hours (per day) ((10+2)×300=3600 hours). 720 hours are calculated in 60 days (12×60=720). And 60 hours result from five days (12×5=60). The total number, which we have referred to above, is 4380 hours (3600+720+60=4380). If you add the same number of nocturnal hours to these, you should be sure to find a total of 8760 hours (2×4380=8760 hours). You should then learn that three hours remain from the division (of this total) by seven (8760mod7=3 hours). You should observe that twelve hours are produced by these every four years. They believe that these are united into a bissextile day after the course of four years has passed.

---

this theory was very popular among Irish computists, as CE and MC illustrate, and in the eighth century it can only be found in texts directly influenced by Irish computistical thought. Cf. BC 39, 40, DIAL. BURG. 14A, DE BISSEXTO I 994C14–995B3; from DE BISSEXTO it made its way into LIB. COMP. 1.3f; the manuscript evidence of this algorithm has still to be fully explored.

92–106 Aliter…putant] Cf. Borst, *Schriften*, 366 and p. XXII, CL–CLII, CLXIII of the introduction.

A year of bissextile days

Annus plenus de bissextis in mille CCCCtisLX continetur. Bissexti totius libri Victorii CXXXIII. Ab initio mundi IIII anni de bissextis pleni sunt usque in pręsentem annum sub consulibus Bero et Bardoa.

<XLII.> DE AETATE

Def. and etym. of *aetas*

Ętas pro uno anno, pro VII annis et VI milibus *dicitur. Et aetas quasi aeuitas, id est similitudo aeui. Nam aeuum aetas perpetua, cuius neque initium neque extremum noscitur. Quod Greci uocant* aeuoas. *Quod apud* nos *aliquando pro seculo, aliquando pro aeterno ponitur.*

Classific. of *aetas*

Aetas aut hominis, ut *infantia, aut mundi,* ut *aetas ab Adam usque ad Noe.*

---

107 in] *add.* <diebus> *Borst.* **108** CXXXIII] *add.* <sunt> *Mac Carthy.* **109** IIII] III *M (corr. according to* BC *and* LIB. COMP.*; also Mac Carthy).* **110** Bero] uero *M* (also Schwartz).*

107–110 Annus…Bardoa] This passage provides one of two dating clauses that reveal that one of MC's sources was a now lost Victorian computus of AD 689 (cf. the parallel passage dealing with the *saltus* in c. 62, ll. 64–67). This Victorian text as transmitted here shows parallels to seventh-century southern Irish computistics: The 133 bissextile years in the Victorian cycle of 532 years are also calculated in EPISTOLA CUMMIANI 225–226: *quem <Victorius, per uicesimas> et oc<tau>as uic<es> cum kalendis DXXXII et bissext<is CXXXIII, i>n id i<psum>, unde ortus est, redire fecit.* and in the SIRMOND MS (Oxford, Bodleian Library, Bodley 309, fol. 95v; the relevant passage transcribed in Ó Cróinín, 'Bede's Irish computus', p. 209; it also appears in Paris, Bibliothèque Nationale, Lat. 16361, p. 241): *Uictorius, in quo ordine scriptorum? ... Cursum paschalem ratione certissima composuit. ... In hoc circulo bissexti sunt CXXXIII.* Moreover, the correct spelling of the two consul names given here was *Vero* and *Bradua* (cf. Krusch, *Studien* II, p. 33), but the Victorian Easter table in the SIRMOND MS (Oxford, Bodleian Library, Bodley 309, fol. 115r) records the same misspelling as *M* for the second name in the year equivalent to AD 689: *XIII, Ivro et Bardua, annus CXXX, Kalendae Ianuarii VI feria luna IIII, pascha XIIII Kalendas Maii luna XXII.* **XLII,2–8** Ętas…Noe] The definition and etymology of *aetas* is copied from ETYM. 5.38.3–4 (cited in RM 96.8–14, PV §§186–187, both reading *eonas* instead of *αἰῶνας*): *Aetas plerumque dicitur et pro uno anno, ut in annalibus, et pro septem, ut hominis, et pro centum, et pro quovis tempore. Vnde et aetas tempus, quod de multis saeculis instruitur. Et dicta aetas, quasi aevitas, id est similitudo aevi. Nam aevum est aetas perpetua, cuius neque initium neque extremum noscitur, quod Graeci vocant αἰῶνας; quod aliquando apud eos pro saeculo, aliquando pro aeterno ponitur*. The classification of *aetas* is an abridged version of ETYM. 5.38.5 (cited in BC 54A9–B1, PV §188; only the first part in RM 96.15–18): *Aetas autem proprie duobus modis dicitur: aut enim hominis, sicut infantia, iuventus, senectus; aut mundi, cuius prima aetas est ab Adam usque ad Noe; secunda a Noe usque ad Abraham; tertia ab Abraham usque ad David; quarta a David usque ad transmigrationem Iuda in Babyloniam; quinta deinde <a transmigratione Babylonis> usque ad adventum Salvatoris in carne; sexta, quae nunc agitur, usque quo mundus iste finiatur.*

107–108 Annus…continetur] The idea outlined in MC here that the 365 bissextile days occurring in the course of 1460 years (4×365=1460) form a year of bissextile days made its way into early ninth-century Frankish computistics: LIB. COMP. 3.8 (cited in LIB. CALC. 45): *In quot annis de bissextilibus diebus unus annus adcrescat ... In mille quadragentis sexaginta annis annum integrum, id est trecentos sexaginta sex dies esse probantur*. Interestingly enough, QUAEST. AUSTR. 2.9D counts the number of days in a four-year period as 1460 and one, i.e. the bissextile day. The number of bissextile days in an Easter-cycle of 532-years, i.e. 133 (532/4=133), is also outlined by Bede: DTR 65.2–7 (cited in LIB.

'A year full of bissextile days' is contained in 1460 (solar) years (365×4=1460 years). The bissextile days of the entire Victorian cycle (amount to) 133 (532/4=133 bissextile days). Four 'years full of bissextile days' (i.e. 4×1460=5840 years) have passed from the beginning of the world to the present year under the consuls Berus and Bardoa (i.e. AM 5889).

42. ABOUT THE *AETAS*

(The term) *aetas* is used for a year, for seven years, and for 6000 (years). And *aetas* (is) just as time unending (*aeuitas*), namely an analogy to eternity. For eternity (is) a perpetual age, of which neither beginning nor end is known. The Greeks call this *aeuoas*. By us this (term) is sometimes taken for a millennium, sometimes for eternity.

(The term) *aetas* is used (in two different ways, referring) either (to a period in the life) of a human being, like childhood, or (to a period in the history) of the world, like the age from Adam to Noe.

CALC. 98, RM 94.2–9): *Circulus paschae magnus est qui ... DXXXII conficitur annis. ... idem circulus magnus ... habeat ... bissextos decies novies septenos, id est CXXXIII.* Similar to MC, BC 119A7–12 counts the exact number of bissextile days from the creation of the world to the *annus praesens* of the author (AM 5988=AD 788): *... quod dies bissextiles ... ab origine mundi. ... Bissextiles autem dies sunt mille CCCCXCVII, in quibus continentur anni IIIIor et dies XXXVII.* DRC 58.4–6 calculates the number of bissextile days from the beginning to the end of the world as 1500 (6000/4=1500), DIAL. NEUSTR. 28aD from the beginning of the world to the cruxifiction as 1307 (5228/4=1307). **XLII,2–8** Ętas...Noe] MC's account of *aetas* is directly copied in ANGERS 477, fol. 10r (gloss to Bede's DNR): *Aetas pro uno anno uel pro VI annis et VI m accipitur. Aetas quasi aeuitas, id est similitudo aeui, nam euus aetas perpetua, cuius neque initium neque extremum noscitur. Quod Greci eonas uocant. Quod aliquando apud nos pro seculo, aliquando pro inifinito* (!) *ponitur. Aetas aut hominis aut mundi, ut prima aetas ab Adam usque Noe*. MC's variants to Isidore's definition of *aetas* can also be found in DRC 59, another clear proof of the interdependency of these two Irish computistical textbooks: *Sciendum nobis quid sit aetas. Grecum nomen; perpetuitas interpretatur. Pro uno anno et pro septem annís et pro sex milibus accipitur, a cuius uero aetas perpetua appellatur et cui initium et finis non cognuscitur*. Moreover, MC has many parallels, especially the mentioning of the 6000 years, to a tract headed *De aetate* (Bern, Burgerbibliothek, 417, fol. 33r; Basel, Universitätsbibliothek, F III 15k, fol. 35r), which probably formed, at least in its original form, one of the final chapters of the now lost Irish *De divisionibus temporum* (cited from the Bern MS); if this was the case, then this text may have been one of MC's sources for this chapter: *DE AETATE. Aetas aliquando pro VII annis dicitur; sicut dicuntur VI aetates unius hominis; aliquando aetas L anni; aliquando tempus uite hominis a natiuitate sua usque ad mortem; aliquando pro VI milibus VI aetates dicuntur. ... Aetas ergo dicitur etiam totum tempus uite praesentis ab initio usque ad finem. Inde Isidorus dicit: Aetas perpetua est, cuius neque initium neque extremum noscitur. Item alia diffinitio: Aetas est quod de multis saeculis instruitur, et dicta est ętas quasi aeuitas, hoc est similitudo aeui; nam aeuus dictus est Greco uocabulo nenas* (*eonas* Basel), *quod apud Grecos aliquando pro saeculo, aliquando pro aeterno ponitur*.

107–110 Annus...Bardoa] Transcribed in Mac Carthy, *Annals of Ulster* 4, p. clxxviii; Schwartz, 'Ostertafeln', 89; only the first sentence transcribed in Graff, 'Thirteenth figure', p. 333 and Borst, *Schriften*, p. 494–5, who mistakenly thinks that the number of days in a four-year period is described here; only the second half transcribed in Krusch, *Studien* I, p. 10; idem, 'Einführung', p. 162; Mommsen, 'Victorii Aquitani cursus', p. 697–8; cf. also Ohashi, 'Sexta aetas', p. 59 and p. XXII, LVII–LVIII, LXXXI–LXXXII, XCIII–XCIV, CXXIV, CLXIX of the introduction. **XLII,2–8** Ętas...Noe] Cf. p. CLXXXVII of the introduction. **3–5** aetas...aeuoas] Transcribed in Borst, *Schriften*, p. 436.

<XLIII.> DE SECULO

Etym. and def. of *saeculum*

Seculum secus colendo dictum, uel sex et colo, quia sex aetates mundi continet. Et pro infinito numero ponitur.

<XLIIII.> *DE MUNDO*

Def. and etym. *mundus*

Mundus est caelum et terra et *mare et* omnia, *quae in eis* sunt. *De quo dicitur*: *mundus per eum factus est.* | fol. 23v *Mundus Latinę a philosofis dictus, quod in sempiterno moto sit, ut caelum, sol, luna, aer,* terra, *maria.* His *enim elementis nulla requies concessa est. Greci uero nomen mundo* inposu*erunt de ornamento, propter diuersitatem elementorum et pulchritudinem siderum. Appellatur enim apud eos cosmos, quod significat ornamentum.* Inde Ysidorus ait: *nihil pulcrius* est *mundo.*

Creation – passion

Rimari *ab* initio huius creaturae per *ann*os *VCCXXVIII* usque ad *pass*ionem.

XLIIII,11 ab initio] ad initium *M (corr. by Schwartz).*

XLIII,2–3 Seculum...ponitur] There is no apparent source for these etymologies and definitions of *saeculum*; only the definition of *saeculum* representing an infinite number of years may ultimately derive from the psalms; cf., e.g., PSALM 47:15: *quoniam hic est Deus Deus noster in aeternum et in saeculum saeculi ipse reget nos in saecula*. Cf. also ETYM. 5.38.3–4, cited in the previous note. MC's account is strikingly different from Isidore's definition and etymology, who considered *saeculum* as a unit of time smaller than *aetas* (and so also MC in c. 1, l. 7). ETYM. 5.38.1 (cited in RM 95.6–7, PV §185): *Saecula generationibus constant; et inde saecula, quod se sequantur.* **XLIIII,1–10** De...mundo] MC's etymology and definition of *mundus* is directly copied from ETYM. 13.1.1–2 (MS C reads *eis* instead of *eius*): *De mundo. Mundus est caelum et terra, mare et quae in eis opera Dei. De quo dicitur* (JOHN 1:10)*: Et mundus per eum factus est. Mundus Latine a philosophis dictus, quod in sepiterno motu sit, ut caelum, sol, luna, aer, maria. Nulla enim requies eius elementis concessa est ... Graeci vero nomen mundo de ornamento adcommodaverunt, propter diversitatem elementorum et pulchritudinem siderum. Appellatur enim apud eos κόσμος, quod significat ornamentum. Nihil enim mundo pulchrius oculis carnis aspicimus*. Cf. also ETYM. 3.29 (cited in PV §475), DNR 9. **11–12** Rimari...passionem] The world-era applied here is directly copied from PROLOGUS VICTORII 9 (Krusch, *Studien* II, p. 24; cited in DRC 89.5–6, SIRMOND MS (Oxford, Bodleian Libray, Bodley 309, fol. 95v; transcribed in Ó Cróinín, 'Early Irish annals', p. 83)): *Passum autem dominum nostrum Iesum Christum peractis ab ortu mundi quinque milibus ducentis viginti et octo annis*. The period from the creation to the passion was essential for Victorius, since his Easter table began with the year of the passion of Christ, so that this passage set his table into context with the world-era. The same number of years is given in a more indirect way for this period from the creation of the world to the passion of Christ in c. 68, ll. 32–33 (10×532−92=5228). Cf. also c. 64.

XLIII,2–3 Seculum...ponitur] MC's etymology and definition of *saeculum* received some attention in early eighth-century Irish computistics, but apparently not beyond. The definition of a *saeculum* containing all six ages of the world is copied almost verbatim in ANGERS 477, fol. 10r (gloss to Bede's DNR): *Seculum sepius colendo dictum, uel sex et colo, quia VI <aetates> mundi continet*. In can also be found in DRC 60.1–2: *Sciendum nobis quid sit seculum. Nomen conpossitum, id est sex et copulum, uel cultus, eo quod sex mundi temporibus*. and DIAL. LANGOB. 2A, 3B (cited in PV §184): *saeculum tempus est, quod sex aetates in se copulat, id est custodiat. ... Saeculum autem dicitur eo, quod copulat in se sex aetates mundi*. The second definition, *saeculum* representing an infinite number, is repeated in DRC

43. ABOUT THE *SAECULUM*

The *saeculum* was so called from cultivating differently (*secus colendo*), or from 'six' (*sex*) and 'to cultivate' (*colo*), since it contains the six ages of the world. Or it is taken for an infinite number.

44. ABOUT THE *MUNDUS*

The *mundus* covers the sky, and the earth, and the sea, and everthing that is in these. It is said about it: The *mundus* was created by him. The *mundus* was so called in Latin by the philosophers, because it appears to be in everlasting motion (*moto*), as the sky, the sun, the moon, the air, the earth, (and) the seas. In fact, no halt was conceded to these elements. The Greeks, however, had given (their) name to the *mundus* from (their term for) ornament, because of the diversity of the elements and the beauty of the stars. In fact, it is named *cosmos* among them, which means ornament. Accordingly Isidore says: Nothings is more beautiful than the *mundus*.

(The period) from the beginning of its (i.e. the *mundus*) creation to the passion is reckoned as 5228 years.

60.2–3: *Et seculum ad infinitatem dicitur.* and it also appears, though based on the psalms rather than MC, in RM 95.12–13: *Aliquando tamen seculum in Propheticis Libris pro aeterno positum repperimus.* For other definitions cf. Bern, Burgerbibliothek, 417, fol. 33r–v; Basel, Universitätsbibliothek, F III 15k, fol. 35r–v. **XLIIII,2–10** Mundus…mundo] Besides MC, the only other printed text that discusses *aetas*, *saeculum*, and *mundus* in successive chapters, DRC, shares some of MC's variants to ETYM.: DRC 61.1–3: *Sciendum nobis quid sit mundus. Congregatio creaturum, quod caelum et terra et mare et omnia quae in eis sunt, ut Isidorus dicit: Mundus est universitas creaturum inter caelum et terram et mare. Sciendum nobis unde mundus nominatus est. Id est de ornamento, ut Isidorus dicit: Greci de ornamento nomen mundo dederunt, propter diuersitatem elimentorum et pulchritudinem siderum. Nam Greci cosmos ornamentum dicunt. Quid enim pulchrius mundo oculis carnalibus aspicientibus.* Cf. also DDT 1B6: *deinde plenitudo dicitur mundus.* and the discussion in Bern, Burgerbibliothek, 417, fol. 33v; Basel, Universitätsbibliothek, F III 15k, fol. 35v, where *aetas*, *saeculum*, and *mundus* are also discussed in this order. **11–12** Rimari…passionem] In general, the Victorian world era (ultimately based on Eusebius' chronicle) was extremely popular right up into the ninth century. The explicit number of years from the creation to the passion is directly copied from Victorius in DRC and the SIRMOND MS (cf. *app. font.*), it occurs in a different passage in DRC 62.1–2 (MS B reads *VCCXXVIIII* before correction): *Sciendum nobis quot anni sunt a principio mundi usque ad passionem Christi. Id est VCCXVIIII.* and then also features prominently in early eighth-century Frankish computistics: DIAL. BURG. 17B: *Fiunt insimul anni usque ad passionem Domini quinque milia ducenti viginti octo.* DIAL. NEUSTR. 30B: *Ab initio mundi usque ad crucem Domini quinque milia ducenti viginti octo.*

XLIII,2–3 Seculum…ponitur] Cf. Borst, *Schriften*, 374 and p. CLXXV, CLXXXVII, CXCIII of the introduction. **XLIIII,11–12** ab…passionem] Transcribed in Schwartz, 'Ostertafeln', p. 90; cf. Ohashi, 'Sexta aetas', p. 59; Borst, *Schriften*, p. 373, 421, 461; and p. XXII of the introduction.

Three comparable years in the course of the world

Tres in hoc mundo Augustinus modos immobiles posuit, hoc est cursus solis et lunae et annos utriusque, in quibus III coetaneos fixit: annum primum mundi et annum egressionis de Aegypto populi et annum resurrectionis Christi ⌐ annumque iudicii futuri ¬ .

---

15 quibus] anni *add. Borst.* | III] *corr. to* IIII *Borst.* **17** annumque…futuri] *I regard this phrase as a later gloss that crept into the text, since the* annus futurus *is the fourth year in a list that should explicitly include only three years, and since it is not discussed later in this chapter (note that Borst (cf. previous notes and app. comm.) has a different opinion on this).*

13–80 Tres…XVII] There are only two pre-MC texts that discuss and compare the chronology of these three years, namely PROLOGUS VICTORII 7–9 (Krusch, *Studien* II, p. 22–25) and PROLOGUS PASCHAE 4–6 (Krusch, *Studien* I, p. 229–232). Whether the Cologne text was known to MC can only be speculated, the Easter reckoning described in it (the *Supputatio Romana*, an 84-year Easter table with 12-year *saltus*), however, was never in use in Ireland (it was known to MC only through the Victorian prologue; cf. c. 65, ll. 2–12). The Victorian prologue, on the other hand, was certainly known to MC and may have served as the model here. Note, however, that in the Victorian prologue the chronology of these three years is not identical. MC's ultimate impetus for applying the same chronology to these three years may have been the traditions (cf. the following notes) that record the same weekday, Julian calendar, and lunar data for the fourth day of creation and the Last Supper, namely Thursday, 25 March, *luna* 14. The weekday-terminology applied in this section, particularly the use of the bilingual term *dies cetene* for Wednesday, follows the advice to apply Christian rather than pagan terminology for the weekdays, especially for Wednesday (if such Christian terminology was available in the respective vernacular of the reader) outlined by AUGUSTINE, *Enarrationes in psalmos*, *Enarratio in psalmum 93, sermon 3* (*CCSL* 39, p. 1302–3): *Quare ergo talem habet titulum: in quarta sabbati? Vna sabbati, dies dominicus est; secunda sabbati, secunda feria, quem saeculares diem Lunae uocant; tertia sabbati, tertia feria, quem diem illi Martis uocant. Quarta ergo sabbatorum, quarta feria, qui Mercurii dies dicitur a paganis, et a multis christianis; sed nollemus; atque utinam corrigant, et non dicant sic. Habent enim linguam suam qua utantur. Non enim et in omnibus gentibus ista dicuntur; multae gentes aliae atque aliae aliter atque aliter uocant. Melius ergo de ore christiano ritus loquendi ecclesiasticus procedit.* The definition of lunar days extending from midday to midday is ultimately based on DRP 5.97–106; cf. also c. 8, ll. 36–38; c. 48, ll. 2–4.

13–17 Tres…futuri] MC's comparison of the chronology of these three most important Biblical years (or MC's ultimate source, the Victorian prologue) may have stimulated the discussion of the first full moon of these three years in DIAL. NEUSTR. 15A–B, which in turn, together with MC, influenced COMP. COL. 5.1, 4C: *De quarta decima <luna> prima mundi in principio, et quarta decima primae paschae filiorum Israhel, et quarta decima, in qua Iesus cum discipulus inmolavit agnum in caena. Quis has consecravit quartas decimas. ... Item tria aiunt esse solemnia similia Paschae initio mundi, luna quarta decima. Id: Primum Pascha mundi, secundum Pascha filiorum Israhel, tertium Pascha manna, quartum Pascha resurrectio Christi. Similes sunt in diebus solis, in aetate lunae, sed tantum grece, non latine.*

13–17 Tres…Christi] Transcribed in Bisagni & Warntjes, 'Latin and Old Irish', p. 11. **14–17** in…futuri] Transcribed (and corrected differently; cf. *app. crit.*) in Borst, *Schriften*, p. 924.

Augustine postulated three immovable measures (to be) in this world (*mundus*), namely the course of the sun and of the moon and the years of both, among which he fixed three corresponding years: the first year of the world (*mundus*), the year of the Exodus of the people from Egypt, and the year of the resurrection of Christ ⌐ and the year of future judgement ¬ .

The year of creation

Prima enim creatio in uere facta est, in XII Kalendas Aprilis, id est VI horę restabant de XII die dominico. A VI hora dominici usque in VI horam diei lunis, XI Kalendas Aprilis; a VI hora diei lunis usque in VI horam diei martis, X Kalendas Aprilis; a VI hora <diei> martis usque in VI horam diei cetene, VIIII Kalendas Aprilis. In initio diei sol in cęlo positus est. Hora et dimidium horae uenit de luce,

---

20 lunis] lunes *M, Mac Carthy, Ó Cróinín (corr. according to* MC*)*. **22** diei] *om. M, Ó Cróinín (add. according to* MC*; also Mac Carthy)*. | usque] utque *Mac Carthy*.

18–36 Prima…XVII] The general idea behind the chronology outlined for the year of creation here, as well as for the following two years, seems to be a reconciliation of the 'Roman' vernal equinox of 25 March with the 'Greek' one of 21 March. This was achieved by placing the fourth day of creation, the day that saw the creation of the sun and the moon, on 25 March, and by reckoning a day from midday to midday (cf. c. 8, ll. 34–43), resulting in the first day of the creation of the world beginning at midday of 21 March. Similarly, MARTIN OF BRAGA, *De Pascha* 5 (Barlow, *Opera omnia*, p. 273; cited in TRACTATUS ADTHANASI 5 (Krusch, *Studien* I, p. 332)) placed the fourth day of creation on 25 March, but since he reckoned a day from midnight to midnight, he calculated 22 March as the first day of creation: *Sed non sine causa maiores nostri super VIII Kal. Apr. tres dies addiderunt, ut primum diem mundi invenirent. XI enim Kal. Apr. primum mensem mundi et diem maiores nostri existimaverunt, quia antequam sol in principatum mundi conderetur triduum ante praecesserat. Refert enim Genesis quarta die facta luminaria solis et lunae*. This chronology for the creation was already outlined in PROLOGUS PASCHAE 4, 6 (Krusch, *Studien* I, p. 231–2); note that in this text the moon has age 15 at its creation rather than MC's 14: *Inde intellegimus in principio creature primam diem factam esse XI kl apr, saecundam diem X kl apr, III diem VIIII kl apr, IIIItam diem VIII kl apr, quando fecit et luminaria in caelo. ... Principium huius mundi XI kl ap facta sit dies prima; quarta autem die VIII kl ap solem et lunam fecit, id est lunam XV*. DE ORDINATIONE FERIARUM PASCHALIUM (*PL* 90, col. 608B) also places the creation of the moon (i.e. the fourth day of creation) on 25 March: *Nunc ergo investigavimus quomodo in principio factus fuerit mundus, id est, die dominico. Verno tempore in aequinoctio, quod est octavo Kalendarum Aprilium, luna plena per ipsum tantummodo tempus et elementa resurgunt*. Note that MC's ultimate source for this chapter, PROLOGUS VICTORII, outlines a different chronology: in c. 7 (Krusch, *Studien* II, p. 23) the first day of creation is set on *VIII Kalendas Aprilis*, and consequently the fourth day with *luna* 14 on *V Kalendas Aprilis* in c. 8 (Krusch, *Studien* II, p. 23; cited in DRC 67.2–3). For related chronological statements about the creation of the world see c. 7, ll. 9–11; c. 11, ll. 37–40; c. 36, ll. 10–15. The question whether the moon takes its light from the sun or from itself is based on DNR 18.1.4–8, 18.3.21–22, 19.1.3–7: *Alii namque dicunt proprium eam habere lumen, globique eius unam partem esse lucifluam, aliam obscuram, et dum mouetur in circulo suo, eandem partem qua lucet paulatim ad terras conuerti, ut uideri a nobis possit ... At contra alii dicunt lunam non suo fulgere lumine, sed a sole accipere lumen ... Quia cum a sole accipiat lumen et ita nobis lucere uideatur, non est dubium eam moueri potius quam stare. Terris autem uicina luna breuiori orbe conuertitur, et iter quod sol in diebus CCCLXV peragit, ista per XXX dies percurrit*. Cf. ETYM. 3.53.1–2, 3.57 (cited in c. 46, ll. 6–8, 10–12, 15): *Lunam quidam philosophi dicunt proprium lumen habere, globique eius unam partem esse lucifluam, aliam vero obscuram, et paulatim se vertendo diversas formas efficere. Alii e contra aiunt lunam non suum lumen habere, sed solis radiis inluminari. ... Luna vicinior est terris quam sole. Inde et breviori orbe celeries peragit cursum suum. Nam iter, quod sol in diebus trecentis sexaginta quinque peragit, ista per triginta dies percurrit*. Note that contrary to Isidore, his main source, AUGUSTINE, *Enarrationes in psalmos, Enarratio in psalmum 10, Sermo 3* (*CCSL* 38, p. 75–6; partially cited in DTR 25.12–21, in turn cited in LIB. CALC. 74), like MC does not express a preference for either opinion. Cf. MC's later statements on this question in c. 46, ll. 6–8, 15; c. 48, ll. 16–18.

18–36 Prima…XVII] Only two texts, one computistical, the other theological, outline the same chronology for the year of creation: The computistical is BC 148B5–C4, which certainly comes from

In fact, the first creation was made in spring, on 21 March, i.e. six out of twelve hours remained (from that day), a Sunday. (The first day of creation extended) from the sixth hour (i.e. noon) of Sunday to the sixth hour of Monday, 22 March; (the second day of creation extended) from the sixth hour of Monday to the sixth hour of Tuesday, 23 March; (the third day of creation extended) from the sixth hour of Tuesday to the sixth hour of Wednesday, 24 April. At the beginning of the (following) day (i.e. the fourth day, which started at that point, the sixth hour of Wednesday) the sun was set in the sky. One and a half hours came from

---

the same intellectual milieu as MC: *Recto ergo tramite dierum, quos praediximus, superius nuper principium et finis illorum numeratur, quod primum XII Kalendas Aprilis in principio die dominicae incipit, quasi fuisset ante in sabbato dimedium ipsius diei et in sexta hora dominici diei finitur, sed paulo post X Kalendas Aprilis consecutus est. Et similiter in sexta hora II feria consummatus est, et X Kalendas Aprilis usque ad dimedium III feriae incipiens a sexta hora secunda feria migrauit, et VIIII Kalendas Aprilis usque ad dimedium IIII feriae apparuit incipiens ante sexta hora V feriae, et in hoc die sol luxit. Initio ipsius et in VIII Kalendas Aprilis usque ad dimedium V feriae morabatur incipiens a sexta hora IIII feria ante, et in VIII Kalendas Aprilis ostensa est luna in caelo in fine illius diei in XIIIImae lunae etate.* Even closer related, and certainly directly dependent on MC, is the Irish reference Bible: PP I §§92, 123, 128: *Quid hic nomine lucis dicitur? Alii dicunt lux pallida diei sine sole, sed uerius inspectio lucida angelorum qua semetipsos et Creatorem suum agnouerunt. ... Huc usque opus tertii diei, X Kl. Aprilis, III feria. Isti sunt tres dies pallidi, ut alii, sed tamen lux angelica excellit lucem corporalem ... Sol in luce diei, id est in VIIII Kl. Aprilis et IIII feria, possitus est in caelo, et precedit eum hora demedia de pallida luce. Luna uero in demedia hora post sextam horam accensa est a sole, et orta est in fine diei incetate XIIII luna.* Accordingly, this chronology appears to be a characteristic feature of Irish computistics of the first half of the eighth century. For other texts referring to the tradition of the world being created on 21 or 22 March cf. c. 11, ll. 37–40. The chronology outlined by MC was then superseded by Bede's theory that the fourth day of creation, i.e. the creation of the heavenly bodies, occurred on the 'Greek' spring equinox, 21 March, and the first day of creation subsequently three days earlier, on 18 March, which also marks the entry of the sun into the sign of Aries: DTR 6.65–69 (cited in PV §76, up to *notandum* in LIB. CALC. 66D): *Nunc admonere contenti XII kalendarum aprilium die occursum aequinoctii at ante triduum, hoc est XV kalendarum earundem, primum saeculi diem esse notandum, cuius ad indicium primatus ibi reor antiquos initium zodiaci circuli voluisse praefigere.* Cf. RM 23.13–19. MC's reckoning of the days in general, the equinox in particular, as extending from midday to midday influenced COMP. COL. 3.4: *Aequinoctium vernum XII Kalendas Aprilis <est>. Tamen adprehendit XI <Kalendas> XII <Kalendas> a mane usque ad sextam horam, quia dies a sexta hora usque ad sextam horam diei sequentis <est>.* Moreover, COMP. COL. 5.2A account of the full moon at the creation of the world is directly based on MC: *Quarta decima <luna> prima mundi facta est in vere, in mense Martio, in quarta feria? ... Tamen secundum numerum in media hora post sextam horam accenditur.* For the quarter-day at the beginning of the world cf. c. 7, ll. 9–11; c. 39, ll. 2–22; c. 41, ll. 38–42; the question of the origin of the light of the moon is discussed in the context of the creation of the moon on the fourth day also in PP I §134: *Alii dicunt quod non a sole lumen accepit, sed a sua natura.* and COMP. COL. 5.2A–B, which is directly based on MC: *Secundum veritatem accensio a sole, etiam si variantur alii de partibus caeli et horis. Tamen secundum numerum in media hora post sextam horam accenditur. Alii interrogant, quomodo fuit in primo die, si a sole vel a se ipsa accensa sit? Utrumque non inpossibile.*

18–24 Prima…est] Transcribed in Mac Carthy, *Annals of Ulster* 4, p. clxxx, save for the last sentence in Ó Cróinín, 'Old Irish gloss', p. 131–2; idem, 'Earliest Old Irish glosses', p. 16. **18–80** Prima…XVII] Cf. Borst, *Schriften*, p. 930; for the weekday-terminology cf. Bisagni & Warntjes, 'Latin and Old Irish', p. 8–12; p. XIX, LXII, LXXV, CIV, CLXXXI–CLXXXII of the introduction. **24–25** Hora…dicunt] Cf. Borst, *Schriften*, p. 902.

ubi materiam quadrantis apparari dicunt, | et finis trium luminum pallide lucis, quia angelorum erat. Inde sol cursum suum incipit. In media hora post VI horam *luna accep*it *lumen* siue *a sole* siue a semetipsa, quia alii dicunt plenam esse lucis, quam celabat usque dum sol prope occasum descenderat. Inde in aetate XIIII oritur, usquedum in tricesima circa terram conuertitur. A VI hora in IIII feria in VI horam in V feria, VIII Kalendas Aprilis et luna XIIII; a VI hora in V feria in VI horam in VI feria, VII Kalendas Aprilis et luna XV; a VI hora in VI feria in VI horam sabbati, VI Kalendas Aprilis et luna XVI; a VI hora sabbati in VI horam dominici, V Kalendas Aprilis et luna XVII. fol. 24r

The year of the Exodus from Egypt

Veni ad annum egressionis de Aegypto: XII Kalendas Aprilis usque in VI horam ante egressionem populi; a VI hora diei dominici in VI horam diei lunis, XI Kalendas Aprilis; a VI hora diei lunis in VI horam diei martis, X Kalendas Aprilis est; a VI hora diei martis in VI horam diei cetene, VIIII Kalendas Aprilis est; a VI hora IIII feriae in VI horam V feriae, VIII Kalendas Aprilis est et luna XIIII, in qua dictum est: occidat *unusquisque agnum* su*per familia*m *sua*m, nisi tantum in hac uice in VI hora diei lunis *agnus immolab*atur. Hinc dicitur per dominum populo: *Nox ista* uener*abilis* uobis, in qua *eduxi uos de terra Aegypti*, id est initium eductionis | in V feria: Initio fol. 24v

---

31 tricesima] uicesima *M (corr. according to* DNR*)*. | conuertitur] comexerunt *M (corr. according to* DNR*)*. **32** in[1]] a *M (corr. according to M; Mac Carthy corr.* a<d>*)*. | horam] hora *M (corr. according to* MC*; also Mac Carthy)*. **33** horam] hora *M (corr. according to* MC*)*. **39** diei[2]] die *M (corr. according to* MC*; also Mac Carthy without indication)*. **42** VIIII] VIII *M (corr. according to* MC*; also Mac Carthy)*. **43** horam] haram *M**. | V] VI *Mac Carthy (contra M)*.

37–54 Veni…XVII] The chronology outlined for the year of the Exodus from Egypt is modelled on the one established for the year of creation in the previous passage; theoretically, this chronology could have originated in a conflation of PROLOGUS VICTORII 9 (Krusch, *Studien* II, p. 24; partially cited in DRC 85.6–7): *in diem, quo filii Israhel paschale mysterium caelesti initiavere mandato et ab clade agni occisione salvati sunt ... quinta feria VIIII Kal. aprl. luna XIII ... VIII Kal. april. luna XIIII noctis initio Hebreos claruit agni sacrificium peregisse*. If *VIIII Kal. aprl. luna XIII* is omitted in this excerpt, the chronology constructed in that way agrees exactly with the one given in MC; yet, it appears more likely that MC once again neglected the Victorian prologue and applied his own chronology here, which he had established for the year of creation, and which he believed to be the same for the year of the Exodus as well as the year of the passion. Cf. also the chronological details given for this year in c. 64; interestingly enough, the observations outlined in that chapter, namely that 3688 years have passed from the creation to the Exodus, 1540 from the Exodus to the passion, should have made MC aware that these years could not have the same solar and lunar chronology, since neither 3688, nor 1540, nor their sum, 5228, is divisible by 532. For the Biblical references used in this passage cf. the following notes. **44–46** occidat…immolabatur] EXODUS 12:3–6: *loquimini ad universum coetum filiorum Israhel et dicite eis decima die mensis huius tollat unusquisque agnum per familias et domos suas sin autem minor est numerus ut sufficere possit ad vescendum agnum adsumet vicinum suum qui iunctus est domui*

the light wherein, they say, the material of the quarter-day is prepared, and (it was) the end of three days of pale light, because it belonged to the angels. At this point, the sun started its course. In the mid-hour after the sixth hour (i.e. half past noon) the moon received (its) light either from the sun or from itself, since some say that it was full of light, (a fact) which (the moon) concealed until the sun had almost descended to (its) setting. Accordingly, (the moon) rose at the age of 14, and it is turned around the earth until (the age of) 30. (The fourth day of creation extended) from the sixth hour of Wednesday to the sixth hour of Thursday, 25 March and *luna* 14; (the fifth day of creation extended) from the sixth hour of Thursday to the sixth hour of Friday, 26 March and *luna* 15; (the sixth day of creation extended) from the sixth hour of Friday to the sixth hour of Saturday, 27 March and *luna* 16; (the seventh day of creation extended) from the sixth hour of Saturday to the sixth hour of Sunday, 28 March and *luna* 17.

Turn to the year of the Exodus from Egypt: 21 March up to the sixth hour (still was) before the Exodus of the people; (the first day of the Exodus extended) from the sixth hour of Sunday to the sixth hour of Monday, 22 March; (the second day of the Exodus extended) from the sixth hour of Monday to the sixth hour of Tuesday, (which) was 23 March; (the third day of the Exodus extended) from the sixth hour of Tuesday to the sixth hour of Wednesday, (which) was 24 March; (the fourth day of the Exodus extended) from the sixth hour of Wednesday to the sixth hour of Thursday, (which) was 25 March and *luna* 14, about which it is said: everybody (presiding) over his family should kill a lamb, unless the lamb was only sacrificed at this (other earlier) turn, the sixth hour of Monday. Thus it is said by the Lord to the people: That night should be holy to you, in which I brought you forth out of the land of Egypt, i.e. the beginning of the Exodus on Thursday:

---

eius iuxta numerum animarum quae sufficere possunt ad esum agni erit autem agnus absque macula masculus anniculus iuxta quem ritum tolletis et hedum et servabitis eum usque ad quartamdecimam diem mensis huius immolabitque eum universa multitudo filiorum Israhel ad vesperam. **47–48** Nox…feria] Cf. EXODUS 12:42,51: *nox est ista observabilis Domini quando eduxit eos de terra Aegypti hanc observare debent omnes filii Israhel in generationibus suis ... et in eadem die eduxit Dominus filios Israhel de terra Aegypti per turmas suas.* The phrase *eduxi vos de terra Aegypti* occurs five times in the Vulgate: LEV 11:45, 19:36, 25:38, 26:16; NUM 15:41. For the Exodus happening in the month of the new fruits, which is equated with March, cf. c. 55, ll. 4–6. **48–50** Initio…sunt] Cf. EXODUS 12:37,41: *profectique sunt filii Israhel de Ramesse in Soccoth sescenta ferme milia peditum virorum absque parvulis ... quibus expletis eadem die egressus est omnis exercitus Domini de terra Aegypti.*

37–54 Veni…XVII] The year of the Exodus from Egypt is hardly ever mentioned in seventh- to ninth-century computistcal literature. Its chronological details are compared to the year of the passion in COMP. COL. 5.4B, a comparison directly influence by MC: *Alii adfirmant initium quartae decimae a medio die quartae feriae, usque in medium sextae feriae quintam decimam fuisse, sextam decimam usque in medium diei sabbati, septimam decimam usque in medium diei dominici in resurrectione Christi. Quod credendum est. Sic fuit in Pascha prima cum filis Israhel et convenit, quomodo Dominus dicit: Ad vesperum comeditis azyma. Id <est>: Ad finem quartae decimae tantum ad vesperum diei solis potest dici, quia vespere dicitur dies solis post sextam horam.*

27–28 In…semetipsa] Transcribed in Borst, *Schriften*, p. 926; partially transcribed in Bisagni & Warntjes, 'Latin and Old Irish', p. 14. **31–32** A…XIIII] Transcribed in Mac Carthy, *Annals of Ulster* 4, p. clxxx. **37–43** Veni…XIIII] Transcribed in Mac Carthy, *Annals of Ulster* 4, p. clxxx.

congregato ęxercitu in *Ramise*, VI hora eiusdem V feriae *egressi* sunt. A VI hora in V feria in VI horam in VI feria, VII Kalendas Aprilis et luna XV; a VI hora in VI feria in VI horam sabbati, VI Kalendas Aprilis et luna XVI; a VI hora sabbati in VI horam dominici, V Kalendas Aprilis et luna XVII.

The year of the passion

Veni ad annum resurrectionis Christi: XII Kalendas Aprilis usque in VI horam dominici, in quo *sedit super pullum asinae*, a *Bethania uenit* in *Hierusalem*, ubi *subuertit mensas nummulariorum*; a VI hora diei dominici in VI horam diei lunis, XI Kalendas Aprilis est, in quo die

---

50 VI^{2}] *an additional superscript* I *above* V *in M.* **52** horam] haram *M**.

55–80 Veni…XVII] The chronological details for the year of the passion as applied here, namely that the last Supper took place on Thursday, 25 March, *luna* 14, that Christ was crucified on Friday, 26 March, *luna* 15, and that he resurrected on Sunday, 28 March, *luna* 17, ultimately stems from PROLOGUS VICTORII 9 (Krusch, *Studien* II, p. 24–5; cited in DRC 89.5–17): *Passum autem dominum nostrum Iesum Christum ... si quidem VIII Kal. aprl., primo mense, luna XIIII vespera prodiente, sicut ab initio a creatore quarto die facta est, coepisse dicitur; ... Dehinc sexta feria subsequente, id est VII Kal. aprl. crucifixus est et sepultus, tertia die, hoc est quinto Kal. aprl., dominica, resurrexit a mortuis*. Only the weekday and / or lunar data is also given in PROLOGUS THEOPHILI c. 4 (Krusch, *Studien* I, p. 225; cited in DTR 61.28–51; the first part partially in COMP. COL. 5.4B), c. 5 (Krusch, *Studien* I, p. 226): *Dehinc quoniam etiam salvator quarta decima luna traditus est, hoc est V feria, quinta decima crucifixus est, tercio die surrexit, hoc est XVIIa luna, que tunc in dominicam inciderat diem ... et XV luna, qua salvator noster crucifixus est, eadem incidat septimanam et septimus decimus dies, quo surrexit*. PROLOGUS CYRILLI 7 (Krusch, *Studien* I, p. 342): *XIIII traditur, XV passus et XVII resurrexit*. EPISTOLA PROTERII 2 (Krusch, *Studien* I, p. 271–2; the first part cited in DRC 86.10–14; the second part in COMP. COL. 5.4B): *quinta sabbatorum XIIII luna mensis primi in cenaculo cum discipulis pascha manducans, paulo post a Juda traditur et sequenti die XV luna crucifigitur, id est sexta feria, et ad inferos discendens ac dispensationes salutis nostrae perficiens, vespere sabbati, luciscente dominico, resurrexit a mortuis, in quo die lunam XVII primi mensis iuxta Hebreos extetisse manifestum est. ... XIIII namque luna primi mensis iuxta Hebreos, ut superius dictum est, Jesus pascha tipicum manducavit, sequenti vero VI ferea XV luna, ut ovis occisionis, cruci pro nobis affixus est et vespere sabbati, luciscente dominico, XVII luna resurrexit a mortuis*. This passage of the EPISTOLA PROTERII is cited in the Irish tract DE COMPARATIONE EPACTARUM 2.4–14 (see Appendix 2) with the noteworthy addition of Julian calendar dates which are in agreement with the once given in MC; this tract may have served as a direct source here: *Hac ratione quoniam ipse, plenitudo legis existens, quando dignatus est homo fieri VIII Kalendas Aprelis, quinta sabbatorum, XIIII luna mensis primi in caenaculo cum discipulis pascha manducans, paulo post a Juda traditur et sequenti die, VII Kalendas Aprelis, VI feria, XV luna, crucifigitur, et ad inferos discendens ac dispensationes salutis nostrae perficiens, uespere sabbati, VI Kalendas Aprelis, luna XVI, luciscente dominico, id est V Kalendas Aprelis resurrexit a mortuis, in quo die luna XVII primi mensis iuxta Haebreos extitisse manifestum est*. For the chronology of the *annus resurrectionis* cf. also c. 52, ll. 3–13, 44–48. For the Biblical references used in this passage cf. the following notes. **55–58** XII…nummulariorum] JOHN 2:15, 12:1,12,14–15: *et cum fecisset quasi flagellum de funiculis omnes eiecit de templo oves quoque et boves et nummulariorum effudit aes et mensas subvertit ... Iesus ergo ante sex dies paschae venit Bethaniam ubi fuerat Lazarus mortuus quem suscitavit Iesus ... in crastinum autem turba multa quae venerat ad diem festum cum audissent quia venit Iesus Hierosolyma ... et invenit Iesus asellum et sedit super eum sicut scriptum est noli timere filia Sion ecce rex tuus venit sedens super pullum asinae*. Cf. MAT 21:1–17; LUKE 19:28–48; MARK 11:1–17. **58–62** a…arente] Cf. MARK 11:12–14,20–22: *et alia die cum exirent a Bethania esuriit*

After the army (i.e. of the people of Israel) had been assembled at the beginning (of the day) in Ramises, they emigrated at the sixth hour of this very same Thursday. (The fifth day of the Exodus extended) from the sixth hour of Thursday to the sixth hour of Friday, 26 March and *luna* 15; (the sixth day of the Exodus extended) from the sixth hour of Friday to the sixth hour of Saturday, 27 March and *luna* 16; (the seventh day of the Exodus extended) from the sixth hour of Saturday to the sixth hour of Sunday, 28 March and *luna* 17.

Turn to the year of the resurrection of Christ: On 21 March, until the sixth hour of (that) Sunday, (Jesus) sat on an ass' colt (and) came from Bethany to Jerusalem, where he overthrew the tables of the moneychangers; (the first day extended) from the sixth hour of Sunday to the sixth hour of Monday, (which) was 22 March, (and) on that day the Lord cursed the fig tree; (the second day

---

cumque vidisset a longe ficum habentem folia venit si quid forte inveniret in ea et cum venisset ad eam nihil invenit praeter folia non enim erat tempus ficorum et respondens dixit ei iam non amplius in aeternum quisquam fructum ex te manducet et audiebant discipuli eius ... et cum mane transirent viderunt ficum aridam factam a radicibus et recordatus Petrus dicit ei rabbi ecce ficus cui maledixisti aruit et respondens Iesus ait illis habete fidem Dei. Cf. also MAT 21:18–22.

55–80 Veni…XVII] The adoption of Victorius' chronology for the year of the resurrection (cf. *app. font.*) appears to be a characteristic of early eighth-century (southern?) Irish computistics. Besides the apparently Irish tract DE COMPARATIONE EPACTARUM, which served MC as source here (cf. *app. font.*), this chronology can also be found DRC 86.5–16 (between the two sentences given here occurs the citation from EPISTOLA PROTERII, quoted in *app. font.*): *Et deinde quoque dominus, quando fuit in carne, hanc regulam seruans, V feria, VIII Kl Aprelis, XIIII luna cum discipulis pascha celebrauit. ... Et per hoc nobis ostenditur XVII luna primi mensis iuxta Ebreos in pascha <luna> resurrectionis fuisse.* This chronology of the year of resurrection favoured by the Irish was not the only tradition, and particularly the fact that the Dionysiac Easter table did not solve this problem (since none of the years AD 30 to 34 had acceptable data; cf. Krusch, *Studien* II, p. 71, *s.a.* 562–566) led to continued speculation (cf. DTR 47.65–75). Bede knew at least three different traditions, namely 25, 27, and 28 March as the day of the resurrection: DTR 61.51–54 (cited in RM 87.3–7): *Ubi autem primum dominicus resurrectionis Christi dies fuerit, varie refertur. Et quidem, ut supra memoravimus, quidam VIII kalendarum aprilium, sed alii VI, nonnulli V, kalendarum earundem die fuisse asseverant.* 28 March as the date of the resurrection is the tradition outlined in MC here; 25 March is recorded in DE ORDINATIONE FERIARUM PASCHALIUM (*PL* 90, col. 607A, cited in DRC 87.9–10, DTR 47.92–94, in turn cited in RM 64.28–30; col. 608D, where it mistakenly reads *septimo* instead of *octavo Kalendas*; cited in DRC 88.2–4, DTR 47.88–90, in turn cited in RM 64.22–24); 27 March was the most popular date for the resurrection of Christ, since it was explicitly noted in the JEROME, *Martyrologium* (De Rossi & Duchesne, *Acta Sanctorum Novembris* 2,1, p. 37, cited in RM 64.7–10), and it is implicit in the statement that the passion took place on 25 March recorded in AUGUSTINE, *De Trinitate* 4.5 (*CCSL* 50, p. 172, cited in RM 64.11–16) as well as in PS-DIONYSIUS, *Argumentum XV* (Krusch, *Studien* II, p. 79–80); this tradition then found its way into PP I §156, DIAL. LANGOB. 18B, EPIST. RAT. 10, 12, CAP. COMP. 2, as well as into numerous calendars (cf. Borst, *Reichskalender*, p. 713–6, 722–3). It seems that one of the continental copyists of MC was a follower of this latter tradition and therefore deliberately omitted the last sentence so that it would appear that Christ resurrected on 27 rather than 28 March. In the tenth century and later, the incongruence of the Dionysiac Easter table with these traditions led to a recalculation of the incarnation era by Abbo of Fleury, Marianus Scotus, Sigibert of Gembloux, and others.

55–64 Veni…XIIII] Partially transcribed in Mac Carthy, *Annals of Ulster* 4, p. clxxx.

*maledix*it dominus *ficum*; a VI hora diei lunis in VI horam diei martis, X Kalendas Aprilis est, in qua die narrat ei discipulus de *fico ar*ente; a VI hora diei martis in VI horam diei cetene, VIIII Kalendas Aprilis est; a VI hora IIII feriae in VI horam V feriae, VIII Kalendas Aprilis est et luna XIIII, cuius accensio in IIII Idus Martii, in die VI feria. In VIII Kalendas Aprilis *accesserunt discipuli ad I*hesum et *di*xerunt: *Ubi uis paremus tibi comedere pascha*? Inde media hora post VI horam *accepit Ihesus panem et benedixit* et reliqua, hoc est in discessu solis. Deinde continentur duo mysteria in uno die, ac inde conuenit *lapis angularis*, ut Paulus ait: In unum *coniunxit nouum et uetus testamentum*. Inde Greci faciunt pascha in XV luna, quia non est | minus pascha passio quam resurrectio. Inde fol. 25r Paulus ait: *etenim pascha nostrum immolatus est Christus*, et *qui* pascha *est propter peccata nostra*, et *resurrexit propter iustificationem nostram*. A VI hora V feriae in VI horam VI feriae, VII Kalendas Aprilis et luna XV; a VI hora in VI feria in VI horam sabbati, VI Kalendas Aprilis et luna <XVI; a VI hora sabbati in VI horam dominici, V Kalendas Aprilis et luna> XVII.

60 dominus] *M gives the abbr.* ds *for* deus *instead of the correct* dns *for* dominus. **65** VI] V *M*. **70** continentur] continent *M, Walsh & Ó Cróinín*. **78** horam] hora *M (corr. according to* MC*)*. **79–80** XVI…luna] *om. M (add. according to* MC*)*.

66–69 accesserunt…reliqua] MAT 26:17,26: *prima autem azymorum accesserunt discipuli ad Iesum dicentes ubi vis paremus tibi comedere pascha ... cenantibus autem eis accepit Iesus panem et benedixit ac fregit deditque discipulis suis et ait accipite et comedite hoc est corpus meum*. **70–72** ac…testamentum] The theory of Christ being the chief corner-stone, joining the New with the Old Testament, is taken from ISIDORE, *Libri differentiarum* 2.7 (*PL* 83, col. 73C): *Lapis angularis, quia Vetus et Novum Testamentum, veluti duo parietes ex adverso venientes, tanquam angulus, sibimet copulavit ac fidei unitate conjunxit*. The attribution to Paul derives from Paul's characterisation of Christ as the *lapis angularis* in his epistle to the Ephesians: EPH 2:20: *superaedificati super fundamentum apostolorum et prophetarum ipso summo angulari lapide Christo Iesu*. **74** etenim…Christus] 1 COR 5:7: *expurgate vetus fermentum ut sitis nova consparsio sicut estis azymi etenim pascha nostrum immolatus est Christus*. **75–76** et[1]…nostram] 1 PETER 2:24: *qui peccata nostra ipse pertulit in corpore suo super lignum ut peccatis mortui iustitiae viveremus cuius livore sanati estis*. ROM 4:25: *qui traditus est propter delicta nostra et resurrexit propter iustificationem nostram*. Cf. JEROME, *Commentarium in Esaiam* 14.53 (*CCSL* 73A, p. 598): *Qui traditus est propter peccata nostra, et resurrexit propter justificationem nostram*. AUGUSTINE, *Sermones*, *Sermo 231* (*PL* 38, col. 1105; likewise, with *enim* instead of *inquit*, in *Sermo 169* (*PL* 38, col. 916)): *Traditus est, inquit, propter peccata nostra, et resurrexit propter justificationem nostram*. AUGUSTINE, *De civitate Dei* 20.6 (*CCSL* 48, p. 707): *qui pro omnibus mortuus est propter peccata nostra et resurrexit propter iustificationem nostram*.

70 continentur…die] Transcribed in Walsh & Ó Cróinín, *Cummian's letter*, p. 24. **73** non…resurrectio] Transcribed in Walsh & Ó Cróinín, *Cummian's letter*, p. 23.

extended) from the sixth hour of Monday to the sixth hour of Tuesday, (which) was 23 March, (and) on that day a disciple told him about the fig tree withering away; (the third day extended) from the sixth hour of Tuesday to the sixth hour of Wednesday, (which) was 24 March; (the fourth day extended) from the sixth hour of Wednesday to the sixth hour of Thursday, (which) was 25 March and *luna* 14; the kindling of this (lunar month occurred) on 12 March, on a Friday. On 25 March the disciples came to Jesus and said: Where wilt thou that we prepare for thee to eat the passover? Then, half an hour after the sixth hour, Jesus took the bread and blessed it, and so on, i.e. at the departure of the sun. Hence, two mysteries are contained in one day, and thereupon the chief corner-stone materialized, as Paul says: He joined the New and the Old Testament together into one. Accordingly, the Greeks celebrate Easter on *luna* 15, since the passion is not less part of Easter than the resurrection. Therefore Paul says: For even Christ our passover is sacrificed (for us), and he is the passover for our sins, and he was raised again for our justification. (The fifth day extended) from the sixth hour of Thursday to the sixth hour of Friday, 26 March and *luna* 15; (the sixth day extended) from the sixth hour of Friday to the sixth hour of Saturday, 27 April and *luna* 16; (the seventh day extended) from the sixth hour of Saturday to the sixth hour of Sunday, 28 March and *luna* 17.

<XLV. DE ANNO SOLIS>

A proof for the length of the solar year

Annus solis CCCLXV diebus constat et sic probatur: a VI Nonas Maii in II Kalendas Iunii XXX dies sunt, Iunius XXX, Iulius XXXI, et faciunt XCI; Augustus, Septimber, et Octimber XCII; Nouimber, December, Ianuarius XCII. V hos dies superfluos coniunge et hii sunt interkalares dies. Februarius XXVIII dies. Adde super Februario II dies, Kalendas Martii et VI Nonas; fiunt XXX dies. Diuide Februarius in ter X et da super XC ter, et faciunt CCC. A VI Nonas in XI Kalendas Aprilis XX dies sunt. Adde interkalares V et faciunt CCCXXV. A <XI> Kalendas Aprilis cum Kalendis Maii XL dies sunt. Inde fiunt CCCLXV dies.

<XLVI.> DE LUNA

Etym. of moon (*luna*)

*Luna dici*tur *quasi Lucina, ablata media syllaba. De qua Virgilius* ait*: casta faue Lucina.* Luna *autem per diriuationem a solis lu*mine *nomen* as*sumpsit.* Deinde dicitur: *Luminare maius et luminare minus.*

XLV,11 CCCXXV] CCCLXV *M (corr. according to* MC*).* | XI] *om. M (add. according to* MC*).* **XLVI,3** Luna] luna- *M.* | autem] *follows* aut *M (mistaken repetition of the preceding* aut-*).*

XLV,2–13 Annus…dies] The method for calculating the number of days of the solar year applied here is later used by MC for the calculation of the number of days of a lunar year (c. 59, ll. 70–105); the basis of the calculation is that the period from 2 May to 22 March consists of 325 days, a fact that is here established in detail, and which is then later taken for granted. Since a lunar year is reckoned from *luna* 15 of the Easter month to the Easter full moon of the following year, the beginning and end of a lunar year fall into the period that is not considered here (22 March to 2 May), so that the number of lunar days from the beginning of the respective lunar year to 2 May, as well as the number of days from 22 March to the end of that lunar year have to be added to the 325 days to establish the total of days of that year; in a solar year, however, which is under consideration here, 40 further days exist between 23 March and 1 May inclusively, leading to a total of 365 days. A very interesting feature of this calculation here is the underlying division of the solar year into four intervals, namely into a) May, June, July; b) August, September, October; c) November, December, January; d) February, March, April; this division of the year agrees with the Irish reckoning of the seasons as recorded by COLUMBANUS, *Regula monachorum* 7 (Walker, *Sancti Columbani opera*, p. 130): *initium hiemis, id est Kalendas Novembris*; cf. c. 29, ll. 17–24; c. 38, ll. 39–43. **XLVI,2–5** Luna…minus] The source for this etymology of *luna* is ETYM. 3.71.2 (cited in BC 142(1354)D13–(1355)A2, CE p. 95 which reads *dicitur* and *kasta febe*, PV §439 which reads *fabe*, RM 37.19–20 which reads *dicitur* and *Virgilius ait*; the first two sentences only in DRC 63.4–6 which reads *dicitur* and *Uirgilius dicit*): *Luna dicta quasi Lucina, ablata media syllaba. De qua Vergilius: Casta fave Lucina. Sumpsit autem nomen per derivationem a solis luce*. The substitution of the term *luce* with *lumine* may derive from EPIT. 11.53–55: *Luna a quadam Diana aestimatur nominari, quae hoc nomine uocitabatur. Sed, quod uerius dicendum est, luna ex ipso lumine uocatur, quod est luminare*. The Biblical quote at the end is from GEN 1:16 (cited in ARN SERM. 1.3B, in a different context in BC 26A7–9): *fecitque Deus duo magna luminaria luminare maius ut praeesset diei et luminare minus ut praeesset nocti.*

45. ABOUT THE SOLAR YEAR

The solar year consists of 365 days and (this) is proven in the following way: There are 30 days from 2 May to 31 May, June (has) 30, July 31 (days), and these sum up to (a total of) 91 (days); August, September, and October (have a total of) 92 (days); November, December, and January (have a total of) 92 (days). Unite these 5 superfluous days (i.e. the days that exceed 90 in each of these three intervals), and these are the intercalary days. February (has) 28 days. Add two days on top of February, (namely) 1 and 2 March, (then these) add up to 30 (days). Divide February into three times ten and add (each of these) on top of the three times 90, (then) they add up to 300 (days). There are 20 days from 2 to 22 March (excluding 2 March itself). Add the five intercalary days, (then) they add up to 325 (days). There are 40 days from 22 March (to and) including 1 May (excluding 22 March itself). Therefore (the days of the solar year) add up to (a total of) 365 days.

46. ABOUT THE MOON

The moon (*luna*) is so called as though it were the goddess Lucina, the middle syllable having been dropped. Virgil says about her: Be clement, pure Lucina. The moon (*luna*), however, received its name by derivation from the light of the sun (*lumine solis*). Hence it is said: The greater light and the lesser light.

XLV,2–13 Annus…dies] The method for calculating the number of days of the solar year applied here is unique to Mc; if an early-medieval computist would have wanted to calculate that total, he would have added the number of days of each Julian calendar month. The division of the year, however, which is the basis for the calculation here, can be found in various texts, apparently all of Irish provenance; in these, it is always related to the seasons: Bc 105: *De principiis quatuor temporum. Si uis scire, qua die septimanae intrant quatuor temporum, hoc est veris, et aestas, authumnus, atque hiemis, sume principium a Kalendis Maii; quacumque enim feria fuerint Kalendae Maii, sequenti die septimane intrant Kalendae Agusti, hoc est principium authumni, tertia die Kalende hiemis, quarta die principium ueris, hoc est Kalendae Februarii, nec bissextus mutat istam rationem.* Dial. Langob. 8B: *Interrogatio: Quomodo secundum consuetudinem rationis? Responsio: Quomodo nos in usu habemus, hoc est a principiis mensium, ut ver a principio Februarii mensis inchoatur, aestas a principio Maii, autumnus a principio Augusti, hiems a principio Novembris. Hoc est consuetudo, ut integri fierent pariter tres menses in unoquoque tempore.* This division of the year is also suggested in Drc 49, Quaest. Austr. 2.7B, and c. 38, ll. 41–43. For divisions of the year into different intervals of the same lengths see c. 38, ll. 17–22, 26–27; for the five intercalary days cf. c. 24, ll. 12–15; c. 38, ll. 23–25; c. 40; c. 41, ll. 80–88; for 365 days as the total of the solar year cf. c. 31, ll. 3–6, c. 46, ll. 11–12; for 365¼ days cf. c. 31, ll. 15–16, c. 41, ll. 29–31. **XLVI,2–5** Luna…minus] Though discussions of the moon feature very prominently in computistical texts, the etymology of *luna* is only rarely considered; the etymology given in the Frankish Dial. Burg. 10B derives directly from Mc: *Luna a quadam dea nominatur, vel a lumine. Dicitur enim: Luminare maius et luminare minus.*

XLVI,2 Luna…Lucina] Transcribed in Borst, *Schriften*, p. 362. **2–5** Luna…minus] Cf. p. CLXXII of the introduction.

Light and shape of the moon

Lunam philosophi dicunt proprium habere lumen; *una par*s *globi eius luciflua*, *al*ter*a obscura, et se uertendo diuersas formas* facit.

The course of sun and moon

Hoc inde | Dominus ait per Moysen: *Fecit dominus duo luminaria magna. Luna uicinior terr*ae *quam sol. Inde breuiori* cursu *orbe peragit cursum suum. Nam iter, quod sol CCCLXV die*s *currit,* luna *XXX dies peragit. Unde annos in sol*e, *menses* uero *in luna* dix*erunt*. fol. 25v

Names

Hericon in Hebreo, silene Aegyptu, gamsia Grece.

Light of the moon

Alii uol*unt* a *radiis solis inluminari*.

---

7 una…obscura] una pars globi eius luciflua-, altera- obscura- *M (corr. according to* Drc*; but cf.* Etym.*,* Dnr*)*. **14** Hericon] hercion *Ó Cróinín (contra M)*.

6–8 Lunam…facit] Concerning the question of the origin of the light of the moon, two general theories existed at the time of Mc, both formulated by Isidore, who took his own information from Augustine. In the first theory, which is the one outlined here, it is argued that the moon has its own light, and Mc takes his information directly from Etym. 3.53.1: *Lunam quidam philosophi dicunt proprium lumen habere, globique eius unam partem esse lucifluam, aliam vero obscuram, et paulatim se vertendo diversas formas efficere*. Cf. Dnr 19.1.4–8: *Alii namque dicunt proprium eam habere lumen, globique eius unam partem esse lucifluam, aliam obscuram, et dum mouetur in circulo suo, eandem partem qua lucet paulatim ad terras conuerti, ut uideri a nobis possit; et ideo prius quasi corniculato lumine fulgit.* The second theory, i.e. the moon taking its light from the sun, is then presented further below, c. 46, l. 15, and later underlies c. 48, ll. 16–18, so that it appears that Mc favoured this second option. In c. 44, l. 28, however, Mc mentions both options without any statement of preference. **9–13** Hoc…dixerunt] The fact that the moon is closer to the earth than the sun and therefore has a shorter orbit is taken from Etym. 3.57: *Luna vicinior est terris quam sole. Inde et breviori orbe celerius peragit cursum suum. Nam iter, quod sol in diebus trecentis sexaginta quinque peragit, ista per triginta dies percurrit. Vnde et antiqui menses in luna, annos autem in solis cursum posuerunt*. Cf. Dnr 19.1.5–8: *Terris autem uicina luna breuiori orbe conuertitur, et iter quod sol in diebus CCLXV peragit, ista per XXX dies percurrit. Vnde et antiqui menses in luna, annos autem in solis cursu posuerunt*. For the terminology (*annus* for the completion of the course of the sun through the zodiac, *mensis* for that of the moon) cf. also c. 53, ll. 2–3. The Biblical citation at the beginning of this passage stems from Gen 1:16, already cited c. 46, l. 5. **14** Hericon…Grece] As in the case of the discussion of the sun (c. 37, ll. 3–5), Mc lists the names for the moon in various languages; there is no apparent source either for this list of names as a whole, nor for each individual term. **15** Alii…inluminari] After these two insertions, Mc returns to the question of the origin of the light of the moon, here outlining the second option, namely that the moon is illuminated by the sun, quoted from Etym. 3.53.2: *Alii e contra aiunt lunam non suum lumen habere, sed solis radiis inluminari*. Cf. Dnr 18.3.21–22 (cited in Drc 63.12–13): *At contra alii dicunt lunam non suo fulgere lumine, sed a sole accipere lumen*. Note that this theory underlies c. 46, ll. 21–23.

6–8 Lunam…facit] The question whether the moon has its own light, or whether it is illuminated by the sun, was not common in computistical literature, but it troubled Insular computists of the early eighth century. Like Mc (especially in c. 44, l. 28), the Irish Drc outlines both options without expressing any preference; interestingly enough, in outlining the first option, Drc 63.8–11 is actually closer to Mc than to the Isidorian original: *Sciendum nobis an proprium lumen habet án a sole accenditur. Isidorus ostendit dicens: Lunam philosophi dicunt proprium lumen habere, globique una pars est eius luciflua, altera obscura, et paulatim se uertendo diuersas formas facit*. For the second opinion cf. c. 46, l. 15. **10–13** Luna…dixerunt] As an explanation for the shorter orbit of the moon when compared to that of the sun, the same Isidorian passage as quoted by Mc is cited (whith notable differences) in Bc 113A5–9: *Luna uicinior est terris quam sol. Unde et breuiore orbe caelerius peragit cursum suum. Nam iter, quod in diebus CCCLXV et quadrante sol peragit, luna currit per XXX dies. Unde et antiqui*

The philosophers say that the moon has its own light; one part of its globe (is) luminescent, the other obscure, and while turning itself it creates different shapes.

Hence, the Lord says this through Moses: The Lord made two great lights. The moon (is) closer to the earth than the sun. Accordingly, it completes its course on a shorter orbit in the sky. For the moon completes the course, which the sun traverses in 365 days, in 30 days. Therefore they call (these courses) years according to the sun, but months according to the moon.

(The moon is called) *hericon* in Hebrew, *silene* in Egyptian, *gamsia* in Greek.

Others argue that (the moon) is illuminated by the rays of the sun.

---

menses in luna annos in solis cursum posuerunt. DRC 68.2–4: *Luna uicinior est terris quam sol. Inde breviore orbe celerius peragit cursum suum. Nam iter, quod in diebus CCCLXV peragit sol, per XXX dies luna currit. Inde mensis in luna annus in sole numeratur*. For 365 days constituting the length of a solar year, cf. c. 31, ll. 3–5, c. 41; for 30 days being the length of a lunar month, cf. c. 24, ll. 12–13, 36–37. **14** Hericon…Grece] DRC 63.1–3 is the only text that gives a comparable list (note the difference in the Greek term; DRC's *mene* stems from Jerome), which impressively illustrates the interdependency of the two texts: *Sciendum nobis quomodo luna uocatur in tribus principalibus linguis. Ita Hericon in Ebreo; Silene apud Aegyptios, unde lapis Silentitis dicitur; Mene apud Grecos, inde mensis dicitur; Luna apud Latinos*. **15** Alii…inluminari] As mentioned above (c. 46, ll. 6–8), DRC shows no preference for either option concerning the origin of the light of the moon. The second option, i.e. the moon taking its light from the sun, reads: DRC 63.11–12 (followed by the quotation from DNR given in *app. font.*): *Siue luna uitria est creatura, et a sole luminatur*. The Irish indecision in this question is particularly remarkable since Isidore, though outlining both theories, clearly favours the opinion that the moon is illuminated by the sun; it may be that Irish computists regarded Isidore's source, Augustine, who does not speak in favour of any of the two theories, as more authoritative (cf. *app. font.* to c. 44, l. 28). Bede, on the other hand, leaves no room for dout that the sun is the origin of the moon's light: BEDE, *De natura rerum* 20 (*CCSL* 123A, p. 211; cited in RM 45.2–4): *Lunam non minui nec crescere dicunt sed, a sole inlustratam, a parte quam habet ad eum paulatim uel ab eo recedendo uel ei proprinquando nobis candidam partem reuoluere uel atram*. DTR 6.33–36, 25.12 (quoting AUGUSTINE, *Enarrationes in psalmos, Enarratio in psalmum 10, Sermo 3* (*CCSL* 38, p. 76); cited in LIB. CALC. 74): *luna autem et stellae, quae non proprio, ut dicunt, sed adventitio et a sole mutuato lumine fulgent, ..., qui, illuminari non illuminare valentes, caelestis gratiae munus accipere sciunt, dare nesciunt. ... non habere lumen proprium, sed a sole illustrari*. Cf. also PP I §134: *Alii dicunt quod non a sole lumen accepit, sed a sua natura*.

6–15 Lunam…inluminari] Cf. Bisagni & Warntjes, 'Latin and Old Irish', p. 14. **14** Hericon…Grece] Transcribed in Walsh & Ó Cróinín, *Cummian's letter*, p. 172; idem, 'A seventh-century Irish computus', p. 105–6.

The position of the moon relative to that of the sun

Et in omni parte accendi nisi in Aquilone. Tribus modis accenditur: Superincensio, id est super solem accenditur in breuitate diei; mediaincensio, id est communiter soli in aequalitate diei et noctis; subincensio, id est sub sole in longitudine diei.

Different shapes of the moon

Inde, *quando* soli proxima, *nihil candoris habe*t. *Cum uero ab* eo *reced*it, *inlustratur*, quia soli coniuncta *obscura*tur.

---

17 Superincensio] *follows* est *M.* **19** communiter] communi *M, Borst.*

16–20 Et…diei] There is no pre-MC text discussing the position of the kindling of the moon relative to the sun; MC himself is rather vague here, only arguing that the moon is kindled above the sun when nighttime exceeds daytime, that it is kindled below the sun when daytime exceeds nighttime, and that it is kindled on the same level as the sun when day and night are of equal length. More details about this theory are given in DRC (cf. *app. comp.*). **21–23** Inde…obscuratur] MC's account of the different shapes of the moon seems to be based not only on DNR, but also on Isidore's source, AUGUSTINE, *Enarrationes in psalmos*. The terminology of the first sentence derives from AUGUSTINE, *Enarrationes in psalmos, Enarratio in psalmum 10, Sermo 3* (*CCSL* 38, p. 75): *Nam et si facias pilam ex dimidia parte candidam, et ex dimidia obscuram, si eam partem quae obscura est ante oculos habeas, nihil candoris uides*. DNR 18.2.9–11: *Nam et si formes pilam ex parte media candidam et ex parte obscuram, tunc eam partem qua obscura est si coram oculis habeas, nihil candoris aspicies*. The theory described in this passage, namely that the illuminated part of the moon grows the further away the moon travels from the sun, presupposes that the light of the moon has its origin with the sun, not in the moon itself; MC's source here is AUGUSTINE, *Enarrationes in psalmos, Enarratio in psalmum 10, Sermo 3* (*CCSL* 38, p. 76; cited in DTR 25.12–16, in turn cited in LIB. CALC. 74): *Alii autem dicunt non habere lunam lumen proprium, sed a sole illustrari; sed quando cum illo est, eam partem ad nos habere qua non illustratur, et ideo nihil in ea lucis uideri; cum autem incipit ab illo recedere, illustrari ab ea etiam parte quam habet ad terram, et necessario incipere a cornibus, donec fiat quinta decima contra solem*. DNR 18.3.22–26 (MSS FS read *inlustratur*; cited in DRC 63.12–17): *At contra alii dicunt lunam non suo fulgere lumine, sed a sole accipere lumen. Sol enim illi loco superior est. Hinc euenit ut, quando sub illo est, parte superiore luceat, inferiore uero, quam habeat ad terras, obscura sit, cum uero ab illo discedere coeperit, inlustretur etiam et a parte quam habet ad terras, incipiens a cornibus*. These passages may also underlie c. 48, ll. 8–12; for the moon's dependence on the sun cf. c. 46, l. 15; c. 48, ll. 16–18; the other option, i.e. that the moon has light of its own, is outlined in c. 46, ll. 6–8; both options are given in c. 44, l. 28.

16–20 Et…diei] The discussion of the position of the kindling of the moon relative to the sun certainly is a characteristic of early eighth-century Irish computistics. It first occurs in MC, and is then found in a more detailed version in DRC 65 (the numbers of Ó Cróinín's emendation are wrong and are here corrected according to the context): *Sciendum nobis quot sunt genéra accensionis lunae. Tres scilicet: subincensio, supraincensio, mediaincensio. Et unumquodque genus de his tribus partem anni, id est CXXI dies et XVI horas, <tenet.> Subincensio in breuitate diei inuenitur, quando sol sub luna sit, quae circa hiemale solstitium retinet <et LX* (*recte CXXI*) *dies et XX* (*recte XVI*) *horae circa solstitium diuiduntur, id est XXX* (*recte LX*) *dies et X* (*recte XX*) *horae solstitium precedunt,> et hic numerus dierum et horarum sequitur solstitium. Et hoc modo diuiditur supraincensio in longitudine diei circa estiuale solstitium. Mediaincensio circa duos equinoctios inuenitur, et LXta dies et XX horae circa aequinoctium uernale diuiduntur, id est XXX diés et X horae equinoctium precedunt et similis numerus dierum et horarum sé subsequitur. Et sic LX dies et XX horae circa equinoctium autumnale diuiduntur*. Note that DRC reverses the terminology of MC, with MC's *superincensio* corresponding to DRC's

And it (the moon) is kindled in every direction save in the North. It is kindled in three ways: the 'above-kindling' (*superincensio*), i.e. it rises above the sun in the shortness of the day (i.e. when daytime is shorter than nighttime); the 'middle-kindling' (*mediaincensio*), i.e. together with the sun in the equality of day and night (i.e. on the same level as the sun when daytime and nighttime are of equal length); the 'below-kindling' (*subincensio*), i.e. below the sun in the length of the day (i.e. when daytime is longer than nighttime).

Therefore, when (the moon is) close to the sun, it has nothing of its brightness. When, however, (the moon) recedes from (the sun), it is illuminated, since it is obscured when it is near the sun.

---

subincensio and vice versa; moreover, DRC assignes explicit Julian calendar periods to each of these three phenomena, namely the moon being kindled below the sun for 121 days and 16 hours around the summer solstice (i.e. the summer solstice constituting the exact middle of this period), it is kindled above the sun for the same period around the winter solstice, while it is kindled more or less on the same level as the sun for 60 days and 20 hours around each of the equinoxes. DRC then seems to have influenced a gloss to DTR (*PL* 90, col. 406C–D): *Tria sunt genera incensionis lunae: subincensio, media incensio, superincensio. Subincensio est, quando luna desubtus accenditur a sole per sexaginta dies, et viginti horas, circa undecimum diem Kalendarum Januariarum, quia tunc luna altior est quam sol. Media incensio est, quando luna a sole accenditur, aequaliter per triginta dies et decem horas, circa undecimum diem Kalendarum Octobrium, quia tunc aequales sunt sol et luna in altitudine. Superincensio est, quando luna desuper accenditur, per sexaginta dies et viginti horas, circa undecimum diem Kalendarum Juliarum, quia tunc altior est quam luna.* which can also be found, in slight variation, as a separat tract in Zürich, Zentralbibliothek, Car C 180, 35r–36r. The terminology was also know to DE SALTU LUNAE I 988A12–B1, which generally draws heavily on Irish computistica: *Illa ergo incensio diei quomodo intelligi potest, utrum in die, an in nocte luna accenditur? vel qua <hora>, vel quo puncto, immo etiam momento luna accendi dicitur? et utrum subincensio, vel media incensio, vel superincensio sit?* MC, for its part, directly influenced COMP. COL. 5.2A: *Superincensione super sole, in brevi? In diei media, comminus soli, aequa longitudine diei et noctis? Subincensione sub sole, in longiore die?* KAL. G 3 notes the end of the *subincensio* on 19 February. **21–23** Inde…obscuratur] Similar statements, based on the same sources as used by MC, can be found in DRC and DTR (cf. *app. font.*)

17–20 Superincensio…diei] Transcribed in Borst, *Schriften*, p. 926; cf. also p. XXII, CLXII, CLXXXIII of the introduction. **21–23** Inde…obscuratur] Transcribed in Bisagni & Warntjes, 'Latin and Old Irish', p. 15.

<XLVII.> *DE QUATTUOR PARTIBUS* MUNDI

The four directions of the world

In *quattuor partes* orbis diuiditur: *Oriens ab ortu solis* uocatur, quia metitur ab ortu solis in VIII Kalendas Iulii usque ad ortum solis in VIII Kalendas Ianuarii. *Meridies* a loco orientis solis in VIII Kalendas <Ianuarii> usque ad locum occidentis eodem die mensuratur. *Occidens* dicitur, *qu*ia *diem occidere facit*; a loco occidentis solis in VIII Kalendas Ianuarii usque in locum occidentis solis in VIII Kalendas Iulii. *Septentrio a VII stellis axis, qu*e *in ipso reuolutae* | *rotantur*; a loco occidentis in VIII Kalendas Iulii usque in locum orientis solis in eodem die. Hęc pars tenebrarum. fol. 26r

---

XLVII,3 solis[1]] *follows* in occasum *M (om. according to* ETYM., MC*)*. **5** Ianuarii] *om. M (add. according to* MC*)*.

XLVII,1–12 De...tenebrarum] This most interesting chapter on the directions of the world is unique to MC. The etymologies are taken from ETYM. 3.42.1–3 (almost verbatim also in 13.1.3–6): *De quattuor partibus caeli. Climata caeli, id est plagae vel partes, quattuor sunt. ... Oriens ab exortu solis est nuncupatus. ... Occidens, quod diem faciat occidere atque interire. ... Septentrio autem a septem stellis axis vocatus, quae in ipso revolutae rotantur. ... Meridies autem vocata, vel quia ibi sol facit medium diem quasi medidies*. The methods for determining the position of the four directions, however, appear to be MC's own composition; MC's phrasing is figurative rather than concise, basically saying that east is the point in the sky where the sun rises on 25 March, south where the sun stands on midday of 24 June, west where the sun sets on 24 September, north the middle between the point of sunset and that of sunrise on 24 June. For the 'Latin' dates of the solstices applied here cf. c. 38, ll. 26–27, 31–38; c. 48, ll. 8–12.

XLVII,1–12 De...tenebrarum] Since the directions of the world are not immediately relevant to computistical theory, this chapter received no reception in later computistical texts.

47. ABOUT THE FOUR DIRECTIONS OF THE WORLD

The world is divided into four directions: The east (*oriens*) is so called from the rise of the sun (*ortu solis*), since it is measured from sunrise on 24 June to sunrise on 25 December. The south (*meridies*) is measured from the place of sunrise on 25 December to the place of (sun)set on the very same day. The west (*occidens*) is so called since it causes the day to fall (*occidere*); (it is measured) from the place of sunset on 25 December to the place of sunset on 24 June. The north (*septentrio*) (is so called) from the seven stars (*septem stellis*) of the North Pole, which are rotating in the process of returning to it; (it is measured) from the place of (sun)set on 24 June to the place of sunrise on the very same day. This direction (belongs) to the shadows.

<XLVIII. CONTINUATIO: DE LUNA>

Place and time of the kindling of the moon

Tribus tantum partibus luna de his currit: Secundum Hebreos et Aegyptios in VI horam accenditur. In VI horam diem habent et in medio austri accenditur. Cum Grecis uero altera uice in septentrione in loco ignoto caeli, altera uice in Meridie. Cum Romanis autem in omnibus horis accenditur.

XLVIII,4 Grecis] Greci *M**.

XLVIII,2–7 Tribus...accenditur] Two important preconditions underlie this passage on the place and time of the kindling of the moon: 1) a lunar day starts at midday (cf. c. 8, ll. 36–38 and especially c. 44); 2) the length of a lunation varies according to different traditions (cf. c. 24, ll. 35–39). Now, if a lunation is reckoned as consisting of a certain number of full days, like 30 or 29 days (or both), then every kindling of the moon is considered to happen at the same time of the day (midday) and in the same place in the sky (south), here ascribed to the Hebrews and Egyptians; if, however, a lunation is reckoned as consisting of 29½ days, then the kindling alternates between midday and midnight, between south and north, here ascribed to the Greeks; if, however, the fact is taken into consideration that such lunations are mathematical constructions rather than an exact reflection of nature (as is conceded by the division of the *saltus* into an increment of 4 *momenta*, 1/12 of a *momentum*, and 1/(12×47) of a *momentum* per lunation; cf. c. 62, ll. 14–61), then the kindling of the moon changes with each lunation, leading to the theory that the moon is actually kindled at every time of the day, here ascribe to the Romans.

XLVIII,2–7 Tribus...accenditur] The question of the place and time of the kindling of the moon was central in eighth-century Insular computistics. The following two Irish texts outline the theory that the kindling of the moon alternates between midday and midnight, between north and south: BC 61D12–B2 (partially cited in QUAEST. AUSTR. 2.4A, in turn cited in COMP. COL. 6.4): *Sed quod dicimus lunaris mensis longitudinem dispares secundum numerum, non secundum naturam omnes lunae mensis aequales sunt. Unde apud Ebreos omnes menses XXVIIII et semis semper habent aetates; uerbi gratia, si accendatur luna media die, et eiusdem dies quatuor ebdomadibus in medio expletis, altera media die XXVIIII luna esse deinde semis usque in medium noctis: una enim uice apud Ebreos media die, altera mensae uice media nocte accenditur luna. Sed parum Romani ab Ebreis discernant, non enim multum distat, si in altera luna diem integrum adicias, an duos menses diuidens equas partes utriusque distribuus. Cum autem media die luna accenditur, locus accensionis medium Austri adfirmatur. Cum in uero media nocte inluminatur, in medio Septentrionis, sed ignito caeli loco inluminari traditur. In duobus horis apud Ebreos luna accenditur, id est in media hora post VIa, et in medio noctis apud Romanos quibus uideatur altera luna XXX, altera XXVIIII in media hora post VIIa accenditur*. DRC 66: *Sciendum nobis in qua hora diei accenditur luna. Anatolius dicit in hore dimedio, post VI <horam> diei accendi, ita dicens: Omnis namque lune conputatione non eodem numero, quo mane iniciatur, ad uesperum finitur. Quia dies, qui mane in luna, id est usque ad VI et dimedium horae horam, XIII adnumeratur, idem ad uesperum XIIII inuenitur. Alii tamen dicunt in media die accendi, et altera uice in media nocte. Ebrei enim XXVIIII dierum et semis lunam esse dicunt. Hinc luna altera uice media die, altera uero media nocte accenditur. Uerbi gratia, quattuor ebdomadibus in medio <die> expletis, iterum medio die XXVIIII luna fit, et in die semis remanens ad mediam noctem tendit eodemque modo iterum <in> medio die accensio repetit*. The following two texts, on the other hand, add that this alternation of the kindling of the moon is the result of mathematical theory; in nature, the kindling of the moon occurs at every time of the day, and in every part of the sky: DE SALTU LUNAE I 988B2–D5: *Secundum aestimationem Hebraeorum luna dicitur accendi post mediam diem, hoc est, post VI horas et dimidiam horam septimae. Proinde Hebraei a media die initium diei sumunt, quia secundum lunam numerant, et in media die lunae aetas commutatur. Inde Anatholius Alexandrinus dicit: Omnis dies in lunae computatione non eodem numero, quo mane initiatur, ad vesperam finitur, quia dies, qui mane*

48. CONTINUATION: ABOUT THE MOON

The moon runs only through three (of these four) directions: According to the Hebrews and the Egyptians it is kindled at the sixth hour (i.e. midday), (since) they commence the day at the sixth hour and (therefore the moon) is kindled in the middle of the south. According to the Greeks, however, it is kindled alternately in the north, in an unknown place of the sky, and in the south. According to the Romans, however, it is kindled in every hour.

---

initiatur usque ad VI horam et dimidiam, XIII adnumeratur, eadem ad vesperum XIIII invenitur. In hoc exemplo ostendit Anatholius, quod lunae aetas commutatur omni die post sextam horam diei in dimidia horae septimae; et sic putatur etiam in illa hora lunam accendi. Sed tamen haec ratio secundum veritatem naturae quomodo probari potest, investigandum est. Si enim luna XXX semper fuisset, aut etiam XXVIIII, fieri potuisset, ut in una hora, et in eodem puncto et momento, eodem diei tempore, quo primum facta est luna, semper accenderetur finita trigesima, vel etiam vigesima nona. Sed cum XXVIIII dies et semis diei, hoc est, XII horas luna in se habere monstratur, illa diei medietas incensionem lunae mutabilem facit: et ideo, si media die lunae accensio dicatur, computa XXVIIII dies a media die incensionis lunae, usque ad mediam diem vigesimae nonae diei. Adde XII horas a media die usque ad mediam noctem, tunc in media nocte illa luna accenditur. Non ergo unis eisdemque articulis temporum vel coeli climatibus lunae accensio celebratur, et ita naturaliter in unaquaque hora et in unoquoque momento, et in omni parte coeli, super terram et sub terra, annua revolutione immo etiam menstruo circulo luna accenditur. QUAEST. AUSTR. 2.9C: *Si quis autem de varietate accensionis diffidat, advertat, ut quid diversum aliquid Anatolius et Hebraei de lunae accensione sentiunt. Ille enim accensionem lunarem in sexta et dimidia diei hora semper defigit, ille vero vicissim media diei, vicissim media nocte definiunt. Conpertum est ergo eorum varietatibus, non in uno eodem caeli cardine lunae accensionem defigi, sed in totius caeli circuitu, quo luna currere consuevit, ubique accendi. Nam si unum eademque certam aetatem summam lunam semper habere videtur, ut aut semper tricesimam, aut semper vicesimam nonam, fortasse firmius locum, quo prima luna proficisceretur, novissima luna repetire possit, ut firma ac stabilis accensio in eodem loco adfigeretur. Sed dum summa aetatum variat, et accensio variabit.* Likewise, Bede mentions the alternation of the kindling of the moon between midday and midnight in his discussion of the tides (DTR 29.19–23, cited in LIB. CALC. 81) for calculation purposes, but is more specific in his discussions of the *saltus lunae* (DT 12.2–9, cited in RM 56.4–13; DTR 42.18–21, cited in LIB. ANN. 50B, PV §292, LIB. COMP. 4.9A, in turn cited in LIB. CALC. 46A) by arguing that the kindling of the moon alternates between two places and times only in mathematically constructed theory, not in nature, where the kindling appears in every part of the sky and at every time of the day.

XLVIII,2–7 Tribus…accenditur] Cf. p. XXII of the introduction.

The facing of the moon

In media parte anni, hoc est ab VIII Kalendas Ianuarii in VIII Kalendas Iulii, lunae in ascensione accenduntur, in plenilunia uero declinantur. Ab VIII Kalendas Iulii in VIII Kalendas Ianuarii lunae in discensione accenduntur, plenilunia uero sublimantur.

The divisions of time in lunar reckoning

Quomodo momenta et minuta et puncta luna numerantur? Cum sint propria solis, id est *momentum dicit*ur *quasi motus mentis*. Inde mens nostra metitur eum. In luna sic momentum a mente nostra ponitur. Siue uerius Augustinus ait: *a parte solis* numerantur, quia luna tomel diem solis.

Epacts in lunar reckoning

Quid primum in lunari anno percunctari debet? Id est ępactę, in quibus maximae cursus lunae intellegitur.

9 accenduntur] accenditur *M (corr. according to* MC*)*. **11** discensione] discension *M*. **16** momentum] minutum *M (corr. according to* MC*)*. | Siue] sine *M (corr. by Ó Cróinín)*.

8–12 In…sublimantur] The impetus for this investigation of the question at what times the waxing and the waning moon face up- or downwards probably was AUGUSTINE, *Enarrationes in psalmos, Enarratio in psalmum 10, Sermo 3* (*CCSL* 38, p. 76; cited in DTR 25.12–21, in turn cited in LIB. CALC. 74): *Alii autem dicunt non habere lunam lumen proprium, sed a sole illustrari; sed quando cum illo est, eam partem ad nos habere qua non illustratur, et ideo nihil in ea lucis uideri; cum autem incipit ab illo recedere, illustrari ab ea etiam parte quam habet ad terram, et necessario incipere a cornibus, donec fiat quinta decima contra solem. Tunc enim sole occidente oritur, ut quisquis occidentem solem obseruauerit, cum eum coeperit non uidere, conuersus ad orientem, lunam surgere uideat; atque inde ex alia parte cum ei coeperit propinquare, illam partem ad nos conuertere, qua non illustratur, donec ad cornua redeat atque inde omnino non appareat, quia tunc pars illa quae illustratur, sursum est ad caelum, ad terram autem illa quam radiare sol non potest*. DNR 18.3.22–31 (cited in DRC 63.12–22): *At contra alii dicunt lunam non suo fulgere lumine, sed a sole accipere lumen. Sol enim illi loco superior est. Hinc euenit ut, quando sub illo est, parte superiore luceat, inferiore uero, quam habeat ad terras, obscura sit, cum uero ab illo discedere coeperit, inlustretur etiam et a parte quam habet ad terras, incipiens a cornibus. Sicque paulatim, sole longius discedente, pars omnis subterior inluminatur, donec efficiatur quinta decima luna. Post dimidium autem mensem, cum coeperit ex alio semicirculo propinquare soli, quanto magis superiore parte inlustratur, tanto magis, ab ea parte quam terris auertit, non potest excipere radios solis et propterea uidetur decrescere*. Again, MC's discussion here presupposes the theory that the moon takes its light from the sun; cf. c. 46, ll. 15, 21–23; for the 'Latin' dates of the solstices cf. c. 38, ll. 26–27, 31–38; c. 47. **13–18** Quomodo…solis] The theory presented here that the same units of time used in solar reckoning should also be applied to lunar reckoning may ultimately derive from Isidore's explanation of the night being nothing but the shadow of sunlight and therefore totally dependent on the sun: DNR 28.2 (note the correspondence between DNR and MC in the phrase *a parte solis*): *Ergo sicut in die cum a parte solis aliquod corpus hominis uel arboris occurrit, ex ea parte qua lumen repercutitur umbra subsistit, sic, cum recedente die sol ad eum locum peruenerit ubi occidere dicitur, ibi montium magnitudine a nobis separatur, sicque terrae obiectu a septentrionali parte odumbratur aer, adeo ut noctem nobis faciat haec ipsa umbra terrarum*. The general idea of tranferring the divisions of time from solar to lunar reckoning, however, probably stems from CE p. 108: *dum de sole hucusque conputauimus, item de luna conputemus*. Moreover, CE p. 85 as well as DDT 3D11–13 make the connection between the *momentum solis* and the *momentum lunae*, and CE p. 85 also is the source for the etymology of *momentum*: *Momentum dictum est quasi motus mentis*. Note that the opposite relation is outlined for the origin of the months above c. 12, ll. 20–23 (i.e. the solar months being modelled on the lunar ones). **19–20** Quid…intellegitur] The epacts being the defining factor in

In one half of the year, i.e. from 25 December to 24 June, the moons are kindled in ascent (i.e. they face upwards), but they are bent down (i.e. they face downwards) from the full moon. From 24 June to 25 December the moons are kindled in descent (i.e. they face downwards), but they are raised (i.e. they face upwards) from the full moon.

How are *momenta*, and *minuta*, and *puncta* reckoned according to the moon? (In the same way) as if they belonged to the sun, i.e. the *momentum* is (so) named as though it were the motion of the mind (*motus mentis*). In fact, our mind measures this. Thus, the *momentum* is assigned to lunar reckoning due to (its connection to) our mind. Or, as Augustine says more accurately: They are reckoned from a part of the sun, since it is the moon which consumes the solar day.

What has to be investigated first in a lunar year? The epacts, by which means the course of the greatest moon can be understood.

---

lunar reckoning, the ultimate characteristic of a lunar year, derives from CE p. 108: *Per quod de luna conputabimus? Id est per ępactas.*

8–12 In...sublimantur] Since this discussion of the question whether the waxing and waning moon face up- or downwards has no immediate relevance for computistical theory in its stricter sense, similar passage can only rarely be found in computistical texts. Bede, for his part, discusses this question in detail in order to counter weather-prognostics related to up- or downwards facing moons: DTR 25.21–34 (after quoting Augustine (cf. *app. font.*) at lenghth; cited in LIB. CALC. 74): *Cum ergo die crescente sol a meridianis plagis ad boreales paulatim partes ascenderit, necesse est luna quae eo tempore nata est ociori transitu solem ad borealia signa praecurrat. Atque ideo cum nova post occasum solis videtur, quae ad septrionem solaris occasus occasura est, nimirum non iuxta sed supra solem sita est, quo inferiora eius illustrante, aequalia pene cornua protendere et instar navis supina ire videtur. At reverso post solstitium aestivum ad inferiora et australia cursu solis, luna quoque illis nata mensibus, ad inferiora cursim tendat necesse est. Unde fit ut quae ad australem partem solis qui occiderat occusara est; absque ulla dubietate cum primo post occasum solis apparet, non iam supra illum, sed iuxta illum, ad meridiem posita videtur. Atque ideo aquilonalia eius latera sole aspectante cernitur erecta progredi. Semper enim luna aversis a sole cornibus, rotundam sui partem pandit ad illum.* **13–18** Quomodo...solis] The question whether the same units of time as in solar reckoning could also be applied in lunar reckoning was only discussed in the late-seventh, early-eighth century Irish intellectual circles. In this contexts DRC 68.23–24 argues concerning the epacts: *Et deinde manifestum est nobis epacta in sole primitus natas, et postea ad lunam translatas*. For the etymology of *momentum* cf. BC 34A5–6: *Momentum pro uelocitate dictum est, quasi motus mentis*.

13–18 Quomodo...solis] Partially transcribed in Bisagni & Warntjes, 'Latin and Old Irish', p. 13 (with a facsimile of the last sentence of this passage); Ó Cróinín, 'Old Irish gloss', p. 132; idem, 'Earliest Old Irish glosses', p. 17; cf. also Bisagni & Warntjes, 'Latin and Old Irish', p. 15–6; p. XIX, LXXV of the introduction.

<XLVIIII.> DE AEPACTIS

Names and def. of epacts

*Aepacta*e *Grec*e *adiectiones lunares*. In *Latin*e quoque nomen insignum numero non inuenitur. Aepactae *adiectiones* dictae sunt, quia XI aetates | recedenti labentis anni aetati *adieciunt*. Quod ideo euenit, quia *solaris ann*us lunari anno prolixior est *XI die*bus. fol. 26v

The number of epacts

Quot aepactę sunt? Tot et solaris *anni* CCCLXV *die*bus. Quot uniuscuiusque diei aepactę? Id est XVIIII. Et in *nonodecim*o *ann*o finito cursu emersum prius *re*uertuntur in ordinem.

XLVIIII,2 Latine] Latina *M*. **7** Quot] quod *M (note that* CE *also reads* quod *instead of* quot*)*. | sunt] ut *M (corr. from* CE*; I suspect that the copyist mistook the Irish abbr. for* sunt *for* ut*)*. **8** Quot] quod *M*. **9** nonodecimo] nodecimo *M**.

XLVIIII,2 Aepactae…Latine] The definition of epacts being a Greek term to be translated as lunar additions seems to stem from ETYM. 6.17.29 (cited in DRC 68.30): *Epactas Graeci vocant, Latini adiectiones annuas lunares*. or PROLOGUS DIONYSII (Krusch, *Studien* II, p. 64): *epactas etiam, quas Greci vocant, id est adiectiones annuas lunares XI*. Dionysius defines *epactae* as *adiectiones lunares* frequently: EPISTOLA DIONYSII (Krusch, *Studien* II, p. 84; cited in DE SALTU LUNAE I, 984C5–6, DTR 42.46, in turn cited in LIB. COMP. 4.9B, in turn cited in LIB. CALC. 46B; PV §297), DIONYSIUS, *Cyclus* (Krusch, *Studien* II, p. 60–74; cited in numerous Easter tables based on the Dionysiac reckoning), DIONYSIUS, *Argumentum III* (Krusch, *Studien* II, p. 75; cited in numerous computistical formularies), DIONYSIUS, *Argumentum IX* (Krusch, *Studien* II, p.77; cited in numerous computistical formularies). Cf. also CE p. 108: *Aepactę ergo adiectiones lunares interpretatur*. This definition is mentioned again in c. 51, l. 4. The explanation of these lunar additions referring to the eleven-day difference between the solar and the lunar year derives from ETYM. 6.29.23–30.1 (cited in DRC 68.31–35): *Quas ideo Aegyptii adiciunt, ut lunaris emensio rationi solis aequetur. Luna enim iuxta cursum suum viginti novem semis dies lucere dinoscitur, et fiunt in anno lunares CCCLIV; remanent ad cursum anni solaris dies undecim, quos Aegyptii adiciunt. Unde et adiectiones vocantur*. Cf. PROLOGUS DIONYSII (Krusch, *Studien* II, p. 66): *Ac per hoc in duodecim lunaribus mensibus CCCLIIII dierum summam colligi, cui epactas Aegyptii annuas, id est undecim dies accommodant, ut ita demum lunaris emensio rationi solis adaequetur*. DNR 1.5.69–73 (cited in DRC 68.18–22, BC 69D8–14, LIB. ANN. 13): *Dies epactarum sunt XI, qui per singulos annos ad cursum lunarem adcrescunt. Nam dum in annum XII lunae CCCLIIII dies habeant, remanent ad cursum anni solaris dies XI, quos epactas Aegyptii uocauerunt, pro eo quod ad inueniendam lunam per totum annum adiciantur*. PROLOGUS CYRILLI, *Addendum* (Krusch, *Studien* I, p. 343): *Nam et sol in anno lucet dies CCCLXV et quadrans, luna vero lucet die CCCLIIII. Reliqui sunt XI, quos supra videtur sol lucere. Et hii sunt XI, qui annis singulis ad lunae cursum adduntur*. For the eleven-day difference between a solar and a common lunar year see also c. 60, ll. 3–6; for the placement of these eleven days within the Julian calendar year cf. c. 51. **7–10** Quot…ordinem] The epacts certainly were the most important feature of lunar reckoning, as MC argues himself (c. 48, ll. 19–20). As outlined in the previous paragraph, the solar year was eleven days longer than the lunar year, so that the lunar age of any given Julian calendar day increased by eleven every year, minus 30 whenever this number exceeded 30 (because of the intercalation of an embolismic month of 30 days). In the 19-year cycle (which is discussed in detail in c. 58), then, the lunar age of any fixed Julian calendar date increase by eleven every year (subtracting 30 whenever that number was exceeded), so that the same lunar age recurred only after the completion of 19 years. Accordingly, each year of a 19-year cycle could unambiguously be defined by the lunar age of any fixed Julian calendar day. The date chosen usually was 1 January or 22 March (cf. the following paragraph). The source for MC in this paragraph here is CE p. 108–9: *Quod (recte quot) sunt ępactę, per quas de luna conputabimus? Hoc sunt dies anni; tot ępactę sunt. Circulus enim ępactarum supra circulum dierum anni est, uel uti dicitur: rota in rota. ... Hic queritur quantum sit temporis quousque in hanc aetatem lunę, qua fuit in Kalendis*

49. ABOUT EPACTS

Epacts in Greek (mean) lunar additions. In Latin this remarkable term does not even stand for a number. Epacts were called additions, because they add eleven (lunar) ages to the receding (lunar) age of the passing year. This happens for the reason that the solar year is longer than the lunar year by eleven days.

How many epacts exist? As many as (there are) in the 365 days of the solar year (i.e. every single Julian calendar day could potentially be considered as the epact of a year). How many epacts belong to every single (Julian calendar) day? 19. And after the completion of a course of 19 years they recur as previously in the renewed order.

Ianuariis aut in qualibet die, eadem luna redeat, id est XIX annis completis.

XLVIIII,2 Aepactae…lunares] The definition of the epacts as lunar additions was extremely common; interestingly enough, it appears that Dionysius' texts were more popular in this respect than Isidore's account of the calculation of Easter, which received its principal reception from late seventh-, early eighth-century Insular computists (cf. *app. font.*). For some variations of this definition cf. PROL. AQUIT. 3D: *Epactae etiam lunares, id est adiectiones*. DTR 50.3–5 (cited in LIB. CALC. 89, PV §214, RM 68.6–8): *Inde epactae graeco vocabulo, id est adiectiones, dictae, quod per annos singulos XI dierum, ut diximus, accumulentur*. LECT. COMP. 2.4A: *epactae, id est adiectiones lunae*. COMP. COL. 4.2A: *Epactae quid sunt? Adiectiones lunares*. This definition appears also (albeit not within a formal discussion of the epacts) in BC *Praefatio* (1282B10), 74D13–14, 76(1312)B1–2, 155(1365)B11–12, DIAL. NEUSTR. 21B, LIB. COMP. 1.9c and many more unpublished computistical formularies. For the following explanation of lunar additions referring to the eleven-day difference between the solar and the lunar year cf. the works referred to in *app. font.* and DTR 50.2–3 (cited in LIB. CALC. 89, PV §214, RM 68.4–6): *Tertia praefati circuli linea continet epactas lunares, quae ad cursum solis annuatim undenis diebus accrescere solent*. LECT. COMP. 2.4A (cited in LIB. COMP. 2.15d): *Si vis scire, unde procedunt epactae, id est adiectiones lunae, tene dies de anno solari et de anno lunari. Annus solaris habet dies trecentos sexaginta quinque, annus lunaris trecentos quinquaginta quattuor. Vide, quantum superat alter ab altero, id est undecim dies. Id sunt undecim adiectiones in capite*. Cf. also BC 14(1285)C4–11, DTR 11.24–27, LECT. COMP. 2.4a. **7–10** Quot…ordinem] Bede describes the increase of the epacts by eleven each year, and the recurrence of each lunar age on the same date after 19 years, in the following terms: DTR 50.7–12 (cited in LIB. CALC. 89, PV §§214–215, RM 68.10–16): *Et quidem per totum anni vertentis orbem suas quaeque dies habet adiectiones lunares XI. Nam, verbi gratia, si hodie cum scribo quinta est luna, isto ipso die post annum decima sexta erit luna, post duos annos vicesima septima, post tres octava; nec eadem huc quae nunc est prius quam XVIIII annorum peracto circulo redit.*

XLVIIII,1–32 De…V] Cf. Borst, *Schriften*, p. 910.

The placement of epacts

Communes sunt totius anni dies aepactis, sed tamen *Greci in* XI *Kalendas Aprilis* legitime, Romani uero *in Kalendis Ianuarii epactas* enumerant.

Change of Greek placement

Unde Romani nec minus Scotti in Kalendis Ianuarii Grecorum obseruationes ępactas rimantur.

---

12 uero] uera *Walsh & Ó Cróinín (contra MS).* **13** epactas] epactis *Walsh & Ó Cróinín (contra MS).* **15** ępactas] epactis *Walsh & Ó Cróinín (contra MS).*

11–13 Communes…enumerant] As already mentioned in the previous paragraph, the epacts could be fixed on any given Julian calendar day, with the lunar age of that date serving as the defining feature of a lunar year. In the Dionysiac reckoning (the reckoning favoured by MC), 22 March was chosen as the *sedes epactarum*, i.e. the placement of the epacts. Interestingly enough, Dionysius himself does not mention any date for the placement of the epacts. Isidore is the first Latin author who explicitly referred to 22 March as the *sedes epactarum* of the Dionysiac (or 'Greek') reckoning: ETYM. 6.17.31: *Istae epactae semper XI Kal. April. reperiuntur in eadem luna quae fuerit eo die.* The older custom in the Latin West, ascribed to the Romans by MC, was to set the *sedes epactarum* at the beginning of the civil year, 1 January. This is the practice followed in the Victorian Easter cycle and the *latercus*: VICTORIUS, *Cyclus* (Krusch, *Studien* II, p. 27–52), *latercus* (this table was tentatively published by Mc Carthy & Ó Cróinín, 'Irish 84-year Easter table', p. 68–9, corrected by McCarthy, 'Easter principles', p. 218–9; it is translated in Blackburn & Holford-Strevens, *Companion to the year*, p. 873–5; a full facsimile is printed in Warntjes, '84 (14)-year Easter reckoning', p. 80–2); note that both of these tables had been in use in the region where MC was compiled before the introduction of the Dionysiac reckoning, first the *latercus*, then the Victorian system. MC's direct source for this passage here probably was CE p. 109: *Item interrogandum est, in quo loco ponunt Greci aepactas et Latini? Hoc est: ponunt Greci in XI Kalendas Aprilis; Latini autem in Kalendas Ianuarii.* **14–15** Unde…rimantur] If Easter reckonings were to be compared, it was obviously absolutely essential to establish comparable data. When first encountering the Dionysiac reckoning in form of the Dionysiac Easter table, the Irish had the problem that none of the data outlined in that table was comparable to the columns of the Easter tables known to them, the *latercus* and the Victorian one; most importantly, the epacts were listed for 1 January in both the *latercus* and the Victorian table (cf. previous note), but for 22 March in the Dionysiac table (Krusch, *Studien* II, p. 69–74). In order to compared the lunar cycles underlying these different systems, it was essential, then, to translate the Dionysiac epacts to 1 January. Mathematically, the Dionysiac epact of 1 January was calculated by adding 9 to the epact of 22 March (and subtracting 30 whenever the sum exceeded that number); alternatively, various algorithms for calculating the Dionysiac epact of 1 January existed, in MC's time predominently PS-DIONYSIUS, *Argumentum XIII*. It must also be presumed that the Dionysiac Easter tables used in Ireland noted the epacts for 1 January rather than (or additional to) 22 March (cf. *app. comp.*); note, in this respect, that Cummian argues that the 19-year cycle known to him contained three columns, one of which being a list of epacts of 1 January. EPISTOLA CUMMIANI 215–220: *Decimo trecentorum decem et octo episcorum decennouennalem cyclum, qui Grece enneacedeciterida dicitur, in quo Kalendae Ianuarii lunaeque eiusdem diei et initia primi mensis ipsiusque XIIIImae lunae recto iure ... sunt adnotatae.* The possibility of translating the epacts of different systems to the same date is also outlined in CE p. 109: *Sed ępactę Grecorum possunt esse in loco, ubi Latini ponunt ępactas. Ut Latinorum ępactae poni possunt in loco, ubi ępactas Greci ponunt.* The subsequent table in CE contains both the Dionysiac epacts of 1 January and 22 March.

11–13 Communes…enumerant] As will be outlined in the following note, the Irish appear to have been the first to translate the Dionysiac epacts from 22 March to 1 January; after this task had been accomplished, computists usually chose either option, but did not refer to both. One of the few texts mentioning both possibilities is QUAEST. AUSTR. 2.10A: *Nunc lunae aetates in Kalendis ostendamus, et adgrediatur a primo ogduadis anno ostendere. Sed iuxta Victorium et Latinos initium in Kalendis Ianuarii ponamus, quamvis novimus non hinc incipere iuxta Aegiptios initium anni, sed ab XI Kalendas*

The days of the whole year are common to the epacts (i.e. the epacts can be reckoned on any Julian calendar day), but, nevertheless, the Greeks lawfully (i.e. according to Dionysius' authority) reckon the epacts on 22 March, the Romans, however, on 1 January.

Therefore, the Romans and not less the Irish convert the epacts (as they appear in) Greek usage (i.e. on 22 March) to 1 January.

Aprilis. Only 1 January is mentioned as the placement of the epacts in COMP. COL. 4.2A; only 22 March in BC 22A15, 26A2–5 (cited in LIB. COMP. 1.9c; partially cited in LIB. ANN. 9, in turn cited in LIB. CALC. 62), PROL. AQUIT. 3D, DTR 42.36–37 (cited in PV §296, LIB. COMP. 4.9B, in turn cited in LIB. CALC. 46B), 50.13 (cited in LIB. CALC. 89A, PV §215, RM 68.17), LIB. COMP. 5.20aA, PV §22, and many calendars (cf. Borst, *Reichskalender*, p. 702). **14–15** Unde…rimantur] In order to be able to compare the Dionysiac 19-year cycle with the Victorian one, it was necessary to translate the Dionysiac epacts from 22 March to 1 January, a task probably first accomplished by Irish computists in the seventh century (cf. *app. font.*). This possibly led to the production of tables listing the Dionysiac epacts of 1 January and 22 March in parallel columns: Cf. St Gall, Stiftsbibliothek, 250, p. 80; St Gall, Stiftsbibliothek, 251, p. 2; London, British Library, Harley 3017, fol. 3r; Paris, Bibliothèque Nationale, Nouvelle acquisition latine 1615, fol. 152v (defective); Paris, Bibliothèque Nationale, Lat. 7530, fol. 252v; Milan, Biblioteca Ambrosiana, H 150 inf, fol. 7v (published in BC *Praefatio*, but not in parallel columns). Both lists also appear, e.g., in a Dionysiac Easter table from AD 798 to 854 (cf. Appendix 8) and in the Dionysiac Easter table from AD 817 to 832 in St Dunstan's classbook (Oxford, Bodleian Library, Auct. F 4 32, fol. 21r).

11–17 Communes…incipimus] Transcribed and translated in Mac Carthy, *Annals of Ulster* 4, p. lxviii–ix, and the most part of it is also translated in Harrison, 'Luni-solar cycles', p. 73; transcribed in Walsh & Ó Cróinín, *Cummian's letter*, p. 84; except for the last sentence in Schwartz, 'Ostertafeln', p. 92; only the middle part transcribed in Krusch, *Studien* I, p. 13; idem, 'Einführung', p. 163; cf. Walsh & Ó Cróinín, *Cummian's letter*, p. 181 and p. LX, LXXI, LXXIII–LXXIV, CLXII, CLXXXVI of the introduction.

[Dionysiac epacts 1 Jan] Hinc ab VIII singulari *in Kalendis Ianuarii* ępactas incipimus: *VIII, XX, I, XII, XXIII, IIII, XV, XXVI, VII, XVIII, XXVIIII, X, XXI, II, XIII, XXIIII, V, XVI, XXVII.*

[Non- Dion. epacts] Haec sunt aetates XI, quae secundum *Gre*cos hanc rationem in Kalendis Ianuarii non inueniuntur: XVIIII, XXX, XI, XXII, III, XIIII, XXV, VI, XVII, XXVIII, VIIII.

[Victorian epacts (from *ogd.*)] *Haec sunt epactae secundum* Latinos in Kalendis Ianuarii: *VIIII, XX, I, XII, XXIII, IIII, XVI, XXVII, VIII, XVIIII, XXX, XI, XXII, III, XIIII, XXV, VI, XVII, XXVIII.*

---

16 singulari] singularis *Walsh & Ó Cróinín (contra MS).* **17** VIII] VIII *corr. to* VIIII *M (since* MC *places the* saltus *in March rather than November of the previous year (cf. c. 49, ll. 15–16; c. 50, ll. 7; c. 52, ll. 9–10, 29; c. 62, ll. 67–86), there cannot be any doubt – contra Mac Carthy, Harrison – that* VIII *was the original number here; the continental copyist, then, changed this number to* VIIII *to suit his own custom of placing the* saltus *in November).* **21** XVII] XVIII *M**. | XXVIII] XXVIIII *M**. **23–24** VIIII…XXVIII] VIIII, XX, I, XII, XXIII, IIII, XV, XXVI, VII, XVIII, XXVIIII, X, XXI, II, XIII, XXIIII, V, XVI, XXVII *M (the list given in M is, save for the first number, identical with the list of Dionysiac epacts given previously in ll. 17–18 of this chapter; I doubt that this list here was compiled only to illustrate a difference in the first epact, which would essentially reflect a difference in the placement of the* saltus *only; the term* Latini *rather implies that the Victorian epacts were originally listed here, and I have corrected the list accordingly; the copyist, then, seems to have changed the Victorian epacts he found in his exemplar to the epacts he followed himself, i.e. the Dionysiac ones; this is supported by the fact that the upper stroke of* I *occurs in M after the numeral* XV, *i.e. after the first number that was corrected by the copyist (cf. the facsimile in Appendix 1): it seems therefore that the copyist started to copy the original number* XVI, *realized that this was Victorian practice, and then changed this and the following numbers to the Dionysiac values by reducing the Victorian epacts by one; cf. also* CE *p. 109 (a facsimile is printed in Appendix 1), where the list of Dionysiac epacts of 1 January is followed by the equivalent Victorian one, and this appears to have been the source here).*

16–32 Hinc…V] This most interesting table is, unfortunately, badly preserved in the only extant copy of MC (for a facsimile and further discussion see Appendix 1). Originally, it appears to have had five columns, listing a) the 19 Dionysiac epacts of 1 January starting with the *ogdoas*, b) the eleven numbers between 1 and 30 that do not feature in column a (obviously compiled by comparing the Dionysiac epacts with the beginning of the Victorian Easter table, since the eleven numbers given here are the first eleven epacts of that table), c) the Victorian epacts of 1 January starting with the *ogdoas*, d) the Victorian epacts of 1 January starting with the *hendecas*, e) the eleven numbers between 1 and 30 that do not occur in columns c and d. The principal purpose of this table was to compare the epacts of the Dionysiac reckoning with the Victorian ones; to seventh-century Irish computist it was, for various reasons, not perfectly clear in what kind of way the Dionysiac epacts were to be aligned with the Victorian ones. Historically correct was the alignment of the two sets of epacts so that both sets started with the *ogdoas*, i.e. comparing column a with column c; this comparison, however, led to Dionysiac data conflicting with the Gospels and other authoritative texts concerning the passion of Christ (cf. c. 52), and therefore a different method was invented, namely an alignement of the Dionysiac *ogdoas* with the Victorian *hendecas*, i.e. column a with d. The source for columns a, c, and d was CE p. 109, which transmits the only extant table (cf. Appendix 1) in which the Dionysiac epacts of 1 January (starting with the *ogdoas*) are paralleled with the two lists of Victorian epacts, namely one starting with the *ogdoas*, the other with the *hendecas* (note in general that the terms *ogdoas* and *hendecas* are retrospectively translated from the Dionysiac to the Victorian reckoning); columns b and e, then, appear to be MC's own additions. The first text to have introduced the unhistorical comparison of the Dionysiac with the Victorian epacts (i.e. aligning Dionysiac *ogdoas* with Victorian *hendecas*) appears to be DE COMPARATIONE EPACTARUM, transmitting a table (1.13–36; see Appendix 2, facsimile in Appendix 3) in which the Dionysiac epacts of 1 January are only compared to the Victorian epacts starting with the

Hence we start the epacts on 1 January from the one-digit number (*singularis*) eight (column a): 8, 20, 1, 12, 23, 4, 15, 26, 7, 18, 29, 10, 21, 2, 13, 24, 5, 16, 27.

The following are the eleven (lunar) ages which, according to the Greeks, are not found on 1 January in this reckoning (column b): 19, 30, 11, 22, 3, 14, 25, 6, 17, 28, 9.

The following are the epacts on 1 January according to the Latins (starting with the *ogdoas*; column c): 9, 20, 1, 12, 23, 4, 16, 27, 8, 19, 30, 11, 22, 3, 14, 25, 6, 17, 28.

---

hendecas; the subsequent explanation of this practice (1.88–90, 2.17–20; see Appendix 2) probably was the immediate source for MC's description of it here: *alium sapientem, qui Grecam magis regulam sequitur, ipsum dicentem ogdoadem melius debere incipere a loco, a quo endicas incipere. ... Sed quamuis ista causa aliquis hunc Victorium et ipsum Dionysium, qui Grecam magis sequitur regulam, ogdoadibus et endicadibus discrepare putet.* Interestingly enough, it seems that this unhistorical comparison of the Dionysiac with the Victorian epacts underlies the two sets of epacts listed in the ANNALS OF ULSTER for the years AD 646–654 (Mac Airt & Mac Niocaill, *Annals of Ulster*, p. 124–8). For this comparison cf. also c. 52, ll. 14–26, 49–60; c. 56, ll. 15–18.

16–32 Hinc…V] Since the unhistorical comparison of the Dionysiac with the Victorian epacts of 1 January proved to be flawed, tables as found in CE, DE COMPARATIONE EPACTARUM, and MC remained a product of seventh- and early eighth-century Irish computistics and found no immitators. Even the translation of the Dionysiac epacts from 22 March to 1 Janury discussed in the previous note became unnecessary at a time when the Dionysiac reckoning became the universally accepted system, and accordingly stopped to be compared to older systems. Only few eighth- or ninth-century texts still list Dionysiac epacts on 1 January (though such a list could still fequently be found as a separate column in certain tables; cf. previous note), namely the HARVARD FRAGMENT (Harvard, Houghton Library, MS Type 613, fol. 7r), probably directly influenced by MC, and these epacts are also mentioned in LECT. COMP. 9.4B, COMP. COL. 4.2A,C (where they are termed *usurpative*). Victorian epacts, which became obsolete with the eventual abandoning of that system towards the end of the eighth century, can still be found in the Victorian DIAL. BURG. 13B (starting with epact 11, i.e. the fourth year of the *hendecas*), as well as in Paris, Bibliothèque Nationale, Lat. 4860, fol. 157v (starting with the *ogdoas*); Munich, Bayerische Staatsbibliothek, Clm 14725, fol. 16r (starting with epact 1, i.e. the third year of the *ogdoas*); ANGERS 477, fol. 43v (starting with epact 12, i.e. the fourth year of the *ogdoas*; note, however, that the preceding epact 1 is missing, so that this may have been the inital epact originally as in the Munich MS). An interesting addendum to the COMPUTUS COTTONIANUS, fol. 80r, uniquely list the Victorian epacts in numerical order, as well as the eleven lunar ages that are not included in that list (otherwise only to be found in MC): *Haec luna, quae kanonica\n/t: I, III, IIII, VI, VIII, VIIII, XI, XII, XIIII, XVI, XVII, XVIIII, XX, XXII, XXIII, XXV, XXVII, XXVIII, XXX. Haec lunae, quae dicuntur quasi non kanonicant: II, V, VII, X, XIII, XV, XVIII, XXI, XXIIII, XXVI, XXVIIII.* Interest in a list of epacts of 1 January, as well as the lunar ages that do not occur on this date, is expressed in QUAEST. AUSTR. 1.10A: *Et quas aetas Ianuarii Kalendae usque ad finem decennovenalis cycli observabunt et quas aetates devitent?*

16–24 Hinc…XXVIII] Cf. Borst, *Schriften*, p. 404. **16–32** Hinc…V] Cf. p. LXXI, LXXXIV XCIV, CII, CXXXIII, CLXXXVI, CXCVIII of the introduction and Appendix 1.

Comp. of Dion. and Vict. epacts

| Sunt tamen quidam, qui endecadem Latinorum contra ogdoadem Grecorum ponunt. Grecos enim in accensione lunae praecedere Latinos ferunt. fol. 27r

Vict. epacts (from *hend.*)

Et *ita Latini* ab endecade incipiunt: *VIII, XVIIII, XXX* et reliqua.

Non- Vict. epacts

Haec sunt XI aetates contrariae Latinis in Kalendis Ianuarii: <XV, XXVI, VII, XVIII, XXVIIII, X, XXI, II, XIII, XXIIII, V>.

<L. DE AETATIBUS LUNARIBUS IN KALENDIS MENSIUM>

Description of the method

Per hunc ordinem *epacta*rum, quę in Kalendis Ianuarii enumerauimus, per XII menses *Grec*e menses et aetates in *Kalend*is *mensium* per *XVIIII anno*s subtilis quisque intellegitur.

---

28 VIII XVIIII] VIIII, XVIII *M (corr. according to* MC, CE, DE COMPARATIONE*; also Schwartz).* **31–32** XV...V] *om. M (add. according to* MC*; I have reconstructed this list presupposing that it was compiled on the basis of a comparison of the Dionysiac with the Victorian epacts, with* XV *being the first Dionysiac epact that does not occur in the Victorian list; a similar technique was used to compile the list of the eleven lunar ages not to be found among the Dionysiac epacts of 1 January above; but note that the only extant list of the eleven lunar ages not to be found among the Victorian epacts, given in the addendum to the* COMPUTUS COTTONIANUS*, is in numerical order).* **L,6** quisque] quique *M, Schwartz.*

L,3–6 Per…intellegitur] In this chapter, MC discusses the lunar ages on the Calends of each month, not only for the Dionysiac reckoning, but also for the Victorian reckoning and the *latercus*, so that the comparative character of the previous chapter is continued here; note that MC's principal source for this chapter, CE, compares only the Dionysiac with the Victorian reckoning, so that the *latercus* data (c. 50, ll. 67–76) is MC's own addition. The information for every single year discussed in this chapter could be calculated by means of the respective list of epacts (which are, save for the *latercus*, given in the previous chapter) and the respective sequence of lunations (given in ll. 30–40, 62–67, 74–78 of this chapter), and this principle of construction is what MC outlines here; it cannot be found in MC's source, as CE p. 109 simply introduces its discussion with the following phase: *Item secundum Grecos XIX annorum aepactę per Kalendas mensium hoc modo explorari desunt.*

L,3–6 Per…intellegitur] MC's rather clumsy method for calculating the lunar age of the Calends of each month of any Julian calendar year was soon replaced by more sophisticated algorithm which made use of the epacts of 22 March and regulars for each month (defined as the lunar age of the Calends of the respective month in the first year of the 19-year cycle); this algorithm can first be found in the COMPUTUS COTTONIANUS of AD 689 (London, British Library, Cotton Caligula A XV, fol. 75v), then in BC *Praefatio* (1282C2–8), 22, DIAL. NEUSTR. 21, before it reached a wider reception through DTR 20 (cited in LIB. CALC. 10, PV §§15–21 (with additional tables in §14), RM 70); interesting enough, LECT. COMP. 2.3 (cited in LIB. COMP. 2.15c) and LIB. COMP. 2.17 (cited in LIB. CALC. 14) use the same method as MC for explaining the regulars that form the basis of this algorithm. Even more basic, a table listing the lunar ages on the Calends of each month for each year of the 19-year cycle, commonly labelled *pagina epactarum*, became extremely popular, since it can be found in almost every computistical MS; note that various versions of this table circulated (with the differences being due to differing placements of the *saltus* and the embolisms), which still have to be classified; printed examples of this table can be found as c. 4.1 in LIB. COMP., as c. 9 in LIB. CALC., while Dicuil (LDA 1.7) gives two differing versions.

There are, however, certain people who place the *hendecas* of the Latins alongside the *ogdoas* of the Greeks. They establish, in fact, that the Greeks precede the Latins in the kindling of the moon.

And the Latins start (the epacts of 1 January) from the *hendecas* in the following way (column d): 8, 19, 30, and so on (i.e. 11, 22, 3, 14, 25, 6, 17, 28, 9, 20, 1, 12, 23, 4, 16, 27).

The following are the eleven lunar ages that are contrary to the Latins on 1 January: 15, 26, 7, 18, 29, 10, 21, 2, 13, 24, 5.

50. ABOUT THE LUNAR AGES ON THE CALENDS OF THE MONTHS

By means of this sequence of epacts, which we have enumerated for 1 January, (and) by means of the 12 (lunar) months (i.e. the sequence of lunations), every astute person understands the (lunar) months and the (lunar) ages on the Calends of the months according to the Greek (reckoning) throughout the 19 years.

25–29 Sunt…reliqua] Transcribed and corrected in Schwartz, 'Ostertafeln', p. 95. **L,1–80** De…continet] Cf. p. LXXI–LXXII, XCIV, CXXXIII, CXCVIII of the introduction. **3–39** Per…conturbet] Cf. p. XCIV, CXC–CXCI of the introduction. **4–36** Grece…Decimbri] Transcribed in Schwartz, 'Ostertafel', p. 97–8.

The first year of the Dionysiac 19-year cycle

VIII luna in Kalendis Ianuarii; *VIIII in Kalendis Februarii*; *VIII luna in Kalendis Martii*; saltu *X in Kalendis Aprilis et Maii*; transilit XI et *XII in Kalendis Iunii*; *XIII in Kalendis Iulii*; *XIIII in Kalendis Augusti*; transilit XV et *XVI in Kalendis Septimbris et Octimbris*; transilit XVII et est *XVIII in Kalendis Nouimbris et Decimbris*; transilit XVIIII, fit XX luna in Kalendis Ianuarii. Hucusque primus annus cum saltu.

The second year of the Dionysiac 19-year cycle

Incipit II annus sine saltu: *XX luna in Kalendis Ianuarii*; *XXI in Kalendis Februarii*; *XX in Kalendis Martii*; *XXI in Kalendis Aprilis et Maii*; transilit XXII et *XXIII* in *Iuni*o; *XXIIII* in *Iuli*o; *XXV* in *August*o; transilit XXVI et fit *XXVII* in *Septimbri et Octimbri*; transilit XXVIII et est *XXVIIII* in *Nouimbri et Decimbri*; transilit XXX et est *I* in *Kalendis Ianuarii*. Hucusque secundus annus.

---

8 saltu] saltu- *M, Schwartz*. **10** XVI] XV *M (corr. by Schwartz)*.

7–14 VIII…saltu] This outline of the lunar ages on the Calends of each month for the first year of the Dionysiac 19-year cycle is copied directly, with few alterations, from CE p. 109–110: *VIII luna in Kalendis Ianuarii; VIIII luna in Kalendis Februarii; VIII luna in Kalendis Martii; X luna in Kalendis Aprilis et Maii; XII luna in Kalendis Iunii; XIII luna in Kalendis Iulii; XIIII luna in Kalendis Augusti; XVI luna in Kalendis Septimbris et Octimbris; XVIII luna in Kalendis Nouimbris et Decimbris*. Note that CE here and in the following two years applies the same peculiar sequence of lunations as MC, in which April is full and May hollow (and therefore the Calends of April and May have the same lunar age), and which is uniquely described in c. 50, ll. 29–39. **15–21** Incipit…annus] Again, CE p. 110 is MC's direct source for this outline of the lunar ages on the Calends for the second year of the Dionysiac 19-year cycle: *XX luna in Kalendis Ianuarii; XXI luna in Kalendis Februarii; XX luna in Kalendis Martii; XXI luna in Kalendis Aprilis et Maii; XXIII Kalendis Iunii; XXIIII Kalendis Iulii; XXV Kalendis Augusti; XXVII Kalendis Septimbris et Octimbris; XXIX Kalendis Nouimbris et Decimbris; I Kalendis Ianuarii.*

7–14 VIII…saltu] This first year of the Dionysiac 19-year cycle is, for various reasons, discussed in quite a number of eighth-century computistical texts: Cf. DRC 73.4–8 (embedded in a discussion of the *luna abortiva*; the emendation is mine; this chapter is copied in St Dunstan's classbook (Oxford, Bodleian Library, Auct. F 4 32, fol. 22r–v) and with considerable variation in the ninth-century Carolingian text DE CONTRARI AEATIBUS edited in Appendix 6): *Ita etiam exploratur, id est in anno in quo fit VIII luna in Kalendis Ianuarii separatur in illo anno geminatione (geminatio MSS) Aprelis a Februario. Uerbi gratia, VIII<I> luna in Kalendis Februarii, X luna in Kalendis Aprelis, causa saltus in XI Kl Aprelis. Et haec est contraria regula huius anni, quam saltus efficit*. DTR 20.2–6 (as an introduction of the regulars for the algorithm mentioned in the previous note; cited in LIB. CALC. 10, 16, PV §15, RM 70.6–13): *Primo decemnovenalis circuli anno, in quo nullae sunt epactae, in kalendas ianuarias VIIII est luna; in kl. februarias, X; in kl. martias, VIIII; in kl. apriles, X; in kl. maias, XI; in kl. iunias, XII; in kl. iulias, XIII; in kl. augustas, XIIII; in kl. septembres, XVI; in kl. octobres, XVI; in kl. novembres, XVIII; in kl decembres, XVIII*. The difference between MC, CE, and DTR is due to the fact that the Irish computists place the *saltus* in March, Bede, however, in the previous November (for these two placements of the *saltus* cf. c. 62, ll. 68–72, 97–110). This difference is nicely considered in COMP. COL. 4.2B (in a discussion of the epacts; note, however, that it should rather read *octo vel novem Martio*): *<Inde aetates lunae in Kalendas primi anni sunt Ianuario> octo vel novem, novem Februario vel decem, novem vel decem Martio, decem Aprili, undecim Maio, duodecim Iunio, tredecim Iulio, quattuordecim Agusto, sedecim Septembri, sedecim Octubri, octodecim (MS) Novembri, octodecim Decembri, viginti in Ianuario*. Most strikingly, only CE and MC apply the peculiar sequence of

(The first year of the Dionysiac 19-year cycle records) *luna* 8 on 1 January; (*luna*) 9 on 1 February; *luna* 8 on 1 March; because of the *saltus* (*luna*) 10 on 1 April and 1 May; it passes over (*luna*) 11 to (*luna*) 12 on 1 June; (*luna*) 13 on 1 July; (*luna*) 14 on 1 August; it passes over (*luna*) 15 to (*luna*) 16 on 1 September and 1 October; it passes over (*luna*) 17 and (*luna*) 18 occurs on 1 November and 1 December; it passes over (*luna*) 19 and *luna* 20 happens to be on 1 January. Thus far the first year (of the Dionysiac 19-year cycle), (which) contains the *saltus*.

Here starts the second year (of the Dionysiac 19-year cycle), (which is) without *saltus*: (*luna*) 20 on 1 January; (*luna*) 21 on 1 February; (*luna*) 20 on 1 March; (*luna*) 21 on 1 April and 1 Mai; it passes over (*luna*) 22 to (*luna*) 23 in June; (*luna*) 24 in July; (*luna*) 25 in August; it passes over (*luna*) 26 and (*luna*) 27 happens to be in September and October; it passes over (*luna*) 28 and (*luna*) 29 occurs in November and December; it passes over (*luna*) 30 and (*luna*) 1 occurs on 1 January. Thus far the second year (of the Dionysiac 19-year cycle).

---

lunations outlined in c. 50, ll. 29–39 to this year, which is obviously a very characteristic feature of a specific Irish computistical school of the late seventh, early eighth centuries. **15–21** Incipit…annus] Since there is no apparent reasons why the second year of the Dionysiac 19-year cycle should be discussed in detail (other than MC's to demonstrate a certain concept or rule), an analysis of it can hardly be found in computistical texts; cf., however, the discussion of this year in the context of the *luna abortiva* in DRC 73.8–10 (for texts citing this passage cf. the previous note), as well as COMP. COL. 4.1A, where this year is considered as the first year of creation.

The 11th year of the Dionysiac 19-year cycle

Quia annus difficilis quidam, quem tertio ferimus, licet non tertius annorum ordine manet: *XXVIIII* in *Kalendis Ianuarii*; XXX in *Kalendis Februarii*; XXVIII in *Martio*; *XXX* in *Kalendis Aprilis*; prima in Kalendis *Maii*; *secunda* in *Iunio*; | *III* in *Iulio*; *IIII* in *Augusto*; transilit V et est *VI* in fol. 27v *Septimbri et Octimbri*; transilit VII et est *VIII* in *Nouimbri et Decimbri*; transilit VIIII et *X* in *Kalendis Ianuarii*.

---

22 Quia] qui *M (cf. Schwartz)*. **23** non] enim *Schwartz (contra M)*. **24** XXVIII] XXVIIII *M, Schwartz (corr. according to* MC, DTR*; if 1 February has* luna *30 (which is the embolism in this case), then 1 March must have* luna *28;* luna *29 on 1 March as given in M would only occur if this year was bissextile and if the extra lunar day was placed as* luna *30 at the end of the February lunation, in this case on 2 March – cf.* MUNICH 18156*; but* MC *does not follow this practice for the lunar bissextile day; yet, it probably was the practice of the copyist, and for that reason he may have changed the numeral here)*.

22–28 Quia…Ianuarii] In the previous two passages, MC has illustrated his method for calculating the lunar age of the Calends of each month by the examples of the first two years of the Dionysiac 19-year cycle. The first of these had already shown irregular features because of the application of the *saltus* in that year. More problems occurred in the 11th year of the Dionysiac 19-year cycle, the main one being, in MC's theory, that in this year 1 January and 1 March do not have the same lunar age, as they normally would, because of the application of an embolism in this year; this embolism, however, could also be placed later in that year, leading to slightly different numbers. This is the case in CE p. 110, and since CE certainly was the source for most part of this capter, it may be presumed that the numbers given in CE may, in fact, have been the ones originally recorded in MC, to be later tampered with by copyists: *Item hoc eodem modo per XIX annos facias. Praeter tertium annum endegdis (!), qui a XXIX in decimam aetatem egreditur in Kalendis Ianuaris, hoc modo: XXIX luna Kalendis Ianuarii; I luna Kalendis Februaris; XXIX luna Kalendis Martis; XXX luna Kalendis Aprilis et Maii; II luna Kalendis Iunii; III luna Kalendis Iulii; IIII luna Kalendis Augusti; VI luna Kalendis Septimbris et Octimbris; VIII luna Kalendis Nouimbris et Decimbris; X luna Kalendis Ianuarii*. Note especially that CE here also employs the sequence of lunations applied in the two years previously discussed, while MC here, contrary to its previous practice, employs the alternating one; this may be another indication that MC's numbers here were considerably changed by later copyists (cf. also *app. crit.*).

22–28 Quia…Ianuarii] This year is also discussed in DRC 73.34–42 (embedded in a discussion of the *luna abortiva*; for texts citing this discussion see ll. 7–14 of this chapter): *In anno quoque in quo fit XXVIIII luna in Kalendis Ianuarii abortiua luna in Ianuario (MS B) XXVIIII inuenitur. Licet per artificalitatem Februario deputatur, deputantes sibi calculatores luna quae in se nec nascitur nec finitur; et XXX a luna fit Apreli in hoc anno, causa XIIII lunae primi mensis, quae in hoc anno in XVII Kl Maii inuenitur. Et in hoc anno abortiua luna XXVIIII inuenitur in Apreli et eadem aetas fit in hoc anno Februario et Maio, id est prima luna. Et in hoc anno separatur geminatione (geminatio MSS) Aprelis a Februario, et haec est contraria regula huius anni, quam abortiva luna et XIIII luna primi mensis efficiunt. Et nihil perturbationis in hoc anno a Maio usque in finem inuenitur*. DRC's numbers vary considerably from the ones given in CE and MC, since DRC applies the alternating sequence of lunations and places the embolism later than MC. The same numbers as in DRC are given in a discussion of this embolismic year in BC 130. The eleventh year of the Dionysiac 19-year cycle is also one of three exceptions to the algorithm for calculating the lunar age of the Calends of each month (cf. ll. 3–6 of his chapter); see DTR 20.30–32 (cited in LIB. COMP. 4.12, in turn cited in PV §283; LIB. CALC. 10, RM 70.33–36): *Item anno XI, quia embolismi pridie nonas decembres accenditur, facit lunam in kl. martias vicesimam esse et octavam, cum hanc ratio argumenti vicesimam nonam tunc existere doceat*. The impact of a bissextile year (with the extra lunar day being added to the end of the February lunation) on Bede's discussion of this year is explicitly expressed in a gloss to DTR 41 in Munich, Bayerische Staatsbibliothek, Clm 18156, fol. 32r: *Verbi gratia, si embolismalis luna II Nonarum Decembris*

Since a certain year (namely the eleventh of the Dionysiac 19-year cycle) is difficult, which we produce (here) in third place, it may be granted that the third example does not remain in the order of years (of the Dionysiac 19-year cycle): (*luna*) 29 on 1 January; (*luna*) 30 on 1 February; (*luna*) 29 in March; (*luna*) 30 on 1 April; (*luna*) 1 on 1 May; (*luna*) 2 in June; (*luna*) 3 in July; (*luna*) 4 in August; it passes over (*luna*) 5 and (*luna*) 6 occurs in September and October; it passes over (*luna*) 7 and (*luna*) 8 occurs in November and December; it passes over (*luna*) 9 to (*luna*) 10 on 1 January.

---

incenditur, XXVIIII in Kalendis Ianuarii erit, ac XXVIII in Kalendis Martii computatur, si bissextus non fuerit. Qui si eue\ne/rit in illo XI anno, eadem ętatem, quam in Kalendis Ianuarii habuit, habebit in Kalendis Martii. In the SIRMOND MS (Oxford, Bodleian Library, Bodley 309, fol. 106v; the relevant passage is transcribed in Ó Cróinín, 'New light on Palladius', p. 29; idem, 'Bede's Irish computus', p. 206) it is argued that this is one of only three years within the Dionysiac 19-year cycle in which April and May have the same lunar age on the Calends (due to the placement of the embolism in April; as in all other texts save MC and CE, the alternating sequence of lunations is obviously applied), which is the exact opposite argument to MC.

22–28 Quia…Ianuarii] Cf. p. XXII of the introduction.

The Dionysiac sequence of lunations

Qua auctoritate praefatas in Kalendis epactas mensium explorant? Hoc est ex aetate lunae mensium. Extincta enim et accensa has praedictas explora epactas. Isto enim ordine lunae feruntur secundum Grecos: XXX luna in Ianuario; XXVIIII in Februario; XXX luna in Martio et Aprili; XXVIIII in Maio et Iunio; XXX in Iulio; XXVIIII Augusto; XXX in Septimbri; XXVIIII in Octimbri; XXX in Nouimbri; XXVIIII in Decimbri. Per has itaque lunae aetates quicumque obseruare uoluerit, ordinem euidenter repperiet et bissextum uigili sensu caueat, ne hanc obseruationem conturbet.

---

29 praefatas] praefata *M, Schwartz (corr. according to* MC*).* | epactas] regulę *M, Schwartz (corr. according to* MC*; it may also have read* regulas *or* regulares *originally).* **30** explorant] explora *M.* **33** in[1]] *follows* k- *for* Kalendis *M (corr. by Schwartz).* **34** XXVIIII[2]] XXXVIIII *M*; add.* in *Schwartz.*

29–39 Qua…conturbet] This sequence of lunations is not outlined in any other text. MC applies it consistently throughout, namely in two of the three previous examples (c. 50, ll. 7–21; the only exception is the third of these examples, c. 50, ll. 22–28), and then again in c. 61, ll. 17–23 and c. 62, ll. 97–110, but argues against it in c. 52, ll. 37–40. It is also implicit in the discussions of the three Dionysiac lunar years in CE discussed in the preceding notes, but the Dionysiac sequence of lunations outlined in CE p. 111 is the alternating one (though it can – and must – be presumed that a later copyist reworked this passage, so that CE, in my opinion, quite certainly was MC's source here also). None of the pre-CE, -MC texts (i.e. neither Dionysius nor Isidore) had discussed the Dionysiac sequence of lunations, so that seventh-century Irish computists were left with the task of reconstructing it from the data available in Dionysiac Easter table. From the lunar data given for the 8th, 11th, and 19th year of the Dionysiac 19-year cycle (namely the epact of 22 March and the Julian calendar date for the Easter full moon, as well as the lunar age of Easter Sunday), Irish computists quite rightly calculated that the lunation ending in April of these years was full, i.e. consisted on 30 days; the question was whether this was a regular feature of the April lunation, or whether in these years embolisms were applied in April (or any other place within the year that would have shifted a full lunation to April). CE and MC decided for the April lunation being full in every year of the Dionysiac 19-year cycle, while every later computist took the Dionysiac sequence of lunations to be alternating between full and hollow lunar months (which made April hollow), and that the three years mentioned above were exceptions due to the application of embolisms.

29–39 Qua…conturbet] The only text, to my knowledge, in which MC's sequence of lunations appears to have been discussed is the HARVARD FRAGMENT (Harvard, Houghton Library, MS Type 613, fol. 7r); though crucial parts of the passage in question have been lost, the statement that May has 29 lunar days survives and therefore the implication is that the sequence was structured in the same way as in MC with April being full, May hollow. It appears, however, that further manuscript research may bring more texts to light containing this feature, as Jones ('Legend of Pachomius', p. 208) draws attention to a passage in Rouen, Bibliothèque Publique, 524 (now I 49), fol. 87r (a MS that I have not examined), which reads: *Apud Grecos XXX luna aprilis semper, et XXVIIII apud Latinos*. DRC 72, then, is the first and only text to argue explicitly against such a sequence of lunations (referring to followers of it as *alii*), with this indicating that DRC read and disagreed with MC and CE: *Sciendum nobis quomodo menses lunares mensibus solis apud Grecos et Latinos secundum hanc regulam iuguntur. Ita etiam, id est XXX luna Ianuario, XXVIIII Februario, XXX Martio, XXVIIII Apreli (etsi alii dicunt lunam Aprelis XXXmam esse semper apud Grecos, quod non sinit natura, nisi in tribus annis tantum, id est quando XXVI uel XXVII uel XXVIIII in Kalendis Ianuarii fit. Nisi enim in his tribus annis XXXma fuerit luna Aprelis, XIIII luna primi mensis, quae contra Maium regulatur, non erit in naturalibus mensis diebus); XXX Maio, XXVIIII Iunio, XXX Iulio, XXVIIII Agusto, XXX Septimbri, XXVIIII Octimbri, XXX Nouimbri,*

According to which authority do they examine the aforesaid epacts on the Calends of the months? According to the age of the moon of the months (i.e. the length of the lunations). Indeed, examine the above-mentioned epacts concerning the extinguished (moon) (i.e. the final moon of a lunation) and the kindled (moon) (i.e. the first moon of a lunation). The moons are, in fact, implemented in the following order according to the Greeks: *luna* 30 in January; (*luna*) 29 in February; (*luna*) 30 in March and April; (*luna*) 29 in May and June; (*luna*) 30 in July; (*luna*) 29 in August; (*luna*) 30 in September; (*luna*) 29 in October; (*luna*) 30 in November; (*luna*) 29 in December. Therefore, whoever would want to calculate the ages of the moon (i.e. the length of the lunations) by means of these (i.e. the above-mentioned epacts) will clearly discover this rule and should beware of the bissextile day with a watchful sense, so that it does not confuse this observation.

XXVIIII Decimbri. In all other texts that mention the Dionysiac sequence of lunation it is given as alternating between full and hollow lunar months (obviously strating with January); cf. BC 64, DT 5.2–4, DTR 11.24–26, DIAL NEUSTR. 6A, RM 28.31–35, NOTKER LABEO, *De quatuor questionibus compoti* (Piper, *Nachträge*, p. 316); it also appears among the tables of some MSS (e.g. Vatican, Biblioteca Apostolica, Reg. Lat. 1260, fol. 16r).

29–39 Qua…conturbet] Cf. Borst, *Schriften*, p. 388 and p. CLXXXVII–CLXXXVIII, CXCVII of the introduction.

The first year of the Victorian 19-year cycle

Secundum uero Latinos ita aepactae Kalendarum mensium obseruandę: *VIIII* in *Kalendis Ianuarii*; *X* in *Kalendis Februarii*; *VIIII* in *Marti*o; *X* in *Aprili*; *XI* in *Mai*o; *XII* in *Iuni*o; *XIII* in *Iuli*o; *XIIII* in *Augusto*; transilit XV et in *Septimbri XVI et* in *Octimbri*; transilit XVII et erit *XVIII* in *Nouimbri et Decimbri*; transilit XVIIII et erit XX in Ianuario.

Two exceptions:

Per hunc ordinem *XVIIII anno*rum ępactae disputandae sunt, excepto iam bino *anno* iam: saltus et annus *XXX* luna in Kalendis Ianuarii.

---

49 annus] *follows* in *M.*

40–46 Secundum...Ianuario] Since the principal purpose of this chapter in MC as well as in its source, CE, is a comparison of Dionysiac and Victorian lunar data, MC now turns to the Victorian reckoning; the fact that this discussion starts with the first year of the Victorian 19-year cycle reveals that both CE and MC were fully aware that the proposed alternative comparison of the two reckonings, namely aligning the Dionysiac *ogdoas* with the Victorian *hendecas* (cf. c. 49, ll. 16–32; c. 52), was not to be followed. MC's direct source here is again CE's description of this Victorian year: CE p. 110: *Hoc autem modo secundum Romanos diverse conputa: IX luna Kalendis Ianuarii; X luna Kalendis Februarii; IX luna Kalendis Martii; X luna Kalendis Aprilis; XI luna Kalendis Maii; XII luna Kalendis Iunii; XIII luna Kalendis Iulii; XIIII luna Kalendis Augusti; XVI luna Kalendis Septimbris et Octimbris; XVIII luna Kalendis Nouimbris et Decimbris.* **47–49** Per...Ianuarii] MC's principal source, CE p. 110, mentions only the 11th year of the Victorian 19-year cycle as an exception to the outlined rule, not the year of the *saltus*: *Item hoc eodem modo facias per annos XIX praeter annum unum, qui a XXX in XI ętatem egreditur.* Yet, the SIRMOND MS (Oxford, Bodleian Library, Bodley 309, fol. 106v; the relevant passage is transcribed in Ó Cróinín, 'New light on Palladius', p. 29; idem, 'Bede's Irish computus', p. 206), which appears to have originated in the same intellectual milieu as MC, describes the same two Victorian years (the sixth and the 11th of the Victorian 19-year cycle) as problematic, whithout adding any further details: *Notandum est quod in XVIIII ciclo quattuor anni contrariae regulae inueniuntur, quorum ratio diligenter arguteque animaduertenda est. ... Item, secundum rationem Latinorum, IIII lunae annus, et XXX.*

40–46 Secundum...Ianuario] Interestingly enough, besides CE and MC there are hardly any other texts from what may be called the transformation period from the Victorian to the Dionysiac reckoning that contain detailed comparisons of the two reckonings; this is particularly surprising since this process took decades in Francia at a time when the production of computistical literature reached its first peak. A noteworthy exception is the colossal comparison of the Dionysiac with the Victorian Easter table in Leiden, Universiteitsbibliotheek, Scaliger 28, fol. 2v–22r. Therefore, descriptions of this first year of the Victorian 19-year cycle can, besides the two Irish computistical textbooks, only be found in texts dealing exclusively with the Victorian reckoning. One of the earliest and most neglected computistical MS, Bern, Burgerbibliothek, 645, gives a full discussion of all 19 years of the Victorian 19-year cycle, in which the first years reads (fol. 55r): *Kalendis Ianuarii luna VIIII; Kalendis Februarii luna X; Kalendis Marcii luna VIIII; Kalendis Aprilis luna X; Kalendis Maii (MS Md-) luna XI; Kalendis Iunii luna XI (recte XII); Kalendis Iulii luna XIII; Kalendis Agusti luna XIIII; Kalendis Septembris luna XVI; Kalendis Octobris luna XVI; Kalendis Novembris luna XVIII; <Kalendis Decembris luna XVIII;> Kalendis Ianuarii luna XX.* The text in the Bern MS probably was the source for QUAEST. AUSTR. 2.11A: *I. Kalendas Ianuarii luna nona. Kalendas Februarii luna decima. Kalendas Martii luna nona. Kalendas Aprilis luna decima. Kalendas Maii luna undecima. Kalendas Iunii luna duodecima. Kalendas Iulii luna tertia decima. Kalendas Agusti luna quarta decima. Kalendas Septembris luna sexta decima. Hicque luna brevis in mense longo, ideo una aetas transilitur. Kalendas Octobris luna sexta decima. Hic aetas ideo duplica<tur>, eo quod Septembris triginta <dies> et luna eius tricesima simul*

According to the Latins, however, the epacts of the Calends of the months are observed in the following way (in the first year of the Victorian 19-year cycle): (*luna*) 9 on 1 January; (*luna*) 10 on 1 February; (*luna*) 9 in March; (*luna*) 10 in April; (*luna*) 11 in May; (*luna*) 12 in June; (*luna*) 13 in July; (*luna*) 14 in August; it passes over (*luna*) 15 to (*luna*) 16 in September and October; it passes over (*luna*) 17 and it will be (*luna*) 18 in November and December; it passes over (*luna*) 19 and it will be (*luna*) 20 in January.

The epacts (of the Calends) of the 19 years are discussed according to this order, except for two years at one time and another: (the year of the) *saltus* (i.e. the sixth year of the Victorian 19-year cycle) and the year (that starts) with *luna* 30 on 1 January (i.e. the eleventh year of the Victorian 19-year cycle).

---

terminantur. Kalendas Novembris luna octava decima. Kalendas Decembris luna octava decima. For a discussion of the lunar data transmitted in these Victorian texts see now Holford-Strevens, 'Lunar calendars', p. 194–6.

40–46 Secundum...Ianuario] Cf. Borst, *Schriften*, p. 403 (but note that MC refers to the Victorian reckoning here, while DIAL. NEUSTR. lists the regulars for the Dionysiac reckoning). **47–49** Per...Ianuarii] Cf. p. XXII of the introduction.

The sixth year of the Victorian 19-year cycle

Nota saltum Latinorum apparere in Kalendis Ianuarii IIII XVI, | qui in XV Kalendas Decimbris inponitur. In quo fol. 28r anno Nouimber and December disiunguuntur: XIII luna in Kalendis Nouimbris et in XV Kalendas Decimbris saltus est et XIIII luna in Kalendis Decimbris, quae accenditur in XIIII Kalendas Decimbris.

The 11th year of the Victorian 19-year cycle

XXX in *Kalendis Ianuarii*; *II* in *Februari*o; *XXX* in *Marti*o; *I* in *Aprili*; *II* in *Mai*o; *III* in *Iuni*o; *IIII* in *Iuli*o; *V* in *August*o; transilit VI et erit *VII* in *Kalendis Septimbris et Octimbris*; transilit VIII et erit *VIIII* in *Kalendis Nouimbris et Decimbris*; transilit X et *XI* erit in *Kalendis Ianuarii*.

---

50 Ianuarii] de *add. Borst.* **51** IIII] in *add. Borst.* | XVI] XV *M (corr. according to* MC*; also Schwartz, Borst; the number may have been corrected by the copyist in accordance with the Dionysiac reckoning)*. **54** et] *follows* in *M.*

50–55 Nota…Decimbris] Since CE does not record the year of the Victorian *saltus* as an exception to the general rule for calculating the lunar age of the Calends of each month, MC had no apparent source for this description of that year; from this perspective, MC's addition here should be regarded as a refinement of CE's account. This sixth year of the Victorian 19-year cycle certainly constituted an exception to the general rule, since the *saltus* in November led to an increase of the lunar age of 1 December by one compared to that of November, whereas normally these two Julian calendar dates would have had the same lunar age. For the Victorian *saltus* see c. 62, ll. 87–95. **56–60** XXX…Ianuarii] The 11th year of the Victorian 19-year cycle is an exception to the general rule for calculating the lunar age of the Calends of each month since the difference between the lunar age of 1 January (and 1 March) and that of 1 February is 2 rather than 1, as it would usually have been; the reason for this is the placement of the embolism on 31 January to 1 March or 2–31 March rather than earlier in the year; if it is placed earlier, however, problems are only shifted, not solved (cf. *app. comp.*). MC, here, is in total agreement with CE p. 110: *XXX luna Kalendis Ianuarii; II luna Kalendis Februarii; XXX luna Kalendis Martii; I luna Kalendis Aprilis; II luna Kalendis Maii; III luna Kalendis Iunii; IIII luna Kalendis Iulii; V luna Kalendis A\u/gusti; VII luna Kalendis Septimbris et Octimbris; IX luna Kalendis Nouimbris et Decimbris; XI luna Kalendis Ianuarii.*

50–55 Nota…Decimbris] Besides MC, this Victorian year of the *saltus* is only discussed in the two texts just mentioned that outline the lunar ages on the Calends of every single year of the Victorian 19-year cycle: Bern, Burgerbibliothek, 645, fol. 55r: *Kalendis Ianuarii luna IIII; Kalendis Februarii (MS Frb-) luna V; Kalendis Marcii luna IIII; Kalendis Aprilis luna V; Kalendis Maii (MS Mad-) luna VI; Kalendis Iunii luna VII; Kalendis Iulii luna VIII; Kalendis Agusti luna VIIII; Kalendis Septembris luna XI; Kalendis Octobris luna XI; Kalendis Novembris luna XIII; Kalendis Decembris luna XIIII; Kalendis Ianuarii luna XVI*. QUAEST. AUSTR. 1.11C, 2.11A: *Sciendum quoque: Cur anno, quo quarta lunae aetas in Kalendis Ianuarii fuerit supputata, eodem anno Kalendae Novembris et Decembris, quae semper eadem aetate conveniunt, hac tantum vice inter se differunt, ut tertia decima luna in Kalendis Novembris et quarta decima <in> Kalendis Decembris supputetur? ... VI. Kalendas Ianuarii luna quarta. Kalendas Februarii luna quinta. Kalendas Martii luna quarta. Kalendas Aprilis luna quinta. Kalendas Maii luna sexta. Kalendas Iunii luna septima. Kalendas Iulii luna octava. Kalendas Agusti luna nona. Kalendas Septembris luna undecima. Kalendas Octobris luna undecima. Kalendas Novembris luna tertia decima. Kalendas Decembris luna quarta decima. Hac vice tanum aetates discordant causa augmenti lunaris, quod in Novembri interponitur. Nam luna, quae tertia decima Kalendas Novembris, vicesima octava est XVI Kalendas Decembris. Deinde aetas transilitur. XV Kalendas <Decembris> tricesima esse definitur. Qui legit, intellegat, utrum peracto an incipienti saltus adiungatur anno. Ex <qua> causa non in quinta decima luna, sed in sexta decima luna Kalendas Ianuarii.* **56–60** XXX…Ianuarii] Compared to CE and MC, the embolism is placed differently in

Note that the *saltus* of the Latins, which is placed on 17 November, appears on 1 January (from *luna*) 4 (to *luna*) 16. In this year November and December differ: *luna* 13 (occurs) on 1 November, the *saltus* is on 17 November, and *luna* 14 occurs on 1 December, which is kindled (i.e. the first moon of that lunation occurs) on 18 November.

(In the eleventh year of the Victorian 19-year cycle *luna*) 30 (occurs) on 1 January; (*luna*) 2 in February; (*luna*) 30 in March; (*luna*) 1 in April; (*luna*) 2 in May; (*luna*) 3 in June; (*luna*) 4 in July; (*luna*) 5 in August; it passes over (*luna*) 6 and it will be (*luna*) 7 on 1 September and 1 October; it passes over (*luna*) 8 and it will be (*luna*) 9 on 1 November and 1 December; it passes over (*luna*) 10 and it will be (*luna*) 11 on 1 January.

Bern, Burgerbibliothek, 645, fol. 55r, namely on 3 December–1 January or 2–31 January, leading to 1 January and 1 March not having the same lunar age as they normally would: *Kalendis Ianuarii luna XXX; Kalendis Februarii (MS Frb-) luna I; Kalendis Marcii luna XXVIIII; Kalendis Aprilis luna I; Kalendis Maii (MS Mad-) luna II; Kalendis Iunii luna III; Kalendis Iulii luna IIII; Kalendis Agusti luna V; Kalendis Septembris luna VII; Kalendis Octobris luna VII; Kalendis Novembris luna VIIII; Kalendis Decembris luna VIIII; Kalendis Ianuarii luna XI.* Interestingly enough, QUAEST. AUSTR. 2.11B notes two possibilities for this 11th year of the Victorian lunar cycle, depending on the placement of the embolism, namely 1) the embolism ending on 1 January as in the Bern MS, or 2) the embolism ending on 1 March, as in MC and CE, so that it appears that the author of this text worked from the Bern text and either of the Irish textbooks: *XI. Kalendas Ianuarii luna tricesima. Haec luna embolismi est. Kalendas Februarii luna prima. Hac vice una luna continet tres menses. Accenditur pridie Kalendas Februarii et totum inlustrat Februarium et deficit in Kalendas Martii. Si byssextus in hoc anno succurrerit, luna Februaria in Kalendas Martii tricesima conputabitur, et quattuor lunae tricesimae in ordine sunt. Embolismus Kalendas Martii. Kalendas Martii luna vicesima nona. Kalendas Aprilis luna prima. Kalendas Maii luna secunda. Kalendas Iunii luna tertia. Kalendas Iulii luna quarta. Kalendas Agusti luna quinta. Kalendas Septembris luna septima. Kalendas Octobris luna septima. Kalendas Novembris luna nona. Kalendas Decembris luna nona.* This year is also referred to in QUAEST. AUSTR. 2.4A (cited in COMP. COL. 6.4) and further below in this chapter (ll. 78–81).

50–51 Nota…inponitur] Transcribed in Schwartz, 'Ostertafel', p. 95; Borst, *Schriften*, p. 367.

The Victorian sequence of lunations

Qua ratione Latinorum ępactas inuestigant? Hoc est e lunarum mensium aetate, quae hoc ordine feruntur: *XXX luna*m *Ianuari*us; *XXVIIII Februari*us; *XXX Marti*us; *XXVIIII Aprili*s; *XXX Mai*us; *XXVIIII Iuni*us; *XXX Iuli*us; *XXVIIII Augustu*s; *XXX Septimbe*r; *XXVIIII Octimbe*r; *XXX Nouimbe*r; *XXVIIII Decimbe*r; XXX Ianuarius.

A year of the *latercus* with epact 8

Apud latercum ępactae in Kalendis mensium: VIII in Kalendis Ianuarii; VIIII in Kalendis Februarii; VIII in Kalendis Martii; X Aprili; XI Maio; XII Iunio; XIII Iulio; XIIII Augusto; XV Septimbri; XVI Octimbri; XVII Nouimbri; XVIII Decimbri; XVIIII Ianuario.

---

61 e] et *M (corr. according to* MC*; also Schwartz).* **67** mensium] ian- *M, Schwartz (corr. according to* MC*; also Warntjes; but note that the* HARVARD FRAGMENT *has the same reading as M).* **67–68** in²...Ianuarii] *om. Schwartz.* **70** XVII] XVIII *M (corr. from* HARVARD FRAGMENT *and according to* MC*; also Schwartz; it seems that the copyist corr. this number in accordance with the Victorian and Dionysiac reckonings, but continued the linear sequence that is the characteristic of the* latercus *afterwards; cf. the detailed discussion of this problem in Warntjes, '84 (14)-year Easter reckoning', p. 46–51).* **71** XVIII] XVIIII *M (corr. from* HARVARD FRAGMENT *and according to* MC*; also Schwartz).* | XVIIII] XX *M (corr. from* HARVARD FRAGMENT *and according to* MC*; also Schwartz).*

61–66 Qua...Ianuarius] As in the case of the Dionysiac sequence of lunations (c. 50, ll. 29–39), seventh-century Irish computist had to reconstruct the Victorian sequence of lunations from the meagre data available in Victorius' Easter table, since Victorius himself did not record this important technical detail in his writings. The first text to outline the Victorian sequence of lunations is MC's source, CE p. 111: *De aetate XII lunarum mensium anni: ... Secundum Romanos ita est: XXX luna Ianuario; XXIX luna Februario; XXX luna Martio; XXIX luna Aprili; XXX luna Maio; XXIX luna Iunio; XXX luna Iulio; XXIX luna Augusto; XXX luna Septimbri; XXIX luna Octimbri; XXX luna Nouimbri; XXIX luna Decimbri.* **67–71** Apud...Ianuario] MC had copied and slightly refined CE's comparative discussion of the Dionysiac and the Victorian 19-year cycles based on the lunar ages of the Calends of each months. In this section here, MC introduces the third Easter reckoning used in Ireland into the discussion, the *latercus*. It seems that he knew the sequence of lunations (cf. the following note) from an oral source, and on that basis reconstructed the lunar data for a year of the *latercus* with epact 8 on 1 January, a *latercus*' year which MC obviously regarded as comparable to the first year of the Dionysiac 19-year cycle, which had the same epact; note, however, that the lunar cycles of these two systems could not sensibly be compared, since the *latercus* was based on an 84-year, not a 19-year lunar cycle.

61–66 Qua...Ianuarius] When listing the Dionysiac sequence of lunations, DRC 72 mentions in passing that this alternating sequence was also applied by the 'Latins', i.e. the followers of the Victorian reckoning: *Sciendum nobis quomodo menses lunares mensibus solis apud Grecos et Latinos secundum hanc regulam iuguntur. Ita etiam, id est XXX luna Ianuario, XXVIIII Februario, XXX Martio, XXVIIII Apreli ... XXX Maio, XXVIIII Iunio, XXX Iulio, XXVIIII Agusto, XXX Septimbri, XXVIIII Octimbri, XXX Nouimbri, XXVIIII Decimbri.* The Victorian QUAEST. AUSTR. 2.4A,B (cited in COMP. COL. 6.4) describes this sequence of lunations in detail: *Menses quoque lunares secundum Latinos et ipsi longitudine dispares sunt, sex longi et sex breves, altrinsecus per omnes annos communes positi. ... In distributione autem lunarium mensium in solaribus mensibus hic ordo servandus est. Verbi gratia, in Ianuario tricesima, in Februario vicesima nona, in Martio tricesima, in Aprile vicesima nona, in Maio tricesima, in Iunio vicesima nona, in Iulio tricesima, in Agusto vicesima nona, in Septembre tricesima, in Octubre vicesima nona, in Novembre tricesima, in Decembre vicesima nona.* while BC 61B1–2 only refers to its alternating character: *apud Romanos quibus uideatur altera luna XXX, altera XXVIIII.* In Bern, Burgerbibliothek, 645, fol. 54v, this sequence of lunations is embedded in a table that records the number of solar days, nones, lunar days, and calends of each month. Since this MS deals exclusively

According to which reckoning do they investigate the epacts of the Latins? According to the age of the moons of the months (i.e. the length of the lunations), which are implemented in the following order: January (has) *luna* 30; February (*luna*) 29; March (*luna*) 30; April (*luna*) 29; May (*luna*) 30; June (*luna*) 29; July (*luna*) 30; August (*luna*) 29; September (*luna*) 30; October (*luna*) 29; November (*luna*) 30; December (*luna*) 29; January (*luna*) 30.

According to the *latercus*, the epacts on the Calends of the months (are): (*luna*) 8 on 1 January; (*luna*) 9 on 1 February; (*luna*) 8 on 1 March; (*luna*) 10 in April; (*luna*) 11 in May; (*luna*) 12 in June; (*luna*) 13 in July; (*luna*) 14 in August; (*luna*) 15 in September; (*luna*) 16 in October; (*luna*) 17 in November; (*luna*) 18 in December; (*luna*) 19 in January.

---

with the Victorian reckoning, there cannot be any doubt that the sequence of lunations included in the mentioned table refers to that reckoning. The COMPUTUS COTTONIANUS, fol. 78v (cited in BC 15B6, LIB. ANN. 58), states – quite mistakenly – that in the Victorian reckoning every lunation consists of 30 days. **67–71** Apud…Ianuario] MC's discussion of this *latercus* year is directly copied in the HARVARD FRAGMENT (Harvard, Houghton Library, MS Type 613, fol. 7r; note the same mistake as in *M* at the beginning, *ian-* instead of the correct *mensium*): *Aput latercum aepact<ae in Kalendis> Ianuarii* (*recte mensium*)*: VIIII* (*recte VIII*) *luna Ianuario, VIII* (*recte VIIII*) *luna in Kalendis Februarii, VIIII* (*recte VIII*) *<luna in Kalendis Martii>, X luna in Kalendis Aprilis, XI luna in Kalendis Maii, XII luna in Kalendis <Iunii>, XIII luna in Kalendis Iulii, XIIII in Kalendis Augusti, XV luna in K<alendis Septembris>, XVI in Kalendis Octembris, XVII luna in Kalendis Novembris, XVIII l<una in Kalendis> Decembris, XVIIII luna in Kalendis Ianuarii.*

61–66 Qua…Ianuarius] Transcribed in Schwartz, 'Ostertafeln', p. 97; cf. Borst, *Schriften*, p. 483. **67–76** Apud…luna[3]] Transcribed in Schwartz, 'Ostertafeln', p. 100; transcribed and translated in Warntjes, '84 (14)-year Easter reckoning', p. 41, 48 (facsimile p. 83); only the sequence of lunations of the *latercus* transcribed in Mc Carthy, 'Easter principles', p. 211. Cf. also Mc Carthy, 'Easter principles', p. 211–3, 220–1, 223; Strobel, *Texte*, p. 17–8; Schäferdiek, 'Osterzyklus', p. 361–2 and p. LXXII, XCIV, XCVII–XCVIII, CXXII–CXXIII, CLV, CLXXXIX of the introduction.

The sequence of lunations of the *latercus*

Apud latercum lunae: Ianuario XXX luna; Februario XXVIIII luna; Martio XXVIIII luna; Aprili XXVIIII luna; Maio XXX luna; Iunio XXVIIII luna; Iulio XXX luna; Augusto XXX luna; Septimbri XXVIIII luna; Octimbri XXX luna; Nouimbri XXVIIII luna; Decimbri XXX luna.

Addition to the 11th year of Vict.

Estque quidam annus, quo unus mensis tres continet lunas, hoc est Ianuarius. Et luna sequens, quae in II Kalendas Februarii accenditur et in Kalendis Martii XXX extinguitur, III menses continet.

---

72 lunae] lunas *M, Schwartz, McCarthy, Warntjes.*

72–76 Apud…luna[3]] As already argued in the previous note, MC probably obtained accurate knowledge of the *latercus*' sequence of lunations from an oral source; at the time of composition of MC, AD 719, the *latercus* had just been abandoned in Iona, while it had presumably been replaced by the Victorian reckoning in the monastery that saw the production of MC ca. 85 years earlier. Note that with the newly discovered HARVARD FRAGMENT (cf. *app. comp.*), MC is the only existing text transmitting information about this crucial technical detail of this Easter reckoning. **77–80** Estque…continet] This passage is an addition to the discussion of the 11th year of the Victorian 19-year cycle (c. 50, ll. 56–60): In that year, according to MC and CE, the January lunation ended on 1 January, the February lunation on 30 January, while the embolism or March lunation stretched from 31 January to 1 March. The reasons for the discussion of the *latercus* separating the two passages dealing with the 11th year of the Victorian 19-year cycle must remain speculative, but it may well be that the *latercus* information had originally been composed on a separate leaf and was then later included slightly too early in the main body of the text.

72–76 Apud…luna[3]] The (heavily corrupted) HARVARD FRAGMENT (Harvard, Houghton Library, MS Type 613, fol. 7r) is more detailed in this passage, since it additionally lists the number of days of the solar months and with this nicely illustrates the correspondence of long solar with long lunar months (as well as short solar with short lunar months) in this sequence of lunations (with the exception of March): *Quia aput latercum <Ianuarius XXXI> diem* (*recte dies*) *habet, lunas XXX; Februarius XXVIII dies habet, <lunas XXVIIII>; Martius XXXI diem* (*recte dies*) *habet, lunas XXVIIII; Aprilis XXX die<s habet, lunas> XXVIIII; Maius XXXI diem* (*recte dies*) *habet, luna eius* (*recte lunas; s* appears to have been misread as the Irish abbr. for *eius*) *XXVIIII* (*recte XXX*)*; Iunius <XXX dies habet, lunas> XXVIIII; Iulius XXXI habet, lunas XXX; Agustus XXXI di<es habet>, luna eius* (*recte lunas*) *XXVIIII* (*recte XXX*)*; September XXX dies habet, luna eius* (*recte lunas*) *XX<VIIII; Octember> XXXI diem* (*recte dies*) *habet, luna eius* (*recte lunas*) *XXX; November XXX dies h<abet, lunas> XXVIIII; December XXXI diem* (*recte dies*) *habet, luna \eius/* (*recte lunas*) *XXX*. This correspondence is explicitly expressed in BC 61B3–5, where *Aquillonares* undoubtedly refers to the northern Irish: *Ideo Aquillonares hanc regulam non observant, quia consequens est apud eos, ut menses XXXI die, luna XXX, mensis XXX die, luna XXVIIII habet*. Cf. also BC 62A8–11: *Hoc quoque non est admittendum, quod in illo solare, in quo luna incipit accendere, si habet dies XXXI, tunc dicitur luna XXXa. In illo uero, qui habet minus quam XXXI, tunc dicitur XXVIIII*. Interestingly enough, this sequence of lunations was ascribed to the Hebrews in the twelfth century; cf. HEIMO OF BAMBERG, *De decursu temporum* 1.2 (Weikmann, *Heimo von Bamberg*, p. 135): *Hebrei quippe seriem temporum per lunares annos estimabant, sed Greci et Latini per solares, et lunationes Aprilis, Iunii, Augusti et cęterorum mensium, quas nos numeramus XXIX dierum, illi numerabant XXX, quas nos XXX, illi XXIX*. The correspondence between the sequence of lunations of the *latercus* and that of the Easter reckoning of Anatolius (?) is neatly expressed in the COMPUTUS COTTONIANUS, fol. 78v–79r (cited in BC 15B6–7, LIB. ANN. 58): *IIII differentiae lunae ... Anatolius et laterculus alius XXX, alius XXVIIII*. **78–80** Et…continet] Of the texts dealing with the Victorian reckoning only QUAEST. AUSTR. 2.11B mentions this phenomenon in its discussion of the 11th year of the Victorian 19-year cycle; the phrasing suggests that this Frankish text was directly influenced by MC:

According to the *latercus*, the moons (i.e. the lengths of the lunations) (are): *luna* 30 in January; *luna* 29 in February; *luna* 29 in March; *luna* 29 in April; *luna* 30 in May; *luna* 29 in June; *luna* 30 in July; *luna* 30 in August; *luna* 29 in September; *luna* 30 in October; *luna* 29 in November; *luna* 30 in December.

And there is a certain year (namely the eleventh year of the Victorian 19-year cycle), in which one month contains three lunations, (and) that is January. And the following lunation, which is kindled on 31 January and is extinguished on 1 March with (*luna*) 30, contains three months.

Hac vice una luna continet tres menses. Accenditur pridie Kalendas Februarii et totum inlustrat Februarium et deficit in Kalendas Martii.

77–80 Estque…continet] Transcribed and translated in Mc Carthy, 'Easter principles', p. 214 (where it is mistakenly connected to the *latercus*); Warntjes, '84 (14)-year Easter reckoning', p. 44 (where it is correctly connected to the Victorian reckoning; facsimile p. 83); cf. p. CLXIX of the introduction.

<LI. DE LOCO ET CALCULATIONE AEPACTARUM>

Comp. of the lunar with the solar year; the epactal period of 11 days

Et illo taliter ępactae decurrunt | in XI Kalendas Aprilis: aepactas, id est adiectiones lunares, omnibus diebus mensis adsistimus, quae a IIII Idus Martii incipiunt et in XI Kalendas Aprilis finiri adsuescunt. Itaque uterque annus lunaris scilicet et solaris ab XI Kalendas Aprilis cursum incipit et in V Idus Martii parem gerīm extendit. In quo finito lunari anno XI dies de solari a IIII Idus Martii in XI Kalendas Aprilis supersunt. Quibus lunares totidem adieciuntur aetates, quae embolismum tertio anno prestant esse. fol. 28v

---

LI,5 mensis] mensium *M.* **11** adieciuntur] adieciunt *M.*

LI,3–12 Et…esse] The idea of trying to physically locate an epactal period of eleven days within the Julian calendar year, as well as taking 22 March as the starting point of this investigation, presumably derived from EPISTOLA CYRILLI 4 (Krusch, *Studien* I, p. 346; partially cited in EW 1.18–20): *Debemus enim investigare epactas lunares in mensibus totius anni, ut caelebremus pascha in luna primi mensis, in principio anni post veris exordium. Sol enim ipse cotidiae terrae marique pariter et cluditur in fine diei et in diei initio apertitur. Et finit sol cursum totius anni in XII kl aprl, et lunaris globi plenitudo vel diminutio digitorum motu, rationis magisterio et calculi supputatione aliqua praevidetur, ut numeremus XII tantum lunas iuxta supputationem legalem Hebreorum in anno communi et XIII in embolismo secundum supputationem dierum.* Taking 22 March as the starting point and knowing that a lunar year is shorter than the solar year by a few days, MC located 11 March of the following Julian calendar year as the date having the same lunar age as 22 March of the previous year, and therefore declared this as the end of the lunar year; consequently, MC argued that the following 11 days, from 12 to 22 March inclusively, constitute the epactal period, the difference between the solar and the lunar year. Note the use of the Old English term *gerīm* in this passage. For the definition of the epacts cf. c. 49, l. 2; for the eleven-day difference between a solar and a common lunar year see also c. 49, ll. 2–6; c. 60, ll. 4–5.

LI,3–12 Et…esse] This theory of MC (but, interestingly enough, not its correction; cf. c. 51, ll. 35–44) found its way into the Angers glosses: ANGERS 477, fol. 76v (gloss to DTR): *Aepactę Grece adiectiones lunares, et plures in fine anni semper fiunt secundum solem, et incipiunt a IIII Idus Martii, licet naturaliter a II* (*recte V*) *Idus Martii, et finiunt in XI Kalendas Aprilis. Sciendum unde factę sunt ab initio mundi, quia facta est mundus in XI Kalendas Aprilis, et exinde incipit annus solis et lunae ab XI Kalendas Aprilis, sed in tribus primis diebus non luxit sol et luna. Ita et annus lunae V* (?) *Idus Martii finitur et restant XI dies de anno solaris a IIII Idus Martii usque in XI Kalendas Aprilis. Deinde adiciuntur* (?) *XI dies de anno lunaris ad suplementum incipienti harum XI dierum. Inde adiectiones lunares uero* (?) *necessitatę* (?) *sunt. Inde semper epactae fiunt in XI Kalendas Aprilis et hę* (?) *dies XI tertio anno praestant embolismum.*

LI,3–44 Et…reliqua] Cf. Borst, *Schriften*, p. 911 and p. CLXIII, CLXXXI of the introduction. **4–12** aepactas…esse] Facsimile, transcription and translation in Warntjes, 'Earliest occurrence', p. 98–9; for the use of Old English *gerīm* in this passage see, besides the article just cited, also p. LXXVI of the introduction; for this passage in general see also p. CLXXXVII of the introduction. **6–8** Itaque…incipit] Transcribed in Borst, *Schriften*, p. 362. **6–12** Itaque…esse] Cf. Borst, *Schriften*, p. 365, 455.

51. ABOUT THE EPACTAL PERIOD AND THE CALCULATION OF EPACTS

And in this (i.e. the month of March) the epacts run down to 22 March as follows: We place the epacts, i.e. the (eleven) lunar additions, on all those days of this month which start on 12 March and usually end on 22 March. Accordingly, both the lunar and the solar year naturally start their course on 22 March and extend over an equal number of days (*gerīm*) up to 11 March. While the lunar year ends on this (day), eleven days of the solar (year) remain from 12 March to 22 March. To these just as many lunar ages are added, which ensure that an embolism occurs in the third year.

The epacts of 22 March, 1 January, 11 March

Quod *ita probauimus* esse numeratum aepactarum *in XI Kalendas Aprilis* et in Kalendis Ianuarii et *V Idus Martii*. *Aetas*, *qua* utriusque *anni* ab *XI Kalendas Aprilis* cursus incipit, aetatibus *in Kalendis Ianuarii* iuniori fieri apparebat. Et una eademque aetate lunaris annus ab XI Kalendas Aprilis egredi innotescit et illa, quae in <V> Idus Martii gressum finit. De X quae aetatibus V Idus Martii senior luna uiget quam ea, quae in Kalendis eius fuisse claret, quod mox in ępactis fieri florebit.

Examples: The final year of the Dionysiac 19-year cycle

Non aliter hinc anno nouissimo epactas saltus uice numeremus. In quo XVIII luna in XI Kalendas Aprilis, VIII luna in Kalendis Ianuarii et Martii, XVIII in V Idus Martii apparet esse. Cum XI aetates additis a IIII Idus Martii, in XI Kalendas Aprilis XXVIIII, in XI quam XXX eueniens. Hoc anno occultus saltus rationem obumbrat. Hinc ab XVIII in XXX | in XI Kalendas Aprilis. fol. 29r

The first year

XXX in XI Kalendas Aprilis, *XX in Kalendis Ianuarii* et *Martii*, *XXX in V Idus Martii*, *XI in XI Kalendas Aprilis*. Hinc a XXX in XI in XI Kalendas Aprilis.

The second year

XI in XI Kalendas Aprilis, *I luna in Kalendis Ianuarii* et *Martii*, *XI luna in V Idus Martii*, *XXII in XI Kalendas Aprilis*. Inde ab XI in XXII in XI Kalendas Aprilis.

---

18 V] *om. M (add. according to* MC*)*. **23** luna] lunas *M (corr. according to* MC*)*. **27** occultus] occulta *M.* | rationem] sartione *M.* **32** I] id est *M (I suspect that the numeral* I *was mistaken for the Irish abbr. for* id est *by the continental copyist)*.

13–34 Quod…Aprilis[2]] MC's theory of the 11-day epactal period outlined in the previous passage was based on certain basic lunar relations within the Julian calendar: The lunar age of 22 March is the same as the lunar age of 11 March of the following Julian calendar year; moreover, the lunar age of 1 January and 1 March is 10 less than the lunar age of 11 March of the same and of 22 March of the previous Julian calendar year. This theory MC demonstrates by three examples, subtracting 10 from the lunar age of 22 March to get the lunar age of 1 January and 1 March of the following year, adding 10 again to get the lunar age of 11 March, which is the same as the lunar age of 22 March of the previous year; if the epactal period of 11 days is then added to the lunar age of 11 March, the result is the epact of that year, the lunar age of 22 March. This theory, as well as the following examples, are described more precisely in MC's source for this passage, CE p. 110–111 (note that MC omits the final two examples of CE, but adds the last year of the Dionysiac 19-year cycle in order to demonstrate the impact of the *saltus* on this theory): *Et ita hoc certum esse probamus: Lunaris enim ętas, quę est in XI Kalendas Aprilis, hęc erit in V Idus Martii finito anno. Cuius probatio hęc est: Etsi non probemus per Kalendas mensium, dum longum est probare sic: De aetate lunae, quę fit in eo anno in XI Kalendas Aprilis, X subtrae and sic de reliquo repperies in Kalendis Ianuarii. Hoc modo probandum est: XXX luna in XI Kalendas Aprilis; X subtrahe, XX luna in Kalendis Ianuarii; XXI luna in Kalendis Februarii; XX luna in Kalendis Martii; X adde, XXX luna in V Idus Martii est; XI luna in XI Kalendas Aprilis; X subtrahe, I luna in Kalendis Ianuarii; II luna in Kalendis Februarii; I luna in Kalendis Martii; X adde, XI luna in V Idus Martii; XXII luna in XI Kalendas Aprilis; X subtrahe, XII in Kalendis Ianuarii; XIII luna in Kalendis Februarii; XII \luna/ in Kalendis Martii; X adde, XXII luna in V Idus Martii; III luna quasi XXXIII; hoc numero subtrahe X, XXIII luna in Kalendis Ianuarii; XXIIII in Kalendis Februarii; XXIII luna in Kalendis Martii; in VIII <Idus> Martii XXX luna est; in V Idus autem Martii III luna est. Sic itidem per XIX annorum ciclum de residuo probabis.*

We have thus proven that the epacts are to be reckoned on 22 March, on 1 January, and on 11 March. The (lunar) age on 22 March, on which (date) the course of both years begin, appeared to be of a younger age on (the following) 1 January. And it is known that a lunar year sets off on 22 March with one and the same lunar age (as the one) which ends the course (of that lunar year) on 11 March (of the following Julian calendar year). This moon on 11 March grows older by ten (lunar) ages compared to the one which shone on the Calends of the same month (i.e. 1 March), since it will soon enter the epacts (i.e. the period from 12 to 22 March).

Hence, we should reckon the epacts in the same manner in the final year (of the Dionysiac 19-year cycle) at the turn of the *saltus*. In this (year) *luna* 18 appears on 22 March, *luna* 8 on 1 January and 1 March, (*luna*) 18 on 11 March. When you add the eleven lunar ages (starting) from 12 March, (*luna*) 29 (occurs) on 22 March, which appears as (*luna*) 30 on 22 (March). In this year the hidden *saltus* obscures the rule. Hence, (the epacts change) from (*luna*) 18 to (*luna*) 30 on 22 March.

(In the first year of the Dionysiac 19-year cycle *luna*) 30 (occurs) on 22 March, (*luna*) 20 on 1 January and 1 March, (*luna*) 30 on 11 March, (*luna*) 11 on 22 March. Hence, (the epacts change) from (*luna*) 30 to (*luna*) 11 on 22 March.

(In the second year of the Dionysiac 19-year cycle *luna*) 11 (occurs) on 22 March, *luna* 1 on 1 January and 1 March, *luna* 11 on 11 March, (*luna*) 22 on 22 March. Accordingly, (the epacts change) from (*luna*) 11 to (*luna*) 22 on 22 March.

13–34 Quod…Aprilis[2]] The relation outlined here between the lunar ages of 22 March, 1 January, 1 March, and 11 March did not find any imitators, for two principal reasons: a) the main objective of this chapter, the determination of an epactal period of eleven days within the Julian calendar, was hardly discussed outside of seventh- and early eighth-century Irish computical circles (cf. the previous and the following note); b) algorithms were developed to easily calculate the lunar age of any given Julian calendar date, so that these basic rules became obsolete.

22–34 Non…Aprilis[2]] Cf. p. XCIII of the introduction.

A different opinion on the epactal period of 11 days

Ita per XVIIII annorum ciclum adsisti: *A V Idus Martii in XI Kalendas Aprilis* deuenit. Sed artificiosae hae epactae subputationis arguti fuerunt, quae in superficie neminem ledunt. Sed inspecto interius sensu minus placent, qui melius disputantes. *Ab XI Kalendas Aprilis* incipiunt et *lunarem ann*um *in VI Idus Martii* finiunt et *a V Idus Martii in XII Kalendas Aprilis epacta*s numerari precipiunt. Ostensa igitur quadam obiectione quam scientiam quaerunt. Epactas *natural*ius ita disputant: a XXX X, ab XI in XXI, a XXII II, a III in XIII, a XIIII in XXIIII, sic et reliqua.

---

35 ciclum] ciclus *M.* | V] VI *M (corr. from* CE*).* **36** artificiosae…epactae] artificiosas has epactas *M.* **40** V] VI *M (corr. according to* MC, CE, DRC, COMP. COL.*).* **42** scientiam] scientia *M.* **44** III] VI *M (corr. according to* MC*).* | XIII] XII *M (corr. according to* MC*).*

35–44 Ita…reliqua] In this final passage of this chapter, MC corrects the theory outlined at the beginning (c. 51, ll. 3–12), where it was argued that the epactal period of 11 days is to be located between 12 and 22 March inclusively; that theory had been based on the fact that 22 March, considered as the beginning of a lunar year, has the same lunar age as 11 March of the following Julian calendar year, which was, for that reason, regarded as the end of that lunar year. Yet, the conclusion drawn was obviously wrong: the Julian calendar date with the same lunar age as the first day of a lunar year was to be taken as the first day of the following lunar year rather than the last day of this one. Hence, if a lunar year started on 22 March, it ended on 10 March of the following Julian calendar year, and the epactal period of 11 days, i.e. the difference between the solar and the lunar year, was to be located between 11 and 21 March inclusively. This is the theory taught by MC's source, CE p. 110: *Item interrogandum est: In quo loco ępactę in anno naturaliter fiunt? Hoc ita soluendum est: A quinto Idus Martii usque in XI Kalendas Aprilis. Egressus enim lunarum XII communis anno ab XI Kalendas Aprilis usque in VI Idus Martii plene perficitur, hoc est diebus CCCLIIII. A V Idus vero Martii usque in XI Kalendas Aprilis ępactę naturales sunt. Sed quare utrum cum V Idus Martii et cum XI Kalendas Aprilis hic ępactarum numerus secundum naturam est? Hoc ita soluendum est: Cum V Idus autem et non cum XI. Licet in XII Kalendas Aprilis finitur.*

35–44 A…reliqua] DRC 68.24–29 only relates the results of the discussions of the epactal period in CE and MC: *Et inciperunt ab inicio mundi, quia, cum factus esset mundus in XII Kl Aprelis, quamuis in tribus primis diebus prima uice non luxit sol et luna, et ita annus lunaris in VI Idus Marcii finitur, et restabant undecim dies de anno solaris á V Idus Marcii usque in XII Kl Aprelis. Et idcirco epactae adiectiones lunares nominatae sunt, eo quo de anno lunari incipienti aetates ad hos solares undecim dies adduntur*. The only non-Irish text that includes a discussion of the physical placement within the Julian calendar of an epactal period of eleven days is COMP. COL. 4.2B–C, a text generally heavily influenced by Irish computistics, as well as interested in theories outdated at its time of compilation (presumably for antiquarian reasons): *Idcirco ostendamus exemplum de eis in uno loco, etiam si exparsim sit per omnes menses. Accensio <lune> primi mensis <est> XII (recte XI? MS X) Kalendas Aprilis. <Inde aetates lunae in Kalendas primi anni sunt Ianuario> octo vel novem, novem Februario vel decem, novem vel decem (recte octo vel novem) Martio, decem Aprili, undecim Maio, duodecim Iunio, tredecim Iulio, quattuordecim Agusto, sedecim Septembri, sedecim Octubri, septendecim (recte MS octodecim) Novembri, octodecim Decembri, viginti Ianuario <secundi anni>, viginti unus in Februario, viginti in Martio. Luna Martii extinguitur in VI Idus, et undecim sunt epactae. Id <est:> Assiti sunt iste semper super V Idus. Deinde enumera dies usque V Idus: Undecim dies sunt, et undecim epactae sunt, quia non addit luna Martii nisi in V Idus Martii. Ubi est adiectio? Id: Luna, que accenditur in V Idus Martii, epacta est et finitur in XII Kalendas Aprilis, etiam si duae lunae vel una luna. Quicquid est in V Idus, et undecim epacta est, quia adiectio lunae <addit> super duodecim dies solis. Sic facias in cyclo novemdecim annorum semper. Alii dicunt esse tribus modis epactas. Id est:*

Therefore, the (epacts) are placed (as follows) throughout the course of 19 years (i.e. throughout the 19-year cycle): It (i.e. the epactal period of eleven days between the end of the lunar and that of the solar year) runs down from 11 March to 22 March. But these artificial epacts, which trouble nobody on the surface, only belonged to this subtle calculation (i.e. the one outlined above). However, after the sense had been examined more closely, they (i.e. these epacts) pleased (those) less, who calculated better: They start on 22 March, they end the lunar year on 10 March, and they teach that the epacts are counted from 11 March to 21 March. After this particular objection had been made, they searched for some theory. They reckon the epacts more naturally like this: From (*luna*) 30 (on 22 March to *luna*) 10 (on 21 March of the following year), from (*luna*) 11 to (*luna*) 21, from (*luna*) 22 (to *luna*) 2, from (*luna*) 3 to (*luna*) 13, from (*luna*) 14 to (*luna*) 24, and so on.

Naturaliter, usurpative, artificaliter. Naturaliter, ut dixi, in undecim; ...; artificaliter ut est septem, octodecim, vel novem, inde viginti.

36–43 artificiosae…naturalius] For the artificial and natural epacts cf. Borst, *Schriften*, p. 439, 912.

<LII. DE ANNO RESURRECTIONIS IN COMPUTATIONE GRAECORUM>

Initial and resurrection year: Victorius, *latercus* and Dionysius

Nobis sciendum est unde Victorius incipit. Id est ab initio endecadis, secundum endecadis annum. Cum eo annus resurrectionis est. Sic enim inuenies in initio cicli eius: *Kalendis Ianuarii V feria, luna XVIIII*. Sic latercus cum Victorio comitatur: Prima uice post resurrectionem, ab VIII luna in XVIIII luna. Grecus uero in initio ogdoadem incipit ab VIII XX. Et ipse est annus resurrectionis cum eo: XIIII luna in VIII Kalendas Aprilis et pascha in V Kalendas Aprilis et luna XVII. | Victorius et latercus XIIII lunam in fol. 29v VII Kalendas Aprilis faciunt et pascha in *V Kalendas, luna XVI*.

LII,4 secundum] *corr. to* secundo *Mac Carthy*. | annum] *corr. to* anno *Mac Carthy*. **8** XVIIII] XXVIIII *M (corr. by Mac Carthy, Schwartz, Schäferdiek, Warntjes)*. | luna[2]] lunam *Schwartz, Schäferdiek*. ogdoadem] *corr. to* ogdoadis *Warntjes*. **9** VIII] in *add. Schwartz, Warntjes*. **11** lunam] luna *M, Schwartz, Schäferdiek (corr. by Mac Carthy, Warntjes)*. **12** in] *om. Schäferdiek (contra M)*. **13** XVI] XIII *M**.

LII,3–13 Nobis…XVI] The general problem discuss in this chapter and introduced in this paragraph here is the discrepancy between the lunar data at the passion and resurrection of Christ transmitted in the Victorian reckoning (and John's Gospel) on the one hand (*luna* 14 and 16 respectively), in authoritative 'Greek' texts (and the Synoptic Gospels), which were traditionally connected to the Dionysiac reckoning, on the other hand (*luna* 15 and 17 respectively). The question asked by seventh-century Irish computists was how these differences could be reconciled. One of the main problems was that only the Victorian Easter table explicitly referred to the *annus passionis*; in the Dionysiac reckoning there was no indication of which year was to be considered as the year of the passion and resurrection of Christ. Accordingly, the Victorian *annus passionis* was taken as the point of orientation; but how could the Dionysiac data be brought into accordance with that year? Since the question was considered to be purely technical, based on nothing but lunar reckoning, the method applied was a comparison of the epactal series of the two reckonings; these had already been aligned in two different ways in c. 49, ll. 16–32, and this table formed the principal basis for this chapter here. The result that the computists were looking for was that the Dionysiac epact exceeded the Victorian one by one in what was regarded as the *annus passionis* (so that Victorian *luna* 16 at the resurrection would correspond to 'Greek' *luna* 17). This is explicitly formulated in MC's main source for this chapter, the tract DE COMPARATIONE EPACTARUM 1.102–106 (see Appendix 2): *Qui diligenter adtendat, inueniet Christum luna XVII secundum Dionisium resurrexisse, XVI uero secundum Victorium. Illud quidem siue Grecorum siue Hebreorum auctoritate confirmatur, istud autem ipsius Victorii et caeterorum consensu Latinorum commendatur*. As for the details of MC's passage here, the information that the first year of the Victorian Easter table had *feria* 5 and *luna* 19 on 1 January was presumably taken directly from such a table (cf. Krusch, *Studien* II, p. 27). Moreover it is possible that the Easter table used marked the first year as the *annus resurrectionis*, as does the Victorian Easter table in the SIRMOND MS (Oxford, Bodleian Library, Bodley 309, fol. 113r), which shows parallels to the Victorian Easter table used in MC (cf. c. 41, ll. 107–110; c. 62, ll. 64–67). If this was not the case, a combination of the PROLOGUS VICTORII 10 (Krusch, *Studien* II, p. 25) with a Victorian Easter table would easily have supplied this information, since Victorius argues that the passion took place in the consulship of the two Gemini, and the first year of the Victorian table is so denoted. Easter Sunday is given as *V kal. aprl., luna XVI* in the first year of that table. The information about the *latercus* was presumably deduced from a list of epacts rather than directly from that 84-year Easter table (for this Easter table, which includes a column of the epacts of 1 January, cf. c. 49, ll. 11–13); note in general that the *latercus* information is only introduced

52. ABOUT THE YEAR OF RESURRECTION ACCORDING TO THE COMPUTATIONS OF THE GREEKS

We have to know where Victorius starts (his Easter table). From the beginning of the *hendecas*, in the second year of the *hendecas*. According to him (this second year of the *hendecas*) is the year of the resurrection. In fact, you will find the following at the beginning of his cycle: Thursday (and) *luna* 19 on 1 January. The *latercus* is connected to Victorius in the following way: In the first turn after the resurrection, from *luna* 8 to *luna* 19 (i.e. the *latercus* and the Victorian reckoning agree in the epact of 1 January of their respective initial year, 19). The Greek (reckoning), however, at (its) beginning, starts the *ogdoas* from (*luna*) 8 (to *luna*) 20 (on 1 January). And the following is (the essential data of) the year of the resurrection according to this (reckoning): *luna* 14 on 25 March and Easter on 28 March and *luna* 17. Victorius and the *latercus* reckon *luna* 14 on 26 March and Easter on 28 March, *luna* 16 (in the year of resurrection).

---

in this chapter to provide for a different perspective (a third opinion so to say) on the question and is the addition of MC (since DE COMPARATIONE EPACTARUM does not refer to the *latercus* at all). As already mentioned, the comparison between the Dionysiac and the Victorian reckoning in this chapter seems to be based on the comparative lists of epacts in c. 49, ll. 16–32. For the Julian calendar dates and the 'Greek' lunar data of the *annus resurrectionis* cf. c. 44, ll. 55–80.

LII,3–13 Nobis…XVI] The question of how to reconcile the differing lunar data given for Christ's passion and resurrection in Victorius' writings on the one hand, 'Greek' authorities on the other, was a problem that appears to have troubled only seventh- and early eighth-century Irish computists. The earliest text dealing with this problem presumably was DE COMPARATIONE EPACTARUM; CE did not mention the question, but knew the two methods of comparing the Victorian with the Dionysiac epacts which underlie this chapter here. An interesting alternative explanation for the difference between 'Greek' and 'Latin' tradition concerning the lunar age of the resurrection is given in the third Irish computistical textbook, DRC 86.21–22: *Sciendum nobis quoque quod XVI luna apud Latinos causa bissexti in die resurrectionis fuit. Nisi enim bissextus, XVII luna esset.* The only non-Irish, post-ca. AD 730 text that includes remnants of this discussion is COMP. COL. 5.4A,C, 5.5A: *Cum Victorio sexta decima in Pascha resurrectionis et cum Latinis; cum Graecis septima decima. Hoc adfirmant Protherius, Theofilus. ... Et artificaliter alii disponunt sextam decimam cum Victorio et septimam decimam cum Graecis in resurrectione Christi idem esse. ... Principium cycli: <Luna> nona decima Kalendas Ianuarii, quinta feria, vicesima cum Graecis. Utrumque proba! Victorius facit quartam decimam sexta feria, quintam decimam in sabbato, sextam decimam in Pascha dominica, prima feria. Congruus <est> Victorius in diebus solis, non congruus in lunae aetate; facit namque quartam decimam in sexta feria. Graeci daciunt quartam decimam in quinta feria, vel quintam decimam in sexta feria, sextam decimam in sabbato; septima decima <est> dominicus dies Pasche Domini. Quod congruum <est> in diebus solis et in aetate lunae.*

LII,3–4 Nobis…annum] Transcribed in Borst, *Schriften*, p. 908. **3–26** Nobis…pascha] Partially transcribed in Mac Carthy, *Annals of Ulster* 4, p. lxxii; Krusch, *Studien* I, p. 10–1; Schäferdiek, 'Osterzyklus', p. 361–3, 369–70; transcribed and translated in Warntjes, '84 (14)-year Easter reckoning', p. 62–3 (facsimile p. 84–5); cf. Krusch, *Studien* II, p. 58; O'Connell, 'Easter cycles', 84–8 and p. LXXII, XCIV–XCV, CXXII–CXXIII of the introduction. **3–60** Nobis…est] Transcribed, corrected and commented in Schwartz, 'Ostertafeln', p. 95–6; cf. p. LI, LXXI, LXXXIV, XCIV, CI–CII, CLIII–CLV, CLXIV, CLXXXI of the introduction.

Comp. of Vic. and Dion. epacts: Dion. ogd. = Vic. hend.

Hinc *apparet*, *qu*od *Grecus praecepit Latinu*m *in* accensione *lunae*. Tribus tantum annis *con*ueniunt in accensione lunę: Quando cum Victorio IIII luna in Kalendis Ianuarii, in anno secundo *XVI luna* pro saltu, cum Grecis a V luna in XVI luna; in quo conueniunt, et in *XXVII* luna et in *VIII* luna. Hic separat *saltus* Grecorum.

Comp. of Victorian, latercus and Dionysiac epacts

Sciendum est nobis, quod Victorius et latercus conueniunt. Id est usque ad annum in quo luna XII fit in Kalendis Ianuarii. In hoc enim anno saltus laterci separat eos, a XII luna in Kalendis Ianuarii in XXIIII. Haec luna conuenit cum Greco, et in V luna. Inde a V luna in XVI, ubi cum Victorio conueniunt. Inde Grecus *precedit* Latinos in XIIII luna et in pascha.

---

17 XVI] XV *M**. **18** Grecis] Grecus *M, Schwartz, corr. to* Greco *Warntjes*. **20** Victorius] Victurius *M*, Krusch*. **21** conueniunt] conuenit *M, Krusch, Schwartz (corr. by Mac Carthy, Warntjes)*. **25** conueniunt] *corr. to* conuenit *Mac Carthy*.

14–26 Hinc…pascha] The first method applied to solve the problem of the discrepancy between the resurrection lunar data as transmitted by Victorius on the one hand, and 'Greek' tradition on the other, is by comparison of the epacts of the Dionysiac *ogdoas* with those of the Victorian *hendecas* (this method has already been defined in more general terms in c. 49, ll. 25–27); the basis for this comparison is the table in c. 49, ll. 16–32 and in CE p. 109. The argument is best illustrated by the tables in Schwartz, 'Ostertafeln', p. 94 and especially Warntjes, '84 (14)-year Easter reckoning', p. 67: In this comparison, the epacts of 1 January of Dionysius and Victorius agree in only three years (epacts 16, 27, and 8), and disagree in the remaining 16 because of the Dionysiac *saltus* being applied in the year with epact 8 (i.e. the first year of the Dionysiac 19-year cycle), leading to the Dionysiac epacts exceeding the Victorian ones by 1 in these 16 years; most importantly, this is the case also in what was regarded as the *annus passionis* in the Victorian reckoning (i.e. the second year of the Victorian *hendecas*), leading to the desired *luna* 17 for the resurrection in the Dionysiac data. The *latercus* information introduced into this discussion can easily be checked with the tables referred to above; note that MC's wrong application of this data here reveals that he was unfamiliar with this reckoning; MC's introduction of the *latercus* into this discussion was purely artificial, and had no relevance to the general argument, as MC's main source, DE COMPARATIONE EPACTARUM 1.37–100 (see Appendix 2) reveals, which does not mention the *latercus* at all: *Ex his quidem aparet, quis sepius precidit in lune aetatibus, quis subsequitur: Dionisius an Victorius, quorum superior Grecus, alter Latinus est. Sed notandum est illos sicut suprascriptis tramitibus manifestum est tunc in ipsis aetatibus concordare, quando luna XVI, aut XXVII, aut VIII Kalendis Ianuarii prouenit. ... Est quidem in annis XVI, quibus Dionisius et Victorius discordant, quedam concordia scilicet singulis. Estque in tribus, quibus concordant, quedam discordia utique singulis. ... Ex illo decennouennali ciclo, quem et secundum Victorium et secundum Dionisium possuimus fere, manifesto apparet, quod quibus lunis ipsi inter se per ogdoadem et endicadem discordant, et quod et quibus sibi concordant: per saltum enim Dionisii incipient discordare, per Victorii uero concordare. ... Unde quidem ut ipse ait: fere in principio ogdoadis Christus passus est. Igitur ab ogdoadis primo anno Dionisius et Victorius per Dionisii saltum, qui est ut supra secundum Grecos, in lune aetatibus sepius discordant. Usque ad nonum endicadis annum XVI enim fere annis sepius discordant, cum utique in hisdem annis Dionysius in ipsis aetatibus precedat, Victorius uero subsequatur. Tribus uero fere annis per saltum Victorii sepius concordant. Finito ergo decennouenalis ciclo per saltum Dionisium iterum discordant.*

It is apparent from these facts that the Greek (reckoning) precedes the Latin one in the kindling of the moon. Only in three years do they agree in the kindling of the moon: When *luna* 4 occurs on 1 January according to Victorius, (and) *luna* 16 in the following year due to the *saltus*, (the epacts of 1 January change) from *luna* 5 to *luna* 16 according to the Greeks; in this (i.e. *luna* 16 on 1 January) they agree, as well as in *luna* 27 and *luna* 8 (on 1 January in the subsequent two years). Here the *saltus* of the Greeks separates (them).

We have to know to what extent Victorius and the *latercus* agree. Up to the year, in which *luna* 12 happens to be on 1 January. In that year, in fact, the *saltus* of the *latercus* separates these, (leaping) from *luna* 12 to (*luna*) 24 on 1 January. This moon (i.e. *luna* 24) agrees with the Greek (reckoning), as (they agree) in *luna* 5 (on 1 January of the following year). Then, (the epacts of 1 January change) from *luna* 5 to (*luna*) 16, where (i.e. in *luna* 16) they agree with Victorius. Therefore, the Greek (reckoning) precedes the Latins in *luna* 14 (i.e. the Easter full moon) and on Easter.

14–26 Hinc…pascha] Again (cf. previous note), the only post-MC text that contains remnants of this discussion is COMP. COL. 5.5C: *Tamen Graeci Latinique verius tertio anno tantum concordant in Kalendas Ianuarii. Idemque dies Paschae <est>, sed non eadem aetas lunae. Nisi in sexta decima et vicesima septima et in octava discrepant aetate.*

Comp. of Victorian and Dionysiac epacts by alignment of the ogdoades

Cum hanc incongruitatem uiderunt, temptauerunt duas ogdoades coniungere. Sic et fecerunt: A VIIII luna in XX cum Latinis, ab VIII uero luna in XX cum Grecis. Et in *VI annis* coeunt, et tres X annis *dis*crepant. Quid separat eos? Id est *saltu*s Latinorum, a IIII luna in XVI luna, cum Grecis uero a IIII luna in XV. Una etas inter eos XIII annis, nisi in duobus annis: in embolismo *fin*iente *ogdoad*e, in quo luna XXVI in Kalendis Ianuarii cum Grecis, cum

---

27 hanc incongruitatem] in hanc congruitatem *M (corr. by Schwartz).* **28** Sic et] et sic *Schwartz (contra M).* | VIIII] VII *M (corr. by Schwartz).* **29** Grecis] *corr. from* Grecus *to* Grecos *M;* Grecis *Schwartz.* **30** VI] II *M (corr. by Schwartz, Bisagni & Warntjes).* | et] *add.* in *Bisagni & Warntjes.* **31** luna[2]] luna- *M, Schwartz.* **32** Grecis] Grecus *M, Schwartz.* | luna] *om. Schwartz (contra M).* | XV] XII *M (corr. by Schwartz).* | etas] aetas *Schwartz (contra M).* **34** Grecis] Grecos *M, Schwartz.*

27–43 Cum…XVIII] Though the comparison of the Dionysiac with the Victorian epacts outlined in the previous passage led to the desired results for the Dionysiac lunar data in what was regarded as the *annus passionis*, computists realized that this comparison was unhistorical; historically and therefore scientifically correct and thus the only valid comparison obviously was by alignement of the two *ogdoades* (and consequently also the two *hendecades*, which is not a different comparison, as Mc makes the reader believe, but only the continuation of this one). In this comparison, the Dionysiac epacts of 1 January agree with the Victorian ones in six years, and disagree in the remaining 13 (cf. the table in Schwartz, 'Ostertafeln', p. 94), with the Victorian epacts exceeding the Dionysiac ones by 1; Mc draws attention to two exceptions, namely the final years of the *ogdoas* and *hendecas* respectively, where, according to Mc, the difference is 2 rather than 1 for one month, since the embolism is placed a month earlier in the Dionysiac compared to the Victorian reckoning (the last two sentences of the first paragraph mean that April being a full lunation in the Dionysiac reckoning, but hollow in the Victorian one, was not a general feature; i.e. the Dionysiac April lunation had generally to be considered as hollow; cf. the contrary argument in c. 50, ll. 29–39). The same argument, though in a different context (namely the discussion of the correlation of the Dionysiac *ogdoas* with the Victorian *hendecas*), is given in Mc's main source for this chapter, DE COMPARATIONE EPACTARUM 1.45–50 (see Appendix 2): *Est quidem in annis XVI, quibus Dionisius et Victorius discordant, quedam concordia scilicet singulis. Estque in tribus, quibus concordant, quedam discordia utique singulis. Quod utrumque per lunam Aprilis fieri probatur, que secundam Victorium XVIIII (recte XXVIIII) secundum Dionysium uero XXX est.* Interestingly enough, the historically correct method discussed in Mc here is ascribed to the adherents of the Victorian reckoning in DE COMPARATIONE EPACTARUM 1.77–87 (see Appendix 2): *Ipsi, qui Victorium et Dionisium fere sex a principio ogdoadis annis in lune aetatibus concordare confirmant, ac deinceps ceteris per saltum Victorii discordare, qui tunc ut aiunt: In lunae aetatibus praecedit, cum Dionisius subsequatur in ipsis. Ita ut etiam in fine ocdoadis et endicadis ipse istum duabus lunae aetatibus superet, quarum una per saltum, altera per quandam causam facta intelligitur, unde nec per saltum soluitur. Finita autem endicade per Dionisi saltum, qui est utique secundum Grecos, iterum concordant. Sed hii, qui ita sentient, in Victorio confidentes audiant.* This ascription clearly demonstrate that adherents of the Victorian reckoning had accused their Dionysiac rivals of following an Easter reckoning in which the lunar data recorded for the resurrection (*luna* 15; cf. the following passage) was uncanonical; the followers of the Dionysiac reckoning, then, appear to have countered this argument by the alternative comparison (correlating the Dionysiac *ogdoas* with the Victorian *hendecas*; cf. previous passage), which led to the desired lunar data for the Dionysiac reckoning.

27–43 Cum…XVIII] This historically correct comparison between the Dionysiac and Victorian epacts is ascribed to Victorius himself in COMP. COL. 5.5B, revealing that the author of this text worked from DE COMPARATIONE EPACTARUM (cf. *app. font.*; this is hardly surprising, since both texts are transmitted

When they realized this incongruence (of comparing the Dionysiac *ogdoas* with the Victorian *hendecas*), they tried to combine the *ogdoades* (of) both (reckonings). And they did this in the following way: (The epacts of 1 January change) from *luna* 9 to (*luna*) 20 (from the first to the second year of the *ogdoas*) according to the Latins, but from *luna* 8 to (*luna*) 20 according to the Greeks. And they agree in six years and differ in (the remaining) 13 years. What separates them? The *saltus* of the Latins, (leaping) from *luna* 4 to *luna* 16 (on 1 January), while (the epacts of 1 January change) from *luna* 4 to (*luna*) 15 according to the Greeks. There is (a difference of) one (lunar) age between them in 13 years, with certain exceptions in two years: in the embolismic year that ends the *ogdoas* (i.e. the eighth year of the 19-year cycle), in which *luna* 26 occurs on 1 January according to the Greeks, but (*luna*) 27 according to Victorius; and in the embolismic year that ends the *hendecas* (i.e. the 19th year of the 19-year cycle),

---

in the same codex): *Oritur item aliud id difficile, quod interrogant alii: Victorius prior <est> in accensione lunae in tredecim annis, in aliis sex annis congruunt Graeci et Latini. Illi hoc habent meritum, quia Victorius iungit occulte duos ogdoados et duos endegados <in> duos ogdoados, id <est> octavum vel nonum. Cum Latinis quasi prius endegado, <cum> Graecis cum endegado cucurrisset, post ogdoadum Graeci, cum ogdoado Latini <posuerunt>.* The historically correct comparison was also briefly mentioned by Bede, and was consequently discussed in some detail in glosses to that text: DTR 42.51–55 (cited in LIB. COMP. 4.9B, LIB. CALC. 46B): *Huiusmodi ergo cogit diversitas sex solum primos circuli decemnovenalis annos aequali nobiscum aetate lunam putare Victorium, nam deinceps inserta mutatione saltus una semper die maiorem nobis eam, usque dum etiam nos finito eam circulo toto inserimus, habere non cessat.* Gloss to DTR 11 (Munich, Bayerische Staatsbibliothek, Clm 18156, fol. 16v): *Hoc igitur concordans Victorius in omni lunę conputatione cum Dionisio. Pergunt simul etiam concordare usque ad finem VI anni istius decennouenalis, in quo uidelicet VIto anno simili utentes ętatę lunę, id est IIII. Saltum facit Victorius. Additis autem ad IIII XI XV Dionisius attingit lunam. In sequenti, id est in VII anno, quam transilit Victorius ac XVI deprehendit. Qui sane ibidem discordare contendentes tendunt sicut ad finem usque cycli peruenire, qua Dionius (!) saltum paret, Victorio semper per hos XIII annos una ętatę illum anticipante. ... Unde scire poterimus, quod VI annis concordant Dionisius et Victorius.* Gloss to DTR 42 (*CCSL* 123B, p. 410): *Sc. dum nos 'de quarta' in quintam decimam conuertimus lunam kl. Ianuarii. Nos enim non fecimus adhuc saltus qui in fine decennouenalis facturi sumus, et ideo minorem aetatem lunae habemus. In eo quod dicit 'de quarta' concordamur quia usque ad illum annum nec ille nec nos fecimus saltum. In eo tamen quod dicit 'in sextam decimam' non concordamur quia ille fecit saltum, nos uero non fecimus, et ideo ibi maiorem aetatem lunae habet quam nos. A primo enim anno usque ad sextum annum decennouenalis concordamur nos et ille: a VIo uero usque ad finem discordamur, et hoc potes probare per argumentum. Accipe enim epactas illius anni et iunge concurrentibus Nouembris, in Nouembri enim facit saltum sicut nos, et inuenies in kl. Nouembris et apud nos et apud illum, quia nec nos nec ille adhuc fecimus saltum, secundam lunam, et computa illa usque ad kl. Ianuarii et inuenies in kl. Ianuarii quartam lunam, id est in praecedenti anno. In sequenti uero ibi saltus fit inuenies in eisdem kl. per idem argumentum XVIam, et ideo dixit 'de quarta in sextam decimam', dum apud nos est de quarta in quintam qui non fecimus adhuc saltum.* Note that COMP. COL. 5.5C disproves (in a similar way as MC at the end of the first paragraph here) of the argument that a full April lunation is the reason for the lunar age of Easter Sunday differing between the Dionysiac and the Victorian reckoning: *discrepant aetate, quia cum tricesima luna Aprilis ipsius <aetatis> est luna vicesima nona Maii? Quod probare potest contra: Vicesima octava, vicesima septima non <est> eadem aetas in Kalendas Ianuarii cum Graecis Latinisque, sed eadem aetas pro causa supradicta lunae.*

29–30 Et…discrepant] Transcribed and translated in Bisagni & Warntjes, 'Latin and Old Irish', p. 26.

Victorio uero XXVII; et embolismo finiente *endecad*e, | id est luna XXVII cum Grecis, XXVIII cum Victorio. Quid addit alteram aetatem? Id est quia *luna Aprilis XXX*, *cum Victori*o *uero XXVIIII*. Hoc tamen <non> uerum est. Si enim una ętas cum utrisque esset in Kalendis Ianuarii, non discreparent in XIIII luna et in pascha. fol. 30r

Comp. of the *hendecades*

Cum uiderunt falsum esse, temptauerunt duas endecades coniungere. Quomodo fecerunt? Cum Victorio a VIII luna XVIIII, cum Grecis uero a VII luna in XVIII.

Implic. on the resurrection year

Inde XIIII luna cum Victorio VII Kalendas Aprilis VI feria, XV sabbato, XVI V Kalendas Aprilis dominica. Cum Grecis uero XIIII luna VI Kalendas Aprilis, luna XV V Kalendas Aprilis dominica. Hoc tamen contra epistulam Proteri et Augusti.

---

35 et] in *add. Schwartz.* **36** Grecis] Grecos *M, Schwartz.* **38** non] *om. M (corr. by Schwartz).* **42** endecades] ogdoades *M (corr. according to* MC*; also Schwartz).* **43** VIII] VIIII *M (corr. by Schwartz); add.* in *Schwartz.* | luna[1]] *om. Schwartz (contra M).* | XVIIII] XXVIIII *M (corr. by Schwartz).* | Grecis] Grecos *M, Schwartz.* | luna[2]] lunu *Schwartz (contra M).* **46** Grecis] Grecos *M, Schwartz.*

45–48 Cum…Augusti] The historically correct comparison by correlation of the two *ogoades* and the two *hendecades* led to the Victorian epacts exceeding the Dionysiac ones by 1 in the second year of the *hendecas*, which was, in the Victorian reckoning, regarded as the lunar year in which the resurrection had occurred; consequently, since the Victorian reckoning recorded *luna* 16 at the resurrection, the Dionysiac equivalent was *luna* 15, which was contrary to the Synoptic Gospels and other authorities that gave *luna* 17 as the lunar age of that event; in this context, MC refers explicitly to the violation of the data outlined in the EPISTOLA PROTERII 2 (Krusch, *Studien* I, p. 271–2; the first part cited in DRC 86.10–14; the second part in COMP. COL. 5.4B): *quinta sabbatorum XIIII luna mensis primi in cenaculo cum discipulis pascha manducans, paulo post a Juda traditur et sequenti die XV luna crucifigitur, id est sexta feria, et ad inferos discendens ac dispensationes salutis nostrae perficiens, vespere sabbati, luciscente dominico, resurrexit a mortuis, in quo die lunam XVII primi mensis iuxta Hebreos extetisse manifestum est. ... XIIII namque luna primi mensis iuxta Hebreos, ut superius dictum est, Jesus pascha tipicum manducavit, sequenti vero VI ferea XV luna, ut ovis occisionis, cruci pro nobis affixus est et vespere sabbati, luciscente dominico, XVII luna resurrexit a mortuis*. For the *annus resurrectionis* cf. also c. 44, ll. 55–80; c. 52, ll. 3–13.

44–48 Inde…Augusti] For other texts discussing the chronological data of the *annus passionis*, though in a different context, cf. c. 44, ll. 55–80.

36–40 Quid…pascha] Cf. p. XCIV of the introduction.

(in which *luna*) 27 (occurs on 1 January) according to the Greeks, (but *luna*) 28 according to Victorius. What (is it that) adds the extra lunar age (i.e. why are the two reckonings exceptionally separated by two lunar ages for a certain period of these two years)? Because the lunation of April (has) 30 (days according to the Greeks), but (only) 29 according to Victorius. This, however, is not true. In fact, if there was one lunar age on 1 January according to both (reckonings), they would not differ in *luna* 14 (i.e. the Easter full moon) and on Easter (i.e. if both reckonings have the same epact on 1 January, they have the Easter full moon on the same day, and the same lunar age on Easter Sunday; ergo, their lunar calendar is identical between 1 January and Easter, and therefore they cannot generally differ in the April lunation).

When they realized that (this theory) is wrong, they tried to combine the two *hendecades*. How did they do this? (The epacts of 1 January change) from *luna* 8 (to *luna*) 19 according to Victorius, but from *luna* 7 to (*luna*) 18 according to the Greeks.

Consequently, (in the year of the resurrection) *luna* 14 (occurred) on Friday, 26 March, (*luna*) 15 on Saturday, (*luna*) 16 on Sunday, 28 March according to Victorius. According to the Greeks, however, *luna* 14 (occurred) on 27 March, *luna* 15 on Sunday, 28 March. Yet, this (is) against the letter of Proterius and Augustus.

Discrep. of one month between Vict. and Dion. Easter Sunday when Dion. *ogd.* = Vict. *hend.*

Si est alius locus, *in quo* Grecos uideretur precedere Victorium? Veni ad finem ogdoadis. In isto enim anno embolismus luna XXVI. Cum Victorio uero XXV et communis est in endecade. Inde *fac*it *communem de embolismo*. Et *du*a *pascha* in hoc anno facitur, quia luna XIIII huius anni cum Grecis XIIII Kalendas Maii, cum Victorio uero luna XIIII XIII Kalendas Aprilis ante ęquinoctium. Sic inuenies primo endecadis: Luna VII cum Grecis in Kalendas Ianuarii et initium endecadis. Cum Victorio uero in hoc anno VI luna in Kalendis Ianuarii. Et embolismum de commune facit. Et hic annus cum eo sine pascha est.

<LIII. DE ANNO LUNAE>

The term *annus* in lunar reckoning

Annus abusiue lune dicitur. Annus enim proprie solis est, menses autem lunę. Sed ut menses solis formati sunt a luna, ita annus lunę a sole formatus est.

---

51 embolismus] embolismu̲ *M, Schwartz.* | luna] lunae *Schwartz.* | XXVI] XV *M (corr. according to* MC*; also Schwartz).* **53** facitur] facit *M, Schwartz.* **54** Grecis] Grecos *M, Schwartz.* **56** endecadis] endecade *M, Schwartz.* **59** commune] cummune *M*.* **LIII,2** lune] lunae *Borst (contra M).*

49–60 Si…est] This discrepancy of one month between the Dionysiac and the Victorian Easter occuring in the (unhistorical) comparison of the Dionysiac *ogdoas* with the Victorian *hendecas*, namely in the final year of the Dionysiac *ogdoas* compared to the eighth year of the Victorian *hendecas*, is also discussed in DE COMPARATIONE EPACTARUM 1.51–62 (see Appendix 2): *In quo etiam anno hii duo in paschae solemnitatibus mense integro discordant. Sed si tunc accederit, ut si secundum Victorium in VIII Kalendas Aprelis pascha celebretur, secundum Dionisium in X Kalendas Maii celebrabitur; si uero secundum ipsum in VII Kalendas Aprelis, secundum istum in Nonis Kalendas Maii; si autem secundum ipsum in VI Kalendas Aprelis, secundum istum in VIII Kalendas Maii; item si secundum ipsum in V Kalendas Aprelis, secundum istum in VII Kalendas Maii. Sed in his, ut in ceteris, errat Victorius in XIII luna prioris anni pascha conponens et communem de embolismo faciens, quod quidem euidens error est.* Note that a discrepancy of one month between the Victorian and the Dionysiac Easter occurs only in this comparison of the two reckonings, and there only in this one year; such a discrepancy did not arise in the historically correct comparison. The final statement of this passage of MC that this year is without Easter in the Victorian reckoning refers to the fact mentioned slightly earlier, namely that the Easter full moon of this year falls before the vernal equinox in the Victorian reckoning; for this reason the Victorian Easter of this year was regarded as unlawful by adherents of the Dionysiac reckoning. Interestingly enough, Pascasinus mentions a similar discrepancy between the *Supputatio Romana* and the Alexandrian reckoning for AD 417: EPISTOLA PASCASINI 2 (Krusch, *Studien* I, p. 249; cited in DTR 43.77–79, in turn cited in LIB. CALC. 51): *Tunc enim cum declinatur, ne X kl mai die pascha teneretur, caelebratum est octavo die kl aprilium, id est pro embolismo communis est annus tentus.* Even more interesting, a discrepancy of one month between the Alexandrian and the Victorian reckoning occurring in AD 607 is discussed in EPISTOLA CYRILLI 3 (Krusch, *Studien* I, p. 345–6): *De pascha vero, ut scripsistis, nunciamus vobis, VIIII kl Mai nos futura Xma indictione caelebrare. Et alia manu: Deus et dominus noster sanctam ecclesiam, id est congregationem, custodiat. Quod optamus carissimi fratres, ut simul pascha caelebremus VIIII kl Mai propter rationem embolismi anni. Quod si facitis VII kl apl*

Is there another place, in which it would seem that the Greeks precede Victorius? Turn to the end of the *ogdoas* (i.e. the final year of the Dionysiac *ogdoas*). In fact, an embolism (and) *luna* 26 (on 1 January occur) in this year. According to Victorius, however, (*luna*) 25 (occurs on 1 January), and it is a common year in the *hendecas* (namely the eighth year of the Victorian *hendecas*). Consequently, (Victorius) turned an embolismic into a common year. And two Easter Sundays are celebrated in this year, since *luna* 14 (i.e. the Easter full moon) of this year (occurs) on 18 April according to the Greeks, but *luna* 14 (i.e. the Easter full moon, occurs) on 20 March, before the equinox, according to Victorius. You will find the following in the first year of the (Dionyisac) *hendecas*: *luna* 7 on 1 January and the beginning of the *hendecas* according to the Greeks. According to Victorius, however, *luna* 6 (occurs) on 1 January in this year (which is the ninth year of the Victorian *hendecas*). And (Victorius) turned a common into an embolismic year. And this year is without Easter according to him.

53. ABOUT THE LUNAR YEAR

The lunar year is improperly so called. In fact, the (term) 'year' (*annus*) belongs exclusively to the sun, the months, however, to the moon. But just as the solar months were formed by the moon, so the lunar year was formed by the sun.

---

luna XXII, ut preparatis, communem annum facitis de embolismo, dum observatis lunam incensam in III non Mar iuxta regulam Latinorum: kl enim ianuarias dominica luna XXVII. But the reason for this was an uncommon interpretation of Victorius' writings (Easter full moon on 18 March, Easter New moon on 5 March; this was explicitly recorded as 'Latin' practice in PROLOGUS VICTORII 4 (Krusch, *Studien* II, p. 19), but these early dates for the Easter new and full moon were avoided by Victorius in his Easter table) rather than the comparison between the Dionysiac *ogdoas* and the Victorian *hendecas*, which underlies MC here. **LIII,2–3** Annus[1]...lunę] The theory that the term 'year' is only appropriate for the course of the sun, the term 'month', on the other hand, only for the course of the moon, derives from Isidore's description of the solar and the lunar course: ETYM. 3.57 (cited c. 46, ll. 10–13): *Luna vicinior est terris quam sole. Inde et breviori orbe celeries peragit cursum suum. Nam iter, quod sol in diebus trecentis sexaginta quinque peragit, ista per triginta dies percurrit. Vnde et antiqui menses in luna, annos autem in solis cursum posuerunt.* DNR 19.1.5–8: *Terris autem uicina luna breuiori orbe conuertitur, et iter quod sol in diebus CCLXV peragit, ista per XXX dies percurrit. Vnde et antiqui menses in luna, annos autem in solis cursu posuerunt.* MC, then, is only consequent in arguing that solar terminology was later applied also to lunar reckoning and vice versa; this theme recurs frequently in MC (cf. c. 12, ll. 20–23; c. 53, ll. 12–15).

49–60 Si...est] Since this unhistorical comparison of the two reckonings hardly found any reception (cf. ll. 14–26 of this chapter), this detail is not mentioned in any post-MC text. **LIII,2–4** Annus[1]...est] The origin of lunar or solar terminology did not generally bother seventh- or eighth-century computists and may therefore well be described as one of MC's own particular interests. Moreover, the term *annus* was hardly ever used for a lunar month; an interesting exception is the Irish DRC 46.6–7, where a lunar month consisting of 29½ days is called *annus brevis*: *Sciendum nobis quid sit annus lunaris: XXVIIII dies et dimedius, et ille nominatus annus breuis.*

LIII,2–4 Annus[1]...est] Transcribed in Borst, *Schriften*, p. 435.

The beginning of the lunar year

| Annus lunę tria habet initia: Ab accensione lunę primi mensis; iterum alterum a XV primi mensis usque in XIIII; iterum, quia in lege omnis tertia decima primi mensis finiuerat, XIIII autem initium anni; in nouo uero omnis XIIII luna primi mensis finis anni, XV autem initium tertium. Prima luna, quae accenditur in X Kalendas Aprilis post saltum, initium anni in preparatione saltus. fol. 30v

8 finiuerat] finierat *M.*

5–11 Annus...saltus] There is no apparent source for the statement that *luna* 1 of the Easter lunation is regarded as the beginning of the lunar year; still, this certainly is one of the beginnings of the lunar year favoured by MC, as c. 57; c. 59, ll. 38–64 illustrate. If the lunar year starts with *luna* 1 of the Easter lunation, this first moon occurs on 23 March in the first year of the Dionysiac 19-year cycle. Since the *saltus* occurs on the previous day according to MC (cf. 62, ll. 68–72 and references there), this first moon coincides with the beginning of the gradual increment (the 'preparation') of the *saltus* throughout the 19-year cycle (for the calculation of this increment per lunation cf. c. 62, ll. 14–61), and this is what is meant in the last sentence of this passage. The second beginning of the lunar year mentioned here, *luna* 15 of the Easter lunation, derives from EPISTOLA DIONYSII (Krusch, *Studien* II, p. 84; cited in c. 59, ll. 4–7, CE p. 113, DRC 83.3–5, DTR 42.40–42, in turn cited in PV §297, LIB. COMP. 4.9B, in turn cited in LIB. CALC. 46B): *A quinta decima luna paschalis festi anni, verbi gratia, praecedentis usque ad XIIII sequentis paschae, si communis annus est, CCCLIIII dies habebit: si embolismus, CCCLXXXIIII <dies>*. As the frequent citations of this passage show, this certainly was the most commonly used beginning of the lunar year, and it is also prominently applied by MC in c. 59, ll. 70–105. The third possible beginning of the lunar year, i.e. the Easter full moon, is purely theoretical, since it was neither applied in MC, nor even mentioned outside of the early eighth-century Irish computistical context (cf. *app. comp.*). The reference to the Old Testament here alludes to the belief that God created the moon in its fullest form (*luna* 14) on the fourth day of creation (GENESIS 1:14–19), as well as the explicit statement that the first pasch and the Exodus from Egypt happend on *luna* 14 (EXODUS 12). Both events were taken as proofs that the year began with *luna* 14 in the Old Testament. This view may have originated more directly from PROLOGUS VICTORII 9 (Krusch, *Studien* II, p. 24; partially cited in DRC 85.4–10): *Rursusque omnibus annis temporibus diebus ac luna maxime, quae iuxta Hebraeos menses facit, rite discussis, a mundi principio secundum praedictae historiae fidem usque in diem, quo filii Israhel paschale mysterium caelesti initiavere mandato et ab Aegyptia clade agni occisione salvati sunt, bissextorum pariter necessitate decursa, quantum fida supputatio vestigavit, anni tria milla sescenti octoginta novem quinta feria VIIII Kal. aprl. luna XIII, incipiente iam vespera, docentur impleti. Cuius sequenti die tertio millesimo scilicet anno ac sexcentesimo nonagesimo procedente mense primo, VIII Kal. april. luna XIIII noctis initio Hebreos claruit agni sacrificium peregisse. Pascha quippe sicut omnimodo traditione cognoscitur anni principio, non fine celebratur*. DRP 12.202–206 (cited in DRC 91.4–7): *cum in ueteri quidem lege hoc solum custodiri debere decernitur, ..., XIIII die mensis primi, qui non diei sed lunae cursibus conputatur, pascha immoletur*. Cf. also MC's discussion of the year of creation and the year of the Exodus from Egypt in c. 44, ll. 18–54. Because of the reference to the Old Testament, MC wanted to make perfectly clear here that the generally advocated beginning of the lunar year, *luna* 15 of the Easter lunation as outlined by Dionysius, is also Biblical and even more authoritative, since it has its origin in the New Testament, namely in the chronology of the Synoptic Gospels, according to which Christ was crucified on *luna* 15. Cf. the discussion of the *annus passionis* in c. 44, ll. 55–80.

5–11 Annus...saltus] Dionyisus' definition of the lunar year extending from *luna* 15 to *luna* 14 of the Paschal month certainly was most popular (cf. the texts that cite this Dionysiac passage in *app. font.*); *luna* 15 of the Easter lunation is the only beginning of the lunar year recorded in DIAL. NEUSTR. 24: *Interrogandum est: Horum initia <et> fines annorum secundum lunam quae sunt? Id est: Initia eorum et fines, ut dignae memoriae Diunisius Exiguus dixit: A quinta decima luna primi mensis semper usque*

The lunar year has three beginnings: from the kindling of the moon of the first month; a different one is from (*luna*) 15 of the first month to (*luna*) 14; yet another beginning, however, (was *luna*) 14 (of the first month), since every (year) had ended on *luna* 13 of the first month in the Old Testament; but in the New Testament *luna* 14 of the first month (i.e. the Easter full moon) (is) the end of every year, (*luna*) 15, then, the third beginning. The first moon, which is kindled on 23 March after the *saltus*, is the beginning of the year according to the preparation of the *saltus*.

---

ad quartam decimam lunam eiusdem lunae annus est. QUAEST. AUSTR. 2.8A, 2.10A (the 1 January mentioned in this passage rather refers to the beginning of the solar year, to which the lunar reckoning was adapted; cf. c. 36, ll. 2–5): *Initium autem lunaris anni iuxta Aegyptios, qui solent a plenilunio mensium initia denumerare, <numeratur> a quinta decima luna paschalis festi usque ad quartam decimam lunam eiusdem primi mensis. ... initium lunaris anni et communium et embulesmorum iuxta eosdem a quinta decima luna usque ad quartam decimam lunam*. Even though Bede quoted this Dionysiac passage in a different context (cf. *app. font.*), he did not consider *luna* 15 of the Easter lunation as a possible beginning of the lunar year, but rather *luna* 1 of that lunation (as MC) as well as *luna* 1 of the January lunation: DTR 36.15–18, 45.11–13 (cited in LIB. CALC. 85, PV §268, RM 59.13–15), 56.4–6 (cited in LIB. CALC. 92, PV §226, RM 77.6–9): *Qui uterque (i.e. annus communis et annus embolismus) apud hebraeos a principio mensis paschalis incipit, ibidemque finitur: apud vero romanos ab incipiente luna mensis ianuarii sumit initium, ibique terminatur. ... Qui utrique (i.e. annus communis et annus embolismus), sicut et supra dictum est, ab exordio primim mensis quem hebraei nisan vocant, hoc est ab accensione lunae paschalis, initium sumunt. ... Nam sicut annus quisque decemnovenalis circuli propter legalem hebraeorum observationem a paschali mense inchoat ibidemque finitur, ita et hic romanorum institutione a luna ianuarii mensis inchoat atque ibi desinit*. Cf. also glosses to DTR (*CCSL* 123B, p. 420). DRC 85.7–8 (framed by the Victorian passage quoted in *app. font.*) ascribes the tradition of *luna* 14 of the first month (i.e. the Easter full moon) being regarded as the beginning of the year to the Hebrews: *Et a XIIII luna primi mensis semper apud Ebreos annus incipiebat, pascha in eo die celebrantes*. In BC 4C14–D1 the lunar years start with the Easter full moon, so that this tradition appears to be prominent in early eighth-century Irish computistics, but hardly beyond. Note also MC's remark in ch. 36, l. 6 that the beginning of the lunar year is dubious. For this question of the beginning of the lunar year cf. also NOTKER LABEO, *De quatuor questionibus compoti* (Piper, *Nachträge*, p. 315).

5–11 Annus…saltus] Cf. Borst, *Schriften*, p. 406.

The 12 lunations of a lunar year

Quid mensurauit annum lunę? Id est cursus lunae, id est XII lunae. Quis primus ordinauit annum lunae in XII lunas? Greci. Annum autem solis Gignus rex Sabinorum in XII menses sub XII lunas ordinauit.

Features underlying the 19-year cycle

Ubi certus lunae cursus XVIIII annorum? Quicquid mensurauit numerum istum: XIIII luna primi mensis et plenitudo saltus; ut bissextus dominatur cursum solis, XXVIII annis usque bisęxtus omnes septimanę dies circumierit.

<LIIII.> DE PASCHA

Etymology and names of Easter (*pascha*)

Pasche uocabulum apud *Grec*os *a passione* sumitur: Pasci per chi f*ati dicitur. Sed* uerius *Hebreo uerbo* hic *pascha a transitu* dicitur, *qu*ia *tunc populus Dei ex Aegypto transi*u*it.* In figuram: populi Christiani transituri de peccatis per transitum Christi. *Unde* Iohannes a*it*: *Cum uidisset Ihesus*, *quia uenit hora* eius, *ut transiret de* hoc *mundo ad patrem*.

---

14 Greci] Grecos *M.* | Sabinorum] Salinorum *M (corr. from* HARVARD FRAGMENT*).* **15** XII[2]] XI *M (corr. according to* MC*).* **19** annis] anni *M (corr. from* HARVARD FRAGMENT*).* | usque] quodum *M (corr. from* HARVARD FRAGMENT*).*

12–15 Quid…ordinauit] For the Greeks having invented the lunar year of 12 months cf. SAT. 1.13.1,9 (cited in COMP. COL. 2.2A): *Sed secutus Numa, quantum sub caelo rudi et saeculo adhuc impolito solo ingenio magistro conprehendere potuit, vel quia Graecorum observatione forsan instructus est, quinquaginta dies addidit, ut in trecentos quinquaginta et quattuor dies, quibus duodecim lunae cursus confici credidit, annus extenderetur. ... nam et Graeci cum animadverterent temere se trecentis quinquaginta quattuor diebus ordinasse annum*. Cf. c. 12, ll. 35–37, where the invention of the 354-day lunar year consisting of 12 months is credited to Numa. For Gignus having divided the solar year into months cf. DNR 4.5.32–33 (almost all MSS read *Cingum* instead of *Sancum*, MS L *Cign-*; cited in DRC 28.9–10 which reads *Gingium*, and *Cignus* in the margin of MS V): *Plerique autem adserunt Sancum Sabinorum regem prius annum in menses diuisisse*. In c. 12, ll. 20–23, c. 53, ll. 2–4 it is also argued that the solar months were modelled on the lunar months; for Gignus being the inventor of the bissextile day see c. 41, ll. 17–20. **16–20** Ubi…circumierit] As is the case with the previous passage (and, in terms of direct sources, the whole chapter) there is no apparent source for this passage, which indicates that the entire chapter appears to have been MC's own composition. For the 19-year lunar cycle cf. c. 1, l. 14; c. 60; for the 19 Easter full moons cf. c. 55, ll. 62–69; for the lunar increment that completes the *saltus* after 19 years cf. c. 62, ll. 13–61; for the 28-year solar cycle cf. c. 1, ll. 14–15; c. 65. **LIIII,2–8** Pasche…patrem] The argument that the term *pascha* may have had either a Greek or a Hebrew origin, the latter being the one to be prefer, is taken from ETYM. 6.17.11 (MS K reads *hora eius ut*; cited in BC 140(1353)B4–10; up to *transierit* cited in CE p. 85, DRC 84.1–4): *Paschae autem vocabulum non Graecum, sed Hebraeum est; nec a passione, quoniam πάσχειν Graece dicitur pati, sed a transitu Hebraeo verbo pascha appellata est, eo quod tunc populus Dei ex Aegypto transierit. Vnde et in Evangelio: Cum vidisset, inquit, Iesus quia venit hora ut transiret de mundo ad patrem*. Similar passages occur frequently in Augustine's writings. Cf. especially AUGUSTINE, *Epistulae, Epistula 55* (*CCSL* 31, p. 235), *Sermones, Sermo 155* (*PL* 38, col. 843). The Biblical citation is from JOHN 13:1, which MC probably consulted when citing Isidore (because of the explicit reference to John and the addition of *eius* and *hoc*): *ante diem autem festum paschae sciens Iesus quia venit eius hora ut transeat ex hoc mundo ad Patrem cum dilexisset suos qui erant in mundo in finem dilexit eos*. The reference to John may also derive from EPISTOLA PASCASINI 2 (Krusch, *Studien* I, p. 249; cited in DRC 84.7–11):

What measured the lunar year ? The course of the moon, i.e. the twelve lunations. Who was the first to arrange the lunar year in twelve lunations? The Greeks. Gignus, King of the Sabines, however, arranged the solar year in twelve month on the basis of the twelve lunations.

Whereby (was) the lunar cycle of 19 years fixed? By whatever measured that number: *luna* 14 of the first month (i.e. the Easter full moon) and the completion of the *saltus*; just as the bissextile day rules the course of the sun (i.e. the solar cycle is not completed) before the bissextile day will have gone through all days of the week in 28 years.

54. ABOUT EASTER

The word Easter (*Pasche*) was derived from (the word for) suffering by the Greeks: 'to suffer' is called *pasci*, with (the syllable) *chi*. Yet, Easter is more truly so called from the Hebrew word for departure (*transitu*), since at that time God's people departed (*transiuit*) from Egypt. In allegory: the Christian people will depart (*transituri*) from (their) sins through the departure (*transitum*) of Christ. Whence John says: At that time Jesus had realized that his hour was come that he should depart out of this world unto the Father.

---

dies passionis ..., a qua, sicut Grecis videtur, pascha nomen accepit, licet Hebreae linguae interpretes tradunt, pascha transitum dici: Quod tamen id esse et Johannis evangelistae dicto firmatur, quo ait, cum de passione loqueretur: Cum venisset hora, ut transiret Jesus de hoc mundo ad patrem. Cf. also AUGUSTINE, *Enarrationes in psalmos, Enarratio in psalmum 68, Sermo 1* (*CCSL* 39, p. 902), *De Trinitate* 2.17 (*CCSL* 50, p. 120). The connection between the Pasch and the change from sins to justice is made in MAXIMUS OF TOURS, *Sermones, Sermo 54* (*CCSL* 23, p. 218): *Pascha enim hebraeice latine transitus dicitur uel profectus, scilicet quia per hoc mysterium de peioribus ad meliora transitur. Bonus igitur transitus est transire de peccatis ad iustitiam.*

12–15 Quid…ordinauit] A copy of this passage can only be found in the HARVARD FRAGMENT (Harvard, Houghton Library, MS Type 613, fol. 7v), which is to be restored according to MC: *<Scien>dum, quid mensurat annum lunae. Id est lu\n/ae cur<sus, id est> XII lunae. Quis primus ordinauit annum in XII <lunas? Greci. An>num autem solis Gingius rex Sabinorum in XII menses <sub XII lun>as ordinauit.* For Gignus having divided the solar year into months cf. DIAL. LANGOB. 13A: *Interrogatio: Quis primus divisit annum in menses? Responsio: Apud antiquos Latinos Rucingus rex Sabinorum primus divisit annum in decem menses.* **16–20** Ubi…circumierit] Again, as in the case of the previous passage, a copy of this passage here can only be found in the HARVARD FRAGMENT (Harvard, Houghton Library, MS Type 613, fol. 7v), which is to be restored according to MC: *Sciendum est, ubi cursus certus <lunae XVII>II annorum. Quid mensurauit hunc numerum: lunę XIIII pri<mi mensis> et plenitudo saltus; ut bissextus dominatur cur<sum solis: XX>VIII annis usque bissextus omnes septimanę <dies circum>ierit.* Cf. also the related passage in DRC 105, with the phrasing revealing that it is certainly based on MC: *Sciendum nobis quid constituit hunc numerum XVIIII annorum in lunari et in decinnouenali ciclo. Saltus uidelicet. Sicut bissextus XXVIII annorum ciclum solis mensurauit, omnes septimane dies in hoc spatio circumiens, sic saltus cursum lunaris XVIIII praebuit. Nisi enim esset saltus in finé, XX anno non esset reuersio. Cum uero saltus <in> XVIIII reuersio in epactis et XIIII luna primi mensis inuenitur, et in fine XVIIII annorum integra preparatio saltus inuenitur.* **LIIII,2–8** Pasche…patrem] The discussion of the etymology of *pascha* on the basis of Isidore's account appears to be a characteristic feature of seventh- and eighth-century Irish computistics (cf. the texts citing ETYM. in *app. font.*) that did not find imitators on the Continent or elsewhere.

12–20 Quid…circumierit] Cf. p. CI, CLXXIII, CLXXVI CLXXXIX, CXCV of the introduction.

Develop. of its pronoun-ciation

Phase in Hebreo, quia Hebrei non habent p; <Petrus et Paulus> Fetrus et Faulus dicuntur. | Sed Grecus posuit fol. 31r pro tanta lenitate uocis suae et extremam syllabam per chi pro eadem tanta lenitate scripsit. Pascha uox uulgi. Inde commune est apud Grecos et Latinos.

Pascha is not inflected

Hoc nomen non declinatur, ut alia nomina dierum festorum apud gentiles, ut Saturnalia, Minerualia, et reliqua. Sed apud ueteres prime declinationis erat.

LIIII,9 p] in *M (corr. from* DRC*)*. **9–10** Petrus…Paulus] *om. M (add. according to* MC, DRC, BEDE*)*. **11** tanta] tantae *M*. **12** tanta] tante *M*.

9–13 Phase…Latinos] The fact that *pascha* is called *phase* in Hebrew may derive from any of the following texts: JEROME, *Commentarii in Esaiam* 10.31 (*CCSL* 73, p. 403): *Pro transeunte, in Hebraico phase. In tribus, praeter LXX, interpretibus ὑπερβαίνον ponitur. Ex quo manifestum est, Pascha, hoc est phase Domini, non passionem significare, sed transitum*. JEROME, *Commentarii in Matthaeum* 4 (*CCSL* 77, p. 245): *Pascha quod hebraice dicitur phase, non a passione ut plerique arbitrantur sed a transitu nominatur*. LEO THE GREAT, *Tractatus septem et nonaginta* 72.6 (*CCSL* 138A, p. 447): *Propter quod ipsa festiuitas quae a nobis Pascha nominatur, apud Hebraeos Phase, id est transitus, dicitur, adtestante euangelista atque dicente: Ante diem autem festum Paschae sciens Iesus quia uenit hora eius, ut transiret de mundo ad Patrem*. For the absence of p from the Hebrew alphabet and its replacement by Greek φ, Latin ph, f in Hebrew words cf. JEROME, *Liber interpretationis hebraicorum nominum* (*CCSL* 72, p. 141): *Sed sciendum est quod apud Hebraeos p littera non habetur nec ullum nomen est quod hoc elementum sonet. Abusiue igitur accipienda quasi per f litteram scripta sint*. JEROME, *Commentarii in Hiezechielem* 8.25 (*CCSL* 75, p. 344–3): *non generali nomine omnium gentium quae non sunt de genere Iudaeorum, sed speciali gentis suae quae nunc dicitur Palaestina, mutata 'phe' littera, iuxta Graecorum consuetudinem, in 'pi', sicut apud nos pro hebraico 'phascha' graecum et latinum pascha celebratur*. JEROME, *Commentarii in Esaiam* 1.2 (*CCSL* 73, p. 31; cited in ETYM. 9.2.58): *quae est hodie gens Palaestinorum, quasi Philisthinorum, quia p litteram sermo Hebraicus non habet; sed pro ea phi Graeco utitur*. JEROME, *Commentarii in Danielem* 4.11 (*CCSL* 75A, p. 935): *notandum autem quod cum 'pe' littera hebraeus sermo non habeat, sed pro ipsa utatur 'phe' cuius uim graecum Φ sonat – in isto tantum loco apud Hebraeos scribatur quidem 'phe' sed legatur 'pe'*. **14–16** Hoc…erat] A text like POMPEIUS (MAURUS), *Commentum artis Donati* (Keil, *Grammatici Latini* 5, p. 177), who discusses the grammatical number of *Pascha* in the context of other Roman feast-days, including the *Saturnalia* and *Minervalia*, may have been MC's source here: *quaerebatur Pascha cuius esset numeri. dies festus est. omnia nomina dierum festorum numeri sunt tantum pluralis, Vulcanalia Compitalia. dicebat ille qui obiciebat etiam hoc numeri esse tantum pluralis. sed sunt causae quae repugnant: primo, quod illa nomina in ia exeunt, Vulcanalia Saturnalia Compitalia, et habent principale suum, unde oriantur, Vulcanal Vulcanalia, Minerval Minervalia, Compital Compitalia, habent principale suum, unde oriantur; hoc non habet, nec ita exit in ia, sed in a. deinde hoc nomen latinum est, a latinitate descendit; illud vero graecum est. et novimus nomina graeca, quae ita exeunt apud Graecos, puta colyma colymata, pegma pegmata, stemma stemmata, ista neutralia quae sunt et sic exeunt, cum coeperint transire in numerum pluralem, necesse habent ut in ta exeant. quo modo ergo vis esse hoc? ut graecam sequaris rationem, aut ut latinam? si graecam vis sequamur rationem, non habet numerum pluralem in ta exeuntem; si latinam, non habet ia. unde constat non esse numeri pluralis. ergo sunt nomina numeri tantum pluralis, ut Minervalia Vulcanalia et similia.*

9–12 Phase…scripsit] The discussion of grammatical features in computistical textbooks generally is a characteristic feature of seventh- and eighth-century Irish computistics, and this phenomenon certainly would deserves a separate study; it may well be that teachers of computistics also taught Latin grammar in early medieval Irish monastic schools. A similar account on the pronounciation of *phase* and *pascha*

(*Pascha* is known as) *Phase* in Hebrew, since the Hebrews do not have the letter p; Peter and Paul are called Feter and Faul. The Greeks, on the other hand, employed (the letter p) because of the great softness of their pronounciation and they expressed the final syllable with *chi* in writing because of this great softness. Pascha (is the word used by) the ordinary people. Therefore, it is common (both) to the Greeks and the Latins.

This term is not inflected, like other names of festive days among the heathens, such as Saturnalia, Minervalia, and so on. Among the ancients, however, it belonged to the first declension.

---

can, in computistical literature, only be found in DRC 84.13–17, which in this was certainly influenced by MC: *Sed hoc nomen, quod fasse dicitur apud Ebreos, leuius sonatur, quod Ebrei 'p' litteram non habent, sed 'fhe' huius littere nobiscum uicem tenet 'p'. Et quaecumque nomina apud nos per 'p' scribuntur, apud Ebreos per 'fhe', ut Falistina, Faulus, Fasse, quod Greci ita dicunt: Palestina, Paulus, Pascha*. Note that the example of Peter and Paul in terms of pronounciation is explicitly referred to in BEDE, *Retractatio in actus apostolorum* 9 (*CCSL* 121, p. 138): *neque enim putandum est Paulum siue Petrum Hebraea et non potius Latina uel Graeca esse uocabula, cum constet Hebraeos P litteram omnimodis non habere*. BEDE, *In Lucae euangelium* 2.6 (*CCSL* 120, p. 133): *Violenter ergo quidam Latino uel Graeco nomini Hebraeam quaerentes aethimologiam dicunt Petrum dissoluentem siue disculciantem uel agnoscentem interpretari cum et expositio Iohannis euangelistae cuius memini et ipsa lingua Hebraea quae P litteram omnino non sonat Hebraeum hoc nomen non esse testetur. Abusiue enim Fetrum pro Petro sicut et Faulum et Filatum pro Pilato et Paulo scribentes uiolenter ficto nomini falsam interpretationem subnectunt*. Bede also describes this phenomenon in *Retractatio in actus apostolorum* 16 (*CCSL* 121, p. 151), *In Ezram et Neemiam* 2 (*CCSL* 119A, p. 320). **14–16** Hoc...erat] The declension of *pascha* is, to my knwoledge, not discussed in any other computistical text; cf., however, the following early ninth-century grammatical tracts: ALCUIN, *Orthographia* (Keil, *Grammatici Latini* 7, p. 306): *Pascha et paschae unum est*. ARS ANONYMA BERNENSIS (Keil, *Grammatici Latini* 8, p. 96): *Neutra nomina primae declinationis in a desinentia sic declinantur: hoc Pascha huius Paschae huic Paschae hoc Pascha o Pascha ab hoc Pascha: sic declinatur hoc manna hoc † iliaria hoc toria, quia pluralem numerum non habent*.

LIIII,9–13 Phase...Latinos] Cf. p. CXCV–CXCVI of the introduction.

<LV.> *DE PRIMI MENSIS REGULIS*

The first month in different traditions

Cauendum est nobis, ne in primi mensis regulis fallamur. Quem Dominus Iudaeis commendauit dicens: *Obserua*bitis *mensem nou*o*rum fruguum* hoc *uerni temp*us,

---

LV,1 De…regulis] *not indicated as a heading M (cf.* CE*).*

LV,1 De…regulis] A chapter with many of the items following in MC here is headed *DE REGULIS PRIMI MENSIS* in CE p. 111, so that it appears that this phrase was originally a heading in MC. **2–12** Cauendum…mensis] Having defined the term Easter in the previous chapter, MC proceeds to outline general rules for the determination of this most important Christian feast. At the beginning of his discussion, MC lists authoritative texts dealing with the question of which Julian calendar period is supposed to be regarded as the Easter month, i.e. the period in which Easter Sunday could lawfully fall. MC sets out with a Biblical citation from DEUT. 16:1 (cited in PROLOGUS DIONYSII (Krusch, *Studien* II, p. 65); only the first part in DTR 35.56–57), 23:4: *observa mensem novarum frugum et verni primum temporis ut facias phase Domino Deo tuo quoniam in isto mense eduxit te Dominus Deus tuus de Aegypto nocte. ... quia noluerunt vobis occurrere cum pane et aqua in via quando egressi estis de Aegypto et quia conduxerunt contra te Balaam filium Beor de Mesopotamiam Syriae ut malediceret tibi.* Note that in earlier computistical literature, DEUT. 16:1 is usually combined with NUM 9:3: PROLOGUS THEOPHILI 1 (Krusch, *Studien* I, p. 222; cited in COMP. COL. 5.11A, which reads *observa* and adds *frugum*): *Serva mense novorum et facies pascha domino tuo quarto decimo die mensis primi.* EPISTOLA PROTERII 2 (Krusch, *Studien* I, p. 271; cited in DRC 85.11–15, which reads *nouarum frugum*): *Olim quidem dominus per Moysen tempus paschale significavit dicens: Custodi mensem novorum, primum hunc esse pronuntians. Sicut iterum dicit: Mensis iste vobis inicium mensium, primus erit in mensibus anni. Et facies pascha domino deo tuo XIIII die mensis primi.* For the Exodus happening in March cf. c. 44, ll. 37–54. The most prominent (and most controversial; cf. *app. comp.*) passage on the different traditions concerning the Easter month is cited next, DRP 2.36–39: *Est ergo in primo anno initium primi mensis, quod est X et VIIII annorum circuli principium, secundum Aegyptios quidem mensis Famenoth XXVI die, iuxta Macedones uero Dystri mensis XXII, secundum Romanos uero Martii mensis XXV die, id est VIII Kal. Aprilis*. For the correspondence of the months mentioned in DRP with March cf. POLEMIUS SILVIUS, *Fasti* (Mommsen, *Corpus inscriptionum latinarum* 1, p. 339; cited in DE ANNO): *Martius ... vocatur apud Hebraeos Nisan, apud Aegyptios Faminod, apud Atheniensis Antesterion, apud Graecos Distros*. But note the difference in the Hebrew month; MC may have taken the correspondence between Roman March and Hebrew *Adar* from the table in DRP 10, which parallels Hebrew and Julian calendar months (though *Adar* is equated with March only in MS A; cf. the facsimile in Mc Carthy & Breen, *De ratione paschali*, p. 59). Most probably, however, MC took the identification of the mentioned months with March from the now lost Irish *De divisionibus temporum*: cf. DDT 17C13–15: *De mense Martio: Adar in Hebraeo, Phamenoth apud Aegytios, Distros apud Macedones, Martios in Graeco, Martius in Latino*. Isidore (ETYM. 5.33.6, cited in COMP. COL. 5.11 with the specification *frugum*; PV §108) equates the *mensis nouorum fruguum* with March: *Idem appellatur et mensis novorum, quia anni initium mensis est Martius*. Note also that March is regarded as the first month in PROLOGUS PASCHAE 8 (Krusch, *Studien* I, p. 232): *in mense primo, hoc est marcio*.

LV,2–12 Cauendum…mensis] Interestingly enough, the same (and very uncommon) combination of DEUT 16:1 with 23:4 as given at the beginning of this passage can also be found in ECGBERT OF YORK, *De institutione catholica dialogus* (*PL* 89, col. 441A–B): *Et iterum Dominus ad Moysen: Observate mensem novarum frugum, quando egressi estis de terra Aegypti, legitimum in generationibus vestris*. As for the passage cited from DRP (cf. *app. font.*) that was so differently interpreted by early medieval computists, MC, by equating the 26th day of Phamenoth with 26 March, suggests here that the Egyptian months started on the Calends as Roman months did; a thorough study of Isidore (DNR 4.7.50; cf. c. 24, ll. 12–14), however, would have made MC aware that the Egyptian equivalent to March started on 25 February; moreover, the Egyptian date in question is explicitly translated to 22 March in VICTOR OF CAPUA's writings (transmitted through EW 8.11–13), the preceding date (the 25th day of Phamenoth) to

55. ABOUT THE RULES OF THE FIRST MONTH

Concerning the rules of the first month (i.e. the Easter month): We have to be cautious, so that we are not deceived in the rules of the first month. The Lord recommended the following to the Jews, saying: You will observe as the month of

---

21 March in EPISTOLA PROTERII 8 (Krusch, *Studien* I, p. 276–7; partially cited in EW 1.13–17). This is taken into account in DRC's (91.10–15) interpretation of this Anatolian passage, where the 26th day of Phamenoth = 22nd day of Dystrios = 11th calends of April (22 March) which constitutes the beginning of the first month among the Egyptians and Greeks, while the Roman beginning of the first month is placed on the 8th calends of April (25 March): *Alii uero, ut Greci et Aegytii, hunc primum mensem secundum solem a XI Kl incipere putant, Latini uero ab VIII Kl Aprelis incipere certant, ut Anatolius dicit: Est ergo in primo anno initium primi mensis, quod est circuli X et VIIII annorum principium, secundum Aegyptios quidem mensis Faminoth XXVI dies, iuxta Macedones uero Distri mensis XXII die, secundum Romanos uero Martii mensis XXV die, id est VIII Kl Aprelis*. Bede interpretes this passage in a way that the given Egyptian, Greek, and Roman dates are supposed to refer to the same Julian calendar date, namely 22 March, following a tradition of DRP in which the Roman date is given as the 11th calends of April (rather than the 8th calends as in the Irish texts). DTR 14.24–37: *est ergo in primo anno initium primi mensis, quod est decem et novem annorum circuli principium, secundum aegyptios quidem mensis phamenoth uicesimo sexto die, secundum macedones distri mensis uicesimo secundo, secundum romanos uero undecimo kl. apriles. Et hic enim vicesimam secundam diem distri mensis aeque vicesimam secundam martii fore commendans, indicat manifeste quia simul uterque mensis initium sumit. Et ne quis dicat quod Anatolius in hac sententia non scripserit undecimo kalendas sed octavo kl. apriles, convicet hoc non ita esse mensis aegyptiorum phamenoth, cuius vicesima sexta dies non octava kalendarum, sed undecima kl. aprilium dies est. Utrique autem, id est et qui octava kalendas et qui undecima kalendas apriles in Anatolio legunt, vicesimam sextam diem aegyptii mensis in eadem sententia habent annotatam, quae absque ulla dubietate in undecimo kl. apriles devenire probatur, iuxta quod superius eorum annalem describentes signavimus*. EW 11.13–22: *Est ergo in primo anno initium primi mensis, quod est XVIIII annorum circuli principium secundum aegyptios quidem mensis famenoth XXVI die, iuxta macedones vero distri mensis XXII, secundum romanos vero martii mensis XXV die, id est octavo kal. aprilis; cum XXVI dies mensis aegyptiorum famenoth et XXII dies mensis macedonum distri non sit VIII kal. aprilium, sed XI kal. aprilium. Nonne verisimillium apparet quam sit infalsata sententia ubi ipse qui VIII kal. aprilium pro XI kal. mutavit oblitus est etiam aegyptiorum vel graecorum mensium statum mutare, sed his quae minus nota erant manere permissis ut fuerunt, illud solummodo quod sibi suisque erat visibile mutavit?* The opinion challenged by Bede, namely that all dates mentioned in this Anatolian (?) passage refer to 25 March, is expressed in DE BISSEXTO I 997A6–B1: *Secundum Anatholium Alexandrinum episcopum Laudatiae iste quadrans in VIII Kal. Aprili positus invenitur, quia illa die, ut tradunt auctores, sol et luna duo luminaria facta sunt, propterea VIII Kal. Aprilis initium anni, et initium cycli decennovenalis secundum Anatholium dicitur: Est ergo in primo anno initium primi mensis, quod est XVIIII annorum circuli principium, secundum Aegyptios quidem mensis Famenoth XXVI die, secundum Macedones vero XXV die, id est, VIII Kal. Aprilis*. The same correspondence of the respective months with March can be found in the lists given in DRC 36, 40, 41 (without Roman equivalents for the Macedonian months), BC 151 (the correlation with the Macedonian months starts with November instead of December, so that *Disthor* is given as the equivalent to February), DTR 11.6–24 (cited in LIB. CALC. 69A; without Roman equivalents in RM 29.7–12), 73–80 (cited in LIB. CALC. 69C; without Roman equivalents in RM 30.7–12), 14.13–17 (cited in LIB. CALC. 72; RM 31.12–17); QUAEST. LANGOB. 4A only lists the Hebrew months, but equates March with *Nisan*; note that in all lists of Hebrew months *Adar* is given as the twelfth, not the first month, in accordance with Esther 3:7,13, 8:12, 9:1. For the beginning of the Hebrew, Egyptian, Macedonian, and Roman year cf. also c. 24, ll. 2–11, 19–26.

quando egressi estis de Aegypto. Primus mensis secundum Hebreos *Adar*, id est *Martius*; *secundum Aegyptios* primus *mensis Faminoth*, id est Martii, *XXVI die*, id est VII Kalendas Aprilis; secundum *uero Macedonas Distri mensis*, id est Martii, *XXII* die, id est XI Kalendas Aprilis; *secundum* autem *Romanos Martii mensis XXV die*, *id est VIII Kalendas Aprilis*. Inde apparet cum omnibus Martius primus esse mensis.

Theophilus on the Easter limits

Theophilus de primo mense: *Omnes XXX dies a domino consecrati sunt*. *Episcopi* respond*erunt principium mundi esse aequinoctium* XII *Kalendas Aprilis*. *Et ab* XI *Kalendas Aprilis in* XI *Kalendas Maii* hos dies *legimus consecratos*. *Theophilus* huic *sacrament*o diuino III menses consecrauit: Mensis agni ab VIII Kalendas Aprilis in VIII Kalendas Maii; mensis passionis ab VIII Kalendas

13–27 Theophilus…XXIII] After outlining his own version of DRP's passage on the beginning of the Easter month according to different traditions, MC here wants to define the entire period in which Easter Sunday could fall, a period that obviously was easily deducable from the Dionysiac Easter table (Krusch, *Studien* II, p. 69–74) or any of its continuations; yet, none of the texts considered as authoritative by MC explicitly mention this period, so that MC had to combined various texts to get the desired result (22 March to 25 April; to MC, it apparently did not suffice to simply cite CE p. 111). At first, DE ORDINATIONE FERIARUM PASCHALIUM (*PL* 90, col. 608D–610A) is cited, which refers to a period of exactly 31 days (a month) for Easter Sunday, 22 March to 21 April: *Omnes ergo triginta dies a Domino consecrati sunt. Episcopi dixerunt: Jam superius dedimus responsum, principium mundi esse aequinoctium octavo Kalendarum Aprilium. Et ab octavo Kalendarum Aprilium usque in octavum Kalendas Maias, legimus esse consecratos. Theophilus dixit: Ecce impium non est ut passio Dominica, tantum sacramenti mysterium, extra limitem excludatur? Passus namque Dominus ab undecimo Kalendarum Aprilium, qua nocte a Judaeis traditus est, et a septimo Kalendarum resurrexit, quomodo ergo tres dies extra terminum excluduntur? Omnes episcopi dixerunt: Nulla ratione fieri debet, ut tantum sacramentum extra limitem excludatur, sed hi tres dies intra terminum inducantur, et de subter retrahere constitutum est. Ergo in illa synodo ab undecimo Kalendarum Aprilium usque in undecimo Kalendarum Maiarum, Pascha debent servare, et nec ante nec postea cuicunque constitutum limitem transgredi liceat. Similiter et de luna praeceptum divinum teneatur. Mandatum est per Moysen: Sit vobis observatum a decima quarta luna usque XXI. Has ergo septem lunas similiter in Pascha tenendo constat fuisse, consecrata sunt. Quando ergo fit intra suum limitem ab undecimo Kalendarum aprilium usque in undecimo Kalendarum Maiarum, dies dominicus, et luna ex illis octava sanctificata, Pascha nobis visum est celebrare*. This period is also outlined by LEO THE GREAT, *Epistola ad Marcium imperatorem* 2 (Krusch, *Studien* I, p. 259), MARTIN OF BRAGA, *De Pascha* 7 (Barlow, *Opera omnia*, p. 273; cited in TRACTATUS ADTHANASI 6 (Krusch, *Studien* I, p. 334)), and condemned in DRP 12. The addition of four days to the end of this period *propter angustiam temporis inminentem*, so that it would extend to 25 April, is implicit in EPISTOLA PROTERII 4–5 (Krusch, *Studien* I, p. 273–4; the first sentence cited in DRC 94.28–30): *et dominicum pascha ... VII kalendas maias, constat esse celebratum propter angustiam temporis inminentem. Item ... dominicum pascha VIIII kl mai, tunc nos caelebrasse meminimus. Necesse est igitur ... VIII kl mai, dominicum caelebremus pascha propter adprehendentem rursus angustiam*. and directly connected to Theophilus (and explicitily referring to the embolism as the reason for this extension) in PROLOGUS CYRILLI 4–5 (Krusch, *Studien* I, p. 339–40): *Quae quidem luna efficitur XIIII a XII kalendarum aprilium die usque in XIIII kalendarum maiarum. Et propterea hos V dies protendere minime dubitavi, quia ipsa luna in VII kalendarum maiarum protenditur, infra primi mensis terminos praenoscitur et impletur et secundum supputationem Hebraeorum, quibus pascha in praefigurationem Christi praeceptum est, post annum embolismum occurrit, qui XIII menses lunares, id*

new corn that time of spring when you came out of Egypt. According to the Hebrews, the first month is *Adar*, i.e. March; according to the Aegyptians, the first month (begins) on the 26th day of *Faminoth*, i.e. of March, that is on the seventh calends of April; according to the Macedonians, however, (it begins) on the 22nd day of the month of *Dystrios*, i.e. of March, that is on the eleventh calends of April; according to the Romans, however, (the first month) begins on the 25th day of March, that is on the eighth calends of April. Accordingly, it appears that March is the first month according to all people.

Theophilus (has the following to say about) about the first month: All 30 days were consecrated by the Lord. The bishops replied that the equinox of 21 March is the beginning of the world. And we read that those days from 22 March to 21 April are sacred. Theophilus consecrated (the following) three months to this divine sacrament: the month of the lamb, from 25 March to 24 April; the month of

---

est CCCLXXXIIII dies, habere. ... Pascha autem, quod ultra XII kl maiarum celebrare formidant, sanctus Theophilus usque VII k mai deferre non dubitat. For the lunar limits of Easter Sunday cf. c. 58, l. 3. The three different lunar limits mentioned here may have been a modification (and adaption to the Dionysiac reckoning) of similar statements in PROLOGUS PASCHAE 8 (Krusch, *Studien* I, p. 232–3): *Cuius observatia perenni voto per singulos dies vicibus congruis a nobis caelebranda, hoc ordine sive dispusitione, ut hii tres dies sacratismi, qui trinitatis gracia sanctificati sunt, passio, requies et resurrectio habeant singulas sibi adscriptas lunae ebdomadas: passio a XIIII luna usque ad XXmam, requies a XV luna, in qua azema, usuqe ad XXI, resurrectio, novi firmenti ingressio, a XVI luna usque ad XXII.* EPISTOLA CUMMIANI 80–85: *Unde orie<ntalis tota aecclesia> tres ebdomadas tribus sacratissimis solennitatibus Domini <nostri Iesu> Christi uenerabiliter, id est passioni, sepulturae, resurrectioni, deputau<it>: passioni a XIIII in XX; sepulturae a XV in XXI; resurrectioni a <X>VI in XXII lunam, septimanam pro reuerentia Dominici diei consecrans. Quia si XIIII luna resurrectioni deputetur, ut uos facitis, XIII in sepultura et XII in passione, prepostero ordine, fiet.*

13–27 Theophilus…XXIII] Very similar to MC, DRC 94.22–31 creates an authoritative account of the Julian calendar limits for Easter Sunday by combining information from ps-Morinus and Proterius. Interestingly enough, Bede and Rabanus Maurus like MC refer to the embolisms as the reason for this period extending over 35 days rather than a month of 31 days: DT 13.17–19: *Septimo dies dominici paschae reperiuntur ab XI kal. aprilium usque in VII kal. maiarum, ob rationem embolismorum licenter extenti.* RM 88.17–27 (RM's source, LIB. ANN. 38 or LIB. CALC. 57, is almost identical, but does not explicitly mention the Julian calendar limits for Easter Sunday, since there this theory is not only applied to the 35-day Dionysiac period mentioned by RM, but also to the 34-day Victorian one): *D. In quot diebus euenit nobis pascha celebrare? M. XXXIIII uel XXXV, id est, ab undecimo kalendis aprilis usque in septimo kalendis maii, et hoc in decennouenali ciclo fit propter concordiam epactarum. Isti autem XXXV dies sunt inde, quia usque in CCCLIIII dies anni cummunes progrediuntur, et anni embolismi usque in CCCLXXXIIII, quia in communibus annis ante CCCLIIII dies pascha non licet celebrare, supra uero licet. In embolismo autem ante CCCLXXXIII non decet celebrari, supra uero decet. Iste ergo numerus dierum in communibus annis et embolismis in celebratione paschae obseruandus est.* For the Julian calendar limits of Easter Sunday in the Dionysiac reckoning (22 March–25 April) cf. DRC 94.22–31, BC 8 (cited in COMP. COL. 6.8C), 12 (cited in LECT. COMP. 6.9C, LIB. ANN. 37, in turn cited in LIB. COMP. 4.18D, in turn cited in LIB. CALC. 55D), PROL. AQUIT. 1A, DTR 45.19–20 (cited in LIB. CALC. 85, PV §269), 51.69–70 (cited in LIB. COMP. 4.21), LECT. COMP. 6.1C, COMP. COL. 6.8B. For the lunar limits of Easter Sunday cf. c. 58, l. 3.

LV,5–6 Primus…Martius] Cf. Borst, *Schriften*, p. 521. **6–11** secundum…Aprilis] Transcribed and correlated with DRP in Mac Carthy, *Annals of Ulster* 4, p. lxix; cf. also p. XXII of the introduction. **11–12** Inde…mensis] Transcribed in Borst, *Schriften*, p. 480. **13–27** Theophilus…XXIII] Cf. Mac Carthy, *Annals of Ulster* 4, p. lxix and also p. LXV of the introduction.

Aprilis in VIII Kalendas Maii; mensis resurrectionis | *ab XI Kalendas Aprilis in XI Kalendas Maii* XXX dies. Huic mensi Theophilus IIII dies addidit *propter angustiam* embolismorum: X Kalendas, VIIII Kalendas, *VIII Kalendas*, *VII Kalendas Maii*. Inde *<quando> ab XI Kalendas Aprili*s *in* VII *Kalendas Maii dies dominicus et luna a* XV *usque XXI*, *pascha obserua*ndum est. Agno autem a XIIII in XX, passioni in XXI, resurrectioni in XXIII. fol. 31v

ps-Jerome on the Pasch

Sed *nunc* abicitur. Hieronimus ait: *Cum Dominus*, *uerus agnus*, et *uerum pascha progreditur*, in mundum *aliqu*a *permanere uolens custodiuit*, *ali*a *non seruar*e *cupiens mutauit*. Nam *XIIII* luna *umbral*em relinquens et *figural*em preuidens *ueritatem* per se expressam in dominicum diem cum luna transiuit.

Three rules for the understanding of Easter

*Tres regul*as *primi mensis*, *qu*ibus *pascha* intellegimus, esse innotescimus. Hoc: *prima* secundum *accension*em *lunae*, *secunda XIIII luna*, *tertia* secundum *dominic*os *die*s *resurrectionis*.

---

24 quando] *om. M (add. from* DE ORDINATIONE FERIARUM*)*. **25** dominicus] domincus *M*. **28** Hieronimus] Hieronymus *Ó Cróinín (contra M)*. **29** progreditur] progredituri *M (corr. from* DE SOLLEMNITATIBUS, BC, DRC*)*. **31** luna] lunam *Ó Cróinín (contra M)*. **32** expressam] et per suam *M, Ó Cróinín (corr. from* DRC*)*.

28–33 Sed…transiuit] Having finished the previous paragraph with a statement on the lunar limits of Easter Sunday, MC directly follows this up with a citation from DE SOLLEMNITATIBUS 2 (Walker, *Sancti Columbani opera*, p. 200–1; cited in BC 150(1362)A5–C15, the second part partially in turn cited in COMP. COL. 5.14B) to the extent that *luna* 14 was, no doubt, unlawful for the celebration of Easter Sunday, a quote directed against followers of the *latercus*, the old 84 (14)-year Easter reckoning: *Quas rationes ipse Dominus, verus agnus, cum ad verum Pascha progreditur, aliquas permanere volens custodivit, aliquas non servari cupiens commutavit. ... et sic etiam nunc a nobis credi debeat. Et hoc etiam intueri debemus, quod non in decima quarta die ad vesperum, ut lex praecepit, ille agnus Dei, qui tollit peccatum mundi, et Pascha nostrum immolatus est Christus, sed decima quinta die, in quo manifestum est diem festum Iudeorum cum suo sacrificio a Domino esse solutum. Sed quid in hoc intelligere debemus, quod prius figuralis agni carnes comedere et postea sui corporis cibo apostolus refecit, et post Iudeorum typicum nostrum Pascha immolatur est Christus? Hoc, ut opinor, ut non veritas figuram, sed figura veritatem praecedet, quia non prius quod spiritale, sed quod animale, deinde quod spiritale. Unde electa et amica sponsa Christi, universalis ecclesia, anatematizat eos, quia cum Iudeis in festivitate Paschali decimam quartam celebrari diffiniunt et sabbata et cetera huiusmodi umbralis observantiae*. Note that MC's and DRC's (cf. *app. comp.*) are the earliest known attributions of this tract to Jerome. This ascription can also be found in the *incipit* of this tract in Cologne, Dombibliothek, 83², fol. 201r, Geneva, Bibliothèque Publique et Universitaire, 50, fol. 121r, Vatican, Bibliotheca Apostolica, Vat. Lat. 642, fol. 89r; the text itself was edited among Jerome's and Columbanus' works, but the content suggests that this tract is of Irish or British origin of ca. AD 650–720 rather than composed by Columbanus or Jerome. For the abandonment of the shadows and the coming of the truth on *luna* 14 cf. also EPISTOLA CUMMIANI 126–129: *Pascha occiditur et uiuificatur. Pars autem diei festi, XIIII luna, non totus dies, in qua seducimur; sed in parte sabbato otioso, et neomenia bucinata, quae sunt umbra, non corpus, Christi. Umbra occiditur, ueritas uiuificatur.* **34–37** Tres…resurrectionis] The general idea of considering the Julian calendar dates for the Easter new and full moon as well as for Easter Sunday as the decisive features of Easter calculation is based on Victorius. In chapter 4 of his prologue, after an introductory statement that set the scene (PROLOGUS

the passion, from 25 March to 24 April; the month of the resurrection, from 22 March to 21 April, (that is) 30 days. To this (latter) month Theophilus added four days on account of the narrowness (*angustiam*) of the embolisms: 22, 23, 24, and 25 April. Therefore, whenever a Sunday (together with) a moon from (*luna*) 15 to the (*luna*) 21 occur between 22 March and 25 April, Easter is to be celebrated. According to the lamb, however, (the lunar limits are) from (*luna*) 14 to (*luna*) 20, according to the passion to (*luna*) 21, according to the resurrection to (*luna*) 23.

But (this view) is now abandoned. Jerome says: When the Lord, the true lamb, proceeded to the true Pasch, he kept some things that he wanted to remain in the world, (and) changed others that he did not wish to preserve. For, abandoning the shadowy truth on *luna* 14 and forseeing a figurative (truth) personified by himself, he departed with the moon on Sunday.

We learn that three rules belong to the first month, by means of which we understand Easter: the first refers to the kindling of the moon (i.e. the Easter new moon), the second to *luna* 14 (i.e. the Easter full moon), the third to the Sundays of the resurrection.

VICTORII 4 (Krusch, *Studien* II, p. 20): *Adcrescit praeter eos etiam aliud erroris genus, dum in regulis primi mensis, quo pascha dominicum celebrari statuunt, magna oritur utriusque contentio.*), Victorius continues to discuss the Julian calendar limits of the Easter new and full moon according to 'Greek' and 'Latin' tradition (cf. the following notes), and finally argues that Easter itself has to be celebrated on the Sunday following the Easter full moon. MC's direct source for this passage here, however, is CE p. 111: *Quot sunt regulę primi mensis? III sunt secundum Grecos: Prima regula accensionis lunę, secunda regula XIIIImae lunae, tertia regula diei dominice resurrectionis.* CE continues to list the Dionysiac Julian calendar limits for all three features mentioned. MC, on the other hand, is only concerned with the Easter new and full moon, since the limits for Easter Sunday are already discussed at length in the preceding passages.

28–33 Sed…transiuit] This passage is directly quoted in DRC 98.22–26 and no other text to my knowledge, once again clearly proving the interdependency between MC and DRC: *Item Hieronimus dicit: Cum dominus, uerus agnus, uerum pascha progreditur, in mundum aliqua permanere uolens custodiuit, aliqua non obseruare cupiens motauit. Nam XIIII lunam umbralem relinquens, et figuralem praeuidens ueritatem per sé expressam, in dominicam diem cum luna transiuit.* **34–37** Tres…resurrectionis] The reference to these three rules, ultimately based on Victorius' account (cf. *app. font.*), is a characteristic feature of late seventh-, early eighth-century Irish computistics, since it can also be found in DRC 94.16–20: *Et haec regula in diebus accensionis lunae primi mensis et in XIIII luna eiusdem et in diebus dominicis paschalibus per legalem conputationem et sécundum angeli ordinationem et secundum lapidem Sileniten et secundum miraculum putei, qui in nocte paschae in Meltina manat, usque ad diem iudicii apud Christianos obseruabitur.* but in no other text. The reason for this discussion not finding any reception outside of these Irish circles presumably lies in the fact that the calculation of Easter can better be defined by different rules (cf. c. 56, ll. 2–14).

28–33 Hieronimus…transiuit] Transcribed in Walsh & Ó Cróinín, *Cummian's letter*, p. 204; cf. Ó Cróinín, 'Computistical works', p. 52; idem, 'A seventh-century Irish computus', p. 118–9 and p. CXXXI, CXCIII of the introduction.

'Greek' and 'Latin' limits for Easter new and full moon

Quarum fines *Theofilus* statuit. Ita accensionis fines *ab VIII Idus Martii in Nonis Aprilis* tenti, et *XIIII luna*rum *a XII Kalendas Aprilis in XIIII Kalendas Maii* curiose *quȩri* ȩxposuit.

Apud *Latino*s uero *a III Nonas Martii in IIII Nonas Aprilis initium primi mensis censu*isse manifestauit, et *XIIII*

---

39 Nonis] VIIII k- *M (corr. from* PROLOGUS VICTORII*)*. **42** III] VI *M (corr. from* PROLOGUS VICTORII*)*.

38–41 Quarum…ȩxposuit] Concerning the Julian calendar limits of the Easter new and full moon, MC cites Victorius' discussion of them directly. First, MC deals with the Dionysiac ('Greek' or Alexandrian) limits, which were the ones to be followed: PROLOGUS VICTORII 4 (Krusch, *Studien* II, p. 20; cited in BC 9B10–C2 which gives *XV Kalendas Aprilis* and *XV Kalendas Maii* as the limits for the Easter full moon, in turn cited in COMP. COL. 6.7B, EPIST. RAT. 9B, LIB. ANN. 12, in turn cited in LIB. COMP. 4.17, all four transmitting the correct limits for the Easter full moon; partially cited with attribution to Victorius in DRC 90.2–4, 92.4–5): *Sanctae memoriae contra Theophilos quondam Alexandrinus antestis ad Theodosium imperatorem datis epistolis, ex primo ipsius et Gratiani quinto consulato, conditum paschale direxit, in quo ab VIII idus martias usque in diem nonarum aprilium, quolibet in medio eiusdem temporis spatio natam perhibet lunam, facere primi mensis exordium, a duodecimo vero kalendarum aprilium usque in quartum decimum kalendarum maiarum lunam quartam decimam solerter inquiri.* Cf. PROLOGUS DIONYSII (Krusch, *Studien* II, p. 65; cited in DRC 94.9–15): *Sed quia mensis hic, unde sumat exordium vel ubi terminetur, evidenter ibi non legitur, praefati venerabilis CCCXVIII pontifices antiqui moris observatiam et exinde a sancto Moyse traditam, sicut in septimo libro ecclesiasticae refertur historiae, sollertius investigantes, ab VIII Idus Martii usque in diem Nonarum Aprilis natam lunam facere dixerunt primi mensis exordium; et a XII kl. april. usque in XIIII Kalendas Mai lunam XIIII sollertius inquirendum.* Interestingly enough, MC preferred Victorius' account to the more concise CE p. 111: *Regula accensionis lunae ab VIIII Idus Martii usque in Nonis Aprilis; regula XIIIIrum lunarum a XII Kalendas Aprilis usque in XIIII Kalendas Maii.* Only the limits for the new moon are also given in ETYM. 6.17.20 (cited in BC 140(1354)A3–5): *Graeci primi mensis lunam ab VIII Id. Mart. usque in diem Non. Apr. observant.* Only the limits for the Easter full moon in c. 56, ll. 3–4 and PROLOGUS CYRILLI 4 (Krusch, *Studien* I, p. 339–40): *Quae quidem luna efficitur XIIII a XII kalendarum aprilium die usque in XIIII kalendarum maiarum.* **42–45** Apud…adseruit] For the 'Latin' Julian calendar limits, MC again cites PROLOGUS VICTORII 4 (Krusch, *Studien* II, p. 19–20; MSS VC read *XVI* instead of *XVII*; cited in BC 6 which reads *luna* and *XVI*, in turn cited in LIB. ANN. 11aA which also reads *XVI*; attributed to Victorius in DTR 51.4–18 which reads (save for MS *M*) *XVI*, in turn cited in LIB. COMP. 4.21): *Latini namque a tertio nonas martias usque ad in quarto nonas apriles, diebus scilicet viginti et novem observandum maxime censuerunt, ut quocumque eorum diae luna fuerit nata, et efficiat primi mensis initium. ... Quartas decimas porro lunas mensis eiusdem a XV kalendarum aprilium usque in XVII kal. mai. diae sancxerunt esse servandum.* Note that *XVII Kalendas Maii* as the upper limit for the Easter full moon corresponds to the given upper limit for the Easter new moon (*IIII Nonas Aprilis*), and therefore appears to have been the original reading; the corruption *XVI Kalendas Maii*, however, seems to have crept in at a very early stage of the transmission of Victorius' work, and was then repeated in almost all subsequent computistical texts citing this Victorian passage. Cf. Victorius' source, QUINTUS JULIUS HILARIANUS, *Expositum de die paschae et mensis* 1 (*PL* 13, col. 1106D–1107A; these rules are repeated in c. 2, 11 of that work): *Cur ergo hodie abs tertio nonarum Martiarum usque in quartum nonarum Aprilium diem luna prima a nobis quaeritur, seu ex quinto decimo kalendarum Aprilium in septimo decimo kalendarum Maiarum die, ut sit quarta decima, confirmatur.* Only the limits for the Easter new moon are given in ETYM. 6.17.19 (cited in BC 140(1353)D14–(1354)A2): *Latini namque a III Non. Mart. in III Non. Apr. primi mensis lunam inquirunt.*

38–41 Quarum…ȩxposuit] The Julian calendar limits of the Easter new and full moon were vital for the calculation of Easter (note, however, that one of the two would obviously have sufficed), so that it does not surprise to find them in almost every computistical text of the period, with some quoting Victorius

Theophilus set the limits of these. Thus he argued that the limits of the kindling (i.e. of the Easter new moon) extend from 8 March to 5 April, and that (the limits) of *luna* 14 (i.e. of the Easter full moon) are to be carefully examined from 21 March to 18 April.

He related, however, that, among the Latins, they determined the beginning of the first month (i.e. the Easter new moon) from 5 March to 2 April, and he

---

like MC (cf. *app. font.*), while most give their own accounts: BC 11C15–D4 (cited in LECT. COMP. 6.9aA, LIB. ANN. 33–34, in turn cited in LIB. COMP. 4.18A–B, in turn cited in LIB. CALC. 55A–B; RM 81.26–30): *Quaerenda est natiuitas lunae XIIII ab VIII Iduum Marciarum usque in Nonis Aprilis, que primi mensis nouorum initium ostendit; a XII uero Kalendas Aprilis usque in XIIII Kalendas Maii, in quacumque die XIIII luna occurrerit ipsa te ad caelebrationem sanctae paschae perducit.* DT 13.13–16: *Sextus XIIII lunas, qua veteres pascha faciebant, quae a duodecimo kal. aprilium usque in XIIII kal. maiarum vario discursu vagantes, tempus incensionis ab VIII iduum martiarum usuque in nonas aprilis accipiunt.* DTR 45.16–19 (cited in LIB. CALC. 85, PV §269), 51.66–69 (cited in LIB. COMP. 4.21): *Unde fit ut ab VIII id. martiarum usque in non. aprilium diem lunaris anni sint quaerenda primordia. Ipsa autem luna quarta decima, cum prium XII kal. aprilium cum novissime XIIII kal. maiarum occurendo, ... initium videlicet primi mensis ab octava iduum martiarum usque in nonarum aprilium diem; quartas decimas vero lunas paschae a duodecima kalendarum aprilium usque in XIIII kalendas maias.* Bede repeats the limits of the Easter full moon in DTR 59.5–8 (cited in LIB. CALC. 93, PV §252, RM 80.10–13). DIAL. NEUSTR. 10A, 11: *Graeci vero ab VIII Idus Martii usque in Nonas Aprilis limites incensionis lunae primi mensis servant. ... Porro Graeci a XII Kalendas Aprilis usque in XIIII Kalendas Maii <limites> praedictae lunae eligerunt.* LECT. COMP. 8.1aA (MS Bo has the correct *VIII* instead of *VII*, but later quite wrongly reads *XII* instead of *XIII*): *prima autem luna primi mensis ab VII Idus Martii usque in Nonas Aprilis variatur, sic etiam quarta decima luna paschalis inter viginti et novem dies, hoc est a XII Kalendas Aprilis usque ad XIII Kalendas Maii.* EPIST. RAT. 9A: *Sancta synodus, ubi trecenti octodecim pontifices, qui apud Nicaeam civitatem Bithiniae convenerunt, inlustrante Spiritu Sancto firmiter sancxerunt, ut ab VIII Idus Martii usque in diem Nonas Aprilis, quod sunt dies viginti novem, <qualiscumque> nata luna fuerit, dixerunt facere primi mensis initium. Et a XII Kalendas Aprilis usque in XIIII Kalendas Maii, diebus scilicet viginti novem, luna quarta decima paschalis festi solertius inquirendum.* Only the limits of the Easter new moon in BC 7 (cited in LIB. ANN. 11aB, EPIST. RAT. 9A (cf. above)); only the limits of the Easter full moon CAP. COMP. 7, 9. **42–45** Apud...adseruit] It is an interesting fact that the 'Latin' limits given by Victorius were copied right into the early ninth century (cf. *app. font.*), which may be explained by the assumption that the Victorian reckoning was still in use in some parts of the Frankish kingdom right to the end of the eighth century. Yet, the 'Latin' limits noted by Victorius are not his own, but those of his source, Quintus Julius Hilarianus (cf. *app. font.*); Victorius never outlined the limits he had followed for the construction of his table, but his table itself reveals, e.g., that the earliest Easter full moon in the Victorian reckoning was 20 March (as against Hilarianus' 18 March). DIAL. NEUSTR. 10, 11 is the only text that points out the difference between the limits ascribed to the Latins by Victorius and the ones applied by the Aquitanian computist himself: *Latini autem limites habent incensionis lune <a> III (MS IIII) Nonas Martii usque in IIII Nonas Aprilis. ... Victorius autem a Nonas Martii usque in IIII (MS III, recte II) Nonas Aprilis lunae primi mensis limites incensionis caute voluit habere. ... Limites quartarum decimarum lunarum primi mensis apud Latinos a XV Kalendas Aprilis usque in XVII Kalendas Maii servantur.* Cf. also COMP. COL. 6.7A: *Latini dicunt: A IIII Nonas Martii usque in IIII Nonas Aprilis, dies scilicet viginti novem, qualiscumque luna nata fuerit, perhibet facere initium primi mensis. Quartas decimas vero a XIIII Kalendas Aprilis usque in XV Kalendas Maii eiusdem mensis, id est primi mensis, adserunt esse servandas.* Only the 'Latin' limits of the Easter new moon are mentioned in DRC 90.11–15 (which may be the source for the wrong *IIII Nonas Martii* given in the Frankish texts above?): *Aliter Latini de regulis primi mensis sentiunt, et ideo a Grecis in diebus accensionis lunae primi mensis et <in> XIIII luna eiusdem primi mensis discripant, quattuor antecedentes inrationabiliter Grecis. Aput ipsos enim incipit a IIII Nonis Marcii et in IIII Nonis

lunas eius a XV Kalendas Aprilis in XVI Kalendas Maii ob*seruar*e adseruit.

Comp. of the Dionysiac with the Victorian Easter full moons

| Sciendum quod, quando endecas Latinorum contra ogdoadem Grecorum ponitur, omnes XIIII lunae mensis primi in libro Victorii incerte sunt, nisi III tamen, quia aut ante aequinoctium sunt aut XV Grecorum. Quando uero coniunguntur duo ogdoades, easdem XIIIImas lunas habent usque in IIII lunam in Kalendis Ianuarii. Ab hoc loco precedit XIIII luna Latinorum. fol. 32r

The Julian calendar dates of the Dionysiac Easter new moons

Dies incensionis primi mensis hii sunt, quos XVIIII dicimus: <X> Kalendas Aprilis, IIII Idus Martii, II Kalendas Aprilis, XIII Kalendas Aprilis, VII Idus Martii, V Kalendas Aprilis, XVI Kalendas Aprilis, Nonae Aprilis. Hucusque ogdoas. Nunc endecas: VIII Kalendas Aprilis, II Idus Martii, IIII Nonas Aprilis, XI Kalendas Aprilis, V Idus Martii, III Kalendas Aprilis, XIIII Kalendas Aprilis, VIII Idus Martii, VI Kalendas Aprilis, XVII Kalendas Aprilis, II Nonas Aprilis.

---

44 lunas] luna *M (corr. from* PROLOGUS VICTORII, DTR, LIB. ANN.*; but note that* BC *also reads* luna*)*. **46** endecas] endecadem *M.* **47** omnes] omnis *M.* **48** incerte] inc&ete *M.* **49** XV] XIII *M (what may have been meant here is that* lunae *13 in the Victorian reckoning correspond to the Greek full moons)*. **53** XVIIII] XVIII *M**. **54** X] *om. M.* | IIII] V *M.* **56** XVI] VI *M.* **57** ogdoas] ogdoades *M.* | endecas] endecades *M.* | Kalendas] id- *M.*

46–52 Sciendum…Latinorum] Both methods of comparing the Dionysiac with the Victorian 19-year cycle are already discussed in c. 49, ll. 17–30 and c. 52. Here, MC is only concerned with a comparison of the Easter full moons. DE COMPARATIONE EPACTARUM 1.13–36 (see Appendix 2) uniquely records a table (a facsimile can be found in Appendix 3) in which the Julian calendar dates of these are compared by paralleling the Dionysiac *ogdoas* with the Victorian *hendecas*. In two years both columns list the same date (MC argues for three years, which is correct for the epacts of 1 January, but not the Easter full moons, since the Dionysiac *saltus* is applied just before the Easter lunation, i.e. on 22 March; this shows that MC worked from the epacts, but did not execute the comparison of the Easter full moons year by year himself); in 16 years the Dionysiac full moons occur a day earlier than the Victorian ones, so that the Victorian full moons correspond to *luna* 15 of Dionysius; in one year a discrepancy of one month occurs between the two full moons, since Victorius placed the Easter full moon on 20 March, i.e. before the vernal equinox, as MC relates; this year is already discussed in detail in c. 52, ll. 49–60. In the historically correct comparison (in which the *ogdoades* of both reckonings as well as their *hendecades* are paralleled), the Easter full moons of both reckonings agree from the beginning of the *ogdoas* up to the sixth year (the year with *luna* 4 on 1 January, as MC states); the Victorian *saltus* then causes the disagreement in the remaining 13 years, in which the Victorian Easter full moons occur a day earlier than the Dionysiac ones. **53–61** Dies…Aprilis] It is entirely possible that MC worked from a Dionysiac 19-year table like the one described by Cummian, which consisted of three columns, namely the epacts of 1 January and the Easter new and full moons: EPISTOLA CUMMIANI 215–220: *Decimo trecentorum decem et octo episcorum decennouennalem cyclum, qui Grece enneacedeciterida dicitur, in quo Kalendae Ianuarii lunaeque eiusdem diei et initia primi mensis ipsiusque XIIIImae lunae recto iure ... sunt adnotatae*. Note especially that this list of Dionysiac Easter new moons does not feature in MC's principal source, CE; it also occurs in a different context in c. 57, ll. 10–17, where it starts with the final date given here (the difference stems from the different purposes of the two lists: the list below

declared that they observe *luna* 14 of this (month) (i.e. the Easter full moon) from 18 March to 15 April.

The fact is to be known that, when the *hendecas* of the Latins is placed alongside the *ogdoas* of the Greeks, all *lunae* 14 of the first month (i.e. all Easter full moons) in the book of Victorius are unreliable save three, since they either fall before the equinox or they correspond to *luna* 15 of the Greeks. When, however, the two *ogdoades* are combined, (the two reckonings) have the same *lunae* 14 (i.e. Easter full moons) up to *luna* 4 on 1 January (i.e. the year in which *luna* 4 occurs on 1 January). From that place *luna* 14 (i.e. the Easter full moon) of the Latins precedes (the Greek Easter full moon, i.e. it occurs a day earlier).

The days of the kindling of the first month (i.e. the Julian calendar dates of the Easter new moon) are those 19 which we specify here: 23 March, 12 March, 31 March, 20 March, 9 March, 28 March, 17 March, 5 April. Thus far the *ogdoas*. Now the *hendecas*: 25 March, 14 March, 2 April, 22 March, 11 March, 30 March, 19 March, 8 March, 27 March, 16 March, 4 April.

---

introduces the beginnings of the lunar years (modelled on the list found in EPISTOLA DIONYSII), while the list here records the Easter new moons corresponding to the Easter full moons as outlined in the Dionysiac Easter table; note that *X* is also omitted in the list below, so that it appears that MC used the same source in both chapters); for the Easter new moon being regarded as the beginning of the lunar year cf. c. 53, ll. 5–6; c. 57, ll. 18–19; c. 59, ll. 38–69.

Aprelis finitur. For a condemnation of the Victorian limits see DRC 51 (cited in LIB. COMP. 4.21), 62 (the relevant passage only partially cited in LIB. CALC. 95). **46–52** Sciendum…Latinorum] Victorian and Dionysiac Easter full moons are hardly ever compare in post-MC texts, and if they are, then only in the historically correct way. Such a comparison is embedded in a table in LIB. COMP. 2.22A, in which four of the six columns record Julian calendar dates for Easter full moons (the remaining two list the Dionysiac epacts of 22 March and the year in the *cyclus lunaris*); these four columns are listed separately in Berlin, Staatsbibliothek, Phillipps 1831, fol. 128r (and, according to Jones, *Bedae opera*, 24, in Paris, Bibliothèque Nationale, Lat. 7296, fol. 111v), where they are headed *Latinorum*, *Grecorum*, *Victorii*, *Eusebii* respectively; the *Latinorum* column records the Dionysiac full moons, the *Victorii* column the Victorian ones (cf. the facsimile in Appendix 3). **53–61** Dies…Aprilis] This list of the Julian calendar dates of the Dionysiac Easter new moons occurs as a separate table primarily in early computistical textbooks. Cf. DRC 90.6–10, HARVARD FRAGMENT (Harvard, Houghton Library, MS Type 613, fol. 7v). From the mid eighth century, it is rather embedded in larger tables, especially those that list the different *termini*, but even then only rarely. Cf. RM 83.49–94 (copied from Basel, Universitätsbibliothek, F III 15k, fol. 44v), Paris, Bibliothèque Nationale, Nouvelle acquisition latine 1615, fol. 153r–v.

46–52 Sciendum…Latinorum] Cf. p. LXXI, LXXXIV–LXXXV, XCIV, CLXIV of the introduction and Appendix 3. **53–69** Dies…Maii] Cf. p. XCIII–XCIV of the introduction.

The Julian calendar dates of the Dionysiac Easter full moons

Deinde XIIIIme lunę primi mensis hae sunt: Nonae Aprilis, VIII Kalendas Aprilis, Idus Aprilis, IIII Nonas Aprilis, XI Kalendas Aprilis, IIII Idus Aprilis, III Kalendas Aprilis, XIIII Kalendas Maii. Hucusque ogdoas. VII Idus Aprilis, VI Kalendas Aprilis, XVII Kalendas Maii, II Nonas Aprilis, VIIII Kalendas Aprilis, II Idus Aprilis, Kalendae Aprilis, XII Kalendas Aprilis, V Idus Aprilis, IIII Kalendas Aprilis, XV Kalendas Maii.

---

65 ogdoas] ogdoades *M.* | VII] VI *M**. **67** II] V *M.*

62–69 Deinde…Maii] The list of Dionysiac Easter full moons could have been taken from a table like the one described by Cummian (cf. previous note); moreover, it constituted a column in the original Dionysiac Easter table (Krusch, *Studien* II, p. 69–74), and may have been copied from it or any of its continuations (the only known Dionysiac Easter table contemporary to MC is Willibrord's: Paris, Bibliothèque Nationale, Lat. 10837, fol. 40v–41v (AD 703–759), 43r–v (AD 760–797), 44r (AD 684–702), and the Périgueux table which could not have been MC's source). Such a list is also part of MC's principal source, CE p. 111, but it does not appear that MC worked from CE in this instance: *Quomodo nominantur lunę XIIIImę in diebus mensis? Nonae, VIII Kalendas, Idus, IIII Nonas, XI <Kalendas>, IIII Idus, III Kalendas, XIIII Kalendas Maii. ogdoas. VII Idus, <VI Kalendas>, XVII Kalendas Maii, II Nonas, VIIII Kalendas, II Idus, Kalendae, XII <Kalendas>, V Idus, IIII Kalendas, XV Kalendas Maii.*

62–69 Deinde…Maii] As in the case of the list of the Julian calendar dates of the Dionysiac Easter new moons, this list of the Julian calendar dates of the Dionysiac Easter full moons occurs as a separate table primarily in early computistical textbooks. Cf. DRC 93.5–9, DIAL. NEUSTR. 9aB, HARVARD FRAGMENT (Harvard, Houghton Library, MS Type 613, fol. 7v). From the mid eighth century, it is rather embedded in tables, especially those that list the different *termini*, or it occurs in poem form, as in the poem *Nonae Aprilis*, which can be found in numerous computistical MSS. Cf. London, British Library, Cotton Caligula A XV, fol. 66v, BC 3(1282)C15–(1283)C3, LECT. COMP. 1.12–15, 6.2B–C, LIB. ANN. 32B–C, LIB. COMP. 1.11E (cited in LIB. CALC. 11B), LIB. CALC. 7A–D, PV §§23, 25, RM 83.49–94 (copied from Basel, Universitätsbibliothek, F III 15k, fol. 44v), 84.6–24.

62–69 Deinde…Maii] Cf. p. C, CLXXXIX–CXC of the introduction.

Accordingly, these are the *lunae* 14 of the first month (i.e. the Easter full moons): 5 April, 25 March, 13 April, 2 April, 22 March, 10 April, 30 March, 18 April. Thus far the *ogdoas*. 7 April, 27 March, 15 April, 4 April, 24 March, 12 April, 1 April, 21 March, 9 April, 29 March, 17 April.

<LVI. DE PASCALI SUBPUTATIONE>

Three essential rules for the understanding of Easter

His *tribus* rebus *pascha* indiget, hoc est *dies*, *luna*, *mensis*, qui *a*b *XII Kalendas Aprilis in XIIII Kalendas Maii* extenditur. Et luna et dies dominicus, quorum termini idem sunt. Et *dies luna VII a XV in XXI* extensi sunt. *Quid* horum trium honorabilius apud *ueter*es extitit? Id est *aetas lunae*. Nam *XIIII luna* | primi mensis integro *trans*gresso uernali *ęquinoctio* quolibet mensis die aut ebdomadis contigisset, in eo *agnu*m *immola*bant. *In nouo* uero testamento *<dies> ebdomadis, hoc est dominicus resurrectionis dies*, amplius his honorari elegunt. Nam licet *XIIII luna* preterierit et dies mensis, nisi *dominic*us

fol. 32v

LVI,3 XII] XI *M (corr. from* CE*; this mistake may derived from* MARTIN OF BRAGA*).* **5** VII] VI *M (corr. from* CE*).* | in] *follows* in *by mistaken repetition M.* | extensi sunt] extensis *M.* **8** uernali] uernalis *M.* **10** dies] *om. M (add. from* CE*).*

LVI,2–14 His…Christiani] The rules outlined here for the determination of Easter Sunday basically follow the generally accepted formula that Easter is to be celebrated on the Sunday (here referred to as *dies*) after the full moon (i.e. on *luna* 15 to 21, here referred to as *luna*) after the vernal equinox (21 March, here referred to by *mensis*, specified as the Julian calendar period in which the Easter full moon was supposed to fall). The general model for these three rules (with MC correcting the Julian calendar limits of the Easter full moon according to the Dionysiac reckoning) seems to have been MARTIN OF BRAGA, *De pascha* 7 (Barlow, *Opera omnia*, p. 274; cited in TRACTATUS ADTHANASI 6 (Krusch, *Studien* I, 333–4)): *Sic Pascha nec ante XI Kal. Apr. nec post XI Kal. Mai. celebrare licebit. Sed cum in hoc mense et luna et dies convenissent, luna scilicet XIV et dies Dominica, tunc Pascha celebraretur. Sane quia rursus frequenter luna XIV cum Dominica die non concurreret, extendi lunam in septem dies maluerunt, dummodo diem Dominicam in resurrectionis laetitia retinerent. Ita quando sic dies venerit usque ad vicesimam primam lunam propter Dominicam diem Pascha differimus semper, ut neque ante XI Kal. Apr. neque post XI Kal. Mai. celebretur. Sic inventum est ut mensis et dies et luna in celebratione Paschae retinerentur*. Very similarly DE ORDINATIONE FERIARUM PASCHALIUM (*PL* 90, col. 609A–610A, cited c. 55, ll. 23–26): *Quando ergo fit intra suum limitem ab undecimo Kalendarum aprilium usque in undecimo Kalendarum Maiarum, dies dominicus, et luna ex illis octava sanctificata, Pascha nobis visum est celebrare*. The ultimate source for the change in emphasis from celebrating the pasch on *luna* 14 to celebrating Easter on a Sunday is DRP 7.115–120: *regulam susceperunt indubitanter omnibus annis quando XIIII luna adfuisset et agnus apud Iudaeos immolaretur, aequinoctio transuadato pascha celebrauerunt, non adquiescentes auctoritati quorundam, id est Petri et Pauli et successorum eorum ... sollemnitatem resurrectionis domini in die tantum dominica posse celebrari docuerunt*. Cf. also ETYM. 6.17.10 (cited in BC 140A12–C4, COMP. COL. 5.9B): *Antiquitus Ecclesia pascha quarta decima luna cum Iudaeis celebrabat, quocumque die occurreret. Quem ritum santi Patres in Nicaena synodo prohibuerunt, constituentes non solum lunam paschalem et mensem inquirere, sed etiam et diem resurrectionis Dominicae observare; et ob hoc pascha a quarta decima luna usque ad vicesimam primam extenderunt, ut dies Dominicus non omitteretur*. The direct source for this passage, however, is CE p. 112: *Hic aliud interrogandum est, quot sunt profecturę, quę cooperantur circa hanc sollepnitatem resurrectionis dominicę. Hoc ita soluendum est, quod III sunt, id est aetas lunę, dies mensis, dies ebdomę. Sed quid de his specialius principatum teneat? Dies utique epdomadis in nouo, hoc est dies dominicus. Nam VII aetates lunę, id est a XVma luna usque in XXIma sustineant hunc dominicum diem per XXXV dies. In ueteri autem aetas lunae principatum tenuisse uisa est. VII enim dies epdomadis unam aetatem lunae, id est XIIIImam per mensem sustinebant unum, id est a XII Kalendas Aprilis usque in XIIII Kalendas Maii*. Close to MC, and written only nine years earlier (in AD 710), is EPISTOLA CEOLFRIDI (transmitted through HE 5.21): *Tres sunt ergo regulae sacris inditae*

56. ABOUT THE CALCULATION OF EASTER

Easter requires these three things, namely the (right) (week)day, the (right) moon, and the (right) month, which extends from 21 March to 18 April (i.e. the Julian calendar limits for the Easter full moon). Likewise, the moon and the Sunday, whose *termini* are the same (i.e. the lunar age and the Julian calendar date of Easter Sunday are both determined by the same *termini*, namely the Easter full moons). And the day (and) the moon extend (through) seven (days), from (*luna*) 15 to (*luna*) 21 (i.e. Easter Sunday and its lunar age fall within a period of seven days following the Easter full moon). Which of these three (rules) stood out more honourably among the elders? The age of the moon. For on whatever day of the month or the week *luna* 14 of the first month would haved occurred, having passed the vernal equinox, on that day they sacrificed the lamb. In the New Testament, (however,) they chose the day of the week, i.e. the Sunday of the resurrection, to be honoured more than the others. For the Christians decided that, after *luna* 14 (i.e. the Easter full moon) will have passed, the days of the month

---

litteris, quibus paschae celebrandi tempus nobis praefinitum, nulla prorsus humana licet auctoritate mutari; e quibus duae in lege Mosi diuinitus statutae tertia in euangelio per effectum dominicae passionis et resurrectionis adiuncta est. Praecepit enim lex, ut pascha primo mense anni et tertia eiusdem mensis septimana, id est a XV die usque ad XXI, fieri deberet; additum est per institutionem apostolicam ex euangelio, ut in ipsa tertia septimana diem dominicam expectare, atque in ea temporis paschalis initium tenere debeamus. Quam uidelicet regulam triformen quisquis rite custodierit, numquam in adnotatione festi paschalis errabit. Cf. also PROLOGUS THEOPHILI 1, 4, 5 (Krusch, *Studien* I, p. 221–2, 225–6), EPISTOLA PROTERII 2 (Krusch, *Studien* I, p. 271–2), PROLOGUS VICTORII 4 (Krusch, *Studien* II, p. 20; the relevant passage cited in DRC 92.4–7, BC 9B17–C2 which gives *XV Kalendas Aprilis* and *XV Kalendas Maii* as the limits for the Easter full moon, in turn cited in COMP. COL. 6.7B, EPIST. RAT. 9B, LIB. ANN. 12, in turn cited in LIB. COMP. 4.17, all four transmitting the correct limits for the Easter full moon), PROLOGUS DIONYSII (Krusch, *Studien* II, p. 65), EPISTOLA CYRILLI 5 (Krusch, *Studien* I, p. 347; the relevant passage cited in DRC 98.13–16, DIAL. NEUSTR. 18A, COMP. COL. 4.9A). For the Hebrew pasch being celebrated on *luna* 14 of the first month cf. the Vulgate citations given in COLOGNE FRAGMENTS 1 (Krusch, *Studien* I, p. 241–2, transmitted through COMP. COL. 6.2), COMPUTUS CARTHAGINIENSIS 2.16 (Krusch, *Studien* I, p. 296–7), PROLOGUS DIONYSII (Krusch, *Studien* II, p. 65) and c. 44, ll. 44–46; c. 55, ll. 3–6. For the Julian calendar limits for the Easter full moon cf. c. 55, ll. 39–41; c. 56, ll. 3–4; for the lunar limits for Easter Sunday cf. c. 58, ll. 2–3.

LVI,2–14 His…Christiani] The rule of celebrating Easter on the Sunday after the full moon of the first month (which is obviously based on the same three criteria explicitly mentioned in CE and MC; cf. *app. font.*) is also outlined in PROL. AQUIT. 1B, DTR 59.1–12 (cited in LIB. CALC. 93, PV §§251–253, RM 80.4–17). Yet, two chapters later Bede (DTR 61.3–27, cited in LIB. CALC. 94, RM 86.11–38, only the first half in PV §259) lists four rather than three criteria for determining Easter, as he adds the vernal equinox, on which he places special emphasis throughout his work: The Old Testament sets out three rules, i.e. that the pasch is to be celebrated after the vernal equinox, in the third week of the first month; these were then supplemented by the observance of the Lord's day, i.e. Sunday, deriving from the New Testament. Bede's focus on the vernal equinox is not shared by his contemporaries, since its observance was understood as part of the rule concerning the first month.

LVI,2–14 His…Christiani] Cf. p. XCIII–XCIV of the introduction.

presto innotescat, *usque ad XXI* dies ebdomadis, qui est *dominic*us, expectari statuunt Christiani.

Method for determin. Easter Sunday

Qualiter ex his pascha exploramus? Hoc est: Scita XIIII luna ebdomadis diem, qua XIIII luna stat, sollicitae quęras; cui in sequenti dominico, in quocumque die mensis et lunae, nisi XXI excedat, pascha facias.

1. Example: AD 719

Quod inminentis anni exemplo monstrabimus: In <V> Idus Aprilis XIIII luna est, cuius ebdomadis diem a Kalendis Ianuarii probemus: Kalendae Ianuarii dominico; hinc primo die <Aprilis> sabbatum, id est cum non sit bissextus; si enim fuerit, eodem die Kalendae Ianuarii et Aprilis ecce apparebit; deinde Kalendis Aprilis sabbatum et VI Idus eius, V Idus dominicus, in quo XIIII luna est. Hinc in XVI Kalendas Maii sequens dominicus dies et XXI luna.

2. Example: AD 720

Itaque secundi anni XIIII luna IIII Kalendas Aprilis; et annus bissexti; Kalendae Ianuarii in II feria; hinc IIIIo die post, qui est V feria, Kalendae Februarii et Martii, VIII Idus Martii et XI Kalendas Aprilis et IIII Kalendas, VI feria mutante bissexto inde XIIII luna, quae est IIII Kalendas Aprilis. | In sequenti uero dominico, id est II Kalendas fol. 33r
Aprilis et luna XVI, pascha celebrari debet.

---

13 XXI] XI *M (corr. from* MARTIN OF BRAGA, ETYM., CE*; it may well rather have read* VII *(also according to* MARTIN OF BRAGA *and* CE*), and would then have to be translated: for up to seven days).* **16** luna[1]] et *add. Schwartz.* | diem] *corr. to* die *Schwartz.* | sollicitae] *corr. to* sollicitate *Mac Carthy.* **19** monstrabimus] monstrauimus *M (corr. by Mac Carthy, Schwartz).* **20** V] *om. M (add. according to* MC*; also Schwartz).* **22** Aprilis] *om. M (add. according to* MC*; cf. Schwartz).* **26** XVI] IIII *M (corr. according to* MC*; also Schwartz).* **30** et Martii] *corr. to* K. Mar. et *Schwartz.* **31** IIII] VII *M (corr. according to* MC*; also Schwartz).*

15–43 Qualiter…licet] The method outlined here for the calculation of Easter Sunday is principally based on the weekday-calculation discussed in c. 29: The preconditions are the weekday of 1 January and the Julian calendar date of the Easter full moon; from this data the weekday of the Easter full moon could be calculated, with Easter falling on the following Sunday. MC took this method from CE p. 112: *Itaque, dum XIIIImas lunas prenotauimus in diebus mensis, necesse est eas ponere in diebus epdomadis, et diem paschę ponere in hoc modo: Kalendis Ianuarii uerbi gratia dominicus, Kalendis Aprilis sabbatum, IIII Nonas dominicus, III Nonas II feria, II Nonas III feria, Nonis IIII feria, id est XIIII luna. Pascha autem V Idus Aprilis, luna XVIIIma. Kalendis Ianuarii II feria, Kalendis Februarii V feria et Kalendis Martii et VIII Idus Martii et Idibus Martii et XI Kalendas Aprilis, X Kalendas Aprilis VI feria, IX Kalendas sabbatum, VIII Kalendas Aprilis dominicus, id est luna XIIIIma. Pascha autem Kalendis Aprilis, luna XXI. Kalendis Ianuarii III feria, Kalendis Aprilis II feria et VI Idus Aprilis, Idibus autem sabbatum, id est luna XIIIIma. Pascha autem XVIII Kalendas Maii, luna XVma. Kalendis Ianuarii IIII feria et Kalendis Aprilis, IIII Nonas Aprilis V feria, luna XIIIIma. Pascha autem Nonis Aprilis, luna XVIma (recte XVII). Kalendis Ianuarii VI feria, Kalendis Februarii II feria, Kalendis autem Martii et VIII Idus Martii et Idibus Martii et XI Kalendas Aprilis II feria, luna XIIIIma. Pascha autem V Kalendas Aprilis, luna XVIIIIma (recte XX). Hęc autem exempli causa ostendimus. Sed si quis in decennouenali ciclo huius numeri ueritatem cognoscere uoluerit, presenti sibi anno indesinenter inuestiget.* Note that the examples given in CE are theoretical, not related to any specific year; yet, CE argues that these rules should be sought until (*luna*) 21 for that day of the week, which is a Sunday, should the Sunday not immediately be known.

How do we determine Easter from these (rules)? In the following way: After *luna* 14 (i.e. the Julian calendar date of the Easter full moon) is known, you should carefully determine the day of the week on which *luna* 14 falls; on the Sunday following this day, on whatever day of the month and the moon (it may fall), as long as it does not exceed (*luna*) 21, you should celebrate Easter.

This we will demonstrate by the example of the imminent year: *luna* 14 (i.e. the Easter full moon) falls on 9 April, whose weekday we should verify from 1 January: 1 January (falls on) a Sunday; consequently, a Saturday (occurs) on the first day of April, i.e. since no bissextile day happens to be (in this year); if, in fact, there was (one), 1 January and 1 April would occur on the same (week)day; moreover, a Saturday (occurs) on 1 April and on 8 April, a Sunday on 9 April, on which *luna* 14 (i.e. the Easter full moon) falls. Therefore, the following Sunday (occurs) on 16 April and on *luna* 21.

Accordingly, *luna* 14 (i.e. the Easter full moon) of the second year (falls) on 29 March; and (this is) a bissextile year; 1 January (falls) on a Monday; consequently, on the fourth (week)day after (this), which is a Thursday, (fall) 1 February and 1 March, (as well as) 8 March, 22 March and 29 March; but the bissextile day changes (these) to Friday, and accordingly (also) *luna* 14 (i.e. the Easter full moon), which falls on 29 March. On the following Sunday, however, i.e. 31 March and *luna* 16, Easter has to be celebrated.

---

could and should easily be applied to the reader's *annus praesens*, and this is exactly what MC does here.

15–43 Qualiter…licet] Various algorithms for the calculation of the Julian calendar date and weekday of the Easter full moon, as well as the Julian calendar date and lunar age of Easter Sunday, were in circulation by the early eighth century; the most prominent among these was PS-DIONYSIUS, *Argumentum XIV*, which certainly was well established by the time of the composition of MC: It was probably invented in the sixth or early seventh century, with formulae matching the first three years of the Dionysiac Easter table (AD 532–534), while two prominent later recensions adapted it to the years AD 684–686 and 776–777 respectively; by the second half of the eighth century, it circulated in numerous versions. Cf. BC 55, 73, LECT. COMP. 6.5, LIB. ANN. 19 (cited in LIB. COMP. 4.19), 29 (cited in LIB. COMP. 4.19, in turn cited in LIB. CALC. 53), 29a, LIB. COMP. 4.19aA, RM 82. For other methods cf. DIONYSIUS, *Argumenta IX* (cited in BC 74; repeated in LECT. COMP. 6.6, RM 90), *X*, LECT. COMP. 6.2, 6.3, LIB. ANN. 11 (cited in LIB. COMP. 4.16, in turn cited in LIB. CALC. 54), 32, LIB. COMP. 4.20 (cited in LIB. CALC. 52), RM 84, PV §§24–25. Accordingly, there certainly was no need to copy or apply CE's and MC's rather simplistic method.

15–41 Qualiter…dubitet] Transcribed in Schwartz, 'Ostertafeln', p. 91; only the beginning in Mac Carthy, *Annals of Ulster* 4, p. lxx; cf. p. LIX, CXLVI–CXLVII of the introduction. **15–57** Qualiter…conprobet] Cf. p. LX, LXIII, XCII–XCIV of the introduction.

3. Example: AD 721

In IIIo quoque finienti endecadi anno XIIII luna XV Kalendas Maii; qui annus bissexti secundum; hinc Kalendis Ianuarii IIII feria; transgressa IIII feria Kalendis Aprilis III feria et VI Idus Aprilis, XVII <Kalendas Maii>; XVI IIII feria; XV V feria, qui XIIII luna. Et pascha sequenti dominico die, hoc est XII Kalendas Maii, luna XVII, inesse nemo dubitet.

Every year likewise

Ita per XVIIII annos pascalis subputatio intellecto occurrenti bissexto IIIIo quoque anno obseruari licet.

Calculation of the Julian calendar date of the Easter full moon

XIIII lunae ita a Kalendis Ianuarii probari debent, ne dubitatio in his oriri inopinate uideatur:

In anno, in quo luna VIII in Kalendis Ianuarii et Martii, XVIII in V Idus Martii, XXX in XI Kalendas Aprilis, una luna in X Kalendas Aprilis; inde XIIII luna in Nonis Aprilis.

In anno, in quo XX luna in Kalendis Ianuarii et Martii, XXX in V Idus Martii, I luna in IIII Idus Martii, XIIII luna in VIII Kalendas Aprilis.

In anno, in quo I luna in Kalendis Ianuarii et Martii, XI luna in V Idus Martii, XXX luna in III Kalendas Aprilis, II luna in Kalendis Aprilis, XIIII luna in Idibus Aprilis.

Non aliter per XVIIII annorum ciclum XIIII lunas sollicitus quisque a Kalendis Ianuarii conprobet.

---

37 IIII] *corr. to* III *Schwartz.* **38** Aprilis[1]] *follows* II id est *M (om. by Schwartz;* id est III *may originally have been an interlinear correction of the wrong number* II*).* | Aprilis[2]] luna *M (corr. according to* MC*; Schwartz corr. to* et*).* | Kalendas Maii] *om. M (add. according to* MC*; also Schwartz).* **39** IIII] VII *M (corr. according to* MC*; also Schwartz).* | V] II *M (corr. according to* MC*; also Schwartz).* **40** Maii] *corr. from* ap- *M.* **45** inopinate] inopinatim *M.* **46** In] *follows* octo *M.* **47** Martii[2]] Mai- *M.* **50** In] *preceded by* XX *M.* **53** anno] *follows* I *M.*

44–57 XIIII…conprobet] With this calculation here MC wants to verify one of the preconditions of the previous calculation, the Julian calendar dates of the Easter full moon; note that this calculation is not found in MC's source, CE. The precondition here is the epact of 1 January; if this calculation here is combined with the previous one, then Easter Sunday could be reckoned from the ferial and lunar data of 1 January. Generally, MC always tried to use nothing but the data of 1 January as preconditions for any kind of calendrical calculation, and it must be presumed that these methods had already been applied to the two reckoning previously followed in southern Ireland, the *latercus* and the Victorian system, since the ferial and lunar data of 1 January feature very prominently in the Easter tables of both reckonings (in the *latercus*, these data are, in fact, the defining features of a year). The method of lunar calculation applied here, namely to take advantage of the fact that 1 January and 1 March have the same lunar age, and that this lunar age +10 equals that of 11 March, is already outlined and demonstrated by the same three examples in c. 51, ll. 12–34.

44–57 XIIII…conprobet] For the most prominent algorithm for calculating the Julian calendar date of the Easter full moon, in MC's time and later, cf. the previous note. Interestingly enough, Bede (DTR 60.1–15, cited in PV §§254–257, RM 81.5–20) does not supply an algorithm to calculate the Julian calendar date of the Easter full moon either, but outlines a method equally basic as MC's by advising to count the Julian calendar days from the epact of 22 March to the following *luna* 14 (or to observe the

In the third year, which also marks the end of the *hendecas*, *luna* 14 (i.e. the Easter full moon) (falls) on 17 April; this year follows a bissextile one; accordingly, a Wednesday (occurs) on 1 January; Wednesday passes into Tuesday on 1 April, 8 April and 15 April; a Wednesday (occurs) on 16 (April); a Thursday on 17 (April), which (is) *luna* 14 (i.e. the Easter full moon). And nobody should doubt that Easter falls on the following Sunday, 20 April, *luna* 17.

The calculation of Easter should be observed in this manner throughout the 19 years, once (the impact on this calculation of) the four-year bissextile cycle is understood.

The *lunae* 14 (i.e. the Easter full moons) have to be examined from 1 January in the following way, so that no doubt in these threatens to arise unexpectedly:

In the year, in which *luna* 8 (occurs) on 1 January and 1 March (i.e. the first year of the Dionysiac 19-year cycle), (*luna*) 18 (occurs) on 11 March, (*luna*) 30 on 22 March, *luna* 1 on 23 March; accordingly, *luna* 14 (i.e. the Easter full moon) (falls) on 5 April.

In the year, in which *luna* 20 (occurs) on 1 January and 1 March (i.e. the second year of the Dionysiac 19-year cycle), (*luna*) 30 (occurs) on 11 March, *luna* 1 on 12 March, *luna* 14 on 25 March.

In the year, in which *luna* 1 (occurs) on 1 January and 1 March (i.e. the third year of the Dionysiac 19-year cycle), *luna* 11 (occurs) on 11 March, *luna* 30 on 30 March, *luna* 2 on 1 April, *luna* 14 on 13 April.

In no other way should a careful person verify (the Julian calendar dates of) *luna* 14 (i.e. the Easter full moon) from 1 January throughout the cycle of 19 years.

Easter full moon on 21 or 22 March if the epact is 15 or 14 respectively). A similarly basic method is described in LECT. COMP. 6.4. These basic methods based on simple finger-counting were soon superseded by more complex algorithms (cf. the previous passage). However, Bede adds a more general rule in the same chapter (DTR 60. 15–19), namely that in common lunar years the Easter full moon falls 11 days earlier than the previous, in embolismic ones 19 days later; this rule is not only cited in PV §258, RM 81.20–25, but also repeated in LECT. COMP. 2.10.

44–57 XIIII…conprobet] Cf. p. CLXI of the introduction

<LVII. DE SPATIO ANNORUM LUNAE>

Julian calendar dates of Dionysiac *lunae* 30 preceding Easter

Nunc XXXmas lunas, in quibus XVIIII lunae anni finiuntur, dicimus: XI Kalendas Aprilis, V Idus Martii, III Kalendas Aprilis, XIIII Kalendas Aprilis, VIII Idus Martii, VI Kalendas Aprilis, XVII Kalendas Aprilis, II Nonas Aprilis, VIIII Kalendas Aprilis, III Idus Martii, Kalendae Aprilis, XII Kalendas Aprilis, | VI Idus Martii, IIII Kalendas Aprilis, XV Kalendas Aprilis, Nonae Martii, VII Kalendas Aprilis, Idus Martii, III Nonas Aprilis. fol. 33v

Julian calendar dates of Dionysiac Easter new moons

Haec sunt initia horum annorum: II Nonas Aprilis, <X> Kalendas Aprilis, IIII Idus Martii, II Kalendas Aprilis, XIII Kalendas Aprilis, VII Idus Martii, V Kalendas Aprilis, XVI Kalendas Aprilis, Nonae Aprilis, VIII Kalendas Aprilis, II Idus Martii, IIII Nonas Aprilis, XI Kalendas Aprilis, V Idus Martii, III Kalendas Aprilis, XIIII Kalendas Aprilis, VIII Idus Martii, VI Kalendas Aprilis, XVII Kalendas Aprilis.

The intervals of lunar years

A II Nonas in XI Kalendas Aprilis, a X Kalendas Aprilis in V Idus Martii, et reliqua.

LVII,2 lunae] lunas *M.* **6** VIIII] VIII *M (corr. according to* MC*).* **7** VI] V *M (corr. according to* MC*).* | IIII] VII *M (corr. according to* MC*).* **8** Martii] Aprilis *M.* **9** Martii] Aprilis *M.* **11** X] *om. M (add. according to* MC*).* **12** Aprilis[2]] p- *M.*

LVII,2–19 Nunc…reliqua] The purpose of this chapter is to outline the Julian calendar intervals for each of the 19 lunar years of the Dionysiac 19-year cycle. Dionysius had done this himself in EPISTOLA DIONYSII (Krusch, *Studien* II, p. 85–6, cited in DRC 82, BC 14(1285)D10–(1287)A14, 124(1333)C7–(1334)B5, DIAL. NEUSTR. 24, and many MSS); Dionysius' intervals extend from *luna* 15 of the Easter lunation to the Easter full moon of the following year, while his first year ends with the first Easter full moon of his Easter table, i.e. *Nonae Aprilis* (5 April). MC, however, considers the Easter new moon as the correct beginning of a lunar year (for the different traditions cf. c. 53, ll. 5–11), so that MC's task here was to translate the years of the Dionysiac list to an Easter new moon beginning. At first, the final dates of each lunar year are listed, followed by the starting dates, and finally these two list are combined to define each lunar year of the Dionysiac 19-year cycle. Note that the list of Easter new moons also occurs in a different context in c. 55, ll. 53–61, where it starts with the second date given here (the difference stems from the different purposes of the two lists: the list here introduces the beginnings of the lunar years modelled on the list found in EPISTOLA DIONYSII, while the previous list records the Easter new moons corresponding to the Easter full moons in the order given in the Dionysiac Easter table; note that *X* is also omitted in the previous list, so that it appears that MC used the same source in both chapters); for the Easter new moon being regarded as the beginning of the lunar year cf. also c. 59, ll. 38–69.

LVII,2–19 Nunc…reliqua] The Julian calendar intervals of the Dionysiac lunar years outlined by MC here are extremely unusual even in early medieval computistics; in fact, I could not locate MC's list of lunar years in any MS. Most common was rather the list of lunar years outlined by Dionysius himself (cf. *app. font.*). Bede (DTR 56.19–56, cited in PV §§229–247, LIB. CALC. 92, and many MSS), on the other hand, lists the 19 lunar years in intervals extending from *luna* 1 of the January lunation to *luna* 29 of the December one; he starts with the first year of the *cyclus lunaris* (which corresponds to the fourth year of the *cyclus decemnovenalis*, the 19-year cycle) because the beginning of the January lunation fell

57. ABOUT THE (JULIAN CALENDAR) INTERVAL OF THE LUNAR YEARS

Now we list the *lunae* 30 (preceding the Easter lunation), in which the 19 lunar years are terminated: 22 March, 11 March, 30 March, 19 March, 8 March, 27 March, 16 March, 4 April, 24 March,13 March, 1 April, 21 March, 10 March, 29 March, 18 March, 7 March, 26 March, 15 March, 3 April.

The following, (then,) are the beginnings of these years: 4 April, 23 March, 12 March, 31 March, 20 March, 9 March, 28 March, 17 March, 5 April, 25 March, 14 March, 2 April, 22 March, 11 March, 30 March, 19 March, 8 March, 27 March, 16 March.

(Therefore, the lunar years are to be reckoned in the following way:) From 4 April to 22 March, from 23 March to 11 March, and so on.

---

on 1 January in that year. Some MSS (like Bc, Basel, Universitätsbibliothek, F III 15k, 34v–36r) have both Dionysius' as well as Bede's lists. In Bc 4C14–D1 the lunar years start with the Easter full moons, the first year with the 19th full moon of Dionysius' Easter table.

LVII,2–19 Nunc…reliqua] Cf. p. XCIII–XCIV of the introduction.

<LVIII. DE INITIO QUADRAGESIMAE>

Lunar limits for Easter Sunday and *initium*

Septem aetates *paschae*, quas prediximus, hae sunt: *Secundum Grecos a XV luna* usque *in XXI*, initii uero *a III* singulari *in VIIII* extenduntur. *Secundum* autem Victorium aetates *pasche a XVI luna in XXII*, initii *a IIII* singulari *in*

LVIII,3 uero] *add.* aetas *Mac Carthy.* | III] VI *M (corr. from* CE, DRC*; also Mac Carthy, Schwartz).* **4** extenduntur] extenditur *M, Mac Carthy, Schwartz.* **5** XVI] XV *M, Krusch (corr. from* CE, DRC*; also Mac Carthy, Schwartz).*

LVIII,2–7 Septem…singularem] After the discussion of Easter Sunday in c. 55, MC here turns to a feature of the liturgical calendar directly connected to Easter Sunday, namely the *initium quadragesimae*, i.e. the beginning of the Lenten fast. The *initium*, according to early medieval computists, was fixed on the sixth Sunday, i.e. 42 days, before Easter Sunday. Since the lunation ending within this 42-day period had 30 days, the *initium* occurred 12 (42–30) lunar days earlier than Easter Sunday (cf. also the table at the end of this chapter); accordingly, lunar age of Easter Sunday−12=lunar age of the *initium*. Interestingly enough, MC applied this algorithm not only to the Dionysiac, but also to the Victorian reckoning and the *latercus*, not realizing that this algorithm was incompatible with *latercus* principles and disagreed with the data given in the Padua table. The origin of this wrong application was, however, not MC, but rather MC's source, CE p. 119: *De paschae figuris. Quot sunt figurę paschę? III sunt: A XIIII luna in XXmam secundum latercum, a XVmam (!) luna in XX\I/ secundum Grecos, a XVI luna in XXIIdam secundum Romanos. Dum autem pro certo nouimus has III formas in pascha esse, aeque III formas in initio fieri oppinamur. Et hoc uerum esse fatemur, hoc est secundum latercum initium fit a IIda luna VIIIuam lunam, secundum Grecum (!) a III luna in IXnam lunam, secundum Romanos a IIII luna in X lunam.* Generally, the *initium* is not discussed in any pre-CE text. As for the lunar limits of Easter Sunday, both the Dionysiac and the Victorian ones are outlined in PROLOGUS VICTORII 4 (Krusch, *Studien* II, p. 19–20; the first part cited in DRC 98.8–12, DTR 51.13–17, in turn cited in LIB. COMP. 4.21; the second part cited in DRC 92.5–7, BC 9B10–C2, in turn cited in COMP. COL. 6.7B, EPIST. RAT. 9B, LIB. ANN. 12, in turn cited in LIB. COMP. 4.17): *Nec minus eiusdem dominicae, quam peragendo mysterio distinarunt, quam sextam decimam neque amplius quam vicensimam et secundam lunam aliqando recipiunt, eligentes potius in vicensimam et secundam lunam diem festi paschalis extendi, quam dominicam passionem ante lunam quartam decimam nullatenus incoari. ... Sanctae memoriae contra Theophilos quondam Alexandrinus antestis ad Theodosium imperatorem datis epistolis .... etiamsi diae inciderit sabbatorum, post quam sequente dominico luna quinta decima celebrandum pascha sine dubitatione conscribit.* Authoritative for the Dionysiac lunar limits of Easter Sunday was PROLOGUS THEOPHILI 4 (Krusch, *Studien* I, p. 225; cited in DTR 61.27–37): *Dehinc quoniam etiam salvator quarta decima luna traditus est, hoc est V feria, quinta decima crucifixus est, tercio die surrexit, hoc est XVII luna, que tunc in dominicam inciderat diem, ut evangeliorum scriptura demonstraet, et tardioris paschae habemus ex necessiate solacium. Et si XIIII dies lune aut sabbato die incederit aut superioris sabbati die, his sine dubio pascha celebranda est. Quod si in dominica die incurrerit, in alteram septimanam celebracio differenda est.* For these cf. also c. 56, l. 5. All three lunar limits for Easter Sunday could easily have been deduced from the respective Easter tables, but if the *latercus* had been consulted, CE and MC would have realized that their *latercus* lunar limits for the *initium* were wrong. Anatolius (?) (DRP 3–5, 8, 13 (the relevant passage cited in DRC 98.3–7)), who presumably was the general source for the *latercus* practice, argued in favour of Easter lunar limits 14 to 20, and against celebrating Easter Sunday on *luna* 21 or 22. This stand was repeated by COLUMBANUS in his *Epistulae* 1.3–5, 2.5 (Walker, *Sancti Columbani opera*, p. 2–9, 16–7), defending the *latercus* practice, and condemning Victorius' principles as followed in Gaul. For arguments against celebrating Easter Sunday on *luna* 14 cf. c. 63.

LVIII,2–7 Septem…singularem] Computistical discussions of the *initium quadragesimae* first occur in Irish computistical textbooks, so that these discussions should be considered as one of the main achievements of seventh-century Irish computistics, and a very influential one, particularly since Bede

58. ABOUT THE BEGINNING OF LENT

The following are the seven lunar ages of Easter, which we have spoken of earlier: According to the Greeks they extend from *luna* 14 to (*luna*) 21, (the ones) of the *initium* (i.e. the beginning of Lent), however, from the third single digit (*singulari*) to the ninth (i.e. from *luna* 3 to *luna* 9). According to Victorius, the (lunar) ages of Easter (extend) from *luna* 16 to (*luna*) 22, of the *initium* from the fourth single digit to the tenth (moon, i.e. from *luna* 4 to *luna* 10). According to

---

did not say a word about this important date in the liturgical calendar. Yet, the exact same information outlined in MC's passage here can only be found in Irish sources, as would have been expected from the fact that Ireland is the only region in which all three Easter reckonings mentioned here were followed at some point in time. Cf. DRC 99: *Sciendum nobis quod, quemadmodum hi uiri in septem paschae aetatibus discrepant, <in septem aetatibus initii idcirco discrepant>. Laterci enim sectatores, qui <a> XIIII luna usque ad XX septem aetates paschae numerant, a II luna usque ad VIII lunam septem aetates initii conputant. Latini uero, septem aetates paschae a XVI luna usque ad XXII conputantes, a IIII luna usque in X lunam septem initii aetates numerant. Greci uero rationalibus, quos nos sequimur, a XV luna usque in XXI lunam septem aetates paschae numerantes, a III luna usque in VIIII lunam septem aetates initii conputant.* The lunar limits of Easter Sunday and the *initium* are given for the Dionysiac reckoning only in BC 50B9–11: *Luna in pascha non potest fieri minor quam XV nec maior quam XXI. In quadragesima uero minor non potest fieri quam IIIa nec maior quam VIIIIa.* LECT. COMP. 6.1aA: *Et sicut dominicus festi paschalis septem tantum aetates lunae observat, hoc est a quinta decima usque ad vicesimam primam, sic initium Quadragesimi septem tantum aetates observare videtur, hoc est a tertia luna usque ad nonam.* Cf. also COMP. COL. 6.8C. Only the lunar limits of Easter Sunday, but of all three reckonings, are discussed in DRC 98, and are also mentioned in the Irish influenced COMP. COL. 5.3B: *Latini solemniter Pascha <celebrant> a quarta decima primi mensis usque vicesimam, Victorius et Latini a sexta decima usque vicesimam secundam vel vicesimam tertiam, Graeci a quinta decima usque vicesimam primam.* These are also discussed at great length in Ceolfrith's letter to the Pictish king Nechtan of AD 710 (transmitted through HE 5.21). The lunar limits of Easter Sunday only of the Victorian and the Dionysiac reckonings are outlined in DIAL. NEUSTR. 13: *De limitibus lune in pascha. Latini et Victorius a sexta decima luna usque in vicesimam secundam sanctum Pascha celebrare faciunt. Qui limites in sacrae scripturae auctoritate minime habentur. Graeci predictam Pascham solemnitatem a quinta decima luna usque in vicesimam primam sine ambiguo celebrare scripserunt.* Only of the *latercus* and the Dionysiac reckoning in Bede's account of the Synod of Whitby (HE 3.25). In his computistica, Bede vehemently argues against the Victorian and the *latercus*' lunar limits for Easter Sunday (DTR 59.10–44, only the Dionysiac and Victorian part cited in LIB. CALC. 93, only the Dionysiac part in RM 80.10–13, only the first sentence in PV §253), while at the same time promoting the Dionysiac ones (cf. DTR 51.70–71 (cited in LIB. COMP. 4.21), 61.6–7 (cited in LIB. CALC. 94, RM 86.14–16, PV §259), 62.2–12 (cited in LIB. CALC. 95, PV §§260–261, RM 89.4–16)). The Dionysiac lunar limits obviously became standard, and therefore they are frequently found in computistica from the eighth century onwards. Cf. BC 140(1353)B2–D8 (partially cited in COMP. COL. 5.9B), which is a copy of ETYM. 6.17 where the Isidorian lunar limits of 14 to 21 are changed to 15 to 21 in three instances; ARG. AQUENS. 3C, 4B.

LVIII,2–7 Septem…singularem] Transcribed in Schwartz, 'Ostertafeln', p. 92; Mac Carthy, *Annals of Ulster* 4, p. lxxiv–lxxv; only the part dealing with the Victorian reckoning and the *latercus* in Krusch, *Studien* I, p. 11; only the part dealing with the *latercus* transcribed and translated in Warntjes, '84 (14)-year Easter reckoning', p. 53 (facsimile p. 84); cf. also Krusch, *Studien* II, p. 58; O'Connell, 'Easter cycles', p. 84–5; Walsh & Ó Cróinín, *Cummian's letter*, p. 204–5; Borst, *Schriften*, p. 394 and p. LXXI, XCIV, CXXII, CXLVIII–CL, CLXV, CXCVI of the introduction.

Xam. Iuxta uero *latercum a XIIII luna in XX*, et *initi*i *a II luna in VIII* singularem.

Disturban. of the lunar relation Easter Sunday – *initium*

Haec est perennis initii regula, quod expressimus, excepto hoc, quod *bissextus* aut *saltus* turbatum interdum ordinem prestat. Si enim in quadragesimo tempore *bissextus* sit, *in* maiorem initium *aetatem* tendit, ut si VI singulari sine bissexto esset, in VII prouehi initium prebet. Et cum inter initium et II singularem, quo initium ut pascha a XIIII luna regulatur, bissextus interseritur, non

---

6 Xam] X anni *M, Krusch (corr. by Schwartz; Mac Carthy corr.* Xmam*).* | II] VI *M (corr. from* CE, DRC; *also Mac Carthy, Schwartz, Warntjes; Krusch corr. to* III*).* **12** VII] IIII *M (corr. according to* MC*).* **13** inter] *om. M**. | quo] quia *M.* **14** interseritur] interserit *M.*

8–32 Haec…saltus] In early medieval Irish computistics, the Dionysiac *saltus* was placed on 22 March (cf. c. 62, ll. 68–72), the lunar bissextile, as is implicit in this passage here, on the solar bissextile day, 24 February. This meant that the *saltus* would always, the lunar bissextile day often fall into the quadragesimal period, which obviously had an impact on the lunar relation between Easter Sunday and the *initium*: The *saltus* reduced the lunation ending in the 42-day quadragesimal period from 30 to 29, so that the lunar day difference between the *initium* and Easter Sunday was 42−29=13 in years of a *saltus* rather than the common 12; on the other hand, the extra lunar day in bissextile years meant that the lunation ending in the quadragesimal period had 31 days rather than 30, leading to a lunar day difference between Easter Sunday and the *initium* of 42−31=11 days. The most detailed discussion of this problem in any computistical text or MS is provided by MC's source, CE p. 119–122 (I only cite the immediately relevant passages here, leaving aside the many illuminating examples; note that CE does not mention the weekday-relation between the *secundum singularis* the Easter full moon here): *Hic quaeritur utrum aliud huic obsistit bissextus utique huic contra uenit et saltus. Hoc est: Bissextus in ętatem ampliorem mutat. Addit enim ętatem lunae, quia una ętas lunę fit in duabus diebus ebdomadis. Saltus uero in minorem ętatem lunae transducit, duę namque aetates lunae in uno die fiunt. ... Ideo enim bissextus mutat hoc anno in ampliorem ętatem, quia bissextus est post initium. Si enim ante esset bissextus, hęc mutatio non esset lunae. ... Item probemus si simul bissextus ac saltus fuerit. Quod uix uno euentu in decennouenali ciclo fieri potest, quomodo nec bissextus in maiorem aetatem, nec saltus in minorem transducit. ... Hinc uidemus, quod neque bissextus mutat in aetatem ampliorem neque saltus in minorem reducit, etiam si simul eodem anno euenerint.*

8–32 Haec…saltus] Note that both CE and MC approach the question of the lunar relation between the *initium* and Easter Sunday not in calendrical sequence, but rather backwards from Easter Sunday. The same perspective is outlined in c. 31 of an unfinished Fulda computus of AD 789 in Basel, Universitätsbibliothek, F III 15k, fol. 44v–45r (cited in Paris, Bibliothèque Nationale, Nouvelle acquisition latine 1615, fol. 150r–v and as part of a computus of AD 825/6 in Bern, Burgerbibliothek, 417, fol. 14v–15r): *Siue ante initio fuerit bissextus siue post. Et in illo uero anno, quando bissextus fuerit, plures sunt dies septimane quam dies solis. In illo uero anno, in quo saltus contigerit, plures sunt ętates lunae quam dies solis et dies septimane. Si autem in uno anno bissextus et saltus euenerint, conplures ętates lunae et dies septimanarum. Et nihil mutant bissextus et saltus in eo anno de regula et concordia initii et pascha, quod saltus retro mouet initium in iuniorem ętatem et in pauciores epactas, bissextus auget in ante lunae ętatem, hoc est in seniorem et in plures epactas*. In BC 50B10–12 it is suggested that the lunar age of the *initium* may extend to ten due to the bissextile day: *In quadragesima uero minor non potest fieri quam III nec maior quam VIIII, nisi bissextus euenerit*. Similarly, the impact of the *saltus* is mentioned in Vatican, Biblioteca Apostolica, Reg. Lat. 1260, 121r: *Et de luna quadragesimae: <nec> ante III nec post VIIII initium quadragesimae fieri potest, nisi saltus euenerit*. The opposite perspective, i.e. approaching this problem from the *initium*, is outlined in DRC 100.12–102.3: *Haec tamen regula bissexto et saltu turbatur. Bissextus enim, si fuerit inter initium et*

the *latercus*, (they extend) from *luna* 14 to (*luna*) 20, and (the ones) of the *initium* from *luna* 2 to the eighth single digit (i.e. to *luna* 8).

What we have (just) outlined is the perpetual rule for the *initium*, except for those cases, in which (either) the bissextile day or the *saltus* occasionally cause the rule to be disturbed. If, indeed, the bissextile day happens to occur in the quadragesimal period (i.e. the 40-day Lenten period), the *initium* passes into a greater (lunar) age, as, (e.g., in the case) when the *initium* falls on the sixth single digit (i.e. *luna* 6) without a bissextile day, it has to be moved to the seventh. And if the bissextile day is inserted between the *initium* and the second single digit (i.e. the *secundum singularis*, *luna* 2 of the March lunation), by which the *initium* is regulated in the same way as Easter Sunday is by *luna* 14 (i.e. the Easter full moon), it relocates the second single digit to the third (i.e. *luna* 2 to *luna* 3), while

---

pascha, transilitionem hanc inter aetatem initii et paschae prohibet. Uerbi gratia: VIII luna initii, XVIIII luna <in> pascha, causa bissexti inter initium et pascha. Nisi enim bissextus, XX luna in pascha esset. Si enim et ante initium bissextus fiat, pascha in iuniorem semper pertrachit, licet transilitionem hanc inter aetatem initii et pascha non prohibet. Ita regulatur initium, quod semper secunda luna Martio mense, uel secunda luna Apreli mense in sequenti dominico, prior in Februario accensa, altera in Martio indubitanter inuenitur. Sciendum nobis utrum ulla est aetas lunae contra quam regularetur initium, sicut regulatur pascha contra XIIIIam lunam primi mensis. Est etiam: secunda luna Martio mense et Apreli. Quia, quacumque die septimane Martio mense secunda luna fuerit uel Apreli, sequenti dominico die initium erit. Hoc nobis sciendum quod, quacumque die septimane secunda luna regulandi initii contigerit, eadem die septimane XIIII luna pascha erit, nisi fuerit bissextus. Si enim inter secundam lunam regulandi initii et XIIIIam lunam primi mensis bissextus intererit, postera die a II luna regulandi initii XIIII luna primi mensis. Uerbi gratia: II feria, II luna regulandi initii, III feria, XIIII luna primi mensis. Sciendum nobis quod bissextus, ut diximus, in iuniorem aetatem pascha pertrachit; saltus uero in maiorem aetatem trudit. Si uero in uno anno bissextus et saltus fuerint, nihil perturbant, sicut unicuique probare licet. More generally DIAL. LANGOB. 24B: *Quaerendum est de die bisexti. Quando venit, conturbatne anni regulam annon? Responsio: Aliquando mutat regulam, aliquando salvat <et> conservat. Interrogatio: Quomodo facit? Responsio: sunt tria loca, in quibus dies bisexti invenitur, hoc est: Initium Quadragesimae et ante secundam lunam initii. Quando sic invenitur, nullam regulam conturbat. Inter initium et Pascha necnon et inter secundam lunam initii et quartam decimam Paschae totas regulas salvas relinquet. Alter locus illius post secundam lunam initii, sed tamen ante initium. De quo loco non conturbat regulas initii et Paschae, sed tantum regulam quartae decimae et secundae lunae commutat, id est diem septimanae. Tertius locus est Quadragesima post initium. Totas regulas commutare facit, sive initii sive Paschae sive secundae lunae et quartae decimae*. All of the above mentioned texts are either Irish or directly influenced by Irish computistical thought. Obviously, the described problems would only occur if the *saltus* was placed on 22 March, and if the lunar bissextile day corresponded to the doubling of the Julian calendar date, i.e. 24 February. This was common theory in Irish computistics of around AD 700, but soon the *saltus* and the lunar bissextile day were placed outside of the quadragesimal period (on 24 November and at the end of the February lunation respectively), most prominently by Bede, precisely to avoid the difficulties outlined here; with this, these Irish discussions lost their currency. But even then new difficulties in the calculation of the *initium* arose, since the extra lunar day at the end of the February lunation led to the *terminus quadragesimi* (which is called *secundum singularis* by MC) falling a Julian calendar date later, if it occurred before 24 February, as Rabanus Maurus relates (RM 83.33–42; cf. also HELPERIC, *De computo* 35 (*PL* 137, col. 46A)). The weekday-correspondence between the Easter full moon and the *terminus quadragesimi* (since they are separated by exactly six weeks) is also referred to in LECT. COMP. 6.2E, RM 83.5–14 (based on Basel, Universitätsbibliothek, F III 15k, fol. 43v), 84.29–33.

8–32 Haec…saltus] Cf. Borst, *Schriften*, p. 458 and p. XLVIII, LXXI, CLXVIII, CLXXVIII of the introduction.

turbato initio II singularem in III remouet, et a die, in qua XIIII luna fit, in alterum mutari prebet. Nisi enim bissextus aut saltus eueniat, in uno die | secundum singularis, quo fol. 34r initium certatur, et XIIII luna, qua pascha regulatur, non dubium esse ratio adfirmat. *Si* autem *bissextus ante* <initium> et II singularem fit, nullam turbationem utriusque addit. Ita per omnes initii aetates: Bissextus in seniores uehit usque VIIII in X mutet. Quod ob hoc fieri uiget, nam plures solis dies quam lune in bissexti anno in quadragesimo haberi innotescunt. *Saltus uero in* iuniorem initii *aetatem mutat*, *II* enim singularem *in I luna* abiecit. Hinc quacumque aetate euenerit: A III singulari in VIIII hanc minutionem facit. Nam si saltu non stante VIIII luna initio caderit, ueniente illo in VIII moueri innotescerit. *Si* uero *bissextus et saltus* in uno anno contigerint, nullam *mutatio*nem paret. Inturbatam enim initii etatem continent, respondenti superfluo solari die bissexto lunari additur diei saltus.

---

17 quo] quia *M.* **20** initium] *om. M (add. according to* MC*).* **27** luna] lunas *M.* **31** bissexto] bissexti *M.*

the *initium* itself remains undisturbed; and (the *secundum singularis*) has to be moved from the (week)day on which *luna* 14 (i.e. the Easter full moon) falls to another (i.e. to the preceding one). If, in fact, neither a bissextile day nor a *saltus* occur, this reasoning affirms that the *secundum singularis*, by which the *initium* is determined, and *luna* 14 (i.e. the Easter full moon), by which Easter Sunday is regulated, fall, without doubt, on the same (week)day. If, however, the bissextile day occurs before the *initium* and the *secundum singularis*, it causes no disturbance of either. The following (holds true, then,) for all (lunar) ages of the *initium*: the bissextile day moves (them) to an older (lunar age) until it may change (*luna*) 9 to (*luna*) 10. This happens for the reason that more solar days than lunar (days) are known to be in the quadragesimal period in a bissextile year. The *saltus*, on the other hand, changes the (lunar) age of the *initium* to a younger (one), since it lowers the *secundum singularis* to *luna* 1. Consequently, (the following) may happen to any given (lunar) age (of the *initium*): (the *saltus*) causes a decrease (by one of the lunar limits of the *initium*, which usually extend) from the third single digit to the ninth (i.e. *luna* 3 to *luna* 9) (in the Dionysiac reckoning). Accordingly, if *luna* 9 falls on the *initium* when the *saltus* does not appear, it will be moved to (*luna*) 8 when the *saltus* comes in. If, however, a bissextile day and a *saltus* occur in one and the same year, it is evident that no change (will appear). They preserve, in fact, the undisturbed (lunar) age of the *initium*, since the *saltus* is added to the lunar day, while the (then) superfluous solar day corresponds to the bissextile day.

Three examples: AD 719: feria of luna 2, lunar age of initium

Quod ita fieri in tribus annis, in quibus paschalem concursum nuper disputauimus, probari poterit.

In quorum primo anno XIIII luna primi mensis V Idus Aprilis dominico die est. Pascha uero in XVI Kalendas Maii luna XXI. Hinc a II luna singularem, a qua initium regulari innotescit, dominico die esse et initium VIIII luna consequi oportet.

Method for calc. Julian calendar date of initium

Igitur a Kalendis Ianuarii reuerte et epactas ibi | fol. 34v sollicite quęre. Quibus inuentis usque ad lunam Martii recurre, cuius <II luna> initium regulat. In rarisque annis II lunam et initium Aprilis mensis luna addunt, in quibus pascha in luna Maii raro ferunt.

AD 719: Julian calendar date of initium

Itaque in illo anno, in quo luna XIIII in V Idus Aprilis manifestauimus, in Kalendis Ianuarii V luna et in Kalendis Martii. Hinc una luna in V Kalendas Martii, II in IIII Kalendas Martii. Kalendae Februarii et VI Idus et XV et VIII Kalendas <Martii IIII feria>, VII Kalendas V feria, VI Kalendas VI feria, V Kalendas sabbato, IIII Kalendas dominico die, et sequenti dominico initium, in III Nonas Martii, nona luna.

AD 720: feria of luna 2, lunar age of initium

In IIo igitur anno XIIII luna IIII Kalendas Aprilis, VI feria. In quo anno bissextus occurit, et XIIII luna a V feria in VI feria mouet. Hinc pascha dominico sequenti, id est II Kalendas Aprilis luna XVI. Deinde II singularis initii VI feria et initium IIII luna fieri debet.

---

38 regulari] regulare- *M.* **41** sollicite] sollicitas *M.* **42** II luna] *om. M (add. according to* MC*).* **43** lunam] luna *M.* | luna] luna- *M.* | addunt] addit *M.* **44** luna Maii] lunam *(lacuna)* aii *M.* **45** illo] llo *M*.* **47** IIII] II *M (corr. according to* MC*).* **49** Martii…feria[1]] *om. M (add. according to* MC*).*

33–80 Quod…conponatur] MC here outlines two methods for determining the calendrical data of the *initium quadragesimae*. First, the lunar age of the *initium* is calculated from the weekday of the Easter full moon: the Easter full moon (*luna* 14) falls on the same weekday as the *secundum singularis* (i.e. *luna* 2 of the preceding, the March lunation), because these two days are separated by exactly six weeks (since 42 lunar days lie between *luna* 2 of the March lunation (which is full) and *luna* 14 of the April lunation); from this *secundum singularis*, then, the number of days to the following Sunday are counted and added to 2, resulting in the lunar age of the *initium*. Second, the Julian calendar date of the *initium* is calculated from the weekday and lunar data of 1 January; on this basis, the Julian calendar date and weekday of the *secundum singularis* is calculated, with the *initium* then falling on the following Sunday. These calculations are illustrated by the same examples as applied to the calculation of the Julian calendar date of Easter Sunday (c. 56, ll. 15–43), AD 719–721. Of the two disturbances discussed in the previous paragraph, only one occurs in these three years, namely a bissextile day in AD 720. Interestingly enough, MC's principal source, CE p. 120–121, discusses an example each for the impact of the lunar bissextile day and the *saltus* on the *initium* calculation, with the chronological data matching the years AD 568 and 570 respectively.

33–80 Quod…conponatur] MC's methods of calculation outlined here found no imitators, since, probably already at the time of MC, more sophisticated algorithms had been established; e.g., the weekday of the *secundum singularis* was calculated by the same method as the weekday of the Easter full moon, namely by adding a given regular (representing the weekday-difference between the Julian

That this happens in this way should be proven by (the examples of) the three years for which we have just recently discussed the Paschal course.

In the first of these years (i.e. AD 719) *luna* 14 of the first month (i.e. the Easter full moon) falls on Sunday, 9 April, Easter Sunday, however, on 16 April, *luna* 21. Consequently, it is proper that the single digit (that stems) from *luna* 2 (i.e. the *secundum singularis*), by which the *initium* is known to be regulated, falls on a Sunday, and that the *initium* follows on *luna* 9.

Now, go back to 1 January and search carefully for the epacts in that place. After these have been established, return to the March lunation, whose *luna* 2 regulates the *initium*. Only in few years, in which they exceptionally celebrate Easter in the May lunation, do they place *luna* 2 and the *initium* in the lunation of the month of April.

Accordingly, in that year, for which we have demonstrated that *luna* 14 (i.e. the Easter full moon) (falls) on 9 April (i.e. AD 719), *luna* 5 (occurs) on 1 January and 1 March. Consequently, *luna* 1 (occurs) on 26 February, (*luna*) 2 on 27 February. 1 February, 8 February, 15 February, and 22 February (fall) on a Wednesday, 23 February on a Thursday, 24 February on a Friday, 25 February on a Saturday, 26 February on a Sunday, and the *initium* (is) on the following Sunday, on 5 March, *luna* 9.

Thus, in the second year (i.e. AD 720) *luna* 14 (i.e. the Easter full moon) (falls) on Friday, 29 March. In this year a bissextile day occurs, so that *luna* 14 (i.e. the Easter full moon) moves from Thursday to Friday. Consequently, Easter (is) on the following Sunday, i.e. 31 March, *luna* 16. Hence, the *secundum singularis* of the *initium* ought to occur on a Friday and the *initium* on *luna* 4 (if this year was not bissextile).

---

calendar date of the *secundum singularis* and the place of the *concurrentes*, i.e. 24 March) to the *concurrentes* of the year in question; the *initium*, then, occurred on the following Sunday. Cf. BC 3(1282)C15–(1283)C12, LECT. COMP. 1.12, 6.2D–E, LIB. COMP. 2.21, LIB. CALC. 7A,B,F, PV §§23–24, RM 83.49–84.33. On the other hand, lists of the Julian calendar dates of the *secundum singularis* (which was commonly termed *terminus quadragesimae*) were soon established, usually set in parallel to a list of the Easter full moons; cf., e.g., London, British Library, Cotton Caligula A XV, fol. 66v, St Gall, Stiftsbibliothek, 251, p. 2, BC 3(1282)C15–D6, DIAL. NEUSTR. 19 (*luna* 3 instead of *luna* 2), LECT. COMP. I 12, LIB. COMP. 2.21, LIB. CALC. 7B, PV §23, RM 83.49–71. The Julian calendar date of the *initium* was later not calculated from the data of the *secundum singularis*, but from the Julian calendar date of Easter Sunday and certain marker days; cf. BC 51 (cited in COMP. COL. 6.8, LIB. ANN. 30, in turn cited in LIB. COMP. 4.14, LIB. CALC. 56A, RM 88.5–10), LIB. ANN. 31 (cited in LIB. COMP. 4.15, LIB. CALC. 56B, RM 88.11–16). The Julian calendar dates of the *initium* were also included (together with their lunar age) in some Victorian and Dionysiac Easter tables (Victorian: cf. Krusch, *Studien* II, p. 8; DIAL. BURG. 16. Dionysiac: PROL. AQUIT. 3F; the Easter table of AD 798–854 in Appendix 8; the Easter table of AD 817–832 in St Dunstan's classbook (Oxford, Bodleian Library, Auct. F 4 32, fol. 21r)), a tradition that may have originated with the *latercus* (cf. *app. font.* at c. 49, ll. 11–13).

33–80 Quod…conponatur] Cf. Schwartz, 'Ostertafeln', p. 91 and p. LIX–LXI, LXXIII, XCII–XCIV, CLXI of the introduction.

Julian calendar date of *initium*

XVI in Kalendis <Ianuarii>, II feria, XVI et in Kalendis Martii. Inde XVI Kalendas Martii I, II luna in XV Kalendas Martii. Hinc Kalendae <Februarii> in V feria et VI Idus <et XV Kalendas Martii>, et initium in proximo dominico die, hoc est in XII Kalendas Martii et luna V. Itaque II in III in IIII uero initii aetatem in V bissextus mutabit, nam in quadragesimo est.

AD 721: *feria* of *luna* 2, lunar age of *initium*

Tertius autem annus, secundum bissexti terminans endecadem, XIIII lunam XV Kalendas Maii habet V feria. Pascha in XII Kalendas Maii luna XVII. Deinde II singularem luna in V feria, initium | V luna esse ratio poscit. fol. 35r

Julian calendar date of *initium*

XXVII et IIII feria in Kalendis Ianuarii, XXVII in Kalendis Martii sabbato, XXVIII in VI Nonas dominico die, XXVIIII in V Nonas II feria, XXX in IIII Nonas III feria, I luna in III Nonas IIII feria, II luna in II Nonas V feria, initium dominico die in VII Idus Martii et luna V.

Every year likewise

Si quis exordio XVIIII annorum sollicite usque in finem disputet, cautus in bissextili anno. Et interdum saltus non equaliter repperietur. Cupidus disputandi quisque annum saltus curiosius probet, et rationem, quam expressimus, non alias inueniat. Et annum, in quo utrumque contineatur, artificiose licet conponatur.

The relation between the lunar age of *initium* and Easter Sunday

Itaque quicumque hoc certius nosse quaerit, a paschali cursu per quadragensimum retro in initium naturam subtilius conprobet, ut uerbi gratia per ebdomadis cursum hanc supradictam doctrinam notet. Nam *hoc* ordine *VII initi*i *aetates* et *pascha*e, nisi bissextus aut saltus uariet, per ascensionem *secundum Greco*s conputationem concurrunt:

---

58 Ianuarii] *om. M (add. according to* MC*).* | XVI] XXVI *M (corr. according to* MC*).* **59** I] id est *M (corr. according to* MC*; I suspect that the numeral* I *was mistaken for the Irish abbr. for* id est *by the continental copyist).* **60** Februarii] *om. M (add. according to* MC*).* **61** et[1]…Martii] *om. M (add. according to* MC*).* **62** XII] XV *M (corr. according to* MC*).* **70** IIII] VII *M (corr. according to* MC*).* **77** repperietur] repperiet *M.* **80** contineatur] contineat *M.* | artificiose] artificios *M.* **83** ebdomadis] ebdomis *M.* **84** supradictam] supradicto *M**.

81–94 Itaque…paschę] This passage repeats in table format for the Dionysiac reckoning what has already been expressed at the beginning of this chapter, namely that the lunar age of the initium is, save for bissextile and *saltus* years, 12 less than the lunar age of Easter Sunday. In *M* (or an earlier copying stage), the table structure was lost at the end of this passage, with the copyist transcribing the columns in lines (cf. *app. crit.* and Appendix 4); originally, the table must have looked like the one given in CE p. 119 (cf. the facsimile in Appendix 4), which was MC's source, and on which the edition is modelled here.

81–94 Itaque…paschę] Cf. p. XCIII–XCIV of the introduction and Appendix 4.

(*Luna*) 16 (occurs) on 1 January, a Monday, and (*luna*) 16 (also occurs) on 1 March. Consequently, (*luna*) 1 (occurs) on 14 February, *luna* 2 on 15 February. Moreover, 1 February (falls) on a Thursday, as well as 8 February and 15 February, and the *initium* (occurs) on the following Sunday, that is on 18 February and *luna* 5. Accordingly, the bissextile day will change (*luna*) 2 to (*luna*) 3, (*luna* 3) to (*luna*) 4, (*luna* 4,) however, (and with it) the (lunar) age of the *initium*, to (*luna*) 5, since it (i.e. the bissextile day) falls in the quadragesimal period.

The third year (i.e. AD 721), however, following a bissextile year and ending the *hendecas*, has *luna* 14 (i.e. the Easter full moon) on Thursday, 17 April. Easter (is) on 20 April, *luna* 17. The computation requires, then, that the *secundum singularis* (stemming from) the moon falls on a Thursday, the *initium* on *luna* 5.

(*Luna*) 27 and Wednesday occur on 1 January, (*luna*) 27 on 1 March, a Saturday, (*luna*) 28 on 2 March, a Sunday, (*luna*) 29 on 3 March, a Monday, (*luna*) 30 on 4 March, a Tuesday, *luna* 1 on 5 March, a Wednesday, *luna* 2 on 6 March, a Thursday, the *initium* on the following Sunday, 9 March and *luna* 5.

If somebody wants to compute this thoroughly, from the beginning of the 19 years to the end, he should be cautious in bissextile years. And in the meantime the *saltus* will prove to be different. Every person, who is eager to compute this, should examine the year of the *saltus* more diligently, and he will find no other computations than the one which we have outlined. And a balance (between the contrary effects of the *saltus* and the bissexile day) may skilfully be achieved in a year, in which both (i.e. the bissextile day and the *saltus*) happen to be contained.

Hence, whoever hoped to have gotten more precise knowledge about this, should more carefully check the natural rule from the course of Easter backwards through the quadragesimal period to the *initium*, so that, e.g., he should notice that the theory outlined above (goes) backwards through the course of the week (i.e. the weekday of the *secundum singularis* is established by means of the weekday of the Easter full moon). For, in the computation according to the Greeks, the seven (lunar) ages of the *initium* and Easter Sunday run parallel in a linearly increasing fashion in the following order, as long as neither the bissextile day nor the *saltus* alters it:

Luna III initii, *XV* luna paschę.
Luna IIII initii, *XVI* luna paschę.
Luna V initii, *XVII* luna paschę.
Luna VI initii, *XVIII* luna paschę.
Luna VII initii, *XVIIII* luna paschę.
Luna VIII initii, *XX* luna paschę.
Luna VIIII initii, *XXI* luna paschę.

<LVIIII. *DE COMMUNIBUS ET EMBOLISMIS*>

The length of a lunar year

Quid pasche *mensurae*? Id est *XIIII lunę* | et XIII lunae. fol. 35v Quid dominatur? Hoc est dierum numerus, qui ita esse innotescit, Dyonisio dicente: *A XV luna paschalis festi anni, uerbi gratia, precedentis usque ad XIIII sequentis, si communis annus est, CCCLIIII dies habebit, si embolismis* annus est, *CCCLXXXIIII* dies. *Si plus minusue contigerit, euidens error est. Excepto uidelicet primo anno, quem a XIIII luna paschae ultimi, id est XVIIII anni, usque ad XIIII eiusdem primi numerare curamus.* Per *quot idem ultimus*

88–94 Luna…paschę] luna III, luna IIII, luna V, luna VI, luna VII, luna VIII, luna VIIII, initii, initii, initii, initii, initii, initii, initii, XV luna, XVI, XVI (recte XVII), XVIII, XVIIII, XX, XXI, paschę, paschę, paschae, paschae, paschae, paschae, paschę *M (corr. from* CE, DRC, KAL. A1*; it is obvious that this passage was in table format originally, which the copyist then transcribed column by column; cf. the facsimile in Appendix 4).* **LVIIII,2** lunae] lunas *M.* **5** XIIII] XII *M (corr. from* EPISTOLA DIONYSII, *etc.).*

LVIIII,1 DE…EMBOLISMIS] CE p. 113 gives this heading to a chapter with the same content as what follows in MC here. **2–13** Quid…adcommodat] As an authoritative statement on the lengths of common and embolismic years MC here cites a famous passage from EPISTOLA DIONYSII (Krusch, *Studien* II, p. 84; MSS 2,5 have *dies*, MSS 1,3 *XVIIII* instead of *X et VIII*; cited in DTR 42.40–47, in turn cited in PV §297, LIB. COMP. 4.9B, in turn cited in LIB. CALC. 46B; save for the last sentence in DRC 83.3–8; the last sentence only in BC 46C12–D1, DE SALTU LUNAE I (984)C5–9; partially in QUAEST. AUSTR. 2.8A): *A quinta decima luna paschalis festi anni, verbi gratia, praecedentis usque ad XIIII sequentis paschae, si communis annus est, CCCLIIII dies habebit: si embolismus, CCCLXXXIIII <dies>. Quod si unus dies plus minusve contigerit, evidens error est. Excepto videlicet primo anno saepe dicti decennovenalis cycli, quem a XIIII luna paschae ultimi, id est noni decimi anni, usque ad XIIII eiusdem primi numerare curamus. Propter quod idem ultimus epactas, id est adiectiones lunares, X et VIII tunc retinens, primo anno non XI, ut in ceteris annis fieri solet, sed XII dies accommodat.* This passage is also cited in CE p. 113, a quote preceded by questions similar to the ones outlined by MC at the beginning of this chapter: *Quaeritur, unde mensurantur lunę XIIIImę? Hoc est a communibus et embolismis.* For the common and embolismic lunar year consisting of 354 and 384 days respectively cf. also EPISTOLA PASCASINI 2 (Krusch, *Studien* I, p. 248–9): *suggerimus com- annos aput Hebreos XII tantum lunas, hoc est dies CCCLIIII, emb. vero lunas habere XIII, hoc est dies CCCLXXXIIII.* EPISTOLA DIONYSII (Krusch, *Studien* II, p. 83): *Communis autem annus XII lunares menses collegit, qui CCCLIIII dies efficiunt. Embolismus autem annus et lunas XIII et dies CCCLXXXIIII habere monstratur.* ETYM. 6.17.21–22 (cited in BC 140(1354)A7–13, DRC 76.3–4, 78.5–6, DIAL. NEUSTR. 23B): *Communis annus dicitur, qui duodecim tantum lunas, hoc est dies CCCLIV habet. ... Embolismus annus est qui tredecim menses lunares, id est CCCLXXXIV dies habere monstratur.* DNR 6.4.25–28: *Annus lunaris communis est, id est qui per duodecim lunares menses decurrit, id est diebus CCCLIIII. Annus embolismus est qui lunas tredecim et dies CCCLXXXIIII habere monstratur.* and c. 59, ll. 70–71 below. For the interval from *luna* 15 of the Easter month to the following Easter full moon defining the lunar year cf. c. 53, ll. 6–10 and c.

Luna 3 (belongs) to the *initium*, *luna* 15 to Easter Sunday.
Luna 4 (belongs) to the *initium*, *luna* 16 to Easter Sunday.
Luna 5 (belongs) to the *initium*, *luna* 17 to Easter Sunday.
Luna 6 (belongs) to the *initium*, *luna* 18 to Easter Sunday.
Luna 7 (belongs) to the *initium*, *luna* 19 to Easter Sunday.
Luna 8 (belongs) to the *initium*, *luna* 20 to Easter Sunday.
Luna 9 (belongs) to the *initium*, *luna* 21 to Easter Sunday.

59. ABOUT COMMON AND EMBOLISMIC YEARS

What (are) the measures of Easter (i.e. the main criteria for the determination of Easter)? The *lunae* 14 and the 13 (possible) lunations. What determines (these)? The number of days, which is known to be like this, as Dionysius says: (A year) will have 354 days from, e.g., *luna* 15 of the Paschal feast of the previous year to (*luna*) 14 (i.e. the Easter full moon) of the following (year), if it is common, (and) 384 days, if it is embolismic. Should it contain more or less, there is clearly an error. Except for the first year, of course, which we tend to reckon from *luna* 14 of Easter of the final, i.e. the 19th, year to *luna* 14 of Easter of the

59, ll. 70–105 below.

88–94 Luna...paschę] A table very similar to MC's can be found in DRC 100.5–11; it seems that MC was the source for the table itself, while the terminology used in the sentence introducing this table suggests that DRC also consulted CE. Quite remarkable is the fact that KAL. A1 gives two such tables, making a distinction in perspective, one from Easter Sunday, the other from the *initium* (see the facsimile in Appendix 4). These tables displaying the relation between the lunar ages of Easter Sunday and the *initium* are a characteristic feature of Irish computistics of the late seventh, early eighth centuries (which then found its way into the calendar tradition); they were replaced by the simple algorithm of subtracting 12 from the lunar age of Easter Sunday to calculate the lunar age of the *initium*; cf. BC 117D3–7, DE ORDINATIONE FERIARUM PASCHALIUM, *Addendum* (*PL* 90, col. 610A; cited in KAL. A1,3, LIB. ANN. 36, RM 91; partially in LIB. CALC. 56C). The opposite perspective, i.e. adding 12 to the lunar age of the *initium* to calculate the lunar age of Easter Sunday, can be found in KAL. A1. Still, the relation between the lunar age of the *initium* and Easter Sunday is explictly outlined for the first instance (3 and 15) in LECT. COMP. 6.1aB, for all seven instances in PV §29 (both additionly listing the lunar age of Rogation Sunday). Interestingly enough, Marianus Scotus includes this data in an unparalleled table comparing the lunar ages of the *dominicus septuagesimae*, of the *initium quadragesimae*, of the Sunday of Mid-Lent, of Maundy Thursday, of Easter Sunday, of Low Sunday, of Rogation Sunday, of Holy Thursday, of Whit-Sunday, and of the sixth Sunday after Whit-Sunday (Vatican, Biblioteca Apostolica, Pal. Lat. 830, fol. 2v–3r). **LVIIII,2–13** Quid...adcommodat] The fact that common lunar years consist of 354 days, embolismic ones of 384, was obviously common knowledge. Besides the texts citing Dionysius (cf. *app. font.*) cf. COMPUTUS COTTONIANUS fol. 79r (cited in 15B11–13, in turn cited in LIB. ANN. 59), BC 14(1285)C4–D2, 16 (cited in LIB. ANN. 38B), 124C8–9, DTR 36.12–15 (cited in LIB. ANN. 1A, in turn cited in LIB. COMP. 3.4, LIB. CALC. 39, RM 36.14–17, PV §§130–131), 45.4–5 (cited in LIB. CALC. 85, PV §266, RM 59.5–9), 56.12–13 (cited in BC 124(1333)B13–15, LIB. CALC. 92, PV §227, RM 77.14–16), QUAEST. LANGOB. 7A,C, VERS. TUR. 5.11–12, 6.14–15, LIB. COMP. 4.8, LIB. CALC. 48B.

LVIIII,4–13 Dyonisio...adcommodat] Cf. Borst, *Schriften*, p. 406. **4–7** A...dies] Cf. Borst, *Schriften*, p. 406.

*annus epactas, id est adiectiones lunares, XVIII tunc retinens, primo non XI, ut in ceteris annis fieri solet, sed XII ad*commodat.

Origin of 19-year cycle

Unde orta est ogdoadis *regul*a et endecadis et *communium* atque *embolismorum*? Id est ab *Hebr*eis, quibus indicauit Deus *XIIII* lunam primi *mensis post egressi*onem *de Ęgypt*o.

Def. and etym. of common year

*Communis hoc est sol*ari *anno*, *qui* et ipse *XII* menses *habet*. Siue *communis*, *quia duo incedunt* simul *coniuncti* anni.

---

11 XVIII] XVIIII *M (corr. from* EPISTOLA DIONYSII, *etc.; but note that MSS 1, 3 also read* XVIIII*)*. **12** non] *follows* ab *M (corr. according to* EPISTOLA DIONYSII, *etc.)*. **16** mensis] mensi *M (corr. by Borst)*.

14–17 Unde…Ęgypto] The Hebrew origin of the diagnostic features of the 19-year cycle (*ogdoas* and *hendeas*, as well as common and embolismic years) is derived from EPISTOLA DIONYSII (Krusch, *Studien* II, p. 82): *Quia vero in scriptis ipsis commonium annorum et embolismorum mentio facta est et a nonullis haec ratio, quae ex Hebraeorum, ut fertur, traditione discendit, magnopere quaeritur, scire volentibus, utrum huic paternae regulae consonare videatur, necessarium duximus et hanc notitiam, ne probetur in aliquo dissidere coata brevitate degerere*. Paschasinus and ps-Cyril (cf. following note) associate the common years explicitly with the Hebrews, while it is implicit throughout their letters that they considered the entire system to originate from Hebrew computations. The fact that *luna* 14 of the first month was revealed to the Hebrews after the Exodus is derived from NUM 9:1–5: *locutus est Dominus ad Mosen in deserto Sinai anno secundo postquam egressi sunt de terra Aegypti mense primo dicens: faciant filii Israhel phase in tempore suo. quartadecima die mensis huius ad vesperam iuxta omnes caerimonias et iustificationes eius. praecepitque Moses filiis Israhel ut facerent phase. qui fecerunt tempore suo quartadecima die mensis ad vesperam in monte Sinai iuxta omnia quae mandaverat Dominus Mosi fecerunt filii Israhel.* **18–20** Communis…anni] The definition of a common year consisting of 12 lunations and its name being derived from the fact that usually two common years occur in a row in the 19-year cycle (cf. ll. 28–35 of this chapter) is taken from ETYM. 6.17.21 (cited in BC 140(1354)A7–11, DRC 76.3–5): *Communis annus dicitur, qui duodecim tantum lunas, hoc est dies CCCLIV habet. Dictus autem communis quia saepe duo ita coniuncti incedunt ut invicem se in paschali sollemnitate sequantur*. MC's etymology of common year deriving from its commonness with the solar year stems from CE p. 113: *Vel communis, eo quod communis est anno solis, utique enim XII mensibus est, quamvis longitudine inpares*. Cf. also Isidore's source EPISTOLA PASCASINI 2 (Krusch, *Studien* I, p. 248): *suggerimus com- annos aput Hebreos XII tantum lunas, hoc est dies CCCLIIII*. EPISTOLA DIONYSII (Krusch, *Studien* II, p. 83): *Communis autem annus XII lunares menses collegit, qui CCCLIIII dies efficiunt*. EPISTOLA CYRILLI 4 (Krusch, *Studien* I, p. 346): *ut numeremus XII tantum lunas iuxta supputationem legalem Hebreorum in anno communi*. DNR 6.4.25–27: *Annus lunaris communis est, id est qui per duodecim lunares menses decurrit, id est diebus CCCLIIII*.

14–17 Unde…Ęgypto] The order of the Easter full moons as well as of the common and embolismic years is also credited to the Hebrews in DRC 94.1–2, which presumably depends on MC here: *Sciendum nobis utrum hic ordo fuit in XIIII luna primi mensis et in commonibus et in embolisimis apud Ebreos primitus*. Similar, and clearly dependent on Irish computistical thought, is COMP. COL. 4.3A, 5.11B: *In novemdecim annis quis vel qui invenerunt communes et embolismos? Et ubi est profectus eorum? Et ubi est exemplum de embolismis in lege divina? Quid nocet, si non sunt communes et embolismi? Hic est profectus eorum et inventio eorum: Moysi divina revelatione ostensum est, ut dixi, embolismus, et postea sapientes Graecorum ogdoados et endecados, communes et embolismos instituerunt. ... Et quaeri debet, si interdum communes, interdum embolismos observabant Ebraei? Vere observabant, ut ostendit*

first (year). This is for the reason that this final year, then retaining an epact, i.e. a lunar increment, of 18, does not add 11 to the first (year), as usually happens in the other years, but 12.

Whence did the rule of the *ogdoas* and *hendecas* and of common and embolismic years originate? From the Hebrews, to whom God revealed *luna* 14 of the first month after the Exodus from Egypt.

A common year is like a solar year, (since) this and that have 12 months. Or (it is called) common (*communis*), since two (such years usually) come along at the same time (i.e. they follow each other) like joined (*coniuncti*) years.

---

eis Moyses divina revelatione, ne in quarta decima luna ante XI Kalendas Aprilis Pascha celebrarent. Bede (DTR 44.1–5; cited in LIB. CALC. 84, PV §203) ascribes the origin of the 19-year cycle to Eusebius of Caesarea, presumably taking this information from JEROME, *De viris illustribus* 61 (Herding, *De viris inlustribus*, p. 41–2). **18–20** Communis…anni] Isidore's account of the common lunar year was very popular in early eighth-century Irish or Irish influenced texts; cf. *app. font.* and DIAL. NEUSTR. 23B: *Communes igitur anni unde dicuntur? Quia duo saepe coniuncti sunt. Qui duodecim lunas tantum habent.* For a common lunar year consisting of 12 months cf. DTR 36.12–13 (cited in LIB. ANN. 1A, in turn cited in LIB. COMP. 3.4, LIB. CALC. 39, RM 36.14–15, PV §130), 56.12 (cited in LIB. CALC. 92, PV §227, RM 77.14–15), DIAL. BURG. 13A, LIB. ANN. 56D, COMP. COL. 4.6C, LIB. COMP. 4.7A, LIB. CALC. 48B. A neat definition of the common year in relation to Easter calculation can be found in HELPERIC, *De computo* 17 (*PL* 137, col. 32A–B).

14–17 Unde…Ęgypto] Transcribed in Borst, *Schriften*, p. 942. **19–20** communis…anni] Transcribed in Borst, *Schriften*, p. 405

Def. of embolismic year

Embolismus est seriunctus annus, ut Augustinus ait: Seriunctus annus, *qui communium annorum <damna> supplet*. Siue *embolismus Grecus*, *longus* siue *superaugmentatus* annus *Latinę*. *Ipse est* qui *Moysi* communis | *reuelatus* est, *in quo iubentur hii, qui longius habitabant*, et coinquinati, *qui in primo mense pascha facere non poterant*, *in secundo mense pascha celebrare*. fol. 36r

---

21 est] et *M (corr. by Borst).* | seriunctus] *corr. to* seiunctus *Borst.* **22** annus] an *(lacuna)* nus *M.* damna] *om. M (add. from* EPISTOLA DIONYSII, *etc.).*

21–27 Embolismus…celebrare] The attribution of the definition of an embolismic year as a composed (*seriunctus*) year (i.e. composed of a common year of 354 days plus an embolism of 30 days) to a certain Augustine may, in fact, point to one of MC's teachers (cf. p. CXXIX–CXXXI of the introduction); the fact that embolismic years make up for the eleven-day difference between common lunar and solar years is copied from EPISTOLA DIONYSII (Krusch, *Studien* II, p. 83; cited in DRC 78.9–11, BC 17C12–13, DTR 45.20–22, in turn cited in LIB. CALC. 85, RM 59.19–22): *Embolismorum autem ista ratio probatur existere, quod annorum communium videtur damna supplere.* Cf. also Dionysius' source QUINTUS IULIUS HILARIANUS, *Expositum de die paschae et mensis* 13 (*PL* 13, col. 1113B): *et damna communium annorum eadem luna annis embolysmis compenasse.* EPISTOLA DIONYSII (Krusch, *Studien* II, p. 83): *Haec est igitur embolismorum, sicut praediximus, ratio, ut incrementis suis communium annorum detrimenta compensent.* CE p. 113: *Embolismus enim ergo superadditus interpretatur, id est annus damna enim communis inplet.* The theory that *embolismus* may have been a Greek term, to be interpreted as *superaugmentatus* in Latin, derives from ETYM. 6.17.23 (cited BC 140(1354)B1–4, DRC 77.4–6, DIAL. NEUSTR. 23B): *Embolismus autem nomen Graecum est, quod interpretatur Latine superaugmentum; eo quod expleat numerum annorum communium, quibus undecim lunares dies deesse cernuntur.* The term *longus annus* used by MC for an embolismic year is extremely uncommon in early medieval computistics and appears to stem from EPISTOLA PASCASINI 2 (Krusch, *Studien* I, p. 249): *Necessitate ergo embolismi cogimur id tenere, quod longum est, ne a vero deviare videamur.* Cf. DNR 6.4.27–28: *Annus embolismus est qui lunas tredecim et dies CCCLXXXIIII habere monstratur; in quo anno longius dies paschae protenditur.* The connection between the Exodus from Egypt and the origin of the embolismic year is copied from ETYM. 6.17.22 (cited in BC 140(1354)A13–15, DRC 78.6–8, which reads *est reuelatus* and *hí*): *Ipse est annus sancto Moysi divinitus revelatus, in quo iubentur hi, qui longius habitabant, in secundo mense pascha celebrare.* For the unclean cf. NUM 9:6–11: *ecce autem quidam inmundi super animam hominis qui non poterant facere pascha in die illo accedentes ad Mosen et Aaron. dixerunt eis inmundi sumus super animam hominis quare fraudamur ut non valeamus offerre oblationem Domino in tempore suo inter filios Israhel. quibus respondit Moses state ut consulam quid praecipiat Dominus de vobis. locutusque est Dominus ad Mosen dicens. loquere filiis Israhel homo qui fuerit inmundus super anima sive in via procul in gente vestra faciat phase Domino. mense secundo quartadecima die mensis ad vesperam cum azymis et lactucis agrestibus comedent illud.* The phrase that characterizes these unclean seems to have been taken from PS-JEROME, *De exodo, in vigilia Paschae* (*CCSL* 78, p. 537): *Praecipitur et in alio loco, ut qui in primo mense pascha facere non potuerit, secundo faciat.*

21–27 Embolismus…celebrare] The fact that the embolisms make up for the difference between the common lunar and the solar years is often cited in eighth-century computistical literature; cf. *app. font.* and BC 14C11–14: *quod etiam in embolissimorum ratione dirigens preseruatur, poteris inuenire, quorum uis talis est, ut pro sui longitudinem (!) communium annorum detrimentam conpensant.* LIB. ANN. 56D: *quia haec est ratio, ut incrementis suis communium annorum detrimenta suppleant.* VERS. TUR. 7.16–16a: *Embolismi incremento magna larga copia anni lunae brevioris suppletur inopia. Ac per istam sui semper incrementi copiam breviorem longiores compensant inopiam.* For *embolismus* being Greek for Latin *superaugmentum* cf. the texts citing Isidore in *app. font.* and BC 4D4–5, 14C14–D1: *Embolismus uero Greci quod in Latinum uertitur augmentum siue maiorem annum. ... Embolismus*

An embolismic is a composed (*seriunctus*) year (i.e. composed of a common year and an embolismic month), as Augustine says: A composed (*seriunctus*) year, which compensates for the deficits of the common years. Or *embolismus* is a Greek (term), (which means) a long or extremely increased year in Latin. This is (the year), which was revealed to Moses as a common year, in which the ones, who dwelled longer (i.e. the ones who could not travel at the same speed), and the unclean, who could not celebrate the pasch in the first month, were told to celebrate the pasch in the second month.

---

etenim Greci, Latine quia superaugmentum dicitur, quia id est annus XIII lunaris mensibus et diebus CCCLXXXIIII impletur. DRC 77.1–4 (followed by the Isidorian citation given in *app. font.*): *Sciendum nobis quid est embolismus. Nomen Grecum. Embolima enim Grece superhabundantia interpretatur, et embolismus Grece superaugmentum interpretatur, eo quod triginta dies, id est XIII lunaris mensis, ad finem anni commonis adduntur*. DTR 36.14–15 (cited in LIB. ANN. 1A, in turn cited in LIB. COMP. 3.4, LIB. CALC. 39, RM 36.15–16, PV §131): *Quartus qui embolismus graece dicitur, id est superaugmentum*. For the unusual term *longus annus* describing an embolism cf. VERS. TUR. 5.11: *Longus annus embolismi supercrescens dicitur*. (in this respect cf. a gloss to DTR (*CCSL* 123B, p. 275): *embolismus, sc. supercrescens, id est qui habet XIII menses*. HELPERIC, *De computo* 17 (*PL* 137, col. 32B): *Et hic annus embolismalis, id est, supercrescens dicitur*.) The theory that the embolismic year originated at the time of the Exodus was very popular in eighth-century Irish and Irish influenced texts, but apparently only in those texts: cf. *app. font.* (note DRC's dependence on MC) and COMP. COL. 4.3A: *Moysi divina revelatione ostensum est, ut dixi, embolismus*. A neat definition of the embolismic year in relation to Easter calculation can be found in HELPERIC, *De computo* 17 (*PL* 137, col. 32B).

21–24 Embolismus...Latinę] Partially transcribed in Borst, *Schriften*, p. 406, proposing *seiunctus* instead of *seriunctus* (cf. *app. crit.*). **24–27** Ipse...celebrare] Cf. Borst, *Schriften*, p. 1046.

The sequence of common and embolismic years

In communibus et embolismis ordo euentus est: per decennouenalem ciclum tertio embolismi occurrunt, nisi in ultimo anno ogdoadis et endecadis, hoc ordine: *Communis, communis, embolismus, communis, communis, embolismus, communis, embolismus, ogdoas; communis, communis, embolismus, communis, communis, embolismus, communis, communis, embolismus, communis, embolismus, endecas.*

Etym. of *ogd.* and *hend.*

Ogdoas Grece Latinae *VIII*, hoc est *anni*. *Endecas* Grece Latinę *XI*. Ena enim Grece Latine unum, deca X.

---

28 est] et *M.* **29** embolismi] embolismum *M.* **31–35** Communis…endecas] *This sequence is arranged in a table-like, vertical fashion in M (cf. facsimile in Appendix 5).*

28–35 In…endecas] An identical vertical list can be found in CE p. 113, and the close similarity in the layout alone makes it obvious that this was MC's source here (cf. the facsimiles in Appendix 5). A column indicating common and embolismic years was part of Isidore's Easter table (ETYM. 6.17.5), and the vertical structure of CE's and MC's lists suggests that this was the source for seventh–century Irish computists. The sequence of common and embolismic years is first described in Latin in EPISTOLA DIONYSII (Krusch, *Studien* II, p. 82–3): *Ogdoas ergo, quae incipit a primo decennovenali cyclo, qui est lunaris decimus septimus, ac ratione peragitur, ut annos primum et secundum communes, id est minores habeat, tertium embolismum, id est maiorem, annum quartum et quintum item communes, sextum embolismum, septimum communem, octavum embolismum. Ac per hoc ogduadis communes anni quinque et tres embolismi iugiter adscribuntur. ... Item endecas hac lege decurrit. Incipit a nono cyclo decennovenali, qui est lunaris sextus, cuius primus et secundus annus communis est, tertius embolismus, quartus et quintus communis, sextus embolismus, septimus et octavus communes, nonus embolismus, decimus communis, undecimus embolismus: sicque endecas communibus anni septem, embolismis IIII terminatur*. Dionysius' source, EPISTOLA PASCASINI 1 (Krusch, *Studien* I, p. 248), only describes the *ogdoas*: *Coepit ergo ogdoas ... . Cuius ratio haec est: Duo sunt priores anni com-, III embo., iterum IIII et V com-, VI embo., VII com-, VIII embo.* **36–37** Ogdoas…X] MC probably copied the fact that *ogdoas* and *hendecas* are the Greek terms for 'eight years' and 'eleven years' respectively from CE p. 113: *Ogdoas VIII anni, endecas XI anni interpretatur*. Cf. also EPISTOLA DIONYSII (Krusch, *Studien* II, p. 82; cited in DRC 81.3–5): *Noverimus itaque, quia idem decennovenalis circulus per ogdoadem et endecadem semper in sese revolvitur. Octo namque et XI ipse numerus explicatur.*

28–35 In…endecas] Cf. DRC 80: *Sciendum nobis quomodo lunares anni in decinnouenali ciclo currunt. Ita etiam, id est duo commones simul et tertio anno embolesmus, excepto fine ogduadis et endicadis, ubi precedit unus commonis embolesmus propter rationem paschalem. ... Sic itaque currunt: Commonis, commonis, embolesmus, commonis, commonis, embolesmus, commonis, embolesmus. Hucusque ogduas. Item endicas: Commonis, commonis, embolesmus, commonis, commonis, embolesmus, commonis, commonis, embolesmus, commonis, embolesmus. Hucusque endicas*. This sequence of common and embolismic years is also listed separately in BC 13A9–12, 14B10–C4, DTR 45.5–11 (cited in LIB. CALC. 85, PV §267, RM 59.7–13), partially 56.8–12 (cited in LIB. CALC. 92, PV §§226–227, RM 77.9–14, NOTKER LABEO, *De quatuor questionibus compoti* (Piper, *Nachträge*, p. 317)), DIAL. NEUSTR. 22A. From the middle of the eighth century, this list appears more commonly as a column in tables of various forms. Cf. BC 3(1283)C5–7, LECT. COMP. 1.13, LIB. ANN. 14, LIB. COMP. 2.22A–B, LIB. CALC. 11B, RM 83.72–94. **36–37** Ogdoas…X] DRC 81.1–3 (followed by the Dionysiac citation given in *app. font.*) copies MC's explanation of *ogdoas* and *hendecas* directly: *Sciendum nobis quod Greca sunt ogduas et endicas. Ogduas enim Grece octo interpretatur. Enna enim Grece unum Latine; deca uero Grece decim Latine*. Similar statements can be found in BC 13A9, 14B8–9: *Sunt autem ogdoaden VIII, endegaden XI ... per ogdoaden et endegaden in semper in se reuoluitur, id est per VIII et XI annis*. DT 11.4–5 (cited in RM 57.8–9): *Qui dividitur in ogdoadas et*

A sequence emerged among common and embolismic years: throughout the 19-year cycle the embolisms occur in every third (year), save for the final year of the *ogdoas* and of the *hendecas* (respectively), in this order: Common year, common year, embolismic year, common year, common year, embolismic year, common year, embolismic year, (thus far) the *ogdoas*; common year, common year, embolismic year, common year, common year, embolismic year, common year, common year, embolismic year, common year, embolismic year, (thus far) the *hendecas*.

Ogdoas in Greek (means) eight in Latin, i.e. (eight) years. *Hendecas* in Greek (means) eleven in Latin. Indeed, *ena* in Greek (means) one in Latin, *deca* (means) ten.

---

hendecadas, hoc est in VIII et XI annos. DTR 46.2–3 (cited in LIB. CALC. 86, PV §272): *Est et alia praefati circuli divisio qua in ogdoadem et hendecadem, id est in VIII et in XI, distinguitur annos*. PROL. AQUIT. 3B: *Nam quia apud Graecos per ogdoadem et endecadem, id est octo et undecim, semper decurrit*. DIAL. NEUSTR. 22B: *In ogdoade namque octo anni sunt, ... In endecade vero undecim anni sunt*. LIB. ANN. 56A: *Ogdoas octo anni, endegas undecim anni*. VERS. TUR. 14.34: *Ogdoadi deputantur octo anni priores, Endecadis una dena conprobatur series*.

28–35 In…endecas] Cf. Borst, *Schriften*, p. 404, p. CXXXIV, CXCIV of the introduction and Appendix 5. **36–37** Ogdoas…X] Cf. p. CXCIV of the introduction.

Determ. com./emb. The first year of the Dionysiac 19-year cycle

Sic ogdoadem et endecadem ab incensione usque ad defectionem eius probamus:

Hic annus incipit prima luna in II Nonas Aprilis. Ista luna primi mensis et luna Maii et initium communis in anno, quo VIII luna in Kalendis Ianuarii, in quo communis XII lunae sunt hoc modo: Luna Maii I, II Iunii, III: In hac uice <Iunius> II lunas continet: XXX in Kalendis eius <ad> naturam et accenditur luna in IIII Nonas eius, id est XXVIIII, quae deficit in II Kalendas Iulii, I luna cadens inter duas Kalendas et sine mense haec luna. Sic reliqua usque luna Martii, quae in XI Kalendas Aprilis finitur, id est finis communis.

---

38 ogdoadem] ogdoadis *M (presumably wrong expansion of the suspension* ogd- *by the copyist).* **40** Nonas] k- *M (corr. according to* MC*).* **42** VIII] VII *M (corr. according to* MC*).* | luna] lunas *M (presumably wrong expansion of the suspension* lu- *by the copyist).* **43** lunae] lunas *M (presumably wrong expansion of the suspension* lu- *by the copyist).* | Luna] lunas *M (presumably wrong expansion of the suspension* lu- *by the copyist).* **44** Iunius] *om. M (add. according to* MC*).* | lunas] luna *M (presumably wrong expansion of the suspension* lu- *by the copyist).* | ad] *om. M.* **49** communis] commmunis *M.*

38–69 Sic…ibis] MC here outlines the first of two methods for proving the sequence of common and embolismic years underlying the Dionysiac 19-year cycle. The proof is rather simple in that MC only counts the number of lunations occurring within the respective lunar years, with 12 lunations indicating a common, 13 an embolismic year. The basis of this proof is formed by the definition of the 19 lunar years in c. 57; for the first moon of the Easter lunation being regarded as the beginning of the lunar year cf. also c. 53, ll. 6–11. There is no apparent source for this proof; the Augustine to whom this proof is ascribed at the end may, in fact, refer to one of MC's teachers (cf. p. CXXIX–CXXXI of the introduction).

38–69 Sic…ibis] Two aspects are noteworthy in this unparalleled proof of the sequence of common and embolismic years in the 19-year cycle: 1) MC is consistent here in defining the lunar years by the lunar age of the 1 January that falls within the given period, and not of the 1 January of the calendar year in which the respective lunar year begins. 2) MC describes instances of lunations falling between two Calends and which are therefore not to be attributed to any Julian calendar month; interestingly enough, MC lacks the terminology for this concept; such a lunation is defined as *luna abortiva* in DRC 70 (copied as a separate tract called DE DUABUS LUNATIONIBUS in the ninth century (edited in Appendix 6); partially in St Dunstan's classbook (Oxford, Bodleian Library, Auct. F 4 32, fol. 22v); cf. also Ó Cróinín, 'A seventh-century Irish computus', p. 125–6, idem, 'Computistical works', p. 54–5): *Sciendum nobis, cum duae lunae in uno mense inueniuntur, quae est de illis ipsius mensis. Quae finitur in mense, ipsa est luna ipsius mensis. Luna uero sequens, si non peringerit Kalendis mensis sequentis, abortiua luna uocabitur, quam alii dicunt lunam esse embolismi, quod non est uerum, ut Palumbus: Illi mensi conputa lunam, quae in eo finitur, non quae incipit. Unde, quae primum prioris aut initium sequentis non tenet, abortiua dicitur*. In c. 73, DRC gives a full disussion of all *lunae abortivae* in the Dionysiac 19-year cycle; the two *lunae abortivae* mentioned in MC can be found (in reversed order, since DRC bases its discussion on Julian calendar years, so that the first nine lunations of the first lunar year in MC are part of the final year discussed in DRC) in DRC 73.8–10, 62–69 (this chapter is copied with minor ellipses in St Dunstan's classbook (Oxford, Bodleian Library, Auct. F 4 32, fol. 22r–v); it appears in an altered version under the title DE CONTRARI AETATIBUS in the ninth century (edited in Appendix 6)): *In anno quoque in quo fit XX luna in Kalendis Ianuarii, abortiua luna XXX in Decimbri inuenitur. ... In anno quoque in quo fit XXVII luna in Kalendis Ianuarii, in illo anno XXX luna Apreli fit, causa XIIII lunae primi mensis, quae in hoc anno in XV Kalendis Maii inuenitur. XXVIIII uero luna Maio et*

Let us examine the *ogdoas* and the *hendecas* from the kindling (of the moon) to (its) disappearance as follows:

This year (i.e. the first year of the Dionysiac 19-year cylce) begins with *luna* 1 on 4 April. This (is) the lunation of the first month and (at the same time) the lunation of May, as well as the beginning of a common year, in the year in which *luna* 8 (occurs) on (the following) 1 January, (and) in which the twelve lunations of a common year occur in the following way: The first lunation (belongs) to May, the second to June, the third: At this turn, (the Julian calendar month of) June contains two lunations: (*luna*) 30 occurs on the first day of that (month) according to nature and the (following) lunation is kindled on the second day of that (month), i.e. (a lunation of) 29 (days), which disappears on 30 June. Hence, one lunation falls between two Calends, and this lunation (is) without month (i.e. a *luna abortiva*). (And) so on up to the lunation of March, which is terminated on 22 March, i.e. the end of (this) common year.

---

econtrario luna XXX Iunio; et eadem aetas in hoc anno, id est XXVIII in Kalendis Februarii et Aprelis et Maii. Et in hoc anno transilicio inter Maium et Iunium inuenitur. XXVIII enim luna in Kalendis Maii et XXX luna in Kalendis Iunii inuenitur, et abortiua luna XXVIIII in Iulio (recte MSS Iunio). Et ab Iulio usque in finem anni nihil perturbationis inuenitur. Interestingly enough, the term *luna abortiva* is also used in a discussion of the year starting with epact 27 of 1 January (the first year in MC, the final year in DRC) in Zürich, Zentralbibliothek, Car C 180, fol. 104v–106v; yet, the *luna abortiva* (of 29 days) is there placed between 3 and 31 May (which means that two hollow lunations occur in a row). For the *luna abortiva* cf. also the SIRMOND MS (Oxford, Bodleian Library, Bodley 309, 106v–107r; the relevant passage transcribed in Ó Cróinín, *Cummian's letter*, p. 178–9, idem, 'Computistical works', p. 55, idem, 'Bede's Irish Computus', p. 206–7), Oxford, Bodleian library, Digby 63, fol. 33v. QUAEST. AUSTR. and CAP. COMP. also struggled with the problem of certain lunations falling between two Calends, apparently not being aware of the Irish concepts dealing with this phenomenon. QUAEST. AUSTR. 1.10C: *Et notandum est de luna, quae tricesima est: <Cur> conputatur in Kalendis, quocumque mense acciderit, ut in eisdem extinguatur, et alia accendatur in IIII Nonas vel in VI Nonas sequentis mensis et in eodem mense extinguatur? Quae luna ex his duabus deputabitur, si dici fas est, esse superflua, id est tertia decima, et cui mensi aptanda erit? <Et an> debent duodecim semper esse tricesimae?* CAP. COMP. 20: *De eo, quod, cum omnis lunatio eius mensis, in quem finierit, esse putatur, si aliqua secunda die mensis aut tertia inciperet et ante ultimum eiusdem mensis diem finita fuerit, cui deputari debeat, interrogandum est. Hoc autem propter embolismi fieri rationem responderunt.*

38–69 Sic…ibis] Cf. p. XXII, LXX, CLXV of the introduction. **38–105** Sic…curabis] Cf. p. XCIII–XCIV of the introduction.

The second year

I luna in X Kalendas Aprilis. Ista luna primi mensis et luna Aprilis et initium communis | in anno, in quo luna fol. 36v
XX in Kalendis Ianuarii, in <quo> communis XII lunae sunt hoc modo: I luna Aprilis, II Maii, sic usque XII luna Martii, quae XXX in V Idus Martii et finis communis.

The third year

Prima luna in IIII Idus Martii. Ista luna primi mensis et luna Aprilis et initium embolismi in anno, in quo luna I in Kalendis Ianuarii, in quo XIII lunae sunt hoc modo: prima luna Aprilis, II luna Maii, III Iunii, sic reliqua usque X. X enim II lunas continet, id est luna XXVIIII, quae defecit in Kalendis eius, et ista est luna sua, et accenditur luna XXX in IIII Nonas eius, extinguitur in II Kalendas Ianuarii, I luna cadens inter duas Kalendas. Sic usque XIII Martii luna, quae XXX in VI Kalendas Aprilis et ⌐ ipsa est luna eius ¬ finis embolismi.

Description of the method: Counting the number of lunations

Augustinus ait: Omnis ratio embolismorum: in fine numeramus. Non aliter per decennouenalem ciclum ab accensione lunae primi mensis usque ad defectionem lunae finientis annum caute discernens communes et embolismos ibis.

---

51 in[2]] om. *M* *. **52** lunae] lunas *M (presumably wrong expansion of the suspension* lu- *by the copyist).* **54** Martii quae] Martiiq: *M (it is obvious that the continental copyist read* Martiique, *but the original reading must have been* Martii quae, *with the Irish abbr. for* quae, *or* Martii que*).* **57** lunae] lunas *M (presumably wrong expansion of the suspension* lu- *by the copyist).* **62** inter] in *M (corr. according to* MC*).* | XIII] XVI *M (corr. according to* MC*).* **63** luna quae] lunaq: *M (it is obvious that the continental copyist read* lunaque, *but the original reading must have been* luna quae, *with the Irish abbr. for* quae, *or* luna que*).* **63–64** ipsa…eius] *This phrase seems to have been a gloss that crept into the text at a later copying stage, since it has no analogy in the previous two passages and does not make much sense here.* **68** discernens] discernes *M.*

(In the second year of the Dionysiac 19-year cycle) *luna* 1 (occurs) on 23 March. This (is) the lunation of the first month and (at the same time) the lunation of April, as well as the beginning of a common year, in the year in which *luna* 20 (occurs) on (the following) 1 January, (and) in which the 12 lunations of a common year occur in the following way: the first lunation (belongs) to April, the second to May, (and) so on until the twelfth lunation (belongs) to March, which (has its *luna*) 30 on 11 March, and (this is) the end of (this) common year.

(In the third year of the Dionyiac 19-year cycle) *luna* 1 (occurs) on 12 March. This (is) the lunation of the first month and (at the same time) the lunation of April, as well as the beginning of an embolismic year, in the year, in which *luna* 1 (occurs) on (the following) 1 January, (and) in which the 13 lunations occur in the following way: The first lunation (belongs) to April, the second to May, the third to June, (and) so on up to the tenth. In fact, the tenth (month) (i.e. December) contains two lunations, i.e. a lunation of 29 (days), which disappears on the first day of that (month), and this is the lunation of that month, and a lunation of 30 (days) is kindled on the second day of that (month), (and) it is extinguished on 31 December. Hence, one lunation falls between two Calends (i.e. a *luna abortiva*). (And) so on until the 13th lunation (belongs) to March, which (has its *luna*) 30 on 27 March ⌐ and this is the lunation of that month ¬ , (and this is) the end of the embolismic year.

Augustine says: (Concerning) every reckoning of the embolisms: we count (the number of lunations of a year) at the end (of the year). In no other way will you go through the 19-year cycle from the kindling of the lunation of the first month to the disappearance of the lunation that ends the year, carefully separating common and embolismic years.

Another method: Calculating the days of the years of the 19-year cycle

In communibus et embolismis numeremus, utrum *communis CCCLIIII* an *embolismus CCCLXXXIIII dies habeat*; probantes naturam solliciti sciamus. Itaque *a XV luna* primi mensis *in XIIII* eiusdem II annos iuxta lunam inuestigemus. Secundum uero solem a XIIII Kalendas Maii in XV Kalendas Maii rimari obseruemus. *Hinc a VI Nonas Maii in XI Kalendas Aprilis* annorum XVIIII ciclum in

---

71 an] et *M.* **73** annos] anni *M.* **74** Kalendas] luna *M (it seems that the copyist mistook the suspension* k- *for* lu-*).* **75** Kalendas] lu- *M (it seems that the copyist mistook the suspension* k- *for* lu-*).*

70–105 In…curabis] MC here outlines an alternative method for proving the order of common and embolismic years, namely by counting the number of days of each lunar year, with 354 days indicating common, 384 days embolismic years. This method MC directly copied from CE p. 113–4: *Hinc in tali conputatione medietatem quandam quasi centrum in medio ponamus, id est ut si aliquid ante vel post addamus. Illa medietas incommutabilis stet, id est a VI Nonas Maii usque ad XI Kalendas Aprilis diebus conputatis CCCXXV. Hinc primum communem ponamus, hoc est a XV Kalendas Maii, quo die nouissimus annus decennouenalis cicli terminatur, usque in Nonis Aprilis. Inde a XIIII Kalendas Maii* (follows mistakenly *dies*) *usque in VI Nonas Maii dies XIIII sunt. A VI Nonas Maii usque in XI Kalendas Aprilis CCCXXV dies sunt. Adde XIIII dies a X Kalendas Aprilis usque in Nonis Aprilis. Diebus omnibus conputatis hunc numerum inuenies CCCLIII. Hinc euidens error est. Et ne hic error sit in numero, dies XV Kalendas Maii diuidendus est in II dies. Et hęc diuisio secundum solem et non secundum lunam est, sed artificaliter. A XV enim Kalendas Maii usque in Nonis Aprilis dies CCCLIIII sunt secundum solem et lunam addito die salti. Hinc secundum Dionisium a XIIIIma luna usque ad XIIIImam hic primus annus conputatur. Communis* (*MS commu*) *secundus incipit ab VIII Idus Aprilis usque in VIII Kalendas Aprilis. Ab VIII Idus autem Aprilis usque in VI Nonas Maii dies* (follows *dies* by mistaken repetition) *XXVI sunt. A VI Nonas Maii usque in XI Kalendas Aprilis CCCXXV. A XI Kalendas usque* (follows *usque* by mistaken repetition) *in VIII Kalendas Aprilis III dies sunt. Hos autem iungens fecisti dies CCCLIIII. Embolismus: A VII Kalendas Aprilis usque in VI Nonas Maii XXXVII dies sunt. A VI Nonas Maii usque in XI Kalendas Aprilis CCCXXV dies sunt. A X Kalendas Aprilis usque in Idibus Aprilis XXII dies sunt. Hos autem coniungendo fecisti dies CCCLXXXIIII. Sic talem numerum per XIIIImas lunas in decennouenali ciclo unum quemque communem et embolismum conputando repperies*. In this method a lunar year is defined as extending from *luna* 15 of the Easter lunation to the following Easter full moon, which reveals that the ultimate model for this proof is EPISTOLA DIONYSII: The general method (Krusch, *Studien* II, p. 84–5; the first passage cited in CE p. 113, DRC 83.3–5, DTR 42.40–42, in turn cited in PV §297, LIB. COMP. 4.9B, in turn cited in LIB. CALC. 46B; only the lengths of the years in QUAEST. AUSTR. 2.8A): *A quinta decima luna paschalis festi anni, verbi gratia, praecedentis usque ad XIIII sequentis paschae, si communis annus est, CCCLIIII dies habebit: si embolismus, CCCLXXXIIII <dies>. ... et ideo, sicut diximus, a XV luna paschae primi cycli, usque ad finem eius in communibus et embolismis annis praefixos dies nos invenire, non dubium est. ... Et ita semper, quotiens dubitatio talis occurrerit, a XV luna transactae festivitatis, usque in decimam quartam paschae, quam quaerimus, dies sollicite conputemus. Et si communis annus est, CCCLIIII dies, si embolismus, CCCLXXXIIII dies inveniemus*. Similarly ETYM. 6.17.24 (cited in BC 140(1354)B5–8, in reversed order BC 14(1285)D4–7, both reading *XV* instead of the first *quarta decima*): *Si enim a quarta decima luna paschae praecedentis usque ad quartam decimam sequentis CCCLXXXIV dies fuerint, embolismus annus est; si CCCLIV, communis <est>*. In detail, however, Dionysius' method of counting the number of lunar days per year differs from the Irish texts, as his example illustrates (Krusch, *Studien* II, p. 84): *Atque ut hoc manifestius possit intellegi, praesentis anni monstremus exemplo. Indictio quippe quarta est, et lunaris ciclus undecimus, decennovenalis cyclus decimus quartus. Et quoniam endecadis sextus annus est, eum embolismum esse necesse est. A XV itaque luna praeteriti festi, usque ad decimam quartam praesentis, quot sunt dies, diligentius inquiramus; et inveniemus procul dubio, quando pascha celebrare debemus. Transacto anno per indictionem tertiam in pascha lunam decimam quartam nono Kalendarum Aprilium*

We should count (the number of days) in common and embolismic years (to verify), whether it has 354 days like a common year, or 384 like an embolismic year; (with this method) we should (be able to) understand the nature (of each year) by examining (each in turn) carefully. Consequently, according to the moon, we should investigate (these) two (types of) years from *luna* 15 of the first month to *luna* 14 of the same (i.e. the Easter full moon of the following year). According to the sun, however, we should reckon (a year) from 18 April (i.e. the latest possible Julian calendar date of the Easter full moon) to 17 April. Hence we should recognize (the fact) that 325 days are found indifferently between 2 May

---

die, id est XXIIII mensis Martii fuisse, quis dubitet, qui curam rei huius habere quantulumcunque cognoscitur? Et ideo ab VII Kalendarum Aprilium die numerandi sumamus exordium: habemus Martii dies VII, Aprilis XXX, Maii XXXI, Iunii XXX, Iulii XXXI, Augusti XXXI, Septembris XXX, Octobris XXXI, Novembris XXX, Decembris XXXI, Ianuarii XXXI, Februarii XXVIII, Martii XXXI, et Aprilis XII, quod est pridie Idus Aprilis. Fiunt simul CCCLXXXIIII. The difficulty of the first year (Krusch, *Studien* II, p. 84; cited in c. 59, ll. 7–13, CE p. 113, DTR 42.42–47, in turn cited in PV §297, LIB. COMP. 4.9B, in turn cited in LIB. CALC. 46B; save for the last sentence in DRC 83.5–8; only the last sentence in BC 46C12–D1, DE SALTU LUNAE I (984)C5–9; partially in QUAEST. AUSTR. 2.8A): *Quod si unus dies plus minusve contigerit, evidens error est. Excepto videlicet primo anno saepe dicti decennovenalis cycli, quem a XIIII luna paschae ultimi, id est noni decimi anni, usque ad XIIII eiusdem primi numerare curamus. Propter quod idem ultimus epactas, id est adiectiones lunares, X et VIII tunc retinens, primo anno non XI, ut in ceteris annis fieri solet, sed XII dies accommodat.* In his list of all 19 years, Dionysius describes the first three (which are the ones discussed by MC here) in the following way (Krusch, *Studien* II, p. 85; cited in DRC 82.3–8 where it is explicitly stressed that the first year extends exceptionally from one Easter full moon to the next, BC 14(1285)D10–(1286)A7, 124(1333)C10–D1, DIAL. NEUSTR. 24, and many MSS): *Anno decennovenalis circuli primo, lunaris XVII a XV Kalendas Maii, usque Nonas Aprilis, quia communis annus est, fiunt dies haud dubium quin CCCLIIII. Anno decennovennalis circuli secundo, lunaris XVIII ab octavo Idus Aprilis usque in octavum Kalendas Aprilis, quia communis est, fiunt dies CCCLIIII. Anno decennovenalis circuli tertio, lunaris XVIIII, a VII Kalendas Aprilis usque in Idus Aprilis, quia embolismus est, fiunt dies CCCLXXXIIII.*

70–105 In...curabis] This method is also described in BC 14(1285)D3–7: *Nam communium et embolissimorum talis conprobatio est: Si enim a XVa luna preteriti festi usque XIIII anni eiusdem sequentes dies CCCLIIII fuerunt, communis annus erit; si autem fuerint dies CCCLXXXVIIII, embolismus erit.* The difficulties occuring in the first year due to the application of the *saltus*, which reduces the number of days of this year by one from 354 to 353, are also noted by Bede (DTR 56.12–13, cited in BC 124(1333)B13–C3, LIB. CALC. 92, PV §227, RM 77.14–19): *Et habent communes anni menses lunares XII, id est dies CCCLIIII, embolismi autem menses tredecim, dies videlicet CCCLXXXIIII (praeter unum dumtaxat septimum decimum cycli huius annum, qui est decemnovenalis primus, in quo unus dies ratione saltus lunaris intercipitur).* Less precise QUAEST. AUSTR. 2.10A: *et initium lunaris anni et communium et embulesmorum iuxta eosdem a quinta decima luna usque ad quartam decimam lunam, anno revoluto, sive communis sive embulesmus sit, excepto uno anno.* Solving this problem by dividing or doubling 17 April, and attributing one part to the final year of the *hendecas*, the other to the first year of the *ogdoas*, is a diagnostic feature of Irish computistics of the late seventh and early eighth centuries; besides the discussions in MC and CE, an entire chapter is devoted to it in DIAL. LANGOB. 20 (too long to be quoted here). This doubling of 17 April led DE SALTU LUNAE I (985)B12–D3 to the assumption that the *saltus* itself had to occur on this date, containing both *luna* 14 and *luna* 15. For the lunar year extending from *luna* 15 to *luna* 14 of the Easter lunation cf. c. 53, ll. 6–10.

70–105 In...curabis] Cf. p. CXXXIV, CLXXVII–CLXXVIII of the introduction.

communibus et embolismis | *CCCXXV dies in*diferenter repperiri intellegamus. fol. 37r

The first year of the Dioynsiac 19-year cycle

A XIIII Kalendas Maii cum Kalendis eius *XIIII dies sunt*, quibus coniunctis *CCCXXV die*bus, quos *a VI Nonas Maii in XI Kalendas Aprilis* diximus, CCCXXXVIIII dies innotescunt. *A X Kalendas in Nonis Aprilis XIIII dies*. *Hinc in primo communi anno CCCLIIII dies inueni*untur uno die minuto. Ideo in *hoc anno a XIIII luna* in *XIIII* lunam *XV Kalendas Maii die*m *sol*is *artific*iose *diuiden*tes, ut integer *numeru*s sit, quem *salt*us secundum solem de commune primo minuit, quidam rimari *cura*nt. Deinde amissis XVIII horis de XV Kalendas Maii cum anno finienti endecadem VI tamen horas cum primo falso enumerant, et CCCLIIII ita inueniunt.

The second year

A VIII Idus itaque in Kalendis Maii *XXVI dies* inuenientes, his unatis numero, <quem> *a VI Nonas Maii in XI Kalendas Aprilis* habent, dies CCCLI esse adfirmant. Et additis X Kalendas, VIIII Kalendas, *VIII Kalendas*, *III die*bus, *CCCLIIII* in illo pro *commun*e repperierunt.

The third year

*Terti*us *ann*us *a VII Kalendas Aprilis*, *qu*i *est embolismus*, incipit. A quo in Kalendis Maii cum ipso die *XXXVII dies* collecti. Ipsique dierum numerum *a VI Nonas Maii in XI Kalendas Aprilis* inuentum amplexi CCCLXII dies efficiunt. Quibus *a X Kalendas* cum Kalendis Aprilis X dies adiecti, sunt CCCLXXII. A IIII Nonas *in Idibus Aprilis* XII *dies* habet, *CCCLXXXIIII dies* inesse commendant.

Every year likewise

Itaque ita *usque* in *finem* endecadis per XVIIII annos rimari curabis.

80 coniunctis] coniuntis *M.* **89** horas] horae *M.* **91** VIII] VI *M (corr. from* CE*).* **99** inuentum] inuento *M (corr. according to* MC*).* | CCCLXII] CCCLII *M (corr. according to* MC*).* **100** efficiunt] efficiuntur *M.* **101** in Idibus] itaque *M (corr. from* CE*).*

82–86 Hinc…sit] Transcribed in Borst, *Schriften*, p. 453.

and 22 March (i.e. the earliest possible Julian calendar date of the Easter full moon) in common as well as embolismic years within the cycle of 19 years.

(Concerning the first year of the Dionysiac 19-year cycle:) From 18 April (i.e. the Julian calendar date of *luna* 15 of the Easter lunation in the first year of the Dionysiac 19-year cycle) to 1 May inclusively are 14 days. Having combined these with the 325 days which we said (are) between 2 May and 22 March, the result is 339 days. From 23 March to 5 April are 14 days. Consequently, in the first common year 354 days with one subtracted day (i.e. 354−1=353 days) are found. Therefore, certain people calculate from *luna* 14 (instead of *luna* 15) to *luna* 14 in that year, so that the number (of days of a common year, i.e. 354), which the *saltus* reduces from the first common year in accordance with the sun (i.e. the *saltus* reduces the number of lunar days of the first common year from 354 to 353 to bring the lunar course in line with the solar one), remains untouched, carefully dividing the solar day of 17 April. For, even though they have left 18 hours of 17 April to the year that ends the *hendecas*, they nevertheless count (the remaining) six hours, (though) wrongly, to the first year, and so they find 354 (days).

(Concerning the second year of the Dionysiac 19-year cycle:) Having found 26 days between 6 April and 1 May, and after combining these with the number (of days) which they have (calculated for the period) from 2 May to 22 March (i.e. 325 days), they affirm that (these) are 351 days (in total). And, after adding the three days 23, 24, and 25 March, they found the 354 days (required) for a common year in that (year).

The third year (of the Dionysiac 19-year cycle), which is an embolismic year, starts from 26 March. From this (date) to 1 May, including that day, 37 days are collected. And these together with the number of days found between 2 May and 22 March (i.e. 325) result in 362 days. To these the 10 days from 23 March to 1 April inclusively are added, (which) are 372 (days in total). (Since this year) has (another) 12 days from 2 to 13 April, they recommend that 384 days are in it.

In this manner you will take care to calculate through the 19 years up to the end of the *hendecas*.

<LX. *DE INPLETIONE INTER SOLEM ET LUNAM PER DECENNOVENNALEM CICLUM*>

Comp. of lunar with solar year

| Itaque solarem cursum lunaremque per decennouenalis cicli tramitem caperemus. Hinc communis annus XI diebus breuior solari, embolismus uero XVIIII diebus quam solaris annus prolixior, qui CCCLXV dies habet. fol. 37v

Proof for the *ogdoas* being cyclic

Deinde IIo solares anni quam communes II XXII diebus quadrantibus duobus prolixiores. *Sol*aris annus tertio, si *communem de* illo *fac*ias, *XI diebus* et quadrante superior. Qui XI dies coniuncti XXII duorum predictis

---

LX,4 cicli] ciclum *M**. | caperemus] eciperemus *M*. **5** annus] anus *M*. **6** annus] anus *M*. **8** IIo] IIos *M (corr. from* BC*)*. | XXII] XXVI *M (corr. from* BC*)*. **9** Solaris annus] solares anni et *M*. **10** et quadrante] *add.* et quadrante *by mistaken repetition M*. **11** XXII] XXV *M (corr. according to* BC*)*. | duorum] duo *M (the copyist may have spelled out the numeral* II *wrongly)*.

LX,1–2 DE...CICLUM] CE p. 114 gives this heading to a chapter with the same content as what follows in MC here. **3–7** Itaque...habet] The facts that a common lunar year is 11 days shorter, an embolismic lunar year 19 days longer than a solar year presumably ultimately derives from EPISTOLA DIONYSII (Krusch, *Studien* II, p. 83; cited in DTR 45.23–27, in turn cited in LIB. CALC. 85, RM 59.23–28; the last two sentences only in QUAEST. AUSTR. 2.8A, LIB. ANN. 56D): *Quamvis enim anni solaris cyclum per singulos menses luna circumeat, tamen eius perfectionem XII suis mensibus implere non prevalet. Denique in annis communibus ad rationem solaris anni XI dies lunae deesse cernuntur. In embolismis vero X et VIIII diebus eundem annum videtur solarem luna transcendere*. Dionysius' source appears to have been QUINTUS IULIUS HILARIANUS, *Expositum de die paschae et mensis* 12 (*PL* 13, col. 1113A–B): *qui anni communes habent dies trecentenos quinquagenos quaternos, aut enim undecim dies, minus habent ratio lunae a ratione solis et facit annum communem, aut X et IX transcendit, et facit embolysmum*. Cf. also COMPUTUS COTTONIANUS fol. 79r (cited in BC 15B11–13, in turn cited in LIB. ANN. 59): *Communis annus habet dies CCCLIIII. Adde undecim, fiunt CCCLXV. Adde supra decem et VIIII, et fit embulismus annus, dies CCCLXXXIIII*. For the 11-day difference between the common lunar and the solar year cf. also c. 49, ll. 3–6; c. 51. **8–21** Deinde...conectiunt] One of the primary problems faced by seventh-century computists was to prove the cyclic character of the 19-year cycle, since such a proof did not exist in the literature available. On the contrary, the only text that dealt with this problem, i.e. the EPISTOLA DIONYSII, only proved the cyclic character of eight years, i.e. of the *ogdoas*, and this underlies MC's account here. EPISTOLA DIONYSII (Krusch, *Studien* II, p. 83): *In ogduade diximus V annos esse communes, tres embolismos. Quinquies ergo trecenteni quinquageni quaterni fiunt ĪDCCLXX et ter trecenteni octuageni quaterni, ĪCLII, ac per hoc simul fiunt ĪĪDCCCCXXII. Similiter octo anni solares, si in summam redigantur, id est, octies trecenteni sexageni quini et quadrantes, faciunt simul ĪĪDCCCCXXII.* Dionysius appears to have been unable to prove the cyclic nature of the 19-year cycle himself, which led him to follow Quintus Iulius Hilarianus' account of the 8-year lunar cycle underlying the Hippolytan reckoning, not realizing that the two cycles are in no way connected. QUINTUS IULIUS HILARIANUS, *Expositum de die paschae et mensis* 13 (*PL* 13, col. 1113B; the corrections are mine): *Inde per annos singulos hac diversitate dies annorum lunae cum diebus solis, qui in trecentis sexaginta quinque et quadran<te> conficit annum, non sibi concordare videntur. Proinde etiam hoc ostendimus, ut appareat omnibus aequales eos invicem dies habere, et ab initio cursus eorum isto ordine cucurrisse, et damna communium annorum eadem luna in annis embolysmis compensasse. Octo annorum rationem, quae perfecta fore videtur, in medium proferamus. In octo scilicet annos luna quinque annos habet communes et tres embolysmos. Ergo quinquies CCCLIV faciunt <M>DCCLXX, et ter CCCLXXXIV faciunt MCLII; fiunt simul dies duo millia DCCCCXXII. Sic deinde et solis octo annorum summam in unum colligamus octies CCCLXV quadran. duo millia DCCCCXXII. Ita igitur congregavit solis ac lunae*

60. ABOUT THE CONGRUENCE BETWEEN THE SUN AND THE MOON THROUGHOUT THE 19-YEAR CYCLE

Therefore, let us turn to the solar and lunar course through the course of the 19-year cycle. A common year (is) eleven days shorter than a solar one, an embolismic (year), however, 19 days longer than a solar year, which has 365 days.

Consequently, two solar years (are) 22 days and two quarter-days longer than two common (years). A solar year (is) by eleven days and a quarter-day superior to the third (lunar year), if you would make a common (year) of it. These eleven days combined with the above-mentioned 22 of the (first) two (years) result in 33 days. These 33 (days) complete an embolism of 30 moons, and three days and

---

cursus cum diebus suprascriptis. MC's direct source here, however, is CE p. 114–5: *Communis: XI diebus antecedit annus solis annum communem. Communis: XI diebus antecedit annus solis communem annum. Fac communem de embolismo, XI diebus sol antecedit. His autem III annis sol supercrescit diebus XXXIII et III quadrantibus. His diebus oppone XXXmam lunam embolismi, remanent III dies et III quadrantes cum sole. Hoc autem itidem per II communes et embolismum unum faciendo, XXXVI dies et VI quadrantes cum sole remanebunt. Postquam opposueris XXX dies embolismi contra solem. Communem annum XI diebus sol antecedit. Fac communem de embolismo, XI diebus sol antecedit. His II annis sol crescit XXII diebus et II quadrantibus. His adde VI dies predictos, unde* (repeated) *XXVIII dies fiunt. VI autem quadrantes et II coniunge* (*coniunge-* MS), *unde* (repeated) *VIII fiunt. Qui iuncti II dies faciunt. Quos, cum addideris XXVIII diebus, XXX dies faciunt. Hucusque ogdoadis perficitur.*

LX,3–7 Itaque…habet] The differences in the total number of days between the two lunar and the solar year obviously were common knowledge; mostly Dionysius and Bede were cited in this matter (cf. *app. font.*), but cf. also VERS. TUR. 5.13, 6.15: *Novem idem embolismus atque decem diebus pervidetur annum solis superare cursibus. ... Uno nimphe atque decem diebus in ordine brevis annus anno solis superatur agmine*. LIB. COMP. 4.8A (cited in LIB. CALC. 49): *Solaris annus, qui habet dies trecentos sexaginta quinque, superat communem annum diebus undecim et superatur ab embolismo diebus novemdecim*. **8–21** Deinde…conectiunt] This proof of the cyclic character of the *ogdoas* by compensation of the remaining two lunar days with eight solar quarter-days is ultimately based on Dionysius' poor treatment of the cyclic character of the 19-year cycle, which he failed to prove (cf. *app. font.*). MC and CE are the only surviving witnesses of how seventh-century computists tried to validate Dionysius' faulty account. Interestingly enough, the exact same modes of calculation are used in BC 17 and some glosses to DTR (cf. Kassel, Landesbibliothek, 4° Ms astron. 1, fol. 76r, Vatican, Biblioteca Apostolica, Vat. Lat 643, fol. 52r, *PL* 90, col. 485B–C), though for a different purpose, namely to calculate the correct difference of two lunar days between the *ogdodas* and eight solar years; cf., e.g., BC 17D2–14: *Communis, communis, et fiunt IIo anni solares, et resedent XXII dies de duobus annis solis. Fit IIIo anno, et supersunt XXXIII dies solis. Fit embolismus III anno lunae et deponit XXX dies, et supersunt III. Et fiunt similiter communis, communis duo anni, singulares III. IIIo anno ideo* (*id MS*, as in the following instances) *embolismus deponit XXX dies, supersunt III. Ideo VI dies de VI annis. Commune ideo XI dies supersunt anni solis. Quia si communis aliter annus, ideo supersunt a sole XXII dies inter II annos, et superiores VI. Ideo necesse est, ut in altero anno fit embolismus. Et non solum embolismus implet dies XXX* (*XVIIII MS*), *sed supersunt II dies a luna* (*lunae MS*). The first part of this calculation (establishing the difference of three solar days between the first three years of the *ogdoas* and three solar years) is also applied in DIAL. LANGOB. 22A.

LX,3–7 Itaque…habet] Cf. Borst, *Schriften*, p. 365. **8–106** Deinde…obseruabis] Cf. p. CXXXIV–CXXXVI, CXLIV–CXLV, CXCVII of the introduction; for the use of the term *dies superflui* in this section cf. Borst, *Schriften*, p. 474.

*XXXIII die*s efficiunt. Qui XXXIII *embolism*um *luna*rum *XXX* supplent, et *III dies et quadrantes cum sole remanent*. Et ceteris sequentibus annis III expletis et VI dies et totidem quadrantibus a sole resedent. Inde gemini sequentis anni, si *communem* anno finiente ogdoade *fec*eris, *XXII diebus* inferiores et totidem *quadrante*s quam *sol*ares *anni II*. Qui XXII et *VI pr*ius memorati *XXVIII dies* parant. Inde in finem ogdoadis bini *dies* solares desunt, quos de *quadrant*ibus *VIII* predictis quidam faciunt. Et ita solarem ac lunarem concursum in fine ogdoadis equiperari posse conectiunt.

Proof for the *hendecas* being cyclic

Ideo ogdoadem illius in finem solis et lunae cursum esse parem putantes ac *endeca*dem ueteribus firmari placuit, | itaque per endecadem lunae et solis congressum fol. 38r *uno die super*fluo cum sole remanente ȩquiperantes et quadrantibus XI. Quem *salt*u lunae quibusdam coniungi placet. Et *quadrant*ibus *XI <III> dies*, *licet quadrans de*sit, effici libet. Quem Dionisius uigilat dicens: *Tantundem pene* idem *repperies*. Quibus *tr*ibus *die*bus *XXX lun*ae aetates *in Kalendis* mensium cadentes, quorum XXVIIII luna manet, falso retrahunt. Supplentis enim lunae quadrantibus effectis diebus Kalendas mensium naturales dies priuare lunari faciunt.

---

12 XXXIII[1]] XXIII *M (corr. from* BC, CE*)*. | XXXIII[2]] XXX *M (corr. according to* BC, CE*)*. | lunarum] luna *M (probably a wrong expansion by the copyist of the suspension* l-*)*. **15** gemini] geminis *M*. **16** communem] communis *M (probably a wrong expansion by the copyist of the suspension* com-*)*. anno] anni *M*. | XXII] XXV *M (corr. according to* MC*)*. **17** quadrantes] quandrantes *M*. **25** ȩquiperantes] ȩquiperandas *M*. **26** saltu] saltus *M*. **27** III] *om. M (add. from* CE*)*. **29** lunae] luna *M (presumably wrong expansion by the copyist of the suspension* l-*)*. **30** mensium] mansium *M**. **31** lunae] luna *M (presumably wrong expansion by the copyist of the suspension* l-*)*. **33** priuare] priuat *M*.

22–40 Ideo…curant] After proving in the previous paragraph that the *ogdoas* is cyclic, MC is left with the task to prove that the *hendecas* is cyclic as well, since otherwise the 19-year cycle itself would not be truely cyclic. However, this obviously was a lost cause, since both 8 and 11 years can only be cyclic if every single year is cyclic (1 is the greatest common divisor of 8 and 11), which was not the case (a lunar year does not have the same number of days as a solar year). Dionysius, who was MC's ultimate model in the previous passage and is referred to here, offered no solution to this question: EPISTOLA DIONYSII (Krusch, *Studien* II, p. 83): *Simili modo et endicadis annos, qui sunt communes VII et quattuor embolismi, si in summam ea, qua diximus, supputatione congresseis, tantundem paene repperies, quantum XI solares anni conficiunt hoc est* $\overline{IIII}$*XIIII*. MC tries to prove the cyclic character of the *hendecas* by connecting the one remaining solar day with the *saltus*, and the eleven remaining solar quarter-days (almost amounting to three days) with the final day (*luna* 30) of three embolisms that happened to fall on the Calends of the respective months (cf. p. CXXXIV–CXXXVI of the introduction). More light on this strange theory is thrown by a parallel, but more detailed passage in CE, which surely was MC's source here; note that explicit dates are given in CE for the three *lunae* 30, namely 1 May in the third year, 1 December in the fifth year, and 1 June in the final year of the *hendecas*. CE p. 115: *Item endegas hoc eodem modo completur: In communi autem pene ultimo XXXI dies supersunt cum sole. Quibus adpone lunam XXXmam embolismi, et diei uno diem salti oppone. VIII*

quater-days remain with the sun. And after the subsequent three years have been completed, six days and just as many quarter-days are left over from the sun. After that, the following two years as a pair (are) by 22 days and just as many quarter-days (as years in this instance, i.e. two) inferior to two solar years, if you would make a common (year) of the year that ends the *ogdoas*. These 22 (days) and the previously memorized six provide 28 days. Accordingly, at the end of each *ogdoas* two solar days are wanting (since the just calculated 28 solar days are two days short of the 30 lunar days of the embolism that is placed at the end of the *ogdoas*); these (two solar days) certain people form out of the eight quarter-days mentioned above. And so they combine the solar and lunar course (in a way) that these can be equalled at the end of the *ogdoas*.

Hence, thinking concerning the *ogdoas* that the course of the sun and the moon are equal at the end of it, it pleased the elders to affirm this concerning the *hendecas* (also), accordingly equalling the course of the sun and the moon throughout the *hendecas*, with one superfluous day as well as eleven quarter-days remaining with the sun (i.e. the number of solar days exceeds the number of lunar days by one full and eleven quarter-days at the end of the *hendecas*). It pleases certain people that this (i.e. the superfluous solar day) (is) linked to the *saltus lunae* (i.e. the *saltus* was supposedly introduced to compensate for the superfluous solar day). And it pleases (them) that three days are produced by the eleven quarter-days, even though one quarter-day is wanting (i.e. 11+1=12 quarter-days=3 days). Dionysius was aware of this when saying: You will likewise find almost as many. Because of these three days they falsely withdraw the lunar ages of 30 that fall on the Calends of the months, of which *luna* 29 remains (that is to say that some computists mistakenly argue that on three occasions, each time on the Calends of a month, *luna* 30 is introduced to compensate for the three solar days by making a hollow lunation full; yet, the instances referred to here are embolism, which already have 30 lunar days). In fact, after the days have been completed by means of the quarter-days of the supplying moon, they make the Calends of the months deprive the natural days of (their) lunar (age) (that is to say that these computists argue that, whenever the lunar quarter-days complete a full lunar day, this is quite artificially introduced on the Calends of a month).

---

vero quadrantes praetermisi et II XI fiunt. Qui in III dies adunantur, licet unus quadrans deesse cernitur. Hi uero III dies lunis quibusdam XXXmis contra naturam insitis adponantur, hoc ad (recte est) XXXmam lunam in Kalendis Maii luna XXIX in Kalendis Ianuarii, et contra XXXmam lunam in Kalendis Decembris luna XXI in Kalendis Ianuarii, et contra XXXmam lunam in Kalendis Iunii luna XXVII luciscente in Kalendis Ianuarii. Quę lunae contra naturam sunt. Sed hoc falsum esse hoc modo opinamur, nam hęc XXXmę embolismorum sunt.

22–33 Ideo…faciunt] Seventh-century Irish computists apparently struggled for quite some time to prove the cyclic character of the Dionysiac 19-year cycle, as this and the previous passage so neatly illustrate. In the end they succeeded in the task (cf. the following passages) so that earlier unsuccessful attempts like the ones outline here and in the previous passage soon became useless and outdated. Accordingly, nothing similar can be found in any computistical textbook, particularly not in DRC, demonstrating that this text belonged to a later stage in the development of Irish computistics than MC.

Different approach to the problem of the quarterdays

Hinc non ualentes ulla ratione hoc firmare, hos *quadrantes* minutim in annos, in quibus sunt, artificiosę diuidunt *tertia*m *par*tem *momenti unicuique diei* anni *exceptis V diebus*, quos fregimus deputantes. Quod si quis curiose quadrantes partire solet, et non aliter esse inueniat. Sed hoc supplimentum solarem cursum ac lunarem XVIIII annorum in fine nominari curant.

Further criticism of the just described proofs

Quibus huius rei notitiam amplius augit dicentes: Saltum ab hoc non esse, ut solis resedenti superfluo diei in fine endecadis coniungit, innotescerit. Etsi quadrantes in ogdoadem, lunam habeant. Ita et in endecade currentes consequi necesse adfirmant. Et quadrantes VIII conuenire in duos dies in ogdoade merito negant. | Nam quadrans, (fol. 38v) quem in fine endecadis deesse cernunt, ab exordio ogdoadis retro. Hii, dicunt hunc, in ogdoadis fine quadrantes VII repperiri ostendunt. Sunt ceterę obiectiones, quas hic necessitas contendendi euolui non sinit.

---

39 lunarem] lunaris *M.* **40** fine] in ab *add. M.* **43** coniungit] coniungi *M.* **45** consequi] sequi *M**. **49** obiectiones] abiectiones *M.* **50** contendendi] conterdii *M.*

34–40 Hinc…curant] The main problem of the previous theory was the handling of the eleven solar quarter-days, to which no lunar equivalent was assigned, and which were therefore considered as superfluous at the end of the *hendecas*. A different solution is envisaged here, namely dividing each quarter-day (consisting of 3 hours=120 *momenta*) into 360×⅓ of a *momentum*, and distributing these 360 equal parts among 360 days of a year. This division was supposed to make the eleven solar quarter-days irrelevant for the calculation (while at the same time hinting at the fact outlined in the following passage that annual bissextile quarter-days had to be assigned to both solar and lunar years). The calculation itself is executed in more detail in CE p. 115: *Aliter quadrantes endegadis comminuuntur hoc modo: Quadrans III horas habet, hora autem momenta XL. Hinc singula momenta in III partes diuiduntur, quę ter XL faciunt. Et ita hi XImi quadrantes endegadis diuiduntur, et particulas CCCLX faciunt. Quę diebus CCCLX adsignantur, ut tertia pars momenti unicuique diei iungatur exceptis V diebus. Qui expertes sunt huius diuisionis, dum fracti sunt intra dies XC trium.* For the division of the quarter-day, as well as the five days that exceed 360 within a year cf. c. 41, ll. 80–87; for a different division of these five days, which may well have been meant here, cf. c. 38, ll. 23–25; c. 40. **41–50** Quibus…sinit] This passage here discusses the main problem of Dionysius' theory and the false proofs outlined above, namely the assignement of an annual quarter-day (the bissexile increment) to the solar year only, even though the lunar year has its equivalent. Similar criticism can be found in CE p. 115–6, MC's ultimate souce here: *Aliter item interrogari decet: Dum hęc inpletio plena est in ogdoade, ut quid endegas fit, utique enim numerus plene conputatus idem semper currit et in eundem reuertitur? Vel uti est ebdoma? Dum autem hęc inpletio plena est in endegade, quamuis se aliter habeat, ut quid ogdoas fit? Quadrantes ogdoadis ex quibus diebus in fine ogdoadis artificaliter fecimus, quomodo addantur contra dies lunae? Nullus enim dies lunae est, sub quo non esset dies solis, neque dies solis est, sub quo non esset dies lunę. Lunę autem XXXmę contra naturam dictę, nonne dies solis habeant? Et si non dicamus dies fractos ex quadrantibus endigadis contra XXXmas lunas, ut quid non eundem effectum habeant quadrantes ogdoadis et quadrantes endigadis? Si enim comminuuntur, omnes comminuantur. Si uero in dies iunguntur omnes quadrantes, omnes iungantur in dies. Hoc ita soluendum est: Eundem effectum habent quadrantes ogdoadis et endegadis, et nec comminuentur, nec dies inplebuntur, nec diebus lunę opponentur, quia inpletio plena est in ogdoadem et endegadem.*

Then, not being able to confirm this by any argument, they skilfully divide these quarter-days in the years in which they occur piecemeal into a third part of a *momentum* for every single day of the year, save for five days, which we have fragmented by estimation. If somebody is accustomed to divide the quarter-days carefully, he will not find it to be (any) different. But they take care that this addition is assigned to the solar and lunar course at the end of 19 years (i.e. the annual bissextile increment, a quarter-day, is part of solar as well as of lunar reckoning and as such is taken into account for the total number of solar and lunar days in the 19-year cycle).

It further increases the knowledge of this matter when they say about these things: It should be known that the *saltus* does not derive from the fact that it combines with the remaining superfluous solar day at the end of the *hendecas*. Furthermore, even though quarter-days (occur) in the *ogdoas*, they should have a lunar equivalent. And so they affirm that (the solar quarter-days) running in the *hendecas* have to be accompanied (by lunar quarter-days). And they justly deny that the eight quarter-days merge in the two (wanting solar) days in the *ogdoas*. For the quarter-day, which they realize to be wanting at the end of the *hendecas*, (derives) retrospectively from the beginning of the *ogdoas* (i.e. the three quarter-days at the end of the *hendecas* will make a full bissextile day when combined with the first quarter day of the *ogdoas*). Those (who) say this object that (then only) seven quarter-days are found at the end of the *ogdoas*. There are further objections, which the necessity to make haste does not allow (us) to elaborate on here.

34–40 Hinc…curant] As the outdated theory outlined in the previous passage found no imitators, the same is obviously valid for its correction here. **41–50** Quibus…sinit] Dionysius' account of the cyclic character of the 19-year cycle (especially his contradictory treatment of the quarter-days in the *ogdoas* and *hendecas*), which had motivated the faulty theories and their corrections in MC, is also criticized in DRC 79.8–12: *Aliter uero Dionisius concordationem in decinnouenali ciclo inter solem et lunam facit, octo quadrantes in ogduade numerans, et de illis duos dies facit. Undecim uero quadrantes in endicade non numerant. Sed in sua epistola ita esse inuenitur*. Bede's criticism, on the other hand, was rather pointed at computists of his life-time, who still believed that eight as well as eleven years were cyclic (probably with arguments similar to the ones described in MC): DT 11.8–11: *Licet quidam violenter hos dies ex bissextis octo annorum supplere nitantur; cum bissextus soleat in mense februario soli superfusus et lunae futuro tempori nihil praeiudicare, et ipsi nullum hendecadi bissextum addant*. DTR 46.14–21: *Verum etiam nunc qui solis octo annos totidem annis lunaribus itemque solis XI totidem putant annis lunaribus dierum aequalitate componi. Ubi primum dicendum quod absque ulla contradictione, si VIII anni utriusque sideris aliquando longitudine concordant semper idem facere habebunt, neque ullum undenorum annorum aequalitas locum inveniet; si autem vel semel XI anni amborum simul incipiunt simulque terminantur, semper ipsium facturi sunt, nec erit quando octonorum concordia redire possit annorum*. Dicuil, for his part, may have referred to 'Irish teachers of old' with the term *veteres*: LDA 3.2.3: *Prima causa, eo quod veteres errando octo annos solares totidem lunaribus annis, et undecim solares similiter lunaribus undecim annis aequari putabant. Quod nunquam fieri posse manifeste videtur. Quasi luna semel in octo annis suum vere ciclum terminasset, nequaquam alterum ciclum veraciter umquam consumpisset, numquam alterum dissimilem possidere valeret*.

Correct comparison of the *odgoas* with eight solar years

Hinc disputantes a *prim*o ogdoadis *ann*o unum diem quęrunt, quem illo sol saltus uice XIImum superflue habet. Nam in ceteris annis plures XI dies quam communis non XII solis annus possedit, qui inter finem lunaris anni, qui uariat, id est XIIII luna, et solaris anni terminum quęri debet. Sol enim *a XIIII Kalendas Maii* in XV Kalendas Maii annualem cursum per decennouenalem ciclum indesinenter euoluit. Hinc a *Nonis Aprilis* primi mensis, *communis* ogdoadis anni fine, in *XV Kalendas Maii*, solaris anni terminum, enumeris, in quo spatio *XII dies* pro certo repperies. Deinde *IIII dies tri*um annorum superfluos in fine cum sole non III permissa hac uice quadrantium ratione inuenies. Si a XIIII luna primi mensis, fine anni lunaris, *in XV Kalendas Maii,* solaris anni finem, curiose disputes, in fine igitur *XXVIIII dies* cum sole esse notes preuento inuento in primo ogdoadis anno die. Quod ita probabis: Nam *si embolism*us *communis* | finiens ogdoadis fol. 39r
foret, XIIII eiusdem *XIIII Kalendas Aprilis esset*. Quod fieri ab *Kalend*arum *Ianuarii epact*is, si quis conputet,

51 disputantes] disputanter *M*. **57** decennouenalem] decennouelem *M**. **63** mensis] in *add. M*. **69** ab] ad *M*.

51–106 Hinc…obseruabis] After refuting the theories about the cyclic character of the 19-year cycle based on Dionysius' faulty account, MC here provides a correct analysis. The method (for which cf. also c. 59, ll. 70–105), i.e. examining the period between *luna* 15 of the Easter month (the beginning of a lunar year) and 17 April to establish the difference in days between the solar and the lunar course, is more systematically applied for every single year of the 19-year cycle in MC's modell, CE p. 116–118; I only quote the years of immediate relevance to MC here (the first, third, and final year of the *ogdoas*; the first, ninth and final two years of the *hendecas*): *Hoc modo conputa: Primus annus est communis, a XV Kalendas Maii computatus usque in Nonis Aprilis CCCLIII dies naturaliter habet. Tamen secundum morem CCCLIIII dies habet, diuidenda enim die XV Kalendas Maii. Duodecim diebus cursum solis non contingit* (add. *lu-* MS), *id est ab VIII Idus Aprilis usque in XV Kalendas Maii. ... Fac communem de embolismo. Qui XI diebus cursum solis non contingit. Per hunc vero triennium sol crescit <XXXIIII diebus. His oppone lunam> XXXmam embolismi, III<I> dies cum sole remanent, hoc est ab Idibus Aprilis, quo die embolismus finit, usque in XV Kalendas Maii. ... Fac communem de embolismo, a II Kalendas Aprilis usque in XIIII Kalendas Maii. Qui XI diebus cursum solis non addit. His II annis XXII diebus sol crescit. Adpone VII dies praetermissos et fiunt dies XXIX. Hinc queritur: Si esset communis hic annus, in quo die finitus esset? Id est in XIIII Kalendas Aprilis. Epacta autem in Kalendis Ianuarii hoc est: XXVI luna, XXVII luna in Kalendis Februarii, XXVI luna in Kalendis Martii, VI luna in V Idus Martii. Inde luna XIIIIma in XIIII Kalendas Aprilis, in quo, si communis esset, finiretur. Hic embolismo ab hinc inplentur XXX dies secundum solem usque in XV Kalendas Maii. Inde VIII anni ogdoadis explentur usque in XV Kalendas Maii iuxta solem. Addita autem una die, hoc est XIIII Kalendas Maii, XXX dies secundum lunam conpleueras. Et VIII anni ogdoadis in XIIII Kalendas Maii secundum lunam sunt pleni, secundum solem uero VIII anni sunt et unus dies est. Hinc per huius diei additamentum endegas fit, et endegas per hunc diem paratur. Item a XIIII Kalendas Maii usque in VII Idus Aprilis communis annus est, CCCLIIII dies habens, XI diebus solis cursum non adiens, hoc est a VI Idus Aprilis usque in XIIII Kalendas Maii. Si autem usque in XV Kalendas Maii X dies sunt atque ad quem diem verius conputabis. ... Fac communem de embolismo. XI diebus cursum solis non adiit. Hoc triennio sol crescit XXXIII diebus. His lunam embolismi XXXmam oppone, remanent III dies cum sole. Quos iungas V*

Consequently, by examining the first year of the *ogdoas* they search for the one day, i.e. the twelfth, which the sun has superfluously in this year due to the turn of the *saltus*. For in the other years the solar year possesses eleven not twelve days more than a common (lunar year), which have to be sought out between the end of a lunar year, which varies, i.e. *luna* 14 (i.e. the Julian calendar date of the end of a lunar year varies, since it is defined by the Easter full moon), and the end of a solar year. For the sun revolves incessantly in an annual course from 18 April to 17 April through the 19-year cycle. Therefore, if you count from 5 April of the first month, (i.e.) from the end of the (first) common (lunar) year of the *ogdoas*, to 17 April, (i.e.) the end of the solar year, you will certainly find 12 days in that interval. Thereafter you will find four days (to be) superfluous with the sun at the end of three years, as the reckoning of the three quarter-days is not permitted on this occasion (i.e. at the end of the first three years of the Dionysiac 19-year cycle there are four solar days more than lunar ones, while the three quarter-days of these years are left out of the count). If you examine carefully from *luna* 14 of the first month (i.e. the Easter full moon), (i.e.) from the end of a lunar year, to 17 April, (i.e.) the end of a solar year, you will note, accordingly, that 29 days are with the sun at the end (of the *ogdoas*), because the (extra solar) day found in the first year of the *ogdoas* had been anticipated. This you will prove in the following manner: For if the embolismic (year) that ends the *ogdoas* was common, its *luna*

---

praemissis diebus, fiuntque VIII dies. Hoc est a IIII Idus Aprilis usque in XIIII Kalendas Maii IX dies sunt, quod non conuenit. Communis a IIII Idus Aprilis usque in IIII Kalendas Aprilis est. XI diebus cursum solis non inplet. Fac communem de embolismo. XI diebus cursum solis non inplet. Hoc autem biennio sol crescit XXII diebus. Ad quos adde VIII dies remanentes, fiuntque XXX dies. Adde lunam XXXmam embolismi et omne numerum ad perfectum deduxisti. Hinc queritur: Si hic embolismus communis esset, in quo finiretur? Hoc est in XV Kalendas Aprilis. Epacta enim in Kalendas Ianuarii hoc est: luna XXVII (followed by *in Kalendis Martii, luna VII in* with indications that it should be deleted)*, XXVIII luna in Kalendis Februarii, luna XXVII in Kalendis Martii, luna VII in V Idus Martii. Unde fit, ut XIIII luna fiat in XV Kalendas Aprilis. Ab hinc sunt dies XXX tam iuxta solem quam iuxta lunam usque XV Kalendas Maii. Huc accidit ut hęc inpletio in ogdoade et endegade plena fiat absque quadrantibus.*

51–106 Hinc…obseruabis] CE's and MC's method certainly represents one of the earliest (if not the earliest) attested successful attempts in the early medieval Latin West to prove the cyclic character of the 19-year cycle. But since it was neither concise nor mathematically suggestive, it did not find imitators once more precise algorithms were invented soon thereafter: DRC 79.1–8 calculates a difference of 209 solar days between 19 common lunar and solar years, and proves that this number equals seven embolisms minus the *saltus* (the same method is applied to the *hendecas* in a gloss to DTR (*CCSL* 123B, p. 424)); Bede gives two methods, in one comparing the total number of lunar and solar days of the 19 years (DTR 46.53–62, cited in LIB. CALC. 86, RM 60.45–57), in the other comparing the total numbers separately for the *ogdoas* and *hendecas*, and with this demonstrating that the two lunar days remaining in the *ogdoas* are compensated for by two solar days remaining in the *hendecas* (DT 11.4–13; DTR 46.21–53, cited in LIB. CALC. 86, partially in RM 60.22–45); the total number of days of the 19-year cycle is also given in DIAL. NEUSTR. 29bA, VERS. TUR. 15.37 (6936); BC 17 and some glosses to DTR calculate only the excess of two lunar days in the *ogdoas* by calculating the difference between lunar and solar days for every year (cf. c. 60, ll. 8–21); NOTKER LABEO, *De quatuor questionibus compoti* (Piper, *Nachträge*, p. 312–5), early in the eleventh century, proves the cyclic character of the 19-year cycle by analyzing the epacts of 1 January (a discussion motivated by DTR 56); other methods still remain unpublished.

euidenter lucebit. Inde *embolism*i *XXX* luna*s* a *XIIII Kalendas Aprilis* in XIIII Kalendas Maii rimeris. In quo XXX dies secundum lunam, XXVIIII secundum solem, *in XV Kalendas Maii* cursum finit, conputator solitus inuenies. *XIIII Kalendas Maii* itaque qui ogdoadem *secundum lunam* finit, *endeca*dem iuxta solis cursum incipit pertrahas, et lunarem in fine superfluam aetatem coniungas, qui superfluus cum sole in endecadis fine obesset, nisi in ogdoade luna suppleri innotescerit.

Correct comparison of the *hendecas* with eleven solar years

Hinc primum *endeca*dis annum si sollicite computes, in eo *X* cum sole *dies* superfluos inuenies. Predicto dies numero minuente XImam ideo VIIII annorum in fine *VIII dies non VIIII* cum sole superesse cernis, a XIIII luna finienti et annum lunae in XV *Kalendas Maii*, solaris anni finem, curiose disputans. *VIII* igitur *dies* predicti et *XXII die*s, quos sequentibus duobus annis endecadem finientibus, si extremum faciamus *commun*em, solis gemini anni superfluos eripiunt, *XXX dies* simul solares efficiunt. Nouissimi itaque endecadis anni, *si communis* efficeretur, XIIII eiusdem *XV Kalendas Aprilis* notaretur esse. Quod ita morari *epacta*s | *in Kalendis Ianuarii* (fol. 39v) disputans probabiliter conperiat: In extremo itaque endecadis anno *XXVII in Kalendis* Ianuarii et *Martii*, XXVIII in VI Nonas, XXVIIII in V Nonas, XXX in IIII Nonas, I luna in III Nonas, II luna in II Nonas, III luna in Nonis, XI in Idibus, XII in XVII Kalendas, XIII in XVI Kalendas, *XIIII in XV Kalendas Aprilis*. Deinde embolismi XXX lunas et solis dies a XIIII Kalendas Aprilis in *XV Kalendas Maii* cautus computes, in quo spatio *XXX dies* secundum utriusque cursum pro certo comperies, nisi qualibet in subtilitate excitatus seduci innotescas. Conpas itaque solis et lunę cursus in fine *endecad*is iuxta hanc probati rationis doctrinam. In XV Kalendas Maii unus terminatus esse innotescit, et XIIII Kalendas Maii *in ogdoade* simul incipit. Nemo ergo arguti sensu occultaeque computationis

---

70 lucebit] *It is possible that the calculation of the Julian calendar date of* luna *14 (19 March) from epact 26 of 1 January followed here originally, as it does in* CE *and for the final year of the* hendecas *in* MC *also below.* | Inde] in *M.* | embolismi] embolismus *M (presumably wrong expansion by the copyist of the suspension* emb-*).* | lunas] lunam *M (presumably wrong expansion by the copyist of the suspension* l-*).* **71** rimeris] rimare *M.* **72** XXVIIII] XXVIII *M (corr. from* MC, CE*).* **73** conputator solitus] conputatos solitos *M.* **77** superfluus] superfluas *M.* **80** dies[2]] die *M.* **88** anni] ani *M.* **95** XIII] XVI *M (corr. according to* MC*).* | XIIII] XVII *M (corr. according to* MC*).* **96** lunas] luna *M (presumably wrong expansion by the copyist of the suspension* l-*).* **100** Conpas] conpar *M.* **101** probati] probate *M.* **102** unus] una *M.*

14 would occur on 19 March. That this happens (in this way) will be evidently clear from the epacts of 1 January, if somebody would calculate it. Then you should reckon the 30 moons of the embolism from 19 March to 18 April. In this (interval) you, (being) a competent calculator, will find 30 days according to the moon, (but only) 29 according to the sun, (since the sun) ends (its) course on 17 April. On that account, you should perceive 18 April, which ends the *ogdoas* according to the moon, (as the date that) begins the *hendecas* according to the sun, and you should combine the superfluous (lunar) age (at the end of the *ogdoas* with the day), which would be in the way (as being) superfluous with the sun at the end of the *hendecas*, if it were not known that a lunar day is supplied in the *ogdoas*.

Consequently, if you carefully reckon the first year of the *hendecas*, you will find ten days (to be) superfluous with the sun in it. Since the above-mentioned number reduces (these) days, you will accordingly realize that eight and not nine days remain with the sun at the end of nine years in the *hendecas* (*XImam*) (i.e. it is outlined in the previous sentence that the first year of the *hendecas* has ten superfluous solar days; the second to ninth year of the *hendecas* have the same general structure as an *ogdoas* without a *saltus*; an *ogdoas* with a *saltus* is described in the previous paragraph as having one superfluous lunar day; an *ogdoas* without a *saltus*, however, has two superfluous lunar days, and therefore the same applies to the second to ninth years of the *hendecas*; subtracting these two (not one) lunar days from the ten solar days of the first year, the result is eight (not nine) superfluous solar days at the end of the ninth year of the *hendecas*), carefully examining from *luna* 14 that ends the lunar year to 17 April, the end of the solar year. Hence, the above-mentioned eight days and the 22 days, which two paired solar years detach as superfluous from the subsequent (i.e. subsequent to the previously-discussed first nine years of the *hendecas*) two (lunar) years that terminate the *hendecas*, if we turned the final (year of the *hendecas*) into a common (year), together produce 30 solar days. It should be noted that (*luna*) 14 of (this) final year of the *hendecas* would be on 18 March, if it was turned into a common (year). (One) may find this out acceptably as follows, examining it (in a way) that the epacts remain on 1 January (i.e. the more recently introduced placement of the epacts on 22 March is not considered in this calculation): In the final year of the *hendecas*, therefore, (*luna*) 27 (occurs) on 1 January and 1 March, (*luna*) 28 on 2 (March), (*luna*) 29 on 3 (March), (*luna*) 30 on 4 (March), *luna* 1 on 5 (March), *luna* 2 on 6 (March), *luna* 3 on 7 (March), (*luna*) 11 on 15 (March), (*luna*) 12 on 16 (March), (*luna*) 13 on 17 (March), (*luna*) 14 on 18 March. Thereafter, you, (being) a careful person, should compute the 30 moons of the embolism and the days of the sun from 19 March to 17 April. In this interval you will certainly find 30 days according to the course of both (i.e. the sun and the moon), as long as you, excited (by the calculation), are not in any way mislead into minuteness (i.e. as long as you do not get lost in details of any sort). Compare, therefore, the courses of the sun and the moon at the end of the *hendecas* according to this instruction of the approved calculation. It is known that one (cycle of 19 years) is terminated on 17 April, while at the same time (another) begins on 18 April in the *ogdoas*. Consequently, nobody with a fine sense and ignorant of obscure computation (i.e. knowledgeable in and approving

ignarus hanc raris inutam ueracis texturae auctoritatis culpauit. Sed agnito legitimo intellectu merito obseruabis.

<LXI.> DE EMBOLISMIS

The reason for embolisms

*Embolism*um ideo fieri adfirmant, *ut communium* atque embolismorum *damna supple*at, et *solis ac lunę cursu*m, nisi esset dispar, XVIIII annorum *in fine ęqu*iperari prebeat.

Number of embolisms

Embolismos in ogdoade III, in endecade uero IIII esse indubitanter dicunt, qui in unum comulati | VII fieri fol. 40r efficiunt.

Length of emb.

Et VII embolismorum luna XXX morari pro certo quidam notant.

Correlation between epacts of 1 Jan. and embolisms

Hoc etiam scias quocumque anno: Singularem *luna*e aetatem aut decimam *usque* ad *XVIIII Kalendis Ianuarii* inuenias, in sequenti anno ne quęras. *A XX* uero *usque* ad XXVIIII *qua*s*cumque epactas* ibi intendas, sequentem annum amplecti embolismum sollicite, quas VII esse lucide colligas.

---

106 agnito] agnita *M*. **LXI,4** dispar] disparem *M*. **6** Embolismos] embolismus *M**. **9** luna] lunas *M (presumably wrong expansion by the copyist of the suspension* l-*)*. **12** decimam] decimum *M*. **14** XXVIIII] XXVII *M (corr. according to* MC*)*. | intendas] intendass *M**.

LXI,2–5 Embolismum…prebeat] The fact that the embolisms compensate for the deficiencies of the common years, and with this balance the course of the sun and the moon at the end of 19 years, ultimately derives from EPISTOLA DIONYSII (Krusch, *Studien* II, p. 83; cited in DRC 78.9–11, BC 17C12–14, DTR 45.20–23, in turn cited in LIB. CALC. 85, RM 59.19–22; partially in c. 59, ll. 22–23): *Embolismorum autem ista ratio probatur existere, quod annorum communium videtur damna supplere: quatenus ad solare tempus lunaris exaequetur excursio*. MC's direct source here is CE p. 118–9: *Has VII lunas embolismorum esse reputa, quę nullius mensis erunt, sed in dampna communium supplenda perfice, ut in fine decennouen\n/alis cicli cursus solis ac lunę ęqualis uidetur*. For the embolisms compensating for the deficiencies of the common years cf. also c. 59, ll. 22–23; for the embolisms ensuring the balance between the solar and the lunar course cf. c. 60. **6–8** Embolismos…efficiunt] The fact that there are three emolisms in the *ogdoas*, four in the *hendecas*, may derive from EPISTOLA DIONYSII (Krusch, *Studien* II, p. 83; cited in DRC 81.6–10, DIAL. NEUSTR. 22B): *Ac per hoc ogduadis communes anni quinque et tres embolismi iugiter adscibuntur. ... sicque endecas communibus annis septem, embolismis IIII terminatur*. The total number of seven embolisms per 19-year cycle is also recorded in DE MIRABILIBUS 2.4 (*PL* 35, col. 2176) and CE (cf. previous note). However, the information given here could also easily have been deduced from the sequence of common and embolismic years listed in c. 59, ll. 28–35. **9–10** Et…notant] The fact that the embolisms consist of 30 days is not explicitly mentioned in any of MC's sources, but it was implicit in many statement (such as a common year having 354, an embolismic 384 days; cf. DRC in *app. comp.*); here, MC probably relied on CE (cf. following note). **11–16** Hoc…colligas] The general rule that embolisms can only be found in years with epact 20 to 29 on 1 January is directly copied from CE p. 118: *Hę XXXmę lunae ... qualibet epactis in Kalendis Ianuaris reuelantur? Hoc ita soluendum est: Has lunas XXXmas in epactis Kalendarum Ianuarii hoc modo explorabis: Sed quandocumque epactas in Kalendis Ianuarii a prima luna usque in XIXmam lunam uideris, nullatenus has lunas embolismorum deprehendes. A XXma autem luna usque in XXXmam lunam in Kalendis scire ualebis, hoc est quando uideris lunam XXmam in*

of clear and precise computation only) criticized this † ... † of rare and true textual authority. But you will rightly observe (this) after the correct meaning has been detected.

61. ABOUT THE EMBOLISMS

Accordingly, they affirm that an embolism occurs to compensate for the deficiencies of the common and embolismic (years), and to ensure that the course of the sun and the moon is balanced at the end of 19 years, lest it should be unequal.

They argue without doubt that there are three embolisms in the *ogdoas*, four, however, in the *hendecas*, which, taken together, result in seven.

And some people declare as certain that (all) seven of the embolisms extinguish on *luna* 30.

You should also know this about any given year: If you find a single digit age of the moon or (*luna*) 10 to 19 on 1 January (i.e. in case of epact 1 to 19 on 1 January), then you should not search for (an embolism) in the following year (i.e. in the Julian calendar year that starts with the mentioned 1 January). If you find, however, any epact from 20 to 29 there (i.e. on 1 January), then you should carefully conclude that the following year contains an embolism, (as) you should clearly conclude that these are seven (since only seven epacts of a value between 20 and 29 exist).

Kalendis Ianuarii aut cuiuslibet ętatis usque in XXXmam lunam, tunc lunam XXXmam, quam superfluam intimabis, in Kalendis mensium fieri posse.

LXI,2–5 Embolismum...prebeat] The facts outlined by MC here were usually quoted from Dionysius (cf. *app. font.*). For the embolisms ensuring the balance between the solar and the lunar course cf. DIAL. LANGOB. 22B. **6–8** Embolismos...efficiunt] The number of embolisms was obviously common knowledge among computists, though hardly ever mentioned explicitly. DIAL. LANGOB. 23A may, in fact, derive from MC: *In ciclo decennovenali septem sunt embolismi, tres sunt in ogdoade et quattuor in endecade*. For the seven embolisms in the 19-year cycle cf. also VERS. TUR. 14.35, LIB. COMP. 4.7A, 4.8A. **9–10** Et...notant] DRC 74.7–8 outlines one possible method of establishing the length of an embolism, namely by subtracting the number of days of a common from that of an embolismic year: *luna uero embolismi XXXa semper inuenitur. Nisi enim XXX sit luna embolismi, non erunt CCCLXXXIIII dies in anno embolismo*. That this question of the length of an embolism was not naive is illustrated by DIAL. LANGOB. 22B, which rightly asks why an embolism, a lunar month by definition, should not have 29 or 29½ days: *Interrogatio: Quare luna embolismi tricesima est, et non vicesima nona fit sicut alterae lunae, qui sunt viginti novem et semis? Responsio: Ideoque habet illa luna triginta dies per se, quibus superfluis diebus solis respondit, qui fiunt triginta*. **11–16** Hoc...colligas] This simple observation that embolisms occur only in years with epact 20 to 29 on 1 January is unique to CE and MC.

LXI,6–8 Embolismos...efficiunt] Cf. Borst, *Schriften*, p. 456. **11–27** Hoc...sunt] Cf. p. LX, XCIV of the introduction.

The first embolism

In hoc itaque anno, in quo *XX luna in Kalendis Ianuarii*, embolismum quęras: XX luna in Kalendis Ianuarii et Martii, XXI in Kalendis Februarii et Aprilis et Maii, XXIII in Iunio, XXIIII in Iulio, XXV in Augusto, XXVII in Septimbri et Octimbri, *XXVIIII in* Nouimbri et *Decimbri*, *XXX in <IIII>* Nonas eius. Ipsa est *embolismi luna* hoc anno.

A list of the remaining 6 embolismic years

Ita a *XXIII in Kalendis Ianuarii* et *XXVI* in ogdoadem embolismi, *in end*a*ecade* autem lunam *XXVIIII in Kalendis Ianuarii*, *luna*m *XXI* ac *XXIIII* et *XXVII*, IIII *embolism*i in sequentibus annis inueniendi sunt.

---

21 XXVII] XXVI *M (corr. according to* MC*)*. | XXVIIII] XXVIII *M (corr. according to* MC*)*. **22** IIII] *om. M (add. from* CE, DIAL. LANGOB.*)*. **25** lunam] in *M (I suspect that the copyist mistook the suspension* lu- *for* in*)*. **26** lunam] in *M (I supsect that the copyist mistook the suspension* lu- *for* in*)*.

17–27 In…sunt] MC's source, CE p. 118 (for which see now Holford-Strevens, 'Lunar calendars', p. 201), does not only list the epacts of 1 January for the years in which an embolism occurs, but also discusses the exact placements of the embolisms within these years for all seven years in question (MC does this only for the first year): *XXma luna in Kalendis Ianuarii, luna XXIX in Kalendis Decembris, luna prima in Kalendis Ianuarii. Itaque in Kalendis Decembris luna XXIX est, de qua lunam embolismi facies XXXmam in IIII Nonas. Luna autem Decembris in III Nonas prima est, et in \II/ Kalendas Ianuarii extinguitur, quę luna XXIX semper est. Item luna XXIII in Kalendis Ianuarii, in Kalendis autem Octimbris luna XXX, quę luna similiter embolismo est. Item luna XXVI in Kalendis Ianuarii, in Kalendis autem Iulii luna XXXma est; postquam luna XXX fieri potest, huius enim mensis luna XXXma est. Item in endegade XXIX in Kalendis Ianuarii, in Kalendis autem Maii luna XXXma est; cuius luna XXIX est. <Item in Kalendis Ianuarii luna XXI, in Kalendis Decembris luna XXXma est; cuius luna XXIX est.> Item in Kalendis Ianuarii luna XXIIII, in Kalendis Augustii luna XXIX, de qua XXXmam lunam facies. Item in Kalendis luna XXVII, in Kalendis Iunii luna XXXma, quam embolismum intimabis*. Note that both CE and MC here again apply the peculiar sequence of lunations they attribute to the Dionysiac reckoning (cf. c. 50, ll. 7–21, 29–39).

17–27 In…sunt] The correlation between the epacts of 1 January and the embolisms for the Dionysiac reckoning is a characteristic feature of late seventh-, early eighth-century Irish computistics. Besides CE and MC, this feature also occurs in the Irish DIAL. LANGOB. and then found its way into the Frankish LECT. COMP.; both texts also discuss the exact placements of the embolisms within the Julian calendar year, which reveal that only LECT. COMP. appears here to be dependent on CE: DIAL. LANGOB. 23: *Initium ogdoadis vicesima vel in Kalendas Ianuarii, vel embolismus in Nonas Novembris initiatur, quia IIII Nonas <Decembris finitur>. Vicesima tertia vel in Kalendas Ianuarii, vel embolismus IIII Nonas Septembris invenitur, quia Kalendas Octobris finitur. Vicesima sexta vel in Kalendas Ianuarii, vel embolismus II Nonas Martii inchoatur, quia in Nonas Aprilis finitur. Initium endicadis vicesima nona vel in Kalendas Ianuarii, vel embolismus V Nonas Martii initiatur, quia Kalendas Aprilis terminatur. Vicesima prima vel in Kalendas Ianuarii, vel embolimsmus IIII Nonas Novembris invenitur, quia Kalendas Decembris finitur. Vicesima quarta vel in Kalendas Ianuarii, vel embolismus IIII Nonas Iulii inchoatur, quia IIII Nonas Augusti finitur. Vicesima septima vel in Kalendas Ianuarii, vel embolismus III Nonas Martii initiatur, quia III Nonas Aprilis terminatur*. LECT. COMP. 6.8a (where the order in the *ogdoas* is mixed up, since the second and third embolisms are followed by the first): *Tres embolismi ogdoadis sunt: Kalendas Ianuarii luna vicesima tertia, embolismus in Octubri invenitur, Kalendas Ianuarii luna vicesima sexta, embolismus in Iulio, Kalendas Ianuarii luna vicesima, embolismus in Decembri. Quattuor embolismi endigadis sunt: Kalendas Ianuarii luna vicesima nona, embolismus in Maio, Kalendas Ianuarii luna vicesima prima, embolismus in Decembri, Kalendas Ianuarii luna vicesima quarta, embolismus in Agusto, Kalendas Ianuarii luna vicesima septima, embolismus in Iunio*

Consequently, you should search for the embolism in that year, in which *luna* 20 (occurs) on 1 January, (in the following way): *luna* 20 on 1 January and 1 March, (*luna*) 21 on 1 February and 1 April and 1 Mai, (*luna*) 23 in June, (*luna*) 24 in July, (*luna*) 25 in August, (*luna*) 27 in September and October, (*luna*) 29 in November and December, (*luna*) 30 on 2 December. This is the moon of the embolism in this year.

In this manner the embolisms have to be found from (*luna*) 23 and (*luna*) 26 on 1 January in the *ogdoas* (i.e. besides the embolism mentioned in the previous passage, there are two more in the *ogdoas*, one occurring in the year with epact 23 on 1 January, the other in the year with epact 26, and the place of the embolisms within these years should be calculated by the above described method, starting with the lunar age of 1 January); in the *hendecas*, however, the four embolisms have to be found in the years following *luna* 29, *luna* 21, (*luna*) 24, and (*luna*) 27 on 1 January.

repperitur vel in Aprili. Interestingly enough, QUAEST. AUSTR. 1.11A asks the question about which lunar ages on 1 January correspond to embolismic years (which is subsequently only implicitly answered in the Victorian lunar table of 2.11): *Sciendum quoque est: Qua aetate lunae Kalende Ianuarii inlustrantur per septem embulesmos annos, qui revertuntur in ogdoade et endicade usque ad nonum decimum annum?* It has to be kept in mind that no consensus existed concerning the exact placements of the Dionysiac embolisms before the Bedan practice became standard from the early ninth century onwards. For the different practices cf. DRC 74.8–10, 82, BC 127–133 (the same placements in DIAL. NEUSTR. 23A), DTR 45.35–37 (cited in LECT. COMP. 1.16a, LIB. CALC. 85, PV §270, RM 59.37–42), LIB. ANN. 52, and many other MSS; similarly, the calendar tradition (and implicitly the *paginae epactarum*) also transmits differing practices; note particularly that only CE, DIAL. LANGOB., and LECT. COMP. record the same placement as MC for the first embolism; yet, one generally has to bear in mind that the difference between some of these practices only lay in terminology, not in the actual sequence of lunations, since it was, strictly speaking, irrelevant which of two successive lunations of 30 days was termed embolismic. QUAEST. AUSTR. 1.11 is the only known explicit source for the placements of the Victorian embolisms, but they seem to be modelled on some Dionysiac practice; otherwise, only the lunar table in Bern, Burgerbibliothek, 645, fol. 55r–56v can shed more light on Victorian practice.

17–27 In…sunt] Cf. Borst, *Schriften*, p. 456, 628 and p. XCII–XCIII, CLXVIII of the introduction.

<LXII.> DE SALTU

Etym. and def. of *saltus*

Saltus non libris hoc nomen inueniri quidam ferunt, sed qualitatem ei inponi tradunt. Tamen Augustinus ait: Saltus quasi saliendo quoddam signum, quia compotus transilit *un*um *die*m, ut *ab VIII in Kalendis Ianuarii in XX* et a XVIII in XXX *in XI Kalendas Aprilis*, *transilit* enim *XXVIIII luna in XXX*. Aut saltus dicitur non saliendo, | sed geminando dicitur duas aetates uno die, ut VIII luna <Kalendae> Ianuarii et Martii. Saltus euentus famosus in luna, cui nulla uis est secundum solem. Et saltus idem sunt, qui cum haec nomina habent alia uero: Cum preparatur per incrementa sua, incrementa lunaria dicuntur. fol. 40v

---

LXII,3 qualitatem] qualitate *M.* **6** XVIII] XVIIII *M (corr. from* EPISTOLA DIONYSII*; also Borst; note that* MC *makes the same mistake l. 72 below).* **7** XXVIIII] a XVIIII *M (corr. from* CE*).* | XXX] XXVIIII *M (corr. from* CE*).* **8** aetates] in *add. Borst.* **9** Kalendae] *om. M, Borst.* **11** qui] quia *M.* | habent] habet *M.*

LXII,2–13 Saltus…dicuntur] The fact that the term *saltus* could not be found in any written source is strictly speaking true for texts available to MC, since especially neither Victorius, nor Dionysius, nor Isidore used this term. This statement is repeated in ll. 81–82, 96, 111–112 of this chapter. The etymology *saltus a saliendo* cannot be found in any of Augustine's works, nor in any other pre-MC text. The reference to the epactal increase by 12 instead of the usual 11 from *luna* 8 to *luna* 20 on 1 January may have been derived directly from the epactal sequence give in c. 49, ll. 17–18; the reference to the epactal increase from *luna* 18 to *luna* 30 on 22 March may have been taken from any Dionysiac Easter table, or from EPISTOLA DIONYSII (Krusch, *Studien* II, p. 84; cited in BC 46C12–D3, DE SALTU LUNAE I (984)C5–10, DTR 42.42–49, in turn cited in PV §297, LIB. COMP. 4.9B, in turn cited in LIB. CALC. 46B; the first sentence only in c. 59, ll. 10–13, CE p. 113): *Propter quod idem ultimus epactas, id est adiectiones lunares, X et VIII tunc retinens, primo anno non XI, ut in ceteris annis fieri solet, sed XII dies accommodat. Et quia XXX dierum fine volvuntur, nulla epacta in principio ipsius cycli ponitur*. That these epacts mentioned by Dionysius actually referred to 22 March was known through ETYM. 6.17.31: *Istae epactae semper XI Kal. April. reperiuntur in eadem luna quae fuerit eo die*. This epactal increase is also mentioned in ll. 70–72 of this chapter. Yet, the epactal information given here MC certainly copied directly from CE p. 124–5: *Item querendum est, quomodo duplex lunaris aetas in una eadem die fieri potest, ut dicatur: luna XXIX et XXX in XI Kalendas Aprilis. ... Extrinsecus quidem ut alii, hoc est ex aepactis in Kalendis Ianuarii, ut dicimus: Ab VIIIua luna in XXmam saltus hoc facit. Aliter in semetipso intus saltandi habitus est, ueluti in uno eodemque die lunam XXmam IXnam et XXXmam diximus, hoc est transilit lunam XXIX in XXXmam aetatem in una die*. Concerning the proposition that the *saltus* was so called from *geminando*, this may ulimately derive from the litteral doubling of the respective Julian calendar date, i.e. *luna XXVIIII in XI Kalendas Aprilis et luna XXX in XI Kalendas Aprilis*, in the sense outlined by AUGUSTINE, *Locutiones in Heptateuchum* 2.116 (*CCSL* 33, p. 417): *Ne de omnibus diceret, de duabus dicit more suo geminando, sicut puteos puteos, aceruos aceruos, generationes et generationes, et similia*. There is no apparent source for the statement that the *saltus* is an *eventus famosus*, but note that the same phrase is ascribed to Augustine in DRC (cf. *app. comp.*) and also in c. 41, ll. 1, 5–7, where it characterizes the bissextile day. For the definition of the *saltus* being a lunar increment cf. c. 62, ll. 112–114.

LXII,2–13 Saltus…dicuntur] Most of MC's account of the definition and etymology of *saltus* is directly copied in DRC 107.7–8, 111 (and cf. ll. 112–118 of this chapter): *Item Agustinus dicit: Saltus est euentus famosus in luna, ut bissextus in sole habetur. ... Sciendum nobis cur saltus dicitur. Agustinus*

62. ABOUT THE *SALTUS*

Some assert that the term *saltus* is not found in books, but they relate that this characteristic (i.e. 'leaping' over a lunar age) is assigned to it. Still, Augustine says: (The term) *saltus* is like a certain image for leaping (*saliendo*), because the calculation passes over one day, just as (the epact increases) from (*luna*) 8 to (*luna*) 20 on 1 January ('leaping' over *luna* 19), and from (*luna*) 18 to (*luna*) 30 on 22 March, (where,) in fact, *luna* 29 passes into (*luna*) 30. Or the *saltus* is so called not from leaping, but from uniting two (lunar) ages in one (Julian calendar) day, just as (conversely the two Julian calendar dates) 1 January and 1 March (are united in the same lunar age, e.g.) in *luna* 8 (in the first year of the Dionysiac 19-year cycle). The *saltus* is a famous feature in the moon (i.e. in lunar reckoning), which has no strength according to the sun (i.e. it does not occur in solar reckoning). And the *saltus* refers to the same (phenomenon as those terms), which have, however, these other names: Since (a *saltus*) is prepared by its increments, they (i.e. the *saltus*) are called lunar increments.

---

ostendit dicens: Saltus, quasi quoddam saliendi signum, quia conpotus transilit unum diem. Quamuis non hic saltus sed geminatio uideatur et in numeri libris hoc nomen non habetur, sed tractatores pro qualitate dant ei. The Angers glosses, then, quote DRC: ANGERS 477, fol. 20v (gloss to DT): *Saltus euentus famosus in luna, ut bissextus in sole habetur. Saltus saliendo dicitur. Augustus: Saltus quasi saliendi quoddam signum, quando compotus transilit diem. Quamuis non hic saltus sed gemminatio uideatur numeri*. The same terminology is used to described the Victorian *saltus* in QUAEST. AUSTR., and by Alcuin when referring to the November placement of the Dionysiac *saltus*. QUAEST. AUSTR. 2.11A: *Nam luna, quae tertia decima Kalendas Novembris, vicesima octava est XVI Kalendas Decembris. Deinde aetas transilitur. XV Kalendas <Decembris> tricesima esse definitur*. ALCUIN, *Epistola* 126 (*MGH Epp*. 4, p. 185–6): *si quis non curaverit unum diem in aetate lunae ratione saltus in Novembrio mense transilire*. Similarly DTR 42.20–21 (cited in LIB. ANN. 50B, LIB. COMP. 4.9A, in turn cited in LIB. CALC. 46A; PV §292): *ut sit unus anno nono decimo transiliatur dies*.

LXII,2–9 Saltus...Martii] Cf. p. CXCIV of the introduction. **4–9** Saltus...Martii] Partially transcribed in Borst, *Schriften*, p. 409; cf. Borst, *Schriften*, p. 904 and for the ascription of the first quote to Augustine p. CXXIX of the introduction. **9–10** Saltus...solem] Cf. p. CXXX, CXCIV of the introduction. **10–13** Et...dicuntur] Cf. Borst, *Schriften*, p. 369.

Calc. of *saltus* increment

Sciendum, quot lunis preparatus. Id est *CCXXXV*. Quomodo? Id *est IIII momenta* uniuscuisque *lunę* usque ad

---

14 lunis] lunas *M (presumably wrong expansion by the copyist of the suspension* l-*)*.

14–61 Sciendum…XXXV] In this passage the 24 hours of a *saltus* are divided by the 235 lunations of the 19-year cycle to establish the lunar increment of the *saltus* per lunation. The result is an increment of 4 *momenta*+1/12 of a *momentum*+1/47 of 1/12 of a *momentum*. The computist deals with these three summands successively, and accordingly this passage is divided into three sections. The general calculations underlying these sections are: 1) 4 *momenta*×235 lunations=940 *momenta*=23½ hours; 2) ½ hour=20 *momenta* are left wanting to complete the full 24 hours of a *saltus*; these 20 *momenta*=240×1/12 of a *momentum*=235×1/12 of a *momentum*+5×1/12 of a *momentum*, i.e. 1/12 of a *momentum* per lunation and a remainder of 5/12; 3) this remainder of 5/12 of a *momentum*=5/12×47/47 of a *momentum*=235/(12×47) of a *momentum*, i.e. 235 times 1/47 of 1/12 of a *momentum*. The exact same modes of calculation as in MC (though more concise and precise) are applied in CE p. 122–3, which is MC's source here: *Incrementum utique lunare per decennouenalem ciclum crescit, hoc est IIII momentis et XII parte momenti et XLma et VIIma parte XIImę partis momenti per singulas lunas, quę sunt CCXXXV decennouennali ciclo. Itaque probemus: IIII momenta per singulas lunas. Veluti est IIII momenta ab una luna, et alia IIII momenta ab altera luna, quę simul VIII sunt. Adde item VIII momenta a duabus lunis, quę XVI momenta faciunt. Item adde ab una luna IIII momenta, quę omnia simul iuncta XX momenta fiunt a lunis V. Item adde ab aliis V lunis XX momenta, quę coniuncta reliquis XL momenta faciunt. Hinc est hora a X lunis. Cum autem perduxeris CC lunas, XX horas fecisti. Adde XXX et V lunas, quę simul faciunt horas XXIII et dimidia horam. Hinc dimidia hora deest, hoc <est> XX momenta. Quibus conpudandis XII lunas iunge, a quibus duces unum momentum, hoc est XIImam partem momenti a singulis lunis. Item ab aliis XII lunis momentum accipe. Hinc II momenta a lunis XXIIII. A lunis vero XLVIII IIII momenta sunt. A lunis autem LX V momenta adduc. Hinc a lunis LX quater ductis XX momenta duces. Si vero sunt lunę LX quater, totam inpleuisti diem. Igitur conputa CCCXXXV (recte CCXXXV) lunas, ut scias, utrum sexagesimum numerum enumerare poteris quater, hoc modo: CC diuide in quinquagenarium numerum, fiuntque L quaterna diuisione. His adde XXX, fiuntque LX ter. Adiece V lunas super L quaternum. V autem lunę desunt. Et V partes, quas diximus esse XII particulas momenti, desunt, id est maá triun laigu leut. Quere igitur diuisionem a lunis CCXXXV. Hoc est XLmam VIImam particulam XIImę partis momenti. Hoc modo diuide lunas: CC, quę faciunt XL quinquies. Adde XXXV lunas, quę diuise fiunt VII quinquies. Hinc singulę partes XLVII sunt. Deinde IIII momenta et XIIma pars momenti et XLma VIIma pars partis XIImę momenti. Ab hanc diem perficiendam a singulis lunis accipiuntur, et hoc modo lunam uelocius accensam dicimus, quam putamus.*

14–61 Sciendum…XXXV] The calculation of this increment of the *saltus* per lunation given in COMP. COL. 4.10B derives directly from MC (since MC's curious description of products is adopted; cf. especially the correspondence in the sentence *Tolle decem de unoquoque sexaginta, remanent quattuor quinquaginta, quinquaginta, quinquaginta, quinquaginta, id est ducenta*; the MS reading of this sentence (Cologne, Dombibliothek, 83², fol. 50v) actually shows the dependency far more clearly: *Tolle X de unoquoque LX, remanent IIII L L L L, id est CC*): *Et sic disponuntur quattuor momenta, adieciuntur, diminuuntur: Centum incensiones portant quadringenta momenta. Aliae centum incensiones portant quadringenta momenta. Octingenta momenta <sunt> cum ducentis incensionibus, qui faciunt horas viginti, quod probare potes. Incensiones triginta, qui faciunt quadraginta, quadraginta, quadraginta momenta, id <est> tres horas, et fiunt horae viginti tres. Quinque incensiones portant viginti momenta, id est dimidium horae. Et fiunt horae viginti tres dimidium. Desunt quinque incensiones, quae fecissent momenta viginti. Quae desunt. Deinde iterum revertamur. Dividamus viginti momenta. Dividitur unumquodque momentum in duodecim partes, et sic potest dividi duodecima pars uniuscuiusque incensionis per incensiones ducentas triginta quinque. Quod probare potes sic: Duo momenta faciunt viginti quattuor, unum momentum facit duodecim, <quinque momenta faciunt> sexaginta. Iterum quinque momenta faciunt sexaginta, quinque momenta faciunt sexaginta, quinque momenta faciunt sexaginta. Tolle decem de unoquoque sexaginta, remanent quattuor quinquaginta, quinquaginta,*

It has to be known by how many lunations (the *saltus*) is prepared. 235. How (i.e. what is the increment per lunation)? Four *momenta* and the twelfth part of a

---

quinquaginta, quinquaginta, id est ducenta. Qui respondent incensionibus ducentis. Quattuor decem, quos dixisti, tolle tres decem, respondent triginta incensionibus triginta. Decem divide, <tolle quinque,> respondent quinque incensionibus. Item quinque remanent, qui iterum dividuntur. Restant igitur quinque, unaquaeque pars duodecima <viginti momentorum>. Iterum revertamur. Deinde omnes incensiones ducentae triginta quinque dividuntur in partes quinque: Quadraginta septem, quadraginta septem, quadraginta septem, quadraginta septem, quadraginta septem. Et quinta pars ponitur uniuscuiusque quadraginta septem, id: Quinta pars dividitur semper in quadraginta septem. As could have been expected, very similar (but more concise) modes of calculating the three summands of this increment were used by Dicuil (LDA 4.3.2). Interestingly enough, DIAL. NEUSTR. 26B calculates 23½ hours from four *momenta* per lunation (which is the sole concern of the first paragraph in MC here), but then argues that the remaining ½ hour cannot be calculated. The three texts just mentioned and MC are the only ones, to my present knowledge, in which the increment of the *saltus* per lunation is calculated in these term (i.e. four *momenta*+1/12 of a *momentum*+1/47 of 1/12 of a *momentum*), which appears to be a diagnostic feature of Irish computistics of the late seventh to early ninth centuries. This increment is, however, also frequently calculated in different terms (e.g. 1/12 being denoted as *uncia*, 1/564 as *atomus*). Cf. BC 137, DE SALTU LUNAE I (986)C3–(988)A7, DE SALTU LUNAE II (cited in LECT. COMP. 7.4), DE SALTU LUNAE III–IV, LIB. COMP. 4.7, LIB. CALC. 48, HELPERIC, *De computo* 18 (*PL* 137, col. 32–33). It is mentioned (without calculation) in the same terms as in MC in DRC 108: *Sciendum nobis quomodo per XVIIII annos hic dies preparatur. Hoc ita est. In XVIIII annis CCXXXV lunationes sunt. Et sic collegitur ab unaquaqe luna saltus materia per IIII momenta et XIImam partem momenti, et per <quadragesimam> VIImam partem XIImae partis momenti et quicumque probauerit, sic inueniet.* and also in some glosses to DTR (e.g., Paris, Bibliothèque Nationale, Nouvelle acquisition latine 1632, fol. 13v; ANGERS 477, fol. 46v); in different terms in RM 57.24–29. Bede mentions and calculates the increment of the *saltus* per year, not per lunation, as do other texts following his example; cf. DT 12.5–9, DTR 42.9–18 (cited in LIB. COMP. 4.9A, LIB. CALC. 46A, PV §§290–291), LECT. COMP. 9.4, DE SALTU LUNAE V–VII, LIB. ANN. 50A (the first part cited in LIB. CALC. 47), ALCUIN, *Epistola* 126 (*MGH Epp.* 4, p. 186–7), RM 57.20–23. For the opinion that the increment of the *saltus* per lunation constitutes the time between the factual and the calculated kindling of the moon cf. DRC 109.1–3 (using the same phrase as MC): *Sciendum nobis quomodo saltus exploratur in naturali lunae cursu. Hoc ita est: Quando dicis accensionem lunae precesserunt IIII momenta <secundum> opinionem nostram.* COMP. COL. 4.9A–B (citing the fact that four *momenta* precede the kindling of the moon): *Et quare fit saltus? Id: Quia quattuor momenta inceditur luna secundum naturam, antequam numeratur. Et quomodo fit saltus? Id: Aliud secundum naturam, aliud secundum numerum, quia in meridie semper accensio lunae numeratur in numeratione. Iterum in unaquaque hora annua revolutione secundum naturam accenditur. Item interrogatur, si quattuor momenta semper in accensione praecedunt? Diminuuntur de die lunae viginti quattuor horarum a sexta hora usque ad sextam horam alterius diei anni. Superflua crescunt cum horis viginti quattuor.* BC 18 (note the closeness to MC in the phrase *in capite XVIIII*): *De augmentum lunae: Incipit augmentum lunae. Augmentum lunae, quare fit et quomodo fit et quid esset, si non fuisset? Quare idem, quia est in naturae lunae tarditas, quod superius ascensio eius in natura, quam putaretur. Addit tarditas eius, donec nox interest. Quod addit in capite XVIIII. In noctem quam adderet id, postquam uenerint XVIIII per XV. Quando putaretur ascensio lunae esse in plenitudine, licet prius depraehenderetur.* Cf. also DE SALTU LUNAE I (986)A6–C2. The argument that the *saltus* is caused by the fact that the factual kindling of the moon occurs earlier than calculated is also mentioned in DT 12.2–5 (cited in RM 56.4–8), DTR 42.2–6 (cited in LIB. COMP. 4.9A, LIB. CALC. 46A, PV §289), QUAEST. AUSTR. 2.9B, ALCUIN, *Epistola* 126 (*MGH Epp.* 4, p. 187).

14–61 Sciendum…XXXV] Cf. Schwartz, 'Ostertafeln', p. 101; Borst, *Schriften*, p. 410, 920–1 and p. XXII, LXXI, CII–CIII, CXXXVI–CXXXVIII, CLXVI, CLXXXIII, CXCIV, CXCVIII of the introduction. **14–110** Sciendum…Ianuarii] Cf. Borst, *Schriften*, p. 915.

caput anni XVIIII, *et XII par*s *momenti <et VII par*s *XLma XIImae partis momenti*>. Quomodo fit? Quando dicis accensionem lunae, id est precederunt IIII momenta pro uelocitate accensionis lunę. In capite XVIIII annorum haec IIII momenta I diem efficiunt horarum XXIIII. Sciendum quomodo crescit. Id est per *octa <momenta* inter> *II luna*s, *XVI momenta* inter IIII lunas, *XX momenta* inter *V luna*s, *XL momenta* inter *X lun*as, LXXX momenta inter XX lunas. Id est II horas coniunge, et fiunt IIII horae in fine lunarum XL. Adde X lunas et fiunt CC momenta in luna L et V horae. CCCC momenta in luna <C> et fiunt *XX hora*e in fine *luna*rum *CC*. *Adde XXX lunas*, per <X> quorum XL momenta crescunt, unam horam coniunge, *hora*e et fiunt *XXIII*. Adde *V* lunas, per quas XX momenta fiunt, id est *dimidia hora*.

Deest alia *dimidia*, | id *est XX momenta*. Tolle *I* fol. 41r *momentum* de his *XX*, diuide illud in *XII part*es. Sic diuide aliud *momentum* in *XII* partes. Coniunge bis XII et efficiunt *XXIIII* partes. Adde alia duo momenta et efficiunt XXIIII. Coniunge hoc et fiunt *XLVIII* inter *IIII momenta*. Remanent tamen XVI momenta. Adde I momentum de XVI, ac diuide illud in XII partes, ad XLVIII partes; <id est adde II> et fiunt L, adde X et fiunt *LX* inter *V momenta*. Supersunt XV momenta. Sic inter V alia momenta LX fiunt partes. Sic in V aliis LX partes. Id est *LX quater* in *XX moment*is. Tolle X quater de IIII LXtis, et remanent *IIII L* L L L, id est *CC*. Cum his CC partibus CCas lunas impleui. Coniunge X quater, quas de <IIII> LXtis tollisti, et efficiunt XL partes. Diuide hanc XL et tolle XXXV. Nunc impleui *CCas lunas* et *XXXV*.

Sunt autem partes *V*, quas de <IIII> LXtis tulisti. ⌐ Diuide <CC in> IIII L L L L. Trahe X de unoquoque L et

---

16–17 et[2]...momenti] *om. M (add. from* CE, MC*)*. **18** accensionem] acensionem *M (corr. from* DRC*)*. **21** momenta inter] *om. M (add. according to* MC*)*. **24** horas] horae *M*. | lunarum] luna *M (presumably wrong expansion by the copyist of the suspension* l-*)*. **25** lunas] luna *M (presumably wrong expansion by the copyist of the suspension* l-*)*. | horae] harae *M**. **26** C] *om. M (corr. from* COMP. COL.*)*. | XX] LX *M (corr. from* CE, COMP. COL.*)*. **27** XXX] XX *M (corr. from* CE, COMP. COL.*)*. | lunas] luna *M (presumably wrong expansion by the copyist of the suspension* l-*)*. | X] *om. M (add. according to* MC*)*. quorum] quas *M*. **28** unam horam] una hora *M*. | horae] horas *M (corr. according to* MC*)*. **32** his] hac *M*. **35** Remanent] *follows* remanen *by mistaken repetition M*. **37** XII] XV *M (corr. according to* MC*)*. ad] per *M*. | id...II] *om. M (add. according to* MC*)*. **41** Cum] in *M*. **43** quas] quae *M (corr. according*

momentum and the 47th part of the twelfth part of a *momentum* from every single lunation up to the head of the 19th year. How does this happen? Whenever you say 'kindling of the moon', i.e. four *momenta* precede this due to the velocity of the kindling of the moon (that is to say that every kindling of the moon is observed or rather calculated four *momenta* late – the computist is here at first not concerned with the minor parts of the increment). At the head of 19 years these four *momenta* complete one day of 24 hours (this is strictly speaking not correct, since the computist neglects the minor parts of the increment here; but the following shows that his concern lies only with the four *momenta* here). It has to be known how it grows. (It grows) by eight *momenta* in the course of two lunations, 16 *momenta* in the course of four lunations, 20 *momenta* in the course of five lunations, 40 *momenta* in the course of ten lunations, 80 *momenta* in the course of 20 lunations. Sum (these) up into two hours (i.e. 80 *momenta* equal two hours), and four hours are produced at the end of 40 lunations. Add ten lunations and 200 *momenta*, (i.e.) five hours, are produced by the 50th lunation. 400 *momenta* are produced by the 100th lunation, and 20 hours at the end of 200 lunations. Add 30 lunations, (with) 40 *momenta* growing in the course of each ten of these, (which) you should sum up into one hour (i.e. each 10 of these 30 lunations produce one hour), and (consequently) 23 hours are produced (at the end of 230 lunations). Add five lunations, by which 20 *momenta* are produced, i.e. half an hour.

Another half an hour, i.e. 20 *momenta*, is left wanting. Take one *momentum* from these 20 (and) divide that into 12 parts. Likewise divide another *momentum* into 12 parts. Combine (these) twice 12 and 24 parts are produced. Add another two *momenta* and (another) 24 (parts) are produced. Combine this and 48 (parts) are produced by four *momenta*. However, 16 *momenta* remain (to be dealt with). Add one *momentum* of these 16, which you should divide into 12 parts, to the 48 parts (produced by the four *momenta* mentioned above); i.e. add two and 50 are produced, (and then) add (the remaining) ten and 60 (parts) are produced by five *momenta*. 15 *momenta* are (still) left over. Likewise 60 parts are produced by another five *momenta*. Likewise 60 parts (are) in another five. Consequently, four times 60 (parts) are in 20 *momenta*. Take four times ten from (these) four times 60, and four times 50 remain, i.e. 200. With these 200 parts I have filled up 200 lunations. Combine the four times ten, which you have taken from the four times 60, and 40 parts are produced. Divide these 40 and take 35. (With these 35) I have now filled up 235 lunations.

However, there are five parts (remaining from the four times ten), which you have taken from the four times 60. ⌐ Divide 200 into four times 50. Take ten from every single one (of these) fifties and combine (these) four times ten and 40 parts

---

to MC*).* | IIII] *om. M (add. according to* MC*).* **45** IIII] *om. M (add. according to* MC*).* **46–48** Diuide…XL³] *I suspect that this passage was a marginal gloss to ll. 54–55 originally, since it presents nothing but the reversed calculation of these lines; it clearly breaks the narrative here, since it is not concerned with the five parts mentioned in the previous and the following sentence; moreover, note that the terminology used here is strikingly different from the rest of the passage, as the (assumed) glossator applies the imperative* trahe *instead of the* tolle *used by* MC *throughout, and he uses* XL XL XL XL *for four times forty, which* MC *would have expressed by* IIII XL XL XL XL. **46** CC in] *om. M.*

coniunge <IIII> X et efficiunt <XL> partes. Inde fiunt XL XL XL XL. ¬ Diuide V in hunc numerum, id <est> I I I I I. Tolle I de his et *diuide* in *partes XLVII*. Sic diuide unum aliud in XLVII; sic diuide I aliud in XLVII; sic diuide I aliud in XLVII; sic diuide I aliud in XLVII. <Id est XLVII quinquies. Tolle VII quinquies de his XLVII quinquies, et remanent XL quinquies.> | Sic diuide I de his XLtis et da X (fol. 41v) quater super XL quater et fiunt IIII L L L L, id est CC. <Supersunt tamen VII quinquies partes. Tolle II quinquies de his, et remanent V quinquies.> Coniunge V quinquies et fiunt XXV. <Supersunt tamen II quinquies partes.> Coniunge II quinquies et fiunt X. Diuide hoc X in bis quinque. Aliud V super XXV et fit XXX. <Aliud V super XXX et fit XXXV.> Sic diuiditur *VII pars XLma XIImae partis momenti* uniuscuiusque lunae de *lunis CC*tis *XXXV*.

saltus a Greek invention

Saltus in *ciclo XCV annorum*, quem *conposuit Quirillus* in noua calculatione, apud Grecos inuentus est.

47 IIII] *om. M.* | XL^{1}] *om. M.* **48** est] *om. M (add. according to* MC*).* | I^{1}...I^{5}] I I I I *M (corr. according to* MC*).* **49** XLVII] XLVIII *M (corr. according to* MC, CE, COMP. COL.*).* **50** XLVII^{1}] XLVIII *M (corr. according to* MC, CE, COMP. COL.*).* | aliud^{2}] *om. M**. | XLVII^{2}] XLVIII *M (corr. according to* MC, CE, COMP. COL.*).* | aliud^{3}] alid *M**. **51** XLVII^{1}] XLVIII *M (corr. according to* MC, CE, COMP. COL.*).* | XLVII^{2}] XLVIII *M (corr. according to* MC, CE, COMP. COL.*);* Sic diuide I aliud in XLVIII. Sic diuide I aliud in XLVIII. Sic diuide I aliud in XLVIII. *follows by mistaken repetition M (corr. from* COMP. COL.). **51–53** Id...quinquies] *om. M (add. according to* MC*).* **54** IIII] I I I I *M (corr. according to* MC*).* | CC] DC *M (corr. according to* MC*).* **55–56** Supersunt...quinquies^{1}] *om. M (add. according to* MC*).* **57** Supersunt...partes] *om. M (add. according to* MC*).* **58** bis] duo *M (corr. according to* MC*).* **59** XXV] XXII *M (corr. according to* MC*).* **59–60** Aliud^{2}...XXXV] *om. M (add. according to* MC*).* **60** XLma] XLmae *M (corr. from* CE*).* **61** partis] partes *M (corr. from* CE*).* **62** ciclo] ciclo *M**, ciiclo *M.*

62–63 Saltus...est] Cf. the statement that the common and embolismic years, as well as the *ogdoas* and *hendecas*, are of Hebrew origin in c. 59, ll. 14–17, which is somewhat contradictory to the argument here that the *saltus* is a Greek invention. The fact that the 95-year Easter table was composed by Cyril of Alexandria is taken from PROLOGUS DIONYSII (Krusch, *Studien* II, p. 63): *sanctus Cyrillus, cyclum temporum nonaginta et quinque annorum componens*. Cf. also ETYM. 6.17.2 (cited in DRC 115.6–9, BC 156, PROL. AQUIT. 4C): *Cuius quidem rationem beatissimus Cyrillus Alexandriae urbis episcopus in nonaginta quinque annos per quinquies decem novies calculans, quoto Kal. vel luna debeat paschalis sollemnitas celebrari, summa brevitate notavit*. The idea that the 19-year *saltus* was a Greek invention may derive from Dionysius' statement that the technical term for the 19-year cycle, *enneacaidecaeteris*, is a Greek word. PROLOGUS DIONYSII (Krusch, *Studien* II, p. 63, 68; the second passage cited in BC 119A7–8): *Immo potius eundem decennovennalem cyclum, qui enneacaidecaeteris graeco vocabulo nuncupatur, sollicite retinentes, paschalem cursum nullus diversitatibus interpolasse monstrantur. ... Incipit cyclus decennovenalis, quem Graeci enneacaidecaeterida vocant*. Cf. also EPISTOLA CUMMIANI 216: *decennouenalem cyclum, qui Grece enneacedeciterida dicitur*.

62–63 Saltus...est] The invention of the *saltus* is ascribed to the Egytians in DRC 106: *Sciendum nobis quis primitus saltum inuenit. Aegyptii uidelicet, a quibus poene omnes didicerunt, ut Dionisius dicit: Addita etiam ab Aegyptis uno die, fiunt XXX, id est luna mensis unius integra, et nihil remanet de*

are produced. Then, four times 40 are produced (i.e. four times 40 remain after four times 10 have been subtracted, so that the overall result is that 200 equals five times 40). ¬ Divide five into this number, i.e. five times one. Take one from these and divide (it) into 47 parts. Likewise divide another one into 47. Likewise divide another one into 47. Likewise divide another one into 47. Likewise divide another one into 47. (Consequently,) these are five times 47. Take five times seven from these five times 47, and five times 40 remain. Then, divide one of these forties and add four times ten on top of four times 40 and four times 50 are produced, i.e. 200. Five times seven parts are still left over. Take fives times two from these, and five times five remain. Combine (these) five times five, and 25 are produced. Five times two parts are still left over. Combine (these) five times two and ten are produced. Divide this ten into two times five. (Add) one of these five on top of 25 and 30 are produced. (Add) the other five on top of 30, and 35 are produced. In this way, the 47th part of the twelfth part of a *momentum* of every given lunation is divided according to the 235 lunations (that is to say that, by this method, this fraction is established as the result per lunation of the division of 5/12 of a *momentum* by 235 lunations).

The *saltus* in the cycle of 95 years, which Cyril composed according to the new reckoning, was invented among the Greeks.

---

epactis. DE SALTU LUNAE I (984)D5–(985)A7: *Qui sunt ergo, qui primum istum diem saltus invenerunt et addiderunt? Aegyptii sine dubio. Inde Dionysius dicit: A primo anno, qui non habet epactas lunares, hoc est, congruas pro eo, quod cum sint numeri decimi inferioris anni X et VIII epactae, addito etiam ab Aegyptiis uno die, fiunt XXX id est, luna mensis unius integra. Illum ergo unum diem, quem dicit Dionysius ab Aegytiis additum, diem saltus sine dubio demonstrat. ... De quo etiam augmento lunari, hoc est, de die saltus Victorius ostendit dicens: Post XVIIII annos Aegyptiorum more continuato ordine hoc augmentum lunare accrescit.*

62–63 Saltus…est] Cf. p. XXII of the introduction.

Number of *saltus* from creation

Anni VDX faciunt annos saltus <decies> XXVIIII. Saltus in libro Victorii a principio mundi numerantur CCLXXXII <usque ad annum> sub Bero et Bardua consulibus.

The 'Greek' placement of the Dionysiac *saltus* on 22 March

Sciendum ubi sit saltus *Greco*rum. Id est in uere, in mense Martio. *In quo* die? Id est *in XI Kalendas Aprilis*. Quid demonstrat illum? Id est luna. Quomodo? A luna XVIII in *luna*m *XXX in XI Kalendas Aprilis*; in *Kalendi*s *Ianuarii* uero a *luna VIII* in XX luna.

---

64 VDX] VCX *M.* | annos] annum *M.* | decies] *om. M.* **66** CCLXXXII] CCLXXX *M, Krusch, Mommsen (corr. by Mac Carthy, Schwartz).* | usque…annum] *om. M, Krusch, Mac Carthy, Schwartz, Mommsen (add. according to* MC*).* | Bero…Bardua] *corr. to* Vero et Bradua *Mac Carthy.* **68** in uere] inuenire *corr. to* in uere *Ó Cróinín (contra M).* **71** XVIII] XVIIII *M (corr. from* EPISTOLA DIONYSII, DE SALTU LUNAE I*; also Schwartz, Ó Cróinín; note that M makes the same mistake l. 6 of this chapter above).* XI] *om. Ó Cróinín (contra M).* **72** XX] XXX *corr. to* XX *Ó Cróinín (contra M).*

64–67 Anni…consulibus] This passage calculates the number of *saltus* from the beginning of the world to the year of the consuls Verus II and Bradua according to Victorius. In Victorius' reckoning, AD 1=AM 5200 (at least in the opinion of seventh-century computists), and the mentioned consuls are listed under the year equivalent to AD 157. Therefore, AD 157=AM 5356. Now, AM 1=BC 5199 according to Victorius, which equals the ninth year of the 19-year cycle (the *cyclus decemnovenalis*); since the Victorian *saltus* is applied in the sixth year of the 19-year cycle, AM 1 is the third year after a Victorian *saltus*. Hence, the first Victorian *saltus* occurred in AM 17 and it recurred every 19th year thereafter. Since AM 17+281×19=AM 5356=AD 157, 282 *saltus* occurred from the beginning of the world to AD 157, and the number in the text is corrected accordingly. This year was apparently chosen since it referred to the *annus praesens* of the compiler of this passage, who could only have written in the second Victorian cycle, i.e. 532 years after AD 157, in AD 689. These facts clearly suggests that this passage was copied by MC from a Victorian computus of AD 689. The correct spelling of the two consul names given here was *Vero* and *Bradua* (cf. Krusch, *Studien* II, p. 33), but note that the Victorian Easter table in the SIRMOND MS (Oxford, Bodleian Library, Bodley 309, fol. 115r) gives the same misspelling as *M* for the second name in the year equivalent to AD 689: *XIII, Ivro et Bardua, annus CXXX, Kalendae Ianuarii VI feria luna IIII, pascha XIIII Kalendas Maii luna XXII.*
68–72 Sciendum…luna²] The placement of the Dionysiac *saltus* on 22 March is ultimately based on an interpretation of EPISTOLA DIONYSII (Krusch, *Studien* II, p. 84; cited in BC 46C12–D3, DE SALTU LUNAE I (984)C5–10, DTR 42.42–49, in turn cited in PV §297, LIB. COMP. 4.9B, in turn cited in LIB. CALC. 46B; the first sentence only in c. 59, ll. 10–13, CE p. 113): *Propter quod idem ultimus epactas, id est adiectiones lunares, X et VIII tunc retinens, primo anno non XI, ut in ceteris annis fieri solet, sed XII dies accommodat. Et quia XXX dierum fine volvuntur, nulla epacta in principio ipsius cycli ponitur.* That these epacts, mentioned by Dionysius here as well as in his Easter table (Krusch, *Studien* II, p. 69–74), actually referred to 22 March was known through ETYM. 6.17.31: *Istae epactae semper XI Kal. April. reperiuntur in eadem luna quae fuerit eo die.* The epactal change on 1 January may have been taken directly from the list of epacts in c. 49, ll. 14–18, and it is also referred to c. 62, ll. 4–7. MC also consulted CE p. 123–4: *Item querendum est, in quo loco ponemus talem diem tam secundum Grecos et quam secundum Latinos. Hoc est: In XI Kalendas Aprilis secundum Grecos. Aepacta autem in illo anno supra Kalendas Ianuarias erit luna VIII, in Kalendis Februariis luna IX, in Kalendis Martiis luna VIII, in V Idus Martias luna XVIIIma, in XI Kalendas Apriles luna XXIX et XXX.*

64–67 Anni…consulibus] A similar calculation of the number of *saltus* from the beginning of the world can be found in BC 119A7–10, which was apparently composed in the second half of the eighth century: *De numero cyclorum et quod dies bissextiles et quod saltus ab origine mundi: Quod ergo cyclos XVIIIIles habet ab origine mundi, hoc est CCCXIIII, tot saltus sunt, id est CCCXIII<I>.*

5510 years incorporate ten times 29 years of a *saltus* (since 5510=290×19). In the book of Victorius 282 *saltus* are counted from the beginning of the world to the year of the consuls Berus and Bardua.

It has to be known where the *saltus* of the Greeks should be placed. In spring, in the month of March. On which day? On 22 March. What indicates this? The moon. How? (The epact changes) from *luna* 18 to *luna* 30 on 22 March; on 1 January, however, from *luna* 8 to *luna* 20.

68–72 Sciendum...luna²] As has so often been repeated by MC in this chapter, not much information about the *saltus* could be found in the literature available, and this was especially true for the placement (i.e. Julian calendar date) of the Dionysiac *saltus*. Before roughly the middle of the ninth century, two principal rival placements were known and used among computists, both mentioned by MC, namely 22 March and 24 November. MC's preferrence for 22 March, the placement under discussion here, is evident from the fact that it is applied various times throughout this computus (cf. c. 50, ll. 7–14; c. 51, ll. 22–28; c. 52, ll. 8–9, 29; c. 53, ll. 10–11; c. 56, ll. 46–49; c. 58, ll. 8–32, 84–87; c. 59, ll. 40–52), while the November placement is not used once. For Dicuil, writing in the early ninth century, these rival concepts had national connotations, since he connects the November placement with Anglo-Saxon traditions (cf. ll. 97–110 of this chapter below), this 22 March placement of the *saltus* explicitly with Irish customs. LDA 1.5.2: *Sed si secundum Grecorum ac Latinorum, quam mea gens in Hibernia in hac ratione semper custodit, praedictum saltum in vigesimo quarto die mensis Martii sequentis, iuxta primum tempus creationis lunae rationabiliter observabis*. This is in line with the evidence from MC and the fact that the 22 March placement is the only Dionysiac one mentioned in both CE (cf. *app. font.*) and DRC (in slight variation, cf. following note). This March placement of the Dionysiac *saltus* is also implicit (since it falls in the quadragesimal period) in the Irish DIAL. LANGOB. 25. Moreover, it is mentioned in various texts influenced by Irish computistica; but note that it does not appear to be the preferred one in these texts (except for DE SALTU LUNAE III): DIAL. NEUSTR. 25A: *In quibus locis sit saltus? ... Secundum autem Diunisium in XI Kalendas Aprilis, quia inicium est mundi.* VERS. TUR. 13.31: *Graeci dicunt et Aegypti de saltu certissima in undenis Kalendarum aepacta trigesima.* COMP. COL. 4.10C: *Alii motationem lunae in XI Kalendas Aprilis dicunt.* DE SALTU LUNAE I (985)D3–(986)A5, (988)D10–989A1: *Item secundum Graecos in epactis dies saltus in XI Kal. April. demonstratur, quia tunc illo anno in principio cycli nullae epactae dicuntur, quia non sunt congruae, oro eo quod superfluae sunt; tunc enim XII epactae additae invenitur de XVIII in XXXII (recte XXX); addito etiam ab Aegyptiis uno die ad suas XI epactas. ... Et hic saltus, sicut diximus, in quatuor locis ponitur. ... Secundum Graecos in XI Kal. Aprilis luna, quae tunc debuit fieri vigesima nona, fit trigesima propter saltum.* DE SALTU LUNAE III (990)B9–11: *Verbi gratia poni debet saltus lunae in XI Kal. Aprilis secundum Grecos.* LIB. CALC. 46C, PV §295 correct to 22 March Bede's statement that the best placement of the *saltus* would be 21 March (cf. following note).

65–67 Saltus...consulibus] Transcribed in Krusch, *Studien* I, p. 10; idem, 'Einführung', p. 162; Mommsen, 'Victorii Aquitani cursus', p. 697; Schwartz, 'Ostertafeln', p. 90; transcribed and translated in Mac Carthy, *Annals of Ulster* 4, p. lxxi; cf. also p. XXII, LVII–LVIII, XCIV, CXXIV of the introduction. **68–72** Sciendum...luna²] Transcribed in Schwartz, 'Ostertafeln', p. 93; Walsh & Ó Cróinín, *Cummian's letter*, p. 181; cf. Borst, *Schriften*, p. 407, 919 and p. XXII, LXXI, LXXXI–LXXXII, CLXVII of the introduction. **68–110** Sciendum...Ianuarii] Cf. p. LXXI, XCIV, CIII of the introduction

Connecting the *saltus* to the creation of the world and placing it on 21 March

Qui annus incipit saltus in materia? Embolismus. III dies tantum restant de communi: X Kalendas, VIIII Kalendas, VIII Kalendas Aprilis. Accensio, quae XIIII in VIII Kalendas Aprilis in IIII feria, in IIII Idus Martii, si esse fit, id est VIIII dies <ante> XII Kalendas Aprilis. Hic quęstio oritur, quia non inuenimus in saltu Grecorum nisi tantum XVIIII annos. Alii dicunt quod VIIII dies ante XII Kalendas Aprilis uicem anni tenet, uel III dies | tamen XII, fol. 42r
XI, X Kalendas Aprilis. Hic saltus non facilis, quia non inuenitur in libris. Nisi tamen copulatur cum initio anni, quando omnia plena facta sunt, hae sunt super XII Kalendas Aprilis: XXI, II, XIII, XXIIII, V, XVI, XXVII, VIII, XVIIII, XXX, XI, XXII, III, XIIII, XXV, VI, XVII, XXVIII, X, pro saltu transilit VIIII lunam.

---

73 Qui] quis *M.* | saltus] *Three light strokes appear on the top left of this word in M, only two on the top right, very similar to modern quotation marks.* **76** IIII^{2}] VII *M (corr. according to* MC*).* **77** ante] *om. M (add. according to* MC*).* **82** cum] in *M.* **83** XII] X *M (corr. according to* MC*).* **84–85** XXI…X] *This list is arranged in column format in M.* **85** XXV] XXVI *M*.*

73–86 Qui…lunam] As the *saltus* was generally not well explained in his sources (cf. c. 62, ll. 2–3, 96, 111–112), MC thought that this feature should possibly be connected to the creation of the world. Did not everything have its origin at that point? So MC constructed the theory that the creation of the world, which began at the end of the second year of the *ogdoas* in his opinion (i.e. the first four days of creation belong to the second year of the *ogdoas*; the following year, the third and embolismic year of the *odgoas*, started with *luna* 15, i.e. the fifth day of creation; cf. c. 36, ll. 14–15; c. 44, ll. 18–36), was also the beginning of the increment of the *saltus*, which would lead to the *saltus* being applied in the 19th year after the creation (i.e. at the end of the second (!) year of the *ogdoas*). The only problem MC struggled with in the construction of this queer theory was that the first day of creation (21 March) should have been *luna* 10 (had the moon been created at that stage), and therefore would not generally represent the beginning of a lunar year (which had either to begin on *luna* 1 or *luna* 15; cf. c. 53, ll. 5–11; c. 57; c. 59, ll. 38–105); still, MC finishes this discussion with a list of epacts for 21 March contructed according to this theory.

73–86 Qui…lunam] MC's rather peculiar theory, which included the construction of an unparaleled list of epacts for 21 March and with this effectively a new 19-year lunar cycle, found no imitators. Nevertheless, the *saltus* was also connected to the creation of the world in the second year of the *ogdoas* in a rather difficult passage of COMP. COL. (4.8B): *in communi ogdoadi <anno> in Graecos facit augmentum. Quare? Id: Primus annus mundi communis secundus ogdoadi lucet omnibus. Victorius* (!) *secundo anno ogdoadi facit augmentum.* Interestingly enough, the placement of the Dionysiac *saltus* on 21 March (albeit in the correct year, i.e. the first of the *ogdoas*, of course) apparently appealed also to other computists, though for slightly different reasons. DRC 112.1–4 places the *saltus* on 21 March (which was consequently supposed to contain both *luna* 28 and *luna* 29) by maintaining that this was the final day of the solar year. Bede argues that, though the epactal change from 18 to 30 on 22 March indicates the *saltus*, the best placement of it would nevertheless be 21 March, since the moon was, according to his own theory, created on that date (DTR 42.34–37, 55–59; cited in LIB. COMP. 4.9B–C, LIB. CALC. 46B–C, PV §§295–296, the latter two correcting 21 March to 22 March; only the second part cited in RM 56.31–34). Cf. also KAL. B2, G1,4. Note that from a technical point it obviously made no difference whether the *saltus* was placed on 21 March (as in DRC and DTR), containing both *luna* 28 and *luna* 29, or on 22 March (cf. previous note), containing both *luna* 29 and *luna* 30.

What kind of year begins at the origin of the *saltus* (this is a question about the year of creation, which was regarded as the origin of everything, including the gradual increment of the *saltus*)? An embolismic (year). Only three days remain from the common (year) (i.e. between the just-mentioned 22 March and 26 March, which is *luna* 15, and therewith the beginning of the following embolismic year, namely the third of the *ogdoas*): 23, 24, 25 March. The kindling (of the moon, i.e. *luna* 1), which (leads to) *luna* 14 on Wednesday, 25 March, (would have occurred) on 12 March, if it existed (this means that since the moon was created as a full moon on 25 March, any lunar day before that is to be regarded as hypothetical, not historical), i.e. nine days before 21 March (which is the first day of creation, a Sunday). Here, a question arises, since we found only (the reference to) 19 years relating to the *saltus* of the Greeks (i.e. in the authoritative literature no information about the Dionysiac *saltus* could be found other than that it recurred after 19 years). Others say that the ninth day before 21 March (i.e. the day of *luna* 1) holds the turn of the year, or, still, the three days 21, 22, 23 March. This *saltus* (i.e. the Dionysiac *saltus*) is not easy, since it is not found in books. Yet, even if it is not linked to the beginning of the (lunar) year, these are (the lunar ages) corresponding to 21 March, when all was made full (on this first day of creation): 21, 2, 13, 24, 5, 16, 27, 8, 19, 30, 11, 22, 3, 14, 25, 6, 17, 28, 10; because of the *saltus* it passes over *luna* 9.

73–86 Qui…lunam] This passage is referred to as 'confuse Speculationen' by Schwartz, 'Ostertafeln', p. 93. **81–82** Hic…libris] Transcribed in Borst, *Schriften*, p. 407.

The placement of the Victorian *saltus* on 17 Nov

Nobis sci*endum* in quo tempore fit saltus Victorii. Id est in hieme, in VIo anno ogdoadis, in mense Nouimbri, in embolismo. *In quo loco*? *In XV Kalendas Decimbris*. Quomodo fit? Id est prima luna in XIIII Kalendas Decimbris. Inde facitur *XIIII luna in Kalendas Decimbris*, *XIII luna in Kalendis Nouimbris*. Inde in anno saltus hos duo menses distat in aetate lunę. Sciendum quid demonstrat saltum. <Id est luna. Quomodo?> Id est a IIII luna in XVI luna in Kalendis Ianuarii.

---

88 VIo] Vo *M (corr. according to* MC, DTR*; also Schwartz, Borst)*. | ogdoadis] endecadis *M, Schwartz, Borst (corr. according to* MC, DTR*)*. **91** facitur] facit *M*. **92** XIII] XII *M (corr. from* MC, CE, QUAEST. AUSTR., BERN 645*)*. | hos] IIII *M*. **94** Id[1]...Quomodo] *om. M (add. according to* MC*)*.

87–95 Nobis...Ianuarii] The epactal increase from 4 to 16 on 1 January due to the *saltus* could derive directly from a Victorian Easter table (cf. Krusch, *Studien* II, p. 27–52); some of these tables were divided into *ogdoades* and *hendecades* (cf. Paris, Bibliothèque Nationale, Lat. 4860, fol. 147v–148r; Vatican, Biblioteca Apostolica, Reg. Lat. 586, fol. 9r–10v), so that the fact that the Victorian *saltus* occurs in the sixth year of the *ogdoas* could also have been taken directly from such a table; note, however, that the Easter table in the SIRMOND MS (Oxford, Bodleian Library, Bodley 309, fol. 113r–120r), which shows parallels to the Victorian Easter table used by the Irish author of MC's source, the Victorian computus of AD 689 (cf. c. 41, ll. 107–110; c. 62, ll. 64–67), does not contain this division. Many of the details outlined by MC here, and especially the Julian calendar date of the Victorian *saltus*, 17 November, probably derive directly from CE p. 123–4: *Item querendum est, in quo loco ponemus talem diem tam secundum Grecos et quam secundum Latinos. Hoc est: ... Item secundum Romanos saltus in XV Kalendas Decimbris fit. Aepacta uero in Kalendis Ianuarii secundum Romanos illo anno ita est: Luna IIII in Kalendis <Ianuarii et Martii, in Kalendis Aprilis> luna V, in Kalendis Maii luna VI, in Kalendis Iunii VII, in Kalendis Iulii luna VIII, in Kalendis Augusti luna IX, in Kalendis Septimbris et Octimbris luna XIma, in Kalendis Nouimbris luna XIIIma, in XV Kalendas autem Decimbris luna XXIX aut XXX, in Kalendis Decimbris luna XIIII, in Kalendis Ianuarii luna XVIma*. Cf. also the very similar earlier discussion of this Victorian year of the *saltus* in c. 50, ll. 50–55. The epactal increase and the (rather implicit) fact that the Victorian *saltus* occurs in the sixth year of the *ogdoas* are also mentioned in c. 52, ll. 16–17, 27–32.

87–95 Nobis...Ianuarii] The November placement of the Victorian *saltus* was not mentioned in any pre-CE text, and therefore it appears to have been popularized by seventh-century Irish computists, whence it was then adopted by eighth-century Frankish computists: DIAL. NEUSTR. 25A: *In quibus locis sit saltus? ... Secundum Victorium XV Kalendas Decembris*. QUAEST. AUSTR. 1.11C, 1.12A, 2.11A: *Sciendum quoque: Cur anno, quo quarta lunae aetas in Kalendis fuerit supputata, eodem anno Kalendae Novembris et Decembris, quae semper eadem aetate conveniunt, hac tantum vice inter se differunt, ut tertia decima luna in Kalendis Novembris et quarta decima <in> Kalendis Decembris supputetur? De lunari mutatione, quam nos secundum consuetudinem saltum vocamus, interrogandum videtur: Quo anno vel in quo mense interponendus sit? Vel que ratio cogit eum iuxta veterem consuetudinem in Novembre luna interesse, quo mense nec annus incipit nec finitur? ... VI. Kalendas Ianuarii luna quarta. ... Kalendas Novembris luna tertia decima. Kalendas Decembris luna quarta decima. Hac vice tantum aetates discordant causa augmenti lunaris, quod in Novembri interponitur. Nam luna, quae tertia decima Kalendas Novembris, vicesima octava est XVI Kalendas Decembris. Deinde aetas transilitur. XV Kalendas <Decembris> tricesima esse definitur. Qui legit, intellegat, utrum peracto an incipienti saltus adiungatur anno. Ex <qua> causa non in quinta decima luna, sed in sexta decima luna Kalendas Ianuarii*. VERS. TUR. 13.32: *Victorius et Latini diem saltus memorant, in quindecim Kalendarum Decembris concelebrant*. DE SALTU LUNAE I (985)A8–10, (988)D10–12: *Interrogandum est igitur, in quo loco hic saltus juxta Latinos ponitur? Sine dubio in XV <Kal.>*

We have to know in which season (*tempore*) the *saltus* of Victorius occurs. In winter, in the sixth year of the *ogdoas*, in the month of November, in an embolismic (year). In which place? On 17 November. How does (this) happen? *Luna* 1 (occurs) on 18 November. Consequently, *luna* 14 is produced on 1 December, *luna* 13 on 1 November. Hence, in (this) year the *saltus* separates these two month in (their) lunar age (on the Calends). It has to be known what indicates the *saltus*. The moon. How? (The epact changes) from *luna* 4 to *luna* 16 on 1 January.

Decembris. ... Et hic saltus, sicut diximus, in quatuor locis ponitur. Juxta Victorium et Latinos in XV Kal. Decembris ponitur. KAL. A4,9–10 also note *hic saltus apud Latinos* under 17 November. The Victorian year in question is discussed without explicit reference to the *saltus* in Bern, Burgerbibliothek, 645, fol. 55r, but from the lunar ages given for 1 November and 1 December it is obvious that it was supposed to fall in November. When touching the Victorian *saltus* in passing, Bede only refers to the year within the 19-year cycle and the epactal increase on 1 January: DTR 42.32–34 (cited in LIB. COMP. 4.9B, LIB. CALC. 46B): *At contra Victorius tertio ante finem ogdoadis anno hanc interserendam censuit, lunam kl. ianuarium de quarta in sextam decimam convertens*. Cf. also glosses to DTR, e.g. Kassel, Landesbibliothek, 4° MS astron. 1, fol. 71r: *At contra Victorius tertio ante finem ogdoadis anno, hoc est VI decennovennolis anno, saltum facit, hoc est quod dicit lunam Kalendarum Ianuarii de IIIIta in XVImam converti. Cum enim in VI anno cicli decennovennalis IIIIta luna in Kalendis Ianuarii existit, sequenti anno in eisdem Kalendis Ianuarii adiectis XI debuit esse XVma. Sed (?) ille propter saltum adiecit XII, ut fieri XVIma*. A different gloss to DTR (*CCSL* 123B, p. 408) more specifically argues that the Victorian *saltus* was to be placed in November, in the sixth year of his 19-year cycle, even giving an explanation for this placement: *Victorius autem in VIo anno decennouenalis cicli, et Anatolius facit saltum in Martio; nos autem et Victorius in Nouembri. Nam December uicesimam nonam lunam habet, et ideo non possumus ibi facere saltum de uicesima nona in primam dimittentes tricesimam, quia non habet tricesimam*. Only the epactal increase in DIAL. BURG. 13B, 15A: *Et quia in quattuor annos unus dies adcrescit, ita et in annis novemdecim una luna augetur, ea tamen racione, ut quandoquidem addendo undecim et tollendo triginta eveniret quinta decima. Ex arbitrio suo unaquaeque augetur secundum sanctum Victurium, et efficitur sexta decima. ... Annus lunaris surgit ad annos novemdecim. Ubi luna in Kalendas Ianuarias fuerit quinta decima, addis sextam decimam, et inde dicitur saltus lunae*. Interestingly enough, DRC 112.4–6 condems 17 November to be the wrong placement of the Victorian *saltus* and instead proposes 31 December, containing both *luna* 14 and *luna* 15, in accordance with the Julian calendar year (and with DRC's placement on 21 March of the Dionysiac *saltus* also one day before the *sedes epactarum*, 22 March in the Dionysiac, 1 January in the Victorian reckoning). QUAEST. AUSTR. 1.12A considers the preceding 26 March as the Victorian placement of the *saltus*, which is obviously modelled on the placement of the Dionysiac *saltus* in March. COMP. COL. 4.8A lists, strangely enough, 18 December as the Victorian placement of the *saltus*, which may be the result of a scribal error (*Ianuarii* instead of *Decembris*). Note that in the Victorian Easter table in the SIRMOND MS (Oxford, Bodleian Library, Bodley 309, fol. 113r–120r) the *saltus* is uniquely placed in the second year of the *hendecas* rather than the sixth year of the *ogdoas*, presumably for the purpose of bringing the lunar age of the resurrection in line with the Synoptic Gospels (17 instead of Victorius' 16).

87–95 Nobis...Ianuarii] The first part transcribed, the rest discussed in Schwartz, 'Ostertafeln', p. 93; part of the beginning transcribed in Borst, *Schriften*, p. 919; cf. Borst, *Schriften*, p. 407–8 and p. XXII, LXXXIV, CXCIX of the introduction.

Term: *saltus*

Notandum nobis, quod hii II saltus non dicuntur.

The 'Egyptian' placement of the Dionysiac *saltus* on 24 Nov.

Inde uerius saltus Aegyptiorum. Sciendum quomodo fit. Id est a XXVII in Kalendis Ianuarii in VIIIIam in Kalendis Ianuarii. <In quo loco?> Id est in VIII Kalendas Decimbris. Sciendum quomodo fit. XXVII in Kalendis Ianuarii et Martii, XXVIII in Kalendis Aprilis et Maii, XXX in Kalendis Iunii, I in Kalendis Iulii, II in Kalendis Augusti, IIII in Kalendis Septimbris et Octimbris, VI in Kalendis Nouimbris, VII in Kalendis Decimbris pro saltu; numeramus VI lunam in Kalendis Nouimbris, X in Nonas, XVIII in Idibus, XX in XVII Kalendas Decimbris, XXVI in XI Kalendas, XXVIIII in VIII Kalendas Decimbris et XXX lunam; luna I in VII Kalendas Decimbris et reliqua sic

---

96 hii] hi *Schwartz (contra M)*. | saltus] saltus *add. Schwartz (contra M)*. **99** In…loco] *om. M (add. according to* MC*; cf. Schwartz, Borst)*. **108** VII] VIII *M (corr. according to* MC *and from* ALCUIN*; also Borst)*.

---

96 Notandum…dicuntur] Cf. similar statements in c. 62, ll. 2–3, 81–82, 111–112. **97–110** Inde…Ianuarii] Even though MC refers to the 24 November placement of the Dionysiac *saltus* as *verius saltus* here, he certainly preferred 22 March, as demonstrated in *app. comp.* of ll. 68–72 of this chapter above. The origin of this 24 November placement, which became very popular from the eighth century onwards, is not quite clear; it seems to be modelled on the Victorian placement just discussed, but then the origin of that is equally obscure. Note that this placement cannot be found in CE and that MC again applied his peculiar Dionysiac sequence of lunations (cf. c. 50, ll. 29–39) here.

97–110 Inde…Ianuarii] Bede, despite arguing that the Dionysiac *saltus* could *non alibi aptius* be placed on 21 March (cf. previous note), nevertheless continues to mention the 24 November placement of the *saltus*, and, in fact, this is the date used by him for calculation throughout his work (cf. DTR 20.41–45, cited in LIB. CALC. 10, LIB. COMP. 4.12, in turn cited in PV §286; DTR 46.51–2 (where *in initio ogdoade* refers to the March, *in hendecade* to the November placement, since Bede's lunar years started with 1 January), cited in LIB. CALC. 86, partially in RM 60.42–44): DTR 42.59–62 (cited in LIB. COMP. 4.9C, LIB. CALC. 46C, RM 56.34–37): *Sed sunt qui hoc nobis in luna novembri mensis agendum magis autument: quatenus huiusmodi impedimentis cum praecedentis anni fine absolutis, novum de caetero annum libero possint computo ingredi*. Supporting the argument that Bede adhered to 24 November as the date of the Dionysiac *saltus*, Dicuil, as mentioned above, ascribed this placement more generally to Anglo-Saxon customs: LDA 1.5.1: *Etsi lunarem saltum in vigesimo quarto die mensis Novembris, secundum Anglos, complere volueris*. This is, moreover, confirmed by Alcuin's preference for this placement: ALCUIN, *Epistola* 126 (*MGH Epp.* 4, p. 185; referring to AD 797): *Nam causa huius praesentis cartulae lunaris saltus effecta est, qui in hac lucenti luna Novembri mensis anno circuli decennovenalis novissimo certissime inspici poterit. Igitur haec luna Novembri, quae per decem et octo annos tricesima conputabatur, in hoc anno praesenti praefati circuli nono decimo, undetricesima conputari debet. Ita ut de vicesima nona, quae fit octavas Kal. Decembri, vertatur septimas Kal. in primam; et sit die Kalendarum Decembrium septima, quae iuxta regulares et epactas sexta conputata est. Et erunt tres lunae pariter – Octobri scilicet Novembri et Decembri mensis – undetricenarium dierum*. This apparently Anglo-Saxon practice soon influence Frankish computistics. DIAL. NEUSTR. 25B (the MS reads *VI* instead of *VII Kalendas Decembris*): *Si vero in Novembri facias saltum luna vicesima octava, in VIII Kalendis Decembris tricesimam fac. Et in VII Kalendas Decembris prima fiat, ut in Kalendis Ianuarii potueris invenire lunam nonam*. LECT. COMP. 2.4C (cited in LIB. COMP. 2.15d;

We have to note that these two are not called *saltus* (i.e. the Dionysiac and the Victorian *saltus* are not referred to under that term in the writings of Dionysius and Victorius).

Therefore the more appropriate *saltus* of the Egyptians. It has to be known how it happens. (The epact changes) from (*luna*) 27 to (*luna*) 9 on 1 January. In which place? On 24 November. It has to be known, how (this) happens. (*Luna*) 27 occurs on 1 January and 1 March, (*luna*) 29 on 1 April and 1 May, (*luna*) 30 on 1 June, (*luna*) 1 on 1 July, (*luna*) 2 on 1 August, (*luna*) 4 on 1 September and 1 October, (*luna*) 6 on 1 November, (and *luna*) 7 on 1 December because of the *saltus*; we count *luna* 6 on 1 November, (*luna*) 10 on 5 (November), (*luna*) 18 on 13 (November), (*luna*) 20 on 15 November, (*luna*) 26 on 21 (November), (*luna*) 29 and *luna* 30 on 24 November; *luna* 1 (occurs) on 25 November and so on up to

---

only the second part in LIB. ANN. 49): *In finem noni decimi anni habes epactas decem et octo. Adde undecim super decem et octo, fiunt viginti novem. Adde saltum lunae, fiunt triginta. Hic apparet, quod non addit, sed salit in retro unum diem. Ista triginta epacta est in capite. Memento, quod anno novissimo circuli decennovenalis luna Novembri mensis vicesima nona propter saltum lunae, qui eo anno inserendus est, computare debes.* COMP. COL. 4.5B, 4.10C: *Non inmerito creditur in omni embolismo ac byssexto luna tricesima numerari. Sed quoniam in omni embolismo ac byssexto, quos pariter in septem lunis ante ostendimus, supputando triginta semper luna plena ex supradicto momentorum numero crescere in fine decennovenalis cycli non potest, cui deesse omni luna minuta decem conprobatur, idcirco luna tricesima, qui eodem anno mense Novembri vicesima nona extinguitur, viginti quattuor horas amittere videtur. ... motationem lunae ... alii in VIII Kalendas Decembris in eodem anno secundum Aegyptiorum auctoritatem; luna in Kalendas Ianuarii non octava, sed nona manifeste deprehenditur. Lege argumenta Graecorum, qua necessitate constringimur, ut illam Novembris mensis ... lunam aliquam tricesimam eodem anno supputemus vicesimam nonam.* ARG. AQUENS. 6A (referring to AD 817; note that the only MS reads *VIII* instead of *VII Kalendas Decembris*): *in Kalendas Novembris sexta, in VIII Kalendas Decembris iterum vicesima nona. Et ita eo anno luna Novembris mensis terminatur in vicesima nona, propter saltum, qui tunc adimitur, non ut semper in aliis annis tricesima. Et erit in VII Kalendas Decembris prima, quae secundum epactam esse debuit tricesima, et est in Kalendas Decembris septima, quae, si sequeretur epactas, fieret sexta.* Gloss to DTR (*CCSL* 123B, p. 348; discussing the differences between the July and the November placement): *Nos uero ponimus saltum in VIIo kl. Decembris*. The *saltus* is also noted under 24 November in KAL. C8, E3, G4, H1–4, under 25 November in KAL. A8, B7, C8, F5, G2, H6. Only the epactal increase on 1 January in LECT. COMP. 9.4B. Then, from roughly the late ninth century, computists started to follow Bede's incidental statements that the Egyptians placed the *saltus* in July (DTR 20.41–45, cited in LIB. CALC. 10, LIB. COMP. 4.12, in turn cited in PV §285; DTR 42.62–63, cited in LIB. COMP. 4.9C, LIB. CALC. 46C, RM 56.37–39), which implies *luna* 29 of the July lunation as the placement of the *saltus*; in the 19th year of the Dionysiac 19-year cycle this occured on 29 July, and, in fact, the *saltus* is noted for that date in KAL. B7,9, and a day later (which means *luna* 1 instead of *luna* 30) in KAL. A3,11, F4,6–8, G1,4, H1–4. The main advocate for this July placement (30 July) and against the November placement was HELPERIC, *De computo* 18, 36 (*PL* 137, col. 33–34, 47C–D). Earlier, DIAL NEUSTR. 25A and DE SALTU LUNAE I (989)A2 proposed (but did not follow) 26 September in accordance with the Egyptian year, which is then also found in KAL. B9; similarly COMP. COL. 4.10C. References to 17 April can be found in DE SALTU LUNAE I (985)B12–D3, (986)B12–C2, (989)A2–5, VERS. TUR. 12.28.

---

96 Notandum…dicuntur] Transcribed in Schwartz, 'Ostertafeln', p. 93. **97–110** Inde…Ianuarii] The first part transcribed, the rest discussed in Schwartz, 'Ostertafeln', p. 93, 98; partially transcribed and discussed in Borst, *Schriften*, p. 407; cf. also p. XCIV of the introduction.

usque ad XXVIIII in X Kalendas Ianuarii, luna I in VIIII Kalendas; inde VIIII luna in Kalendis Ianuarii.

Names for and definitions of *saltus*

| Saltus non inuenitur in libris. Hoc nomen ei, sed abusiue, dicitur. Nomina enim naturalia ei *augmentum lunare*, *increment*um *lunar*e; haec nomina, cum preparatur, ei conueniunt. Una luna, cum apparet. *Adiectio lunae*, cum addit aetatem, ut *XXX* facit in *XI Kalendas Aprilis*. Mutatio lunę, cum transilit aetatem, id est XXVIIII, in XXX. fol. 42v

109 VIIII] VIII *M (corr. according to* MC*)*. **111** in libris] *follows* in libris *by mistaken repetition M (cf. Ó Cróinín)*. | Hoc nomen] *om. Ó Cróinín (contra M)*. **112** augmentum] agmentum *M, Ó Cróinín (corr. from* PROLOGUS VICTORII, DRC, DE SALTU LUNAE I*)*. **116** transilit] trasilit *M**. | XXVIIII] XXVIII *M (corr. from* DRC*; also silently Ó Cróinín)*.

111–117 Saltus…XXX] Since MC could not find the term *saltus* in the autoritative literature available to him (cf. also ll. 2–3, 81–82, 96 of this chapter), he felt the need here to list the terminology that was used instead: *augmentum lunare* and *incrementum lunare* (for which see also ll. 11–13 of this chapter) derive from PROLOGUS VICTORII 3 (Krusch, *Studien* II, p. 19; the second sentence is cited in DRC 113.4–6; only the second half of that sentence in DE SALTU LUNAE I (985)A6–7): *Porro ii, qui post centum duodecim annos in id cyclum, unde orsus fuerat, reverti confirmant, ea ipsa incrementa lunaria, quae superius memoravi, post annos sedecim subnecti debere constituunt. Ii vero qui anno quinto et nonegensimo cycli observantiam conprehendunt post decem et novem annos Aegyptiorum more continuato ordine, quod est verius, hoc augmentum lunare subiciunt*. For *augmentum lunare* cf. also EPISTOLA CUMMIANI 206–207: *alium in aucmento lunari, quod uos saltum dicitis*. For *incrementum lunare* cf. DE MIRABILIBUS 2.4 (*PL* 35, col. 2176): *incrementum lunare, quod computatores saltum nominant*. The term *una luna* may also stem from PROLOGUS VICTORII 3 (Krusch, *Studien* II, p. 18–9; partially cited in QUAEST. AUSTR. 2.9B): *Tum deinde ii, qui cyclum annorum octoginta et quattuor ediderunt, duodecim peractis annis lunam unam, quam per idem tempus certis annua revolutione minutiis adserunt calculandi lege subcrescere, adiciendam legitimo cursui esse praecipiunt*. The definition of *saltus* as *adiectio lunae* is based Dionysius' description of the epactal increase of 12 rather than 11 due to the *saltus*: EPISTOLA DIONYSII (Krusch, *Studien* II, p. 84; cited in BC 46C12–D3, DE SALTU LUNAE I (984)C5–10, DTR 42.42–49, in turn cited in PV §297, LIB. COMP. 4.9B, in turn cited in LIB. CALC. 46B; the first sentence only in c. 59, ll. 10–13, CE p. 113): *Propter quod idem ultimus epactas, id est adiectiones lunares, X et VIII tunc retinens, primo anno non XI, ut in ceteris annis fieri solet, sed XII dies accommodat. Et quia XXX dierum fine volvuntur, nulla epacta in principio ipsius cycli ponitur*. That these epacts mentioned by Dionysius actually referred to 22 March was known through ETYM. 6.17.31: *Istae epactae semper XI Kal. April. reperiuntur in eadem luna quae fuerit eo die*. The origin of the last term, *mutatio lunae*, is the most difficult to establish. Interestingly enough, the term itself can also be found, though in a different context, in MAXIMUS OF TOURS, *Sermones, Sermo 30* (*CCSL* 23, p. 118): *Grauior ergo tua quam lunae mutatio est; luna defectum luminis patitur tu salutis*.

111–117 Saltus…XXX] The exact same names for and definitions of *saltus* can be found in the Irish DRC 110 (the sole MS B reads *XVIIII* instead of *XXVIIII*), which is certainly based on MC: *Sciendum nobis quot nomina saltus habet. Sex videlicet, id est naturalia duo, id est augmentum lunare et incrementum lunare; et haec duo nomina dum preparatur sibi conueniunt. Quattuor uero in adnumeratione: luna una cum apparet, adiectio lunae cum addit aetatem, ut XXX facit in XI Kl Aprelis, motatio lune cum transilit XXVIIII in XXX. Et sic saltus dicitur, quo nomine nos utimur*. With the sole omission of the dubious *adiectio lunae* DE SALTU LUNAE I (985)A2–4, (988)A2, D6–10: *Hic ergo saltus lunae, incrementum vel augmentum dicitur lunare, eo quod crescit per XVIII (recte XVIIII) annos unus dies in luna. ... qui dicitur saltus, sive incrementum, vel augmentum. ... Interrogandum est igitur, quot nomina habet hic saltus? Hoc est, quinque. Dicitur incrementum lunare; dicitur etiam augmentum

(*luna*) 29 on 21 December; *luna* 1, (then, occurs) on 22 (December); consequently, *luna* 9 (occurs) on 1 January.

The (term) *saltus* is not found in books. This name is assigned to it, but by improper use. In fact, the natural names (assigned) to it are 'lunar augmentation' and 'lunar increment'; these names suit it, when it is prepared. 'One moon', when it appears. 'Lunar addition', when it adds a (lunar) age, as it does with 30 on 22 March. 'Change of the moon', when it passes over a (lunar) age, i.e. 29, to 30.

---

lunare; dicitur mutatio lunaris; dicitur etiam una luna crescens per omnes annos; dicitur etiam saltus. DIAL. NEUSTR. 26A (the MS reads *mutacio*; cited in COMP. COL. 4.9A): *De nominibus saltus. Incrementum lunae, quando crescit. Augmentum lunae, quando facitur. Item motacio lune. Et una luna.* The term *augmentum lunare* for *saltus* is used frequently in computistical literature of the seventh and eighth centuries: BC 18, DE SALTU LUNAE II, QUAEST. AUSTR. 2.9B, 2.11A, LECT. COMP. 7.4, VERS. TUR. 12.30. Bede applies the term *mutatio* in DTR 42.31, 54 (cited in LIB. COMP. 4.9B, LIB. CALC. 46B), and it is also neatly explained in QUAEST. AUSTR. 1.12A: *De lunari mutatione, quam nos secundum consuetudinem saltum vocamus.* Cf. also DIAL. BURG. 16. Certainly, some of these terms will also be found in still unpublished tracts hidden in the MSS. Cf., e.g., Leiden, Universiteitsbibliotheek, Scaliger 28, fol. 33v: *Saltus est mutacio lune aut adieccio lunaris, ut saltus est diei nam peractum XVIIII annorum cursum, si non lunaris mutacio esset, cum prima luna putaris, secunda esset.*

111–117 Saltus…XXX] Transcribed in Walsh & Ó Cróinín, *Cummian's letter*, p. 210; partially transcribed in Borst, *Studien*, p. 409; cf. Ó Cróinín, 'A seventh-century Irish Computus', p. 124 and p. XXII, CIII, CXCIV–CXCV of the introduction.

The *saltus* of the *latercus*

Saltus laterci per XIIII annos paratur. In IIII anno ogdoadis sit, in communi, in Nouembrio mense. Sic fit: A XII luna in XXIIII in Kalendis Ianuarii primus; secundus saltus a XVII in XXVIIII; tertius a XXII in IIII; quartus a XXVII in VIIII; quintus a II in XIIII in Kalendis Ianuarii.

Impact of bissextile day on the course of the moon, of *saltus* on the course of the sun and time

Sciendum si bissextus aliquid mutat in cursum lunae. Id est, mutat lunam, commutat Kalendas. Sciendum si saltus mutat aliquid in cursu solis. Id est, mutat rationem mensium in anno saltus. Saltus non mutat cursum temporum, quia dimittunt omnes XVIIII anni partem cum aliis XVIIII annis. In anno saltus non fiunt ępacte in numeratione, fiunt tamen secundum naturam, quia finis cicli XVIIII annorum.

---

118 XIIII] XII *M (corr. by Krusch, Schwartz, Warntjes; corr. to* XIII *Mac Carthy, Mc Carthy).* **119** sit] *corr. to* fit *Krusch, Mac Carthy, Schwartz.* | communi] *add.* anno *Mac Carthy, Mc Carthy.* | Nouembrio] Novembrie *Schäferdiek (contra M).* **121** XXII] XV *M (corr. by Krusch, Mac Carthy, Schwartz, Mc Carthy, Warntjes).* **122** II] III *Mac Carthy (contra M).* | XIIII] XIII *M (corr. by Krusch, Mac Carthy, Schwartz, Mc Carthy, Warntjes).*

118–122 Saltus…Ianuarii] It appears that MC had a list of all 84 *latercus* epacts of 1 January at his disposal, or reconstructed it from the basic information that this reckoning had initial epact 19 and employed a 14-year saltus (cf. the column of epacts in the Padua *latercus* in Mc Carthy, 'Easter principles', p. 218–9 (edition); Warntjes, '84 (14)-year Easter reckoning', p. 80–2 (facsimile)); this list of epacts was already applied in the (wrong) comparison of the *latercus* epacts with those of Victorius and Dionysius c. 52, ll. 3–26; from this comparison with the Victorian reckoning derives the statement that the *saltus* occurs in the fourth year of the *ogdoas*: the first *latercus saltus* occurs, in that wrong comparison, in the fourth year of the Victorian *ogdoas*, while the following obviously would not, since the *latercus* employed a 14-year *saltus*, the Victorian reckoning, on the other hand, one of 19 years. Moreover, the statement that the *latercus saltus* was placed in November is nothing but a translation of Victorian practice (cf. ll. 87–95 of this chapter) to the *latercus*. Now, from a list of all 84 epacts of the *latercus* it was a simple task to note the epactal changes which indicate the occurrence of a *saltus*, and the use of such a list of exactly 84 years also explains that the final *saltus*, from the 84th to the first year, from *luna* 7 to *luna* 19 on 1 January, was omitted. **123–130** Sciendum…annorum] The impact of the solar bissextile day on the reckoning of time is that it adds an extra Julian calendar day, since *VI Kalendas Martii* are doubled; moreover, an extra lunar day is implemented, which changes the common sequence of lunations. The *saltus*, on the other hand, also changes the sequence of lunations, since a full lunar month is reduced to a hollow one; additionally, the reckoning of the epacts is disturbed by the *saltus*, since in a year of a *saltus* the epactal increase is 12 rather than the common 11. Yet, and this is explicitly stressed by MC, though these two technical devices are artificially implemented, they develop naturally over a period of four and 19 years respectively. Cf. also the related passage in the chapter about the bissextile day (c. 41, ll. 32–37, 50–71). For the fact that the application of the *saltus* is caused by the divergence between the natural and the calculated course of the moon cf. c. 62, ll. 17–20.

118–122 Saltus…Ianuarii] The only other text that refers to the *saltus* of the *latercus* is COMP. COL. 4.8A. Intriguingly, this reference is followed by exactly five different Julian calendar dates. The entire passage, however, is difficult to interpret, and it may be noted that none of these five dates could be reconciled with the theory of the *latercus*. **123–130** Sciendum…annorum] The fact that the bissextile day and the *saltus* are natural phenomena in their respective cycles, but artificially applied, is mentioned in DRC 107.1–4: *Sciendum nobis utrum saltus in naturali lunae <cursu> deprehenditur, sicut bissextus*

The *saltus* of the *latercus* is prepared in the course of 14 years. It should occur in the fourth year of the *ogdoas*, in a common (year), in the month of November. It happens in the following way: The first (*saltus* changes the epact of 1 January) from *luna* 12 to (*luna*) 24 on 1 January; the second *saltus* from (*luna*) 17 to (*luna*) 29; the third from (*luna*) 22 to (*luna*) 4; the fourth from (*luna*) 27 to (*luna*) 9; the fifth from (*luna*) 2 to (*luna*) 14 on 1 January.

It has to be known, whether the bissextile day changes anything in the course of the moon. It changes the moon and it changes the calends entirely (since it adds an extra lunar day and an extra sixth calends). It has to be known, whether the *saltus* changes anything in the course of the sun. It changes the reckoning of the months in the year of the *saltus* (i.e. it changes the sequence of lunations in such a year, since one lunation is reduced from a full to a hollow one). The *saltus* does not change the course of time, since all 19 years leave a part with the other 19 years (i.e. the same lunar increment grows in every year; this natural and gradual increment, neglected throughout the 19-year cycle, results in one day, i.e. the *saltus*, which is therefore, though artificially implemented, in agreement with the natural course of time). Even though the epacts do not happen to be in (line with) the usual counting (i.e. the epactal increment is 12 instead of the usual 11) in the year of the *saltus*, they still happen to be according to nature, since it is the end of the cycle of 19 years (in which the mentioned natural lunar increment leads to the artificial implementation of the *saltus*, which in turn results in the unusual epactal increase).

---

in naturali cursu solis fit, án in ultimo anno circuli subito per artem efficitur. Non per artem, sed ex certis minutis in naturali lunae cursu crescit, licet uarie preparatur. DIAL. NEUSTR. 26A (cited in COMP. COL. 4.9A) tries to illustrate the difference between the natural increment and the resulting artificial application of the *saltus* by analogy to a pregnant woman (which is also used by MC when explaining the bissextile day; cf. c. 41, ll. 36–37) and to a boat journey. For this question cf. also DIAL. LANGOB. 21B, QUAEST. AUSTR. 1.12B, 2.9B–C. The reduction of one lunation from a full to a hollow one due to the application of the *saltus*, and the resulting change in the sequence of lunations, are described by Bede: DTR 42.21–24, 63–65 (cited in LIB. COMP. 4.9B–C, LIB. CALC. 46B–C, PV §§292–293; only the second part in RM 56.39–41): *Diversis autem locis circuli decemnovenalis, diversi quique calculatores hunc eundem lunae saltum interponendum, id est mensem lunarem, qui caeteris annis tricenos dies habere consueverat, undetriginta dierum esse faciendum putavere. ... Sed sive hic sive illic sive alibi feceris, necesse est ibidem, ni fallor, tres pariter menses undetricenorum computare dierum*. The reduction of the November lunation is explicitly mentioned in ARG. AQUENS. 6A (referring to AD 816), and the impact on the sequence of lunations, if the *saltus* is applied in November, is outlined by ALCUIN, *Epistola* 126 (*MGH Epp.* 4, p. 185; referring to AD 797): *Et erunt tres lunae pariter – Octobri scilicet Novembri et Decembri mensis – undetricenarium dierum*. But note that, according to MC's practice, the phenomenon of three successive hollow lunations would not occur, since MC placed the *saltus* at the end of the March lunation (c. 62, ll. 68–72), and applied a different sequence of lunations, in which the April lunation was full (c. 50, ll. 29–39).

118–122 Saltus…Ianuarii] Transcribed in Krusch, *Studien* I, p. 11; Schwartz, 'Ostertafeln', p. 94; Mc Carthy, 'Easter principles', p. 215; transcribed and translated in Mac Carthy, *Annals of Ulster* 4, p. lxxiii; Warntjes, '84 (14)-year Easter reckoning', p. 39 (facsimile p. 83); the beginning is partially transcribed and discussed in Schäferdiek, 'Osterzyklus', p. 361, 370–1; cf. also Krusch, *Studien* I, p. 14; O'Connell, 'Easter cycles', p. 84–5; Borst, *Schriften*, p. 919 and p. LXXII, XCIV, CXXII–CXXIV, CLXXXI of the introduction. **126–127** Saltus…temporum] Transcribed in Borst, *Schriften*, p. 454. **128–129** non…naturam] Transcribed in Borst, *Schriften*, p. 439.

<LXIII. DE PROHIBITIONE CELEBRANDI PASCHA IN LUNA XIIII>

Arguments for rejecting Easter on *luna* 14

Post resurrectionem domini et *ascensionem, quomodo pascha apostoli obseruare*nt non quęrebant, *quia* discipuli *per uniuersum mundum ad predicandum* | fueru*nt* fol. 43r *dispersi. Sed quacu*m*que die* esset *XIIII luna, pasca celebrabant. Post transitum* uero *omnium apostolorum de hoc mundo, diuersa ieiunia tenebant per singulas*

LXIII,3 resurrectionem] *the* c *in the middle of this word is composed of the first and the third stroke of* x *in M; it seems that the copyist intended to write* x, *but after the first stroke realized that it should rather read* c*; consequently, he only added a slightly rounder third stroke of* x *to the first, since these two strokes do, in fact, resemble a* c.

LXIII,3–24 Post…mandatum] In this passage, MC supplies an interesting account of the origin of celebrating Easter on *luna* 14 and the condemnation of that practice ever since, in MC's opinion, the Synod of Caesarea. By the early middle ages, the ancient practice of celebrating Easter *luna* 14 irrespective of the weekday was long gone. However, Easter Sunday could fall on *luna* 14 in one of the three Easter reckonings of the early middle ages, namely the 84 (14) (or *latercus*) followed in the Ireland and Britain (for the lunar limits of Easter Sunday of these three reckonings cf. c. 58, ll. 2–7). This whole chapter on the interdiction of celebrating Easter on *luna* 14, then, is obviously directed against followers of the *latercus* and probably is a remnant from the height of the paschal controversy in the middle of the seventh century; in fact, it probably formed part of the now lost Victorian computus of AD 689 (cf. introduction p. CXXIIII–CXXVI) originally. The first authority cited in this respect are the so-called Acts of the Council of Caesarea in the version known as DE ORDINATIONE FERIARUM PASCHALIUM (*PL* 90, col. 607A–C; cited up to *tenebant* with attribution to Theophilus in DRC 87.2–12, which reads *ad praedicandum fuerant occupati, quacumque, omnium apostolorum de hoc mundo*): *Post resurrectionem vel ascensionem Domini Salvatoris, apostoli quomodo Pascha deberent observare nihil ordinare potuerunt, quia dispersi erant per universum mundum, et ad praedicandum occupati. Sed quacunque die decima quarta luna mense Martio fuisset, Pascha celebrabant. Post transitum ergo de hoc mundo omnium apostolorum, per singulas provincias diversa tenebant jejunia. Nam Galli quacunque die octavo Kalendarum Aprilium fuisset, quando Christi resurrectio tradebatur, semper Pascha celebrabant: in Italia vero alii XX dies jejunabant, alii septem. Orientales vero sicut apostolos viderunt, ut supradictum est, decima quarta luna, mense Martis Pascha tenebant. Cum ergo hae tales observationes per singulas provincias tenerentur, unde moeror erat sacerdotibus, eo quod a quibus una fides recte tenebatur, eorum dissentirent jejunia. Tunc papa Victor Romanaeque urbis episcopus direxit auctoritatem ad Theophilum Caesariensis Palaestinaeque antistitem, ut quomodo Pascha recto jure a cunctis Catholicis celebraretur Ecclesiis, inibi fieret ordinatio, ubi Dominus et Salvator mundi fuerat in carne versatus. Percepta itaque auctoritate praedictus episcopus non solum de sua provincia, sed etiam de diversis regionibus omnes episcopos evocavit*. The statement that Victor was the 13th bishop of Rome since Peter is an interesting addition to this excerpt from DE ORDINATIONE FERIARUM PASCHALIUM. First of all, it is not apparent if this counting is supposed to be inclusive or exclusive, i.e. if the count includes Peter (which would make Victor the 13th bishop of Rome) or not (which would make Victor the 14th bishop of Rome). But since the counting in MC usually is inclusive, it should be presumed here that Victor is regarded as the 13th bishop of Rome. Whence MC obtained this information is not immediately obvious. It is noteworthy that Eusebius refers to Victor with the exact same phrase as MC as 'the 13th bishop of Rome since Peter' in the Greek original of his Church History (5.28.3: *Τῶν Βίκτορος χρόνων, ὃς ἦν τρισκαιδέκατος ἀπὸ Πέτρου ἐν Ῥώμῃ ἐπίσκοπος*; Schwartz, *Kirchengeschichte* 2.1, p. 500), while this detail is omitted in Rufinus' Latin translation of that work (5.28.3: *usque ad Victoris tempora*; Mommsen, *Kirchengeschichte* 2.1, p. 501). Yet, MC could not have worked from the Greek text. The *liber pontificalis* does not agree with MC, since it lists Victor as the 15th pope (Duchesne, *Liber pontificalis*, p. 48, 61, 137). However, in the fifth- to

63. ABOUT THE INTERDICTION OF CELEBRATING EASTER ON THE 14TH MOON

After the resurrection and ascension of the Lord, the apostles did not inquire how they should observe Easter, since the disciples had been dispersed throughout the whole world on account of their preaching. But they celebrated Easter on whatever day *luna* 14 (i.e. the Easter full moon) happened to be. After the departure of all apostles from this world, however, they held diverse Lenten fasts throughout the various provinces. For the Gauls celebrated Easter on whatever

---

seventh-century lists of Roman bishops, Victor takes 14th place (Duchesne, *Liber pontificalis*, p. 13–41; but note that Victor is listed as the 13th pope in two lists (I and VIII; Duchesne, *Liber pontificalis*, p. 14, 28)), which may be regarded as the source if an exclusive count is assumed. On the other hand, JEROME refers to Victor as the 13th bishop of Rome: JEROME, *Chronicon* (Helm, *Chronik des Hieronymus*, p. 210): *Romae episcopatum suscipit XIII Uictor ann. X, cuius mediocria de religione extant uolumina. De viris illustribus* 34 (Herding, *De viris inlustribus*, p. 29): *Victor, tertius decimus Romae urbis episcopus, super quaestione paschae, et alia quaedam scribens opuscula rexit ecclesiam sub Severo principe annis decem*. Since MC cites Jerome's chronicle elsewhere (directly or through an intermediary: c. 13, ll. 3–4), I regard it as the source here too. This is confirmed by the fact that the ANNALS OF TIGERNACH and BEDE, *Chronica* §339 (*MGH AA* 13, p. 288) transmits the same information, presumably drawing on the same or a similar source (cf. *app. comp.*). The end of this passage here is a summary of the results of the synod, i.e. of the rest of DE ORDINATIONE FERIARUM PASCHALIUM (*PL* 90, col. 607C–610A), too long to be quoted here. For the three rules determining Easter cf. also c. 56, ll. 2–14. The final phrase is attributed to Isidore further below (ll. 28–29 of this chapter).

LXIII,3–50 Post…tradiderunt] The essence of MC's criticism here (or rather that of his source) was more directly formulated by early eighth-century computists. DRC 98.13–26, and more vehemently Bede in DTR 59.36–67, point their criticism explicitly against followers of Easter lunar limits of *luna* 14 to *luna* 20. For contemporary non-computistical statements against these lunar limits and the historical background of this debate cf. HE 2.2; 3.3–4, 25, 28; 5.21–22, and, in view of MC's discussion here, especially ALDHELM, *Epistola ad Geruntium* (*MGH AA* 15, p. 483–4): *Porro isti secundum decennem novennemque Anatolii computatum aut potius iuxta Sulpicii Severi regulam, qui LXXXIIII annorum cursum descripsit, quarta decima luna cum Iudeis paschale sacramentum celebrant, cum neutrum ecclesiae Romanae pontifices ad perfectam calculi rationem sequantur; sed nec Victorii paschalis laterculi curriculum, qui DXXXII annorum circulis continetur, posteris sectandum decreverunt. Erat namque genus quoddam hereticorum aput orientales, quod tessereskaedecaditae vocatur id est quartadecimani, eo quod quarta decima luna cum Judeis Christus blasphemantibus et margaritas evangelii ritu porcorum calcantibus paschae solemnitatem peragunt et ob hoc alieni a beata orthodoxorum sodalitate inter scismaticorum conciliabula infeliciter reputantur, quos beatum Augustinum in libro de nonaginta heresibus scripto commemorasse memini*. As would have been expected, this debate did not concern eighth-century continental computists, as they were completely unfamiliar with the 84 (14). Consequently, the works that copied DTR 59 (LIB. CALC. 93, PV 2.8, RM 80) omitted the second part of that chapter which condemns the celebration of Easter on *luna* 14. Similar passages can only be found in DIAL. NEUSTR. and COMP. COL., both heavily influenced by Irish computistics, while the latter also drew on Bede. In the early tenth century, Helperic only argued that Easter Sunday should not be celebrated with the Jews, and therefore not on *luna* 14. For parallels in detail between MC and other texts cf. the following notes and *app. font.*

LXIII,3–23 Post…preterire] Cf. Mac Carthy, *Annals of Ulster* 4, p. lxix; Schwartz, 'Ostertafeln', p. 101; Krusch, *Studien* II, p. 58; Strobel, *Texte*, p. 86; Walsh & Ó Cróinín, *Cummian's letter*, p. 37; Ó Cróinín, 'A seventh century Irish Computus', p. 103; Machielsen, *CCSL Clavis Patristica* 3A, p. 189 and also p. LXV of the introduction. **3–50** Post…tradiderunt] Cf. Borst, *Schriften*, p. 938 and p. XCV, CXXIV–CXXV of the introduction.

*prouintias. Nam Galli quacumque die VIII Kalendas Aprilis fuisset pascha celebrabant, qu*ia in eo *Christ*um *resurre*xisse credebant, quod falsum est. *In Italia alii XX dies, alii uero VII* dies *ieiunabant* ante pascha. *Cum* uero *hae* diuerse *obseruationes* essent, *papa Victor, urbis Romę episcopus* a Petro XIIImus, *direxit auctoritatem ad Theophilum, Cesariensis Palestine* episcopum, *ut pasca recto iure a cunctis catholicis celebraretur, ubi dominus et saluator mundi* con*uersatus fuerat in carne.* Inde Theophilus *omnes episcopos non solum de sua prouintia, sed diuersis euocauit.* Inde omnes ad principium mundi reuersi sunt, et rectam rationem in obseruatione paschę ab XI Kalendas Aprilis cum auctoritate librorum legis statuerunt, hoc est diem et uernum tempus et lunae VII dies, ut nulli liceat hos terminos preterire. Quicumque enim transgreditur hos terminos, transgreditur mandatum.

Quirillus ait: *Non faciamus pascha in XIIII luna cum Iudęis et hereticis, qui dicuntur Thesserescedecaditę.* Inde
Iohannes consiliarius ait: XIIII dies lunae | ad umbras fol. 43v
pertinebant. Ysidorus ait: Quicumque ante XV luna pascha celebrari iubet, transgreditur mandatum.

---

11 alii] a luna *M (corr. from* DE ORDINATIONE, DRC*; it seems that the copyist mistook* alii *for* a lu-*).* **12** dies[1]] diebus *M (corr. according to* MC, DE ORDINATIONE, DRC*).* **18** episcopos] episcopi *M (corr. from* DE ORDINATIONE*).* | sua] sur *M (corr. from* DE ORDINATIONE*).* **26** Thesserescedecaditę] *an* s *(?) is scratched out at the end of this word in M.* **28** pertinebant] pertinebat *M*, Ó Cróinín, Harrison.* | luna] lunam *Ó Cróinín, Harrison.*

25–29 Quirillus…mandatum] In this passage MC lists the citations most direct in condemning the practice of celebrating Easter on *luna* 14. The first of these is from EPISTOLA CYRILLI 5 (Krusch, *Studien* I, p. 347; MS L reads *thessariscedecaditę*; cited in COMP. COL. 5.9A, with attribution to Cyril in DRC 98.13–17 which reads *Thesserescaidechadite*; only from *ut non fallamur* in EPISTOLA CUMMIANI 226–229 which reads *thesserescedecadite*; only the phrase that is quoted in MC with attribution to Dionysius Exiguus in DIAL. NEUSTR. 13B, the rest with attribution to Cyril in DIAL. NEUSTR. 18A): *Scruteminique diligentissime, quod ordinavit synodus Nicena, lunas XIIII omnium annorum per decennovenalem cyclum, ut non fallamur in luna primi mensis et caelebremus pascha in sequenti dominico et non faciamus in luna XIIII cum Judaeis et hereticis, qui dicuntur teserescedecaditae.* The, no doubt, most interesting quote in the whole chapter is the second one here, as it constitutes an otherwise unattested fragment from the letter of John *consiliarius*, among others, to the northern Irish clergy in AD 640 (the only other passages surviving from this letter can be found in HE 2.19). For a similar statement of *luna* 14 belonging to the shadows cf. c. 55, ll. 31–33. Finally, the sentence attributed to Isidore cannot be found among the bishop of Seville's extant works; note that this phrase is also used above (c. 63, ll. 23–24), and that it is also ascribed to Isidore in DRC (cf. *app. comp.*).

13–14 papa…XIIImus] The fact that Victor was the 13th bishop of Rome is also mentioned in ANNALS OF TIGERNACH (Stokes, *Annals of Tigernach*, p. 9), BEDE, *Chronica* §339 (*MGH AA* 13, p. 288): *Uictor tercius decimus Rome episcopus datis late <li>bellis constituit Pasca die dominico celebrari.* For this question cf. *app. font.* **25–28** Quirillus…pertinebant] The connection between the celebration of Easter on *luna* 14 and Jewish customs, as well as it representing shadows rather than light, is also made by HELPERIC, *De computo* 33 (*PL* 137, col. 43D–44A): *Porro si terminus, id est, XIV luna Dominica die*

day 25 March happened to be, since they believed that Christ resurrected on that day, which is wrong. In Italy, some fasted for 20 days, others, however, for seven days before Easter. However, because these customs were diverse, Pope Victor, the 13th bishop of the city of Rome since Peter, gave the instruction to Theophilus, bishop of Caesarea (and) Palestine, where the Lord and Saviour of the world had lived in the flesh, that Easter was to be celebrated according to (one) legitimate rule by all Christians. Thereupon, Theophilus summoned all bishops, not only from his own province, but (also) from other (provinces). Thereupon, all (bishops) reconsidered the beginning of the world, and they established, with (reference to) the authority of the books of Law, the legitimate doctrine concerning the observation of Easter from 22 March, i.e. the (week)day (i.e. Sunday) and springtime and the seven days of the moon, so that it is not lawful for anyone to ignore these terms. In fact, whosoever violates these terms, violates the commandment.

Cyril says: We shall not celebrate Easter on *luna* 14 with the Jews and (those) heretics, who are called *Thesserecedecaditae*. Accordingly, the *consiliarius* John says: The 14th day of the moon belonged to the shadows. Isidore says: Whosoever approves of Easter being celebrated before *luna* 15 violates the commandment.

---

evenerit, Paschalis festi laetitiam in aliam Dominicam transferimus, ut quidam dicunt, ne unquam cum Judaeis Pascha celebremus, neve cum eis umbrae legali deservire videamur. **28–29** Ysidorus…mandatum] The quote ascribed to Isidore by MC occurs with the same attribution in DRC 98.21–22: *Et Isidorus dicit: Quicumque ante XV lunam pascha celebrari iubet, transgreditur mandatum domini.*

23–29 Quicumque…mandatum] Transcribed in Ó Cróinín, 'A seventh-century Irish Computus', p. 103; Harrison, 'Letter', p. 228; for the quote from ps-Cyril cf. Borst, *Schriften*, p. 394; for the quote from *Iohannes consiliarius* cf. also Ó Cróinín, 'A seventh-century Irish computus', p. 103–4 (including translation), 126; idem, 'New heresy', p. 89, 93–4 (including transcription in both instances) and p. LXXXVI of the introduction; for the citation ascribed to Isidore cf. also Ó Cróinín, 'Irish computus', p. 118–9 and p. CXXVIII, CXCV of the introduction.

Item idem ait: N*ulla ciuitas* dicitur *uel eclesia* transgredi debet, *quae in Niceno concilio statuta sunt. Verum agnum ueri Israhelite, inmaculatum inmaculati comedamus, quia in una domo agnu*s *comede*batur. *Priscam consuetudinem* a *domino credimus predicari in eclesiis unam fidem, unam sol*l*emnitatem paschalem omnibus Christianis*. Atque *soluere ieiunium* non oportet ante *luna*m *XIIII*. Inde Christus <*pascha*> fecit in *XIIII luna* antequam suum corpus transfiguraret. Inde, si in *XIIII luna* feceris pascha, *solue*s *ieiunium* in *XIII*, quod *non oportet*.

---

33 domo] *follows* x *M*. **36** Atque] usque *M*. **37** pascha] *om. M (add. according to* EPISTOLA PROTERII*)*.

30–39 Item…oportet] This passage, combining quotes from Cyril, Proterius and Theophilus, pleads for the unity of the church by urging all Christians to follow the same practice in determining Easter, namely the one allegedly outlined at the Council of Nicaea of celebrating Easter after *luna* 14. Note that the first citation here, taken from Cyril's letter, is introduced with *idem*, which indicates that it directly followed the previous Cyril quote in an earlier draft, and that the sentences attributed to John and Isidore (cf. previous note) were later added (maybe they were first attached to the text on a separate leaf). The first authority cited is EPISTOLA CYRILLI 5 (Krusch, *Studien* I, p. 347–8; cited in EPISTOLA CUMMIANI 229–232, with attribution to Cyril in DRC 98.18–21, DIAL. NEUSTR. 18A), 6 (Krusch, *Studien* I, p. 348; MS C reads *Israhelite*, MS L *Israelite*; up to *comedamus* cited in EPISTOLA CUMMIANI 238–239): *Et constitutum est in omnibus synodis praeter synodum Gangrensem et Caesariensem, ut non faceret ulla ecclesia vel civitas et omnis regio contraria his, quae statuta sunt de pascha in Niceno concilio. ... ut in unitate ecclesiae catholicae verum agnum veri Israhelitae, inmaculatum inmaculati comedamus, quia in una domo preceptum est Israheli carnali agnum anniculum comedere*. This is followed by EPISTOLA PROTERII 9 (Krusch, *Studien* I, p. 278): *Quod etiam nunc iuxta priscam consuetudinem credimus in domino predicari in ecclesiis unam fidem, unum baptisma et unam solemnitatem sacratissimam paschalem ab omnibus Christianis ubique caelebrari in Christo Jesu domino nostro, quia in ipso vivimus et movemur et sumus*. The end of this passage is then based on PROLOGUS THEOPHILI 3 (Krusch, *Studien* I, p. 223–5): *Sed quia sepę propter quartum decimum diem lune eiusdem primi mensis aliquantos contingit errare, cum quartus decimus dies in ipso nonnumquam die dominico incidat, sabbatum solent ieiunium solvere, cum sabbato certe tercius decimus dies esse videatur, hoc autem contra legem est divinam; ergo diligenter oportet observari, ut, si aliquando contingat in dominica quartum decimum diem lune incidere, melius est differri in dein septimanam, propter duas scilicet raciones: primo, ut non sabbato die, quo tercia decima luna est, ieiunare desinamus, quod legis auctoritatę prohibitur. Sed et illud accedit, ut lune circulus adhuc inperfectus esse videatur, dehinc et dominica die et quarta decima luna ieiunare cogamur, quod Manicheorum proprium est. Igitur quoniam nec quarta decima luna si in diem dominicam incidat ieiunare nos oportet, nec consequens est tercia decima luna die sabbato ieiunium solvere, necessaria est omni racione dilacio nec circa observacionem a pasche dilacionem aliquid innovetur. Ut enim decimus numerus etiam nonum in se continet, ita et si quarta decima luna in dominicam incidat diem, in quo ieiunare non oportet, et dilacio fiat, non est inminucio pasche, quoniam in sequentibus diebus etiam superior continetur*. EPISTOLA PROTERII 2 (Krusch, *Studien* I, p. 272): *Quia vero lunae circulus ad solis cursum inaequalis est, et XIIII luna paschalis in die dominico sepe contingit, non est autem possibile tunc festum caelebrare, sed nec pridie sabbato lun- XIII ieiunium solvere, in septimanam sequentem deferendum est, maxime quod habeamus intra eam XV lunam, quando, sicut scripsit apostolus, pascha nostrum immolatus est Christus. XIIII namque luna primi mensis iuxta Hebreos, ut superius dictum est, Jesus pascha tipicum manducavit, sequenti vero VI ferea XV luna, ut ovis occisionis, cruci pro nobis affixus est et vespere sabbati, luciscente dominico, XVII luna resurrexit a mortuis*.

Likewise the same person (i.e. Cyril) says: It is said that no community or church must violate (the rules), which were established at the Council of Nicaea. We shall eat the true lamb of the true Israelite, the immaculate of the immaculate, because the lamb was consumed in one house. We believe in the ancient custom (transmitted) from God that in the churches one faith, one paschal celebration is pronounced by all Christians. And one shall not break the fast before *luna* 14. For Christ celebrated the Pasch on *luna* 14 before his transsubstantiation. Accordingly, if you celebrated Easter on *luna* 14, you would break the fast on (*luna*) 13, which is unacceptable.

36–39 Atque…oportet] The interdiction of breaking the fast on *luna* 13 is also discussed by Bede and Helperic, both with reference to the same source as used by MC (PROLOGUS THEOPHILI): DTR 59.44–67 (partially cited in COMP. COL. 5.9B): *Quos inter alios fidei et actionis catholicae defensores, etiam beatus Theophilus alexandrinae antistes ecclesiae manifesta ratione devincit, scribens imperatori Theodosio maiori: sed quia accidere, inquit, interdum solet, ut occasione quartae deciame lunae mensis primi, nonulli in errorem cadant, si quarta decima ipsa luna in dominicum diem veniat. Quo facto necesse est iam ieiunium solvi sabbati die, quando tertia decima luna venire monstratur, et incipimus legi contraria facere. Convenit itaque diligenter advertere ut, quoties quarta decima luna in dominicum incurrit diem, in sequentem septimanam paschalem diem potius differamus, duplici hoc modo: primum quidem ne tertia decima luna inventa in sabbati die solvamus ieiunium quod consequens non est quod nec ipsa lex praecipit, maxime cum et lumen ipsius lunae imperfectum adhuc in proprio globo esse videatur; deinde ne dominica die luna quarta decima constituta ieiunare cogamur – indecentem rem illicitamque facientes, hoc enim manichaeorum sectae consuetudo possedit. Quoniam igitur nec quarta decima luna veniente in dominicum diem ieiunare debemus, neque consequens est ut si in sabbati die tertia decima eveniat solvamus ieiunium, necessario adserimus hoc in septimanam sequentem deberi differri, sicut paulo superius comprehendi – non tamen ex hac dilatione praevaricatione aliqua circa paschalem calculum perpetrata. Quemadmodum enim decimus numerus complectitur nonum, sic et quoties quarta decima luna in dominicum incurrerit diem, eo quod in eo ieiunare non liceat, in proximam septimanam differri necesse est diem paschae. Nulla enim ex hoc imminutio fieri paschae videtur quia ipsi sequentes dies complectuntur et reliquos.* HELPERIC, *De computo* 33 (*PL* 137, col. 44A): *Ut vero Theophilus Alexandrinus episcopus scribit, ne vel XIII luna Sabbato occurrente finire cogamur jejunium contra legem facientes, cum necdum luna plenum luminis sui orbem exhibeat, vel ne in Dominico die XIV luna existente auctoritate legis, quae hac die agnum ad vesperum immolare jubet, jejunare cogamur contra Ecclesiae facientes traditionem, secundum quam Dominica die jejunare fas non est.*

Item *Iudei ignorantes* d*ominum tempus quoque pasche ignorauerunt. Unde sepius a primo mense recedunt.* Item *nos Christiani non solum XIIII* non *requirimus in pascha, hoc enim facientes sine festiuitate sunt.* Item *XIIII lunas paschalis obseruantiae stabiles in*muta*tas*, quas *non tam peritia seculari quam spiritus sancti inlustratione* sancti patres *uenerabilium CCCXVIII pontificum sanxerunt. Qua*s *postmodum* alii *arrogantię dispicientes uel transgredientes inscientia, Iudaicis* s*educti fab*o*lis, diuersam* | fol. 44r *contrariam formam festiuitatis unicae tradiderunt.*

---

44 lunas] luna *M (corr. from* Prologus Dionysii*)*. **45** quas] quae *M*. **49** unicae] untae *M (corr. from* Prologus Dionysii *)*.

40–50 Item…tradiderunt] Finally, Mc higlights the differences between Jewish and Christian customs. The first two quotes are from different chapters of the letter of Proterius: Epistola Proterii 7 (Krusch, *Studien* I, p. 276; MSS LS read *dominum*; cited with attribution to Proterius in Drc 94.5–7 which om. *namque* and reads *dominum, pasche*): *Judaei namque ignorantes deum, tempus quoque paschae ignoraverunt. Unde sepius a primo mense recedunt.* Epistola Proterii 2 (Krusch, *Studien* I, p. 271; cited in Drc 89.19–21): *Nos ergo Christiani non solum XIIII luna in pascha requirimus, hoc enim Judaei facientes sine festivitate sunt, sed etiam resurrectionis diem redemptoris nostri.* The remainder of this passage comes from Prologus Dionysii (Krusch, *Studien* II, p. 63; MSS 3, 5 read *dispicientes*; parts of this are also cited in Dial. Neustr. 15C): *sequentes per omnia venerabilium CCCX et octo pontificum, qui apud Niceam, civitatem Bithiniae, contra vesaniam Arrii convenerunt, etiam rei huius absolutam veramque sententiam; qui XIIII lunas paschalis observantiae per decem et noveni annorum redeuntem semper in sese circulum stabiles inmotasque fixerunt, quae cunctis eculis eodem, quo repetuntur, exordio sine varietatis labuntur excursu. Hanc autem regulam praefati circuli non tam peritia saeculari quam sancti spiritus inlustratione sancxerunt et velut anchoram firmam ac stabilem huic rationi lunaris dimensionis apposuisse cernuntur. Quam postmodum nonnulli vel arrogantia despicientes, vel transgredientes inscientia, Iudaicis inducti fabulis, diversam atque contrariam formam festivitatis unicae tradiderunt.*

The Jews, being ignorant of the Lord, also were ignorant of the time of the Pasch. For they regularly move backwards (in the calendar) from the first month (to the previous month, i.e. the 12th month). Not only do we Christians demand that (*luna*) 14 (is) not (to fall) on Easter, since the ones who do so are, in fact, without a feast. The firm and unchanged *lunae* 14 of the paschal observance (i.e. the Easter full moons), these the holy fathers among the 318 venerable bishops (i.e. the bishops attending the Council of Nicaea) have established through the illumination of the Holy Ghost rather than through secular knowledge. Thereafter, others of arrogance, disregarding or violating these (Easter full moons) due to (their) ignorance, (which resulted from them being) seduced by Jewish gossip, transmitted a different and contradictory idea of this unique feast.

<LXIIII. DE ANNIS IN TEMPORIBUS>

Creation – Exodus – *annus resurrectionis*

Moyses primum *pascha* in *Aegypto domino* precipiente *peregit*, *annis a principio mundi peractis I̅I̅I̅DCLXXXVIII*, *V feria*, *VIII Kalendas Aprilis*. A primo pascha in Aegypto facto usque ad primum pascha resurrectionis anni sunt mille DXXXX.

LXIIII,6 mille DXXXX] mille DCVIIII *M (corr. according to* MC, *MSS VBT of* PROLOGUS VICTORII, BC*)*.

LXIIII,2–6 Moyses…DXXXX] This passage, dealing with the number of years from the creation to the Exodus, and then from the Exodus to the year of Christ's resurrection, seems to be a conflation of PROLOGUS VICTORII 9 (Krusch, *Studien* II, p. 24; MSS VBT read *I̅I̅I̅DCLXXXVIII*, MSS VCS*T *VIII Kal.* instead of *VIIII Kal.*; partially cited in DRC 85.4–10 (cf. *app. comp.*), 89.5–7, BC 77C2–8, which reads *I̅I̅I̅DCLXXXVIII*): *Rursusque omnibus annis temporibus diebus ac luna maxime, quae iuxta Hebraeos menses facit, rite discussis, a mundi principio secundum praedictae historiae fidem usque in diem, quo filii Israhel paschale mysterium caelesti initiavere mandato et ab Aegyptia clade agni occisione salvati sunt, bissextorum pariter necessitate decursa, quantum fida supputatio vestigavit, anni tria milla sescenti octoginta novem quinta feria VIIII Kal. aprl. luna XIII, incipiente iam vespera, docentur impleti. Cuius sequenti die tertio millesimo scilicet anno ac sexcentesimo nonagesimo procedente mense primo, VIII Kal. april. luna XIIII noctis initio Hebreos claruit agni sacrificium peregisse. Pascha quippe sicut omnimodo traditione cognoscitur anni principio, non fine celebratur. Passum autem dominum nostrum Iesum Christum peractis ab ortu mundi quinque milibus ducentis viginti et octo annis, eadem chronicorum relatione monstratur*. The conflation appears to consist of the omission of the part from *VIIII Kal.* to *mense primo*, which then agrees with the chronology of the Exodus from Egypt outlined in c. 44, ll. 37–54 (for a similar conflation of this Victorian passage cf. DRC in *app. comp.*). The number 1609 given in *M* is corrected to 1540 on the basis of *M*, MSS VBT of PROLOGUS VICTORII and BC: The number of years between the Exodus from Egypt and Christ's resurrection equals the number of years from the creation of the world to Christ's resurrection minus the number of years from the creation of the world to the Exodus from Egypt; consequently, according to MSS VBT, 5228−3688=1540. If 3688 was corrected to 3689 in MC according to the other MSS of the PROLOGUS VICTORII, then the number in question would need to be emended to 1539=5228−3689. This calculation is outlined in detail in the SIRMOND MS (the relevant passage is transcribed in Ó Cróinín, 'Early Irish annals', p. 83; idem, 'Irish provenance', p. 177; idem, 'Virgilius Maro Grammaticus', p. 194; idem, 'Bede's Irish computus', p. 209): *Omnibus annis temporibus diebus ac luna maxime, que iuxta Hebraeos menses facit, rite discussis, a mundi principio usque in diem, quo filii Israhel paschale mysterium initiauere anni sunt I̅I̅I̅DCLXXXVIIII praecedente primo mense VIII Kl. Aprilis, luna XIIII, VI feria. Passum autem dominum nostrum Ihesum Christum peractis VCCXX & VIII annis ab exortu mundi, eadem cronicorum relatione monstratur, VIII Kl. Aprilis primo mense, luna XIIII, VI feria. Inter primum pascha in Aegypto et passionem domini anni sunt I̅DXXXVIIII*. 5228 is given twice elsewhere in MC as the number of years from the creation of the world to the passion; cf. c. 44, ll. 11–12; c. 68, ll. 32–33.

LXIIII,2–6 Moyses…DXXXX] A similar passage, counting the number of years from the creation to the Exodus, is given in DRC 85.4–7: *Moyses uidelicet primus celerauit cum filis Israhel in Aegypto, annis a principio mundi peractis I̅I̅I̅DCXLX (recte I̅I̅I̅DCXC), tertio milissimo scilicet anno ac sexcentissimo nonagentissimo procedente, Uictorio dicente: Primo mense, V feria, VIII Kl Aprelis, luna XIIII noctis initio, Ebreos claruit agni sacrificium peregisse*. Note that DRC has the same chronological details as MC for the date of the first pasch in Egypt, which contraditcs (due to conflation) Victorius' original account (since there *V feria* refers to the previous day, *VIIII Kalendas Aprilis*, *luna XIII*). Generally, it appears that once the Victorian paschal tract was outdated by the unanimous acceptance of the Dionysiac reckoning, computists and chronologists lost interest in the study of that work; consequently, the number of years between the creation of the world and the Exodus from Egypt

64. ABOUT THE YEARS IN TIME

Moses celebrated the first pasch in Egypt, as the Lord set it forth, on Thursday, 25 March, after 3688 years had passed from the beginning of the world. There are 1540 years from the first pasch, celebrated in Egypt, to the first pasch of the resurrection.

---

explicitly mentioned in the PROLOGUS VICTORII, as well as the implicit number of years from the Exodus to Christ's resurrection, did not receive any attention outside of seventh- and early eighth-century Irish computistics.

LXIIII,2–6 Moyses…DXXXX] Cf. p. XXII, LXXXI, XCIII–XCIV, CXXIV–CXXV of the introduction.

Comp. of Easter cycles with Victorian 532-year cycle: 1. 84 (12)

<LXV. DE COMPARATIONE VARIORUM CICLORUM CUM CICLO VICTORII>

Alii Latinorum *ciclum LXXXIIII annorum* conposu*erunt*. In quo ciclo XVIIII quater inueniuntur anni et VIII anni superflui sunt. XXVIII uero anni ter in eo continentur et nihil superfluum habet. In quo ciclo saltus in *XII*o *anno* fit. In quo saltus sol lunam VII annis superat. Et hic saltus incongruus est, quia non XIImam partem diei

---

LXV,5 anni[1]] dies *M, Schwartz (corr. from* MC*; it seems that a later copyist systematically changed all references to years into references to days whenever cycles are compared in this passage; however, he overlooked one instance, which preserves the correct reading).* | XXVIII] XX *M (corr. according to* MC*; also Schwartz).* **6** superfluum] superflui *Schwartz (contra M).* **7** lunam] luna *M, Schwartz (corr. from* MC*).* | annis] diebus *M, Schwartz (corr. from* MC*).*

LXV,3–38 Alii…conuertit[2]] In this passage the basic structure of all Easter tables known to MC (or rather to the author of this chapter, which MC appears to have copied in full) is compared to the Victorian Easter cycle. Each of these tables is analyzed concerning four criteria: 1) The relation to the 19-year lunar cycle: The hypothetical number of full 19-year lunar cycles within the extent of each table is determined, as well as the number of years remaining; if such a remainder does not exist, the Easter table in question is regarded as congruent to the 19-year lunar cycle, if it exists, it is regarded as incongruent. 2) The same is then done for the 28-year solar cycle. 3) The place of the *saltus* within each table relative to the 19-year *saltus*: The number of years between two *saltus* in the respective Easter table is given, followed by a calculation of the number of years remaining to the 19-year *saltus*. 4) The relation to the 532-year Victorian Easter cycle: The number of times that each Easter table would recur in full within the 532-year period is determined, as well as the number of years remaining. All the information necessary for such an analysis of the first three Easter reckonings (i.e. the 84 (12), the Hippolytan 112-year table, and the Alexandrian 95-year table) derive from PROLOGUS VICTORII 3 (Krusch, *Studien* II, p. 18–9): *Discrepare inter se disputatores praecipuae festivitatis primum omnium necessitas illa conpellit, qua paschalis cycli, sive qui post octoginta et quattuor, seu qui post quinque et nonaginta annos, vel etiam qui post centum et duodecim annorum spatia propagati, eodem quo praeterierant ordine in principium sui adque originem redire creduntur, in aliquibus quidem dierum sunt parilitate concordes, sed in lunae varietate dissentiunt, hac rursum, cum in lunari supputatione reperiuntur stabiles, dierum mutabilitate confligunt. Neque solum alius alio est propria definitione dissimilis, verum singuli quique eorum in semetipsis, alternatione praedicta, lunari scilicet ac dierum, nunc veritate subsistunt, nunc adsertione deficiunt. Huc accedit, quod lunaris ipsa demensio ita cunctorum est varietas sententiis, ut, cum isti verbi gratia eandem kalendis ianuariis primam fuisse pronuntiant, alii tricensimam, nonulli secundam provenisse contendant. Tum deinde ii, qui cyclum annorum octoginta et quattuor ediderunt, duodecim peractis annis lunam unam, quam per idem tempus certis annua revolutione minutiis adserunt calculandi lege subcrescere, adiciendam legitimo cursui esse praecipiunt. Item sunt, qui hanc eandem quinto decimo incipiente anno magis adnumerari definiunt. Porro ii, qui post centum duodecim annos in id cyclum, unde orsus fuerat, reverti confirmant, ea ipsa incrementa lunaria, quae superius memoravi, post annos sedecim subnecti debere constituunt. Ii vero qui anno quinto et nonagensimo cycli observantiam conprehendunt post decem et novem annos Aegyptiorum more continuato ordine, quod est verius, hoc augmentum lunare subiciunt.* Since this chapter of MC is favourable to the Victorian reckoning it must be presumed that MC copied it from his now lost Victorian computus of AD 689 (cf. p. CXXIV–CXXVI of the introduction); the author of this exemplar, then, appears to have known the *latercus*, the 84 (14), in the form of a 100-year table, now also lost; note that this author does not use the term *latercus*, which clearly indicates that the discussion here is not related to the other *latercus* passages in MC. The fact that the Victorian cycle consisted of 532-years was evident from the Victorian Easter table itself, as well as from PROLOGUS VICTORII 10 (cf. Krusch, *Studien* II, p. 25, 27–52); the knowledge that the lunar cycle (underlying the Victorian

65. ABOUT THE COMPARISON OF VARIOUS CYCLES WITH THE VICTORIAN CYCLE

Others among the Latins constructed a cycle of 84 year. In this cycle four times 19 years are found and eight years are remaining (i.e. 4×19+8=84). However, three times 28 years are contained in this and it has no remainder (i.e. 3×28=84). In this cycle the *saltus* occurs in the twelfth year. In this (year) of the *saltus* the sun exceeds the moon by seven years (i.e. compared to the 19-year cycle, the *saltus* occurs seven years earlier). And this *saltus* is incongruent, since it does not run up to the twelfth part of a day with the night (i.e. the twelfth part of a lunar day is not the natural annual increment of a *saltus* in an 84-year cycle; this

---

reckoning) consisted of 19 years, the solar cycle of 28 years, and that the combination of these results in a 532-year Easter cycle may have derived from DE MIRABILIBUS 2.4 (*PL* 35, col. 2176): *quorum unusquisque uniformi statu, peractis quingentis triginta duobus annis in semetipsum, id est, in sequentis initium revolvitur, completis videlicet in unoquoque solaribus octovicenis nonodecies, et in lunaribus decemnovenalibus vicies octies circulis*. Cf. also EPISTOLA CUMMIANI 225–226. Authors of Easter tables were explicitly named in a passage of ETYM. that proved to be very popular in the early middle ages; in this passage, Isidore also mentions the fact that Cyril's 95-year Easter table consists of five 19-year lunar cycles. ETYM. 6.17.1–2 (cited in DRC 115.2–8, PROL. AQUIT. 4C; only the list of authors with the additon of Dionysius Exiguus in DIAL. NEUSTR. 19aB, in turn cited with the addition of Gillitanus in COMP. COL. 5.12B): *Paschale cyclum Hippolytus episcopus temporibus Alexandri imperatoris primus conscripsit. Post quem probatissimi auctores Eusebius Caesariensis, Theophilus Alexandrinus, Prosper quoque natione Aquitanus atque Victorius, amplificatis eiusdem festivitatis rationibus, multiplices circulos ediderunt. Cuius quidem rationem beatissimus Cyrillius Alexandriae urbis episcopus in nonaginta quinque annos per quinquies decem novies calculans*. This fact is also implicit in PROLOGUS DIONYSII as well as in the Dionysiac 95-year Easter table, which is arranged in five times 19 years (cf. Krusch, *Studien* II, p. 63–4, 69–74). Interestingly enough, Anatolius (?) (DRP 1.10–11; cited in PROL. AQUIT. 4A) and JEROME, *De viris illustribus* 61 (Herding, *De viris inlustribus*, p. 41; cited in BEDE, *Chronica* §339 (*MGH AA* 13, p. 299)) explicitly linked the 16-year cycle to Hippolytus, a detail that apparently was of no interest to MC.

LXV,3–38 Alii...conuertit[2]] The analysis outlined in this chapter is unique to MC. The relation between the 19-year lunar cycle, the 28-year solar cycle, and the 532-year Easter cycle can, however, also be found in other texts. Cf. DT 13.7–13: *Quartus concurrentes septimanae dies quae ... necessario XXVIII annis implentur; quarum ratio cogit cyclos decemnovenales XXVIII describi ut singulae concurrentes singulos cyclos inchoent, totaque summa paschalis calculi DXXXII annis explicitur*. DIAL. NEUSTR. 29bB: *Et quot in ciclo lunae, qui novemdecim annos habet, quot in ciclo solis, qui viginti octo annos habet, quot in ciclo Victorii, qui <quingentos> triginta duos annos habet?* DTR 53.24–30 (cited in LIB. CALC. 90, PV §218, RM 71.24–31) 65.2–7 (cited in LIB. CALC. 98, RM 94.2–8): *Notandum sane quod huius gyri solaris, qui XXVIII annis peragitur, causa facit decemnovenales circulos XXVIII debere compleri priusquam idem per omnia paschalis observantiae cursus in seipsum redeat, ut omnis nimirum huius circuli annus caput circuli decemnovenalis instituat. Itemque annus quisque circuli decemnovenalis huius caput adsequatur, ac per hoc tota paschalis observantiae series non minus quingentis XXX duobus annis explicetur. ... Circulus paschae magnus est qui, multiplicato per invicem solari, ac lunari cyclo, DXXXII conficitur annis. Sive enim decies novies viceni et octoni seu vicies octies deni ac noveni multiplicentur, DXXXII numerum complent. Unde fit ut idem circulus magnus decemnovenales lunae circulos XXVIII, solis autem (qui vicenis octonisque consummari solent annis) X et VIIII habeat circulos*. Cf. also c. 1, ll. 14–16; VERS. TUR. 15–17.

LXV,3–38 Alii...conuertit[2]] Transcribed in Schwartz, 'Ostertafeln', p. 101; cf. Mac Carthy, *Annals of Ulster* 4, p. lxxi; Borst, *Schriften*, p. 365 and p. XCIII–XCIV, CXXIV–CXXV of the introduction; for discussions of the individual Easter tables mentioned in this chapter cf. following notes.

cum nocte currit. Deinde hic ciclus secundum solem congruus est, incongruus secundum lunam. Hic ciclus sexies conuertit in ciclo Victorii, et VII uice incipit currere, et XXVIII anni superflui in ciclo Victorii.

2. Hippolytan 112-year cycle

Alii *ciclum CXII anno*rum faciunt. In quo ciclo luna sexies, nisi II anni desunt, conuertit. Sol uero <quater> conuertit. Deinde congruus est iuxta solem et <iuxta> lunam incongruus. In hoc ciclo saltus in *XVI anno* habetur. In quo saltus sol <lunam> III annis superat. Hic ciclus quater in ciclo Victorii conuertit, | fol. 44v et V uice incipit currere, et distant per LXXXIIII annos.

3. Alexandrian 95-year table

Alii *ciclum XCV anno*rum faciunt. In quo ciclo luna quinquies conuertit. <Sol uero ter conuertit> et XI anni superflui. Dehinc iuxta lunam congruus, iuxta solem uero incongruus. Saltus huius cicli in *XVIIIIo anno* fit. In quo saltus quot dies habet sol tot et luna. Hic ciclus quinquies conuertit in ciclo Victorii, tamen anni superflui in ciclo Victorii LVII inueniuntur.

4. 84 (14) in the shape of a 100-year table

Alii ciclum annorum C conponunt. In quo ciclo luna quinquies conuertit, et quinque superflui anni sunt in illo. In eo sol conuertit tér, et XVI anni superflui sunt. Dehinc iuxta solem et lunam incongruus est. In quo ciclo saltus XIIIIo anno fit. In quo numero sol lunam V annis superat. <Hic ciclus quinquies in ciclo Victorii conuertit, et VI uice incipit currere, et distant per XXXII annos.>

---

12 XXVIII] XXXII *M (corr. according to* MC*; also Schwartz).* **13** CXII] CXI *M (corr. according to* MC*; also Schwartz).* | ciclo] cicli *M (corr. from* MC*; silently corr. Schwartz).* **14** anni] dies *M, Schwartz (corr. from* MC*).* | quater] *om. M, Schwartz (add. according to* MC*).* **15** conuertit] nisi II dies superflui sunt *add. M (corr. according to* MC*; also Schwartz).* | iuxta[2]] *om. M (add. from* MC*; also Schwartz).* **16** incongruus] congruus *M (corr. according to* MC*; also Schwartz).* **17** lunam] *om. M (add. according to* MC*; also Schwartz).* | III annis] VII diebus *M (corr. according to* MC*; the number corr. by Schwartz).* **17–19** Hic…annos] *In full, this sentence reads* Hic ciclus quinquies in ciclo Victorii conuertit, et VI uice incipit currere, et distant per XXX annos *in M. If the last number is emended to* XXXII, *this sentence would fit perfectly into the context of a different cycle discussed below, where such a sentence is, in fact, left wanting (cf. Schwartz). It appears, therefore, that the sentence given here is misplaced, and I have reproduced it in this emended version in its appropriate place below (ll. 32–34 of this chapter). Here, however, I have corrected this sentence to fit into the discussion of the Hippolytan 112-year table (cf. the following notes), since there cannot be any doubt that such a sentence was originally in this place.* **18** quater] quinquies *M.* | V] VI *M.* **19** LXXXIIII] XXX *M.* **20** annorum] annos *M, Schwartz (corr. according to* MC*).* **21** Sol…conuertit[2]] *om. M (add. according to* MC*; also Schwartz).* | anni] dies *M, Schwartz (corr. from* MC*).* **23** huius] huiuus *M (silently corr. Schwartz).* **28** quinque] quinquies *M, Krusch, Schwartz (corr. by Warntjes).* **29** anni] dies *M, Krusch, Schwartz (corr. from* MC*; also Warntjes).* **31** lunam] luna *Krusch (contra M).* | annis] diebus *M, Krusch, Schwartz (corr. from* MC*; also Warntjes).* **32–33** Hic…annos] *om. M (add. from M; cf. ll. 17–19 of this chapter).*

would rather be the 14th part, since six *saltus* have to be inserted in an 84-year cycle, and 84/6=14). Therefore, this cycle is congruent with the sun, incongruent with the moon. This cycle revolves six times in the cycle of Victorius (which consists of 532 years), and it starts to run in (its) seventh turn, and 28 years (are) remaining in the cycle of Victorius (i.e. 6×84+28=532).

Others composed a cycle of 112 years. In this cycle the moon revolves six times, except for two years that are wanting (i.e. 6×19−2=112). The sun, however, revolves four times (i.e. 4×28=112). Therefore, (this cycle) is congruent with the sun and incongruent with the moon. In this cycle the *saltus* is known (to occur) in the 16th year. In this (year) of the *saltus* the sun exceeds the moon by three years (i.e. compared to the 19-year cycle the *saltus* occurs three years earlier). This cycle revolves four times in the cycle of Victorius, and it starts to run in (its) fifth turn, and they differ by 84 years (i.e. 4×112+84=532).

Others composed a cycle of 95 years. In this cycle the moon revolves five times (i.e. 5×19=95). The sun, however, revolves three times and (there are) eleven years remaining (i.e. 3×28+11=95). Accordingly, (this cycle is) congruent with the moon, but incongruent with the sun. The *saltus* of this cycle occurs in the 19th year. In this (year) of the *saltus* the sun has as many days as the moon (i.e. the 19-year lunar cycle is completed in that year). This cycle revolves five times in the cycle of Victorius, but 57 years are still found to be remaining in the cycle of Victorius (i.e. 5×95+57=532).

Others constructed a cycle of 100 years. In this cycle the moon revolves five times and five years are remaining in this (i.e. 5×19+5=100). In it the sun revolves three times and there are 16 years remaining (i.e. 3×28+16=100). Accordingly, (this cycle) is incongruent with the sun as well as the moon. In this cycle the *saltus* occurs in the 14th year. In this number the sun exceeds the moon by five years (i.e. compared to the 19-year cycle the *saltus* occurs five years earlier). This cycle revolves five times in the cycle of Victorius, and it starts to run in (its) sixth turn, and they differ by 32 years (i.e. 5×100+32=532).

20–26 Alii…inueniuntur] Cf. p. XCI–XCII of the introduction. **27–31** Alii…superat] Transcribed in Krusch, *Studien* I, p. 11–2, partially 14; transcribed and translated in Warntjes, '84 (14)-year Easter reckoning', p. 55 (facsimile p. 84); cf. also Krusch, *Studien* II, p. 58; O'Connell, 'Easter cycles', 84–5; Schäferdiek, 'Osterzyklus', p. 361, 371; Warntjes, '84 (14)-year Easter reckoning', p. 61; Mosshammer, *Easter Computus*, p. 219–22 (wrong conclusion); and p. XCIV, CXXII–CXXII of the introduction.

5. Victorian 532-year cycle

Ciclus Victorii *DXXXII* annis constat. Hic ciclus certus est et ultra hunc numerum non est certus ciclus, quia in hoc numero *ad originem* suam *reuertitur* secundum solem et lunam. *Lunar*is *circul*us in ciclo Victorii *uigies octies* conuertit. *Solar*is uero undeuigies tantum conuertit.

<LXVI. DE CURSU TEMPORUM>

Disruptions of the course of time: 1. The first three days of creation

*Tres dies ante sol*em facti, in quibus numerantur ordines: Initio anni, *in medio*, uel *in fine*. Cum dies dicuntur, et sol et luna et sidera in ipsis non luciscebant. Id est initio anni poni debent, ut *Theophilus* ait. Et alio modo cognouimus | hos III dies initium esse, ut in Genesi: fol. 45r
Factum est uespere et mane dies I.

---

34 annis] anni *M, Schwartz*. **36** suam] sui *M, Schwartz*. **37** octies] septies *M (corr. from* DE MIRABILIBUS*; also Schwartz)*. **38** undeuigies] uigies *M (corr. according to* MC*; also Schwartz)*. **LXVI,2** quibus] quo *M*.

LXVI,2–7 Tres…I] In this chapter MC discusses the various disruptions of time recorded in the Bible. Note that since the bigger part of this chapter is based on DE MIRABILIBUS SACRAE SCRIPTURAE, a text written in AD 654 and based on Victorian chronology, this chapter probably was part of the now lost Victorian computus of AD 689 originally. At the beginning of this analysis, MC (or rather his source) quite naturally wants to define the starting point of time: Did time begin with the first day of creation, or with the creation of the heavenly bodies (which form the basis for any reckoning of time) on the fourth day? This question may have been motivated by AUGUSTINE, *De genesi ad litteram* 2.14 (*CSEL* 28,1, p. 54): *quis igitur animo penetrat, quomodo illi tres dies transierunt, antequam inciperent tempora, quae quarto die dicuntur incipere, uel utrum omnino transierint dies illi?* MC's answer is unmistakenly that time began with the first rather than the fourth day of creation, even though time could not have been reckoned on the first three days, since the heavenly bodies did not exist. The texts consulted for this question seem to have been DE ORDINATIONE FERIARUM PASCHALIUM and MARTIN OF BRAGA's *De Pascha*. These two texts, however, have a different focus in that they discuss the question whether the creation took place at the beginning, in the middle, or at the end of a season, i.e. spring; MARTIN OF BRAGA explictily argues that the fourth day of creation (25 March) fell right in the middle of spring, and to this the three preceding days had to be added if one was to determine the first day of creation; MC, however, seems rather to be concerned with the question whether the first three days constituted the beginning, the middle, or the end of a year in order to determine whether these three days are to be regarded as the beginning of time (which would then obviously coincide with the beginning of the first year). DE ORDINATIONE FERIARUM PASCHALIUM (*PL* 90, col. 607D): *Theophilus episcopus dixit: In quo loco caput mundi esse creditis? In principio temporis, an in medio, aut in fine? Episcopi responderunt: In aequinoctio octavo Kalendarum Aprilium*. MARTIN OF BRAGA, *De Pascha* 5 (Barlow, *Opera omnia*, p. 272–3; cited in TRACTATUS ADTHANASI 4–5 (Krusch, *Studien* I, p. 331–2)): *Inchoasse mundum veris tempore Genesis docet. ... In quo germinare omnia videmus, atque ita in eo esse principium mundi non dubitamus. Sed cum tres menses vernum tempus habeat, horum trium medius est qui initium mundo dedit, nec solum mensis medius, sed etiam dies mensuum medii. Ex V enim Id. Febr. (veris est inchoatio) in V Id. Mart. unus est mensis. Ex V Id. autem Mart. in VIII Kal. Apr. quindecim dies sunt, id est, medietas mensis. Ita unus et dimidius mensis subsequitur. VIII autem Kal. Apr. aequalis est nox et dies, ... Ita in quo aequalitatem noctis et diei invenimus, in eo initium mundi constitutum intellegamus. Sed non sine causa maiores nostri super VIII Kal. Apr. tres dies addiderunt, ut primum diem mundi invenirent. XI enim Kal. Apr. primum mensem mundi et diem maiores nostri existimaverunt, quia antequam sol in principatum mundi conderetur triduum ante praecesserat. Refert enim Genesis quarta die facta luminaria solis et luna*. For the fact that the first three days were termed *dies*, even though the

The cycle of Victorius consists of 532 years. This is a true cycle and beyond this number (of years) there is no true cycle, because in this number (of years) it returns to its origin according to the sun and the moon. The lunar cycle revolves 28 times in the cycle of Victorius (i.e. 28×19=532). The solar cycle, however, revolves only 19 times (i.e. 19×28=532).

66. ABOUT THE COURSE OF TIME

Three days (were) created before the sun, on which (different) periods (of a year) are reckoned: At the beginning of the year, in the middle, or at the end. They (i.e. the first three days of creation) are called days, though the sun and the moon and the stars did not shine on those (days). Accordingly, they have to be placed at the beginning of a year, as Theophilus says. And we (also) knew from a different source that these three days are the beginning, since (it is said) in Genesis: It was made from evening and morning the first day.

---

heavenly bodies did not shine on them, cf. GEN 1:1–19. The citation at the end of this passage in MC is from GEN 1:5: *appelavitque lucem diem et tenebras noctem factumque est vespere et mane dies unus*. Cf. DE ORDINATIONE FERIARUM PASCHALIUM (*PL* 90, col. 607C): *Dixerunt ergo episcopi: Quem credimus factum fuisse in mundum primum, nisi Dominicum diem? Theophilus episcopus dixit: Probate quod dicitis. Responderunt episcopi: Secundum Scripturae auctoritatem factum est vespere et mane dies primus*.

LXVI,2–12 Tres…reliqua] The only computistical text, to my knowledge, explicitly dealing with the disruptions of time recorded in the Bible, is COMP. COL. (cf. following note), and the question of the first three days of creation, as well as the impact of the Flood on the course of time, are not mentioned there.

34–38 Ciclus…conuertit[2]] Cf. Borst, *Schriften*, p. 419. **LXVI,2–23** Tres…peragunt] Cf. p. LXXVIII, CII, CXXIV–CXXV, CLXXXIII of the introduction.

2. The Flood

Quid in his diebus metendum est? Id est mora *diluu*io *fact*a, *luminari*a non erant. Quomodo cursus illorum fuit? Auctoritate librorum legis in hoc habemus: Qui certo numero statuerunt dies, quibus *pluuia super* illos uenit et reliqua.

3. Josuah

*Sol*e *st*ante *contra Gabaon et luna contra uallem* Aqu*ilon*.

8–12 Quid…reliqua] MC proceeds to a discussion of the Flood here. DE MIRABILIBUS SACRAE SCRIPTURAE also discussed the impact of the Flood on the course of the heavenly bodies, but its verdict was that their courses remained unaltered by this event. DE MIRABILIBUS SACRAE SCRIPTURAE 1.8 (*PL* 35, col. 2160): *De cursu solis et lunae in diluvio. De terminis, de recursu solis et lunae in diluvio, quamvis famosa plurimorum mentes quaestio pulsat; nihil in luminarium aut siderum consuetis ministrationibus diluvium commutasse, ipsius anni quo diluvium est factum, cursus manifestat. ... Ex quibus indiciis manifestissime ostenditur, quod in iis quae ad anni circulum peragendum pertinebant, in luminarium ministrationibus dierum ac noctium plenitudine decursa, nihil decedere videtur. Si enim luminarium, solis scilicet et lunae, cursus in aliquo titubaret; dierum ac noctium primitus vicissitudo, et deinde totius anni circulus impeditus appareret: dum vero nullo modo turbatus in se anni circulus revertitur, manifestum est quod in anno praecedente luminarium cursus non turbatur*. MC, on the other hand, argues that the heavenly bodies were 'delayed' by the Flood, presumably due to the fact that the flood gates of heaven were open for the forty days of continuous rain, which made it impossible for the heavenly bodies to continue their course. Cf. GEN 7:4,11,12,17: *adhuc enim et post dies septem ego pluam super terram quadraginta diebus et quadraginta noctibus et delebo omnem substantiam quam feci de superficie terrae ... anno sescentesimo vitae Noe mense secundo septimodecimo die mensis rupti sunt omnes fontes abyssi magnae et cataractae caeli apertae sunt et facta est pluvia super terram quadraginta diebus et quadraginta noctibus ... factumque est diluvium quadraginta diebus super terram et multiplicatae sunt aquae et elevaverunt arcam in sublime a terra*. **13–14** Sole…Aquilon] The next disruption of time mentioned in the Bible occurs in JOSHUA, where it is related that the sun and the moon stood still at the command of Joshua, and this is the passage referred to here in MC. JOS 10:12–14: *tunc locutus est Iosue Domino in die qua tradidit Amorreum in conspectu filiorum Israhel dixitque coram eis sol contra Gabaon ne movearis et luna contra vallem Ahialon steteruntque sol et luna donec ulcisceretur se gens de inimicis suis nonne scriptum est hoc in libro Iustorum stetit itaque sol in medio caeli et non festinavit occumbere spatio unius diei non fuit ante et postea tam longa dies oboediente Domino voci hominis et pugnante pro Israhel*. This miracle is also discussed at length in DE MIRABILIBUS SACRAE SCRIPTURAE, which forms the basis of a more detailed discussion of this event further below (ll. 20–21 of this chapter). Here, however, the phrasing indicates that MC's source was the Vulgate rather than DE MIRABILIBUS. Cf. DE MIRABILIBUS SACRAE SCRIPTURAE 2.4 (*PL* 35, col. 2175): *Sic enim prescribitur: Sol stetit contra Gabaon et luna ad vallem Hailon, obediente Domini voci hominis*. Interestingly enough, this event is mentioned in connection to the bissextile day in PS-DIONYSISUS, *Argumentum XVI* (Krusch, *Studien* II, p. 80; cited in BC 42A15–B2, QUAEST. AUSTR. 2.9A). Cf. also the following note.

13–23 Sole…peragunt] The three events discussed here (i.e. the standing still of the sun at the command of Joshua, the backwards movement of the shadow on the sun dial of Ahaz, and the solstice occuring during the crucifixion) are also analyzed in COMP. COL. concerning their impact on the course of time; for this analysis, COMP. COL. certainly consulted MC (cf. especially the phrases *in velociore cursu postea autem more reddiderunt* and *item tenebrae iuxta crucem*) among other sources. The final verdict of this Cologne text, then, is that none of these three events changed the course of time in any way. COMP. COL. 5.7 (cited in COMP. COL. 6.1B; the MS reads *more*): *De statione solis ad Gabaon et lunae contra vallem Achilon. Si inmutat aliquid in numero statio luminarium? Vel reversio solis decem lineis in horologio Achaz? Vel si mutant tenebrae aliquid in hora crucis? Non inmutat aliquid statio luminarium, quia aliqui dicunt: Duo luminaria in velociore cursu, postea autem morem reddiderunt et*

What has to be gathered from these days? A delay occurred due to the Flood, and there were no heavenly bodies. How was the course of these? According to the authority of the books of law we found (the following) concerning this matter: These days, in which the rain came over those, and so on, constituted a certain number.

At the same time as the sun stood still (on its way) towards Gabaon, the moon (stood still) opposite to the Aquilon valley.

---

iterum concurre temporaliter. Haec statio tantum efficiebat longiores horas diei. Aliter Agustinus dicit, quod verius: Non addit longiores horas, quia interim non currebat tempus. Item alii male putant hanc stationem prestare byssextum. Item reversio solis fuit pro signo. Interrogatur, si reversa est <*et*> *luna, quia sermo tacuit? Tamen reversa est sic et luna, dum simul steterunt, et nihil mutavit in numero. Item tenebrae iuxta crucem non mutaverunt numerum, quia interim sol cursum suum currebat. Nihil de his omnibus mutat cursum suum.*

13–23 Sole…peragunt] Cf. Borst, *Schriften*, p. 937.

4. The promise of the Lord to Hezekiah

Auctoritate sancti Augustini in hoc habemus dicentis quod moraretur <tempora *Deus*> *Dauid*, quod uelocitas super traxit, et *X lineas horologi* pro *sign*o regi *sol*e *reuer*tente. Quomodo interim luna fuit? Cui non reputetur reuersio; forsitan reuersa est, uel pergebat itinere suo.

Continuat.: Josuah

Iterum ait: *Sole et luna stantibus* cum *Iosue*, uelociori cursu postea uicem more suo reddiderunt.

5. Solstice at passion

Ita iuxta crucem *tenebrae sol*e uelato cursum suum peragunt.

---

15 Auctoritate] auctoritate- *M (corr. according to* MC*)*. **16** tempora Deus] *om. M (add. according to* ISA*; a different possible emendation here would be* in throno *according to* DE MIRABILIBUS 2.28*, to be translated as 'that he (i.e. Hezekiah) remained on the throne of David (for another 15 years)').* **21** suo] sue *M**. **23** peragunt] peragit *M.*

15–19 Auctoritate…suo] Next in the list of Biblical disruptions of time is an event described in 4 KINGS and also in ISAIAH, where God made the shadow of the sundial of Ahaz move backwards by ten lines. ISA 38:5–8: *vade et dic Ezechiae haec dicit Dominus Deus David patris tui audivi orationem tuam vidi lacrimam tuam ecce ego adiciam super dies tuos quindecim annos et de manu regis Assyriorum eruam te et civitatem istam et protegam eam hoc autem tibi erit signum a Domino quia faciet Dominus verbum hoc quod locutus est ecce ego reverti faciam umbram linearum per quas descenderat in horologio Ahaz in sole retrorsum decem lineis et reversus est sol decem lineis per gradus quos descenderat.* 4 KINGS 20:12: *invocavit itaque Esaias propheta Dominum et reduxit umbram per lineas quibus iam descenderat in horologio Ahaz retrorsum decem gradibus.* MC's attribution to Augustine refers to the Irish ps-Augustinian text DE MIRABILIBUS SACRAE SCRIPTURAE, even though MC's phrasing suggests that the Vulgate rather than DE MIRABILIBUS was used; in the ps-Augustinian text the question of the moon's reaction is not discussed, but it is argued that this event in no way disrupted the course of time. DE MIRABILIBUS SACRAE SCRIPTURAE 2.28 (*PL* 35, col. 2187): *Sed dum Isaias duas prophetias sibimet invicem contrarias eadem hora protulisset, necesse erat ut alteram quae facienda esset, signi ostensione firmaret. Unde rex ut quod dicebatur indubitanter crederet, solis in ortum ab occasu per decem horas cursum recidivo Deus tramite retorquet, quantenus inde veluti mane iterum festinare inciperet, qui diei totius, exceptis duabus horis, spatium consummasset. Per quas duas horas, ut multi aestimant, celeri cursu de occasu in ortum revertitur, ut dies integra rursum sine interveniente nocte in ortu innovaretur, ut scilicet dies duos, etiam in hac vice, in una luce concluderet.* Note, however, that this miracle, as well as the standing still of the sun, are also mentioned in AUGUSTINE, *De civitate Dei* 21.8 (*CCSL* 48, p. 772): *Nos autem in diuinis libris legimus etiam solem ipsum et stetisse, cum hoc a Domino deo pertiuisset uir sanctus Iesus Naue, donec coeptum proelium uictoria terminaret, et retrorsum redisse, ut regi Ezechiae quindecim anni ad uiuendum additi hoc etiam prodigio promissioni Dei significarentur adiuncto.* Concerning the two possibilities outlined for the moon's reaction in this instance, it seems that MC later (c. 66, ll. 22–23) favours the second option, i.e. that the moon continued on its course while the sun was drawn back. **20–21** Iterum…reddiderunt] MC here returns to the question whether the standing still of the sun and the moon described in JOSUAH had an impact on the general course of time (cf. ll. 13–14 of this chapter); MC, as well as its source, DE MIRABILIBUS SACRAE SCRIPTURAE, deny any such impact. DE MIRABILIBUS SACRAE SCRIPTURAE 2.4, 2.28 (*PL* 35, col. 2175, 2187): *De sole et luna stantibus ad imperium Josue. ... Jesus* (*recte Josue*) *filius Nun princeps populi Israel soli in medio die praecepit ut ne se moveret, et lunae ubi fuerat staret, donc se Dei populus de inimicis vindicaret, quod et factum est. Nam sol duos dies in uno conclusit, et luna diei spatio non occurrit. ... Haec luminarium mora nihil novum in natura commisit, etsi in ministerio aliquid varium ostendit. Sed et illa varietas nihil in anni cursu et reliquorum dierum commovit, dum pariter sol et luna unumquodque in suo ordine requievit. Si enim unum luminare curreret, dum alterum interim*

According to the authority of St Augustine we found concerning this matter that he argues that the God of David delayed time, that the velocity (of time) retarded as the sun moved backwards ten lines of the sundial as a sign for the king. How did the moon act in the meantime? A (similar) backwards movement shall not (necessarily) be contemplated for it; perhaps it moved backwards, or it continued on its way.

But then (Augustine) states (more specifically concerning a different occasion): Because the sun and the moon stood with Josuah, they subsequently returned to the initial point (*vicem*) in a swifter course according to their manner.

Hence, darkness completed its course at the time of the crucifixion, after the sun had been covered.

---

requiesceret, dierum et mensium et annorum assuetum cursum conturbaret. Dum autem utrumque moram hanc habuit, quasi post consuetum diem in occasus sui limitem perrexit. ... Quomodo et sub Jesu (recte Josue) filio Nun, in illo praedicto bello moram fecisset, indubitanter et hac vice reverti luna minime dubitatur; quatenus nequicquem, ut praediximus, in cyclorum cursibus conturbaretur. **22–23** Ita...peragunt] This sentence refers to the solstice that occurred for three hours during the crucifixion of Christ according to Luke. LUKE 23.44–45: *erat autem fere hora sexta et tenebrae factae sunt in universa terra usque in nonam horam et obscuratus est sol et velum templi scissum est medium.* Cf. also MAT 27:45: *a sexta autem hora tenebrae factae sunt super universam terram usque ad horam nonam.* MARK 15:33: *et facta hora sexta tenebrae factae sunt per totam terram usque in horam nonam.* This event is also described in DE MIRABILIBUS 3.13 (*PL* 35, col. 2199): *De solis eclipsi in passione Domini. Solis defectio juxta carnem, non ut solet, quando luna accenditur, facta erat; sed in plena luna, scilicet quinta decima die mensis, ne casu putaretur fuisse, quod evenerat: et tantum lucem solis media tenebrae obscuraverunt, ut per tres illas horas defectionis, stellae in caelo visae sint.* MC appears to argue here that this solstice compensated for the one disruption that was recorded for the sun only, namely the backwards movement of the shadow on the sundial of Ahaz (c. 66, ll. 15–19).

<LXVII. DE CICLO GRECORUM XCV ANNORUM>

95-year table not cyclic in solar data

Ciclus Grecorum, XCV anni, quinque uicibus conuertitur in ciclo Victorii. Sed sunt alii anni superflui in eo, probemus: Id est IIII feria in fronte kalendarum Dyonisii super VIIII Kalendas Aprilis, et haec IIII feria post bissextum. Et numera IIII feriam, V feriam, VI feriam, sabbatum. Inde transtulit dominicum in II feria, sic reliqua. Prope finem cicli, id est V feria, cum bissexto transtulit VI feriam in sabbatum. Inde sabbatum, dominicum, II feria in fine cicli. In III feria in hoc loco incongruus est secundum

---

LXVII,3 anni] *corr. to* annorum *Mommsen.* **5** fronte] flonte *M*.* | kalendarum] kali *M, Krusch, Mommsen (corr. by Mac Carthy; or should it read* concurrentium*?).* **9** bissexto] bissextum *M (corr. according to* MC*).*

LXVII,3–26 Ciclus…contrarietatem] At the beginning of this passage, MC declares that its purpose is to demonstrate that five 95-year periods are less than the 532 years of a true Easter cycle (5×95=475<532). The following proof, however, does more than this by illustrating that the 95-year Dionysiac Easter table, or any multiple thereof (up to 28), is not cyclic in its weekday data, which follow a 28-year cyclic structure, and therefore Dionysius' Easter table cannot be regarded as a true cycle. The discussion of this problem must have been a lively one in the seventh century, especially since Isidore implies that the 95-year Easter table is truly cyclic, stating at the end of his table of 95 years: ETYM. 6.17.9: *post cuius expletionem ad primum exordium recurrendum*. This may have motivated the composition of this passage in MC. MC's analysis consists of a comparison of the *concurrentes* of the second, third, fourth, and fifth 95-year period with those of the first, i.e. the orginal Dionysiac one. This comparison is based on blocks of four years between the occurrences of bissextile days; when these blocks are compared to the equivalent years of the first 95-year period, the pattern is that the *concurrentes* of only one of these four years differ from the first 95-year period in the second 95-year period, of two years in the third, of three years in the fourth, and of all four years in the fifth; in the following 95-year periods, this disagreement with the first period only increases (cf. the table provided in Appendix 7). The reasons for this gradually increasing disagreement between the first and the subsequent four 95-year periods are: 1) from one 95-year period to the next the *concurrentes* change only in what have been bissextile years in the first period; 2) the bissextile years occur a year later in every subsequent period. The method applied by MC presumably had its origin in Dionysius' instructions for his Easter table. There he explicitly argues that, after the expiration of a 95-year period, all data of the previous period also applied to the following one, except for the solar data (and consequently also the lunar data for Easter Sunday) of every fourth year, namely in what used to be bissextile years in the previous period; the *concurrentes* of those years could be recalculated for the new 95-year period by simply reducing the *concurrentes* of the previous one by one (as MC demonstrates, even though this is not mentioned in Dionysius' account). PROLOGUS DIONYSII (Krusch, *Studien* II, p. 64): *Hoc praeterea putavimus lectorem admonendum, quod cyclus iste nonaginta et quinque annorum, quem fecimus, cum, finito tempore, in id ipsum reverti coeperit, non per omnia propositam teneat firmitatem. ... Concurrentium autem hebdomadum ratio, quae de solis cursu pervenit, septeno annorum iugi circuitu terminatur. In quo per annos singulos unum numerare curabis; in eo tantummodo anno, in quo bissextus fuerit, duos adicies. Quae causa etiam facit, ut non per omnia circulus iste XCV annorum suo recursui concordare videatur. Nam cum in ceteris annis non dissentiat, in his aliis solis, in quibus se bissextus interserit, pascha dominicum cum sua luna vario modo rationis occurrit. Sed hii, qui ordine fixo per omnia decurrunt tempora, mobilium casum sua stabili circuitione sine ulla possunt difficultate dirigere. Et ideo post expletionem XCV annorum cum harum rerum diligens ad exordium redire voluerit, non ad quintum cyclum sancti Cyrilli, quem nobis necessarium proposuimus, sed ad nostrum*

67. ABOUT THE GREEK CYCLE OF 95 YEARS

The cycle of the Greeks, (consisting of) 95 years, revolves in five turns in the cycle of Victorius. But there are years remaining in this, (as) we shall prove (in the following): Wednesday (occurred) at the beginning of the Dionysiac Easter table (*kalendarum / kali*) under (the rubric) '24 March' (i.e. in the column of the *concurrentes*), and this Wednesday (occurred) after a bissextile day (since the first year of the Dionysiac Easter table was AD 532, a bissextile year, in which the bissextile day occured on 24 February). Count Wednesday, Thursday, Friday, Saturday (i.e. the *concurrentes* of the years AD 532 to 535). Then, it passed over Sunday to Monday (i.e. the *concurrentes* of AD 536), and so on. Near the end of the cycle, i.e. (the fourth year from the end, AD 623, having) Thursday (as *concurrentes*), it (then) passed over Friday to Saturday due to the bissextile day (in AD 624). Then, (the *concurrentes* of AD 624 to 626 are) Saturday, Sunday, Monday at the end of the cycle. In this place it is incongruent with the sun in (providing) Tuesday (as the *concurrentes*) at the recurrence (of the 95-year period, i.e. in the first year of the subsequent 95-year period, AD 627), while it did not

---

primum vigilanter excurrat; et ordine, quo diximus, per eos, qui firmum cursum retinent, eorum progressum, qui videntur titubare, sustentet. Now, since this chapter is clearly directed against the Dionysiac table and is in favour of the Victorian 532-year Easter cycle, it appears likely that it formed part of the now lost Victorian computus of AD 689 cited by MC elsewhere (cf. introduction p. CXXIV–CXXV).

LXVII,3–26 Ciclus…contrarietatem] This chapter is unique to MC (though it probably formed part of the now lost Victorian computus of AD 689). It can be argued that from the early eighth century onwards, an analysis of the Dionysiac 95-year period was unnecessary, since these tables were replaced by 532-year cycles based on Dionysiac principles at that stage, according to Bede and his abbot Ceolfrith (DTR 65.16–21 (cited in LIB. CALC. 93), HE 5.21). Whether the same expertise as Bede outlines for his home territory can be taken for granted everywhere in the British Isles must, however, remain doubtful. Moreover, in the pre-Bedan period, particularly in the second half of the seventh century and in the very early eighth century, computists certainly struggled with the application of the 95-year period. In this respect, the reference to the Victorian table in this passage of MC here may indicate that the 532-year period was not yet transferred to the Dionysiac table. Further evidence for an analysis of successive 95-year periods is provided by a text uniquely transmitted in London, British Library, Cotton Caligula A XV, fol. 110r–117v, dateable to ca. AD 703; this is the only text to my knowledge that shows similarities to the passage of MC in question here. The structure of this text is that it outlines the first year of five successive 19-year cycles on the first page, starting with AD 703, the second year of these cycles on the following page, and so on; the principal purpose of this text is the calculation for these years of the weekday of the Easter full moon and the data of the subsequent Easter Sunday by means of *Argumentum XIV*. The interesting aspect, however, is the fact that every single one of these years is compared to the subsequent two 95-year equivalents; e.g. AD 703 is compared to AD 798 and 893 (note that the AD date for the last of these equivalents exceeds the correct date by one throughout; e.g. AD 893 is given as 894). Part of this comparison is the correlation of the weekdays of 1 January, and it is in this feature that this text resembles MC.

LXVII,1–26 De…contrarietatem] Cf. p. LX, LXXIII, XCII, CXXIV–CXXV of the introduction and Appendix 7. **3–7** Ciclus…bissextum] Transcribed in Krusch, *Studien* I, p. 13 (wrongly commenting on *fronte kali Dionysii*); Mommsen, 'Chronographus', p. 34; Mac Carthy, *Annals of Ulster* 4, p. lxxi (apparently did not understanding this chapter).

solem in reuersione, | dum non in IIII feria uenit, sed in V feria cum bissexto. Inde transilit IIII feriam in V feria, in VI feria, sabbatum, dominicum. Cum bissexto transilit II feriam in III feria. Nota in prima uice transilit dominicum, in hac uice II feriam transilit. Et sic currit usque ad finem cicli: III anni congrui, IIIIus annus incongruus. In IIIa reuersione duo anni congrui, II incongrui. Id est VI feria, cum bissexto transilit sabbatum in dominicum, id est congruus dominicus et II feria, incongrua III feria et IIII feria. In secunda reuersione, in fine sabbatum. Cum bissexto transilit dominicum in II feria. Inde in hac uice unus annus congruus, III incongrui. In alia uice IIII incongrui. Sic currit usque ad finem, et in fine inuenies II feria. Cum bissexto inde transilit III feriam in IIII feria. Et a II feria incipit. Tunc uenit in contrarietatem. fol. 45v

12 reuersione] reuelatione *(corr. according to* MC*)*. | V] VI *M (corr. according to* MC*)*. **13** bissexto] bissextum *M (corr. according to* MC*)*. **14** bissexto] bissextum *M (corr. according to* MC*); follows* quia *M (corr. according to* MC*)*. **20** II] VI *M*. **21** reuersione] reuelatione *M (corr. according to* MC*)*. **21–22** sabbatum…bissexto] bissexti in sabbatum *M (corr. according to* MC*)*. **25** feria[2]] *follows* et IIII feria *M*. **26** II] III *M*.

turn into Wednesday (in the following year, AD 628), but into Thursday due to the bissextile day. Therefore, it passes over Wednesday to Thursday (in AD 628), to Friday (in AD 629), (and) Saturday, Sunday (are the *concurrentes* of AD 630 and 631). Due to the bissextile day it passed over Monday to Tuesday (in AD 632). Note that in the first turn (i.e. the first occurrence of a bissextile day in the first 95-year period) it (i.e. the *concurrentes*) passed over Sunday, in this turn (i.e. the second occurrence of a bissextile day in the second 95-year period) it passed over Monday. And it continues to the end of (this second) cycle according to the following pattern: Three (of four) years (are) congruent, the fourth year (is) incongruent (with the equivalent years of the first 95-year period). In the third recurrence (of the 95-year period) two (of four) years (are) congruent, (the other) two incongruent (with the equivalent years of the first 95-year period). I.e. (the *concurrentes* of the third year from the end of the second cycle, AD 719, is) Friday, it passes over Saturday to Sunday (in AD 720) due to the bissextile day, i.e. Sunday and Monday (i.e. the *concurrentes* of the last two years of the second 95-year period, AD 720 and 721) (are) congruent (with the last two years of the first 95-year period, AD 625 and 626), Tuesday and Wednesday (i.e. the *concurrentes* of the first two years of the third 95-year period, AD 722 and 723) (are) incongruent (with the first two years of the first 95-year period, AD 532 and 533). In the following (i.e. the fourth) recurrence (of the 95-year period), at the end (of the third 95-year period) Saturday (is the *concurrentes* of the second year from the end of that period, AD 815). Due to the bissextile day it passes over Sunday to Monday (in AD 816). Therefore, in this turn (i.e. the fourth 95-year period) one year (is) congruent, three (of four years) incongruent (with the equivalent years in the first 95-year period). In the remaining turn (i.e. the fifth 95-year period) four (of four years are) incongruent (with the equivalent years in the first 95-year period). According to this pattern it continues to the end, and at the end (i.e. in the first year of the sixth 95-year period, AD 1063) you will find Monday (as the *concurrentes*). Due to the bissextile day it passes accordingly over Tuesday to Wednesday (in AD 1064). And from (the *concurrentes*) Monday (of AD 1063 the sixth 95-year period) starts. Thenceforth it leads into contradiction.

<LXVIII. DE XII CICLIS MUNDI>

The 12 cycles of the world: 1st cycle

De ciclo I

Primus ciclus *a mundi principio* incipiens *quingent*esimo *trig*esimo secundo *ann*o post ortum mundi defecit.

2nd cycle

De ciclo II

Inde secundus spatia cursus sui usque ad millesimum et sexagesimum quartum mundi etatis extendit.

3rd cycle

Ciclus III

Post hunc tertius exoritur, qui seculi noui DLXXXXVI anno finitus.

4th cycle

De ciclo IIII

Post tertium quartus suum incipit cursum | donec $\overline{\text{II}}$CXXVIII ab origine consumantur. fol. 46r

5th cycle

Ciclus V

*Quint*us *post diluuium quadringentesimo octauo decimo anno de*sinit.

6th cycle

De ciclo VI

Sextus in primo *aetatis* ab arche *anno finitur*.

7th cycle

Ciclus VII

Septimus quinquennio ante mortem Moysi concluditur.

LXVIII,8 extendit] extenditur *M*. **10** DLXXXXVI] XXVI *M (corr. according to* MC*)*. **11** anno] anni *M*. **14** $\overline{\text{II}}$CXXVIII] $\overline{\text{II}}$CXVIII *M**. **16** diluuium] diluium *M (corr. from:* DE MIRABILIBUS*)*. quadringentesimo] quadrintesimo *M (corr. from* DE MIRABILIBUS*)*. **17** desinit] definit *M*.

LXVIII,2–39 De…ignoratur] This chronicle forming the final chapter of MC outlines a correlation between 532-year (Victorian) Easter cycles and Victorian-Eusebian chronology (which MC had already used in c. 44, ll. 11–12, c. 64). It is based on c. 2.4 of the ps-Augustine tract DE MIRABILIBUS SACRAE SCRIPTURAE (apparently composed in southern Ireland in AD 654), in which the discussion starts with the fifth 532-year cycle from the beginning of the world, since the end of the first age (i.e. the year of the Flood, AM 2242) occurs in that cycle. DE MIRABILIBUS SACRAE SCRIPTURAE 2.4 (*PL* 35, col. 2175–6): *Ut enim hoc manifestis approbationibus pateat, cyclorum etiam ab initio conditi orbis recursus in se breviter digeremus, quos semper post quingentos triginta duos annos, sole ut in principio, et luna per omnia conventibus, nullis subvenientibus impedimentis, in id unde coeperant, redire ostendemus. Quinto namque cyclo a mundi principio, anno centesimo quarto decimo, generale totius mundi diluvium sub Noe venit, qui post diluvium quadringentesimo decimo octavo anno defecit: et inde alius incipiens, id est, sextus, in octavo aetatis Abrahae anno finitur. Et nono ejus anno septimus incipiens, trigesimo quinto anno egressionis filiorum Israel de Aegypto, quinquennio ante mortem Moysi concluditur. Post quem octavus, in quo est illud signum in sole et luna factum, trigesimo sexto anno egressionis Israel de Aegypto incipiens, in trigesimum primum annum Asae regis Juda incidit. Cujus trigesimo secundo anno nonus exordium capiens, in quo et aliud signum in sole, Ezechiae regis tempore, de quo paulo post dicemus, factum legitur, centesimo octavo anno post templi restaurationem, quae sub Dario facta est, sui cursus spatium consummavit: donec decimus inde oriens nonagesimo secundo anno post passionem Salvatoris, Alia et Sparsa consulibus, peractis cursibus consummatur. Post quem undecimus a consulatu Paterni et Torquati ad nostra usque tempora decurrens, extremo anno Hiberniensium moriente Manichaeo inter caeteros sapientes, peragitur. Et duodecimus nunc tertium annum agens ad futurorum scientiam se praestans, a nobis qualem finem sit habiturus*

68. ABOUT THE TWELVE CYCLES OF THE WORLD

About the first cycle

The first cycle, which had started at the beginning of the world, ended in the 532nd year after the creation of the world.

About the second cycle

Thereafter, the second (cycle) extended the intervals of its course to the 1064th (year) of (this first) age of the world.

The third cycle

After this the third came forth, which (was) terminated in the 596th year of the new (i.e. second) millennium.

About the fourth cycle

After the third the fourth started its course (and it continued) until 2128 (years) had elapsed from the beginning (of the world).

The fifth cycle

The fifth came to an end in the 418th year after the Flood.

About the sixth cycle

The sixth was ended in the first year of the age from the ark (i.e. in the first year of the third age, which follows the age that starts with the Flood).

The seventh cycle

The seventh was concluded in the five years before the death of Moses.

---

ignoratur. MC completes this discussion by providing the number of years from the beginning of the world to the final year of each of the first five cycles. There are two notable differences between MC and DE MIRABILIBUS SACRAE SCRIPTURAE: 1) In MC the sixth cycle ends in the first year of the third age, not in the eighth year of that age as in DE MIRABILIBUS; the difference could be the result of a copying mistake, but it may be more significant in that the second age may have been considered as consisting of 949 instead of 942 years by MC. 2) The dating clause of DE MIRABILIBUS (the third year of the 12th cycle) was changed into a more general phrase in MC. Since, again, the bias of this minor chronicle is clearly Victorian, it probably was already part of the now lost Victorian computus of AD 689 (cf. p. CXXIV–CXXV of the introduction).

LXVIII,2–39 De…ignoratur] MC is the only Irish computistical textbook that ends with a short chronicle (though in this he probably just followed his direct source, the Victorian computus of AD 689). Appending a chronicle to the main body of a computistical texbook was, however, not uncommon. Most prominently, the final chapters of Bede's DT and DTR constitute chronicles; in these, Bede compared Eusebian-Septuagint with Vulgate chronology, but evidently favoured the latter. In DTR, the chronicle begins in c. 66 with a condensed summary of the six ages of the world, and this forms the chronicle of Rabanus Maurus' *De computo* (RM 96.21–59). A variation of Bede's DT chronicle, recording only the Vulgate chronology, was composed in AD 807 (SER. NOV.); it circulated widely in the ninth century and was also included in LIB. COMP. (as c. 1.5) and then copied in LIB. CALC. (as c. 1.1). The correlation between 532-year cycles and linear world chronology as outlined in MC, however, did not find any imitators.

LXVIII,2–39 De…ignoratur] Cf. Borst, *Schriften*, p. 461 and p. LXXVIII, XCIV, CXXIV–CXXV of the introduction.

8th cycle De ciclo VIII

Octauus, in quo signum in sole et luna fuit, *in XXXI anno Ase regis Iuda incidit.*

9th cycle Ciclus VIIII

*Nonus, in quo et*iam *signum aliud in sole* fuit, Are*thiae regis tempore* usque *CVIII ann*i *post restaurationem templi, quae sub Dario facta est, sui cursus spatium consumauit.*

10th cycle De ciclo X

Decimus XCII anno post passionem domini *consumatur.*

11th cycle Ciclus XI

Undecimus in *tempor*ibus *nostr*is *currens Hibernensium* doctore *Manchiano moriente peragitur.*

12th cycle De ciclo XII

Duodecimus sua tempora *nunc agens a nobis qualem finem hab*uerit *ignoratur.*

---

24 XXXI] XXI *M (corr. from* DE MIRABILIBUS*).* **35** currens] turrens *M (corr. according to* DE MIRABILIBUS*; also Docen, Mac Carthy, Schwartz, silently Esposito, Grosjean).* **36** doctore Manchiano] doctore- Anchiano *M (corr. according to* DE MIRABILIBUS*; also Docen, Schwartz, Mac Carthy, silently Esposito, Grosjean, the latter three arguing that M transmits the correct spelling of the name, i.e.* Manchiano*, as against* DE MIRABILIBUS*'s* Manichaeo*; for this cf. also Mac Ginty, Ó Cróinín).*

32–39 Decimus…ignoratur] This passage is partially transcribed and paralleled with DE MIRABILIBUS in Schwartz, 'Ostertafeln', p. 90; from the eleventh cycle only in Docen, 'Annales Ratisponensis', p. 516–7; Mac Carthy, *Annals of Ulster* 4, p. lxx; Esposito, '*De mirabilibus*', p. 200 (197–8 for *Manchianus*); Grosjean, 'Quelques exégètes irlandais', p. 74 (73–4 for *Manchianus*); cf. also Schwartz, 'Ostertafeln', p. 92; Krusch, *Studien* II, p. 58; Mac Ginty, 'Irish Augustine', p. 76–8; Ó Cróinin, 'Early Irish annals', p. 82; idem, 'A seventh-century Irish computus', p. 127; idem, 'Old Irish gloss', p. 131; idem, 'Irish provenance', p. 182–3; idem, *Early medieval Ireland*, p. 188; idem, 'Earliest Old Irish glosses', p. 16; Machielsen, *CCSL Clavis Patristica* 3A, p. 189.

About the eighth cycle

The eighth, in which a sign occurred in the sun and the moon, discontinued in the 31st year of Asa, king of Juda.

The ninth cycle

The ninth, in which yet another sign occurred in the sun, completed the length of its course from the time of king Arethia (i.e. Ezechias) to (the time) of the 108th year after the restoration of the temple, which was done under Darius.

About the tenth cycle

The tenth was completed in the 92nd year after the passion of the Lord.

The eleventh cycle

The eleventh, which passes into our time, is brought to an end (at the time) when Manchianus, a teacher of the Irish, died.

About the twelfth cycle

What kind of end the twelfth will have, which happens now in its own time, is unknown to us.

APPENDICES

APPENDIX 1 (MC 49.16–32): COMPARISON OF THE DIONYSIAC WITH THE VICTORIAN EPACTS

The following table, the only one in the entire Munich Computus, is a good example of the extremely corrupt state of certain passages of this text. As it stands, it has the following three columns: a) the epacts of 1 January of the Dionysiac 19-year cycle, starting with epact 8 (altered to 9 by the scribe of this MS by the subsequent addition of an I, an alteration that is clearly evident from the shape of the final I differing sharply from the previous Is) due to the March placement of the *saltus*; b) the eleven lunar ages that do not occur on 1 January in the Dionysiac 19-year cycle; c) the epacts of 1 January of the Dionysiac 19-year cycle, starting with epact 9 instead of 8 due to the November placement of the *saltus*.

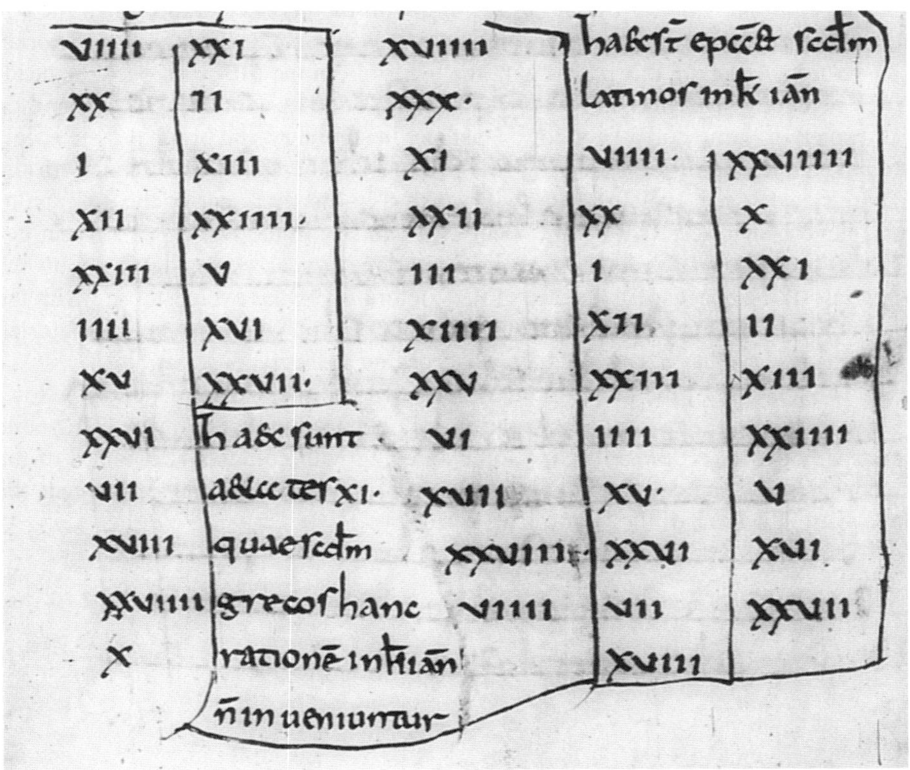

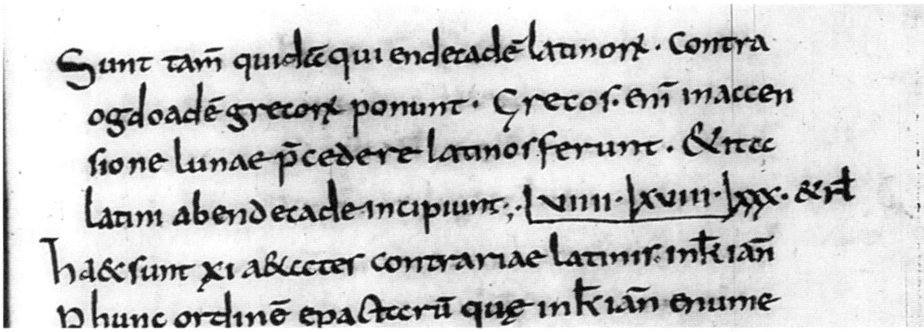

Munich, Bayerische Staatsbibliothek, Clm 14456, fol. 26v–27r.

As I have already argued in the edition, it appears unlikely that originally the only difference between column a and c was in the first epact, and consequently only in the placement of the *saltus*. The term *Latini* in the heading of column c rather suggests that the epacts of the Victorian 19-year cycle were originally listed here. Moreover, the text immediately following the table appears to have originally been part of that table, listing in two additional columns d) the epacts of 1 January of the Victorian 19-year cycle, but starting with the *hendecas* rather than the *ogdoas*, and e) the eleven lunar ages that do not occur on 1 January in the Victorian 19-year cycle; again, this part is extremely corrupt in the MS, with only the headings revealing what has to be regarded as the original content, which was then corrupted or omitted at a later copying stage. In summary, the original table seems to have consisted of the following five columns: a) the epacts of 1 January of the Dionysiac 19-year cycle, starting with epact 8 in the first year of the *ogdoas*; b) the eleven lunar ages that do not occur on 1 January in the Dionysiac 19-year cycle; c) the epacts of 1 January of the Victorian 19-year cycle, starting with the *ogdoas*; d) the epacts of 1 January of the Victorian 19-year cycle, starting with the *hendecas*; e) the eleven lunar ages that do not occur on 1 January in the Victorian 19-year cycle; it has been edited above according to this reconstruction.

This reading of the table in the Munich MS and the text immediately following is confirmed by a similar and fortunately uncorrupted table in *CE*. This table consists of four columns, the first three agreeing with columns a, c, and d of the reconstructed Munich table; the last column of the Einsiedeln table then lists the epacts of 22 March of the Dionysiac 19-year cycle.

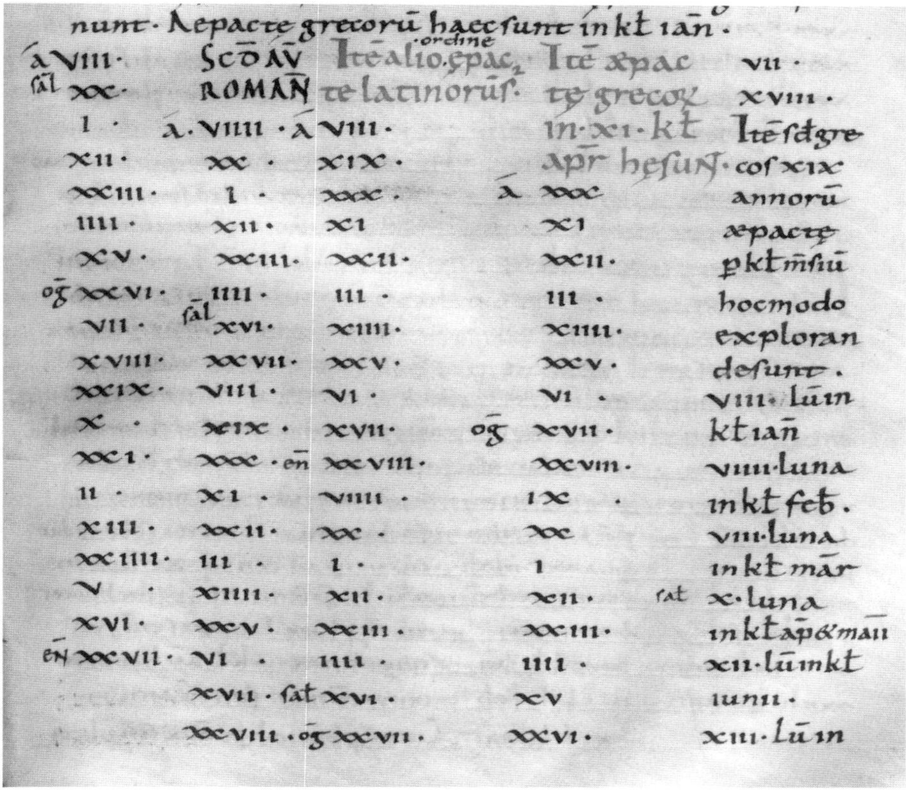

Einsiedeln, Stiftsbibliothek, 321 (647), p. 109.

Hence, it appears that the Munich computist worked from the Einsiedeln table, omitted the final column and instead added two further columns he thought were of interest, i.e. b and e. The unusual correlation of the Dionysiac epacts, starting with the *ogdoas*, with the Victorian epacts, starting with the *hendecas*, can also be found in a table in Cologne, Dombibliothek, 83², fol. 176v (cf. Appendix 3).

APPENDIX 2 (MC 52): *DE COMPARATIONE EPACTARUM DIONYSII ET VICTORII*

This text is here edited for the first time from the sole MS witness Cologne, Dombibliothek, 83², fol. 176v–178v (designated as K in *app. crit.*); for a discussion of this text cf. p. CLII–CLVIII of the introduction.

fol. 176v <I.> | Princeps Dionisius Alaxandrinus episcopus Grece scripserat: *exordium spei nostre*. In uno se iungit cum his, qui ante *aduentum domini nostri Ihesu Christi* fuerant: *antiqui moris obseruantiam ac a sancto Moyse traditam* idem est. Unde exinde inuestigantes coniungendum est. Aliter: antiqui moris, id est Aegyptiorum, in decennouenali, qui de cyclo quod dies sol habet tot et luna, hoc est V$\overline{\text{I}}$DCCCCXXXVI dies, qui possunt diuidi hoc ordine: $\overline{\text{III}}$, $\overline{\text{III}}$; CCC, CCC; CL, CL; VVV, VVV; III, III.

In secundo ogdoadis anno mundus coepit, quod XIIII luna primi mensis tunc occurente confirmatur.

I,7 Aegyptiorum] Aegyptiuorum *K*. **8** V$\overline{\text{I}}$DCCCCXXXVI] *corr. from* V$\overline{\text{I}}$DCCCCXXVI *K*. **9** CL CL] VI, VI *K (corr. according to K)*. **9–10** III III] *follows* r *K*.

I,2 exordium…nostre] PROLOGUS DIONYSII (Krusch, *Studien* II, p. 64): *quantinus exordium spei nostrae notius nobis existeret et causa reparationis humanae, id est, passio redemptoris nostri, evidentius eluceret.* **3** aduentum…Christi] 2 THES 2:1: *rogamus autem vos fratres per adventum Domini nostri Iesu Christi et nostrae congregationis in ipsum*; 1 TIM 6:14: *ut serves mandatum sine macula inreprehensibile usque in adventum Domini nostri Iesu Christi.* **4–5** antiqui…traditam] PROLOGUS DIONYSII (Krusch, *Studien* II, p. 65): *Tanta hac auctoritate divina claruit, primo mense, XIIII die mensis ad vesperum usque ad XXI festivitatem paschalem debere celebrari. Sed quia mensis hic, unde sumat exordium vel ubi terminetur, evidenter ibi non legitur, praefati venerabiles CCCXVIII pontifices antiqui moris observantiam et exinde a sancto Moyse traditam, sicut in septimo libro ecclesiasticae refertur historiae, sollertius investigantes, ab VIII Idus Martii usque in diem Nonarum Aprilis natam lunam facere dixerunt primi mensis exordium; et a XII kl. april. usque in XIIII Kalendas Mai lunam XIIII sollertius inquirendum.*

Kal. Ian. Dionisii	**Luna XIIII**	**Kal. Ian. Victurii**	**Luna XIIII**
VIIII	Non Apr	VIII	VIII Id Apr
XX	VIII Kl Apr	XVIIII	VII Kl Apr
I	Id Apr	XXX	XVIII Kl Mai
XII	IIII Non Apr	XI	III Non Apr
XXIII	XI Kl Apr	<XXII>	X Kl Apr
IIII	IIII Id Apr	III	III Id Apr
XV	III Kl Apr	XIIII	II Kl Apr
XXVI	XIIII Kl Mai	XXV	<XIII Kl Apr>
hucusque ogdoas <Dionisii>.			
VII	VII Id Apr	VI	VII Id Apr
XVIII	VI Kl Apr	XVII	V Kl Apr
<XXVIIII>	XVII Kl Mai	XXVIII	XVI Kl Mai
		<hucusque endecas Victurii>	
X	II Non Apr	VIIII	Non Apr
XXI	VIIII Kl Apr	XX	VIII Kl Apr
II	II Id Apr	I	Id Apr
XIII	Kl Apr	XII	<IIII Non Apr>
XXIIII	XII Kl Apr	XXIII	XI Kl Apr
V	V Id Apr	IIII	IIII Id Apr
XVI	IIII Kl Apr	XVI	IIII Kl Apr
XXVII	XV Kl Mai	XXVII	XV Kl Mai
hucusque endicas <Dionisii>, in anno saltus.		<hucusque ogdoas Victurii>	

13 Kal[1]…XIIII[2]] *the first line of the table, which varies considerably from the edition here (cf. following notes and Appendix 3), is written in red ink K.* **13–36** Kal[1]…saltus] *The table in K is extremely corrupt, its purpose, however, fully evident: it compares the Dionysiac and Victorian epacts on 1 January and Easter full moons by placing, quite artificially, the Dionysiac ogdoas alongside the Victorian hendecas; the original table has accordingly been reconstructed here. For reasons of space, the columns noting* luna XIIII *after every single Julian calendar date of the Easter full moons are omitted and replaced by a heading. A red line separates the first from the second column in K. Cf. the facsimile of this table in Appendix 3.* **13** Luna XIIII[1]] *The notation* luna XIIII *occurs as a separate column in K (specifying every single of the previously listed dates as* luna XIIII*) rather than as a heading (cf. the facsimile in Appendix 3).* | Ian[2]] *follows* V *K.* | Victurii] Victurius *K.* | Luna XIIII[2]] *the notation* luna XIIII *occurs as a separate column in K (specifying every single of the previously listed dates as* luna XIIII*) rather than as a heading (cf. the facsimile in Appendix 3); the first entry, however, notes the Julian calendar date and lunar age of the final moon of the March lunation of the first year of the Victorian hendecas, namely* **X K- Apr-, XXX**. **14** VIII[2]] *VII K.* **15** XVIIII] XVIII *K.* | VII] IIII *K.* **17** III…Apr[2]] *follows* cl- *after the* luna XIIII *entry K.* **18** XXII] *om. K.* **20** XV] XII *K.* | XIIII] X *K.* **21** XXV] XIII *K.* | XIII…Apr] *om. K.* **22** Dionisii] *om. K.* **23** VII[3]] V *K.* **25** XXVIIII] *om. K.* | XVI] XVII *K.* **26** hucusque…Victurii] *om. K.* **29** II[1]] *corr. from* XII *(?) K.* | I] *add. K.* **30** IIII…Apr[2]] *om. K.* **32** IIII[2]] V *K.* **33** XVI[2]…Apr[2]] *follows a blank space in the following row K.* **34** XV[2]] XVI *K.* **35** hucusque[1]] hucus *K.* | hucusque[2]…Victurii] *om. K.*

fol. 177r | Ex his quidem aparet, quis sepius precidit in lune aetatibus, quis subsequitur: Dionisius an Victorius, quorum superior Grecus, alter Latinus est. Sed notandum est illos sicut suprascriptis tramitibus manifestum est:

Tunc in ipsis aetatibus concordare, quando luna XVI, aut XXVII, aut VIII Kalendis Ianuarii prouenit, nec non, uerbi gratia et in octauo ogdoadis anno concordant prima luna die Nonarum Aprilis et deinceps usque ad XXVIIII die V Nonas Maii. Est quidem in annis XVI, quibus Dionisius et Victorius discordant, quedam concordia scilicet singulis. Estque in tribus, quibus concordant, quedam discordia utique singulis. Quod utrumque per lunam Aprilis fieri probatur, que secundum Victorium XXVIIII, secundum Dionisium uero XXX est.

In quo etiam anno hii duo in paschae solemnitatibus mense integro discordant. Sed si tunc accederit, ut si secundum Victorium in VIII Kalendas Aprilis pascha celebretur, secundum Dionisium in X Kalendas Maii celebrabitur; si uero secundum ipsum in VII Kalendas Aprilis, secundum istum in nono Kalendas Maii; si autem secundum ipsum in VI Kalendas Aprilis, secundum istum in VIII Kalendas Maii; item si secundum ipsum in V Kalendas Aprilis, secundum istum in VII Kalendas Maii. Sed in his, ut in ceteris, errat Victorius in XIII luna prioris anni pascha conponens et communem de embolismo faciens, quod quidem euidens error est.

Quid plura? Ex illo decennouennali ciclo, quem et secundum Victorium et secundum Dionisium possumus, fere manifeste apparet, quod quibus lunis ipsi inter se per ogdoadem et endicadem discordant, et quod et quibus sibi concordant. Per saltum enim Dionisii incipiunt discordare, per Victorii uero concordare, excepto quod etiam uerbi gratia in octauo ogdoadis anno, sicut paulo ante diximus, concordabant.

Haec autem decennouennalis cicli ratio ab ipsi passionis domini anno et paulo ante coepisse non dubitatur, sicut et fine cicli Victorii probari potest hoc modo: Kalendis Ianuarii IIII feria, luna VIII. Quem ciclum qui se intellegere putant, dominum nostrum Ihesum Christum fere in principio endicadis passum esse dicunt. Ipsi, qui Victorium et Dionisium fere sex a principio ogdoadis annis in lune aetatibus concordare confirmant, ac

---

39 Latinus] Ltinus *K**. **49** XXVIIII] XVIIII *K*. **54** secundum] secundu *K*. **64** possumus] possuimus *K*. **67** Dionisii] Dionisi *K*. **71** cicli] ciclo *K**. **73** fine] sine *K*.

deinceps ceteris per saltum Victorii discordare, qui tunc ut ai|unt: <Victorius> in lunae aetatibus praecedit, cum Dionisius subsequatur in ipsis. Ita ut etiam in fine ocdoadis et endicadis ipse istum duabus lunae aetatibus superet, quarum una per saltum, altera per quandam causam facta intelligitur, unde nec per saltum soluitur. Finita autem endicade per Dionisii saltum, qui est utique secundum Grecos, iterum concordant. fol. 177v

Sed hii, qui ita sentiunt in Victorio confidentes, audiant alium sapientem, qui Grecam magis regulam sequitur, ipsum dicentem ogdoadem melius debere incipere a loco, a quo endicas incipit. Unde quidem ut ipse ait: Fere in principio ogdoadis Christus passus est. Igitur ab ogdoadis primo anno Dionisius et Victorius per Dionisii saltum, qui est ut supra secundum Grecos, in lune aetatibus sepius discordant. Usque ad nonum endicadis annum XVI enim fere annis sepius discordant, cum utique in hisdem annis Dionisius in ipsis aetatibus precedat, Victorius uero subsequatur. Tribus uero fere annis per saltum Victorii sepius concordant. Finito ergo decennouennalis ciclo per saltum Dionisium iterum discordant.

Quodsi igitur haec ita sunt, immo ut supra dictus adfirmat ita sunt, qui diligenter adtendat, inueniet Christum luna XVII secundum Dionisium resurrexisse, XVI uero secundum Victorium. Illud quidem siue Grecorum siue Hebreorum auctoritate confirmatur, istud autem ipsius Victorii et caeterorum consensu Latinorum commendatur.

Post haec itaque omnia pascha precedens istius anni in ciclo Victorii non querendum est in loco, in quo stimatum est posse designari saltus, ut Kalendae Ianuarii VI feria, luna IIII, pascha XIIII Kalendas Maii, luna XXII. Sed in alio ante non retro, ubi scriptum inuenitur: Kalendae Ianuarii, VI feria, luna III, pascha XIIII Kalendas Maii, luna XXI.

Ecce quidam sapiens XIIII lunam paschę, quod ipse Christus more Iudaeorum comedit V feria de media fere hora, post sextam IIII feriae coepisse dicit. Nam, ut ipse ait, non XIIII, | sed in XV luna eadem V feria ad uesperum fol. 178r

80 Victorius] *om. K.* **81** fine] sine *K.* **85** Dionisii] Dionisi *K.* **103** XVII] *corr. from* XIIII *K.* **106** caeterorum] coeterorum *K.* **108** ciclo] *a dot is placed above* i *K.* **111** in] m *K.* | scriptum] *follows* est *erased K.* **112** Ianuarii] Ianuarias *K.* **116** dicit] dicunt *K.*

nouae officia faciens Christ*us panem accepit et calicem* ipsum et istum *benedi*cens reliqua.

<II.> In nomine Dei summi incipit.

Quidam quidem secundum Grecos dominum nostrum Ihesum Christum tertio ogdoadis anno passum esse et resurrexisse credunt. Hac ratione quoniam ipse, *plenitudo legis existens, quando dignatus est homo fieri* VIII Kalendas Aprilis, *quinta sabbatorum, XIIII luna mensis primi in caenaculo cum discipulis pascha manducans, paulo post a Iuda traditur et sequenti die,* VII Kalendas Aprilis, *VI feria, XV luna, crucifigitur, et ad inferos discendens ac dispensationes salutis nostrae perficens uespere sabbati,* VI Kalendas Aprilis, luna XVI, *luciscente dominico,* id est V Kalendas Aprilis, *resurrexit a mortuis, in quo die luna XVII primi mensis iuxta Haebreos extitisse manifestum est.*

Quidam uero ipsum eodem fere endicadis anno passum et resurrexisse dicunt, quod etiam per ordinem cicli Victorii facile probatur. Sed quamuis ista causa aliquis hunc Victorium et ipsum Dionisium, qui Grecam magis sequitur regulam, ogdoadibus et endicadibus discrepare putet. Quibusdam obiectibus turbamur et econtra: Si alius eos eisdem concordare adfirmet, quaedam difficilia offendet. Turbatur, ut tertii et regis uultis tertio et regis uulto, Probi Iunioris Probo Iuniore; offendet, ut V Kalendas Aprilis luna XV. Cum ergo illud turbari et istud offendere accedat, utrique supperit.

---

118 officia] offitia *K.* **II,1** In…incipit] *written in red ink K.* **3** esse] *a vertical thick line at the end of the final* e *K, which may be a splash of ink, but makes the verb look like* esset. **8** post] potest *K (corr. from* EPISTOLA PROTERII*).* **12** V] VI *K.* **22** tertii] &tii *K.* | tertio] &tio *K.*

118–119 Christus…reliqua] MAT 26:26–27: *cenantibus autem eis accepit Iesus panem et benedixit ac fregit deditque discipulis suis et ait accipite et comedite hoc est corpus meum et accipiens calicem gratias egit et dedit illis dicens bibite ex hoc omnes.* **II,4–14** plenitudo…est] This is a direct quotation from EPISTOLA PROTERII 2 (Krusch, *Studien* I, p. 271; MS A reads *caenaculo* and *aebreos*, MS C *perficens*, MSS C L *luna XVII*) with the addition of Julian calendar dates to complement the chronology of Christ's passion and resurrection: *Sed qui haec per Moysen locutus est dominus, plenitudo legis existens, quando dignatus est homo fieri, quinta sabbatorum XIIII luna mensis primi in cenaculo cum discipulis pascha manducans, paulo post a Juda traditur et sequenti die XV luna crucifigitur, id est sexta feria, et ad inferos discendens ac dispensationes salutis nostrae perficiens, vepere sabbati, luciscente dominico, resurrexit a mortuis, in quo die lunam XVII primi mensis iuxta Hebreos extetisse manifestum est*. Note that the dates given here agree with the chronology outlined in MC 44.55–80.

APPENDIX 3 (MC 55.46–52): COMPARISON OF THE VICTORIAN WITH THE DIONYSIAC EASTER FULL MOONS

The Munich computist compares the Victorian with the Dionysiac full moons by using the two methods he already applied for the correlation of the epacts (cf. Appendix 1), namely a) the historically correct correlation, starting with the *ogdoades* of the two reckonings, and b) the unusual alignment of the Dionysiac *ogdoas* with the Victorian *hendecas*. In fact, the conclusion drawn by the Munich computist from comparing the two sets of Easter full moons according to method b, i.e. that the two reckonings agree in 3 years, but disagree in the remaining 16, is based on the comparison of the epacts rather than the full moons, as the full moons agree, according to this comparison, in only two years. Only one text, to my knowledge, also discussed the comparison of the Dionysiac and Victorian Easter full moons according to method b; the text in question, totally neglected by modern scholarship, is now published, for the first time, as Appendix 2, and a brief introduction to this text can be found on p. CLII–CLVIII of the introduction. On the first page of this text (Cologne, Dombibliothek, 83², fol. 176v), this comparison of the Easter full moons by correlation of the Dionysiac *ogdoas* with the Victorian *hendecas* is given in table form; the table also includes two further columns, listing the Dionysiac epacts of 1 January starting with the *ogdoas*, and the Victorian epacts of 1 January starting with the *hendecas* (for this comparison of the epacts cf. Appendix 1). Thus, the columns of this table are the following: a) the Dionysiac epacts of 1 January, starting with the *ogdoas*; b) the Julian calendar dates of the Dionysiac Easter full moons, starting with the *ogdoas* (followed by a descriptive column noting that these dates refer to *luna* 14); c) the Victorian epacts of 1 January, starting with the *hendecas*; d) the Julian calendar dates of the Victorian Easter full moons, starting with the *hendecas* (followed by a descriptive column noting that these dates refer to *luna* 14). The alignment of the columns is not well executed, and some of the data are corrupt; the facsimile is provided here, for a reconstruction of the table see its edition in Appendix 2.

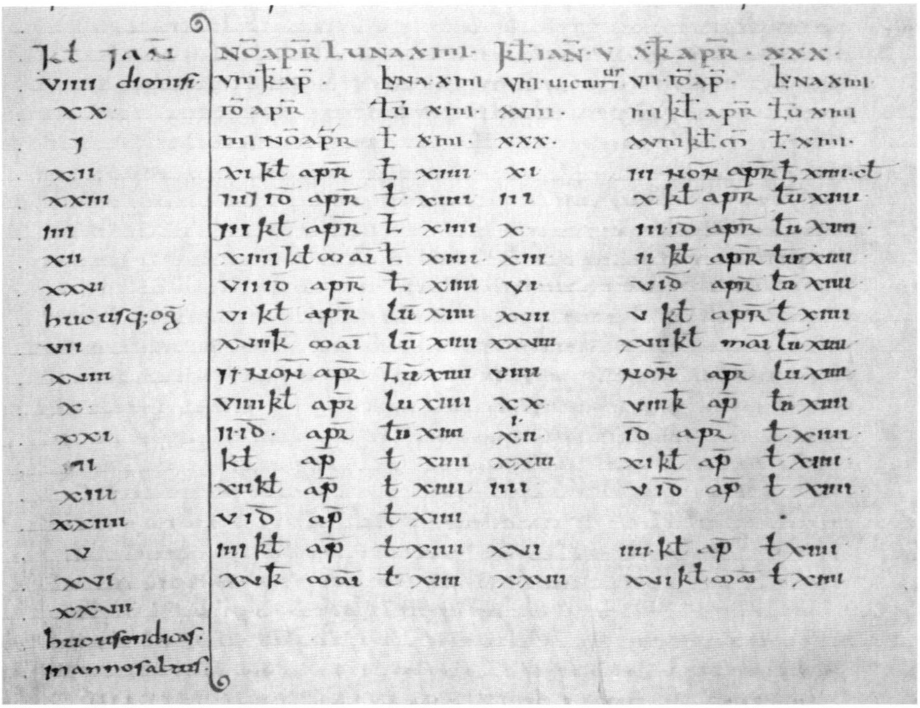

Cologne, Dombibliothek, 83², fol. 176v.

As for the historically correct method (a) of comparing the Dionysiac and the Victorian Easter full moons, even this is rarely found in early medieval manuscripts. It features in a table in Berlin, Staatsbibliothek, Phillipps 1831, fol. 128r that compares four sets of Easter full moons, headed a) *Latinorum*, b) *Grecorum*, c) *Victorii*, d) *Eusebii*. Column a lists the Dionysiac full moons, starting with the *ogdoas*, column c the Victorian ones, starting with the *ogdoas*; columns b and d are yet to be explained; column d is identical with column a except for years 1 and 5; column b is identical with column c except for years 1, 3, 11, 15, 19.

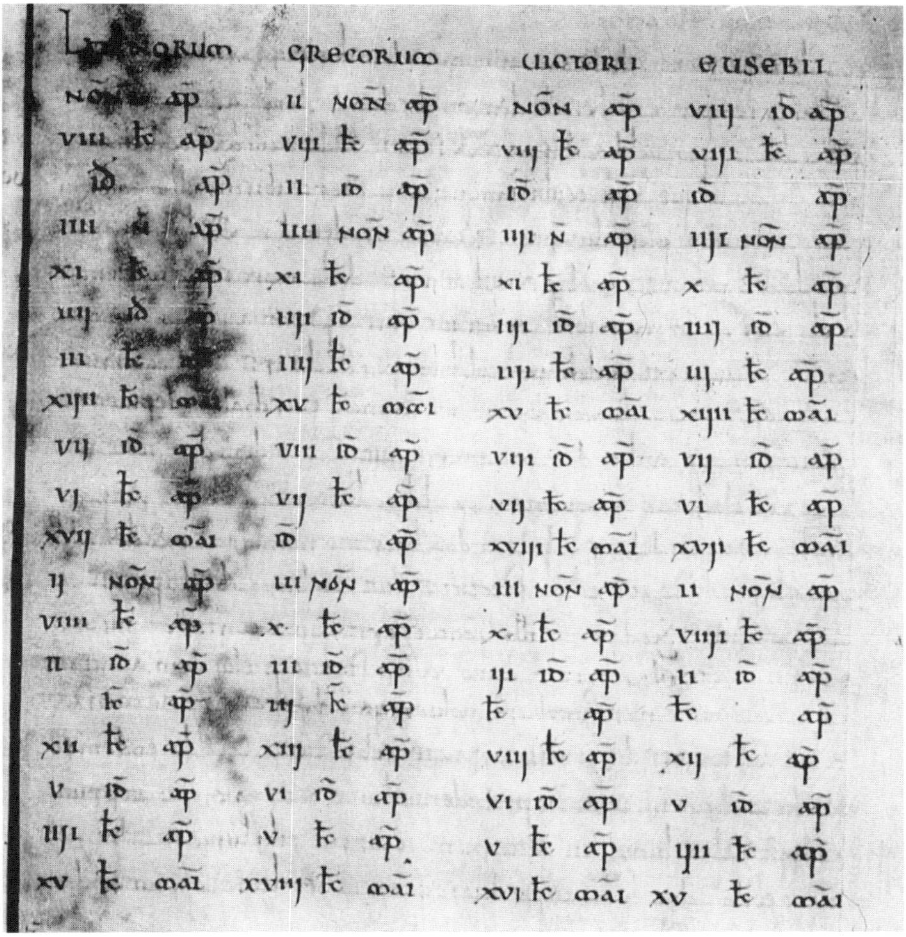

L[illegible]ORUM	GRECORUM	VICTORII	EUSEBII
NON AP	II NON AP	NON AP	VIII ID AP
VIII KL AP	VIIII KL AP	VIII KL AP	VIIII KL AP
ID AP	II ID AP	ID AP	ID AP
IIII [illegible] AP	IIII NON AP	IIII N AP	IIII NON AP
XI KL AP	XI KL AP	XI KL AP	X KL AP
IIII ID AP	IIII ID AP	IIII ID AP	IIII ID AP
III KL AP	IIII KL AP	IIII KL AP	III KL AP
XIIII KL MAI	XV KL MAI	XV KL MAI	XIIII KL MAI
VII ID AP	VIII ID AP	VIII ID AP	VII ID AP
VI KL AP	VII KL AP	VII KL AP	VI KL AP
XVII KL MAI	ID AP	XVIII KL MAI	XVIII KL MAI
II NON AP	III NON AP	III NON AP	II NON AP
VIIII KL AP	X KL AP	X KL AP	VIIII KL AP
II ID AP	III ID AP	IIII ID AP	II ID AP
KL AP	IIII KL AP	KL AP	KL AP
XII KL AP	XIII KL AP	VIII KL AP	XII KL AP
V ID AP	VI ID AP	VI ID AP	V ID AP
IIII KL AP	V KL AP	V KL AP	IIII KL AP
XV KL MAI	XVIII KL MAI	XVI KL MAI	XV KL MAI

Berlin, Staatsbibliothek, Phillipps 1831, fol. 128r.

The dates of the Dionysiac full moons in column a are recorded without mistake; as for the Victorian Easter full moons of column c, note the interesting correction in the date of the 16th Easter full moon by the compiler of this table, who records *VIII Kalendas Aprilis* instead of Victorius' *XIII Kalendas Aprilis*; this is not a scribal error, but a deliberate correction by which the computist tried to obscure the fact that the Victorian Easter full moon of this year fell before the vernal equinox of *XII Kalendas Aprilis*. The Munich computist had already referred to this year of the Victorian 19-year cycle as 'having no Easter', precisely because of this uncanonical Easter full moon (see MC 52.49–60). This deliberate corruption then found its way into the copy of this table in *Lib. comp.* 2.22 (Borst, *Schriften*, 1172–5), where *all* MSS read *VIII Kalendas Aprilis* instead of *XIII Kalendas Aprilis*.

APPENDIX 4 (MC 58.88–94): THE RELATION BETWEEN THE LUNAR AGE OF THE *INITIUM QUADRAGESIMAE* AND THAT OF EASTER SUNDAY

In MC, as in the other two Irish computistical textbooks, the relation between the lunar age of the *initium quadragesimae* and that of Easter Sunday features prominently. In order to illustrate the argument, a table correlating the corresponding lunar ages is outlined in both *CE* and *DRC*. In the transmission of MC, however, this table was so heavily corrupted that it even lost its table format; apparently, at one of the various copying stages this table was transcribed column by column (turning columns into lines) rather than faithfully reproduced in column format:

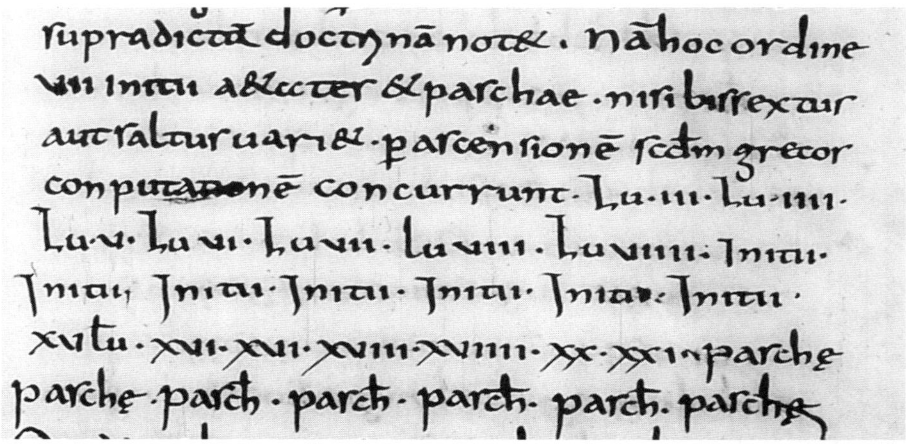

Munich, Bayerische Staatsbibliothek, Clm 14456, fol. 35r.

The original format of this table was quite certainly very similar to the corresponding table in *CE*, which was MC's source here:

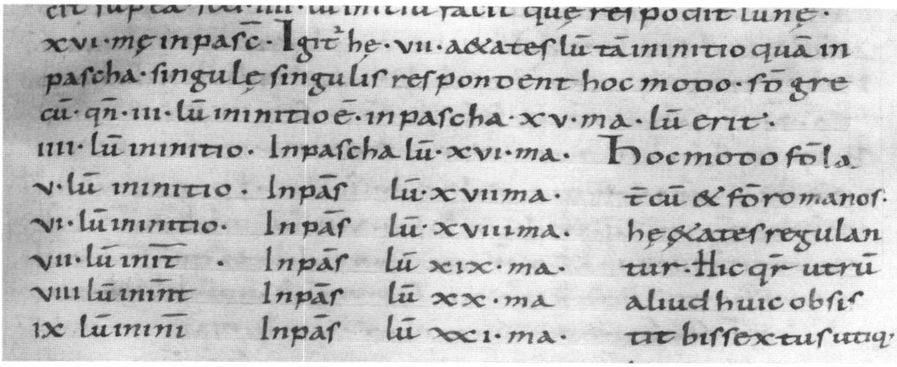
... que rei podit lune·
xvi·me inpasc· Igit he·vii·aetates lu tā ininitio quā in
pascha· singule singulis respondent hoc modo· sō gre
cū· qn·iii·lū ininitio ē·in pascha·xv·ma·lū erit·
iiii·lū ininitio· Inpascha lū·xvi·ma· Hoc modo solā
v·lū ininitio· Inpās lū·xvii ma· tcū & sō romanos·
vi·lū ininitio· Inpās lū·xviii ma· he aetates regulan
vii·lū init · Inpās lū xix·ma· tur· Hic qr utrū
viii lū inint Inpās lū xx·ma aliud huic obsis
ix lū inini Inpās lū xxi·ma· tit bissextus utq;

Einsiedeln, Stiftsbibliothek, 321 (647), p. 119.

Most remarkable is the treatment of this relation in the principal MS of the *Reichskalender*, where an algorithm and a table is given for both perspectives of the question, i.e. a) the calculation of the lunar age of the *initium* from the lunar age of Easter Sunday (by a −12 algorithm) and b) the calculation of the lunar age of Easter Sunday from the lunar age of the *initium* (by a +12 algorithm). The equivalent to the original table of MC is b, the one on the right in the facsimile:

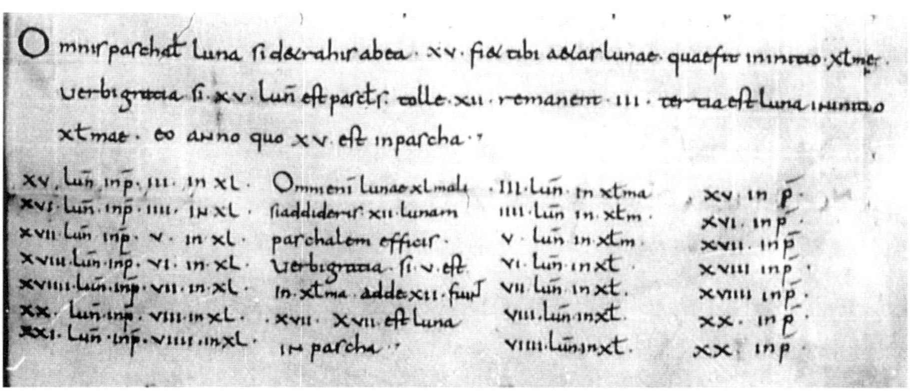
Omnis paschalis luna si detrahis ab ea ·xv· fiat tibi aetas lunae quae fit in initio xlme·
verbi gratia si ·xv· lun est pascha tolle ·xii· remanent ·iii· tertia est luna in initio
xlmae· eo anno quo xv est in pascha·

xv lun inp· iii· in xl·	Omnem lunae xlmalis	iii·lun in xtma	xv in p
xvi lun inp· iiii· in xl	si addideris xii lunam	iiii·lun in xtm	xvi inp
xvii lun inp· v· in xl	paschalem efficis·	v· lun in xtm	xvii inp
xviii lun inp· vi· in xl	verbi gratia si v est	vi lun in xt	xviii inp
xviiii lun inp· vii in xl	in xtma adde xii fuit	vii lun in xt	xviiii inp
xx lun inp· viii in xl	xvii· xvii est luna	viii lun in xt	xx inp
xxi lun inp viiii in xl	in pascha	viiii lun in xt	xx inp

Kal. A1 (Berlin, Staatsbibliothek, Phillipps 1869, fol. 2r).

Finally, note that Marianus Scotus, in the 11th century, even extended this correlation, listing the corresponding lunar ages of the the *dominicus septuagesimae*, of the *initium quadragesimae*, of the Sunday of Mid-Lent, of Maundy Thursday, of Easter Sunday, of Low Sunday, of Rogation Sunday, of holy Thursday, of Whit-Sunday, and of the sixth Sunday after Whit-Sunday (Vatican, Biblioteca Apostolica, Pal. Lat. 830, fol. 2v–3r).

APPENDIX 5 (MC 59.28–35): THE SEQUENCE OF COMMON AND EMBOLISMIC YEARS

The parallels between MC and its main source, *CE*, are not only of a textual, but also of a graphical nature. One example of graphical correspondence between these two textbooks is the sequence of common and embolismic years of the Dionysiac 19-year cycle, which is integrated in both texts in column format rather than pressing it into the line format of the rest of the text (as it was commonly done in other computistical texts).

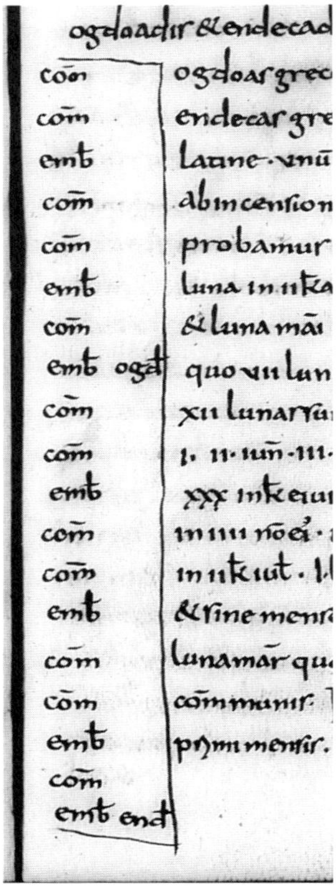

Munich, Bayerische Staatsbibliothek, Clm 14456, fol. 36r.

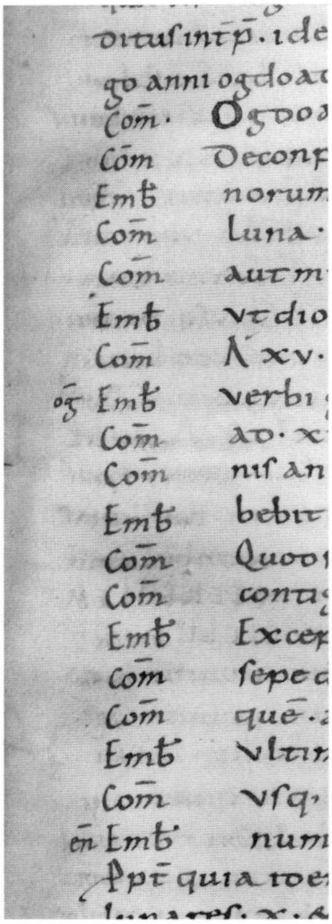

Einsiedeln, Stiftsbibliothek, 321 (647), p. 113.

APPENDIX 6 (MC 59.38–68): *DE CONTRARI AETATIBUS VII CYCLI LUNARIS ET DE DUABUS LUNATIONIBUS IN UNO MENSE ET DE LUNA ABORTIVA*

The following two texts deal with the phenomenon of a lunation falling between the Calends of two months, termed *luna abortiva*; since a lunation is usually called after the Julian calendar month in which it ends, the concept of a *luna abortiva* is introduced to avoid the confusion of two lunations being attributed to the same Julian calendar month; for the historical context of these texts cf. p. LXX–LXXI, CLXV–CLXVI of the introduction and MC 59.38–69. They are principally based on two chapters of *De ratione conputandi* (c. 73, 70), the earliest text to describe this concept in detail. Yet, especially the first text adds considerably to its Irish exemplar (with direct quotations from *DRC* being marked in italics), adapting the theory to ninth-century Carolingian computistics; these two texts are here edited (for the first time) to illustrate the transmission and adaption of this Irish concept in ninth-century continental schools (Fleury, in this case).

The three manuscript witnesses show a direct line of transmission: This ninth-century recension of the two chapters of *De ratione conputandi* originated in Fleury in the mid-ninth century (Paris, Bibliothèque Nationale, Lat. 5543, fol. 123v–124v – F); it was copied from a MS closely related to (not identical with, as this edition here demonstrates) F in Limoges some 80 years later (Paris, Bibliothèque Nationale, Lat. 5239, fol. 108r–109r – L); L's copy then was the basis for another copy in the Limoges region around AD 1000 (Strasbourg, Bibliothèque Nationale et Universitaire, 326, fol. 149r–150r – S; for the relation of these three MSS cf. Borst, *Schriften*, 269–71, 294–5). Hence, F serves as the *Leithandschrift* here, and the line of transmission can easily be checked in the *app. crit.*

DE CONTRARI ETATIBUS VII CYCLI LUNARIS

Hanc autem regulam et cursum epactarum species cycli decennouenalis *luna XIIII primi mensis et abortiua luna*, quam quidam embolismam uocant, ualde perturbant uel commutant. *Veniamus nunc ad principium cycli, ut uideamus in quibus annis contraria regula* con*uenitur, ita etiam exploratur*:

In anno quando XXXs epactas, tunc Kalendis Septembris et Octobris IIII est luna, quod Aegyptii V computant causa saltus illorum, quod in II Kalendas

---

1 DE…LUNARIS] No heading visible on the microfilm of *S.* | VII] septem *L.* **2** epactarum] aepactarum *L, S.* **3** primi] prini *L.* **4** quidam] quidem *F.* | embolismam] embolismum *L, S.* perturbant] conturbant *L, S.* **5** principium] principatum *L, S.* **7** exploratur] id est *add. F.* **8** anno] quoque *add. L, S.* | epactas] aepactas *L, S.* **9** V] Vam *F.* **10** computant] conputant *S.* | saltus] saltuis *L.* | II] pridie *L, S.*

Augusti deputantes, huic diei lunam I et II, tertiam uero in Kalendis Augusti; Kalendis Nouembris et Decembris luna VI, quem ipsi VII computant; II Kalendas Ianuarii saltus Latinorum deputantes, huic diei lunam VII et VIII, *in Kalendis Ianuarii* VIIII, quem Greci *VIII* dicunt, quia *saltu*m lunae *in XII Kalendas Aprilis* ponunt, computantes huic diei XXVIII et XXVIIII, et XXX in XI Kalendas Aprilis, ubi sedes est epactarum. Et haec est contraria regula huius anni, quam saltus efficitur.

In anno quoque, in quo fit XX luna in Kalendis Ianuarii per epactam XI, *abortiua luna*, quam quidam embolismam uocant, *XXX in Decembrio inuenitur*; inchoatur I in IIII Nonas Decembris et finitur in II Kalendas Ianuarii. *Et* nihil *contrariam regulam* huius *ann*i *efficit*.

In anno quoque, in quo fit XXIII in Kalendis Ianuarii per epactam XIIII, *abortiua luna* XXX *in mense Septembri inuenitur*, quam quidam embolismam primam et II ocdoadis uocant; inchoatur luna I in IIII Nonas Septembris et finitur XXX in Kalendis Octobris. Et deinceps nihil perturbat. Sed si qui aliqui aiunt *ualde perturbat*ionem facit, quia mense Septembrio abortiua *XXVIIII* computant et *geminatione*m *Septemb*rii *ab Octobrio separa*nt ita: *XXX enim lunam in Kalendis Septembris et prima*m *in Kalendis Octobres* esse dicunt, *et XXX in Octobri*o *cum semper* sit *XXVIIII, XXVIIII lunam Nouembri*o, *et econtraria XXX Decembri*o esse, quod ita conuenit.

In anno quoque, in quo fit XXVI luna in Kalendis Ianuarii per epactam XVII, *XXVII in Kalendis Febroarii et Aprilis et Maii inuenitur horum trium mensium. Et in hoc anno XXX luna fit Aprili*o, *nisi enim XXX fit, non erit XIIII luna primi mensis in XIIII Kalendas Maii, ut angelus ordinauit*. Quidam hanc lunationem embolismam IIIam ogdoadis esse dicunt, incipiens in II Nonas Martii et finitur XXX luna in II Nonas Aprilis. Haec lunatio contra regulam

11 diei] die *F.* | tertiam] terciam *L, S.* **13** II] pridie *L, S.* **14** Latinorum] latronum *L, S.* **16** lunae] lune *L, S.* **17** XXVIII] XXVIIII *S.* | et[1]] *om. L, S.* | XXVIIII] XXVIII *S.* | XXX] XX *L, S.* **18** epactarum] aepactarum *L, S.* | haec] hęc *S.* **21** epactam] aepactam *L, S.* | XI] XX *L, S.* | abortiua] auortiua *F.* **22** embolismam] embolisma *L, S.* | in] *om. L, S.* **23** inchoatur] incoatur *L, S.* | II] pridie *L, S.* **26** epactam] aepactam *L, S.* | abortiua] auortiua *F.* | Septembri] Febroario *L*, Feb-rio *S*, Febr-o *F.* **27** quam] quo *L, S.* | quidam] quidem *L, S.* **28** ocdoadis] octoadis *L, S.* | inchoatur] incoatur *L, S.* Nonas] *om. L, S.* **29** Octobris] Octub- *L, S.* **30** si] *om. L, S.* | aliqui] aliquid *S.* | aiunt] agunt *L, S.* ualde] uel de *F.* **31** XXVIIII] XXXIIII *S.* **32–34** separant…Octobrio] *om. L, S.* **33** lunam] luna *F.* **35** lunam] lun- *F*, luna *L, S.* **36** Decembrio] Decebrio *L, S.* **37** luna] *om. L, S.* **38** epactam] aepactam *L*, ępactam *S.* | Febroarii] Sept-bris *L, S*, Sepb- *F.* **41** Maii] Ian- *L, S.* **43** ogdoadis] hocdoadis *L, S.* **44** lunatio] lunacio *S.*

epactarum facit *XX uero nonam Maio, et econtrario XXX Iunio, et XXVIIII Iulio. Et in hoc anno inuenitur transilitio inter Maio et Iulio, eadem aetas in Kalendis Iunii et Iulii, id est XXVIIII, et XIIII luna primi mensis efficit omnem perturbationem hanc. In hoc anno abortiua luna XXX in Iulio inuenitur, et ab Augusto usque in finem anni contraria regula non inuenitur.*

In anno quoque, in quo fit XXVIIII luna in Kalendis Ianuarii per epactam XX, *abortiua luna XXVIIII in Ianuario inuenitur*. Et contra epactam I in Kalendis Februarii *licet per artificalitatem Febroarii deputatur. Et XXX luna fit Aprilio*, quam quidam lunam embolismam primam endecadis uocant; inchoatur I in V Nonas Martii et finitur XXX in Kalendis Aprilis *causa XIIII luna primi mensis, quae in hoc anno in XVII Kalendas Maii inuenitur. Et* iterum *abortiua luna XXVIIII in Aprilio inuenitur. Et eadem aetas fit hoc anno Febroario et Maio, id est prima. Et in hoc anno separatur geminatio Aprili a Febroario, et haec est contraria regula huius anni, quae abortiua luna et XIIII luna primi mensis efficiunt. Nihil*que *perturbationis in hoc anno a Maio usque in finem inuenitur.*

In anno quoque, in quo fit XXI luna in Kalendis Ianuarii per epactam XII, *abortiua luna* XXX *in Nouembrio inuenitur*, quam quidam lunam embolismam secundam endecadis uocant, quae ita inchoat: I in IIII Nonas Nouembris, et finitur XXX in Kalendis Decembris, quamuis alii dicunt ipsam lunationem abortiuam esse *XXVIIII*, ut *XXX fit in Kalendis Nouembris* et I *in Kalendis Decembris* et ascendere usque in *XXX*.

In anno quoque, in quo fit XXIIII luna in Kalendis Ianuarii per epactam XV, *abortiua luna XXX in Augusto inuenitur*, quam quidam embolismam III endecadis dicunt; inchoatur I in IIII Nonas Augusti et finitur XXX in II Kalendas Septembris; *II in Kalendis Octobris*, *III in*

---

45 epactarum] aepactarum *L*, *S*. **49** abortiua] auortiua *F*. **53** epactam] aepactam *L*, *S*. **54** epactam] ępactam *S*. **57** inchoatur] incoatur *L*, *S*. **59** quae] que *L*, quę *S*. | XVII] VII X *L*. **61** prima] I *L*, *S*. **62** haec] hęc *S*. **63** quae] que *L*, *S*. | abortiua] auortiua *F*. | XIIII] XVII *L*, *S*. **65** in] ad *F*. **66** XXI] XX *F*. | in[2]] *om. L*, *S*. **67** epactam] epacta *F*, aepactam *L*, *S*. | abortiua luna] luna auortiua *L*, *S*. | luna] *add. F*. **67–68** in…inuenitur] inuenitur *(half erased)* in Nouembrio *F*. **69** quae] que *L*, *S*. | inchoat] inquoat *L*, *S*. | I] prima *L*, *S*. | IIII] IIIIor *L*, *S*. **71** abortiuam] auortiuam *L*, *S*. **72** ut] u *L*, II *S*. **73** in] *om. F*. **74** XXIIII] XXVII *L*, *S*. | luna] *om. S*. **75** epactam] aepactam *L*, *S*. | abortiua] auortiua *L*, *S*. **76** dicunt] dicant *F*. **77** inchoatur] inquoatur *L*, *S*. | IIII] IIIIor *L*, *S*. **78** Septembris] Decembris *(first syllable erased) L*. | II] IIa *F*. | Kalendis Octobris] Octb-ro *L*, *S*. | III] IIIa *F*.

Kalendis Nouembris, IIII *in Kalendis Decembris*, V *in Kalendis Ianuarii*, quod non ita conuenit.

In anno quoque, in quo fit XXVII *luna in Kalendis Ianuarii* per epactam XVIII, *illo anno* XXX *luna Aprili* erit; *fit causa* XIIII *luna primi mensis*, *quae in hoc anno in* XV *Kalendas Maii inuenitur*, quam quidam lunam embolismam nouissimam endecadis uocant; inchoatur I in III Nonas Martii et finitur XXX in III Nonas Aprilis; XX uero VIII *luna in Kalendis Maii*, quod per epactam XXVIIII esse debuit. Sic erit mense *Maio luna*tio XXVIIII *et Kalendis Iunii* XXX *et abortiua luna* XXVIIII *in* ipso mense *Iunio* erit. Aliam quoque perturbationem *ab Iulio usque in finem* in principio cycli posuimus. *Et deinde manifestum est nobis esse* VII *anni contrariae regulae in circulo* XVIIII *annorum per saltum et* XIIII *lunam primi mensis et abortiuam lunam.*

DE DUABUS LUNATIONIBUS IN UNO MENSE ET DE LUNA ABORTIVA

*Cum duae luna*tiones *in uno mense inueniuntur, quae est de illis lunationibus ipsius mensis*? *Quae finitur in mense, ipsa est luna*tio *ipsius mensis*. *Luna uero sequen*tis, *non pertingerit Kalendas mensis sequentis, abortiua luna uocabitur, quam alii dicunt luna esse embolism*a, *quod non est uerum, ut Palum*pbus: *Illi*us *mensi com*puta *lunam, quae in eo finitur, non quae incipit. Unde, quae primum prioris aut initium sequentis non tenet, abortiua dicitur.*

---

79 IIII] IIIIa *F.* | V] Va *F.* **80** conuenit] conueniat *L, S.* **82** epactam] aepactam *L, S.* | XVIII] XVIIIam *F.* **83** quae] que *L, S.* **85** inchoatur] in *L, S.* | I] Ia *F*, prima *L, S.* **86–87** XX…VIII] XXVIII *L, S.* **87** epactam] aepactam *L*, ępactam *S.* **90** Iunio] Iun- *F*, Iunii *L, S.* **92** contrariae] contrarię *S.* | regulae] regule *L*, regulę *S.* **93** annorum] decem et nouem *L, S.* **94** abortiuam lunam] abortiua luna *L, S.* **1–2** DE…ABORTIVA] No heading visible on the microfilm of *S.* **3** duae] due *L, S.* | quae] que *L, S.* **4** Quae] que *L* quę *S.* **8** Palumpbus] Palumbus *S.* | Illius] ipsius *F.* **9** quae[1]] qui *L, S.* | quae[2]] que *S.* quae[3]] que *S.* **10** initium] inicium *S.*

APPENDIX 7 (MC 67): PROOF OF A 95-YEAR PERIOD NOT BEING CYCLIC IN RESPECT OF ITS SOLAR DATA

In chapter 67, the Munich computist copied, from the now lost Victorian computus of AD 689 (for this text cf. especially p. CXXIV–CXXVI above), a proof of the fact that a 95-year period is not cyclic according to its solar data. The method applied is a comparison of the *concurrentes* (i.e. the weekday of 24 March) of successive 95-year periods with the original Dionysiac Easter table (covering the 95 years from AD 532 to 626), which led to the following results: If any given block of four years (the interval between two bissextile days) of the second 95-year period is compared to the first, the *concurrentes* agree in three, but disagree in one year; if the same comparison is made between the third and the first 95-year periods, the *concurrentes* agree in two, and disagree also in two years, while they disagree in three and agree in only one year between the fourth and the first period; if, then, the subsequent 95-year periods are compared to the first 95-year period (i.e. the one originally outlined by Dionysius), the *concurrentes* do not agree in any year. The way in which this proof is described in MC suggests that the original author of this passage work from a table comparing the *concurrentes* of five successive 95-year periods, starting with the Dionysiac one covering the years AD 532–626; for a proper understanding of this proof and as an illustration of the argument, such a table is provided in the following:

1st period			2nd period			3rd period			4th period			5th period			6th period		
by	AD	cc	by	AD	cc	by	AD	cc	by	AD	cc	by	AD	cc	by	AD	cc
B	532	4		627	3		722	3		817	3	B	912	3		1007	2
	533	5	B	628	5		723	4		818	4		913	4	B	1008	4
	534	6		629	6	B	724	6		819	5		914	5		1009	5
	535	7		630	7		725	7	B	820	7		915	6		1010	6
B	536	2		631	1		726	1		821	1	B	916	1		1011	7
	537	3	B	632	3		727	2		822	2		917	2	B	1012	2
	538	4		633	4	B	728	4		823	3		918	3			
	539	5		634	5		729	5	B	824	5		919	4			
B	540	7		635	6		730	6		825	6	B	920	6			
	541	1	B	636	1		731	7		826	7		921	7			
	542	2		637	2	B	732	2		827	1		922	1			
	543	3		638	3		733	3	B	828	3		923	2			
B	544	5		639	4		734	4		829	4	B	924	4			
	545	6	B	640	6		735	5		830	5		925	5			
	546	7		641	7	B	736	7		831	6		926	6			
	547	1		642	1		737	1	B	832	1		927	7			
B	548	3		643	2		738	2		833	2	B	928	2			
	549	4	B	644	4		739	3		834	3		929	3			
	550	5		645	5	B	740	5		835	4		930	4			
	551	6		646	6		741	6	B	836	6		931	5			
B	552	1		647	7		742	7		837	7	B	932	7			
	553	2	B	648	2		743	1		838	1		933	1			
	554	3		649	3	B	744	3		839	2		934	2			
	555	4		650	4		745	4	B	840	4		935	3			
B	556	6		651	5		746	5		841	5	B	936	5			
	557	7	B	652	7		747	6		842	6		937	6			
	558	1		653	1	B	748	1		843	7		938	7			
	559	2		654	2		749	2	B	844	2		939	1			
B	560	4		655	3		750	3		845	3	B	940	3			
	561	5	B	656	5		751	4		846	4		941	4			
	562	6		657	6	B	752	6		847	5		942	5			

	563	7		658	7		753	7	B	848	7		943	6			
B	564	2		659	1		754	1		849	1	B	944	1			
	565	3	B	660	3		755	2		850	2		945	2			
	566	4		661	4	B	756	4		851	3		946	3			
	567	5		662	5		757	5	B	852	5		947	4			
B	568	7		663	6		758	6		853	6	B	948	6			
	569	1	B	664	1		759	7		854	7		949	7			
	570	2		665	2	B	760	2		855	1		950	1			
	571	3		666	3		761	3	B	856	3		951	2			
B	572	5		667	4		762	4		857	4	B	952	4			
	573	6	B	668	6		763	5		858	5		953	5			
	574	7		669	7	B	764	7		859	6		954	6			
	575	1		670	1		765	1	B	860	1		955	7			
B	576	3		671	2		766	2		861	2	B	956	2			
	577	4	B	672	4		767	3		862	3		957	3			
	578	5		673	5	B	768	5		863	4		958	4			
	579	6		674	6		769	6	B	864	6		959	5			
B	580	1		675	7		770	7		865	7	B	960	7			
	581	2	B	676	2		771	1		866	1		961	1			
	582	3		677	3	B	772	3		867	2		962	2			
	583	4		678	4		773	4	B	868	4		963	3			
B	584	6		679	5		774	5		869	5	B	964	5			
	585	7	B	680	7		775	6		870	6		965	6			
	586	1		681	1	B	776	1		871	7		966	7			
	587	2		682	2		777	2	B	872	2		967	1			
B	588	4		683	3		778	3		873	3	B	968	3			
	589	5	B	684	5		779	4		874	4		969	4			
	590	6		685	6	B	780	6		875	5		970	5			
	591	7		686	7		781	7	B	876	7		971	6			
B	592	2		687	1		782	1		877	1	B	972	1			
	593	3	B	688	3		783	2		878	2		973	2			
	594	4		689	4	B	784	4		879	3		974	3			
	595	5		690	5		785	5	B	880	5		975	4			
B	596	7		691	6		786	6		881	6	B	976	6			
	597	1	B	692	1		787	7		882	7		977	7			
	598	2		693	2	B	788	2		883	1		978	1			
	599	3		694	3		789	3	B	884	3		979	2			
B	600	5		695	4		790	4		885	4	B	980	4			
	601	6	B	696	6		791	5		886	5		981	5			
	602	7		697	7	B	792	7		887	6		982	6			
	603	1		698	1		793	1	B	888	1		983	7			
B	604	3		699	2		794	2		889	2	B	984	2			
	605	4	B	700	4		795	3		890	3		985	3			
	606	5		701	5	B	796	5		891	4		986	4			
	607	6		702	6		797	6	B	892	6		987	5			
B	608	1		703	7		798	7		893	7	B	988	7			
	609	2	B	704	2		799	1		894	1		989	1			
	610	3		705	3	B	800	3		895	2		990	2			
	611	4		706	4		801	4	B	896	4		991	3			
B	612	6		707	5		802	5		897	5	B	992	5			
	613	7	B	708	7		803	6		898	6		993	6			
	614	1		709	1	B	804	1		899	7		994	7			
	615	2		710	2		805	2	B	900	2		995	1			
B	616	4		711	3		806	3		901	3	B	996	3			
	617	5	B	712	5		807	4		902	4		997	4			
	618	6		713	6	B	808	6		903	5		998	5			
	619	7		714	7		809	7	B	904	7		999	6			
B	620	2		715	1		810	1		905	1	B	1000	1			
	621	3	B	716	3		811	2		906	2		1001	2			
	622	4		717	4	B	812	4		907	3		1002	3			
	623	5		718	5		813	5	B	908	5		1003	4			
B	624	7		719	6		814	6		909	6	B	1004	6			
	625	1	B	720	1		815	7		910	7		1005	7			
	626	2		721	2	B	816	2		911	1		1006	1			

Comparison of the concurrentes *of five successive 95-year periods, starting with the period originally outlined by Dionysius, AD 532–626 (by = indication of bissextile years; AD =* annus domini*; cc =* concurrentes*)*

APPENDIX 8: EASTER TABLE AD 798–854

In three Vatican MSS (Vatican, Biblioteca Apostolica, Reg. Lat. 1260, fol. 112r–114r; Vatican, Biblioteca Apostolica, Pal. Lat. 1447, fol. 19v–22r; Vatican, Biblioteca Apostolica, Pal. Lat. 1448, fol. 13v–16r) can be found an Easter table for the years AD 798–854. As Dionysiac Easter tables were generally structured in 19-year sections, being model on Dionysius' own table which recorded five 19-year intervals (95 years), it should be presumed that this Easter table, consisting of three 19-year intervals, was compiled after the beginning and before the end of the first of these three intervals, i.e. between 798 and 817. The earliest MS witness (Vatican, Biblioteca Apostolica, Pal. Lat. 1448) dates from 810 and originated in Trier, a date and provenance that may therefore also apply to this Easter table, which, in all likelihood, should be connected to the intellectual milieu that produced *Lib. comp.* (for which see Borst, *Schriften*, 1054–65). This Easter table is the most comprehensive Easter table of its time, with its colums recording: 1) the AD date; 2) the weekday (*feria*) of 1 January; 3) the epact (lunar age) of 1 January; 4) the indiction; 5) the Julian calendar date of the *terminus quadragesimae* (i.e. *luna* 2 of the lunation preceding the Easter lunation); 6) the Julian calendar date of the *initium quadragesimae* (the beginning of the Lenten fast); 7) the lunar age of that date; 8) the epact (lunar age) of 22 March; 9) the *concurrentes* (the weekday of 24 March); 10) the year in the *cyclus lunaris*; 11) the Julian calendar date of the Easter full moon; 12) the weekday of that date; 13) the Julian calendar date of Easter Sunday; 14) the lunar age of that date.

The Munich Computus (at least in its original form), as most other computistical textbooks, was not accompanied by an Easter table, but since the table in question here is frequently referred to throughout the book (and since a modern account on Easter calculation seems incomplete without the reproduction of an Easter table), a facsimile of the first 19-year interval of this Easter table of AD 798–854 is reproduced here from Vatican, Biblioteca Apostolica, Pal. Lat. 1447, fol. 19v–20r.

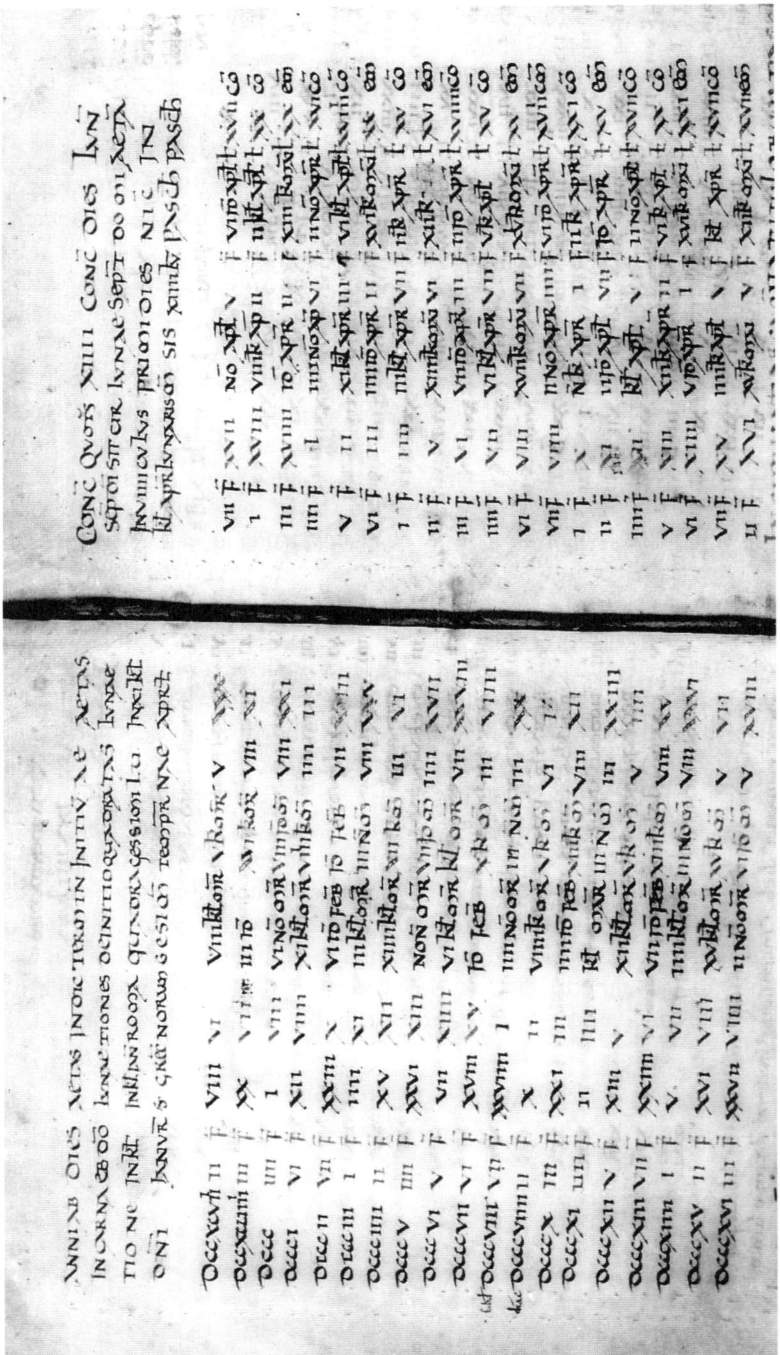

Vatican, Biblioteca Apostolica, Pal. Lat. 1447, fol. 19v–20r.

BASIC COMPUTISTICAL GLOSSARY

As in every exact science, it was and is (in medieval and modern times) paramount to use generally accepted and well-defined terminology. For this reason, many recent publications on the computistics provide brief and concise glossaries, a development that is to be welcomed by both the novice and the professional in the field. The novice, however, often finds it difficult to follow clear-cut definitions without further explanations. And since even professionals sometimes define certain terms differently, some justification is needed for the application of these terms as well as some explanation for the meanings assigned to them. The following, therefore, is designed to define and explain, in its early medieval context, the technical terminology used in the present book. '→' highlights terms found elsewhere in the glossary; since many terms are best explained in a wider context, only the generic terms are listed alphabetically.

bissextile day (*bissextus*): A bissextile day can be defined as the sum of four annual quarter-days, each of which exceeding the integer number of days of a → solar year (as reckoned by medieval computists). It was one of the main achievements of seventh-century computistics, then, to realize that the solar bissextile day had a lunar equivalent.

- **solar:** A → solar year, as calculated by antique astronomers and then applied by medieval computists, consisted of 365¼ days. In the → Julian calendar, the annual quarter-day (*quadrans*) was not applied in the year in which it occurred, but rather four such quarter-days were summed up to constitute a full day, which was then introduced in four-year intervals. Accordingly, three out of four Julian calendar years had 365 days, the fourth, however, 366 days. Though this extra day every four years was, by the eighth century, generally accepted to be introduced as the doubling of the sixth calends of March (therefore the name, *bis sextus*; in modern notation, 24 February was doubled; which of these two 24 Februaries was to be considered as the intercalated days was, however, disputed), other possible dates for the bissextile day had been discussed in the seventh century (2 March, 21 March).
- **lunar:** One of the central computistical questions of the seventh century was whether the bissextile day was a solar phenomenon only, or whether it had also to be applied to a → lunar year. The answer depended on the → lunar cycle followed by these computists, since, e.g., the → 8-year lunar cycle underlying the Hippolytan Easter reckoning and transmitted through Quintus Julius Hilarianus, explicitly denied the existence of lunar bissextile days; lunar bissextile days were, however, a central feature of all → lunar cycles underlying the → Easter reckonings used in the early middle ages. In the → Dionysiac reckoning, by the late eighth century the placement of the lunar bissextile day

was generally accepted as the final day of the February lunation (turning this → hollow lunation into a → full one), while late seventh- and early eighth-century (particularly Irish) computists preferred to set it on the same day as the → solar bissextile day (24 February), though this placement regularly interfered with the → quadragesimal period of the Lenten fast.

calendar: In early medieval computistics, two types of calendars were of paramount importantance, namely the Julian and the lunar calendar.

– **Julian:** The Julian calendar, thus named because it was introduced under Julius Caesar, became the authoritative ordering of a → solar year, calculated as consisting of 365¼ days, throughout the Roman Empire and then, in the early middle ages, in the Christian regions of Western Europe. The Julian calendar year was divided into 12 months, which are the same (and of the same length) as today. The naming of the Julian calendar dates, however, varied considerably from our modern consecutive count (which was, though rarely, also applied by some early medieval computists); each month had three marker-days, the → Calends, → Nones, and → Ides, and from these the preceding days were counted backwards (e.g. the second day before the Nones, the third day before the Nones, etc.).

– **lunar:** A lunar calendar can be defined as the correlation of all Julian calendar dates with their corresponding lunar ages. Since this lunar data is different for every year of a lunar cycle, a full lunar calendar had to cover the entire period of the underlying lunar cycle. For the → Dionysiac 19-year cycle, such a calendar was composed under the auspices of Charlemagne and is printed in Borst, *Reichskalender*, 1645–1727; two manuscripts (Bern, Burgerbibliothek, 645, fol. 55r–56v; Cologne, Dombibliothek, 83^2, fol. 211v–213r, the latter printed as *Quaest. Austr.* II 11 in Borst, *Schriften*, 496–508) transmit a Victorian lunar calendar, which lists, however, only the lunar age on the Calends of each month; for the Dionysiac reckoning, tables transmitting exactly this information (i.e. the lunar ages on the Calends of each month for all 19 years) became standard from the eighth century (commonly referred to as *pagina epactarum*), though, it should be noted, these tables sometimes show considerable variations due to differing placements of the embolisms and the *saltus* (a comprehensive study of these tables still is a desideratum); nothing comparable is transmitted for any 84-year Easter reckoning. For lunar calendars of late antiquity and the early middle ages see now especially Holford-Strevens, 'Lunar calendars'.

Calends/calends (*Kalendae, Kal, Kl, K*): In the → Julian calendar, the Calends of a month referred to the first day of this month. The days preceding this day were counted backwards as 'the second day before the Calends' (*ante diem secundum Kalendas*; abbr.: *a.d. II Kalendas*), 'the third day before the Calends' (*ante diem tertium Kalendas*), etc. up to the → Ides of the preceding month. In the early middle ages, the words *ante diem* (*a.d.*) in this formula were abandoned, so that 'the second day before the Calends of January', e.g., were typically noted as *II Kl Ian* ('the second calends of January'). For sake of precision, in the present book the marker day (i.e. the first day of a month) is written with capital C (Calends), the days preceding, individually or collectively, as calends; note that early medieval computists, when referring to all calends occurring within a specific Julian calendar month, also termed them the calends of that month, even though all except the Calends technically constituted the calends of the following month (e.g., there were 19 calends within the Julian calendar month of January, the Calends proper and the final 18 days of that month, i.e. the calends of February); the context, however, makes the use unambiguous; the number of calends per month varied between 16, 17, 18, and 19, depending on the length of the month and the placement of the → Ides within the respective month.

***concurrentes*:** The *concurrentes* can be defined as the weekday (*feria*) of a fixed Julian calendar date, and in the early middle ages it was traditionally understood to refer to 24 March. Since the weekday data is dependent on the 28-year → solar cycle, the *concurrentes* recur in exactly the same order every 28 years. Note that the term itself was introduced only with the → Dionysiac reckoning.

- **placement of (*sedes concurrentium*):** In the → Easter tables of the → *latercus* and the → Victorian reckoning, the weekday data was noted for 1 January; in the → Easter tables of → Dionysiac reckoning, however, it was rather noted for 24 March. As in the case of the placement of the → epacts, converts from the older two reckoning to the Dionysiac translated the 24 March data to 1 January, so that both dates had currency from the eighth century onwards, though the term *concurrentes* was almost exclusively used for the weekday of 24 March.

cycle: Cycles were of primary importance to early medieval computists as they made full use of recurring data; in fact, particularly lunar and Easter cycles were explicitly constructed to meet this premise.

- **19-year (*cyclus decemnovenalis*):** The → 19-year lunar cycle was the basis of computistics throughout the middle ages. In the → Victorian reckoning, the years of the underlying → 19-year lunar cycle were defined by their characteristic → epact of 1 January. Easier still for the discussion of a → 19-year lunar cycle was to set these 19-years in a specific order. This was done in the → Dionysiac reckoning, with this specific order being termed *cyclus decemnovenalis*.
    - **Dionysiac:** The order set for the → 19-year lunar cycle underlying the → Dionysiac reckoning began with the year of → epact 30 on 22 March (or 8 or 9 on 1 January, depending on the placement of the → *saltus*), so that this first year was equivalent to BC 1 and every 19th year thereafter (i.e. AD 19, 38, etc.). This order starting with the mentioned year was termed *cyclus decemnovenalis* (19-year cycle). Originally, as outlined by Dionysius himself in his letter to Boniface and Bonus (Krusch, *Studien*

II, 85–6), a year of the *cyclus decemnovenalis* started on the day following the Easter full moon (*luna* 15) and ended on the Easter full moon of the following year; the epacts defining these years (either of 1 January or especially of 22 March) occurred, accordingly, towards the end of these years. In this structure, the → *saltus* was introduced in the first year of the Dionysiac 19-year cycle, the → embolisms in the 3rd, 6th, 8th, 11th, 14th, 17th, and 19th year. Note, however, that Dionysius nowhere mentioned the exact placements of either the → *saltus* or the → embolisms, so that those had to be reconstructed by seventh-century computists, which led to differing customs. Many computists (medieval and modern) then preferred to think in terms of Julian calendar years starting with 1 January (and therefore setting the beginning of a year of the *cyclus decemnovenalis* more than eight months later than originally envisaged), which obviously led to a change of the general structure (e.g., the *saltus* may then have been introduced in the final rather than the first year of the cycle, if it was placed at the end of the November lunation)

- **Victorian:** In the course of the seventh and eighth centuries, adherents of the → Victorian reckoning were faced with the introduction of the → Dionysiac one. In order to be able to compare the two systems, the terminology found in the newly introduced Dionysiac system was translated to the Victorian system also, so that the order of years of the Victorian 19-year cycle is the same as that of the Dionysiac one (i.e. the first year of the Dionysiac 19-year cycle corresponds to the first year of the Victorian one). Note, however, that the structure (placement of the → *saltus* and of the → embolisms) is completely different, with the → *saltus*, e.g., being placed in the seventh rather than the first year – or the sixth rather than the final year, if the January beginning of a year of the 19-year cycle is preferred (as it usually is in connection to the → Victorian reckoning, since 1 January was the natural beginning of a year in that reckoning) –, resulting in differing sets of → epacts and with this obviously also in differing → lunar calendars.

– **Easter:** As early medieval computists had difficulties in making a distinction between an Easter cycle and an → Easter table, both terms are briefly defined in this glossary, though trivial as they are. An Easter cycle is a → luni-solar cycle, so that both the same weekday and lunar data recur on any given Julian calendar date after a specific number of years, the length of the cycle. Accordingly, the → *latercus* is an 84-year Easter cycle, the → Victorian and the → Dionysiac reckonings, on the other hand, constitute 532-year Easter cycles, as both are based on a → 19-year lunar cycle, while the → solar cycle consists of 28 years (19×28=532).

– **lunar:** A lunar cycle can be defined as the period of the first recurrence of a fixed sequence of lunar ages on any given Julian calendar date. As the → lunar year was shorter than the → solar year by 11 or 11¼ days (depending on whether a → bissextile day was believed to constitute a part of the lunar year or not), lunar cycles of various lengths could be constructed, primarily differing in the applications of the → embolisms and the → *saltus*. In the early middle ages, the two principal lunar cycles consisted of 19 and 84 years respectively, with the → 19-year lunar cycle underlying both the → Victorian and → Dionysiac reckoning, while the → 84-year lunar (and → luni-solar) cycle, was the basis of the → *latercus* or 84 (14)-year → Easter reckoning. Furthermore, the basic construction of the → 8-year lunar cycle underlying the Hippolytan Easter reckoning was known to the Munich computist, Bede, and other late seventh- and early eighth-century computists. The constructions of these three cycles will be briefly outlined in the following:

- **19-year:** 19 Julian calendar years of 365¼ days consisted of 6939¾ days; 19 lunar years of 354¼ days, on the other hand, consisted of only 6730¾ days. The difference of 209 days was compensated by the application of seven → embolisms of 30 days minus the → *saltus lunae* (6730¾+7×30−1=6939¾).
- **84-year:** 84 Julian calendar years of 365¼ days consisted of 30681 days; 84 lunar years of 354¼ days consisted of only 29757 days. The difference of 924 days was compensated by the application of 31 → embolisms of 30 days minus six → *saltus lunae* (29757+31×30−6=30681).
- **8-year:** The main and essential difference between the 8-year lunar cycle and the two others described above is the fact that no bissextile day (or rather annual increment of ¼ day) was assigned to a lunar year. Consequently, eight Julian calendar years of 365¼ days consisted of 2922 days; eight lunar years of 354 days consisted of 2832 days. The difference of 90 days was compensated by three → embolisms of 30 days (2832+3×30=2922).

– **lunar (*cyclus lunaris*):** A distinction needs to be drawn between a general → lunar cycle and the specific cycle known as *cyclus lunaris* in early medieval computistical literature. The *cyclus lunaris* can be defined as a specific order of the → 19-year lunar cycle underlying the → Dionysiac reckoning, alternative to the → 19-year cycle (*cyclus decemnovenalis*). The *cyclus lunaris* started three years later than the → 19-year cycle (*cyclus decemnovenalis*), so that the first year of the *cyclus lunaris* was equivalent to the fourth year of the *cyclus decemnovenalis* (as is obvious from the → Dionysiac Easter tables). Irish computists of the late seventh, early eighth centuries were not concerned with the *cyclus lunaris*, but it became popularized with Bede's discussion of it (*DTR* 56), where the beginning of the years of the *cyclus lunaris* is set as the beginning of the January lunation (note that Bede's list of years of the *cyclus lunaris* starts a year early, at least from an epactal point of view). In order not to confuse the general → lunar cycle with the specific *cyclus lunaris*, the Latin term stands strictly for this specific, the English term strictly for the general lunar cycle throughout this book.

– **luni-solar:** In a → lunar cycle, it is only essential that the lunar data recurs on any given Julian calendar date after a certain number of years, but not necessarily the weekday data also. A luni-solar cycle, however, is constructed in a way that both the weekday and the lunar data recurs, in the same order, on any given Julian calendar date after a certain number of years, with this number of years defining the length of this luni-solar cycle. Hence, a luni-solar cycle can be defined as a combination of a lunar with the → solar cycle; e.g., an → 84-year lunar cycle constitutes a luni-solar cycle, as does every → Easter cycle, while a → 19-year lunar cycle does not.

- **solar:** A solar cycle can be defined as the period of the first recurrence of the same sequence of weekdays on any given Julian calendar date. A Julian calendar year consisted of 365 days in common, of 366 days in bissextile years so that the weekday data of any given Julian calendar date increased by one in common (365=52×7+1), by two in bissextile years (366=52×7+2). Accordingly, the weekday-cycle of seven days had to be brought in line with the four-year bissextile cycle in order to create a working solar cycle. As seven and four have no common divisor, a solar cycle consisted of 28 (7×4) years.

Easter new moon (*luna I*): The first moon of the Easter → lunation, recorded as *luna I* by medieval computists, is termed the Easter new moon; note that though some modern scholars of computistics discuss whether *luna* 30 or *luna* 1 or the time in-between was regarded as new moon, there is no ambivalence in early medieval sources, which unanimously refer to *luna* 1 as the new moon of a → lunation.

Easter full moon (*luna XIIII*; *terminus paschalis*): The Easter full moon, the 14th moon of the Easter → lunation, probably was the most important feature in medieval computistics, since the Julian calendar date and lunar age of Easter Sunday could easily be calculated once the Julian calendar date and weekday of the Easter full moon was determined; in medieval terms, the Easter full moon 'ruled' Easter Sunday, and was therefore often called *terminus paschalis*. Note that, in the → Dionysiac reckoning, a corresponding lunar date falling on the same weekday as the Easter full moon, namely *luna* 2 of the lunation preceding the Easter lunation (called → *secundum singularis* by the Munich computist, more appropriately *terminus quadragesimae* later), was introduced for the determination of the → *initium quadragesimae*, the Sunday of the beginning of the → Lenten fast.

Easter reckoning: In the early middle ages, three Easter reckonings were followed in the Latin West, of which the main features are outlined in the following.

- ***latercus* (84 (14)-year Easter reckoning):** The *latercus* was arguably invented by Sulpicius Severus, quite certainly early in the fifth century. It constitutes a → luni-solar cycle of 84 years with the six → *saltus* being placed in 14-year intervals (a characteristic feature of this reckoning only, which is accordingly termed 84 (14)-year Easter reckoning). The lunar limits for Easter Sunday were *luna* 14 to 20 inclusively, the Julian calendar ones 26 March to 23 April inclusively, while the Easter full moon could fall as early as 21

March. The only known → Easter table of this reckoning consists of a full cycle of 84 years, each year being defined and characterized by the lunar and weekday data of 1 January, as in the → Victorian reckoning (which may be due to both reckonings being invented in the same intellectual environment, that of fifth-century Aquitaine, a supposition that would further speak for Sulpicius' authorship of the *latercus*).

- **Victorian:** The Victorian reckoning was invented by Victorius of Aquitaine in AD 457 upon a request by the papal curia; despite the bad press which Victorius received by modern scholars, he surely was one of, if not the most competent computists in the Latin West of his day. His system is based on a → 19-year lunar cycle, so that a full → Easter cycle of this reckoning consisted of 532 years (though in manuscripts excerpts of this table are found more often than complete versions). Accordingly, Victorius composed an → Easter table of this number of years, each year being defined, up to Victorius' lifetime, by the names of the two Roman consuls appointed for those years and additionally (and in the years following Victorius' year of writing only) by the lunar and weekday data of 1 January, as in the *latercus*. The lunar limits for Easter Sunday were the old Roman ones, *luna* 16 to 22 inclusively, the Julian calendar limits stretched from 22 March to 24 April inclusively, while the Easter full moon was allowed to fall before the vernal equinox, as early as 20 March (in which case Easter Sunday, however, would always occur after the vernal equinox). In certain years, Victorius left the choice of the right date of Easter to the pope by noting two Julian calendar dates for Easter Sunday (ascribing one to Roman, the other to Alexandrian tradition).
- **Dionysiac:** As the → Victorian reckoning did not solve all difference between Alexandria and Rome in respect of the correct method of calculating the most important Christian feast, Dionysius Exiguus was asked by the papal curia, in AD 525, to make the Alexandrian reckoning, which obviously was composed in Greek, available to the Latin West. Accordingly, Dionysius' task was one of translation rather than calculation, and the evidence suggests that this suited his abilities perfectly, being an excellent translator, but a rather mediocre mathematician. The Alexandrian system which Dionysius translated (and since Dionysius made it, for the first time, available to the Latin West, it is here termed the Dionysiac reckoning) was based on a → 19-year lunar cycle, with lunar limits for Easter Sunday from *luna* 15 to 21 inclusively, Julian calendar limits from 22 March to 25 April inclusively, while the earliest date for the Easter full moon was the vernal equinox of 21 March. The → Easter table itself consisted (for mathematical reasons) of 95-years (though the → Easter cycle of this reckoning obviously had 532 years), each year being defined by the AD-date, which Dionysius may have first introduced to the Latin West (though this era was hardly invented nor its years calculated by him).

Easter table: An Easter table can be defined as a list consisting of, at least, the essential data for Easter Sunday, i.e. its Julian calendar date and lunar age (cf. the Isidorian table (*Etym.* 6.17), which is the most minimalistic of medieval

Easter tables); generally, however, it contained far more data, though the data included in these tables varied considerably, even when representing the same reckoning (cf., e.g., the columns of the Dionysiac Easter table in Appendix 8 with Dionysius' original composition (Krusch, *Studien* II, 69–74)). Likewise, the length of these tables varied: The only known Easter table of the *latercus* covers its entire cyclic period, 84 years; Victorius also had originally composed his Easter table to cover the entire 532 years of his → Easter cycle, but later computists only copied the section that was immediately relevant for their and the following one or two generations, usually not covering more than 100 years; Dionysius had drawn up an Easter table of 95 years (a practise he had copied from his Alexandrian model), with only minimal recalculation becoming necessary after its completion; Isidore, however, as well as other seventh-century computists, then, mistook this 95-year period of the Dionysiac Easter table for an → Easter cycle.

epacts (*epactae*): In a → lunar cycle, the lunar ages recur in a fixed sequence on any given Julian calendar date only after a certain period of years (i.e. the length of the → lunar cycle in question); in the → 19-year lunar cycle, this sequence obviously consisted of 19 differing lunar ages, i.e. the first recurrence of the same lunar age on any given Julian calendar date was after exactly 19 years. Consequently, in the → Dionysiac and → Victorian reckonings, the lunar age of a specific Julian calendar date was a sufficient parameter to unambiguously define every year of the respective 19-year lunar cycle; the lunar age of the date chosen was called the epact of the year in question. From year to year, the epact regularly increased by 11 (due to the 11-day difference between a → solar and a → lunar year), exceptionally by 12 in years that contained a → *saltus*; whenever this value exceeded 30, that number was subtracted (since the maximum length of a → lunar month, and particularly that of an → embolismic lunar month, was 30).

– **placement of (*sedes epactarum*):** In the → Easter reckonings followed in the early middle ages, the epacts were set either on 1 January (→ *latercus* and → Victorian reckoning) or on 22 March (→ Dionysiac reckoning); the latter placement of the epacts was chosen since this was the first possible Julian calendar date for the celebration of Easter. Converts from the older two reckonings to the Dionysiac one, then, translated the Dionysiac epacts of 22 March to 1 January, so that in this reckoning both of these placements were followed by the eighth century.

***hendecas*:** A → 19-year cycle is, due to its specific structure, divided into two main parts: Generally, two → common lunar years were followed by an → embolismic lunar year in a → 19-year cycle; in the 8th and the 19th year, however, an → embolismic lunar year followed a → common one directly, marking the end of each of these two parts respectively. The latter part, i.e. the final eleven years of a → 19-year cycle, are called *hendecas*.

- **Dionysiac:** The final eleven years of the → Dionysiac 19-year cycle constitute the Dionysiac *hendecas*.
- **Victorian:** Retrospectively modelled on the → Dionysiac reckoning, the final eleven years of the → Victorian 19-year cycle constitute the Victorian *hendecas*.

Ides/ides (*Idus*, *Id*): In the → Julian calendar, the Ides were usually placed on the 13th day of the respective months, with four exceptions (March, May, July, October), in which they were set two days later, on the 15th day. The days between the → Nones and the Ides were counted backwards from the Ides, so that the days preceding the Ides were called 'the second day before the Ides' (*ante diem secundum Idus*; abbr.: *a.d. II Idus*), 'the third day before the Ides' (*ante diem tertium Idus*), etc. As in the case of the → calends, medieval scribes preferred a short form of these expressions as *II Id* ('the second ides'), *III Id* ('the third ides'), etc. In the present book the marker day (i.e. the 13th or 15th day of the respective month) is written with capital I (Ides), the days preceding as ides (including the marker day, if ides refers to the entire block of days – e.g., all eight ides of January); the number of ides is eight for every Julian calendar month.

lunar month/lunation: Lunar months, called lunations when referring to a lunar month within the → Julian calendar (→ calendrical lunar month), occur in various forms in early medieval computistical texts, where they are often not clearly defined (though the context, in most instances, provide for an exact interpretation):

- **calendrical:** A lunar month modelled on Julian calendar months may be termed calendrical lunar month or lunation. Since a lunar year was shorter than a solar year by 11 days, calendrical lunar months were not identical with the months of the → Julian calendar. Still, as a → lunar year was also divided into 12 months, each of these months was ascribed to a Julian calendar month. The → naming of a calendrical lunar month was, generally, based on the Julian calendar month in which the lunar month ended. As the calculation with fractions posed some difficulties for medieval mathematicians and computists, preference was given to integer numbers, especially in terms of calendrical calculations. Accordingly, rather than applying calendrical lunar months of the → synodic period of ca. 29,5 days (as reckoned by early medieval computists), six calendrical lunar month were fixed as consisting of 29 days (→ hollow lunations), the remaining six as consisting of 30 days (→ full lunations), which were arranged in specific → sequences.
- **hollow:** A → calendrical lunar month of 29 days is termed a hollow lunation.
- **full:** A → calendrical lunar month of 30 days is termed a full lunation.
- **embolismic (*embolismus*):** Since the → lunar year was by eleven days shorter than the Julian calendar year, extra lunar months were intercalated at appropriate places of the respective → lunar cycle to keep the lunar year in line with the Julian calendar. These extra lunar months were termed embo-

lisms. They consisted of 30 days, so that they additionally balanced the difference between a → synodic lunar month of ca. 29.53 days and a → calendrical lunar month of 29.5 days.

- **abortive (*luna abortiva*):** A *luna abortiva* is defined as a lunation falling between the Calends of two successive Julian calendar months; a *luna abortiva* may be → full or → hollow. It is a specifically early medieval Irish concept and has serious implications on the → naming of the lunar months.
- **naming of:** As mention under → calendrical lunar months, a lunation was generally named after the Julian calendar month in which the lunation ended. Yet, there were exceptions to this rule, since → embolisms had to be intercalated at certain places; these embolisms were not attributed to any Julian calendar month (being simply term *embolismus*) and sometimes led to a shift of the following lunations, so that in these instances, e.g., the May lunation may have ended in June. Note that in this system the → sequence of lunations remained unaffected (except obviously for the fact that embolisms were introduced). Early medieval Irish computists, on the other hand, did not allow exceptions to the rule that lunations were named after the Julian calendar month in which they ended. If two lunations ended in the same Julian calendar month, the latter one of these had to fall between the Calends of two Julian calendar months; this lunation was not attributed to the Julian calendar month but it was simply termed → *luna abortiva*. Accordingly, whenever an embolism did not coincide with a *luna abortiva*, the naming of lunations varied considerably between the two systems.
- **sequence of:** The arrangement of → full and → hollow lunar months within the → Julian calendar (six each) is termed the sequence of lunations. The sequence of lunations differed between the three → Easter reckonings followed in the early middle ages.
    - ***latercus*:** The *latercus* (or 84 (14)-year Easter reckoning) had a very specific sequence of lunations, in which → full lunations were applied to Julian calendar months of 31 days, → hollow ones to Julian calendar months of 30 days, with March being the only exception (i.e. the March lunation was hollow, as was the February one).
    - **Victorian:** Victorius had not outlined any sequence of lunations in his writings, but early medieval computists considered the Victorian sequence of lunations as alternating between → full and → hollow lunations, starting with a full January lunation.
    - **Dionysiac:** Like Victorius, Dionysius gave no information concerning the sequence of lunations applied in his reckoning; early medieval computists, then, reconstructed this sequence as alternating between → full and → hollow lunations; note, however, that both the Munich and the Ensiedeln computists attributed a different sequence of lunations to the → Dionysiac reckoning, in which the April lunation was considered full, the May one hollow.
- **sidereal:** A sidereal lunar month can be defined as the period of the course of the moon through the zodiac, ca. 27.32 days.

– **synodic:** A synodic lunar month can be defined as the period from one new moon to the next, ca. 29.53 days; a synodic lunar month constitutes the basis of → lunar cycles.

Nones/nones (*Nonae*, *Non*, *No*, *N*): The Nones take their name from being placed nine days before the → Ides in every month of the → Julian calendar. Depending on the placement of the → Ides, then, the Nones are set on the fifth or the seventh day of a month. The days preceding up to the → Calends were originally termed 'the second day before the Nones' (*ante diem secundum Nonas*; abbr.: *a.d. II Nonas*), 'the third day before the Nones' (*ante diem tertium Nonas*), etc., which by the early middle ages was expressed in short as *II Non* ('the second nones'), *III Non* ('the third nones'), etc. In the present book, the marker day (i.e. the fifth or seventh day of the respective month) is written with capital N as Nones, the days preceding as nones (including the marker day, if nones refers to the entire block of days – e.g., the four nones of January); the number of nones of a month depends on the placement of the → Ides within a month, so that it varies between four and six.

***ogdoas*:** As discussed under → *hendecas*, a → 19-year cycle is divided into two main parts, covering the first eight and the final eleven years. The first part of eight years is called *ogdoas*.

– **Dionysiac:** The Dionysiac *ogdoas* is defined as the first eight years of the → Dionysiac 19-year cycle.
– **Victorian:** The Victorian *ogdoas* was retrospectively modelled on the Dionysiac one, so that the first eight years of the → Victorian 19-year cycle (corresponding to the first eight years of the → Dionysiac 19-year cycle) can be defined as the Victorian *ogdoas*.

quadragesimal period (*quadragesima*): The fourty-day period of the Lenten fast was called *quadragesima*.

– ***initium quadragesimae*:** The beginning of this period, *initium quadragesimae* in Latin, was generally accepted by early medieval computists to be the sixth Sunday before Easter Sunday (according to the → *latercus* as reconstructed from the Padua table, however, the *initium quadragesimae* appears to have been set on the following Wednesday, i.e. exactly 40 days before Easter Sunday, reckoned inclusively). The Julian calendar date of the *initium quadragesimae* obviously varied, as did its lunar age, depending on the Julian calendar date and lunar age of Easter Sunday (the lunar limits for the *initium quadragesimae* were *luna* 3 to 9 in the → Dionysiac, *luna* 4 to 10 in the → Victorian reckoning); it was often calculated from *luna* 2 of the lunation preceding the Easter lunation (termed → *secundum singularis* by the Munich computist), which had the same weekday (*feria*) as the → Easter full moon and therefore the same weekday and lunar day difference to the *initium quadragesimae* as the Easter full moon had to Easter Sunday.

***saltus lunae*:** In most → lunar cycles, the → embolismic lunar months introduced to compensate for the eleven-day difference between a → solar and a → lunar year provided too many lunar days compared to the number of solar days in the same period. Accordingly, one lunar day had to be subtracted in the → 19-year lunar cycle, six days in the → 84-year lunar cycle. This subtraction, which was generally accomplished by reducing a → full to a → hollow lunation, was called a *saltus lunae* (since the reckoning 'leaps over' lunar age 30).

- **placement of:** Depending on the → Easter reckonings, the *saltus* were introduced in differing intervals and on differing Julian calendar dates.
    - ***latercus*:** Since the → *latercus* was based on an → 84-year lunar cycle, it contained six *saltus*. These were characteristically (since it is the only known → Easter reckoning in which the *saltus* was distributed in these intervals) introduced in regular, 14-year intervals (and for this reason this reckoning is termed 84 (14)-year Easter reckoning). Unfortunately, nothing else is known about these *saltus*, particularly not about their exact placement within the Julian calendar; it is fairly assumed, however, that the *saltus* reduced either the October or, more likely, the December → lunation from a → full to a → hollow one.
    - **Victorian:** The → Victorian reckoning, being based on a → 19-year lunar cycle, had only one *saltus*, occurring on the same Julian calendar date every 19 years. Victorius' writings, especially his Easter table, revealled that the *saltus* was introduced in the sixth year of the → Victorian 19-year cycle; the exact placement at the end of the November lunation, i.e. on 17 November, was not outlined by Victorius himself, but as such accepted by most early medieval computists.
    - **Dionysiac:** Like the → Victorian reckoning, the → Dionysiac reckoning was based on a → 19-year lunar cycle and therefore had only one *saltus*. Contrary to the Victorian reckoning, this *saltus* was placed either at the end or the beginning of the cycle, in the 19th or the first year (at least if the years of this cycle are reckoned to begin on 1 January; if they are reckoned to beginn with *luna* 15 of the Easter lunation, all placements discussed in the following fall in the first year), depending on customs: While particularly Irish computists of the late seventh and early eighth centuries understood Dionysius' writings as suggesting that the *saltus* had to be introduced at the end of the March lunation (22 March), Anglo-Saxon and then Carolingian computists preferred the end of the previous November lunation (24 November); based on Bede, a July placement became popular in later centuries (though this placement was not favoured by Bede himself).

***secundum singularis* (*terminum quadragesimae*):** *Luna* 2 of the → lunation preceding the Easter lunation was termed *secundum singularis* (lit. 'second single digit')by the Munich computist. This lunar date served the same function for calculating the → *initium quadragesimae* as the → Easter full moon did for Easter Sunday, since both the *secundum singularis* and the → Easter full moon fell on the same weekday and therefore the weekday and lunar day difference between the *secundum singularis* and the *initium quadragesimae* on the one hand, between the Easter full moon and Easter Sunday on the other, was identical. Consequently, the *secundum singularis* of the Munich Computus was more appropriately termed *secunda luna regulandi quadragesimae* or simply *terminus quadragesimae* in other early medieval computistical texts.

year: As the calculation of Easter, the most important Christian feast, depended directly on the course of the sun as well as the moon, the principal types of year in medieval computistics were the solar and the lunar year.

– **solar:** A solar year was defined as the course of the sun through the zodiac. At the time of Julius Caesar, this was calculated as 365¼ days, divided into 12 months of differing lengths in the → Julian calendar (named after Caesar); this calendar was used throughout the middle ages.

– **lunar:** Modelled on the → Julian calendar, a lunar year was defined as consisting of 12 months plus an annual bissextile increment of a quarter-day. Since a → synodic lunar month was calculated as 29½ days, a common (*communis*) lunar year of 12 months consisted of 354¼ days. In order to compensate for the 11-day difference between a → solar year and a common lunar year, an → embolismic lunar month of 30 days was implemented in certain years, leading to an embolismic (*embolismus*) lunar year of 13 months equalling 384¼ days.

BIBLIOGRAPHY OF CITED LITERATURE

SOURCES

Acts of the Council of Caesarea,

Recension A, ed. Baluze, *Nova collectio conciliorum*, 13–5.

Recension B (*De ordinatione feriarum paschalium*), ed. J.-P. Migne, *Patrologiae cursus completus, series latina XC* (Paris 1850), 607–10.

Recension C, ed. Krusch, *Studien* I, 306–10.

Recension D, ed. Wilmart, 'Nouveau texte', 20–7.

Adbreviatio librorum sancti Emmerammi, ed. Bischoff, *Mittelalterliche Bibliothekskataloge* 4,1, 144–6.

Admonitio generalis, ed. Georg Heinrich Pertz, *Monumenta Germaniae Historica, Leges I* (Hannover 1835), 53–67.

Alcuin, *Epistola 126*, ed. Ernst Dümmler, *Monumenta Germaniae Historica, Epistolae IV: Epistolae Karolini aevi II* (Berlin 1895), 185–7.

Alcuin, *Interrogationes et responsiones in Genesin*, ed. J.-P. Migne, *Patrologiae cursus completus, series latina C* (Paris 1863), 515–66.

Alcuin, *Orthographia*, ed. Heinrich Keil, *Grammatici Latini VII: Scriptores de orthographia* (Leipzig 1880, repr. Hildesheim 1961), 295–312.

Ps-Alcuin, *De bissexto I–II*, ed. J.-P. Migne, *Patrologiae cursus completus, series latina CI* (Paris 1863), 993–9.

Ps-Alcuin, *De saltu lunae I–VIII*, ed. J.-P. Migne, *Patrologiae cursus completus, series latina CI* (Paris 1863), 984–93.

Ps-Alcuin, *Disputatio puerorum*, ed. J.-P. Migne, *Patrologiae cursus completus, series latina CI* (Paris 1863), 1097–144.

Aldhelm, *Epistula ad Geruntium*, ed. Rudolf Ehwald, *Monumenta Germaniae Historica, Auctores antiquissimi XV: Aldhelmi opera* (München 1919), 480–6.

Ambrose, *De excidio urbis Hierosolymitanae libri quinque*, ed. J.-P. Migne, *Patrologiae cursus completus, series latina XV* (Paris 1887), 2061–310.

Ambrose, *Exameron*, ed. Karl Schenkl, *Sancti Ambrosii opera, pars prima, Corpus Scriptorum Ecclesiasticorum Latinorum XXXII, 1* (Leipzig 1897), 1–261.

Anatolius (?), *De ratione paschali*, ed. Mc Carthy & Breen, *De ratione paschali*, 44–53.

Annales sancti Emmerammi Ratisponensis, ed. Georg Heinrich Pertz, *Monumenta Germaniae Historica, Scriptores I* (Hannover 1826), 91–4.

Annales Cambriae, ed. John Morris, *Nennius. British history and the Welsh annals* (London 1980), 44, 85–91.

Annals of Inisfallen, ed. Seán Mac Airt, *The Annals of Inisfallen* (Dublin 1944, repr. Dublin 1988).

Annals of Tigernach, ed. Whitley Stokes, *The Annals of Tigernach, vol. 1* (repr. from *Revue celtique* 16–17 (1895–1896), Felinfach 1993).

Annals of Ulster, ed. Seán Mac Airt & Gearóid Mac Niocaill, *The Annals of Ulster (to A.D. 1131)* (Dublin 1983).

Anonymus ad Cuimnanum, ed. Bischoff & Löfstedt, *Anonymus ad Cuimnanum*, 1–160.

Arg. Aquens.: 'Der Aachener Vorbehalt von 816', ed. Borst, *Schriften*, 1356–66.

Arn Serm.: 'Die Predigt Arns von Salzburg um 802', ed. Borst, *Schriften*, 833–84.

Ars anonyma Bernensis, ed. Heinrich Keil & Hermann Hagen, *Grammatici Latini VIII: Supplementum* (Leipzig 1870, repr. Hildesheim 1961), 63–142.

Augustine, *Confessionum libri XIII*, ed. Lucas Verheijen, *Sancti Augustini opera, Confessionum libri XIII, Corpus Christianorum Series Latina XXVII* (Turnhout 1990).

Augustine, *Contra Faustum libri XXXIII*, ed. Joseph Zycha, *Sancti Aureli Augustini ... Contra Faustum, Corpus Scriptorum Ecclesiasticorum Latinorum XXV* (Leipzig 1891), 249–797.

Augustine, *De civitate Dei libri XXII*, ed. Bernhard Dombart & Alfons Kalb, *Sancti Aurelii Augustini De civitate Dei, Corpus Christianorum Series Latina XLVII–XLVIII* (Turnhout 1955).

Augustine, *De Genesi contra Manichaeos libri II*, ed. J.-P. Migne, *Patrologiae cursus completus, series latina XXXIV* (Paris 1887), 172–220.

Augustine, *De Genesi ad litteram libri XII*, ed. Joseph Zycha, *Sancti Aureli Augustini De Genesi ad litteram duodecim ..., Corpus Scriptorum Ecclesiasticorum Latinorum XXVIII, 1* (Leipzig 1894), 1–456.

Augustine, *De trinitate libri XV*, ed. William J. Mountain, *Sancti Aurelii Augustini De trinitate libri XV, Corpus Christianorum Series Latina L–LA* (Turnhout 1968).

Augustine, *Enarrationes in psalmos*, ed. Eligius Dekkers & Iohannes Fraipont, *Sancti Aurelii Augustini Enarrationes in psalmos, Corpus Christianorum Series Latina XXXVIII–XL* (Turnhout 1956).

Augustine, *Epistula 36*, ed. Klaus-Detlef Daur, *Sancti Aurelii Augustini Epistulae I–LV, Corpus Christianorum Series Latina XXXI* (Turnhout 2004), 130–53.

Augustine, *Epistula 55*, ed. Klaus-Detlef Daur, *Sancti Aurelii Augustini Epistulae I–LV, Corpus Christianorum Series Latina XXXI* (Turnhout 2004), 234–65.

Augustine, *Epistula 205*, ed. Alois Goldbacher, *Sancti Aureli Augustini Hipponensis episcopi Epistulae IV: ep. CLXXXV&CCLXX, Corpus Scriptorum Ecclesiasticorum Latinorum LVII* (Leipzig 1911), 323–39.

Augustine, *Locutionum in Heptateuchum libri VII*, ed. Iohannes Fraipont, *Sancti Aurelii Augustini ... Locutionum in Heptateuchum libri VII..., Corpus Christianorum Series Latina XXXIII* (Turnhout 1958), 379–465.

Augustine, *Sermones*, ed. J.-P. Migne, *Patrologiae cursus completus, series latina XXXVIII–XXXVIIII* (Paris 1865).

Bede, *Chronica*, parallel edition of Bede's two chronicles (*DT* 16 and *DTR* 66) by Theodor Mommsen, *Monumenta Germaniae Historica, Auctores antiquissimi XIII* (Berlin 1898), 247–354.

Bede, *De natura rerum*, ed. Charles W. Jones, *Bedae Venerabilis opera, pars VI, 1, Corpus Christianorum Series Latina CXXIIIA* (Turnhout 1975), 173–234.

Bede, *De temporibus*, ed. (including only the beginning of the chronicle (c. 16)) Jones, *Bedae opera*, 295–303 [abbr. *DT*].

Bede, *De temporum ratione*, ed. (without chronicle (c. 66)) Jones, *Bedae opera*, 175–291 [abbr. *DTR*].

Bede, *Epistola ad Wicthedum*, ed. Jones, *Bedae opera*, 319–25 [abbr. *EW*].

Bede, *Historia ecclesiastica gentis Anglorum*, ed. Plummer, *Baedae opera* 1, 1–360 [abbr. *HE*].

Bede, *Homiliae subdititiae*, ed. J.-P. Migne, *Patrologiae cursus completus, series latina XCIV* (Paris 1850), 267–516.

Bede, *In Ezram et Neemiam libri III*, ed. David Hurst, *Bedae Venerabilis opera, pars II, 2A, Corpus Christianorum Series Latina CXIXA* (Turnhout 1959), 235–392.

Bede, *In Lucae evangelium expositio*, ed. David Hurst, *Bedae Venerabilis opera, pars II, 3, Corpus Christianorum Series Latina CXX* (Turnhout 1960), 1–425.

Bede, *In Marci evangelium expositio*, ed. David Hurst, *Bedae Venerabilis opera, pars II, 3, Corpus Christianorum Series Latina CXX* (Turnhout 1960), 427–648.

Bede, *Retractatio in actus apostolorum*, ed. Max Ludwig Wolfram Laistner, *Bedae Venerabilis opera, pars II, 4, Corpus Christianorum Series Latina CXXI* (Turnhout 1963), 101–63.

Ps-Bede, *De officiis*, ed. J.-P. Migne, *Patrologiae cursus completus, series latina XCIV* (Paris 1850), 531–40.

Ps-Bede, *De sex dierum creatione*, ed. J.-P. Migne, *Patrologiae cursus completus, series latina XCIII* (Paris 1850), 207–34.

Bobbio Computus, ed. J.-P. Migne, *Patrologiae cursus completus, series Latina CXXIX* (Paris 1853), 1275–372 (checked against the sole MS witness, Milan, Biblioteca Ambrosiana, H 150 inf) [abbr. BC].

Boethius, *De arithmetica*, ed. Henrik Oosthout & Johannes Schilling, *Anicii Manlii Severini Boethii De arithmetica, Corpus Christianorum Series Latina XCIVA* (Turnhout 1999).

Boethius, *Liber de diffinitione*, ed. J.-P. Migne, *Patrologiae cursus completus, series latina LXIV* (Paris 1891), 891–910.

Boniface, *Epistola 75*, ed. Michael Tangl, *Monumenta Germaniae Historica, Epistolae seletae I: Die Briefe des Heiligen Bonifatius und Lullus* (Berlin 1916), 156–8.

Boniface, *Epistola 76*, ed. Michael Tangl, *Monumenta Germaniae Historica, Epistolae seletae I: Die Briefe des Heiligen Bonifatius und Lullus* (Berlin 1916), 158–9.

Boniface, *Epistola 91*, ed. Michael Tangl, *Monumenta Germaniae Historica, Epistolae seletae I: Die Briefe des Heiligen Bonifatius und Lullus* (Berlin 1916), 206–9.

Caesarius of Arles, *Sermones*, ed. Germain Morin, *Sancti Caesarii Arelatensis Sermones, Corpus Christianorum Series Latina CIII–CIV* (Turnhout 1953).

Capitulare duplex in Theodonis villa promulgatum, ed. Georg Heinrich Pertz, *Monumenta Germaniae Historica, Leges I* (Hannover 1835), 131–6.

Cap. comp.: 'Das Aachener Verhör von 809', ed. Borst, *Schriften*, 1040–53.

Cassiodorus, *Chronica*, ed. Theodor Mommsen, *Monumenta Germaniae Historica, Auctores antiquissimi IX* (Berlin 1894), 120–61.

Cassiodorus, *Exposito psalmorum*, ed. Marc Adriaen, *Magni Aurelii Cassiodori Expositio psalmorum, Corpus Christianorum Series Latina XCVII–XCVIII* (Turnhout 1958).

Cassiodorus, *Institutiones*, ed. (a reprint of R.A.B. Mynors 1963 edition) Wolfgang Bürsgens, *Cassiodor. Institutiones divinarum et saecularium litterarum. Einführung in die geistlichen und weltlichen Wissenschaften, 2 vols* (Freiburg 2003).

Cassiodorus (?), *Expositio in cantica canticorum*, ed. J.-P. Migne, *Patrologiae cursus completus, series latina LXX* (Paris 1863), 1055–1106.

Cologne fragments, ed. Krusch, *Studien* I, 241–4.

Columbanus, *Epistolae* 1–2, ed. Walker, *Sancti Columbani opera*, 2–22.

Columbanus, *Regula monachorum*, ed. Walker, *Sancti Columbani opera*, 122–42.

Comp. Col.: 'Das Kölner Lehrbuch von 805', ed. Borst, *Schriften*, 891–950.

Computus Carthaginiensis, ed. Krusch, *Studien* I, 279–97.

Computus Cottonianus: London, British Library, Cotton Caligula A XV, fol. 73r–80r.

Computus Digbaeanus: Oxford, Bodleian Library, Digby 63, fol. 72v–79r (ed. Jan, *Historia cycli dionysiani*, 79–94 (repr. in J.-P. Migne, *Patrologiae cursus completus, series latina LXVII* (Paris 1865), 497–508) and Krusch, *Studien* II, 75–81).

Computus Einsidlensis: Einsiedeln, Stiftsbibliothek, 321 (647), p. 82–125 [abbr. *CE*].

Concilium Avrelianense, ed. Charles De Clercq, *Concilia Galliae A. 511–A. 695, Corpus Christianorum Series Latina CXLVIIIA* (Turnhout 1963), 131–45.

Cummian, *Epistola* (*Epistola Cummiani*), ed. Walsh & Ó Cróinín, *Cummian's letter*, 55–97.

De anno, ed. Mountford, 'De mensium nominibus', 115–6.

De computo dialogus, ed. J.-P. Migne, *Patrologiae cursus completus, series latina XC* (Paris 1850), 647–52.

De divisionibus temporum, ed. J.-P. Migne, *Patrologiae cursus completus, series latina XC* (Paris 1850), 653–64 [abbr. DDT].

De flexibus digitorum (*Romana computatio*), ed. Charles W. Jones, *Bedae Venerabilis opera, pars VI, 3, Corpus Christianorum Series Latina CXXIIIC* (Turnhout 1980), 669–72.

De hebdomadibus, ed. Jones, *Bedae opera*, 394–5.

De mirabilibus sacrae scripturae, ed. J.-P. Migne, *Patrologiae cursus completus, series latina XXXV* (Paris 1864), 2149–200.

De ordinatione feriarum paschalium, cf. Acts of the Council of Caesarea, Recension B.

De ratione conputandi, ed. Dáibhí Ó Cróinín in Walsh & Ó Cróinín, *Cummian's letter*, 113–213 [abbr. *DRC*].

De sollemnitatibus, ed. Walker, *Sancti Columbani opera*, 198–206.

Dial. Burg.: 'Das burgundische Lehrgespräch von 727', ed. Borst, *Schriften*, 353–74.

Dial. Langob.: 'Das langobardische Zwiegespräch um 750', ed. Borst, *Schriften*, 433–61.

Dial. Neustr.: 'Das neustrische Streitgespräch von 737', ed. Borst, *Schriften*, 381–423.

Dicuil, *Liber de astronomia*, ed. Esposito, 'An unpublished astronomical treatise', 378–446.

Dionysius Exiguus, *Argumenta*, ed. (including the ps-Dionysiac *argumenta*) Krusch, *Studien* II, 75–81.

Dionysius Exiguus, *Cyclus*, ed. Krusch, *Studien* II, 69–74.

Dionysius Exiguus, *Epistola ad Bonifatium et Bonum* (*Epistola Dionysii*), ed. Krusch, *Studien* II, 82–7.

Dionysius Exiguus, *Prologus* (*Prologus Dionysii*), ed. Krusch, *Studien* II, 63–8.

Diplomata Merovingica, nr. 179–189, ed. Theo Kölzer, *Monumenta Germaniae Historica, Diplomata I: Diplomata regum Francorum e stripe Merovingica – Die Urkunden der Merowinger I* (München 2001), 444–73.

Disputatio Chori et Praetextati: cited from the Sirmond MS (Oxford, Bodleian Library, Bodley 309, fol. 101r–105v); an edition by Leofranc Holford-Strevens is forthcoming.

Disputatio Morini, ed. Pitra, *Spicilegium Solesmense*, 31–2.

Donatus, *Ars maior*, ed. Holtz, *Donat*, 603–74.

Durand, Guillaume: *Rationale diuinorum officiorum*, ed. Anselme Davril & Timothy M. Thibodeau, *Guillelmi Duranti Rationale divinorum officiorum, Corpus Christianorum Continuatio Mediaevalis CXL–CXLB* (Turnhout 1995–2000).

Ecgbert of York, *De institutione catholica dialogus*, ed. J.-P. Migne, *Patrologiae cursus completus, series latina LXXXIX* (Paris 1850), 435–42.

Epistola Cyrilli, ed. Krusch, *Studien* I, 344–9.

Epist. Rat.: 'Der Regensburger Protestbrief von 809', ed. Borst, *Schriften*, 1027–53.

Eusebius, *Historia ecclesiastica* (translated into Latin by Rufinus), Greek and Latin parallel ed. Eduard Schwartz & Theodor Mommsen, *Eusebius Werke 2: Die Kirchengeschichte, Teil 1: Die Bücher I bis V* (Leipzig 1903).

Felix Gillatanus, *Prologus*, ed. Krusch, *Studien* II, 87.

Fides Romanorum, ed. Vinzenz Bulhart, *Corpus Christianorum Series Latina LXIX* (Turnhout 1967), 265–8.

Fragmenta epistolarum Fuldensium, ed. Ernst Dümmler, *Monumenta Germaniae Historica, Epistolae V: Karolini aevi III* (Berlin 1899), 517–33.

Heimo of Bamberg, *De decursu temporum*, ed. Hans Martin Weikmann, *Heimo von Bamberg. De decursu temporum* (Hannover 2004).

Heimo of Halberstadt, *Homiliae de tempore*, ed. J.-P. Migne, *Patrologiae cursus completus, series latina CXVIII* (Paris 1880), 11–746.

Helperic of Auxerre, *Liber de computo*, J.-P. Migne, *Patrologiae cursus completus, series latina CXXXVII* (Paris 1853), 17–48.

Quintus Julius Hilarianus, *Expositum de die paschae et mensis*, ed. J.-P. Migne, *Patrologiae cursus completus, series latina XIII* (Paris 1845), 1105–14.

Isidore, *Chronica*, ed. Theodor Mommsen, *Monumenta Germaniae Historica, Auctores antiquissimi IX* (Berlin 1894), 424–81.

Isidore, *De natura rerum*, ed. Jacques Fontaine, *Isidore de Séville, Traité de la nature* (Bordeaux 1960), 163–327.

Isidore, *Differentiarum, sive de proprietate sermonum, libri II*, ed. J.-P. Migne, *Patrologiae cursus completus, series latina LXXXIII* (Paris 1862), 9–98.

Isidore, *Etymologiae*, ed. Wallace M. Lindsay, *Isidori Hispalensis episcopi etymologiarum sive originum libri XX* (Oxford 1911).

Jerome, *Chronicon* (translation and expansion of Eusebius' chronicle), ed. Rudolf Helm, *Eusebius Werke 7: Die Chronik des Hieronymus* (second edition Berlin 1956).

Jerome, *Commentarii in Ionam*, ed. Marc Adrien, *Sancti Hieronymi presbyteri opera, pars I, 6, Corpus Christianorum Series Latina LXXVI* (Turnhout 1969), 377–419.

Jerome, *Commentariorum in Danielem libri III*, ed. François Glorie, *Sancti Hieronymi presbyteri opera, pars I, 5, Corpus Christianorum Series Latina LXXVA* (Turnhout 1964).

Jerome, *Commentariorum in Esaiam libri XVIII*, ed. Marc Adriaen, *Sancti Hieronymi presbyteri opera, pars I, 2–2A, Corpus Christianorum Series Latina LXXIII–LXXIIIA* (Turnhout 1963).

Jerome, *Commetariorum in Hiezechielem libri XIV*, ed. François Glorie, *Sancti Hieronymi presbyteri opera, pars I, 4, Corpus Christianorum Series Latina LXXV* (Turnhout 1964).

Jerome, *Commentariorum in Matthaeum libri IV*, ed. David Hurst & Marc Adriaen, *Sancti Hieronymi presbyteri opera, pars I, 7, Corpus Christianorum Series Latina LXXVII* (Turnhout 1969).

Jerome, *De viris illustribus*, ed. Wilhelm Herding, *Hieronymi De viris inlustribus liber* (Leipzig 1879).

Jerome, *Expositio in Evangelium secundum Matthaeum*, ed. J.-P. Migne, *Patrologiae cursus completus, series latina XXX* (Paris 1846), 534–60.

Jerome, *Liber interpretationis hebraicorum nominum*, ed. Paul de Lagarde, *Sancti Hieronymi presbyteri opera, pars I, 1, Corpus Christianorum Series Latina LXXII* (Turnhout 1959), 57–161.

Jerome, *Martyrologium*, ed. Giovanni Baptista de Rossi & Louis Duchesne, *Acta Sanctorum Novembris II 1: Martyrologium Hieronymianum* (Bruxelles 1894).

Ps-Jerome, *De exodo, in vigilia Paschae*, ed. Germain Morin, *Sancti Hieronymi presbyteri opera, pars II, Corpus Christianorum Series Latina LXXVIII* (Turnhout 1958), 536–41.

Jonas, *Vitae Columbani abbatis discipulorumque eius libri II*, ed. Bruno Krusch, *Monumenta Germaniae Historica, Scriptores rerum germanicarum in usum scholarum XXXVII* (Hannover 1905), 1–294.

Kal.: The various calendars on which Borst based his edition of the *Reichskalender* (Borst, *Reichskalender*, 399–1644).

Lect. comp.: 'Die rheinische Anleitung von 760/792', ed. Borst, *Schriften*, 544–659.

Leo the Great, *Epistola ad Marcianum imperatorem*, ed. Krusch, *Studien* I, 257–60.

Leo the Great, *Tractatus septem et nonaginta*, ed. Antoine Chavasse, *Sancti Leonis Magni Romani pontificis Tractatus septem et nonaginta, Corpus Christianorum Series Latina CXXXVIII–CXXXVIIIA* (Turnhout 1973).

Lib. ann.: 'Das Veroneser Jahrbüchlein von 793', ed. Borst, *Schriften*, 676–722.

Lib. calc.: 'Die Salzburger Enzyklopädie von 818', ed. Borst, *Schriften*, 1383–451.

Lib. comp.: 'Die Aachener Enzyklopädie von 809', ed. Borst, *Schriften*, 1087–334.

Liber pontificalis, ed. Louis Duchesne, *Le Liber pontificalis. Texte, Introduction et commentaire, tome 1* (Paris 1886, repr. Paris 1981), 47–523.

Libri ecclesie sancti Emmerammi Ratisbonensis, ed. Bischoff, *Mittelalterliche Bibliothekskataloge* 4,1, 154–61.

Lists of popes from the fifth to seventh centuries, ed. Louis Duchesne, *Le Liber pontificalis. Texte, Introduction et commentaire, tome 1* (Paris 1886, repr. Paris 1981), 13–41.

Macrobius, *Commentarii in somnium Scipionis*, ed. James Willis, *Macrobius II: Ambrosii Theodosii Macrobii Commentarii in somnium Scipionis* (Leipzig 1970).

Macrobius, *Saturnalia*, ed. James Willis, *Macrobius I: Ambrosii Theodosii Macrobii Saturnalia* (Leipzig 1970).

Martianus Capella, *De nuptiis Philologiae et Mercurii*, ed. James Willis, *Martianus Capella* (Leipzig 1983).

Martin of Braga, *De pascha*, ed. Barlow, *Opera omnia*, 269–75.

Maximus Confessor, *Computus ecclesiasticus*, ed. J.-P. Migne, *Patrologiae cursus completus, series graeca XIX* (Paris 1857), 1217–80.

Maximus of Tours, *Sermones*, ed. J.-P. Migne, *Patrologiae cursus completus, series latina LVII* (Paris 1862), 529–772; ed. Almut Mutzenbecher, *Maximi episcopi Taurinensis Sermones, Corpus Christianorum Series Latina XXIII* (Turnhout 1962). The *PL* and *CCSL* editions differ in content; the *PL* edition has been used for sermons that are not included in the *CCSL* edition.

Missale mixtum secundum regulam B. Isidori dictum mozarabes, ed. J.-P. Migne, *Patrologiae cursus completus, series graeca LXXXV* (Paris 1862), 169–1036.

Muretach, *In Donati artem maiorem*, ed. Louis Holtz, *Grammatici Hibernici Carolini aevi, pars I, Corpus Christianorum Continuatio Mediaevalis XL* (Turnhout 1977).

Notker Labeo, *De quatuor questionibus compoti*, transcr. Paul Piper, *Nachträge zur älteren deutschen Litteratur* (Stuttgart 1898), 312–8.

Notker Labeo, *Translatio barbarica psalterii*, ed. Petrus W. Tax, *Notker der Deutsche, Der Psalter, 3 vols* (Tübingen 1979–83).

Pacificus of Verona, *Computus* (including the *Opus excerptum ex libro compoti*), ed. Meersseman & Adda, *Manuale di computo*, 53–166 [abbr. PV].

Padua table, ed. Mc Carthy, 'Easter principles', 218–9; a full facsimile is printed in Warntjes, '84 (14)-year Easter reckoning', 80–2.

Pascasinus, *Epistola* (*Epistola Pascasini*), ed. Krusch, *Studien* I, 247–50.

Paschale Campanum, ed. Theodor Mommsen, *Monumenta Germaniae Historica, Auctores antiquissimi IX* (Berlin 1892), 744–50.

Pauca problesmata de enigmatibus ex tomis canonicis: 'The Irish reference Bible', ed. Gerard Mac Ginty, *Corpus Christianorum Continuatio Mediaeualis CLXXIII* (Turnhout 2000) [abbr. *PP*].

Périgueux table, ed. Angelus Mai, *Scriptorum veterum nova collection V* (Roma 1831), 69; a photographic reproduction of this table can be found in Cordoliani, 'Table pascal de Périgueux', facing p. 57.

Pliny, *Naturalis historiae libri XXXVII, liber VI,* ed. Kai Brodersen, *C. Plini Secundi Naturalis historiae libri XXXVII, Liber VI. C. Plinius Secundus d. Ä., Naturkunde, Lateinisch-deutsch, Buch VI, Geographie: Asien* (Zürich 1996).

Polemius Silvius, *Fasti*, ed. Theodor Mommsen, *Corpus inscriptionum latinarum I* (Berlin 1863), 335–57.

Pompeius (Maurus), *Commentum artis Donati*, ed. Heinrich Keil, *Grammatici Latini V: Artium scriptores minores* (Leipzig 1868, repr. Hildesheim 1961), 95–312.

Priscian, *Institutionum grammaticarum libri XVIII*, ed. Heinrich Keil & Martin Hertz, *Grammatici Latini II–III: Prisciani grammatici Caesariensis Institutionum grammaticarum libri XVIII* (Leipzig 1855/9, repr. Hildesheim 1961).

Prol. Aquit.: 'Das aquitanische Vorwort zur Ostertafel von 721', ed. Borst, *Schriften*, 337–47.

Prologus Cyrilli, ed. Krusch, *Studien* I, 337–43.

Prologus paschae: 'Der Kölner Prolog', ed. Krusch, *Studien* I, 227–35.

Prologus Theophili, ed. Krusch, *Studien* I, 220–26.

Proterius, *Epistola* (*Epistola Proterii*; original Greek version is lost, only Dionysius' Latin translation survives), ed. Krusch, *Studien* I, 269–78.

Quaest. Austr.: 'Die austrasische Abhandlung von 764', ed. Borst, *Schriften*, 466–508.

Quaest. Langob.: 'Die langobardische Abhandlung um 780', ed. Borst, *Schriften*, 514–26.

Rabanus Maurus, *De computo*, ed. Wesley M. Stevens in John McCulloh & Wesley M. Stevens, *Rabani Mauri Martyrologium, De computo, Corpus Christianorum Continuatio Medievalis XLIV* (Turnhout 1979), 199–321 [abbr. RM].

Rabanus Maurus, *Versus ad Barturicum episcopum ex persona Isanberti*, ed. Ernst Dümmler, *Monumenta Germaniae Historica, Poetae latini aevi carolini II* (Berlin 1884), 173–4.

Registrum sive inventarium librorum bibliothece monasterii Sancti Emmerammi episcopi et martyris, ed. Bischoff, *Mittelalterliche Bibliothekskataloge* 4,1, 188–385.

Repertorium patris Conradi Pleystayner prioris super Bibliothecam, ed. Bischoff, *Mittelalterliche Bibliothekskataloge* 4,1, 164–76.

Saltair na Rann, ed. Whitley Stokes, *The Saltair na Rann. A collection of Early Middle Irish poems* (Oxford 1883).

Sanas Cormaic, ed. Kuno Meyer, *Anecdota from Irish manuscripts IV: Sanas Cormaic. An Old-Irish glossary* (Halle 1912).

Sedulius Scottus, *In Matthaeum*, ed. Bengt Löfstedt, *Sedulius Scottus, Kommentar zum Evangelium nach Matthäus, 2 vols* (Freiburg 1989, 1991).

Sedulius Scottus, *In Donati artem maiorem*, ed. Bengt Löfstedt, *Grammatici Hibernici Carolini aevi, pars III, 1, Corpus Christianorum Continuatio Mediaevalis XLB* (Turnhout 1977).

Ser. nov.: 'Die ostfränkische Ahnentafel von 807', ed. Borst, *Schriften*, 971–1008.

Sergius (?), *De littera, de syllaba, de pedibus, de accentibus, de destinctione commentarius*, ed. Heinrich Keil, *Grammatici Latini IV: Probi Donati Servii qui feruntur De arte grammatica libri* (Leipzig 1864, repr. Hildesheim 1961), 473–85.

Smaragdus, *Collectiones*, ed. J.-P. Migne, *Patrologiae cursus completus, series latina XCIV* (Paris 1850), 13–552.

Stephen of Ripon, *Vita sancti Wilfrithi*, ed. Bertram Colgrave, *The life of bishop Wilfrid by Eddius Stephanus* (Cambridge 1927).

Tractatus Adthanasi, ed. Krusch, *Studien* I, 329–36.

Vers. Tur.: 'Das Tourer Lehrgedicht um 800', ed. Borst, *Schriften*, 804–19.

Victorius, *Calculus*, ed. Peden, *Abbo of Fleury*, 1–62.

Victorius, *Cyclus*, ed. Krusch, *Studien* II, 27–52.

Victorius, *Prologus* (*Prologus Victorii*), ed. Krusch, *Studien* II, 17–26.

Virgilius Maro Grammaticus, *Epitomae*, ed. Bengt Löfstedt, *Virgilius Maro Grammaticus, Opera omnia* (Leipzig 2003), 103–245.

Willibrord, Calendar: Paris, Bibliothèque Nationale, Lat. 10837, fol. 34r–40r (H.A. Wilson, *The Calendar of St. Willibrord. A facsimile with transcription, introduction, and notes* (London 1918, reprint 1998)).

Willibrord, Easter table: Paris, Bibliothèque Nationale, Lat. 10837, fol. 40v–41v (AD 703–759), 43r–v (AD 760–797), 44r (AD 684–702).

Zeitz table (*Liber paschalis codicis Cicensis a. CCCCXXXXVII*), ed. Theodor Mommsen, *Monumenta Germaniae Historica, Auctores antiquissimi IX* (Berlin 1892), 501–9, with additions by Krusch, 'Bruchstücke', 996; a new edition is announced in Overgaauw & Steving, *Zeitzer Ostertafel*, 30.

SECONDARY LITERATURE

Anscombe, Alfred: 'The paschal canon attributed to Anatolius of Laodicea', *English Historical Review* 10 (1895), 515–35.
Anspach, August Eduard: 'Das Fortleben Isidors im VII. bis IX. Jahrhundert', in *Miscellanea Isidoriana* (Roma 1936), 323–56.
Arweiler, Alexander: 'Zu Text und Überlieferung einer gekürzten Fassung von Macrobius Saturnalia I,12,2–I,15,20', *Zeitschrift für Papyrologie und Epigraphik* 131 (2000), 45–57.

Bach, Joseph: *Die Osterfestberechnung in alter und neuer Zeit* (Straßburg 1907).
Baluze, Etienne: *Nova collectio conciliorum* (Paris 1683).
Barlow, Claude W.: *Martini episcopi Bracarensis opera omnia* (New Haven 1950).
Barret-Kriegel, Blandine: *Jean Mabillon* (Paris 1988).
Bergkamp, Joseph Urban: *Dom Jean Mabillon and the Benedictine historical school of Saint-Maur* (Washington 1928).
Bergmann, Werner: *Innovationen im Quadrivium des 10. und 11. Jahrhunderts* (Stuttgart 1985).
Betten, Francis S.: 'The adoption of the Roman Easter calculation by the Island Celts', *The Catholic Historical Review* 14 (1928), 485–99.
Bezzel, Irmgard: *Bayerische Staatsbibliothek München* (München 1967).
Bieler, Ludwig: 'Towards a Hiberno-Latin dictionary', *Archivum Latinitatis Medii Aevi* 38 (1972), 248–55.
Bierbrauer, Katharina: 'Die Regensburger Buchmalerei des 8. und 9. Jahrhunderts', in Mütherich & Dachs (eds), *Regensburger Buchmalerei*, 15–22.
Bierbrauer, Katharina: *Katalog der illuminierten Handschriften der Bayerischen Staatsbibliothek in München, Bd. 1: Die vorkarolinigischen und karolingischen Handschriften der Bayerischen Staatsbibliothek* (Wiesbaden 1990).
Bisagni, Jacopo & Warntjes, Immo: 'Latin and Old Irish in the Munich Computus: a reassessment and further evidence', *Ériu* 57 (2007), 1–33.
Bisagni, Jacopo & Warntjes, Immo: 'The Early Old Irish material in the newly discovered Computus Einsidlensis (c. AD 700)', *Ériu* 58 (2008), 77–105.
Bischoff, Bernhard: 'Literarisches und künstlerisches Leben in St. Emmeram (Regensburg) während des frühen und hohen Mittelalters', *Studien und Mitteilungen zur Geschichte des des Benediktiner-Ordens und seiner Zweige* 51 (1933), 102–42, repr. in idem, *Mittelalterliche Studien 2*, 77–115.
Bischoff, Bernhard: *Die südostdeutschen Schreibschulen und Bibliotheken in der Karolingerzeit 1: Die bayerischen Diözesen* (Leipzig 1940).
Bischoff, Bernhard: 'Ostertagtexte und Intervalltafeln', *Historisches Jahrbuch* 60 (1940), 549–80, repr. in idem, *Mittelalterliche Studien 2*, 192–227.
Bischoff, Bernhard: 'Das griechische Element in der abendländischen Bildung des Mittelalters', *Byzantinische Zeitschrift* 44 (1951), 27–55, repr. in idem, *Mittelalterliche Studien 2*, 246–75.

Bischoff, Bernhard: 'Studien zur Geschichte des Klosters St. Emmeram im Spätmittelalter', *Studien und Mitteilungen zur Geschichte des des Benediktiner-Ordens und seiner Zweige* 65 (1953–4), 152–94, repr. in idem, *Mittelalterliche Studien 2*, 115–55.

Bischoff, Bernhard: 'Wendepunkte in der Geschichte der lateinischen Exegese im Frühmittelalter', *Sacris Erudiri* 6 (1954), 189–279, repr. in idem, *Mittelalterliche Studien 1*, 205–73.

Bischoff, Bernhard: 'Die europäische Verbreitung der Werke Isidors von Sevilla', in Manuel C. Diaz y Diaz (ed.), *Isidoriana* (Leon 1961), 317–44, repr. in idem, *Mittelalterliche Studien 1*, 171–94.

Bischoff, Bernhard: *Mittelalterliche Studien, vols 1–2* (Stuttgart 1966–7).

Bischoff, Bernhard (ed. Christine Elisabeth Ineichen-Eder): *Mittelalterliche Bibliothekskataloge Deutschlands und der Schweiz 4,1: Bistümer Passau und Regensburg* (München 1977).

Bischoff, Bernhard: *Kalligraphie in Bayern. Achtes bis zwölftes Jahrhundert* (Wiesbaden 1981).

Bischoff, Bernhard: *Katalog der festländischen Handschriften des neunten Jahrhunderts (mit Ausnahme der wisigothischen), Teil 1: Aachen–Lambach* (Wiesbaden 1998).

Bischoff, Bernhard: *Katalog der festländischen Handschriften des neunten Jahrhunderts (mit Ausnahme der wisigothischen), Teil 2: Laon–Paderborn* (Wiesbaden 2004).

Bischoff, Bernhard & Löfstedt, Bengt: *Anonymus ad Cuimnanum, Corpus Christianorum Series Latina CXXXIII* (Turnhout 1992).

Blackburn, Bonnie & Holford-Strevens, Leofranc: *The Oxford companion to the year* (Oxford 1999).

Boll, Franz: 'Ludwig Traube', in Traube, *Vorlesungen 1*, XI–XLVII.

Borst, Arno: 'Computus. Zeit und Zahl im Mittelalter', *Deutsches Archiv für Geschichte des Mittelalters* 44 (1982), 1–82.

Borst, Arno: *Computus. Zeit und Zahl in der Geschichte Europas* (Berlin 1990, repr. München 1999).

Borst, Arno: 'Alkuin und die Enzyklopädie von 809', in Butzer & Lohrmann (eds), *Science*, 53–78.

Borst, Arno: *Das Buch der Naturgeschichte. Plinius und seine Leser im Zeitalter des Pergaments* (Heidelberg 1994).

Borst, Arno: *Die karolingische Kalenderreform* (Hannover 1998).

Borst, Arno: *Der karolingische Reichskalender und seine Überlieferung bis ins 12. Jahrhundert, 3 vols* (Hannover 2001).

Borst, Arno: *Der Streit um den karolingischen Kalender* (Hannover 2004).

Borst, Arno: *Schriften zur Komputistik im Frankenreich von 721 bis 818, 3 vols* (Hannover 2006).

Boschen, Lothar: *Die Annales Prumiensis* (Düsseldorf 1972).

Bresslau, Harry: 'Geschichte der Monumenta Germaniae Historica', *Neues Archiv der Gesellschaft für ältere deutsche Geschichtskunde* 42 (1921), 1–769.

Breysig, Theodor: *Jahrbücher des fränkischen Reiches. 714–741. Die Zeit Karl Martells* (Leipzig 1869).

Bucherius, Aegidius: *De doctrina temporum. Commentarius in Victorium Aquitanum ... aliosque antiques canonum paschalium scriptores* (Antwerpen 1634).

Butzer, Paul Leo (with the assistance of Helga Butzer Felleisen): 'Scholars of the Mathematical Sciences in the Aachen – Liège – Maastricht region during the past 1200 years: an overview', in Butzer et al. (eds), *Karl der Große*, 43–90.

Butzer, Paul Leo & Butzer, Karl W.: 'Mathematics at Charlemagne's court and its transmission', in Catherine Cubitt (ed.), *Court culture in the early middle ages* (Turnhout 2003), 77–89.

Butzer, Paul Leo et al. (eds): *Karl der Große und sein Nachwirken. 1200 Jahre Kultur und Wissenschaft in Europa, 2: Mathematical Arts* (Turnhout 1998).

Butzer, Paul Leo & Lohrmann, Dietrich (eds): *Science in Western and Eastern civilizations in Carolingian times* (Basel 1993).

Byrne, Francis John et al. (eds): *A new history of Ireland IX: Maps, genealogies, lists* (Oxford 1984).

Cadwell, John: 'The De institutione artthmetica and the De institutione musica', in Gibson (ed.), *Boethius*, 135–54.

Cantor, Moritz: *Vorlesungen über Geschichte der Mathematik 1: Von den ältesten Zeiten bis zum Jahre 1200 n. Chr.* (Leipzig 1907).

Chaîne, Marius: *La chronologie des temps chrétiens de l'Égypte et de l'Éthiope* (Paris 1925).

Christ, Wilhelm: 'Über das argumentum calculandi des Victorius und dessen Commentar', *Sitzungsberichte der Bayersichen Akademie der Wissenschaften, philosophisch-philologische Classe* 1 (München 1863), 100–52.

Chroust, Anton: *Monumenta palaeographica, Denkmäler der Schreibkunst des Mittelalters, Erste Abteilung: Schrifttafeln in lateinischer und deutscher Sprache 1, 1 (Lieferung 1–8)* (München 1902).

Cordoliani, Alfred: 'Les traités de comput du haut moyen age (526–1003)', *Archivum Latinitatis Medii Aevi* 17 (1942), 51–72.

Cordoliani, Alfred: 'Une encyclopédie carolingienne de comput: Les Sententiae in laude compoti', *Bibliothèque de l'école des chartes* 104 (1943), 237–43.

Cordoliani, Alfred: 'Les computistes insulaires et les écrits pseudo-Alexandrins', *Bibliothèque de l'école des chartes* 106 (1945–6), 5–34.

Cordoliani, Alfred: 'Un manuscrits de comput ecclésiastique mal connu de la Bibliothèque Nationale de Madrid', *Revista de archivos, bibliotecas y museos* 57 (1951), 5–35.

Cordoliani, Alfred: 'Contribution à la littérature du comput ecclésiastique au moyen âge', *Studi medievali, Serie 3,* 2 (1961), 169–208.

Cordoliani, Alfred: 'La table pascal de Périgueux', *Cahiers de civilisation médiévale* 4 (1961), 57–60.

David, Pierre: 'Saint Martin de Braga: est-il l'auteur d'un traité de comput paschal ?', *Bulletin des études portugaises* 14 (1950), 283–99.

Declercq, Georges: *Anno Domini. The origins of the Christian era* (Turnhout 2000).

Declercq, Georges: 'Dionysius Exiguus and the introduction of the Christian Era', *Sacris Erudiri* 41 (2002), 165–246.

Dekkers, Eligius & Gaar, Émile: *Corpus Christianorum Series Latina, Clavis patrum latinorum* (third edition Turnhout 1995).

Docen, Bernhard Joseph: 'Anzeige der, die ältere deutsche Geschichte betreffenden Handschriften der vormaligen Abtei St. Emmeram in Regensburg', *Archiv der Gesellschaft für ältere deutsche Geschichtskunde* 1 (1820), 425–9.

Docen, Bernhard Joseph: 'Einige Notizen über die Handschrift der sogenannten Annales Ratisponenses (748–823)', *Archiv der Gesellschaft für ältere deutsche Geschichtskunde* 5 (1824), 515–9.

Dreyer, Johan Ludvig Emil: 'Medieval astronomy', in Charles Singer (ed.), *Studies in the history and method of science, vol. 2* (Oxford 1921), 102–20.

Du Cange, Charles: *ΠΑΣΧΑΛΙΟΝ seu chronicon paschale a mundo condito ad Heraclii imperatoris annum vicesimum*, Paris (1688).

Duchesne, Louis: 'La question des la pâque au concil de Nicée', *Revue des questiones historiques* 14 (1880), 5–42.

Dugmore, Clifford William: 'A note on the quartodecimans', *Studia Patristica* 4 (1961), 411–21.

Eastwood, Bruce: 'Astronomy in Christian Latin Europe c.500–c.1150', *Journal for the History of Astronomy* 28 (1997), 235–58, repr. in idem, *Revival*, article I.

Eastwood, Bruce: 'Plinian astronomy in the middle ages and renaissance', in Roger French & Frank Greenaway (eds), *Science in the early Roman Empire: Pliny the Elder, his sources and influence* (London 1986), 197–251, repr. in Eastwood, *Revival*, article II.

Eastwood, Bruce: 'Plinian astronomical diagrams in the early middle ages', in Edward Grant & John Emery Murdoch (eds), *Mathematics and its applications to science and natural philosophy in the middle ages* (Cambridge 1987), 141–72, repr. in Eastwood, *Revival*, article III.
Eastwood, Bruce: 'The astronomy of Macrobius in Carolingian Europe: Dungal's letter of 811 to Charles the Great', *Early Medieval Europe* 3 (1994), 117–34, repr. in idem, *Revival*, article V.
Eastwood, Bruce: 'The astronomies of Pliny, Martianus Capella and Isidore of Seville in the Carolingian world', in Butzer & Lohrmann (eds), *Science*, 161–180, repr. in Eastwood, *Revival*, article VI.
Eastwood, Bruce: *The revival of planetary astronomy in Carolingian and post-Carolingian Europe* (Aldershot 2002).
Eastwood, Bruce & Graßhoff, Gerd: *Planetary diagrams for Roman astronomy in medieval Europe ca. 800–1500* (Philadelphia 2004).
Englisch, Brigitte: *Die Artes liberales im frühen Mittelalter (5–9. Jh.)* (Stuttgart 1994).
Englisch, Brigitte: *Zeiterfassung und Kalenderproblematik in der frühen Karolingerzeit. Das Kalendarium der Hs. Köln DB 83-2 und die Synode von Soisson 744* (Stuttgart 2002).
Englisch, Brigitte: 'Karolingische Reformkalender und die Fixierung der christlichen Zeitrechnung', in Warntjes & Ó Cróinín (eds), *Computus and its cultural context*, 238–58.
Esposito, Mario: 'An unpublished astronomical treatise by the Irish monk Dicuil', *Proceedings of the Royal Irish Academy* 26C (1907), 378–446, repr. in idem (ed. by Michael Lapidge), *Irish books and learning in medieval Europe* (Aldershot 1990), article VII.
Evans, Claude: 'The Celtic church in Anglo Saxon times', *Études celtiques* 19 (1981), 215–28.
Eves, Howard: *An introduction to the history of mathematics* (New York 1953, fourth edition New York 1976).
Ewig, Eugen: *Die Merowinger und das Frankenreich* (third edition Stuttgart 1997).

Ferrari, Michele Camillo: *Sancti Willibrordi venerantes memoriam. Echternacher Schreiber und Schriftsteller von den Angelsachsen bis Johann Bertels. Ein Überblick* (Luxembourg 1994).
Ferrari d'Occhieppo, Konradin: 'Die Osterberechnung als Kalenderproblem von der Antike bis Regiomontanus', in Günther Hamann (ed.), *Regiomontanus-Studien, Sitzungsberichte der Österreichischen Akademie der Wissenschaften, philosophisch-historische Klasse* 364 (Wien 1980), 93–108.
Fleckenstein, Josef: *Die Hofkapelle der deutschen Könige 1: Grundlegung. Die Karolingische Hofkapelle* (Stuttgart 1959).
Fleuriot, Léon: *Dictionnaire des glosses en vieux Breton* (Paris 1964).
Folkerts, Menso: *„Boethius" Geometrie II* (Wiesbaden 1970).
Folkerts, Menso: 'Anonyme lateinische Euklidbearbeitungen aus dem 12. Jahrhundert', *Denkschriften der Österreichischen Akademie der Wissenschaften, mathematisch-naturwissenschaftliche Klasse* 116 (1971), 5–42.
Folkerts, Menso: 'The importance of the pseudo-Boethian geometria during the middle ages', in Michael Masi (ed.), *Boethius and the liberal arts: a collection of essays* (Bern 1981), 187–209, repr. in Folkerts, *Essays*, article VII.
Folkerts, Menso: 'Die Altercatio in der Geometrie I des Pseudo-Boethius. Ein Beitrag zur Geometrie im mittelalterlichen Quadrivium', in Gundolf Keil (ed.), *Fachprosa-Studien. Beiträge zur mittelalterlichen Wissenschaftsgeschichte* (Berlin 1982), 84–114, repr. in Folkerts, *Essays*, article VIII.
Folkerts, Menso: 'Euclid in medieval Europe', *Quaestio de rerum natura* 2 (1989), 1–63.
Folkerts, Menso: 'The importance of the Latin middle ages for the development of mathematics', in idem, *Essays*, article I.
Folkerts, Menso: 'The geometry II ascribed to Boethius', in idem, *Essays*, article IX.
Folkerts, Menso: *Essays on early medieval mathematics. The Latin tradition* (Aldershot 2003).
Frede, Hermann Josef: *Kirchenschriftsteller. Verzeichnis und Sigel* (Freiburg 1981).

Freise, Eckhard: 'Kalendarische und annalistische Grundformen der Memoria', in Karl Schmid & Joachim Wollasch (eds), *Memoria. Der geschichtliche Zeugniswert des liturgischen Gedenkens im Mittelalter* (München 1984), 441–562.

French Anderson, William: 'Arithmetical computations in Roman numerals', *Classical Philology* 51 (1956), 145–9.

Freund, Stephan: *Von den Agilolfingern zu den Karolingern. Bayerns Bischöfe zwischen Kirchenorganisiation, Reichsintegration und Karolingischer Reform (700–847)* (München 2004).

Fried, Johannes: *Der Weg in die Geschichte. Die Ursprünge Deutschlands bis 1024* (Berlin 1994).

Friedlein, Gottfried: 'Victorii calculus ex codice Vaticano editus', *Bullettino di Bibliografia e di Storia delle Scienze Matematiche e Fisiche* 4 (Roma 1871), 443–63.

Friedlein, Gottfried: 'Der Calculus des Victorius', *Zeitschrift für Mathematik und Physik* 16 (1871), 42–79.

Fuhrmann, Horst: 'Goethe, Frankfurt und die Anfänge der Monumenta Germaniae Historica', *Jahrbuch des Freien Deutschen Hochstifts* 1995, 3–21.

Fuhrmann, Horst: *"Sind eben alles Menschen gewesen". Gelehrtenleben im 19. und 20. Jahrhundert. Dargestellt am Beispiel der Monumenta Germaniae Historica und ihrer Mitarbeiter* (München 1996).

Gentz, Günter: 'Ostern', in Georg Wissowa et al. (eds), *Paulys Real-Encyclopädie der classischen Altertumswissenschaft 18.1* (Stuttgart 1942), 1647–53.

Germann, Nadja: *De temporum ratione. Quadrivium und Gotteserkenntnis am Beispiel Abbos von Fleury und Hermanns von Reichenau* (Leiden 2006).

Gibson, Margaret (ed.): *Boethius. His life, thought and influence* (Oxford 1981).

Ginzel, Friedrich Karl: *Handbuch der mathematischen und technischen Chronologie 3* (Leipzig 1914).

Goldscheider, Franz: 'Über die Einführung des neuen Kalenders in Dänemark und Schweden', *Wissenschaftliche Beilage zum Jahresbericht des Luisenstädtischen Realgymnasiums zu Berlin, Ostern 1898* (Berlin 1898), 1–39.

Gougaud, Louis: *Christianity in Celtic lands* (London 1932, repr. Dublin 1992).

Graff, Eberhard Gottlieb: *Althochdeutscher Sprachschatz oder Wörterbuch der althochdeutschen Sprache 1* (Berlin 1834, repr. Hildesheim 1963).

Graff, Eric: 'The thirteenth figure in the Munich Computus zodiac', *Journal for the History of Astronomy* 36 (2005), 321–34.

Graff, Eric: 'The recension of two Sirmond texts: Disputatio Morini and De divisionibus temporum', in Warntjes & Ó Cróinín (eds), *Computus and its cultural context*, 112–42.

Grosjean, Paul: 'Recherches sur les debuts de la controverse pascale chez les Celtes', *Analecta Bollandiana* 64 (1946), 200–44.

Grosjean, Paul: 'Sur quelques exégètes irlandais du VIIe siècle', *Sacris Erudiri* 7 (1955), 67–98.

Grumel, Venance: 'Le problème de la date paschale aux IIIe et IVe siècles', *Revue des études byzantines* 18 (1960), 163–78.

Gundlach, Wilhelm: 'Über die Columban-Briefe', *Archiv der Gesellschaft für ältere deutsche Geschichtskunde* 15 (1890), 497–526.

Gundlach, Wilhelm: 'Zu den Columban-Briefen', *Archiv der Gesellschaft für ältere deutsche Geschichtskunde* 17 (1892), 425–9.

Gundlach, Wilhelm: 'Columbae sive Columbani abbatis Luxoviensis et Bobbiensis epistolae', *Monumenta Germaniae Historica, Epistolae III* (Berlin 1892), 154–90.

Haller, Klaus: *Die Bayerische Staatsbibliothek in historischen Beschreibungen* (München 1992, second edition München 1998).

Halm, Karl et al. (eds): *Catalogus codicum latinorum Bibliotheca Regiae Monacensis 2, 2* (München 1876).

Hamel, Jürgen: *Geschichte der Astronomie* (Basel 1998).
Harrison, Kenneth: 'Easter cycles and the equinox in the British Isles', *Anglo-Saxon England* 7 (1978), 1–8.
Harrison, Kenneth: 'Luni-solar cycles: Their accuracy and some types of usage', in King & Stevens (eds), *Saints*, 65–78.
Harrison, Kenneth: 'A letter from Rome to the Irish clergy, AD 640', *Peritia* 3 (1984), 222–9.
Harrison, Kenneth: 'Epacts in Irish chronicles', *Studia Celtica* 12/13 (1977/8), 17–32.
Haskins, Charles Homer: *Studies in the history of medieval science* (Cambridge 1924).
Hauck, Albert: *Kirchengeschichte Deutschlands 2* (ninth edition Berlin 1958).
Hauke, Hermann: 'Die Bedeutung der Säkularisation für die bayerischen Bibliotheken', in Josef Kirmeier & Manfred Treml (eds): *Glanz und Ende der alten Klöster. Säkularisation im bayerischen Oberland 1803* (München 1991), 87–97.
Hausberger, Karl: *Geschichte des Bistums Regensburg 1: Mittelalter und frühe Neuzeit* (Regensburg 1989).
Hemmerle, Josef: *Die Benediktinerklöster in Bayern* (München 1951) [abbr.: *Benediktinerklöster* I].
Hemmerle, Josef: *Die Benediktinerklöster in Bayern* (Augsburg 1970) [abbr.: *Benediktinerklöster* II].
Henel, Heinrich: *Studien zum altenglischen Computus* (Leipzig 1934).
Herren, Michael: *The Hisperica Famina: I. The A-Text* (Toronto 1974).
Herren, Michael: 'The pseudonymous tradition in Hiberno-Latin: an introduction', in John Joseph O'Meara & Bernd Naumann (eds), *Latin script and letters A.D. 400–900* (Leiden 1976), 121–31, repr. in Herren, *Latin letters*, article V.
Herren, Michael: 'Some new light on the life of Virgilius Maro Grammaticus', *Proceedings of the Royal Irish Academy* 79C (1979), 27–71, repr. in idem, *Latin letters*, article VII.
Herren, Michael: 'On the earliest Irish acquaintance with Isidore of Seville', in Edward James (ed.), *Visigothic Spain: new approaches* (Oxford 1980), 243–50, repr. in Herren, *Latin letters*, article III.
Herren, Michael W. (ed.): *Insular Latin studies. Papers on Latin texts and manuscripts of the British Isles: 550–1066* (Toronto 1981).
Herren, Michael: 'Sprachliche Eigentümlichkeiten in den hibernolateinischen Texten des 7. und 8. Jahrhunderts', in Heinz Löwe (ed.), *Die Iren und Europa im früheren Mittelalter* (Stuttgart 1982), 425–33, repr. in Herren, *Latin letters*, article III.
Herren, Michael: *Latin letters in early Christian Ireland* (Aldershot 1996).
Heymann, Ernst: 'Bruno Krusch', *Deutsches Archiv für Geschichte des Mittelalters* 4 (1941), 504–18.
Holder-Egger, Oswald: 'Handschriften der Königlichen Bibliothek zu München II', *Neues Archiv der Gesellschaft für ältere deutsche Geschichtskunde* 9 (1884), 541–600.
Holford-Strevens, Leofranc: *The history of time* (Oxford 2005).
Holford-Strevens, Leofranc: 'Paschal lunar calendar up to Bede', *Peritia* 20 (2008), 165–208.
Holford-Strevens, Leofranc: 'Marital discord in Northumbria: Lent and Easter, his and hers', in Warntjes & Ó Cróinín (eds), *Computus and its cultural context*, 143–58.
Holtz, Louis: *Donat et la tradition de l'enseignement grammatical. Étude sur l'Ars Donati et sa diffusion (IVe-IXe siècle) et édition critique* (Paris 1981).
Hoskins, Michael & Gingerich, Owen: 'Medieval Latin astronomy', in Michael Hoskins (ed.), *The Cambridge concise history of astronomy* (Cambridge 1999), 68–93.
Hoskins, Michael: *The history of astronomy* (Oxford 2003).
Hughes, Kathleen: 'The Welsh Latin chronicles: Annales Cambriae and related texts', *Proceedings of the British Academy* 59 (1973), 233–58.

Ideler, Ludwig: *Handbuch der mathematischen und technischen Chronologie, 2 vols* (Berlin 1825–6).

Jadart, Henri: *Dom Jean Mabillon (1632–1707)* (Reims 1879).

Jaffé, Philipp & Wattenbach, Wilhelm: *Ecclesiae metropolitanae Coloniensis codices manuscripti* (Berlin 1874).

Jan, Wilhelm: *Historia cycli dionysiani cum argumentis paschalibus et aliis eo spectantibus* (Wittenberg 1718, repr. J.-P. Migne, *Patrologiae cursus completus, series latina LXVII* (Paris 1865), 453–520).

Janner, Ferdinand: *Geschichte der Bischöfe von Regensburg, vol. 1* (Regensburg 1883).

Jaritz, Gerhard & Moreno-Riaño, Gerson (eds): *Time and eternity. The medieval discourse* (Turnhout 2003).

Jones, Charles W.: 'The Victorian and Dionysiac Paschal tables in the West', *Speculum* 9 (1934), 408–21, repr. in idem, *Bede*, article VIII.

Jones, Charles W.: 'The 'lost' Sirmond manuscript of Bede's computus', *English Historical Review* 52 (1937), 204–19, repr. with omission of the crucial final pages in idem, *Bede*, article X.

Jones, Charles W.: 'Two Easter tables', *Speculum* 13 (1938), 204–5, repr. in idem, *Bede*, article IX.

Jones, Charles W.: *Bedae pseudepigrapha* (Ithaca 1939).

Jones, Charles W.: *Bedae opera de temporibus* (Cambridge 1943).

Jones, Charles W.: 'A legend of St. Pachomius', *Speculum* 18 (1943), 198–210, repr. in idem, *Bede*, article VII.

Jones, Charles W.: *Bedae Venerabilis opera, pars VI, Corpus Christianorum Series Latina CXXXIIIA–C* (Turnhout 1975–80).

Jones, Charles W. (ed. by Wesley M. Stevens): *Bede, the schools and the computus* (Aldershot 1994).

Jones, Leslie Weber: *The script of Cologne from Hildebald to Hermann* (Cambridge 1932).

Jülicher, Adolf: 'Dionysius Exiguus', in Georg Wissowa et al. (eds), *Paulys Real-Encyclopädie der classischen Altertumswissenschaft 5 (9. Halbband)* (Stuttgart 1903), 998–9.

Kalesch, Hans: *Tag und Jahr. Die Geschichte unseres Kalenders* (Zürich 1970).

Kaltenbrunner, Ferdinand: 'Die Vorgeschichte der Gregorianischen Kalenderreform', *Sitzungsberichte der Kaiserlichen Akademie der Wissenschaften, philosophisch-historische Classe* 82 (1876), 289–414.

Kehnel, Annette: *Clonmacnois – the Church and lands of St. Ciarán. Change and continuity in an Irish monastic foundation (6th to 16th century)* (Münster 1997).

Kellner, Stephan & Spethmann, Annemarie: *Historische Kataloge der Bayerischen Staatsbibliothek München, Münchener Hofbibliothek und andere Provenienzen* (Wiesbaden 1996).

Kenney, James F.: *The sources for the early history of Ireland 1: Ecclesiastical. An introduction and guide* (New York 1929, repr. Dublin 1997).

Ker, Neil Ripley: *Catalogue of manuscripts containing Anglo-Saxon* (Oxford 1957, repr. 1990).

King, Margot H. & Stevens, Wesley M. (eds): *Saints, scholars and heroes. Studies in medieval culture in honour of Charles W. Jones, vol. 2* (Minnesota 1979).

Kline, Morris: *Mathematics in Western culture* (London 1954, repr. New York 1964).

Köhler, Wilhelm: *Die karolingischen Miniaturen, Bd. 3* (Berlin 1960).

Krusch, Bruno: *Studien zur christlich-mittelalterlichen Chronologie. Der 84jährige Ostercyclus und seine Quellen* (Leipzig 1880) [abbr.: *Studien* I].

Krusch, Bruno: 'Die Einführung des griechischen Paschalritus im Abendlande', *Neues Archiv der Gesellschaft für ältere deutsche Geschichtskunde* 9 (1884), 99–169.

Krusch, Bruno: 'Über eine Handschrift des Victurius', *Neues Archiv der Gesellschaft für ältere deutsche Geschichtskunde* 9 (1884), 269–81.

Krusch, Bruno: 'Chronologisches aus Handschriften', *Neues Archiv der Gesellschaft für ältere deutsche Geschichtskunde* 10 (1885), 81–94.

Krusch, Bruno: *Ionae vitae sanctorum Columbani, Vedastis, Iohannis. Monumenta Germaniae Historica, Scriptores rerum germanicarum XXXVII* (Hannover 1905).

Krusch, Bruno: 'Das älteste fränkische Lehrbuch der dionysianischen Zeitrechnung', in *Mélanges offerts à M. Émile Chatelain* (Paris 1910), 232–42.

Krusch, Bruno: 'Neue Bruchstücke der Zeitzer Ostertafel vom Jahre 447', *Sitzungsberichte der preußischen Akademie der Wissenschaften, Jahrgang 1933, philosophisch-historische Klasse* (Berlin 1933), 981–97.

Krusch, Bruno: 'Studien zur christlich-mittelalterlichen Chronologie. Die Entstehung unserer heutigen Zeitrechnung', *Abhandlungen der preußischen Akademie der Wissenschaften, Jahrgang 1937, philosophisch-historische Klasse, Nr. 8* (Berlin, 1938) [abbr.: *Studien* II].

Kühnel, Bianca: *The end of time in the order of things. Science and eschatology in early medieval art* (Regensburg 2003).

Kunitzsch, Paul: *Der Almagest. Die Syntaxis Mathematica des Claudius Ptolemäus in arabisch-lateinischer Überlieferung* (Wiesbaden 1974).

Lambert, Pierre-Yves: 'Les commentaires celtique a Bède Le Vénérable I', *Études celtiques* 20 (1983), 119–43.

Lambert, Pierre-Yves: 'Les commentaires celtique a Bède Le Vénérable II', *Études celtiques* 21 (1984), 185–206.

Lambert, Pierre-Yves: ''Thirty' and 'Sixty' in Brittonic', *Cambridge Medieval Celtic Studies* 8 (1984), 29–43.

Lapidge, Michael: 'Theodoros', in Norbert Angermann et al. (eds), *Lexikon des Mittelalters, Bd. 8* (München 1997), 636.

Lapidge, Michael: 'Theodore', in Michael Lapidge et al. (eds), *The Blackwell encyclopaedia of Anglo-Saxon England* (Oxford 1999), 444–6.

Lapidge, Michael & Sharpe, Richard: *A bibliography of Celtic-Latin literature 400–1200* (Dublin 1985).

Law, Vivien: *Wisdom, authority and grammar in the seventh century. Decoding Virgilius Maro Grammaticus* (Cambridge 1995).

Law, Vivien: *Grammar and grammarians in the early middle ages* (London 1997).

Lechner, Anton: *Mittelalterliche Kirchenfeste und Kalendarien in Bayern* (Freiburg 1933).

Leclercq, Henri: *Dom Mabillon, tom. 1* (Paris 1953).

Lehmann, Paul: 'Die alte Klosterbibliothek Fulda und ihre Bedeutung', *Landesbibliothek Fulda* 2 (1928), 5–12, repr. in idem, *Erforschung des Mittelalters. Ausgewählte Abhandlungen und Aufsätze 1* (Leipzig 1941), 213–31.

Lehner, Albert & Berschin, Walter: 'Nachwort', in idem, *Lateinische Kultur im VIII. Jahrhundert. Traube-Gedenkschrift* (St. Ottilien 1989), 243–4.

Lejbowicz, Max: 'Computus. Le nombre et le temps altimédiévaux', in Bernard Ribémont (ed.), *Le temps, sa mesure et sa perception au Moyen Âge* (Caen 1992), 151–96.

Lejbowicz, Max: 'Des tables pascales aux tables astronomique et retour', *Methodos* 6 (2006), 1–67.

Leonardi, Claudio: 'I codici di Marziano Capella', *Aevum* 33 (1959), 443–89 [abbr.: 'Codici I'].

Leonardi, Claudio: 'I codici di Marziano Capella', *Aevum* 34 (1960), 1–99 [abbr.: 'Codici II'].

Lindhagen, Arvid: 'Der schwedische Kalender seit dem Jahr 1700', *Arkiv för matematik, astronomi och fysik* 5 (1909), article 18.

Linn, Charles F.: *The ages of mathematics, vol. 2: Mathematics East and West* (New York 1977).

Löfstedt, Bengt: *Der hibernolateinische Grammatiker Malsachanus* (Uppsala 1965).

Löfstedt, Bengt: 'Spät- und Vulgärlateinisches in der Sprache des Virgilius Maro Grammaticus', *Latomus* 40 (1981), 121–6.

Löfstedt, Bengt: 'Zum Wortschatz des Virgilius Maro Grammaticus', *Philologus* 1982 (126), 99–110.

Lowe, Elias Avery: *Codices Latini Antiquiores. A palaeographical guide to Latin manuscripts prior to the ninth century, vol. 2: Great Britain and Ireland; vol. 5: France: Paris; vol. 8: Germany: Altenburg–Leipzig* (Oxford 1935, 1950, 1959).

Mabillon, Jean: *Veterum analectorum IV, complectens iter germanicum* (Paris 1685, only the description of the tour through Switzerland and Germany is republished under the title *Iter germanicum* in Hamburg 1717).

Machielsen, John: *Corpus Christianorum Series Latina, Clavis patristica pseudepigraphorum medii aevi, vol. 3A: Artes liberales* (Turnhout 2003).

Mac Carthy, Bartholomew: *The codex Palatino-Vaticanus, No. 830, Todd Lecture Series 3* (Dublin 1892).

Mac Carthy, Bartholomew: *Annals of Ulster, vol. 4* (Dublin 1901).

Mac Ginty, Gerard: 'The Irish Augustine: De mirabilibus sacrae scripturae', in Próinséas Ní Chatháin & Michael Richter (eds), *Irland und die Christenheit. Ireland and Christendom. Bibelstudien und Mission. The Bible and the missions* (Stuttgart 1987), 70–83.

Manitius, Max: *Geschichte der lateinischen Literatur des Mittelalters 1: Von Justinian bis zur Mitte des zehnten Jahrhunderts* (München 1911).

Masi, Michael: *Boethian number theory: a translation of the De institutione arithmetica (with introduction and notes)* (Amsterdam 1983).

Mc Carthy, Daniel: 'Easter principles and a fifth-century lunar cycle used in the British Isles', *Journal for the History of Astronomy* 24 (1993), 204–24.

Mc Carthy, Daniel: 'The origin of the *latercus* paschal cycle of the Insular Celtic churches', *Cambrian Medieval Celtic Studies* 28 (1994), 25–49.

Mc Carthy, Daniel: 'The lunar and Paschal tables of De ratione paschali attributed to Anatolius of Laodicea', *Archive for History of Exact Sciences* 49 (1996), 285–320, repr. in Mc Carthy & Breen, *De ratione paschali*, 148–80.

Mc Carthy, Daniel: 'The chronology of the Irish annals', *Proceedings of the Royal Irish Academy* 98C (1998), 203–55.

Mc Carthy, Daniel P.: 'The emergence of anno domini', in Jaritz & Moreno-Riaño (eds), *Time and eternity*, 31–53.

Mc Carthy, Daniel P.: *The Irish Annals. Their genesis, evolution and history* (Dublin 2008).

Mc Carthy, Daniel P. & Breen, Aidan: *The ante-Nicene Christian Pasch. De ratione paschali. The Paschal tract of Anatolius, bishop of Laodicea* (Dublin 2003).

Mc Carthy, Daniel & Ó Cróinín, Dáibhí: 'The 'lost' Irish 84-year Easter table rediscovered', *Peritia* 6–7 (1987–8), 227–42, repr. in Ó Cróinín, *Early Irish history*, 58–75.

McCluskey, Stephen C.: 'Astronomies in the Latin West from the fifth to the ninth centuries', in Butzer & Lohrmann (eds), *Science*, 139–60.

McCluskey, Stephen C.: *Astronomies and cultures in early medieval Europe* (Cambridge 1998).

McGurk, Patrick: 'Computus Helperici: Its transmission in England in the eleventh and twelfth century', *Medium Aevum* 43 (1974), 1–5.

Meersseman, Giles G. & Adda, Edvige: *Manuale di computo con ritmo mnemotecnico dell'arcidiacono Pacifico di Verona* (Padua 1966).

Meier, Gabriel: 'Die sieben freien Künste im Mittelalter 2', *Jahresbericht über die Lehr- und Erziehungs-Anstalt des Benediktiner-Stiftes Maria-Einsiedeln der Studienjahre 1886/87* (Einsiedeln 1887), 3–36.

Mohrmann, Christine: 'Le conflit pascal au IIe siècle. Note philologique', *Vigiliae Christianae* 16 (1962), 154–71.

Molinier, Auguste: 'Manuscrits de la Bibiothèque d'Angers', in *Catalogue genéral des manuscrits des bibliothèques publiques de France, Départments, tom. 31* (Paris 1898), 189–618.

Mommsen, Theodor: 'Zeitzer Ostertafel vom Jahre 447', *Abhandlungen der Königlichen Akademie der Wissenschaften zu Berlin, Jahrgang 1862, philosophisch-historische Klasse* (Berlin 1963), 537–66.

Mommsen, Theodor: 'Chronographus anni CCCLIIII', *Monumenta Germinae Historica, Auctores antiquissimi IX* (Berlin 1892), 13–148.

Mommsen, Theodor: 'Liber paschalis codicis Cicensis a. CCCCXXXXVII', *Monumenta Germaniae Historica, Auctores antiquissimi IX* (Berlin 1892), 501–9.

Mommsen, Theodor: 'Victorii Aquitani cursus paschalis annorum DXXXII ad Hilarum archidiaconum ecclesiae Romae A. CCCCLVII', *Monumenta Germinae Historica, Auctores antiquissimi IX* (Berlin 1892), 667–735.

Mommsen, Theodor: 'Paschale Campanum A. 464–599', *Monumenta Germaniae Historica, Auctores antiquissimi IX* (Berlin 1892), 744–50.

Mommsen, Theodor: 'Die Papstbriefe bei Beda', *Neues Archiv der Gesellschaft für ältere deutsche Geschichtskunde* 17 (1892), 387–96.

Mordek, Hubert: 'Dionysius Exiguus', in Robert-Henri Bautier et al. (eds), *Lexikon des Mittelalters, Bd. 3* (München 1986), 1088–92.

Mosshammer, Alden A.: *The Easter Computus and the origins of the Christian era* (Oxford 2008).

Mountford, James Frederick: 'De mensium nominibus', *The Journal of Hellenic Studies* 43 (1923), 102–16.

Mütherich, Florentine: 'Die Buchmalerei am Hofe Karls des Grossen', in Wolfgang Braunfels (ed.): *Karl der Grosse. Lebenswerk und Nachleben 3: Karolingische Kunst* (Düsseldorf 1965), 9–53.

Mütherich, Florentine: 'Die Bilder des Leidener Aratus', in Bernhard Bischoff et al. (eds), *Aratea. Kommentar zum Aratus des Germanicus Ms. Voss. Lat. Q. 79. Bibliotheek der Rijksuniversiteit Leiden* (Luzern 1989), 31–68, repr. in Mütherich, *Studies*, 147–265.

Mütherich, Florentine: 'Die Erneuerung der Buchmalerei am Hof Karls des Großen', in Christoph Stiegemann & Matthias Wemhoff (eds), 799. *Kunst und Kultur der Karolingerzeit* (Mainz 1999), 560–609, repr. in Mütherich, *Studies*, 1–62.

Mütherich, Florentine: *Studies in Carolingian manuscript illumination* (London 2004).

Mütherich, Florentine & Dachs, Karl (eds): *Regensburger Buchmalerei von frühkarolingischer Zeit bis zum Ausgang des Mittelalters* (München 1987).

Mütherich, Florentine & Gaehde, Joachim E.: *Karolingische Buchmalerei* (München 1976).

Neugebauer, Otto: *Ethiopic Astronomy and Computus* (Wien 1979).

Neugebauer, Otto: 'On the Computus paschalis of "Cassiodorus"', *Centaurus* 25 (1982), 292–302.

Neuß, Wilhelm: 'Eine karolingische Kopie antiker Sternzeichenbilder im Codex 3307 der Biblioteca Nacional zu Madrid', *Zeitschrift des deutschen Vereins für Kunstwissenschaft* 8 (Berlin 1941), 113–40.

Nicklin, T: 'The date and origin of pseudo-Anatolius De ratione paschali', *The Journal of Philology* 28 (1903), 137–51.

Noris, Heinrich: *Annus et epochae Syromacedonum ... Additis fastis consularibus anonymi omnium optimis. Accesserunt nuper dissertationes De paschali Latinorum cyclo annorum LXXXIV, ac Ravennate annorum XCV* (Leipzig 1696).

North, John: *The Fontana history of astronomy and cosmology* (London 1994).

Obrist, Barbara: 'The astronomical sundial in Saint Willibrord's calendar and its early medieval context', *Archives d'histoire doctrinale et littéraire du moyen âge* 67 (2000), 71–118.

Obrist, Barbara: 'La representation carolingienne du zodiaque. À propos du manuscrit de Bâle, Universitätsbibliothek, F III 15a', *Cahiers de civilisation médiévale* 44 (2001), 3–33.

O'Connell, D.J.: 'Easter cycles in the early Irish church', *Journal of the Royal Society of Antiquaries of Ireland* 66 (1936), 67–106.

Ó Cróinín, Dáibhí: 'An Old Irish gloss in the Munich Computus', *Éigse* 18 (1981), 289–90, repr. in idem, *Early Irish history*, 131–2.

Ó Cróinín, Dáibhí: 'A seventh-century Irish computus from the circle of Cummianus', *Proceedings of the Royal Irish Academy* 82C (1982), 405–30, repr. in idem, *Early Irish history*, 99–132.

Ó Cróinín, Dáibhí: 'Mo Sinu maccu Min and the computus at Bangor', *Peritia* 1 (1982), 281–95, repr. in idem, *Early Irish history*, 35–47.

Ó Cróinín, Dáibhí: 'Hiberno-Latin *calcenterus*', *Peritia* 1 (1982), 296–7, repr. in idem, *Early Irish history*, 56–7.

Ó Cróinín, Dáibhí: 'Early Irish annals from Easter tables: a case restated', *Peritia* 2 (1983), 74–86, repr. in idem, *Early Irish history*, 76–86.

Ó Cróinín, Dáibhí: 'The Irish provenance of Bede's computus', *Peritia* 2 (1983), 229–47, repr. in idem, *Early Irish history*, 173–90.

Ó Cróinín, Dáibhí: 'Rath Melsigi, Willibrord, and the earliest Echternach manuscripts', *Peritia* 3 (1984), 17–49, repr. in idem, *Early Irish history*, 145–65.

Ó Cróinín, Dáibhí: ''New heresy for old': Pelagianism in Ireland and the papal letter of 640', *Speculum* 60 (1985), 505–16, repr. in idem, *Early Irish history*, 87–98.

Ó Cróinín, Dáibhí: 'Early Echternach manuscript fragments with Old Irish glosses', in Georges Kiesel & Jean Schroeder (eds), *Willibrord. Apostel der Niederlande, Gründer der Abtei Echternach* (Luxembourg 1989), 135–43, repr. in Ó Cróinín, *Early Irish history*, 133–44.

Ó Cróinín, Dáibhí: 'The date, provenance and earliest use of the works of Virgilius Maro Grammaticus', in Günter Bernt et al. (eds), *Tradition und Wertung: Festschrift Franz Brunhölzl* (Sigmaringen 1989), 13–22, repr. in Ó Cróinín, *Early Irish history*, 191–200.

Ó Cróinín, Dáibhí: 'The Irish as mediators of antique culture on the Continent', in Butzer & Lohrmann (eds), *Science*, 41–52.

Ó Cróinín, Dáibhí: *Early medieval Ireland 400–1200* (London 1995).

Ó Cróinín, Dáibhí: 'The computistical work of Columbanus', in Michael Lapidge (ed.), *Columbanus. Studies on the Latin writings* (Woodbridge 1997), 264–70, repr. in Ó Cróinín, *Early Irish history*, 48–55.

Ó Cróinín, Dáibhí: 'The earliest Old Irish glosses', in Rolf Bergmann et al. (eds), *Mittelalterliche volkssprachliche Glossen* (Heidelberg 2001), 7–31.

Ó Cróinín, Dáibhí: 'Bede's Irish computus', in idem, *Early Irish history*, 201–12.

Ó Cróinín, Dáibhí: *Early Irish history and chronology* (Dublin 2003).

Ó Cróinín, Dáibhí: 'Hiberno-Latin literature to 1169', in idem (ed.): *A new history of Ireland I: Prehistoric and early Ireland* (Oxford 2005), 371–404.

Ó Cróinín, Dáibhí: 'Dionysius Exiguus in the classroom', in Joseph W. Dauben et al. (eds), *Mathematics celestial and terrestrial. Festschrift für Menso Folkerts zum 65. Geburtstag* (Halle 2008), 253–74.

Ó Cróinín, Dáibhí: 'The continuity of the Irish computistical tradition', in Warntjes & Ó Cróinín (eds), *Computus and its cultural context*, 324–47.

Ohashi, Masako: 'Sexta aetas continent annos praeteritos DCCVIIII (Bede, De temporibus, 22): A scribal error?', in Jaritz & Moreno-Riaño (eds), *Time and eternity*, 55–61; now revised as 'The annus domini and the sexta aetas: problems in the transmission of Bede's De temporibus', in Warntjes & Ó Cróinín (eds), *Computus and its cultural context*, 190–203.

Ogg, George: 'Hippolytus and the introduction of the Christian era', *Vigiliae Christianae* 16 (1962), 2–18.

Oulton, J.E.L.: 'The epistle of Cummian, De controversia paschali', *Studia Patristica* 1 (1957), 128–33.

Overgaauw, Eef & Steving, Frank-Joachim: *Die Zeitzer Ostertafel aus dem Jahre 447* (Petersberg 2005).

Pannekoek, Anton: *A history of astronomy* (New York 1961).

Peden, Alison: *Abbo of Fleury and Ramsey: Commentary on the Calculus of Victorius of Aquitaine* (Oxford 2003).

Pedersen, Olaf: *A survey of the Almagest* (Odense 1974).

Pederson, Olaf: 'The ecclesiastical calendar and the life of the church', in George V. Coyne, Michael A. Hoskin & Olaf Pedersen (eds), *Gregorian reform of the calendar* (Vatican 1983), 17–74.

Pedersen, Olaf: *Early physics and astronomy* (London 1974, revised edition Cambridge 1993).

Petavius, Dionysius: *Opus de doctrina temporum, 2 vols* (Paris 1627, second edition Antwerp 1703).

Pillonel-Wyrsch, Roland-Pierre: *Le calcul de la date de Pâques au Moyen Âge. Analyse et commentaires sur 'De Temporum Ratione' de Bède* (Fribourg 2003).

Pingree, David: 'Boethius' geometry and astronomy', in Gibson (ed.), *Boethius*, 155–61.

Piper, Ferdinand: *Karls des Grossen Kalendarium und Ostertafel* (Berlin 1858).

Pitra, Jean-Baptiste: *Spicilegium Solesmense complectens sanctorum patrum scriptorumque ecclesiasticorum anecdota hactenus opera* (Paris 1852).

Plotzek, Joachim M. & Surmann, Ulrike (eds): *Glaube und Wissen im Mittelalter, Die Kölner Dombibliothek* (München 1998).

Plummer, Charles: *Venerabilis Baedae opera historica, 2 vols* (Oxford 1896).

Poole, Reginald Lane: 'The earliest use of the Easter cycle of Dionysius', *English Historical Review* 33 (1918), 57–62, 210–3.

Reeves, William: 'On Augustin, an Irish writer of the seventh century', *Proceedings of the Royal Irish Academy* 7 (1861), 514–22.

Reifferscheid, August: *Bibliotheca partum latinorum Italica 2* (Wien 1871, repr. Hildesheim 1976).

Reindel, Kurt: 'Vom Beginn des Quadriviums', *Deutsches Archiv für Geschichte des Mittelalters* 15 (1959), 516–22.

Richard, Marcel: 'Notes sur le comput de cent-douze ans', *Revue des études byzantines* 24 (1966), 257–77.

Richard, Marcel: 'Le comput pascal par octaétéris', *Le muséon* 87 (1974), 307–39.

Rissel, Maria: *Rezeption antiker und patristischer Wissenschaft bei Hrabanus Maurus* (Frankfurt 1976).

Rissel, Maria: 'Hrabans Liber de computo als Quelle der Fuldaer Unterrichtspraxis in den Artes Arithmetik und Astronomie', in Winfried Böhne (ed.), *Hrabanus Maurus und seine Schule* (Fulda 1980), 138–55.

Roger, Maurice: *L'enseignement des lettres classique d'Ausone à Alcuin* (Paris 1905, repr. Hildesheim 1968).

Rose, Valentin: *Die Handschriften-Verzeichnisse der Königlichen Bibliothek zu Berlin, Bd. 12: Verzeichnis der lateinischen Handschriften 1: Die Meerman-Handschriften des Sir Thomas Phillipps* (Berlin 1893).

Rühl, Franz: *Chronologie des Mittelalters und der Neuzeit* (Berlin 1897).

Ruinart, Thierry: *Mabillon* (Paris 1933).

Ryan, John: 'The abbatial succession at Clonmacnois. From the foundation of the monastery to the coming of the Norse (A.D. 545–799)', in idem (ed.), *Féil-sgríbhinn Eóin Mhic Néill. Essays and studies presented to Professor Eoin MacNeill* (Dublin 1940), 490–507.

Sanftl, Colomann: *Catalogus veterum codicum manuscriptorum ad S. Emmeram Ratisbonae, Pars II–IV* (Regensburg 1809); these second to fourth parts of Sanftl's hand-written catalogue are now kept in the Bayerische Staatsbibliothek in Munich with the shelf-marks Cbm Cat. 14(2), Cbm Cat. 14(3), and Cbm Cat. 14(4) respectively.

Saxl, Fritz: 'Beiträge zu einer Geschichte der Planetendarstellungen im Orient und im Okzident', *Der Islam* 3 (1912), 151–77.

Schäferdiek, Knut: 'Columbans Wirken im Frankenreich (591–612)', in Heinz Löwe (ed.), *Die Iren und Europa im früheren Mittelalter* (Stuttgart 1982), 171–201, repr. in Knut Schäferdiek (ed. by Winrich A. Löhr & Hanns Christof Brennecke), *Schwellenzeit. Beiträge zur Geschichte des Christentums in Spätantike und Frühmittelalter* (Berlin 1996), 349–79.

Schäferdiek, Knut: 'Der irische Osterzyklus des sechsten und siebten Jahrhunderts', *Deutsches Archiv für Erforschung des Mittelalters* 39 (1983), 357–78.

Scherrer, Gustav: *Verzeichniss der Handschriften der Stiftsbibliothek von St. Gallen* (Halle 1875).

Schissel, O. & Ellend, M.: 'Berechnung des Sonnen-, Mond- und Schaltjahrszirkels in der griechisch-christlichen Chronologie', *Byzantinische Zeitschrift* 42 (1943/9), 250–7.

Schmid, Joseph: 'Zur Geschichte der Gregorianischen Kalenderreform', *Historisches Jahrbuch der Görres-Gesellschaft* 3 (1882), 388–415, 543–95.

Schmid, Joseph: *Die Osterfestberechnung auf den britischen Inseln vom Anfang des vierten bis zum Ende des achten Jahrhunderts* (Regensburg 1904).

Schmid, Joseph: *Die Osterfestberechnung in der abendländischen Kirche* (Freiburg 1907).

Schmitz, Gerhard: 'Zur Entstehungsgeschichte der *Monumenta Germaniae Historica*', in Heinrich Beck et al. (eds), *Zur Geschichte der Gleichung „germanisch–deutsch"* (Berlin 2004), 503–22.

Schüling, Hermann: 'Die Handbibliothek des Bonifatius', *Archiv für Geschichte des Buchwesens* 4 (1961–3), 286–349.

Schwartz, Eduard: 'Christliche und jüdische Ostertafeln', *Abhandlungen der königlichen Gesellschaft der Wissenschaften zu Göttingen, philologisch-historische Klasse, Band 8, Nr. 6* (Berlin 1905).

Schwartz, Eduard: 'Osterbetrachtungen', *Zeitschrift für neutestamentliche Wissenschaften* 7 (1906), 1–33, repr. in idem, *Gesammlte Schriften, 5. Bd.: Zum neuen Testament und zum frühen Christentum* (Berlin 1963), 1–41.

Seebass, Otto: 'Über die Handschriften der Sermonen und Briefe Columbas von Luxeuil', *Neues Archiv der Gesellschaft für ältere deutsche Geschichtskunde* 17 (1892), 243–59.

Seebass, Otto: 'Columba der Jüngere', in Albert Hauck (ed.), *Realencyklopädie für protestantische Theologie und Kirche, Bd. 4* (third revised edition Graz 1969), 241–7.

Sickel, Theodor: 'Die Lunarbuchstaben in den Kalendarien des Mittelalters', *Sitzungsberichte der Kaiserlichen Akademie der Wissenschaften, philosophisch-historische Classe* 38 (1862), 153–201.

Simonetti, Manlio: 'De mirabilibus sacrae scripturae. Un trattato irlandese sui miracoli della Sacra Scrittura', *Romanobarbarica* 4 (1979), 225–51.

Singer, Charles: *A short history of scientific ideas to 1900* (Oxford 1959).

Smyth, Marina: 'Isidore of Seville and early Irish cosmography', *Cambridge Medieval Celtic Studies* 14 (1987), 69–102.

Smyth, Marina: *Understanding the universe in seventh-century Ireland* (Woodbridge 1996).

Sotheby & Co: *Bibliotheca Phillippica Medieval manuscripts: New Series: Sixth part: Catalogue of manuscripts on papyrus, vellum and paper of the 7th century to the 18th century* (London 1971).

Spilling, Herrad: 'Die frühe Phase karolingischer Minuskel in Fulda', in Gangolf Schrimpf (ed.): *Kloster Fulda in der Welt der Karolinger und Ottonen* (Frankfurt 1996), 249–84.

Springsfeld, Kerstin: *Alkuins Einfluß auf die Komputistik zur Zeit Karls des Großen* (München 2003).

Springsfeld, Kerstin: 'Rechnen', in Wolfgang Dressen, Georg Minkenberg & Adam C. Oellers (eds): *Ex oriente. Isaak und der weiße Elefant, Bd. 1: Die Reise des Isaak. Bagdad: 800 und heute* (Mainz 2003), 224–33.

Springsfeld, Kerstin: 'Eine Beschreibung der Handschrift St. Gallen, Stiftsbibliothek, 225', in Warntjes & Ó Cróinín (eds), *Computus and its cultural context*, 204–37.

Stahl, William H.: 'Dominant traditions in early medieval Latin science', *Isis* 50 (1959), 95–124.

Stahl, William H.: *Roman science. Origins, development, and influence to the later middle ages* (Madison 1962, repr. Westport 1978).

Steinmeyer, Elias & Sievers, Eduard: *Die althochdeutschen Glossen, Bd. 4* (Berlin 1898, repr. Zürich 1969).

Stenton, Frank: *Anglo-Saxon England* (Oxford 1943, repr. Oxford, 2001).

Stevens, Wesley M.: 'Rabani Mongontiacensis episcopi De computo', in John McCulloh & Wesley M. Stevens, *Rabani Mauri Martyrologium, De computo, Corpus Christianorum Continuatio Medievalis XLIV* (Turnhout 1979), 163–323.

Stevens, Wesley M.: 'Scientific instruction in early Insular schools', in Herren (ed.), *Insular Latin*, 83–111, repr. in Stevens, *Cycles of time*, article IV.

Stevens, Wesley M.: 'Bede's scientific achievement', *Jarrow Lecture* 1985, 1–43.

Stevens, Wesley M.: 'Compotistica et astronomica in the Fulda School', in King & Stevens (eds), *Saints*, 27–63, repr. in Stevens, *Cycles of time*, article IX.

Stevens, Wesley M.: 'Cycles of time: calendrical and astronomical reckonings in early science', in Julius T. Fraser & Lewis Rowell (eds), *Time and process* (Madison 1993), 27–51, repr. in Stevens, *Cycles of time*, article I.

Stevens, Wesley M.: *Cycles of time and scientific learning in medieval Europe* (Aldershot 1995).

Stevens, Wesley: 'A present sense of things past: Quid est enim tempus?', in Jaritz & Moreno-Riaño, *Time and eternity*, 9–28.

Stevens, Wesley: 'Karolingische Renovatio in Wissenschaft und Literatur', in Christoph Stiegemann & Matthias Wemhoff (eds), 799. *Kunst und Kultur der Karolingerzeit* (Mainz 1999), 662–80.

Störmer, Wilhelm: 'Eine Adelsgruppe um die Fuldaer Äbte Sturmi und Egil und den Holzkirchener Klostergründer Troand. Beobachtungen zum bayerisch-alemannisch-ostfränkischen Adel des 8. bis 9. Jahrhunderts', in Richard van Dülmen (red.), *Gesellschaft und Herrschaft. Forschungen zu sozial- und landesgeschichtlichen Problemen vornehmlich in Bayern. Eine Festgabe für Karl Bosl zum 60. Geburtstag* (München 1969), 1–34.

Störmer, Wilhelm: *Früher Adel. Studien zur politischen Führungsschicht im fränkisch-deutschen Reich vom 8. bis 11. Jahrhundert, 2 vols* (Stuttgart 1973).

Stotz, Peter: *Handbuch zur lateinischen Sprache des Mittelalters, Bd. 1* (München 2002).

Strobel, August: *Ursprung und Geschichte des frühchristlichen Osterkalenders* (Berlin 1977).

Strobel, August: *Texte zur Geschichte des frühchristlichen Osterkalenders* (Münster 1984).

Struik, Dirk J.: *A concise history of mathematics* (New York 1948, third revised edition New York 1967).

Stubbs, William: 'Theodorus of Tharsus', in William Smith & Henry Wace (eds): *A dictionary of Christian biography, literature, sects and doctrines during the first eight centuries, vol. 4* (London 1888), 926–32.

Swarzenski, Georg: *Denkmäler der süddeutschen Malerei des frühen Mittelalters, Bd. 1* (Leipzig 1901).

Tannery, Paul: 'Notes sur la Pseudo-Géométrie de Boèce', *Bibliotheca Mathematica* 1 (1900), 39–50.

Teres, Gustav: 'Time computations and Dionysius Exiguus', *Journal for the History of Astronomy* 15 (1984), 177–88.

Thorndike, Lynn & Kibre, Pearl: *A catalogue of incipits of mediaeval scientific writings in Latin* (Cambridge 1937, revised edition London 1963).

Traube, Ludwig: 'Virgilius Maro grammaticus', *Hermes* 24 (1889), 647–9, repr. in Traube, *Vorlesungen 3*, 157–8.

Traube, Ludwig: 'Computus Helperici', *Neues Archiv der Gesellschaft für ältere deutsche Geschichtskunde* 18 (1893), 73–105, repr. with editional notes in Traube, *Vorlesungen 3*, 128–156.

Traube, Ludwig (ed. by Gabriel Silagi): *Rückblick auf meine Lehrthätigkeit* (originally dated to 11/12/1901, printed München 1988).

Traube, Ludwig (ed. by Paul Lehmann et al.): *Vorlesungen und Abhandlungen, 3 vols* (München 1909–1920, repr. München 1965).

Ussher, James: *Britannicarum ecclesiarum antiquitates* (Dublin 1639, second edition London 1687).

van der Hagen, Johannes: *Observationes in Prosperi Aquitani chronicon integrum ejusque LXXXIV annorum cyclum, et in anonymi cyclum LXXXIV annorum a Muratorio editum, nec non in anonymi laterculum paschalem centum annorum a Bucherio editum* (Amsterdam 1733).

van der Hagen, Johannes: *Observationes in veterum patrum et pontificium, prologos et epistolas paschales, aliosque antiquos de ratione paschali scriptores* (Amsterdam 1734).

van der Hagen, Johannes: *Observationes in Heraclii imperatoris methodum paschalem; ut et in Maximi monachi computum paschalem; nec non in anonymi chronicon paschale* (Amsterdam 1736).

van der Hagen, Johannes: *Dissertationes de cyclis paschalibus ..., ut et de enneadecaeteridis ..., nec non de computo solari* (Amsterdam 1736).

van de Vyver, André: 'L'évolution du comput alexandrin et romain du IIIe au Ve siècle', *Revue d'histoire ecclésiastique* 52 (1957), 5–25.

van Euw, Anton: 'Kat. Nr. 23: Beda Venerabilis: Naturlehre, historiographische und zeitrechnerische Werke (Dom Hs. 103)', in Plotzek & Surmann (eds), *Glaube und Wissen*, 129–35.

van Euw, Anton: 'Kat. Nr. 24: Kompendium der Zeitrechnung, Naturlehre und Himmelskunde (Dom HS 83^{II})', in Plotzek & Surmann (eds), *Glaube und Wissen*, 136–56.

van Wijk, Walter E.: 'Le computu emendé de Reinherus de Paderborn (1171), publié d'après le Ms V.P.L. 191-E de la Bibliothèque de l'Université de Leiden', *Verhandelingen der Koninklijke Nederlandse Akademie van Wetenschappen, afd. letterkunde 59, No. 3* (Amsterdam 1951).

Vogtherr, Thomas: *Zeitrechnung. Von den Sumerern bis zur Swatch* (München 2001).

von den Brincken, Anna-Dorothee: *Historische Chronologie des Abendlandes. Kalenderreformen und Jahrtausendrechnungen* (Stuttgart 2000).

Walker, George Stuart Murdoch: *Sancti Columbani opera* (Dublin 1957, repr. Dublin 1997).

Wallis, Faith: *Bede: The reckoning of time* (Liverpool 1999).

Walsh, Maura: 'Some remarks on Cummian's paschal letter and the commentary on Mark ascribed to Cummian', in Próinséas Ní Chatháin & Michael Richter (eds), *Irland und die Christenheit. Ireland and Christendom. Bibelstudien und Mission. The Bible and the missions* (Stuttgart 1987), 216–229.

Walsh, Maura & Ó Cróinín, Dáibhí: *Cummian's letter De controversia paschali, together with a related Irish computistical tract, De ratione conputandi* (Toronto 1988).

Warntjes, Immo: 'A newly discovered Irish computus: Computus Einsidlensis', *Peritia* 19 (2005), 61–4.

Warntjes, Immo: 'The Munich Computus and the 84 (14)-year Easter reckoning', *Proceedings of the Royal Irish Academy* 107C (2007), 31–85.

Warntjes, Immo: 'The earliest occurrence of Old English gerīm and its Anglo-Irish computistical context', *Anglia* 127 (2009), 91–105.

Warntjes, Immo: 'The argumenta of Dionysius Exiguus and their early recensions', in Warntjes & Ó Cróinín (eds), *Computus and its cultural context*, 40–111.

Warntjes, Immo & Ó Cróinín, Dáibhí (eds): *Computus and its cultural context in the Latin West, AD 300–1200* (Turnhout 2010).

Wesche, Markus: 'Der erste Band der Monumenta Germaniae Historica: Scriptores I (1826)', in Alfred Gawlik (red.), *Zur Geschichte und Arbeit der Monumenta Germaniae Historica* (München 1996), 17–21.

White (now Peden), Alison: 'Boethius in the medieval quadrivium', in Gibson (ed.), *Boethius*, 162–205.

Widemann, Josef: *Die Traditionen des Hochstifts Regensburg und des Klosters S. Emmeram* (München 1943).

Wiesenbach, Joachim: 'Wilhelm von Hirsau. Astrolab und Astronomie im 11. Jahrhundert', in Klaus Schreiner (ed.), *Hirsau. St. Peter und Paul 1091–1991. Bd. 2: Geschichte, Lebens- und Verfassungsformen eines Reformklosters* (Stuttgart 1991), 109–54.

Wiesenbach, Joachim: *Sigebert von Gembloux. Liber decennalis* (München 1998).

Williams, Burma P. & Williams, Richard S.: 'Finger numbers in the Greco-Roman world and the early middle ages', *Isis* 86 (1995), 587–608.

Wilmart, André: 'Catéchèses celtiques', in idem, *Analecta Reginensia, Studi e Testi 59* (Vatican 1933), 29–112.

Wilmart, André: 'Un nouveau texte du faux concile de Césarée sur le comput pascal', in idem, *Analecta Reginensia, Studi e Testi 59* (Vatican 1933), 19–27.

Wilmart, André: *Codices reginenses latini 2: Codices 251–500* (Vatican 1945).

Wislicenus, Walter Friedrich: *Der Kalender in gemeinverständlicher Darstellung* (Leipzig 1905, second edition Leipzig 1914).

Wußing, Hans et. al. (eds): *4000 Jahre Algebra* (Berlin 2003).

Zaitsev, Evgeny A.: 'The meaning of early medieval geometry: From Euclid and surveyors' manuals to Christian philosophy', *Isis* 90 (1999), 522–53.

Ziegler, Walter: *Das Benediktinerkloster St. Emmeram zu Regensburg in der Reformationszeit* (Kallmünz 1970).

Zinner, Ernst: *Verzeichnis der astronomischen Handschriften des deutschen Kulturgebietes* (München 1925).

INDICES

Every publication incorporating an edition deals, first and foremost, with texts, many of which are published, while others still remain unpublished. Accordingly, the following indices list sources and manuscripts cited in the present book. Other indices, of persons, places, and terms, were not felt to be helpful to the reader, as only very few names and places are mentioned throughout the book, most of which being easily traceable through the two indices provided in the following; as for discussions of computistical terms and concepts, a guide is provided by the Munich Computus itself and the table of contents of this text given at the beginning of the present book may serve as the easiest way of accessing the relevant discussions within this book (e.g., if a reader is interested in the *saltus*, he should turn to the chapter dealing with the *saltus* in the Munich Computus (chapter 62) and will there find a thorough commentary on the Munich computist's discussion and context of this computistical feature, as well as cross-references to the relevant passages in the introduction).

As for the indices of sources and manuscripts, these are provided in as much detail as possible. Texts are listed by author and title, with the name of the author being in small capitals, the title in italics; if a text is anonymous, it appears under its title, which is additionally in small capitals; if it is still unpublished, the MS details are additionally provided. The texts edited by Borst in his monumental *Schriften zur Komputistik* are listed for ease of reference under the abbreviated Latin title assigned to them by Borst. For each text, the chapters cited in the present book are explicitly listed, with the referencing system applying only Arabic numerals, which will be easily understood when the editions of these texts are consulted; it is hoped that these details may prove helpful to scholars of texts cited in the present book; if, e.g., a scholar is particularly interested in Isidore's discussion of Easter calculation, which constitutes book 6, chapter 17 of the *Etymologiae*, he will only need to turn to ISIDORE, *Etymologiae*, 6.17 in the index. Throughout the indices, references to pages within this book are in normal font and appear first, references to the commentary provided in the apparatus of the edition proper are by chapter and line and additionally printed in bold (e.g., if a certain text is discussed in the commentary to chapter 50, lines 29–39, **50.29–39** will be found in the index).

BIBLICAL REFERENCES

Genesis:
1:1–19: **66.2–7**
1:3: **11.2–3**
1:4: **8.2–10**
1:5: CC; **8.2–10 / 66.2–7**
1:14–19: **53.5–11**
1:16: **46.2–5,9–13**
7:4: **66.8–12**
7:11: **66.8–12**
7:12: **66.8–12**
7:17: **66.8–12**

Exodus:
12: **53.5–11**
12:3–6: **44.44–46**
12:37: **44.48–50**
12:41: **44.48–50**
12:42: **44.47–48**
12:51: **44.47–48**

Leviticus:
11:45: **44.47–48**
12:3: **24.2–11**
19:36: **44.47–48**
23:5: **24.2–11**
25:38: **44.47–48**
26:16: **44.47–48**

Numbers:
9:1–5: **59.14–17**
9:3: **55.2–12**
9:6–11: **59.21–27**
15:41: **44.47–48**

Deuteronomy:
16:1: **55.2–12**
23:4: **55.2–12**

Joshua:
10:12–14: **66.13–14**

4 Kings:
20:12: **66.15–19**

Esther:
3:7: **55.2–12**
3:13: **55.2–12**
8:12: **55.2–12**
9:1: **55.2–12**

Psalms:
47:15: **43.2–3**
118:62: **8.38–43**

Isaiah:
38:5–8: **66.15–19**

Matthew:
5:26: **7.2–11**
21:1–17: **44.55–58**
21:18–21: **44.58–62**
26:17: **44.66–69**
26:26: 326; **44.66–69**
26:27: 326
27:45: **66.22–23**

Mark:
11:1–17: **44.55–58**
11:12–14: **44.58–62**
11:20–22: **44.58–62**
15:33: **66.22–23**

Luke:
2:21: **24.2–11**
19:28–48: **44.55–58**
23:44–45: **66.22–23**

John:
1:10: **44.1–10**
2:15: **44.55–58**
12:1: **44.55–58**
12:12: **44.55–58**
12:14–15: **44.55–58**
13:1: **54.2–8**

Romans:
4:25: **44.75–76**

1 Corinthians:
5:7: **44.74**
15:51–52: **3.6–10**

Ephesians:
2:20: **44.70–72**

2 Thessalonians:
2:1: 322

1 Timothy:
6:14: 322

1 Peter:
2:24: **44.75–76**
4:5: **8.38–43**

SOURCES

ACTS OF THE COUNCIL OF CEASAREA (DE ORDINATIONE FERIARUM PASCHALIUM): LXII, LXV–LXVI; **11.37–40 / 31.26–27 / 44.18–36,55–80 / 54.13–27 / 56.2–14 / 58.88–94 / 63.3–24 / 66.2–7**

ADMONITIO GENERALIS OF AD 789: XLVIII, CCXI
79: XXXIII

ALCUIN:
Epistolae:
126: **62.2–61,97–110,123–130**

Interrogationes et responsiones in Genesin: LXXX

Orthographia: **8.14–16 / 54.14–16**

PS-ALCUIN:
De bissexto: XXII, XLVII, CXII
I:
col. 993: **41.3–7**
col. 994: CLII, CLXII; **31.2–13 / 41.2–16,29–31,92–106**
col. 995: CXXXII, CLXII, CLXIX; **8.38–43 / 41.25–28,72–79,92–106**
col. 996: **41.25–28,38–42**
col. 997: CXLIV; **36.2–5 / 41.2–16, 2–20,38–49 / 55.2–12**
col. 998: **24.12–18 / 41.2–20,43–49,80–88**
II: **41.80–88**

De saltu lunae: XXII, XLVII, CXII
I:
col. 984: **49.2 / 59.4–13,70–105 / 62.2–13, 62–63,68–72,111–117**
col. 985: CLXVII; **59.70–105 / 62. 62–63, 68–72,111–117,87–95,97–110, 111–117**
col. 986: CLXVI–CLXVII; **62.14–61,68–72,97–110**
col. 987: CLXVI; **62.14–61**
col. 988: CLXII, CLXVI–CLXVII, CLXXXI; **46.16–20 / 48.2–7 / 62.14–61,68–72,87–95,111–117**
col. 989: CLXVII; **62.97–110**
II: CLXVI; **62.14–61,111–117**
III: CLXVI, CLXVII; **62.68–72**
IV: CLXVI; **62.68–72**
V–VIII: **62.14–61**

Disputatio puerorum:
6: **35.2–5**

ALDHELM:
Epistola ad Eahfridum: CL

Epistola ad Geruntium: XXXVII; **63.3–50**

AMBROSE:
Exameron:
4.4: **3.6–10**

De excidio urbis Hierosolymitanae:
5.49: **36.7–9**

AMBROSIASTER:
Liber Questionum: LV

ANATOLIUS (?):
De ratione paschali: XLIV, LII, LXII, LXIV–LXVII
2: **11.37–40 / 36.2–5 / 55.2–12**
5: **8.34–39 / 44.13–80**
7: **56.2–14**
10: LXVII, CCXV; **55.2–12**
12: **53.5–11 / 55.13–27**
13: LXVI, CLXI; **4.2–8 / 24.2–11 / 38.26–27 / 39.2–17,23–55 / 40.10–20**
14: **38.39–41**

ANGERS GLOSSES: CIV–CV, CLXXXIII–CLXXXVII, CCIX *et vide* Angers, Bibliothèque Municipale, 477 (461)

ANNALES CAMBRIAE: XL–XLI

ANNALES RATISPONENSIS: XV–XVI, CCXXI

ANNALS OF INISFALLEN: LXXX
§73: **13.2–7**

ANNALS OF TIGERNACH: LXXIX–LXXX; **12.37–40 / 63.3–24**

ANNALS OF ULSTER: LXXIX, LXXXIII–LXXXIV, CLVII; **49.16–32**

ANONYMUS AD CUIMNANUM: XXI
7: **11.2–3**
18: **1.2–3**

ARG. AQUENS. (*'Der Aachener Vorbehalt von 816'*):
3–4: **58.2–7**
6: **62.97–110,123–130**

ARN. SERM. (*'Die Predigt Arns von Salzburg um 802'*):
1: **46.2–5**

ARS ANONYMA BERNENSIS: **8.14–16 / 54.14–16**

AUGUSTINE:
Confessiones:
11.14: **1.2**
11.23: **3.6–10**

Contra Faustum:
18.5: **13.2–7 / 14.2–8**

De civitate Dei:
7.7: **14.2–8**
15.12: **24.12–18**
18.19: **13.2–7**
20.6: **44.75–76**
21.8: **66.15–19**

De Genesi ad litteram:
2.14: **31.2–13 / 66.2–7**

De Genesi contra Manichaeos:
1.4: **41.2–16**
1.14: **8.14–20**

De Trinitate:
2.17: **54.2–8**
4.4: CLXIX; **41.56–61,72–79**
4.5: **44.55–80**

Enarrationes in psalmos:
Enarratio in psalmum 10:
Sermo 3: **44.18–36 / 46.15,21–23 / 47.8–42**
Enarratio in psalmum 68:
Sermo 1: **54.2–8**
Enarratio in psalmum 93:
Sermo 3: LXXV; **11.13–15 / 44.13–80**

Epistulae:
36: **11.13–15**
55: **54.2–18**
205:
§14: **3.6–10**

Locutiones in Heptateuchum:
2.116: **62.2–13**

Sermones:
155: **54.2–18**
169: **44.75–76**
231: **44.75–76**

BEDE:
Chronica:
§72: **13.2–7**
§339: **63.3–24,3–38,13–14**
et vide De temporibus 16; *De temporum ratione* 66

De natura rerum: CLXXX, CLXXXIII
20: **46.15**

De temporibus: LIII–LIV, CIII, CVIII–CX, CLXIV, CLXXXIII, CLXXX, CLXXXIII
1: **1.4–16 / 3.2–5 / 6.2–7**
2: **8.2–10,17–20,34–39**
3: **9.11–27**
4: **10.2–6 / 11.13–15,27–36**
5: **24.12–18 / 50.29–39**
6: **12.24–31,37–40 / 13.2–6 / 14.7–8 / 15.2–6 / 17.2–9 / 18.2 / 19.2–5 / 21.1–23.3 / 25.2–10 / 26.2–7 / 27.2–12**

7: **38.26–27**
9: **24.2–11 / 31.14–16,22–25 / 36.2–5**
10: CXLIV; **31.2–13 / 41.3–7,25–31**
11: **59.26–37 / 60.41–106**
12: **48.2–7 / 62.14–61**
13: **1.9–16 / 55.13–27,38–41 / 65.3–38**
16: CXXI

De temporum ratione: XX, XLVII–LI, LIII–LIV, CX–CXII, CLXIV, CLXXX, CLXXXIII, CCV
2: LXIX; **1.4–16 / 30.2–6**
3: CXLII, CLX; **1.4–16 / 2.2–7,22–28 / 3.2–10**
4: CXLVII; **7.2–5**
5: LXIX; **8.2–10,34–39 / 9.28–34**
6: **11.37–40 / 44.18–36 / 46.15**
7: LXIX; **8.17–20 / 9.2–3,11–27**
8: **10.2–6 / 11.4–8,13–26 / 11.27–36**
11: LXIX; **12.2–14 / 24.2–26,32–39 / 52.27–43 / 41.80–88 / 49.2 / 50.29–39 / 55.2–12**
12: **12.24–37 / 13.6–7 / 14.7–8 / 15.2–6 / 16.2–6 / 17.2–9 / 18.2 / 19.2–5 / 20.2–6 / 21.1–23.3 / 24.2–18 / 36.2–5 / 41.2–20,86–91**
13: **25.2–10 / 26.2–7 / 27.2–12**
14: **24.19–26 / 55.2–12**
16–19: LIII
20: XLVIII–XLIX; **50.3–14,22–28**
22: LXVII
25: **44.18–36 / 46.21–23 / 48.8–12**
29: **48.2–7**
30: **38.9–16, 28–38**
35: LXIX; **30.2–5 / 31.26–27 / 32.2–4 / 33.2–3 / 34.2–4 / 38.39–41 / 55.2–12**
36: LXIX; **24.2–11,32–39 / 31.2–16,21–25 / 36.2–5 / 53.5–11 / 59.2–13,18–27**
38: CXLIV, CLXII; **41.3–7,25–31,54–61,72–79,43–49**
39: CLXIX; **7.12–13 / 41.3–7,72–79**
40: CXLIV
41: XLVIII, CXXXV; **50.22–28**
42: XLVIII–XLIX; **48.2–7 / 49.11–13, 19–20 / 52.27–43 / 53.5–11 / 59.4–13,70–105 / 62.2–13,68–72,111–117**
43: XLIX; **52.49–60**
44: **59.14–17**
45: XLVIII, LXX; **53.5–11 / 55.13–27,38–41 / 58.2–13 / 59.2–13,21–35 / 60.3–7 / 61.2–5,17–27**
46: CXXXV; **59.36–37 / 60.41–106**
47: LXV, CI; **44.55–80**
48: **1.9–16**
49: LIV
50: **49.2,7–13**
51: XLIX, LXXXVII; **55.13–27,38–45 / 58.2–7**
52: **49.19–20**
53: **65.3–38**
54: LIV
56: XLIX; **53.5–11 / 57.2–19 / 58.2–13 / 59.2–13,18–20,28–35,70–105 / 60.51–106**
59: XLIX, CLXV; **55.38–41 / 56.2–14 / 58.2–7**
60: **56.44–57**
61: CI; **44.55–80 / 56.2–14 / 58.2–7**
62: XLIX, LXXXVII; **58.2–7**
65: LXXXI, XCI; **41.107–108 / 65.3–38**
66–71: CXI, CXXI

Epistola ad Wicthedum: CLXXX
1: **38.9–16 / 51.3–12 / 55.2–12**
3: **41.25–28**
8: **55.2–12**
11: **55.2–12**

Historia ecclesiastica gentis Anglorum:
2.2: **63.3–50**
2.19: XXXIX–XL, LXXXI–LXXXII; **63.25–29** *et vide* PAPAL LETTER TO THE NORTHERN IRISH CLERGY OF AD 640
3.3: LXXXI, LXXXIII; **63.3–50**
3.4: LXXXIII; **63.3–50**
3.5: LXXXIII, LXXXIV
3.14–17: LXXXIII
3.25: XL; **63.3–50**
3.28: **63.3–50**
4.2: CLI
5.15: LXXXVIII, CLVII
5.19: XL
5.21: XL, LXXXVIII, XCI, CLVII, CLXV; **56.2–14 / 58.2–7 / 63.3–50 / 67.3–26**
5.22: LXXXIII; **63.3–50**
5.24: LXXXIII

Homiliae subdititiae:
53: LV

In Ezram et Neemiam:
2: **54.9–12**

In Lucae euangelium:
2.6: **54.9–12**
6.6: LV

In Marci evangelium:
4.14: LV

Retractatio in actus apostolorum:
9: **54.9–12**
16: **54.9–12**

PS-BEDE:
De sex dierum creatione: **9.28–34**

De officiis: LV

BOBBIO COMPUTUS (Milan, Biblioteca Ambrosiana, H 150 inf): XXIV, XLVII, L, LVII–LVIII, LXV, LXVIII, LXXIV, LXXXIV, CXXXII, CLXI–CLXII, CLXIX
Praefatio: CLXIII; **49.2,14–15 / 50.3–6**
1: **28.14–22**
2: CLXI; **29.25–35**
3: **55.62–69 / 58.33–80 / 59.28–35**
4: **53.5–11 / 57.2–19 / 59.21–27**
6: **55.42–45**
7: **55.38–41**
8: **55.13–27**
9: **55.38–41 / 56.2–14 / 58.2–7**
11: **55.38–41**
13: **59.28–37**
14: **49.2 / 57.2–19 / 59.2–13,21–37,70–105**
15: **24.12–18 / 50.61–66,72–76 / 60.3–7**
16: **59.2–13**
17: **59.21–27 / 60.8–21,51–106 / 61.2–5**
18: **62.14–61,111–117**
19–20: **25.2–10 / 26.2–7 / 27.2–12**
22: **49.11–13 / 50.3–6**
26: **46.2–5 / 49.11–13**
27: **31.26–27 / 38.17–25**
28: **30.2–6 / 31.26–30 / 32.2–4 / 33.2–3 / 34.2–4 / 35.2–5**
29: **38.9–16,28–38**
30: CLXI; **31.19–20 / 38.2–8,26–27 / 39.2–17,30–42**
31: CLXI; **38.17–25 / 39.2–17,30–55**
32: **38.39–43**
33: CXLII; **4.2–4 / 39.2–17**
34: **48.13–18**
35: **3.2–5**
36–37: **41.80–88**
39–40: CLII, CLXII; **41.92–106**
40: **41.92–106**
41: **41.80–88**
42: **41.2–16,43–49,80–88 / 66.13–14**
43: **41.92–106**
44: **41.25–31,72–79**
45: **38.9–16,28–38**
46: **49.4–13 / 59.70–105 / 62.2–13,68–72,111–117**
49: **12.18–19 / 13.2–7 / 14.2–7 / 15.2–6 / 16.2–6 / 17.2–9 / 18.2 / 19.2–5 / 21.1–23.3**
50: CLXVIII; **58.2–32**
51: **58.33–80**
54: **42.2–6**
55: **56.15–43**
56: **10.2–6 / 11.2–11,16–31**
57: **10.2–6 / 11.13–15,27–31**
58: **12.2–3,8–14,24–31 / 24.19–26**
59: **12.2–14,15–19,24–40 / 13.2–7 / 14.2–8 / 15.2–6 / 16.2–6 / 17.2–9 / 18.2 / 19.2–5 / 21.2–4 / 24.2–11 / 25.2–10 / 36.2–5**
60: **12.20–23 / 24.12–18**
61: **24.32–39 / 48.2–7 / 50.61–66,72–76**
62: **24.32–39 / 50.72–76**
64: **50.29–39**
66: LXXII; **8.11–13**
67: **31.18–20,23–25**
69: **8.6–9,14–22,34–39 / 11.4–11 / 24.12–18 / 26.2–7 / 49.2**
70: CXXXII; **8.2–3,6–9,21–22,23–43 / 9.21–24**
71: **9.2–3,11–20,25–27**
73: **56.15–43**
74: **49.2 / 56.15–43**
76: **49.2**
77: **64.2–6**
82: **11.37–40 / 36.10–15**
89: **1.4–8 / 2.2–31**
92: **36.7–9**
95: **36.7–9**
96: CXXVII; **3.2–10 / 4.6–8 / 5.2 / 6.2–7**
97: CXLVII; **7.2–11**
105: CLXI; **29.17–24 / 38.39–43 / 45.2–13**
113: **46.10–13**
114: CLXI; **39.2–17,23–55**
116: CLXXXVI; **37.2–7**
117: **58.88–94**
119: **41.107–108 / 62.62–67**
124: **59.2–19,70–105**

127–133: **61.17–27**
130: **50.22–28**
137: CLXVI; **62.14–61**
140: LIII; **6.2–14 / 54.2–8 / 41.2–16,43–49 / 55.38–45 / 58.2–7,18–27 / 59.4–13,21–27,70–105**
142: **46.2–5**
148: **11.37–40 / 36.10–15 / 44.18–36**
150: **55.28–33**
151: **24.19–26,32–39 / 55.2–12**
155: **49.2**
156: **62.62–63**

BOETHIUS:
De arithmetica: XXXI

Liber de diffinitione: **7.2–11**

PS-BOETHIUS:
Geometria: XXXI

BONIFACE:
Epistolae:
75: CLXIV
76: CLXIV
91: CLXIV

BRAULIO OF ZARAGOZA:
Epistolae:
22: LXXII

CAESARIUS OF ARLES:
Sermones:
192:
1: **36.2–5**

CALCIDIUS: XXX

CAP. COMP. (*'Das Aachener Verhör von 809'*):
2: **44.55–80**
7: **55.38–41**
9: **55.38–41**
20: **59.38–69**

CAPITULARE DUPLEX IN THEODONIS VILLA PROMULGATUM (AD 805): XXXIII

CASSIODORUS: XXXI
Expositio psalmorum:
Expositio in psalmum 103:
§22: CXCIII; **37.2–7**

Expositio in psalmum 104:
§16: **8.14–20**

Institutiones:
2.3: **7.2–11**

CASSIODORUS (?):
Chronica:
§§78–79: **12.39–40**

Expositio in cantica canticorum:
4: **8.14–20**

COLOGNE FRAGMENTS:
1: **56.2–14**

COLUMBANUS:
Regula monachorum:
7: **38.26–27,39–43 / 45.2–13**

Epistulae:
1: LXXXVI; **58.2–7**
2: LXXXIV–LXXXVI; **58.2–7**

COMP. COL. (*'Das Kölner Lehrbuch von 805'*): CII, CXII, CLXX–CLXXI, CLXXIX–CLXXXIII
1.1–2: CIV
2.1: LXIX, CIV; **12.2–14**
2.2: **41.17–20 / 53.12–15**
2.3: CIII, CIV, CLXXX–CXXXII; **11.37–40 / 40.2–20**
2.4: CIV; **24.12–18**
2.5: CIV
2.6: CLXXXII; **41.80–88**
3.1: CIV, CLXII, CLXIX; **41.17–20,32–37,54–61,72–79**
3.2: LXVI; **39.2–17 / 41.25–28**
3.3: CXLIII; **36.2–5 / 41.38–49**
3.4: **39.18–22 / 44.18–36**
3.5: CIV, CXLIII, CLXII; **41.3–7,43–49,54–61**
3.7: **41.3–7**
3.8: CXLIII; **41.43–49**
4: LXXII
4.1: CLXXXII; **36.10–15 / 50.15–21**
4.2: CLXIII, CLXXXI; **49.2,11–13,16–32 / 50.7–14 / 51.35–44**
4.3: **59.14–17,21–27**
4.5: **62.97–110**
4.6: LXVII; **38.17–25 / 39.2–17 / 59.18–20**
4.7: **36.10–15**

4.8: CLXXXII; **62.73–95,118–122**
4.9: CIII, CIV, CLXXXIII; **56.2–14 / 62.14–61,111–117,123–130**
4.10: CII, CIV, CLXVI–CLXVII, CLXXXI, CLXXXIII; **62.14–61,68–72,97–110**
5: LXXII
5.1: CIV, CLXXXII; **44.13–17**
5.2: CLXII, CLXXXII; **44.18–36 / 46.16–20**
5.3: CLXV, CLXXXII; **58.2–7**
5.4: LI, CII, CLIV, CLXIV, CLXXXI–CLXXXII; **44.13–17,37–80 / 52.3–13,45–48**
5.5: LI, CII, CLIV, CLXIV, CLXXXI; **52.3–43**
5.6: LI
5.7: CII, CLXXXII; **66.13–23**
5.9: **56.2–14 / 58.2–7 / 63.25–29,36–39**
5.11: **55.2–12 / 59.14–17**
5.12: CIV; **65.3–38**
5.14: **55.28–33**
6.1: CII, CLXXXII; **66.13–23**
6.2: CLXXX; **56.2–14**
6.3: CLXXX
6.4: **24.32–39 / 48.2–7 / 50.56–66**
6.6: LXXII
6.7: **55.38–45 / 56.2–14 / 58.2–7**
6.8: **55.13–27 / 58.2–7,33–80**

COMPUTUS CARTHAGINIENSIS: XVII, XXXVI, LII
2.16: **56.2–14**

COMPUTUS COTTONIANUS OF AD 688/9 (London, British Library, Cotton Caligula A XV, 73r–80r): CLVII
78v: **50.61–66**
78v–79r: **24.32–39 / 50.72–76**
79r: **24.12–18 / 59.2–13 / 60.3–7**
80r: **49.16–32**

COMPUTUS DIGBAEANUS OF AD 675 (London, British Library, Digby 63, 72v–79r): CLII, CLXII

COMPUTUS EINSIDLENSIS (Einsiedeln, Stiftsbibliothek, 321 (647), 82–125): XIX, XLIX–LI, LIV–LV, CVII–CX, CXII, CXXII–CXXIII, CXXXIII–CLII, CLIX, CLXX, CCVII–CCVIII
p. 82: CIII, CVIII
pp. 83–84: LXXII, CIII, CVIII, CXLIX
p. 85–95: CXLIII
p. 85: CIII, CVIII, CIX, CXLIII, CLXXVIII; **1.4–16 / 2.2–7 / 3.6–10 / 5.7–8 / 48.13–18 / 54.2–8**
p. 86: CXLII, CLXXVIII; **1.4–16 / 3.2–10 / 4.2–8 / 5.7–8**
p. 87: **5.2–6**
p. 88: CXXVII; **5.2–6 / 6.2–8**
p. 89: CXXVIII, CXLVII, CLXXXV; **7.2–11 / 8.2–10**
p. 90: CXXXIII, CCI; **8.2–10,14–20,23–39 / 9.28–34**
p. 91: CXL; **8.34–39 / 9.11–27**
p. 92: CXL, CLXXXV; **1.4–16 / 8.14–22 / 9.2–3,26–27 / 10.2–6 / 11.2–3,13–36 / 12.2–14**
p. 93: CXXXIII, CXLI, CLXXV, CCI; **1.4–16 / 12.2–19 / 13.2–7 / 14.2–8**
p. 94: **14.2–8 / 15.2–6 / 16.2–6 / 17.2–9 / 18.2 / 19.2–5 / 20.2–6 / 21.2–4 / 30.2–6**
p. 95: CXLI, CLXXV; **1.2 / 12.20–23 / 24.40–41 / 25.2–10 / 30.2–6 / 31.2–13,26–27 / 32.2–4 / 33.2–3 / 34.2–4 / 37.2–7 / 46.2–5**
p. 96: CLXXIV, CLXXVIII–CLXXIX; **25.2–10 / 26.2–7 / 27.2–12 / 28.1–22 / 31.2–13**
p. 97: LXXIII, CXXXIII, CXXXIX–CXL, CLXI, CLXIX, CCI; **28.6–13 / 29.2–12,17–24**
p. 98: LXXIII, CXXXIX–CXL, CLXI, CLXXIV; **29.13–35 / 38.1,2–4,9–27,31–38**
p. 99: LXVI, CIII, CLXI; **38.26–27,31–38 / 39.2–22**
p. 100: LXVI, CIII, CLXI; **39.18–42**
p. 101: LXVI, CIII, CXLVI, CLXI; **39.23–42**
p. 102: LXVI, CIII, CXLVI, CLXI; **39.43–55**
p. 103: LXVI, CIII, CXLV–CXLVI, CLXI, CLXXX–CLXXXI; **24.12–18 / 40.2–9**
p. 104: CIII, CXLIV, CLXXX–CLXXXI; **40.10–20**
p. 105: CIII, CXLIV; **41.2–16,32–49**
p. 106: LXXIII, CIII, CXLIV, CLXII, CLXX; **29.2–12 / 41.32–37,43–49,54–71**

p. 107: CLXII, CLXIX, CLXX; **41.25–28,54–79,89–91**
p. 108: LXXI, CL–CLI, CLXII; **41. 25–28,89–106 / 48.13–18 / 48.19–20 / 49.2, 7–10**
p. 109: LXXI, LXXIII, LXXXIV, LXXXIX, XCIV, CXXXIII, CLIII, CXCVIII, CXCIX, 321; **49.7–32 / 50.3–14 / 52.14–26**
p. 110: LXXI, LXXXIX, XCIV, CXXXIII, CLXIII, CLXXXI, CXCVIII, CXCIX; **50.7–28,40–49,56–60 / 51.13–44**
p. 111: LXXI, LXXXIX, CLXIII, CLXXXI, CXCVIII, CXCIX; **50.29–39,61–66 / 51.13–34 / 55.1,13–27,34–41,62–69**
p. 112: LXXIII, CXLVI; **56.2–43**
p. 113: CXXXIV, 332; **53.5–11 / 59.1,4–13,18–37,70–105 / 62.2–13,68–72,111–117**
p. 114: CXXXIV, CXXXV, CLXXVIII, CXCVI–CXCVII; **59.70–105 / 60.1–2,8–21**
p. 115: CXXXV, CXCVI–CXCVII; **60.8–50**
p. 116: CXXXV, CXCVI–CXCVII; **60.41–106**
p. 117: CXXXV, CXCVI–CXCVII; **60.51–106**
p. 118: CLXX, CXCVI–CXCVII; **60.51–106 / 61. 2–5,11–27**
p. 119: XLVIII, LXXI–LXXIII, LXXXIX, CXLVIII, CXLIX, CLXV, CLXVIII, CLXXVIII, 331; **58.2–32,81–94 / 61.2–5**
pp. 120–121: XLVIII, LXXI, LXXIII, CXLI, CLXVIII, CLXXVIII; **58.8–80**
p. 122: XLVIII, LXXI, CIII, CXXXVII–CXXXVIII, CLXVI, CLXVIII, CLXXVIII, CXCVII; **58.8–32 / 62.14–61**
p. 123: LXXXIX, CIII, CXXXIII, CXXXVII–CXXXVIII, CLXVI, CLXVII, CXCVII, CXCVIII, CCI; **62.14–61,68–72,87–95**
p. 124: LXXXIX, CIII, CLXVII, CXCVIII; **62.2–13,68–72,87–95**
p. 125: **62.2–13**

COMPUTUS RHENANUS OF 775 (Cologne, Dombibliothek, 103, 184v–190v; Wolfenbüttel, Herzog-August-Bibiothek, Weißenburg 91, 169r–173v): CXI, CLXXIX–CLXXX

CONCILIUM AURELIANENSE A. 541:
1: XXXVIII

CUMMIAN:
Epistola: XXXIX, LV, LXXXI, CXCI–CXCII
ll. 10–11: LXXXVIII
ll. 75–79: LXXXVIII
ll. 80–85: LXXXVIII; **55.13–27**
ll. 126–129: **55.28–33**
ll. 206–207: **62.111–117**
l. 213: LXV
ll. 215–220: **49.14–15 / 55.53–61/ 62.62–63**
l. 225: LXXXI; **41.107–110 / 65.3–38**
l. 226: LXXXI; **41.107–110 / 63.25–29 / 65.3–38**
ll. 227–229: **63.25–29**
ll. 230–232: **63.30–39**
ll. 238–239: **63.30–39**
ll. 273–285: LXXXI–LXXXIII

CYRIL OF ALEXANDRIA:
Easter table: XXXIX

PS-CYRIL:
Epistola: LXII, LXIV–LXV
3: **52.49–60**
4: **51.3–12 / 59.18–20**
5: **56.2–14 / 63.25–39**

DE ANNO: **24.2–11 / 55.2–12**

DE COMPARATIONE EPACTARUM DIONYSII ET VICTORII (ed. in Appendix 2): L–LI, CII, CLII–CLVIII, CLXXIX–CLXXXI, CCVII–CCVIII, 322–6
1: **36.10–15 / 49.16–32 / 52.3–43,49–60 / 55.42–45**
2: **44.55–80 / 49.16–32**

DE COMPUTO DIALOGUS: XXII, CVII–CIX; **36.7–9**

DE CONTRARI AETATIBUS (ed. in Appendix 6): LXXI, 333–6; **50.7–15 / 59. 38–69**

DE DIVISIONIBUS TEMPORUM: XXII, XXVIII–XXIX, XLVII, L, CVII, CVIII–CIX, CXIII–CXV, CXXVIII, CXXX, CLX;
1: CXLII; **1.4–16 / 44.2–10**
2: **2.2–31**
3: **3.2–10 / 48.13–18**
4: CXXVII, CXLII; **4.2–4,6–8 / 5.7–8**
5: CXXVII; **5.3–5 / 6.8**
6: **6.2–7**
7: CXLVII; **7.2–7**
8: CXXVIII, CXXXII, CLXXI; **8.2–17,21–33,34–39,38–43 / 9.2–3,28–34**
9: **11.37–40**
10: **11.2–11,32–36**
11: **8.17–20 / 9.4–10**
12: CLXI; **39.2–17,23–29,32–40**
13: **10.2–6**
14: **12.7–8,2–14 / 41.80–88**
15: **13.2–7 / 24.32–39**
16: **14.2–8**
17: **15.2–6 / 55.2–12**
18: **16.2–6**
19: **17.2–9**
20: **18.2**
21: **19.2–5**
22: **20.2–6**
23: **21.1–23.3**
24: **21.1–23.3 / 24.2–11**
25: **21.1–23.3 / 24.19–26**
26: **21.1–23.3**
27: **12.24–39 / 14.7–8 / 24.12–18 / 25.2–10 / 26.2–7 / 27.2–12 / 28.2–5,14–22**

DE DUABUS LUNATIONIBUS (ed. in Appendix 6): LXXI, 333, 336; **59. 38–69**

DE HEBDOMADIBUS: XXVIII; **11.2–11,13–15,27–36**

DE MIRABILIBUS SACRAE SCRIPTURAE: LXIX, LXXVIII–LXXX, XCV, CXXV, CXXIX–CXXX
1.8: CII; **66.8–12**
2.4: LXXVIII, CII; **61.6–8 / 62.111–117 / 65.3–38 / 66.13–14,20–21 / 68.2–39**
2.28: CII; **66.15–21**
3.13: CII; **66.22–23**

DE ORDINATIONE FERIARUM PASCHALIUM: *vide* ACTS OF THE COUNCIL OF CEASAREA

DE RATIONE CONPUTANDI: XXII, XLVII, L–LI, LIV–LV, LXIII, CVI–CXII, CLXXXIV, CLXXXVII, CXCI–CI
1: CLXXXVI, CXCII–CXCIII; **37.2–7**
2–12: CIII
3–4: LXXII
9: **36.7–9**
13: CX, CXLII, CXCII; **1.4–8**
14: CIX; **1.9–16**
15: **1.9–16**
16: CXCII; **2.2–7,141.9–16 / 18**
17: **3.2–5**
18: CXXVII, CXLII, CXLIII; **4.2–4,6–8 / 5.2–8**
19: **5.2–5**
20: **6.2–7 / 36.7–9**
21: CXLVII; **7.2–7**
22: CXXVIII, CLXXXV, CC; **8.2–13,16–17**
23: **8.6–9,21–22**
24: **8.6–9,16–17 / 9.28–34**
25: **8.14–20 / 9.2–3,11–25,26–27**
26: CXXXII; **8.34–43**
27: CC; **10.2–6 / 11.2–8.13–15**
28: **12.2–14,24–39 / 28.14–22 / 53.12–15**
29: CXCII, CC; **12.15–19 / 13.2–8 / 14.2–8 / 15.2–6 / 16.2–7 / 17.2–9 / 19.2–5 / 20.2–6 / 21.1–23.3**
30: **24.2-10,40–41 / 25.2–10 / 26.2–7 / 27.2–12**
31: CLXXIV, CLXXIX; **28.2–5,14–22**
32: **28.6–13**
33: LXXIII, CXXXVIII, CLXI; **29.25–35**
34: LXXIII, CXXXVIII, CLXI, CLXXIV; **29.13–16**
35: LXXIII, CXXXVIII, CLXI, CLXIX, CLXXIV, CXCVIII; **29.2–12,17–24**
36: **24.19–26,32–39 / 55.2–12**
37: **24.12–18**
39: **24.2–11**
40: **24.19–26 / 55.2–12**
41: **24.32–39 / 55.2–12**
42: CLXXXV; **30.2–5 / 31.26–27 / 32.2–4 / 33.2–3 / 34.2–4**
43: CLXXXV; **1.2–3 / 31.26–27 / 32.2–4 / 33.2–3 / 34.2–4**
44: **31.2–13**
45: **24.2–11 / 36.2–5**
46: CXCV; **31.14–25 / 53.2–4**

47: **38.2–16,26–27**
48: **38.39–43**
49: **38.39–43 / 45.2–13**
50: LXVI, CIII, CLXI, CLXXXI; **38.17–27,39–43 / 39.2–17,23–45 / 40.10–20**
51: CIII; **55.42–45**
52: CLXXIV, CXCVIII; **41.2–24,43–49**
53: **41.3–7**
55: **31.2–13 / 41.80–88**
56: CLXIX; **41.72–79**
57: CXXIX, CXCVIII; **41.2–16,43–49,62–71**
58: **41.107–108**
59: **42.2–8**
60: CXCIII; **43.2–3**
61: **44.2–10**
62: **44.11–12 / 55.42–45**
63: **46.2–8,14,15,21–23 / 46.6–8 / 48.8–12**
65: CLXII, CLXXXI; **46.16–20**
66: **8.34–39 / 48.2–7**
67: **44.18–36**
68: CLXIII, CLXXXI; **11.37–40 / 24.12–18 / 46.10–13 / 48.13–18 / 49.2 / 51.35–44**
70: LXX, LXXI, CLXVI, 333, 336; **59.38–69**
72: LXXXIX, XCIV, CXCIX; **50.29–39,61–66**
73: LXX, LXXI, LXXIII, CLXVI, 333–6; **50.7–28 / 59.38–69**
74: LXX; **61.9–10,17–27**
76: **59.4–13,18–20**
77: **59.21–27**
78: **59.4–13,21–27 / 61.2–5**
79: CXXXV, CXCVII; **60.41–106**
80: CXCIII; **59.28–35**
81: CXCIII; **59.36–37 / 61.6–8**
82: **57.2–19 / 59.70–105 / 61.17–27**
83: **53.5–11 / 59.4–13,70–105**
84: CXCV; **54.2–12**
85: **44.37–54 / 53.5–11 / 55.2–12 / 64.2–6**
86: LXXXIX; **44.55–80 / 52.3–13,45–48**
87: **44.55–80 / 63.3–24**
88: **44.55–80**
89: **44.11–12,55–80 / 63.40–50 / 64.2–6**
90: LXXXIX; **55.38–45,53–61**
91: **53.5–11 / 55.2–12**
92: **56.2–14 / 58.2–7**
93: **55.62–69**
94: **55.13–27,38–41 / 59.14–17 / 63.40–50**
98: LXXXIX, CXXVIII, CXXXI, CXCIII, CXCIV, CXCVIII; **55.28–37 / 56.2–14 / 58.2–7 / 63.3–50,25–39**
99: LXXII, LXXXIX, CXLVIII, CXLIX, CLXV, CXCVIII; **58.2–7**
100–102: XLVIII, LXXI, CLXVIII, CLXXVIII; **58.8–32,88–94**
103: LXXIII, CLXXVI
104: LXXIII
105: CXCV; **53.16–20**
106: **62.62–63**
107: CVI, CXXX, CXCIV; **61.2–13**
108: CIII, CLXVI, CXCVII; **62.14–61**
109: CXCIV; **62.14–61**
110: CIII, CXCIV; **62.110–117**
111: CVI, CXXIX, CXCIV; **61.2–13**
112: CLXVII, CXCVIII; **62.73–95**
113: **62.111–117**
115: **62.62–63 / 65.3–38**

De sollemnitatibus: LXVIII–LXIX, CXXXI
2: LXVIII, CXCIII; **55.28–33**

Dial. Burg. (*'Das burgundische Lehrgespräch von 727'*): XXIV–XXV, XLIV, XLVII, LV, CIV, CXI, CLXVIII, CLXXI–CLXXII, CC
1: LXXII
4: CLXXII; **11.37–40 / 24.2–11**
5: CLXXI, CC; **8.2–10 / 10.2–6 / 11.16–26,37–40**
6: LXIX, CC; **12.2–3,18–19 / 13.2–7 / 14.2–7 / 15.2–6 / 16.2–6 / 17.2–9 / 18.2 / 19.2–5 / 21.1–23.3**
7: CLXXII; **24.40–41 / 25.2–10 / 27.2–12**
8: LXIX; **30.2–6 / 34.2–4**
9: LXIX; **31.2–13,26–27 / 36.2–5**
10: LXIX, CLXXII; **37.2–7 / 46.2–5**
11: LXIX, CXXVIII, CLXXI; **8.2–3,11–13 / 9.2–3,28–34**
12: **7.12–13 / 31.15–16**
13: **1.9–16 / 49.16–32 / 59.18–20 / 62.87–95**
14: CLII, CLXII; **41.92–106**
15: **62.87–95**
16: LXXXIV, CLXXI, CC; **58.33–80 / 62.111–117**
17: LXXIX, CLXXI, CC; **44.11–12**

DIAL. LANGOB. (*'Das langobardische Zwiegespräch um 750'*): L, LV, CXI, CLXIX, CLXXIV–CLXXIX
1: CXXXI, CLXXV, CLXXIX; **1.2–3**
2: LXIX, CLXX, CLXXV; **30.2–5 / 43.2–3**
3: CLXXVIII, CLXXIX; **1.4–16 / 3.6–10 / 43.2–3**
4: CLXXIX; **1.9–16**
5: **31.2–13 / 31.2–13 / 31.26–27**
6: LXIX, CLXXV; **32.2–4 / 33.2–3 / 34.2–4**
8: **38.39–43 / 45.2–13**
9: CLXXVI; **12.15–17 / 12.20–23**
10: CLXX, CLXXV; **16.2–6 / 17.2–9 / 12.18–19 / 13.2–7 / 14.2–7 / 15.2–6 / 17.2–9 / 18.2–9 / 19.2–5 / 20.2–6 / 21.1–23.3**
11: LXIX; **12.2–3**
12: CXXXVIII, CLXI, CLXX, CLXXIV, CLXXVIII; **28.2–5,14–22 / 29.13–16**
13: CLXXVI; **12.24–31,37–40 / 53.12–15**
14: **24.40–41 / 28.6–13**
15: **27.2–12 / 26.2–7 / 25.2–10 / 26.2–7 / 27.2–12**
16: LXXIII, CLXI, CLXXIV; **29.17–24 / 29.17–24**
17: CLXXV, CLXXVIII; **31.19–20 / 38.2–8**
18: CLXXVIII; **6.37–40 / 24.19–26 / 38.9–16 / 38.28–38 / 44.55–80**
19: LXVI, CLXI, CLXXVIII, CLXXXI; **38.17–25 / 39.2–17,23–55**
20: CLXXVII; **59.70–105**
21: **62.123–130**
22: **60.8–21 / 61.2–5,9–10**
23: LXXIII, CLXX; **61.6–8,17–27**
24: LXXI, CLXVIII, CLXXVIII; **41.3–7,80–88 / 58.8–32**
25: CLXVII, CLXXVIII; **62.68–72**
26: L, CLXXVII; **1.9–16**
27a: LXXIII
28: LXVI

DIAL. NEUSTR. (*'Das neustrische Streitgespräch von 737'*): XXV, XLVII, LV, CIII–CIV, CXI, CLXVIII–CLXIX, CLXXII–CLXXIV
1: CIII
2: **12.15–17 / 31.2–13**
3: **24.12–18**
5: **28.14–22**
7: **28.2–5**
9: **55.62–69**
10: **55.38–45**
11: **55.38–45**
13: CLXV; **58.2–7 / 63.25–29**
14: **12.2–14**
15: CIII, CLXXXII; **44.13–17 / 63.40–50**
18: **56.2–14 / 63.25–39**
19: **58.33–80 / 65.3–38**
21: **49.2 / 50.3–6**
22: **59.28–37 / 61.6–8**
23: **59.4–13,18–27 / 61.17–27**
24: **53.5–11 / 57.2–19,70–105**
25: CIII, CLXVII; **11.37–40 / 62.68–72,87–95,97–110**
26: CIII, **62.14–61,111–117,123–130**
27: **41.2–16,80–88**
27a: CIII, CXLIV, CLXXIV; **41.17–20,43–49**
28: LXVI, CIII, CLXI, CLXXIV; **39.23–55**
28a: LXVI, CIII, CLXI, CLXII; **39.2–17,43–55 / 41.3–7,32–37,54–61,72–79,107–108**
29: CLXXIII; **38.2–4,9–27**
29a: CIII; **3.6–10 / 11.12 / 37.7–9 / 40.2–9**
29b: **60.51–106/ 65.3–38**
30: LXXIV, CIII, CLXXII; **11.37–40 / 31.26–27 / 44.11–12**

DICUIL:
Liber de astronomia:
1.5: LXXIV, CLXVII; **62.68–72,97–110**
3.2: **60.41–50**
3.4: **36.2–5**
4.1: **41.80–88**
4.3: CLXVI

DIONYSIUS EXIGUUS:
Argumenta: XIX, XLIV–XLVI, LII, LXXII–LXXIII, CLX–CLXI
III: **49.2**
IX: **49.2**
XIII (pseudo): **49.14–15**
XIV (pseudo): XXVIII, XCIX, CCXII, CCXIV; **56.15–43 / 67.3–26**
XV (pseudo): **24.2–11 / 38.31–38 / 44.55–80**
XVI (pseudo): CLII, CLXII, CCXV; **41.2–16,43–49,80–88,92–106 / 66.13–14**

Cyclus: XXXIII, XXXVIII–XXXIX, XLIV–XLVI, LXXI–LXXIII, LXXXIX–XCIII,

CXI, CLXXVI; **44.55–80 / 49.2,14–15 / 55.13–27,53–69 / 57.2–19**

Epistola (*ad Bonifatium et Bonum*): XXXIX, XLV–XLVI, CXXXIII–CXXXIV; **49.2 / 53.5–11 / 55.53–61 / 57.2–19 / 59.4–37,70–105 / 60.3–40 / 61.2–8 / 62.2–13,68–72,111–117**

Prologus (*Libellus; Epistola ad Petronium*): XXXIX, XLV, XCII, 322; **1.14 / 24.12–18 / 38.9–16 / 49.2 / 55.2–12,38–41 / 56.2–14 / 62.62–63 / 63.40–50 / 65.3–38 / 67.3–26**

DISPUTATIO CHORI ET PRAETEXTATI: LXVII, CXLIII *et vide* MACROBIUS, *Saturnalia*

DISPUTATIO MORINI: LXIV–LXV; **11.37–40 / 36.2–5 / 38.9–16**

DONATUS:
Ars maior:
1.2: **2.22–28**

DURAND, GUILLAUME:
Rationale divinorum officiorum:
8.1: XXXIII

ECGBERT OF YORK:
De institutione catholica dialogus: **55.2–12**

EPIST. RAT. (*'Der Regensburger Protestbrief von 809'*): XCIX, CCXXI
9: **55.38–41 / 56.2–14 / 58.2–7**
10: **38.31–38 / 44.55–80**
12: **44.55–80**

EPISTOLA CUMMIANI: *vide* CUMMIAN, *Epistola*

EPISTOLA CYRILLI: *vide* PS-CYRIL, *Epistola*

EPISTOLA DIONYSII: *vide* DIONYSIUS, *Epistola*

EPISTOLA PASCASINI: *vide* PASCASINUS, *Epistola*

EPISTOLA PROTERII: *vide* PROTERIUS, *Epistola*

EPISTOLA VICTORII: *vide* VICTORIUS, *Epistola*

EUCLID:
Elements: XXXI–XXXII

FIDES ROMANORUM:
7: **8.38–43**

FULDA COMPUTUS OF AD 789 (Basel, Universitätsbibliothek, F III 15k, 36r–49r): XLVIII, CI, CLXVIII, CCVIII, CCXI
c. 28: XLVIII
c. 31: LXXI, CLXVIII; **58.8–32**
et vide Basel, Universitätsbibliothek, F III 15k

HARVARD FRAGMENT (Harvard, Houghton Library, MS Type 613, fol. 7r–v): XCIX–CI, CLXXXVII–CXCI, CCIX; **49.16–32 / 50.29–39, 67–76 / 53.12–15 / 55.53–69**

HAYMO OF HALBERSTADT:
Homiliae de tempore:
66: LV

HEIMO OF BAMBERG:
De decursu temporum:
1.2: **50.72–76**

HELPERIC:
De computo: XXI, LV–LVI, CVI, CXII–CXIII
1: CXLVII; **7.2–5 / 41.43–49,54–61**
17: **59.18–27**
18: **62.14–61,97–110**
30: **38.28–38**
33: **63.25–28,36–39**
35: **58.8–32**
36: **62.97–110**

QUINTUS JULIUS HILARIANUS:
Expositum de die paschae et mensis: XLV, CXXXIV
1: **55.42–45**
2: **55.42–45**
11: **55.42–45**
12: **60.3–7**
13: XLV; **59.21–27 / 60.8–21**

HIPPOLYTUS:
Cylus: XXXV, XLV, CXXXIV; **65.3–38**

ISIDORE: XXXI
Chronica:
§§95–96: **13.2–7**
§§152–153: **12.39–40**

De natura rerum: LIII, CXIV
1: **8.2–10,14–22 / 9.28–34 / 11.4–11 / 24.12–18 / 49.2**
2: **8.11–13,34–39 / 9.2–3,11–27 / 10.2–6 / 11.2–11,16–31**
3: CXLII
4: CCXVI; **12.2–19,24–40 / 13.2–7 / 14.2–8 / 15.2–6 / 16.2–6 / 17.2–9 / 18.2 / 19.2–5 / 21.2–4 / 23.2–3 / 24.2–18 / 25.2–10 / 26.2–7 / 31.22 / 36.2–5 / 53.12–15**
6: **24.2–11 / 31.2–21,23–25 / 36.2–5/ / 41.43–49 / 59.4–13,18–27**
7: CXLII; **12.24–39 / 30.2–6 / 31.26–30 / 38.17–25**
8: CLXXV; **31.19–20 / 38.2–16,26–27**
9: CXLII; **44.1–10**
10: CCXIX
18: **44.18–36/ 46.15,21–23/ 48.8–12**
19: **44.18–36 / 46.6–13**
23: **11.16–31 / 31.23–25**
26: **11.16–31**
28: **48.13–18**
37: CCXVII

Etymologiae: LII–LIII, CXLII, CLXXXVII
3.24–71: LIII
3.29: CXLII; **43.1–10**
3.42: **47.1–12**
3.53: **44.18–36 / 46.6–13**
3.57: **44.18–36 / 46.9–13 / 53.2–3**
3.71: CXCIII; **9.16–20 / 37.2–7 / 46.2–5**
5.28–39: LIII
5.29: CXLII, CCXIII; **1.4–16 / 3.2–5 / 6.2–7**
5.30: CXLII; **8.2–10,14–39 / 11.2–11,13–31**
5.31: CXLII; **9.2–24,26–27**
5.32: CXLII; **10.2–6**
5.33: **12.2–23 / 13.2–7 / 14.2–8 / 15.2–6 / 16.2–6 / 17.2–9 / 19.2–5 / 21.2–4 / 23.2–3 / 25.2–10 / 26.2–7 / 55.2–12,38–45 / 46.2–14**
5.34: CLXXV, CLXXXVI; **38.2–16,26–27**
5.35: CXLII, CCXIII; **30.2–6 / 31.26–30 / 32.2–4 / 33.2–3 / 34.2–4 / 35.2–5**
5.36: CXLII; **31.2–13,15–16,19–20,23–25**
5.37: CXLII
5.38: **42.2–6 / 43.2–3**
6.17: LIII, XC–XCI, CXXXIV, CXLIII; **41.2–16,43–49 / 49.11–13,19–20 / 54.2–8 / 58.2–7 / 59.4–13,18–35,70–105 / 62.2–13,62–63,68–72,111–117 / 65.3–38 / 67.3–26**
6.29–30: **49.19–20**
7.1: CCXIII
7.5: CCXIII
9.2: **54.9–13**
13.1: CXLII; **44.1–10**
13.2: CXLII, CXCII; **2.2–33**
14.4: **9.16–20**
16.25: CXLII; **7.2–11**
18.6: **5.2–6**

Libri differentiarum:
2.7: **44.70–72**

JEROME:
Chronicon: **12.39–40 / 13.2–7 / 63.3–24**

Commetarii in Danielem:
2.7: **36.7–9**
4.11: **54.9–13**

Commetarii in Hiezechielem:
1.1: **24.2–11**
4.16: **6.2–7**
8.25: **54.9–13**
9.29: **24.32–39**

Commentarii in Ionam:
2.1: CLXXI; **9.28–34**

Commentarii in Matthaeum:
1: **7.5–7**
4: **54.9–13**

Commentarium in Aggaeum: LV

Commentarium in Esaiam:
1.2: **54.9–13**
10.31: **54.9–13**
14.53: **44.75–76**

De viris illustribus:
34: **63.3–24**
61: **59.14–17 / 65.3–38**

Expositio in Evangelium secundum Matthaeum: **7.2–11**

Liber interpretationis hebraicorum nominum: **54.9–13**

Martyrologium: **44.55–80**

PS-JEROME:
De exodo:
In vigilia Paschae: **59.21–27**

JONAS OF BOBBIO:
Vitae Columbani abbatis discipulorumque eius libri II:
1.4–5: CV
1.20–22: CIV

KAL. (*'Reichskalender'*):
A1: **31.15–16,21,22 / 36.7–9 / 58.88–94**
A2: **36.7–9**
A3: **58.88–94 / 62.97–110**
A4: **36.7–9 / 62.87–95**
A5: **36.7–9**
A7: **36.7–9**
A8: **36.7–9 / 62.97–110**
A9: **36.7–9 / 62.87–95**
A10: **62.87–95**
A11: **62.97–110**
a11: **17.2–9**
a14: **36.7–9**
a17: **13.2–7 / 17.2–9 / 18.2 / 19.2–5 / 20.2–6 / 21.2–4**
a19: **14.7–8 / 36.7–9**
a24: **36.7–9**
B2: **62.73–86**
B7: **62.97–110**
B9: **62.97–110**
C6: **36.7–9**
C8: **36.7–9 / 62.97–110**
c3: **36.7–9**
c6: **36.7–9**
c11: **14.2–8 / 15.2–6 / 16.2–6 / 17.2–9 / 18.2 / 19.2–5 / 21.1–23.3**
d2: **13.2–7 / 14.2–8 / 15.2–6 / 16.2–6 / 17.2–9 / 18.2 / 19.2–5 / 21.1–23.3**
E3: **36.7–9 / 62.97–110**
e4: **13.2–7 / 14.2–8 / 15.2–6 / 16.2–6 / 18.2 / 19.2–5 / 21.1–23.3 / 24.2–11**
F4: **12.39–40 / 13.2–7 / 14.2–8 / / 15.2–6 / 16.2–6 / 17.2–9 / 18.2 / 19.2–5 / 20.2–6 / 21.2–4 / 24.2–11 / 62.97–110**
F5: **19.2–5 / 62.97–110**
F6: **13.2–7 / 14.2–8 / 15.2–6 / 16.2–6 / 17.2–9 / 18.2 / 19.2–5 / 20.2–6 / 21.1–23.3 / 24.2–11 / 62.97–110**
F7: **13.2–7 / 14.7–8 / 15.2–6 / 16.2–6 / 17.2–9 / 18.2 / 19.2–5 / 21.1–23.3 / 24.2–11 / 62.97–110**
F8: **62.97–110**
f2: **13.2–7 / 14.2–8 / 17.2–9 / 18.2 / 19.2–5 / 20.2–6 / 21.2–4**
f3: **13.2–7 / 14.7–8 / 17.2–9 / 18.2 / 19.2–5 / 20.2–6 / 21.2–4**
f9: **36.7–9**
G1: **13.2–7 / 14.2–7 / 15.2–6 / 16.2–6 / 17.2–9 / 18.2 / 19.2–5 / 20.2–6 / 21.1–23.3 / 62.73–86,97–110**
G2: **13.2–7 / 14.2–7 / 15.2–6 / 16.2–6 / 17.2–9 / 18.2 / 19.2–5 / 20.2–6 / 21.1–23.3 / 62.97–110**
G3: **46.16–20**
G4: **62.73–86,97–110**
g1: **21.1–23.3**
g3: **36.7–9**
g13: **36.7–9**
H1: **12.37–40 / 13.2–7 / 4.7–8 / 16.2–6 / 17.2–9 / 18.2 / 19.2–5 / 20.2–6 / 21.1–23.3 / 62.97–110**
H2: **12.37–40 / 13.2–7 / 4.7–8 / 16.2–6 / 17.2–9 / 18.2 / 19.2–5 / 20.2–6 / 21.1–23.3 / 62.97–110**
H3: **12.37–40 / 13.2–7 / 4.7–8 / 16.2–6 / 17.2–9 / 18.2 / 19.2–5 / 20.2–6 / 21.1–23.3 / 62.97–110**
H4: **12.37–40 / 13.2–7 / 4.7–8 / 16.2–6 / 17.2–9 / 18.2 / 19.2–5 / 20.2–6 / 21.1–23.3 / 62.97–110**
H5: **12.37–40 / 13.2–7 / 14.7–8 / 16.2–6 / 17.2–9 / 18.2 / 18.2 / 19.2–5 / 20.2–6 / 21.1–23.**
H6: **13.6–7 / 14.7–8 / 16.2–6 / 17.2–9 / 18.2 / 19.2–5 / 20.2–6 / 21.1–23.3 / 62.97–110**
h9: **13.6–7 / 14.2–8 / 17.2–9 / 18.2 / 19.2–5 / 20.2–6 / 21.1–23.3**

LATERCULUS OF AUGUSTALIS: XVII, XXXVI

LATERCUS (84 (14)-year Easter reckoning): XVI–XIX, XXXVI–XXXVII, XLIV, XLIX, LXXII–LXXIII, LXXXI, LXXXIV, LXXXVIII–LXXXIX,

XCIII–XCV, CXXII–CXXIV, CXLVIII–CL,CLV, CLVII–CLVIII, CLX, CLXXXII, CXCV–CXCVI; **29.2–12 / 36.2–5 / 49.11–15 / 50.3–6,67–80 / 52.3–26 / 55.28–33 / 56.44–57 / 58.2–7 / 62.118–122 / 63.3–24 / 65.3–38**

LECT. COMP. (*'Die rheinische Anleitung von 760/792'*): CXI, CLXX, CLXXIX; **61.17–27**
1.1: LXVII, CCXX
1.3: CCXV
1.4: CCXV
1.9: **28.14–22**
1.11: CCXV
1.12: CCXX; **55.62–69 / 58.33–80**
1.13: CCXX; **55.62–69 / 59.28–35**
1.14: CCXX; **55.62–69**
1.15: **55.62–69**
1.16: **61.17–27**
2.3: **50.3–6**
2.4: **49.2 / 62.91–110**
2.10: **56.44–57**
6.1: **55.13–27 / 58.2–7,88–94**
6.2: **55.62–69 / 56.15–43 / 58.8–80**
6.3: **56.15–43**
6.4: **56.44–57**
6.5: **56.15–43**
6.6: **56.15–43**
6.8: CLXX; **61.17–27**
6.9: **55.13–27,38–41**
7.1: LXVI, CLXI; **39.2–17,23–29**
7.1a: LXVI; **39.30–42**
7.2: **41.3–7,29–31,80–88**
7.4: CLXVI; **62.14–61,111–117**
7: **41.3–7,29–31,80–88 / 62.14–61,111–117**
8a: **41.80–88**
8.1a: **55.38–41**
9.3: **36.7–9 / 41.80–88**
9.4: **49.16–32 / 62.14–61,97–110**
9.8: **31.15–16**

LEO THE GREAT:
Epistola ad Marcium imperatorem:
2: **55.13–27**

Tractatus septem et nonaginta:
72.6: **54.9–13**

LIB. ANN. (*'Das Veroneser Jahrbüchlein von 793'*): XXV–XXVI, CXI
1: **24.32–39 / 31.2–16,21–25 / 59.2–13,18–27**
5: **41.2–16,80–88**
9: **49.11–13**
11: **55.38–45 / 56.15–43**
12: **55.38–41 / 56.2–14 / 58.2–7**
13: **49.2**
14: **59.28–35**
19: **56.15–43**
22: **8.34–39**
23: **24.12–18**
30: **58.33–80**
31: **58.33–80**
32: **55.62–69**
33–34: CCXVI; **55.38–41**
35: CCXVI
36: **58.88–94**
37: **55.13–27**
38: **55.13–27 / 59.2–13**
49: **62.97–110**
50: **48.2–7 / 62.2–61**
52: **61.17–27**
54: LXVI–LXVII, CLXI; **39.2–17**
56: **59.18–27,36–37 / 60.3–7**
58: **24.12–18,32–39 / 50.61–66,72–76**
59: **59.2–13 / 60.3–7**
60: **36.7–9**
64: **36.7–9**
69: LIII

LIB. CALC. (*'Die Salzburger Enzyklopädie von 818'*): XXVI–XXVII, CXII
3: **39.23–29**
7: **55.62–69 / 58.33–80**
10: **50.3–14,22–28 / 62.97–110**
11: **55.62–69 / 59.28–35**
13: **24.12–18**
14: **50.3–6**
16: **50.7–14**
36–38: LIII
39: **24.32–39 / 31.2–16,21–25 / 59.2–13,18–27**
40: **36.7–9**
41: CXLIV; **41.3–7,25–28,43–49,54–61,72–79**
42: CXLIV, CLXIX; **7.12–13 / 41.3–7,29–31,72–79**
45: **41.107–108**

46: **48.2–7 / 49.2,11–13 / 52.27–43 / 53.5–11 / 54.4–13 / 59.70–105 / 62.2–61,68–95,97–117,123–130**
47: **62.14–61**
48: CLXVI; **59.2–13,18–20**
49: **60.3–7**
51: **52.49–60**
52–54: **56.15–43**
55: **55.13–27**
56: **58.33–80,88–94**
57: **55.13–27**
62: **49.11–13**
64: **9.28–34**
66: **11.37–40 / 44.18–36**
67: **10.2–6 / 11.4–8,13–36**
69: **12.2–3,7–14 / 24.2–26,32–39 / 55.2–12**
70: **12.24–39 / 13.6–7 / 15.2–6 / 17.2–9 / 16.2–6 / 17.2–9 / 18.2 / 19.2–5 / 20.2–6 / 21.1–23.3 / 24.2–18 / 36.2–5 / 41.2–20,43–49**
71: **25.2–10 / 26.2–7 / 27.2–12**
72: **24.19–26 / 55.2–12**
74: **44.18–36 / 46.15,21–23 / 48.8–12**
81: **48.2–7**
82: **38.9–16,28–38**
83: **30.2–5 / 31.26–27 / 32.2–4 / 33.2–3 / 34.2–4**
84: **59.14–17**
85: LXX–LXXI; **53.5–11 / 55.13–27,38–41 / 59.2–13,21–35 / 60.3–7 / 61.2–5,17–27**
86: **59.36–37 / 60.51–106 / 62.97–110**
88: **1.9–16**
89: **49.2,7–13**
90: **65.3–38**
92: **53.5–11 / 57.2–19/ 59.2–13,18–20,28–35,70–105**
93: CLXV; **55.38–41 / 56.2–14 / 58.2–7 / 63.3–50 / 67.3–26**
94: **56.2–14 / 58.2–7**
95: **55.42–45/ 58.2–7**
98: LXXXI; **41.107–108 / 65.3–38**

LIB. COMP. (*'Die Aachener Enzyklopädie von 809'*): XXVI–XXVII, CXII
1.3a: **31.26–27**
1.3b: **38.9–16,28–38**
1.3f: CLII, CLXII; **41.92–106**
1.4: CCXV
1.7: CCXX
1.9: **9.2,11–13**
1.11: **55.62–69**
2.14: **28.14–22**
2.15: **49.2 / 50.3–6 / 62.97–110**
2.16: **24. 12–18**
2.17: **50.3–6**
2.21: CCXIII; **58.33–80**
2.22: LXXII, CLXIV, 329; **55.46–52 / 59.28–35**
3.2: **36.7–9**
3.4: **31.2–16,21–25 / 59.2–13,18–27**
3.6: CXLIV; **41.3–7,25–41,43–49,72–79**
3.8: **41.107–108**
3.10: **41.2–16,80–88**
3.14: LXVI–LXVII, CLXI; **39.2–17,23–42**
4.1: CCXVI; **50.3–6**
4.4: **24.32–39**
4.7: CLXVI; **59.18–20 /61.6–8 / 62.14–61**
4.8: **59.2–13 / 60.3–7 /61.6–8**
4.9: **48.2–7 / 49.2,11–13 / 52.27–43 / 53.5–11 / 59.4–13,70–105 / 62.2–61,68–95,97–117,123–130**
4.12: **50.22–28 / 62.97–110**
4.14: **58.33–80**
4.15: **58.33–80**
4.16: **56.15–43**
4.17: **55.38–41 / 56.2–14 / 58.2–7**
4.18: CCXVI; **55.13–27,38–41**
4.19: **56.15–43**
4.20: **56.15–43**
4.21: LXXXVII; **55.13–27,38–45 / 58.2–7**
5.20: **49.11–13**

MACROBIUS: XXX, LII
Commentarii in somnium Scipionis:
2.11: **36.2–5**

Saturnalia:
1.3: **9.16–20**
1.12: **12.24–40 / 15.2–6 / 16.2–6 / 17.2–9 / 18.2 / 19.2–5 / 20.2–6 / 21.2–4 / 22.2–4**
1.13: **12.24–39 / 14.2–8 / 24.2–11 / 26.2–7**
1.14: CXLIII; **28.2–5,14–22**
1.15: **24.12–18 / 25.2–10 / 26.2–7 / 27.2–12 / 28.2–5**
et vide DISPUTATIO CHORI ET PRAETEXTATI

MARTIN OF BRAGA:
De pascha:
4: **11.37–40**
5: **11.37–40 / 38.39–43 / 44.18–36 / 66.2–7**
6: **11.37–40**

7: **55.13–27 / 56.2–14**

MARTIANUS CAPELLA: XXX
De nuptiis Philologiae et Mercurii:
3: CCXIV

MAXIMUS CONFESSOR:
Computus ecclesiasticus: XLVI

MAXIMUS OF TOURS (?):
Sermones:
54: **54.2–8**
60: **38.31–38**

MISSALE MIXTUM SECUNDUM REGULAM B. ISIDORI DICTUM MOZARABES:
In dominica ante epiphaniam domini: **41.25–28**

MURETACH:
In Donati artem maiorem: **11.2–3 / 25.2–10 / 26.2–7**

NOTKER LABEO:
De quatuor questionibus compoti: LVI; **50.29–39 / 59.28–35 / 60.51–106**

Translatio barbarica psalterii:
Psalmus 93: LXXV

PACIFICUS OF VERONA:
Computus: LV, CLXXIX
§5: **28.14–22**
§14: **50.3–6**
§§15–20: **50.7–14**
§21: **50.3–6**
§22: **49.11–13**
§23: **55.62–69 / 58.33–80**
§24: **6.15–43 / 58.33–80**
§25: **6.15–43 / 55.62–69**
§29: **58.88–94**
§41: CXLVII
§50: **8.2–3**
§51: **8.2–9**
§53: **8.34–39**
§54: **9.28–34**
§55: **8.2–10**
§§55–58: **11.16–31**
§59: **11.13–15**
§60: **11.4–11**
§62: **11.2–3**
§63: **8.21–22**
§64: **8.23–25**
§65: **8.26–30**
§66: **8.31–33**
§72: **11.37–40**
§76: **11.37–40 / 44.18–36**
§78: **9.2–10**
§79: **9.4–10**
§81: **9.11–13**
§82: **9.16–20**
§83: **9.14–15**
§84: **9.21–22**
§85: **9.23–24**
§87: **9.23–24**
§§88–90: **9.26–27**
§91: **10.2–6**
§100: **11.32–36**
§§101–102: **11.4–8**
§§103–104: **12.2–14**
§105: **13.2–7**
§106: **14.2–8**
§107: **15.2–6**
§108: **55.2–12**
§109: **16.2–6**
§110: **17.2–9 / 18.2–9**
§111: **18.2–9**
§112: **19.2–5**
§113: **21.2–4**
§§115–116: **41.80–88**
§§119–120: **25.2–10**
§121: **25.2–10 / 26.2–7**
§122: **27.2–12**
§124: **31.2–13**
§126: **31.2–13**
§127: **24.32–39 / 31.14**
§128: **31.14**
§129: **24.32–39 / 31.14**
§130: **31.14,22 / 59.2–13,18–20**
§131: **31.14,22 / 59.2–13,21–27**
§132: **31.14**
§133: **31.2–14,15–16**
§134: **31.14,21**
§§135–143: **31.14**
§144: **31.14,23–25**
§145: **24.2–11 / 36.2–5**
§§176–177: CLXXIX; **1.2–3**
§§180–184: CLXXIX
§§184–185: **43.2–3**
§§186–188: **42.2–6**
§203: **59.14–17**
§214: **49.2**
§214: **49.7–10**
§215: **49.7–10 / 49.11–13**

§218: **65.3–38**
§226: **9.28–35 / 53.5–11**
§227: **9.28–35 / 59.2–13,18–20,70–105**
§§229–247: **57.2–19**
§251: **56.2–14**
§252: **56.2–14 / 55.38–41**
§253: **56.2–14 / 58.2–7**
§§254–258: **56.44–57**
§259: **56.2–14 / 58.2–7**
§260–261: **58.2–7**
§266: **59.2–13**
§267: **59.28–35**
§268: **53.5–11**
§269: **55.13–27,38–41**
§270: LXX; **61.17–27**
§271: LXX
§272: **59.36–37**
§283: **50.22–28**
§§285–286: **62.97–110**
§§289–291: **62.14–61**
§292: **48.2–7 / 62.2–13, 123–130**
§293: **62.123–130**
§295: **62.68–72**
§295–296: **62.73–86**
§296: **49.11–13**
§297: **49.19–20 / 53.5–11 / 59.4–13,70–105 / 62.68–72,111–117**
§299: **41.72–79**
§300: **41.29–31**
§301: CXLIV; **41.3–7,25–28,43–49**
§302: CLXII; **41.54–61**
§304: **41.3–7**
§310: **38.28–38**
§311: **38.9–16**
§313: **38.9–16**
§314: **30.2–5 / 31.26–27**
§318: **32.2–4 / 33.2–3 / 34.2–4**
§§320–336: LIII
§438: **37.2–7**
§439: **46.2–5**
§475: **44.1–10**

PAPAL LETTER TO THE NORTHERN IRISH CLERGY OF AD 640: XXXIX–XL, LXIX, LXXVII, LXXXII–LXXXVII, XCV

PAUCA PROBLESMATA (*'The Irish reference Bible'*):
I:
§14: **11.37–40**
§64: **11.37–40**
§91: **7.9–11**
§92: **44.18–36**
§94: CXXVIII; **8.2–4**
§99: **9.2–3**
§123: **44.18–36**
§126: **37.2–7**
§128: **44.18–36**
§129: **9.4–10**
§131: **24.12–18 / 31.2–13 / 36.2–5**
§134: **44.18–36/ 46.15**
§156: **44.55–80**
II:
§14: **11.37–40**

PASCASINUS:
Epistola:
1: **59.28–35**
2: **52.49–60 / 54.2–8 / 59.4–13,18–27**

PASCHALE CAMPANUM: LXXXIV

PÉRIGUEUX TABLE: XCI; **55.62–69**

PLINY: XXX
Naturalis historia:
6: **24.32–39**

POLEMIUS SILVIUS:
Fasti: **12.15–19 / 14.2–8 / 17.2–9 / 18.2–8 / 19.2–5 / 20.2–6 / 21.2–4 / 24.32–39 / 25.2–10 / 26.2–7 / 27.2–12 / 28.6–13 / 55.2–12**

POMPEIUS (MAURUS):
Commentum artis Donati: **25.2–10 / 54.14–16**

PRISCIAN:
Institutiones grammaticae: **2.22–28 / 7.2–11**

PROL. AQUIT. (*'Das aquitanische Vorwort zur Ostertafel von 721'*):
1: **55.13–27 / 56.2–14**
3: **49.2,11–13 / 58.33–80 / 59.36–37**
4: LXVI; **62.62–63 / 65.3–38**

PROLOGUS CYRILLI: LXVII–LXVIII
4–5: **55.13–27,38–41**
6: **24.32–39**
7: **44.55–80**
8: **24.32–39**

Addendum: **24.12–18,32–39 / 31.15–16,26–27 / 38.31–38 / 49.2**

PROLOGUS DIONYSII: *vide* DIONYSIUS, *Prologus*

PROLOGUS PASCHAE:
4: **44.13–80,18–36**
5: **44.13–80**
6: **44.13–80,18–36**
8: **55.2–27**

PROLOGUS THEOPHILI: LXVII
1: **55.2–12 / 56.2–14**
3: **63.30–39**
4: **44.55–80 / 56.2–14 / 58.2–7**
5: **44.55–80 / 56.2–14**

PROLOGUS VICTORII: *vide* VICTORIUS, *Prologus*

PROTERIUS:
Epistola: LXVII–LXVIII, 326
2: **44.55–80 / 52.45–48 / 55.2–12 / 56.2–14 / 63.30–50**
4: **55.13–27**
5: **55.13–27**
7: **44.55–80 / 63.40–50**
8: **38.9–16**
9: **63.30–39**

PTOLEMY:
Almagest: XXX

QUAEST. AUSTR. (*'Die austrasische Abhandlung von 764'*): XLIV, CXII, CLXIX–CLXXI
1.1: CLXIX; **24.19–31**
1.2: CLXIX; **24.19–26**
1.5: **24.40–41**
1.10: **49.16–32 / 59.38–69**
1.11: **50.50–55 / 61.17–27 / 62.87–95**
1.12: **62.87–95,111–117,123–130**
2.1: CLXIX; **12.15–17 / 24.19–31**
2.2: CLXIX; **18.2 / 19.2–5 / 20.2–6 / 24.19–26**
2.4: **24.32–39 / 48.2–7 / 50.56–66**
2.5: **27.2–12 / 28.2–5,14–22**
2.6: **28.2–5**
2.7: CXXXVIII, CLXI, CLXIX; **29.2–12,17–24 / 45.2–13**
2.8: CLXXVIII; **31.15–16,19–22 / 36.7–9 / 38.17–27 / 53.5–11 / 59.4–13,70–105 / 60.3–7**
2.9: LXVI–LXVII, CIII, CLXII, CLXIX; **41.2–16,32–37,43–49,54–88,107–108 / 48.2–7 / 62.111–117,123–130 / 66.13–61**
2.10: **36.2–5 / 49.11–13 / 53.5–11 / 59.70–105**
2.11: CLXIX; **50.40–46,50–60,78–80 / 61.17–27 / 62.2–13,87–95,111–117**
2.17: LXXIII

QUAEST. LANGOB. (*'Die langobardische Abhandlung um 780'*): CXII, CLXVIII, CLXX
1: **11.4–11,13–26**
2: CLXX; **25.2–10 / 26.2–7 / 27.2–12 / 28.2–5**
3: CLXX; **12.18–19 / 13.2–7 / 14.2–7 / 15.2–6 / 16.2–6 / 17.2–9 / 18.2 / 19.2–5 / 20.2–6 / 21.1–23.3**
4: **24.32–39 / 55.2–12**
5: **36.7–9**
7: **59.2–13**

RABANUS MAURUS:
De computo: XXIV, LV, CI, CXII, CXXVIII, CLIX–CLX
8: CXLVII; **7.2–5**
9: **1.4–16 / 30.2–6**
10: CLX; **1.4–8**
11: **2.2–31**
13: **3.6–10**
15: CXXVII; **4.6–8**
16: CXXVII; **5.3–5,7–8**
17: CXLII; **6.2–7,8**
18: CXLVII; **7.2–5 / 8.16–17**
19: **8.2–10,14–20**
20: **8.21–33**
21: **9.2–3, 4–10**
22: **9.11–34**
23: **8.34–39 / 44.18–36**
24: **10.2–3**
26: **11.13–31**
27: **11.2–8,32–36**
28: **12.7–14 / 41.80–88 / 50.29–39**
29: **24.32–39 / 55.2–12**
30: **24.12–26,32–39 / 26.2–7 / 55.2–12**
31: **24.19–26 / 55.2–12**
32: **12.18–19,24–37 / 13.2–6, 6–7 / 14.7–8 / 15.2–6 / 16.2–6 / 17.2–9 / 18.2 / 19.2–**

5 / 20.2–6 / 21.1–23.3 / 24.2–11 / 36.2–5 / 41.2–20,43–49
33: **25.2–10 / 27.2–12**
34: **30.2–5 / 31.26–27 / 32.2–4 / 33.2–3 / 34.2–4 / 38.17–25,39–41**
35: **31.2–13**
36: **24.32–39 / 31.2–16,21–25 / 36.7–9 / 59.2–13,18–27**
37: **37.2–7 / 46.2–5**
39–42: LIII
45: **46.15**
53: **38.9–27**
54: CXLIV; **31.2–13 / 41.3–7,25–31,43–49**
55: CXLIV
56: **48.2–7 / 62.14–61,73–86,97–110,123–130**
57: CLXVI; **59.36–37 / 62.14–61**
59: **53.5–11 / 59.2–13,21–27 / 60.3–7,28–35 / 61.2–5,17–27**
60: **60.51–106 / 62.97–110**
64: **44.55–80**
66: **1.9–16**
68: **49.2,7–13**
70: **50.3–14,22–28**
71: **65.3–38**
77: **53.5–11 / 59.2–13,18–20,28–35,70–105**
80: **55.38–41 / 56.2–14 / 58.2–7 / 63.3–50**
81: **55.38–41 / 56.44–57**
82: **56.15–43**
83: XLVIII; **55.53–69 / 58.8–80 / 59.28–35**
84: **56.15–43**
86: **56.2–14 / 58.2–7**
87: **44.55–80**
88: **55.13–27 / 58.33–80**
89: **58.2–7**
90: **49.2 / 56.15–43**
91: **58.88–94**
94: LXXXI; **41.107–108 / 65.3–38**
95: **43.2–3**
96: CXXI; **42.2–6**

Versus ad Baturicum episcopum ex persona Isanberti: XCVIII

SALTAIR NA RANN: XXXIII

SEDULIUS SCOTTUS:
In Donati artem maiorem: **11.2–3 / 25.2–10 / 26.2–7 / 27.2–12**
In Matthaeum: LV

SENTENTIAE S. AUGUSTINI ET ISIDORI IN LAUDE COMPUTI: XXIX, L

SERGIUS (?):
De littera, de syllaba, de pedibus, de accentibus, de destinctione commentarius: **2.22–28**

SIRMOND MS: *vide* Oxford, Bodleian Library, Bodley 309

SMARAGDUS:
Collectiones: LV

STEPHEN OF RIPON:
Vita sancti Wilfrithi:
5: XL
7: XL
10: XL
47: XL

SULPICIUS SEVERUS: XXXVII

SUPPUTATIO ROMANA: XXXV–XXXVII, XLIX, CLXXX; **44.13–80 / 52.49–60 / 65.2–12**

TRACTATUS ADTHANASI: LXIV–LXV
4: **11.37–40 / 38.39–43 / 66.2–7**
5: **11.37–40 / 44.18–36 / / 66.2–7**
6: **55.13–27 / 56.2–14**

VERS. TUR. (*'Das Tourer Lehrgedicht um 800'*):
3: **41.17–20**
5: **59.2–13,21–27 / 60.3–7**
6: **59.2–13 / 60.3–7**
7: **59.21–27**
12: **62.97–117**
13: CLXVII; **62.68–72,87–95**
14: **59.36–37 / 61.6–8**
15: **60.51–106 / 65.3–38**

VICTORIAN COMPUTUS OF AD 689: LVIII, LXXXII, LXXXVII,LXXXIX, CVIII, CXX–CXXI, CXXIII–CXXVI, CLII; **41.107–110 / 62.64–67 / 63–68**

VICTORIUS OF AQUITAINE:
Calculus: XXXI

Cyclus: XXXVIII, XLIV, XLIX, LVII–LVIII, LXXXIII–LXXXVI, LXXXIX, CLV–CLVIII, CLIX; **49.11–13 / 50.61–66 / 52.3–13**

Prologus: XXXVIII, XLIV, XLIX
3: **62.111–117 / 65.3–38**
4: **52.49–60 / 55.34–45 / 56.2–14 / 58.2–7**
7: **11.37–40 / 44.13–80,18–36**
8: **44.13–80 / 44.13–80**
9: **24.2–11 / 44.13–80,37–54 / 53.5–11 / 69.2–6**
10: **52.3–13 / 65.3–38**

VIRGILIUS MARO GRAMMATICUS:
Epitomae: XXII, LXVIII–LXIX, CXXVI, CXXIX
11: LXIX, CXCIII; **8.2–10 / 9.2–3 / 12.2–14 / 30.2–6 / 31.2–13 / 32.2–4 / 34.2–4 / 37.2–7 / 46.2–5**

WILLIBRORD: CLXIV
Easter table (Paris, Bibliothèque Nationale, Lat. 10837, fol. 34r–40r): XCI; **12.15–19 / 28.6–13 / 55.62–69**

ZEITZ TABLE: XXXVI–XXXVII, CCXI

MANUSCRIPTS

Angers, Bibliothèque Municipale, 477 (461):
10r: CLXXXVI–CLXXXVII; **42.2–8 / 43.2–3**
13v: CLXXXVI; **37.2–7**
19r: CXXVIII, CLXXXV; **8.2–3,9–10 / 11.2–3,9–11**
19v: CXXXII, CLXXXV–CLXXXVI; **9.2–7 / 31.2–13 / 38.5–8**
20r: CLXXXV–CLXXXVI; **32.2–4 / 34.2–4 / 35.2–5 / 41.3–7,17–24,32–49,72–79**
20v: CVI, CXXIX–CXXX
46v: CLXVI, CLXXXV; **2.6–10 / 5.2,5–6 / 6.2–7**
49v: CXLVII, CLXXXV; **7.2–9**
76v: CLXIII, CLXXXI, CLXXXVII
et vide ANGERS GLOSSES

Basel, Universitätsbibliothek, F III 15a: XCVIII

Basel, Universitätsbibliothek, F III 15k: CCXI
24v–35v: CXIII
31r: **31.14**
34v–36r: **57.2–19**
35r: **42.2–8**
35r: **43.2–3**
35v: **43.2–3 / 44.2–10**
36r–49r: *vide* FULDA COMPUTUS OF AD 789
43v: **58.8–32**
43v–44r: XLVIII
44v: **55.53–69**
44v–45r: LXXI, CLXVIII
45r: **58.8–32**
56v–57r: CLXI

Berlin, Staatsbibliothek, Phillipps 1831:
128r: LXXII, CLXIV–CLXV, 328–9; **55.46–52**
129r–134r: CLXXIX *et vide* DIAL. LANGOB.
134r: CLXXVI

Berlin, Staatsbibliothek, Phillipps 1869:
2r: 331

Bern, Burgerbibliothek, 417:
14v–15r: LXXI, CLXVIII; **58.8–32**
33r: CXIII; **42.2–8 / 43.2–3**
33v: CXIII; **43.2–3 / 44.2–10**
47r–61v: CXIII
52v–53v: XXVIII et *vide* DE HEBDOMADIBUS

Bern, Burgerbibliothek, 645: CCXI
41r–48v: LXXXIV
41r–71v: XLIV
54v: **50.61–66**
55r: **50.40–46,50–60 / 61.17–27 / 62.87–95**
56v: **61.17–27**
59r–67v: LXXXIV
66r: LXXXV

Brussels, Bibliothèque Royale, 5413–22:
77v–107v: CVI

Cologne, Dombibliothek, 83[2]: CLXXIX–CLXXX
50v: **62.14–61**
176v–178r: LI, CCVII *et vide* DE COMPARATIONE EPACTARUM DIONYSII ET VICTORII
176v: CLII, 321, 327–8
201r: CXXXI; **55.28–33**

Cologne, Dombibliothek, 102: CCXI

Cologne, Dombibliothek, 103: CLXXIX–CLXXX
184v–190v: *vide* COMPUTUS RHENANUS OF 775
187v–188r: XCIX

Einsiedeln, Stiftsbibliothek, 321 (647):
82–125: XLVIII, L, CCVII *et vide* COMPUTUS EINSIDLENSIS

Gotha, Forschungsbibliothek, 75:
77v–106r: LVII, LXXXIV
84v: LVII

Geneva, Bibliothèque Publique et Universitaire, 50:
121r: CXXXI; **55.28–33**
139r–v: CLX; **1.4–8**

Harvard, Houghton Library, MS Type 613: *vide* HARVARD FRAGMENT

Karlsruhe, Landesbibliothek, 229:
33r: CLX; **1.4–8**

Karlsruhe, Landesbibliothek, 442:
63v–64r: CLX; **1.4–8**

Kassel, Landesbibliothek, 4° Ms astron. 1:
71r: **62.87–95**
76r: **60.8–21**

Leiden, Universiteitsbibliotheek, Scaliger 28: CCXI
2v–22r: LVII, LXXXIV; **50.40–46**
6r: LV, LXXXV
23v: CLXI
33v: **62.111–117**
40r: CLXI
40r–v: CLCV; **1.4–8**

London, British Library, Cotton Caligula A XV: CCXI
66r: **49.14–15**
73r–80r: *vide* COMPUTUS COTTONIANUS

London, British Library, Harley 3017:
3r: CLXI; **49.14–15**
50r–51r: XC–XCI

Milan, Biblioteca Ambrosiana, H 150 inf:
7v: CLXIII; **49.14–15**
127v: LXXIV
129v: LXXXIV
130r–132r: LVII, LXXXIV
et vide BOBBIO COMPUTUS

Munich, Bayerische Staatsbibliothek, Clm 14456: CCXIII–CCXXI
8r–46r: XXVII, CCVI
46v–47v: XCVIX
48r–53v: XCVIX
65v–67r: XXVIII
70v–73r: XCVIII

Munich, Bayerische Staatsbibliothek, Clm 14725: CCXI
16r: **49.16–32**

Munich, Bayerische Staatsbibliothek, Clm 18156:
12v: **11.2–3**
16v: **52.27–43**
32r: **50.22–28**

Oxford, Bodleian Library, Auct. F 4 32 (St Dunstan's classbook):
21r: XCIII, CLXIII; **49.14–15 / 58.33–80**
22r–v: LXXI, CLXIII, CLXVI; **50.7–14 / 59.38–69**

Oxford, Bodleian Library, Bodley 309 (SIRMOND MS): XXI–XXII, LXXX–LXXXII, XCV, CXXV
62r–v: XXVIII
64v–73v: CXIII–CXIV
73v: XXVIII
95v: LXXXI; **41.107–110 / 44.11–12**
95v–95*bis*r: LXXIII, LXXX
106v: CLXV; **50.22–28,47–49 / 59.38–69**
107r: CLXV; **59.38–69**
113r: **52.3–13**
113r–120r: LVII, LXXXIV; **62.87–95**
114r: LXXXV, LXXXVI
115r: LVII–LVIII, LXXXII, CLVIII; **41.107–110 / 62.64–67**

Oxford, Bodleian Library, Digby 63:
33v: LXXI, CLXVI; **59.38–69**
33v–34r: CLXI
72v–79r: *vide* COMPUTUS DIGBAEANUS

Padua, Biblioteca Antoniana, I 27: CCXI
76r–77v: XLIV
77v–78r: XIX, XXII

Paris, Bibliothèque Nationale, Lat. 2341:
9r: CLX; **1.4–8**

Paris, Bibliothèque Nationale, Lat. 4860:
147v–148r: LXXXIV; **62.87–95**
151r–v: XCIX
157v: **49.16–32**

Paris, Bibliothèque Nationale, Lat. 5239:
108r–109v: 333
145r–162r: LIII

Paris, Bibliothèque Nationale, Lat. 5543:
123v–124v: 333

125r–126r: CLX; **1.4–8**
126r: CLX; **1.2–8**
145r–v: XCI
148–157v: LIII

Paris, Bibliothèque Nationale, Lat. 7296:
111v: LXXII, CLXV; **55.46–52**

Paris, Bibliothèque Nationale, Lat. 7418:
89v–90r: CLX; **1.4–8**

Paris, Bibliothèque Nationale, Lat. 7530:
252v: CLXIII; **49.14–15**

Paris, Bibliothèque Nationale, Lat. 9528:
201: XCI

Paris, Bibliothèque Nationale, Lat. 10837:
34r–40r: **12.15–19 / 28.6–13**
40v–41v: XCI; **55.62–69**
43r–44r: XCI; **55.62–69**

Paris, Bibliothèque Nationale, Lat. 16361:
241: **41.107–110**

Paris, Bibliothèque Nationale, Nouvelle acquisition latine 456:
189v–190v: CVI

Paris, Bibliothèque Nationale, Nouvelle acquisition latine 1203:
124v–126r: XCII

Paris, Bibliothèque Nationale, Nouvelle acquisition latine 1615:
10r: CLXI
150r–v: LXXI, CLXVIII; **58.8–32**
152v: CLXIII; **49.14–15**
153r–v: **55.53–61**

Paris, Bibliothèque Nationale, Nouvelle acquisition latine 1632:
13v: CLXVI; **62.14–61**

Rouen, Bibliothèque Publique, I 49 (524):
87r: **50.29–39**

Schaffhausen, Stadtbibliothek, 61:
22r–v: CIX, CXLIII

St Gall, Stiftsbibliothek, 250:
80: CLXIII; **49.14–15**

St Gall, Stiftsbibliothek, 251: CCXI
2: **49.14–15 / 58.33–80**
13: **36.10–15**
14: **12.18–19 / 36.10–15**

St Gall, Stiftsbibliothek, 878: CXIV

Strasbourg, Bibliothèque Nationale et Universitaire, 326:
108r–109r: 333
164r: **1.4–8**
164v: CLX; **1.2–8**

Turin, Biblioteca Nazionale, F VI 2: LXXXV

Vatican, Biblioteca Apostolica, Pal. Lat. 830:
2v–3r: 331; **58.88–94**

Vatican, Biblioteca Apostolica, Pal. Lat. 1447:
3r: CIX, CXLIII; **1.4–8**
19v–22r: XCII, CLXIII, 339–40
23r: CLXI

Vatican, Biblioteca Apostolica, Pal. Lat. 1448:
1v–2r: CIX, CXLIII; **1.4–8**
13v–16r: XCII–XCIII, CLXIII, 339

Vatican, Biblioteca Apostolica, Reg. Lat. 49: **8.38–43**

Vatican, Biblioteca Apostolica, Reg. Lat. 586:
9r–10v: LXXXIV; **62.87–95**
117r–v: XCVIX

Vatican, Biblioteca Apostolica, Reg. Lat. 1038:
126r: CLX; **1.2–8**

Vatican, Biblioteca Apostolica, Reg. Lat. 1260:
15v: CLXI
16r: **50.29–39**
87r–99v: CV *et vide* DE RATIONE CONPUTANDI
112r–114r: XCII, CLXIII, 339
117r: CLX; **1.2–8**
121r: CLXVIII; **58.8–32**

Vatican, Biblioteca Apostolica, Reg. Lat. 1855:
47r: CLXI
84r: CLX; **1.4–8**

Vatican, Biblioteca Apostolica, Urbs Lat. 290:
34v–35r: CLX; **1.4–8**

Vatican, Biblioteca Apostolica, Vat. Lat. 642:
9r: CXXXI; **55.28–33**

Vatican, Biblioteca Apostolica, Vat. Lat. 643:
52r: **60.8–21**

Wolfenbüttel, Herzog-August-Bibliothek, Weißenburg 91:
169r–173v: *vide* COMPUTUS RHENANUS OF 775

Würzburg, Universitätsbibliothek, M.p.th.f. 6:
inserted fol. 29r (Gospel fragment): LXXXVIII

Zürich, Zentralbibliothek, Car C 180: CCXI
35r–36r: **46.16–20**
104–106v: LXXI, CLXVI; **59.38–69**

Zürich, Zentralbibliothek, Rh. 83:
14r–17r: XC